Moment Resistant Connections of Steel Frames in Seismic Areas

Moment Resistant Connections of Steel Frames in Seismic Areas

Design and reliability

Edited by Federico M. Mazzolani

CRC Press
Taylor & Francis Group
Boca Raton London New York

CRC Press is an imprint of the
Taylor & Francis Group, an informa business
A TAYLOR & FRANCIS BOOK

CRC Press
Taylor & Francis Group
6000 Broken Sound Parkway NW, Suite 300
Boca Raton, FL 33487-2742

First issued in paperback 2019

ISBN-13: 978-0-415-23577-8 (hbk)
ISBN-13: 978-0-367-44736-6 (pbk)

Publisher's Note
This book has been prepared from camera-ready copy provided by the editor.

British Library Cataloguing in Publication Data
A catalogue record for this book is available from the British Library

Library of Congress Cataloging in Publication Data

A catalog record for this book has beeen requested

Visit the Taylor & Francis Web site at
http://www.taylorandfrancis.com

and the CRC Press Web site at
http://www.crcpress.com

LIST OF CONTENTS

LIST OF CONTRIBUTORS

Anthimos ANASTASIADIS
INCERC, Timisoara, Romania

Jean Marie ARIBERT
Institut National des Sciences Appliquees, Laboratoire de Structures et Mecanique Appliquee, Rennes, France

Darko BEG
University of Ljubljana, Faculty of Civil Engineering and Geodesy, Institute of Structures and Earthquake Engineering, Ljubljana, Slovenia

Luis CALADO
University of Lisbon, DECivil - Instituto Superior Tecnico, Lisbon, Portugal

Adrian CIUTINA
Politehnica University of Timisoara, Civil Engineering Faculty, Department of Steel Structures and Structural Mechanics, Timisoara, Romania

Gaetano DELLA CORTE
University of Naples Federico II, Department of Structural Analysis and Design, Naples, Italy

Gianfranco DE MATTEIS
University of Naples Federico II, Department of Structural Analysis and Design, Naples, Italy

Florea DINU
Politehnica University of Timisoara, Civil Engineering Faculty, Department of Steel Structures and Structural Mechanic, Timisoara, Romania

Dan DUBINA
Politehnica University of Timisoara, Civil Engineering Faculty, Department of Steel Structures and Structural Mechanics, Timisoara, Romania

Ciro FAELLA
University of Salerno, Department of Civil Engineering, Salerno, Italy

Peter FAJFAR
University of Ljubljana, Faculty of Civil Engineering and Geodesy, Institute of Structures and Earthquake Engineering, Ljubljana, Slovenia

Ognian GANCHEV
University. of Architecture and Civil Engineering of Sofia, Division of Steel and Timber Structures, Sofia, Bulgaria

Tzvetan GEORGIEV
University. of Architecture and Civil Engineering of Sofia, Division of Steel and Timber Structures, Sofia, Bulgaria

Victor GIONCU
INCERC, Timisoara, Romania

Daniel GRECEA
Politehnica University of Timisoara, Civil Engineering Faculty, Department of Steel Structures and Structural Mechanics, Timisoara, Romania

Alain. LACHAL
Institut National des Sciences Appliquees, Laboratoire de Structures et Mecanique Appliquee, Rennes, France

Raffaele LANDOLFO
University of Chieti G. D'Annunzio, Pescara, Italy

Graziella MATEESCU
INCERC, Timisoara, Romania

Federico M. MAZZOLANI
University of Naples Federico II, Department of Structural Analysis and Design, Naples, Italy

Jordan MILEV
University. of Architecture and Civil Engineering of Sofia, Division of Steel and Timber Structures, Sofia, Bulgaria

Damian MARUSIC
University of Ljubljana, Faculty of Civil Engineering and Geodesy, Institute of Structures and Earthquake Engineering, Ljubljana, Slovenia

Rosario MONTUORI
University of Salerno, Department of Civil Engineering, Salerno, Italy

Dana PECTU
INCERC, Timisoara, Romania

Iztok PERUS
University of Ljubljana, Faculty of Civil Engineering and Geodesy, Institute of Structures and Earthquake Engineering, Ljubljana, Slovenia

Zdravko B. PETKOV
University. of Architecture and Civil Engineering of Sofia, Division of Steel and Timber Structures, Sofia, Bulgaria

Vincenzo PILUSO
University of Salerno,, Department of Civil Engineering, Salerno, Italy

André PLUMIÈR
University of Liège - Insitut du Genie Civil, Liège, Belgium

Nikolay RANGELOV
University. of Architecture and Civil Engineering of Sofia, Division of Steel and Timber Structures, Sofia, Bulgaria

Crt REMEC
University of Ljubljana, Faculty of Civil Engineering and Geodesy, Institute of Structures and Earthquake Engineering, Ljubljana, Slovenia

Gianvittorio RIZZANO
University of Salerno, Department of Civil Engineering, Salerno, Italy

Luis SANCHEZ RICART
University of Liège - Insitut du Genie Civil, Liège, Belgium

Peter SOTIROV
University. of Architecture and Civil Engineering of Sofia, Division of Steel and Timber Structures, Sofia, Bulgaria

Aurel STRATAN
Politehnica University of Timisoara, Civil Engineering Faculty, Department of Steel Structures and Structural Mechanics, Timisoara, Romania

Lucia TIRCA
INCERC, Timisoara, Romania

Ioannis VAYAS
National Technical University of Athens, Laboratory of Steel Structures, Athens, Greece

INTRODUCTORY REMARKS

Federico M. Mazzolani

SCOPE

An unexpected brittle failure of connections and, in same cases, of members occurred during the last earthquakes of Northridge and Kobe. This behaviour was not foreseen in the current engineering practice (Mazzolani, 1995; Mazzolani and Piluso, 1996) and the bad performance of steel joints and members seriously compromised the image of steel structures, which before was considered the most suitable material for seismic resistant structures (Mazzolani, 1995, 1998 & 1999; Bruneau et al, 1998).

For this reason, the international scientific community and, in particular, the earthquake prone Countries of the Mediterranean area and of the Eastern Europe were aware of the urgent need to investigate new topics (such as the influence of the strain rate on the cyclic behaviour of beam-to-column joints) and to improve the current seismic provisions consequently. In addition, the whole background of the modern seismic codes deserves to be completely reviewed in order to grasp the design rules, which failed during the last terrible seismic events. This revision is aimed at the up-dating of seismic codes and in particular at the improvement of Eurocode 8, whose application will be widespread in the next years during the so-called conversion phase.

In this perspective the European research project dealing with the **"Reliability of moment resistant connections of steel building frames in seismic areas"** (RECOS) has been sponsored in 1997 by the European Community within the INCO-Copernicus joint research projects of the 4th Framework Program. The aim of the research was to examine the influence of joints on the seismic behaviour of steel frames, bringing together knowledge and experience of different specialists from several Countries. It has been developed through the co-operation of the following partners coming from eight different European Countries.

Belgium: University of Liège (Prof. A. Plumier)

Bulgaria: University of Sofia (Prof. P. Sotirov)

France: Institut National de Science Appliqué of Rennes (Prof. J.M. Aribert)

Greece: National Technical University of Athens (Prof. I. Vayas)

Italy: University of Naples (Prof. Federico M. Mazzolani, convenor)
University of Catania (Prof. A. Ghersi)
University of Chieti (Prof. R. Landolfo)
University of Salerno (Prof. V. Piluso)

Portugal: Instituto Superior Tecnico of Lisbon (Prof. L Calado)

Romania: Polytechnic University of Timisoara (Prof. D. Dubina)
INCERC, Timisoara (Prof. V. Gioncu)
Romanian Academy (Prof. D. Mateescu)

Slovenia: University of Ljubljana (Proff. P. Fajfar & D. Beg)

The goal has been accomplished through the following objectives:

a) analysis and synthesis of research results, including code provisions, in relation with the evidence of the Northridge and Kobe earthquakes;
b) identification and evaluation by experimental tests of the structural performance of beam-to-column connections under cyclic loading;
c) setting up of sophisticated model for interpreting the connection response;
d) numerical study on the connection influence on the seismic response of steel buildings;
e) assessment of new criteria for selecting the behaviour factor for different structural schemes and definition of the corresponding range of validity in relation to the connection typologies.

In particular, the working program has been subdivided through the following tasks:

TASK 1 "Analysis of design criteria and seismic hazard"
TASK 2 "Evaluation of the material behaviour in extreme conditions"
TASK 3 "Ductility of members and connections"
TASK 4 "Reliable models for the cyclic behaviour of beam-to-column connections"
TASK 5 "Evaluation of global seismic performance"
TASK 6 "Failure mode and ductility demand "
TASK 7 "Evaluation of the q-factor"
TASK 8 "Preparation of an Application Document"

DEVELOPMENT OF ACTIVITY

The project activity started in June 1997 and has finished in November 1999 for a total of 30 months of extension. The activity work plan has been developed according to Table 1. In particular, two progress technical reports (TR) and a final report (FR) have been carried out. Six meetings (M) have been held in different Countries, where partial results and progress reports have been presented and discussed. The project has favoured a strong co-operation among partners, which has been supported by several exchanges of personnel and co-authored papers.

In the following the main developed activities are briefly reported, sub-dividing the dealt topics treated by each involved Country.

Belgium

The activity of the University of Liège in the Copernicus "RECOS" project is devoted to experimental tests for the evaluation of material behaviour in extreme condition, with particular reference to the influence of strain rate. The work involves two main groups of test: tests on small specimens and tests on full-scale specimens.

The *small specimen tests (Task 2)* concern four series of 5 specimens with the size of fillet welds being respectively 6 mm, 8 mm, 10 mm and 12 mm. These analyses are made to obtain information on both the effects of strain rate on the behaviour of welded connections made with fillet welds and the critical size of fillet welds in dynamic cyclic conditions. In each series, a standard tension test has been run for different values of strain rate.

TABLE 1

ACTIVITY WORKING PLAN

TASKS	*First year*		*Second year*		*Six-month extension*	*PARTNERS*
	June 1997 - May 1998		*June 1998 - May 1999*		*Until November 1999*	
Task 1						INCERC Timisoara Univ. Timisoara
Task 2						Univ. Liège Univ. Ljubljana
Task 3						INCERC Timisoara Univ. Naples Univ. Salerno Univ. Timisoara
Task 4						I.N.S.A. Rennes I.S.T. Lisbon Univ. Liège Univ. Ljubljana Univ. Sofia Univ. Timisoara
Task 5						N.T.U. Athens Romanian Academy Univ. Timisoara
Task 6						Univ. Catania Univ. Chieti Univ. Ljubljana Univ. Naples Univ. Salerno Univ. Timisoara
Task 7						INSA Rennes Univ. Naples Univ. Sofia Univ. Timisoara
Task 8						All Partners

Technical Report			TR1		TR2		FR
Meetings	M1 Kyoto	M2 Naples	M3 Rennes	M4 Sofia	M5 Ljubljana	M6 Timisoara	
Time scale	June97	Jan.98	June98	Nov.98	May 99	Sept. 99	

The *full scale specimen tests (Task 4)* refer to different types of typical European steel connections. The aim is to study the influence of the strain rate on the capacity to dissipate energy by plastic mechanism under cyclic loading. Tests have been done at constant deformation amplitude with two different loading speeds, the first with a period of 40 seconds and the second in order to simulate real earthquake condition, which corresponds to a period of 2.5 seconds. Three typologies of beam-to-column connections, namely extended end plate, welded dog-bone and bolted partial strength connections, have been tested for a total of 12 specimens. Different structural steel grades for beams have been used

Bulgaria

The activities of the University of Sofia are implemented into two main directions. The *Experimental study of moment-resisting connections (Task 4)* refers to tests on beam-to-column assemblies "extracted" from an especially designed two-bay four-storey regular moment-resisting frame. Two different beam-to-column joints have been considered. In both of them reduced flanges have been used for the improvement of the dissipative behaviour of the beams, according to the so-called "column tree" technique. A total of 6 specimens have been tested. The testing programme has been based on the ECCS recommendations. The hysteretic behaviour and the energy dissipation capacity of both connection types is investigated and compared to each other. Tests results have emphasised the effect of "haunching" of beams in the beam-to-column connections.

The *Evaluation of the behaviour factor of moment-resisting steel frames (Task 7)* focuses on the evaluation of q-factor by accounting for connection properties. The energy concept has been considered as the basis of this research. In particular, an energy approach to evaluate q-factor, which is a refined version of the one previously proposed in Japan by Akiyama and Kato, has been presented. A wide numerical analysis has been carried out showing the reliability of such a method. Then, another parametrical study has been devoted to investigate the differences in estimating q-factor through the most widely adopted methods for steel structures, based on various criteria. This task has been developed in co-operation with the University of Naples. The influence of strength, rigidity and ductility of rigid and semi-rigid connections and their mutual effect on the evaluation of q-factor for steel frames has been also evaluated. Both static pushover analyses and dynamic inelastic time history analyses, the latter by using three different earthquake records, have been performed.

Greece

The task under investigation at the National Technical University of Athens includes the *Investigation between local and global ductility for moment resisting frames (Task 5)*, developed in co-operation with the Romanian Academy and the Politehnica University of Timisoara. Appropriate numerical tools have been developed allowing for the study of the behaviour of frames subjected to monotonic and seismic loading. Then, such tools have been used for performing parametric studies with reference to a number of frame typologies. The frame response has been investigated by means of pushover monotonic analysis, pushover dynamic analysis that corresponds to monotonic analysis with up-dated dynamic characteristics of the structure, nonlinear dynamic analysis with structural evaluation with respect to serviceability and ductility criteria, nonlinear dynamic analysis with structural evaluation according to low-cycle-fatigue criteria. The effects of frame geometry, level of local ductility, level of vertical loading, joint flexibility, rotation capacity of connections and the susceptibility to low-cycle-fatigue have been considered.

France

The INSA of Rennes is involved into two different tasks. The *Experimental program on composite connections (Task 4)* is dedicated to the experimental investigation of the risk of degradation of the slab and the shear connection in composite joints under cyclic repeated loads. This program considers joints currently used for buildings, where beams are connected to column by means of bolted flush end plates, with solid or composite slabs cast in profiled steel sheeting connected in shear by welded headed studs or cold formed angles. The aim has been to assess first the damage of the shear connection and of the slab; then to evaluate the consequent effect on rotational stiffness, moment resistance, rotational capacity and dissipative capacity of the joint. To allow an easier interpretation of the experimental global results of the joints, complementary push-pull tests have been also performed. The complete experimental program involves 12 cycle tests on bare steel and composite beam-to-column bolted joints and 30 push-pull tests. Experimental results provide information about the actual cyclic moment-rotation behaviour of the joints and the degree of damage of shear connectors adjacent to the joints.

The *base shear approach (Task 7)* is a proposal of a new method for the evaluation of the q-factor. It has been developed in co-operation with the University of Timisoara. The main purpose of the research has been to propose a more significant general method based on column shear forces obtained from a dynamic inelastic analysis, which can determine a value of q-factor without any ambiguity, for a given structure and for any type of accelerogram. A wide parametrical numerical study has been performed, using the DRAIN-2DX computer code. Different types of building steel frames have been considered with various types of accelerogram records. Moreover, the new method has been also applied to structures with semi-rigid and partial-strength connections with different realistic characteristics.

Italy

Besides to the co-ordination of all the activities, the University of Naples, together with the other associated Italian Universities (Catania, Chieti and Salerno), has been mainly engaged in the development of theoretical and numerical aspects dealing with tasks 3 and 6. In particular, the following topics (sub-tasks) have been investigated.

The *design criteria and methods for semi-rigid steel frames (Task 6)* are devoted to state an analytical method in order to design seismic resistant structures able to promote a global-type collapse-mechanism also in case of semi-rigid beam-to-column joints. A sophisticated theoretical procedure, based on the extension of the kinematic theorem of plastic collapse to the concept of mechanism equilibrium curve, together with a wide parametrical analysis, is presented. The main related activities have been developed at University of Salerno.

The *evaluation of available ductility (Task 3)* investigates the prediction of plastic rotation supply of beam-to-column connections from a theoretical point of view by means of the component method. The analysis of bolted T-stubs is developed as the first step towards the prediction of the plastic deformation capacity of whole bolted joints. Then, results of an experimental program devoted to the validation of the theoretical model are presented and discussed. The activity concerned with this task has been carried out in co-operation with University of Salerno.

The *influence of hysteretic model of connections on the evaluation of ductility demand (Task 3)* is focussed on the assessment of the seismic response of steel frame building accounting for the actual hysteretic behaviour of beam-to-column connections. Reliable cyclic models for connections have been set up and implemented into computer programs for global structural analysis. The proposed model has been also calibrated on the basis of experimental results carried out at the University of Lisbon within the activity of this project. A wide numerical study has been carried out, emphasizing joint phenomenological aspects affecting the global response of the structure. The activity related to this task has been carried out in co-operation with the Universities of Chieti and Catania.

The *influence of the structural typology on the seismic performance of steel buildings (Task 6)* is devoted to study the effect of joint behaviour in terms of stiffness and resistance on the global building response. Several structural typologies, including dual frame structures, have been considered. For each typology, a number of joint characteristic levels have been accounted for. A wide parametric study has been therefore carried out. The obtained results focus on the effect of earthquake type and structural overstrength as well. The activity concerned with this task has been carried out in co-operation with the Politehnica University of Timisoara and the University of Chieti.

The *influence of building asymmetry (Task 6)* and the *evaluation of q-factor (Task 7)* have been performed in co-operation with the Universities of Ljubljana and Sofia, respectively. The corresponding main concerns are reported in the developed activity summary of these Countries.

Portugal

The Instituto Superior Tecnico of Lisbon has performed an *Experimental analysis on the effect of column size on the cyclic response of beam-to-column joints (Task 4)*. Specimens consisted of a beam attached to a column by means of either full-penetration welds or bolted cleat angles, which represent frequent solutions adopted in steel construction for beam-to-column connections. Specimens have been fabricated in a manner simulating field conditions, according to the procedures of workmanship and quality control as used by applicable Portuguese standards. Tests have been aimed at determining the effect of column member size on the hysteretic response of joint. Several types of loading histories have been considered, including constant amplitude deformation and ECCS testing methodology.

The Instituto Superior Tecnico of Lisbon also performed an unitary *re-elaboration of experimental data* carried out by all the partners involved in this project, in order to allow for a direct comparison of the obtained results. Such a comparison is mainly based on low-cycle fatigue endurance assessment. Therefore, fatigue strength lines are determined through a pre-defined procedure.

Romania

The Romanian research team, together with its collaborators, has been involved in the following tasks. *Analysis of design criteria and seismic hazard (Task 1)* comprises two sub-tasks: (1) *Comparison among codes*, developed in co-operation by the Polytechnic University and the Romanian Academy, focus on general concepts and rules provided by recent codes for the evaluation of base shear seismic force as well as on design criteria for MR frames. Comparison among different provisions and numerical examples are reported. Five different design codes have been considered for this comparison: EC8, PS-92 (French Code), UBC 1997, AISC 1997, AIJ 1990 and 1993. It has been evidenced that, if compared with the other existing codes, American seismic code (UBC 97) provides important and new provisions, particularly concerning the design criteria for MR frames. (2) *Influence of type of seismic motions and vertical components*, developed by INCERC, analyses the effects of earthquake type on the structural behaviour of steel frames. Particular emphasis is paid to the difference between near-source and far-source earthquakes. Artificial ground motions have been generated and their effects on the designed frames have been investigated under both horizontal and vertical components.

The *prediction by means of plastic yield line method (Task 3)* has been developed by INCERC. A special software has been elaborated allowing the available ductility of members and joints to be determined. The new format of this program is called DUCTROT M&J (DUCTility of ROTation for Members and Joints). The designer has the possibility to determine the ductility of both members and joints. The program is based on collapse plastic mechanism methods, where plastic zones and yielding lines are properly combined to predict the ultimate ductility of nodes, also accounting for the degradation due to the cyclic behaviour. Besides, the program gives the possibility to classify joints from the point of view of strength, rigidity and ductility, and to identify its weakest part. Comparisons with available experimental tests are presented.

The *experimental analysis on the effect of loading asymmetry (Task 4)* refers to several typologies of welded and bolted beam-to-column connections tested in bending in order to study the effect of loading type (symmetrical versus anti-symmetrical) under cyclic actions. The experimental work is based on a total number of 12 full scale specimens. In particular, it comprises 2 symmetric and 2 anti-symmetric cruciform extended end-plate bolted connections, 2 symmetric and 2 anti-symmetric cruciform welded connections, 2 symmetric and 2 anti-symmetric cruciform welded flanges and welded web plate (bolted for erection). The experimental activity has been developed at the Laboratory of the Polytechnic University of Timisoara.

The *ductility demand for semi-rigid joint frames (Task 5)* is developed in co-operation with the Romanian Academy. Two different MR frames, of 3-bay 6-storey and 3-bay 3storey with 3 different beam-to-column joints corresponding of those tested in the Timisoara Laboratory, have been parametrically analysed. In total there are 15 frames with homogenous and dual joint configurations, which have been investigated under the action of 19 different accelerograms corresponding to relevant records from Romania, Italy, Greece, Montenegro, Japan (Kobe) and USA (Northridge). The results expressed in terms of accelerogram multipliers, corresponding to different failure criteria, q-factor values, global and local damage index and the type of collapse mechanism are provided.

The *interaction between local and global properties (Task 5)* has been studied by the co-operation among the Polytechnic University of Timisoara, the Romanian Academy and the University of Athens. The *influence of structural typology (Task 6)* is developed in co-operation with the University of Naples. The *evaluation of q-factor by means of the base shear force approach (Task 7)* is developed in co-operation with INSA of Rennes. The corresponding main results are reported in the developed activity summary of these Countries.

Slovenia

The research team at the University of Ljubljana has been involved in full-scale tests on moment resistant connections, in order to assess the *Behaviour of welded and unsymmetrically bolted connections under dynamic loading (Task 4)*. Loading velocity has been fixed so that applied strain rate has been as much as possible close to that recorded during actual seismic conditions. Two test specimens with welded connections and two test specimens with unsymmetrical bolted connections have been analysed. One specimen with symmetric bolted connection with end-plate extended on both sides and one specimen with unsymmetrical bolted connection and thinner end-plate (semi-rigid connection) have been prepared additionally. Constant amplitude deformations have been imposed with a frequency of 0.5 Hz. At unsymmetrically bolted connections a preloading moment has been applied, it being calibrated on the effects of gravity loading which produces compression at the weaker side of the connection. The specimen with unsymmetrically bolted semi-rigid connection has been tested at higher loading frequency of 0.75 Hz and, in the last stage, even at 1 Hz. Also the magnitude of displacements has been increased during the test.

The experimental activity at the University of Ljubljana has also concerned the *Evaluation of strain rate effect on welds (Task 2)*. In fact, in order to get the information on the behaviour of welds at higher strain rates, a series of standard tensile tests on small speciemns containing welded joints within the gauge length have been executed. The following parameters have been investigated for different strain rate values: type of notch (notch in the weld, notch in HAZ, notch at the edge of the filled weld); temperature (20° C, 0° C, -20° C, -40° C, -60° C); filler metal and corresponding welding procedures.

As far as the numerical studies are concerned, the *Influence of building asymmetry on structural behaviour of steel frame buildings (Task 6)* has been investigated. The aim of the research has been to better understand the inelastic torsional behaviour of asymmetric buildings and to develop a relatively simple and reasonably accurate method for their non-linear analysis. The method represents an extension of the N2 method, which combines pushover analysis and response spectra approach and can provide an estimate of damage of structural elements. Inelastic seismic response of multi-storey steel moment-resisting frame buildings has been studied. The basic frame structures have been designed at the University of Naples. The reference structure is torsionally rigid. A torsionally flexible variant has been also studied. The symmetric structures have been designed according to Eurocode 8. Asymmetry has been introduced in plan by taking into account 5% accidental eccentricity, as proposed by Eurocode 8. In addition, larger mass eccentricity (10 and 15 %) has been also applied. The seismic response has been determined by non-linear dynamic analyses of spatial structural models subjected to six recorded ground motions. Simultaneous excitation in two horizontal directions has been applied. For comparison, uni-directional excitation has been also considered. Four different levels of ground motion intensity have been used. In addition, non-linear static ("pushover") analyses were performed. The obtained results allow several important conclusions to be drawn.

Finally, in order to allow a homogenisation among chapters dealing with the evaluation methods for q-factor, a special chapter has been devoted to the clarify the *basic concepts and main definitions (Task 7)*.

OBTAINED RESULTS

All the above activities have been developed under an effective co-ordination among all the partners, in order to finalise each activity within a given task, avoiding possible overlapping and lacks of homogeneity. Besides this introduction, nine main parts have been planned to organise the research report and to illustrate the obtained results. More or less, these chapters correspond to the project tasks previously mentioned, but some variants have been decided during the process in order to better finalise the scope of the project. For each one of these parts or sub-parts, one person is assumed to be responsible. In the following, a general list of contents is reported, together with the responsible persons (in brackets). This list practically corresponds to the chapters and sub-chapters of the present volume, which fulfil the object of task 8, devoted to the *preparation of an Application Document.*

1. SEISMIC INPUT AND CODIFICATION
   - 1.1 Analysis of design criteria and seismic hazard: Comparison among methods *(Dubina)*
   - 1.2 Influence of the type of seismic ground motions *(Gioncu)*

2. DUCTILITY OF MEMBER AND CONNECTIONS
   - 2.1 Prediction of available ductility by means of local plastic mechanism method: DUCTROT Computer Program. *(Gioncu)*
   - 2.2 Plastic deformation capacity of bolted T-stubs: theoretical analysis and testing *(Piluso)*

3. CYCLIC BEHAVIOUR OF BEAM-TO-COLUMN BARE STEEL CONNECTIONS
   - 3.1 Influence of strain rate *(Beg & Plumier)*
   - 3.2 Influence of column size *(Calado)*
   - 3.3 Influence of "haunching" *(Sotirov)*
   - 3.4 Influence of connection typology and loading asymmetry *(Dubina)*

4. CYCLIC BEHAVIOUR OF BEAM-TO-COLUMN COMPOSITE CONNECTIONS *(Aribert)*
   - 4.1 The present situation
   - 4.2 Research results from literature
   - 4.3 Cyclic behaviour of shear connectors
   - 4.4 Behaviour of full scale joints
   - 4.5 Conclusions

5. RE-ELABORATION OF EXPERIMENTAL RESULTS *(Calado)*

REFERENCES

Bruneau, M., Uang, C.M. and Whittaker, A. (1998) *Ductile Design of Steel Structures*, McGraw-Hill, New York.

Mazzolani F. M. (1995). Some simple considerations arising from the Japanese presentation of damages caused by the Hanshin earthquake. *Proc. of Int. Colloquium on "Stability of Steel Structures"*, Budapest, September.

Mazzolani F. M. (1998). Design of steel structures in seismic regions: the paramount influence of connections. *Proc. of COST C1 Int. Conference on "Control of the semi-rigid behaviour of civil engineering structural connections"*, Liège (Belgium), 17-19 Sept., 371-384.

Mazzolani, F.M. (1999). Principles of Design of Seismic Resistant Steel Structures. *Proc. of 4th National Conference on Steel Structures*, Ljubljana (Slovenia), May, 20th, 27-42.

Mazzolani, F.M. and Piluso, V. (1996) *Theory and Design of Seismic Resistant Steel Frames*, E & FN Spon, London.

Chapter 1

Seismic Input and Codification

1.1

ANALYSIS OF DESIGN CRITERIA AND SEISMIC HAZARD: COMPARISON AMONG CODES

Dan Dubina, Daniel Grecea, Aurel Stratan

INTRODUCTION

Seismic design provisions from six documents are compared in this research work:

- *Eurocode 8 – Design provisions for earthquake resistance of structures*. CEN European Committee for Standardisation, October 1994 (**EC8-94**).
- *Norme Francaise. Regles de construction parasismique PS 92*. Association Francaise de Normalisation (AFNOR), 1995 (**PS 92**).
- *Uniform Building Code, Volume 2, Structural Engineering Design Provisions*. International Conference of Building Officials, Whittier, California, USA (**UBC-97**).
- *Seismic Provisions for Structural Steel Buildings*. American Institute of Steel Construction, Inc. Chicago, Illinois, USA, 1997 (**AISC-97**).
- *Standard for Limit State Design of Steel Structures (draft)*. Architectural Institute of Japan, 1990 (**AIJ_{LSD}-90**).
- *Recommendations for Loads on Buildings*. Architectural Institute of Japan, June 1993 (**AIJ_L-93**).

The documents considered herein were chosen such as to capture as good as possible the "geographical" spread of codification in the field of earthquake engineering around the world. Unfortunately, it was not always possible to use the latest versions of specific codes, due to the fact that translations were not available at this time, as is the case of Japanese codes. Conversely, the US seismic design provisions are most up-to-dated and, more important, are issued after the recent devastating earthquakes of Northridge (1994) and Hyogoken-Nanbu (1995). These are expected to incorporate in some degree the considerable amount of information that is gathered after such events.

All the codes considered here make use of the limit state design philosophy. Anyway, they use not only different names when designating similar procedures (LRFD for US codes and Limit State Design in the other cases), but also different verification formats. For example, the safety format of Eurocode 8 is based on the partial safety factor method (level 1 format), while that for Japan is based on the first order second moment method (level 2 format), Kato (1995). Therefore, a direct comparison of various codes is quite difficult. But it will be of interest to observe the differences that will arise when applying design criteria of different codes.

It is to be noted here that AISC-97 and AIJ_{LSD}-90 are intended only as (seismic) design provisions for steel structures. The definition of seismic action is treated scarcely (AIJ_{LSD}-90), or directly referred as being given by specific provisions (AISC-97). In this way, UBC-97 and AIJ_L-93 are somehow complementary to the previous two. UBC-97 adopts the seismic design provisions for steel structures provided by AISC-94, with only minor changes. Therefore, in the case of American codes, for this study, UBC-97 was used when determining the seismic action, while AISC-97 provided specific design provisions for steel structures.

The Japanese codes are also complementary in respect to the problems covered, but are not as "synchronised" as the American ones, even if issued by the same institution. In fact, AIJ_{LSD}-90 treats the seismic action in a superficial way, further details the Building Standard Law (BSL) being referred. AIJ_L-93 is intended for determination of loads on buildings (including seismic loads) to be used in allowable stress design, ultimate strength design, and limit state design for various types of constructions. The specific values for the appropriate frequency of occurrence or load factors estimation are left open to individual design frame-works and only a few examples are given in the commentaries.

PURPOSE OF EARTHQUAKE CODES

It is important to understand the expressed or implied purpose of a particular design document in order to fully understand its provisions. Of course, the primary purpose of any earthquake code is to protect life. However, the way that this purpose, as well as any additional purposes, is presented, can provide additional insight into the reasons for the presence of specific provisions in the body of the document and its intended audience.

EC8–94 applies to the design and construction of buildings and civil engineering works in seismic regions. Its purpose is to ensure, that in the event of earthquakes:

- human lives are protected,
- damage is limited,
- structures important for civil protection remain operational.

The purpose also states that the random nature of the seismic events and the limited resources available to counter their effects are such as to make the attainment of these goals only partially possible and only measurable in probabilistic terms. The extent of the probabilistic protection that can be provided to different categories of buildings is a matter of optimal allocation of resources and is therefore expected to vary from country to country, depending on the relative importance of the seismic risks of other origin and on the global economic resources. To provide the necessary flexibility in this respect, Eurocode 8 contains a set of safety elements whose values are left to the National Authorities to decide so that they can adjust the level of protection to their respective optimal value. Special structures with increased risks for the population such as nuclear power plants and large dams are beyond the scope of EC8.

The main objective of PS 92 is to protect the human lives, with a small probability of structural collapse for a nominal level of the seismic action. A second important objective is the limitation of material damages, but in this second objective large incursions into the material plastic range are admitted. A proportion of less important structures can be unrepairable after a nominal earthquake. The provisions state that their objective is to design civil engineering works to severe ground motions, in regions of high seismicity, for conferring a satisfactory behaviour globally in order to assure the safety of human lives. They also try to limit the economic damages.

UBC–97 is a complete model building code. The purpose of its seismic provisions is to safeguard against major structural failures and loss of life not to limit damage or to maintain function.

The provisions of AISC–97 are intended for the design and construction of structural steel members and connections in the Seismic Force Resisting Systems in buildings for which the design forces resulting from earthquake motions have been determined on the basis of the various levels of energy dissipation in the inelastic range of response.

AIJ_{LSD}-90 are the only considered specifications that do not contain separate provisions for earthquake design. However, design to seismic action is present in the specification, which introduce the limit state design of steel structures. Anyway, no specific scope of these design provisions is presented within the body of the document.

AIJ_L-93 applies to the estimation of loads on ordinary buildings and similar structures, or parts thereof. Buildings must have a sufficient degree of safety against various loads. The required degree of safety and serviceability must be determined based on social and economical requirements.

Both AISC-97 and AIJ_{LSD}-90 are intended to provide design criteria for earthquake-resistant steel structures. The intensity of the seismic action is to be determined elsewhere. The rest of the codes incorporate (or are dedicated to) determination of earthquake load.

As mentioned earlier, their primary scope is to protect human life. All the codes assume the possibility of some degree of damage to structures under the design earthquake. UBC-97 is the less restrictive in what regards damage limitation. In fact, it states that its scope (beyond protection of life) is to safeguard against major structural failures only, and not to limit damage or to maintain function. PS 92 states the limitation of damage among its scopes, but at the same time recognises the probability that some structures will collapse under the nominal earthquake. Also EC8-94 states damage limitation, but reminds that attainment of its goals is measurable in probabilistic terms only. The Japanese AIJ_L-93 is the least specific in its statements, referring in a quite general way to the "social and economical requirements".

METHODS OF ANALYSIS

General

Equivalent static method

The equivalent static method of analysis may only be used to analyse simple regular buildings. The dynamic response of such systems is only used to distribute the total base shear within each frame (usually with an inverse triangulated load pattern), together with an amplification factor to allow for torsional effects. The system is then designed to resist the static loads induced from this distributed load pattern superimposed upon those arising from dead load and long term live loads. Members may be sized and detailed according to the resulting actions, or, if capacity design is required, as a basis for determining the flexural member capacity and the overstrength capacity that is used to ensure the appropriate hierarchy of strength attained.

Multi-modal analysis method

When more accurate representations of the actual excitation modes of the structure are required, either because of vertical irregularity or for other reasons, the multi-modal method of analysis can be employed. This method determines the actual period and mode shapes of the building that are used, in conjunction with the acceleration response spectra, as the basis for distributing base shear. The contribution from each mode is combined usually using the square root of sum of squares (SRSS) technique.

Time-history analysis

Time-history analysis techniques, although still only being used for major structures, are more widely available with the advent of more powerful, cheaper computing. The process involve developing a computer model of the structure, usually including both the elastically and the post-elastic characteristics of each element. The resulting model, together with the appropriately distributed seismic mass and the ground motion considered, is subjected to a step-by-step integration of the equations of motion.

Issues which require to be included in any such analysis are: the dynamic character and scaling effects of ground motion, the dynamic interaction between the interconnected lateral load resisting elements, and the actual inelastic response of the structural components being modelled. Techniques such as these are most commonly used as checking tools once the structural form has been proportioned in accordance with other methods.

Code provisions

EC8–94

The reference method for determining the seismic effects is the modal response analysis, using a linear-elastic model of a structure and the design spectrum $S_d(T)$. Depending on the structural characteristics of he building, one of the following two types of analysis may be used:

- the "simplified modal response analysis"
- the "multi-modal response spectrum analysis"

The simplified modal response analysis can be applied to buildings that can be analysed by two planar models and whose response is not significantly affected by contributions from higher modes of vibration. These requirements are deemed to be satisfied by buildings which meet the criteria for regularity in plan and in elevation and have fundamental period of vibration T_l in the two main directions less than the following values: $T_l \leq 4T_C$; $T_l \leq 2.0$s.

The multi-modal response spectrum analysis is applicable to all types of buildings. For buildings complying with the criteria for regularity in plan, the analysis may be performed using two planar models, one for each main direction. The response of all modes of vibration contributing significantly to the global response shall be taken into account. This can be demonstrated by: demonstrating that the sum of the effective modal masses for the modes considered amounts to at least 90% of the total mass of the structure, or by demonstrating that all modes with effective modal masses greater than 5% of the total mass are considered. Whenever all relevant modal responses can be regarded as independent from each other ($T_j \leq 0.9T_i$), the maximum value E_E of a seismic action effect may be taken as:

$$E_E = \sqrt{\sum E_{Ei}^2}$$

where:
E_E – seismic action effect under consideration.
E_{Ei} – value of this seismic action effect due to the vibration mode i.

In addition to actual eccentricity, in order to cover uncertainties in the location of masses and the spatial variation of the specific motion, the calculated centre of mass at each floor i shall be considered displaced from its nominal location in each direction by an additional accidental eccentricity:

$$e_{1i} = \pm 0.05\ L_i$$

where:
e_{1i} – accidental eccentricity of storey mass i from its nominal location, applied in the same direction in all floors.
L_i – floor dimension perpendicular to the direction of the seismic action.

As alternative to these basic methods, other methods of structural analysis are allowed:

- power spectrum analysis
- (non-linear) time history analysis
- frequency domain analysis

PS 92

Structures that are not presenting significant irregularities are conventionally classified as regular structures and structures of intermediate regularity.
Regular structures can be calculated by considering a deformed shape based on simple analytical expression. These structures have to respect the following:

- There must be not significant coupling between horizontal and vertical degrees of freedom.
- For each of the two principal vertical planes of the structure, it must be possible to be reduced to a system with a single mass on each storey.
- The structure has to have at least three non-concurrent braced planes.
- The horizontal diaphragms or floors must posses sufficient rigidity in their plane in order to be considered undeformable in that plane.
- The structure must be regular in plan and in elevation.

In the case of structures of intermediate regularity, the deformed shape must be effectively computed or determined approximately. If no other method is available, the deformed shape may be determined by considering the infinitely elastic structure and applying to each mass appropriate (horizontal or vertical) force proportional to its weight.

All other structures that do not meet the above criteria must be computed using modal analysis on a three-dimensional model. For each direction, the computation of modes of vibration must be accomplished up to a frequency of 33 Hz (a period of 0.03 s). The computation of modes can be stopped if the sum of effective modal masses $\sum M_i$ for the considered direction reached 90% of the total vibrating mass of the structure. In no case the number of modes considered must be inferior to three. If for a frequency of 33 Hz (a period of 0.03 s) the sum of effective modal masses is inferior to 90% of the total mass, account must be taken for the neglected modes of vibration. This can be done by multiplying all the variables of interest (forces, displacements, etc.) obtained by the combination of modal responses by the factor $M/\sum M_i$ .Two modes of vibration are considered independent if:

$$\frac{T_i}{T_j} \leq \frac{10}{10+\sqrt{\zeta_i \cdot \zeta_j}}$$

where ζ_i and ζ_j are the relative damping ratios, expressed in percentage, of the two modes.

If the modal responses can be considered as independent, the combination can be accomplished according to the formula:

$$S = \pm\sqrt{\sum S_i^2}$$

where S is the value to be computed and S_i is its maximum value in the i-th mode.

In order to take account for the torsional effects of the vertical axis, masses must be introduced in a manner that will translate the centre of gravity from the theoretical one in both directions by the values:

$$e_x^{'} = 0.05 \cdot L_{rx}$$
$$e_y^{'} = 0.05 \cdot L_{ry}$$

UBC-97

Any structure may be and certain structures shall be designed using dynamic lateral-force procedures.

The simplified static lateral-force procedure may be used for the following structures of Occupancy Category 4 or 5:

- Buildings of any occupancy (including single-family dwellings) not more than three stories in height excluding basements, that use light-frame construction.
- Other buildings not more than two stories in height excluding basements.

The static lateral force procedure may be used for the following structures:

- All structures, regular or irregular, in Seismic Zone I and in Occupancy Categories 4 and 5 in Seismic Zone 2.
- Regular structures under 73 m in height.
- Irregular structures not more than five stories or 20 m in height.
- Structures having a flexible tipper portion supported on a rigid lower portion

The dynamic lateral-force procedure shall be used for all other structures, including the following:

- Structures 73 m or more in height.
- Structures having a stiffness, weight or geometric vertical irregularity of Type 1, 2 or 3.
- Structures over five stories or 20 m in height in Seismic Zones 3 and 4 not having the same structural system throughout their height.
- Structures, regular or irregular, located on Soil Profile Type S_F that have a period greater than 0.7 seconds. The analysis shall include the effects of the soils at the site.

Response spectrum analysis is an elastic dynamic analysis of a structure utilising the peak dynamic response of all modes having a significant contribution to total structural response. Peak modal responses are calculated using the ordinates of the appropriate response spectrum curve, which correspond to the modal periods. Maximum modal contributions are combined in a statistical manner to obtain an approximate total structural response.

The corresponding response parameters, including forces, moments and displacements, may be reduced for purpose of design, such that the corresponding design base shear is not less than 80-100% of the base shear determined by the static force procedure.

The requirement that all significant modes be included may be satisfied by demonstrating that for the modes considered, at least 90 percent of the participating mass of the structure is included in the calculation of response for each principal horizontal direction.

The peak member forces, displacements, story forces, story shears and base reactions for each mode shall be combined by recognised methods. When three-dimensional models are used for analysis, modal interaction effects shall be considered when combining modal maxima.

The time-history analysis is an analysis of the dynamic response of a structure at each increment of time when the base is subjected to a specific ground motion time history.

Where diaphragms are not flexible, the mass at each level shall be assumed to be displaced from the calculated centre of mass in each direction a distance equal to 5 percent of the building dimension at that level perpendicular to the direction of the force under consideration. The effect of this displacement on the story shear distribution shall be considered.

The torsional design moment at a given story shall be the moment resulting from eccentricities between applied design lateral forces at levels above that story and the vertical-resisting elements in that story plus an accidental torsion.

Where torsional irregularity exists, the effects shall be accounted for by increasing the accidental torsion at each level by an amplification factor A_x, determined from the following formula:

$$A_x = \left[\frac{\delta_{max}}{1.2\delta_{avg}} \right]^2$$

where:
δ_{avg} - the average of the displacements at the extreme points of the structure at Level x.
δ_{max} - the maximum displacement at Level x.

The value of A_x need not exceed 3.0.

AIJ_{LSD}–90

For the ultimate limit state design, as a rule, the design of braced and unbraced frames that are classified into S-I or S-II shall be based on plastic analysis. The design of braced and unbraced frames that are classified into S-III or S-IV is to be based on elastic analysis.

The serviceability limit state design is to be based on elastic analysis.

In the design of structures with torsional and/or structures whose vertical stiffness distribution is not uniform, considerations shall be given to the influence of these effects on structural performance in the design process.

AIJ_L–93

Load effects in terms of the stress or deformation are obtained from structural analysis based on estimated loads. Structural analysis methods and procedures are not specified in recommendations, but loads are estimated, in principle, for static analysis. Dynamic loads caused by strong winds or earthquake ground motions are evaluated for design as equivalent static loads.
Seismic loads acting on buildings are evaluated according to the following procedure:

1. The building is idealised as an appropriate vibrational system.
2. The natural periods and vibration modes of the system are computed.
3. The horizontal seismic loads are evaluated by the response spectrum method.

In the response spectrum method, the maximum response can be estimated by the superposition method of "square root of sum of squares", taking the predominant vibration modes into account. The number of vibration modes taken into account may usually be takes as three or less, even if the number of degrees of freedom is greater than three.

This procedure may be simplified for low-rise buildings of small scale with regular geometry, mass distributions and rigidity distributions.

For specific buildings described below, the dynamic analysis shall be used in addition to the static one:

- where damage tends to concentrate on particular stories because of non-uniform vertical distributions of rigidity and strength.
- where torsional effect can not be ignored because of non-uniform horizontal distributions of mass and rigidity.
- where there are horizontal long-span structural members and the effect of vertical vibration must be taken into account.
- where the building has no simple multi-stories but has a special configuration, such as a shell structure.
- where large number of people assemble, such as in high-rise buildings.
- where the building is especially important and the response of installed equipment must be computed, such as for a nuclear power plant.
- where damage to the building affects the environment to a great degree, such as buildings for storing dangerous materials.
- where a newly-developed special construction method is used, such as a base-isolation device.

Comparative Remarks

The three analysis methods specified above, and namely, the equivalent static, multi-modal and time-history analyses are present in all the codes, except in AIJ_{LSD}–90, where only the equivalent static method is specified. PS 92 does not mention the time-history type of analysis, while EC8–94 allows also for other two methods, power spectrum analysis and frequency domain analysis, but no specific descriptions are made. Both PS 92 and UBC-97 allow for a simplified analysis method for certain categories of structures, that make use of arbitrary distribution of the base shear force.

The choice of the analysis method generally depends on structural regularity, possibility to neglect the influence of higher modes and sometimes importance of the building. The default method of analysis is the equivalent static for UBC-97 and AIJ_{LSD}–90, and the modal analysis for EC8–94, PS 92 and AIJ_L–93. The US code requires that certain structures be analysed by dynamic methods, included here are both multi-modal and time history. The only provision that requires a time-history analysis in addition to the multi-modal one is the Japanese AIJ_L–93, though specific features of this type of analysis such as choice of accelerograms, modelling requirements and interpretation of results, are not treated at all. Some provisions of this kind are present in Eurocode 8 and the US code.

EC8–94 chooses the simplified analysis for structures regular in plan and in elevation and with a limitation on the fundamental period of vibration. In order to adopt the equivalent static method, structural regularity and possibility to reduce the model of the structure to a MDOF cantilever model is required by PS 92. Seismic zone, height, and structural irregularities are the main parameters for adoption of simplified static analysis in UBC-97. The equivalent static procedure is the only one covered by AIJ_{LSD}–90. The choice of the analysis procedure in AIJ_L–93 is governed by structural regularity, building importance category, structure type, and need to take account for vertical component.

A specific characteristic of the Japanese provisions is that structural analysis for ULS for certain types of structures is based on plastic analysis.

EQUIVALENT SEISMIC FORCES

General

The implicit method of analysis in UBC-97 and AIJ_{LSD}–90 is the equivalent static one, also EC8–94 and PS 92 easily defaults to it. Only AIJ_{L}–93 has the multi-modal analysis as the default one. Consequently, the equivalent static formulation was used in order to compare the equivalent shear forces provided by the codes, except for AIJ_{L}–93, where the modal analysis was considered.

Generally speaking there are two methods used to determine the equivalent seismic forces in a structure:

- the direct method, when seismic forces are determined directly for each storey (each DOF), and
- the indirect method, when the base shear force is determined first and is distributed afterwards to all the storeys (all DOF) of the structure.

It has to be noted that these two methods differ only in formulation, having the same result.

The equivalent seismic forces are determined quite imperatively in the most of seismic codes. The principles used by the code drafters and their considerations are usually presented in little details or are not presented at all to the code user. Also, the large variety of concepts and formulations in different codes makes the comparison more difficult. Frequently, very different aspects are presented using similar coefficients. Because of these, access to the theoretical bases of the codes should be possible.

The equivalent seismic force F_{tot} is determined in the most of the seismic design codes in the following general form:

$$F_{tot} = \alpha_1 \cdot \alpha_2 \cdot \alpha_3 \cdot \alpha_4 \cdot \alpha_5 \cdot \alpha_6 \cdot \alpha_7 \cdot \alpha_0 \cdot M$$

where the coefficients have the following meaning:

α_1 – ***Zone factor***; most countries are divided into seismic zones with specific coefficients. By multiplying this coefficient with the reference value of the peak ground acceleration, the peak design ground acceleration is obtained.

α_2 – ***Dynamic amplification coefficient***; takes into consideration the fact that the oscillation response of the building is generally higher than the base excitation. It depends on the period of vibration of the building and considers the general shape of the design spectrum. For conditions of California, for example, and for a 5% damping, the dynamic coefficient has an average value of 2.12, in the period range from 0.14 to 0.5 s.

α_3 – ***Soil factor***; the foundation soil can be engaged into an independent oscillation, so that the amplitudes and frequencies at the base of the building are considerably different than those at the base rock. Therefore, the acceleration at the base of the building will be affected by a coefficient of the foundation soil, which depends on its dynamic characteristics (soil stiffness, thickness of soil layers, soil period of vibration) and the period of vibration of the building.

α_4 –***Damping coefficient***; takes into consideration the difference between the values of damping coefficient considered as being generally 5% of critical damping and the one of the building.

α_5 –***Seismic action reduction factor***; a ductile behaviour of the structural system leads to a reduction of the equivalent horizontal forces. In some codes, the reciprocal value of the coefficient is used ($1/\alpha_5$).

α_6 –***Risk factor***; the estimated average of damages suggests ways of total costs reduction (initial plus repairing). When the estimated damage is too large, the equivalent seismic force is increased through the risk factor, and in this way the design return period is increased (e.g. theatres, etc).

α_7 –***Importance factor***; depending on the importance of the building, for example buildings whose integrity during earthquakes is of vital importance, the equivalent seismic force is increased through the importance factor. Due to the fact that the importance factor has the same effect as the risk factor, many codes associate them.

α_0 – ***Ground acceleration***; is considered as the basic value of the peak ground acceleration at the reference soil.

M – ***Total mass of the building***; is the mass corresponding to the building dead and variable loads W ($M=W/g$, where g is the gravitational acceleration).

Depending on the design methods and code, some of these coefficients are partially or totally superposed, other are used as unique terms. It is important that using some of the coefficients (such as the importance and risk factors), the return period is modified, without being evident.

Code provisions

Most of the parameters used to describe the evaluation of the seismic shear forces in the equivalent static analysis are also valid for multi-modal and time-history analyses. A brief overview of the analysis procedures and the coefficients α_0 - α_7, for the computation of the seismic shear forces, is presented in TABLE 1.1.10. Aspects of the design considered here are:

- Design response spectra, as characterising the earthquake ground motion.
- Formulations of design seismic shear forces, directly or from the response spectrum.
- Seismic coefficients α_0 - α_7 as defined above.
- Return period of the design earthquake.
- Simple rules for determination of the fundamental period of vibration for simplified analysis.
- Topographic amplification effects, if present.
- Torsional effects to be taken into account for the simplified analysis.
- Evaluation of displacements in the inelastic system under seismic forces of the design earthquake.
- Vertical component of the seismic action.
- Account for second-order (P-Δ) effects.
- Combination of horizontal and vertical components of the seismic action.
- Determination of the seismic mass
- Seismic combination of loads for safety checks

EC8–94

The seismic hazard is described in terms of a single parameter, the value of a_g of the effective peak ground acceleration in rock or firm soil. The concept of "effective peak ground acceleration" is an attempt to compensate for inadequacy in general of the actual single peak to describe the damaging potential of the ground motion in terms of maximum acceleration and/or velocity induced to the structures. There is not a unique established definition and corresponding techniques for deriving a_g from the ground motion characteristics.

The design ground acceleration corresponds to a reference return period of 475 years. To this reference period an importance factor γ_I is assigned. Seismic zones with a design ground acceleration a_g not greater than $0.1g$ are low seismicity zones, for which reduced or simplified seismic design procedures for certain types or categories of structures may be used.

In seismic zones with a design ground acceleration a_g not greater than $0.04g$ the provisions of Eurocode 8 need not be observed.

The earthquake motion at a given point of the surface is generally represented by an elastic ground acceleration response spectrum called elastic response spectrum. The horizontal seismic action is described by two orthogonal components considered as independent and represented by the same response spectrum. The vertical component of the seismic action should be represented by the response spectrum as defined for the horizontal seismic action but with the reduced ordinates (see TABLE 1.1.10).

For special conditions more than one spectrum may be needed to adequately represent the seismic hazard over an area. This may be necessary when the earthquakes affecting the area are generated by sources differing widely in distance, focal mechanism or travel path geology, as in the case of shallow depth and intermediate depth earthquakes. In such circumstances, different values of a_g as well as different shapes of the response spectrum for each type of earthquake would normally be required.
For important structures in high seismicity zones it is recommended to consider topographic amplification effects.

Alternative representations of the earthquake motion, e.g. power spectrum or time history representation, may be used.

The elastic response spectra $S_e(T)$ for the reference return period and different subsoil conditions are shown in Figure 1.1.1. Values of the parameters that define the elastic spectrum are selected so that the ordinates of the elastic response spectrum have a uniform probability of exceedance over all periods (uniform risk spectrum) equal to 50%.

The influence of local ground conditions on the seismic action is accounted for by considering the three subsoil classes A, B, C, described by the following stratigraphic profiles (see TABLE 1.1.1):

TABLE 1.1.1
CLASSIFICATION OF SUBSOIL CONDITIONS IN EC8-94

Subsoil class	Description
A	- Rock or other geological formation characterised by a shear wave velocity v_s, of at least 800 m/s, including at most 5 m of weaker material at the surface. - Stiff deposits of sand, gravel or overconsolidated clay, at least several tens of m thick, characterised by a gradual increase of the mechanical properties with depth and by v_s values of at least 400 m/s at a depth of 10 m.
B	- Deep deposits of medium dense sand, gravel or medium stiff clays with thickness from several tens to many hundreds of m, characterised by values of at least 200 m/s at a depth of 10 m, increasing to at least 350 m/s at a depth of 50 m.
C	- Loose cohesionless soil deposits with or without some soft cohesive layers, characterised by v_s values below 200 m/s in the uppermost 20 m. - Deposits with predominant soft-to-medium stiff cohesive soils, characterised by v_s values below 200 m/s in the uppermost 20 m.

Additions and/or modifications to this classification may be necessary to better conform to special soil conditions.

For sites with ground conditions not matching the three subsoil classes A, B, C, special studies for the definition of the seismic action may be required.

The capacity of structural systems to resists seismic action in the non-linear range generally permits their design for forces smaller then those corresponding to a linear elastic response. To avoid explicit non-linear structural analysis in design the energy dissipation capacity of the structure through mainly ductile behaviour of its elements and/or other mechanisms is taken into account by performing a linear analysis based on a response spectrum reduced with respect to the elastic one, henceforth called design spectrum. This reduction is accomplished by introducing the behaviour factor q. Additionally, modified exponents k_{d1} and k_{d2} are generally used. The behaviour factor q is an approximation of a ratio of the seismic forces that the structure would experience if its response was completely elastic with 5% viscous damping to the minimum seismic forces that may be used in design – with a conventional linear model - still ensuring a satisfactory response of the structure. The values of the behaviour factor q, which also accounts for the influence of the viscous damping being different from 5% are given for the various materials and structural systems and according to various ductility levels. The non-dimensional design response spectra of EC8-94 are shown in Figure 1.1.2, for a behaviour factor q=6.

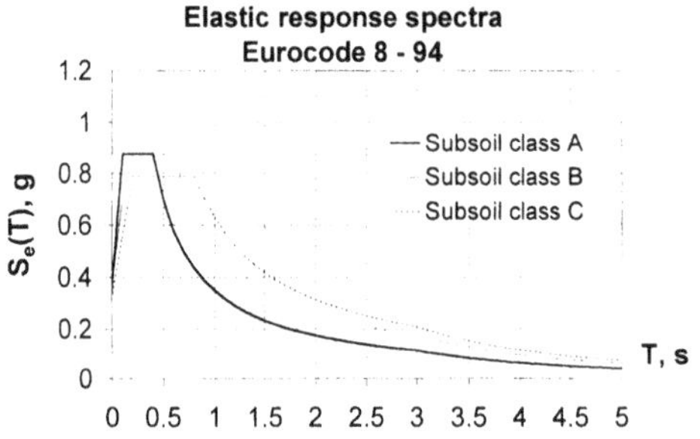

Figure 1.1.1: Elastic response spectra of EC8-94. a_g =0.35g

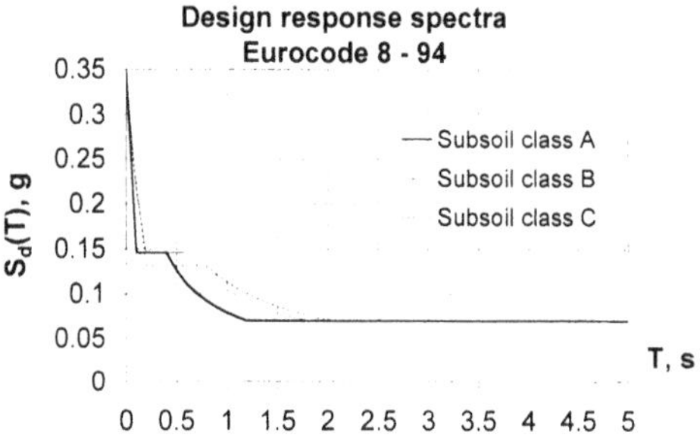

Figure 1.1.2: Design response spectra of EC8-94. a_g =0.35g; q=6

The aspect of seismic hazard shall be taken into consideration in the early stages of the conceptual design of the building. The guiding principles governing this conceptual design against seismic hazard are:

- structural simplicity
- uniformity and symmetry
- redundancy
- bi-directional resistance and stiffness
- torsional resistance and stiffness
- diaphragmatic action at storey level
- adequate foundation

For the purpose of seismic design, building structures are distinguished as regular and non-regular. This distinction has implications on the following aspects of the seismic design:

- the structural model, which can be either a simplified planar or spatial one
- the method of analysis, which can be either a simplified modal or a multi-modal one
- the value of the behaviour factor q, which can be decreased depending on the type of non-regularity in elevation

With regard to the implications of structural regularity on the design, separate consideration is given to the regularity characteristics of the building in plan and in elevation.

Buildings are generally classified into 4 importance categories that depend on the size of the building, on its value and importance for the public safety and on the possibility of human losses in case of a collapse. The importance categories are characterised by different importance factors γ_I.

The definitions of the importance categories and the related importance factors are given in TABLE 1.1.2. Different values of γ_I may be required for various seismic zones of a country.

TABLE 1.1.2
IMPORTANCE CATEGORIES AND IMPORTANCE FACTORS FOR BUILDINGS IN EC8-94

Importance category	Buildings	Importance factor γ_I
I	Buildings, whose integrity during earthquakes is of vital importance for civil protection, e.g. hospitals, fire stations, power plants, etc.	1.4
II	Buildings, whose seismic resistance is of importance in view of the consequences associated with the collapse e.g. schools, assembly halls, cult, institutions, etc.	1.2
III	Ordinary buildings, not belonging to the other categories	1.0
IV	Buildings of minor importance for public safety, e.g. agricultural buildings, etc.	0.8

PS 92

The minimum level of seismic hazard to be used in the design process of a structure is conventionally specified by means of a unique parameter a_N (nominal acceleration). For a given nominal acceleration, seismic hazard, in relation to the macroseismic intensity, depends on the form of the associated normalised spectrum (i.e. nature of the soil). For the same spectrum and the same acceleration level, seismic hazard of a real earthquake depends on the duration of the seismic motion, which is taken into account here in simplified manner by a_N. Interim values of the nominal accelerations are:

TABLE 1.1.3
NOMINAL ACCELERATIONS, IN M/S^2

Seismic zone	Building importance class			
	A	B	C	D
0	/	/	/	/
Ia	/	1.0	1.5	2.0
Ib		1.5	2.0	2.5
II	/	2.5	3.0	3.5
III	/	3.5	4.0	4.5

Seismic provisions shall not be applied nor to zone 0, neither to buildings of class A. Except for special need, structures must not be constructed in the immediate vicinity of active faults.

The soil motion is considered as being the resultant of:

- a concomitant translational movement, when all the points of the soil surface are engaged in the same instant motion;
- differential movement, when distance between the specified points is considered.

The translational movement is defined by three components: two orthogonal horizontal components, and a vertical component. Each component is characterised by a response spectrum in terms of acceleration, from which a design response spectrum is derived. The same spectrum is used for the two horizontal components. Vertical component, if otherwise specified, is considered as of an intensity equal to 70% of the horizontal one. The non-dimensional elastic response spectra for the horizontal components are given in Figure 1.1.3.

Soils are classified into four categories, function of their mechanical properties:

- rock
- category **a**: soils of good and very good resistance
- category **b**: soils of intermediate resistance
- category **a**: soils of weak resistance

Four site classes are defined from the above soil categories, according to TABLE 1.1.4:

TABLE 1.1.4

CLASSIFICATION OF SUBSOIL CONDITIONS IN PS 92

Site class	Description
S0	rock sites (reference site) soils of category **a** less than 15m depth
S1	soils of category **a** more than 15m depth soils of category **b** less than 15m depth
S2	soils of category **b** of depth between 15m and 50m soils of category **c** less than 10m dept
S3	soils of category **b** more than 50m depth soils of category **c** of depth between 10m and 100m

The non-dimensional design response spectra are given for a 5% damping and unit nominal acceleration. These are shown in Figure 1.1.4. Design response spectrum $R_D(T)$ is obtained from the elastic one by adopting a plateau corresponding to the maximum amplification of ground acceleration from zero up to the control period T_C, and by amplifying the ordinates of the descending branches of the spectrum. The design response spectrum is an elastic one, i.e. it is not divided by the structural behaviour factor q.

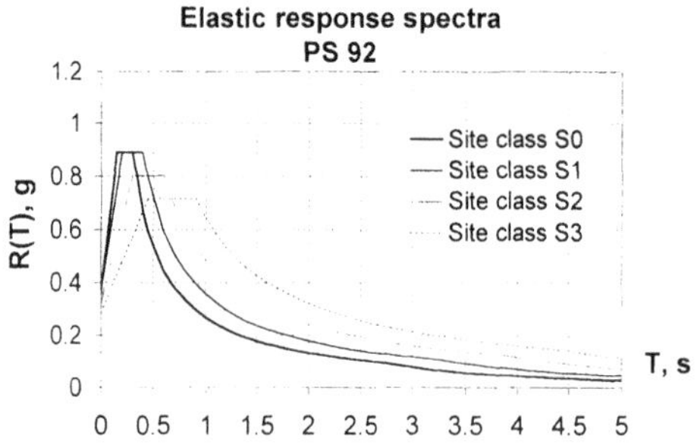

Figure 1.1.3: Elastic response spectra of PS 92. $a_N = 0.36g$

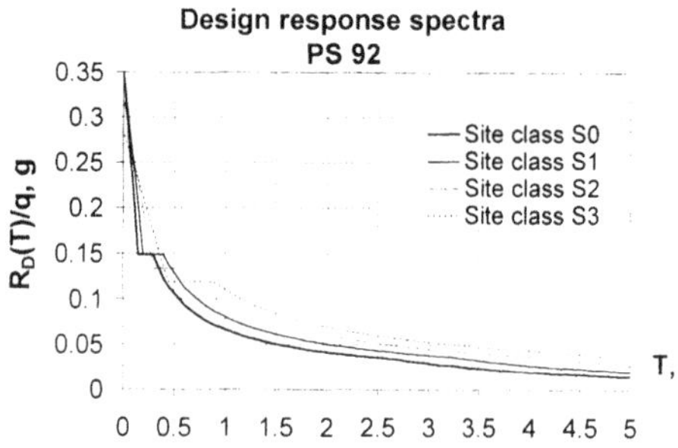

Figure 1.1.4: Design response spectra of PS 92. $a_N = 0.36g$

Account is taken of a coefficient τ of topographic amplification, for buildings constructed near ridges. Seismic action acting on a structure may be considered as composed of:

- dynamic forces due to inertia of masses subjected to the translational movement of the soil beneath the structure
- displacements directly imposed to the structure or its foundation due to differential movement of the soil, considered to be applied statically
- forces developed by torsional oscillations of vertical axis due to differential movement of the soil
- dynamic pressures on the structure coming from earth or water, or from solid or liquid materials it may contain

Soil-structure interaction may be modelled by springs and dampers in relation to nature and number of DOF chosen, and with respect to the position of the foundation relative to the soil surface. In a simplified manner, when a building has an infrastructure, a design height may be considered, which is defined by the heights of the superstructure, infrastructure, and the nature of the foundation soil. If denoting by H_0 the height of the superstructure, and by H_I the height of the infrastructure, the design height H is given by:

- $H=H_0$ in the case of category **a** foundation soil
- $H=H_0+H_I/2\leq 1.5H_0$ in the case of category **b** foundation soil
- $H=H_0+H_I\leq 2.0H_0$ in the case of category **c** foundation soil

In the design process the structure is considered infinitely elastic, whichever the intensity of the seismic action. Displacements of the structure are considered equal to those computed for the fictive elastic model, using the design spectrum. The design forces are obtained from those computed on the elastic model, dividing by a behaviour coefficient q. This coefficient global for a structure, is chosen on:

- the nature of the materials used,
- type of construction,
- possibility to redistribute forces in the structure, and
- the deformation capacity of elements in the post-elastic range.

Values of the behaviour coefficient q are given for different regularity classes and structural materials used. A reduced behaviour factor is used in the case of short periods (inferior to the control period T_B), and for the vertical component of the seismic action (see TABLE 1.1.11).

Buildings are classified into four risk classes, according to TABLE 1.1.5.

TABLE 1.1.5

CLASSES OF RISK FOR BUILDINGS IN PS 92

Risk classes	Buildings
Class A	Structures that represent a minimum risk for persons or economical activity
Class B	Structures and equipment of "current" risk for persons
Class C	Structures of high risk for persons due to their great repeatability or social and economical importance
Class D	Structures and equipment whose security is of primary importance due to civil security, public order, defence, and survive of the region

UBC-97

The procedures and the limitations for the design of structures shall he determined considering seismic zoning, site characteristics, occupancy, configuration, structural system and height in accordance with this section. Structures shall be designed with adequate strength to withstand the lateral displacements induced by the design basis ground motion, considering the inelastic response of the structure and the inherent redundancy, overstrength and ductility of the lateral-force resisting system. The minimum design strength shall be based on the design seismic forces. For purpose of earthquake-resistant design, structures shall be placed in one of the occupancy categories listed in TABLE 1.1.6.

TABLE 1.1.6
OCCUPANCY CATEGORIES OF STRUCTURES IN UBC-97

Occupancy category	Occupancy or function of structure	Seismic importance factor, *I*
1. Essential facilities	Occupancies having surgery and emergency treatment areas	1.25
2. Hazardous facilities	Occupancies and structures housing or supporting toxic or explosive chemicals or substances	1.25
3. Special occupancy categories	Occupancies used for college or adult education	1.00
4. Standard occupancy categories	All structures housing occupancies or having functions not listed in category 1, 2, 3 and 4	1.00
5. Miscellaneous structures	Group U occupancies except for towers	1.00

Each site shall be assigned a soil profile type based on properly substantiated geotechnical data using the site categorisation procedure. Soil profile types S_A, S_B, S_C, S_D and S_E are defined in TABLE 1.1.7 and soil profile type S_F is defined as soils requiring site-specific evaluation as follows:

- Soils vulnerable to potential failure or collapse under seismic loading, such as liquefiable soils, quick and highly sensitive clays, and collapsible weakly cemented soils.
- Peats and/or highly organic clays, where the thickness of peat or highly organic clay exceeds 3 m
- Very high plasticity clays with a plasticity index, PI > 75, where the depth of clay exceeds 7 m.
- Very thick soft/medium stiff clays, where the depth of clay exceeds 36 m.

TABLE 1.1.7
SOIL PROFILE TYPES IN UBC-97

Soil profile type	Generic description	Shear wave velocity, v_s, m/s
S_A	Hard rock	>1500
S_B	Rock	760 to 1500
S_C	Very dense soil and soft rock	360 to 760
S_D	Stiff soil profile	180 to 360
S_E	Soft soil profile	<180
S_F	Soil requiring site-specific evaluation	

Seismic hazard characteristics for the site shall he established based on the seismic zone and proximity of the site to active seismic sources, site soil profile characteristics and the structure's importance factor. Each site shall be assigned a seismic zone and a seismic zone factor Z, in accordance with TABLE 1.1.8. In seismic zone 4, each site shall be assigned a near-source factor and the seismic source type.

TABLE 1.1.8
SEISMIC ZONE FACTOR *Z* IN UBC-97

Zone	1	2A	2B	3	4
Z	0.075	0.15	0.20	0.30	0.40

Each structure shall be designated as being structurally regular or irregular. Structures shall be designed for ground motion producing structural response in any horizontal direction. The elastic design response spectrum for 5% equivalent viscous damping to be used in dynamic (multi-modal) analysis only is given in Figure 1.1.5.

The following earthquake loads shall be used in the load combinations:

$$E=\rho E_h+E_v$$
$$E_m=\Omega_0 E_h$$

where:
E - the earthquake load on an element of the structure resulting from the combination of the horizontal component, E_h, and the vertical component, E_v.
E_h - the earthquake load due to the base shear, V.
E_m - the estimated maximum earthquake force that can be developed in the structure.
E_v - the load effect resulting from the vertical component of the earthquake ground motion and is equal to an addition of $0.5C_aID$ to the dead load effect, D.
Ω_0 - the seismic force amplification factor that is required to account for structural overstrength.
ρ - reliability/redundancy factor as given by the following formula:

$$\rho = 2 - \frac{6.1}{r_{\max}\sqrt{A_B}}$$

where:
r_{max} - the maximum element-story shear ratio. For a given direction of loading, the element-story shear ratio is the ratio of the design story shear in the most heavily loaded single element divided by the total design story shear. For any given story level i, the element-story shear ratio is denoted as r_i. The maximum element-story shear ratio r_{max} is defined as the largest of the element story shear ratios, r_i, which occurs in any of the story levels at or below the two-thirds height level of the building.
For moment frames, r_i shall be taken as the maximum of the sum of the shears in any two adjacent columns in a moment frame bay divided by the story shear. For columns common to two bays with moment-resisting connections on opposite sides at Level i in the direction under consideration, 70 percent of the shear in that column may be used in the column shear summation.
ρ shall not be taken less than 1.0 and need not be greater than 1.5, and A_B is the ground floor area of the structure in square meters. For special moment-resisting frames, except when used in dual systems, ρ shall not exceed 1.25. The number of bays of special moment-resisting frames shall be increased to reduce r, such that ρ is less than or equal to 1.25.

When calculating drift, or when the structure is located in seismic zone 0, 1 or 2, ρ shall be taken equal to 1.

The ground motion producing lateral response and design seismic forces may be assumed to act nonconcurrently in the direction of each principal axis of the structure. In seismic zones 2, 3 and 4, provisions shall be made for the effects of earthquake forces acting in a direction other than the principal axes if the structure has certain plan irregularities or a column forms part of two or more intersecting lateral-force-resisting systems.

The inelastic response spectrum is derived from base shear force formulation and is used only for the static force procedure. It is reduced with respect to the elastic design response spectrum by a numerical coefficient R and is presented in Figure 1.1.6. The coefficient R represents the inherent overstrength and global ductility capacity of lateral-force-resisting systems. For undefined structural systems, the coefficient R shall be substantiated by approved cyclic test data and analyses. The following items shall be addressed when establishing R:

- Dynamic response characteristics,
- Lateral force resistance,
- Overstrength and strain hardening or softening,
- Strength and stiffness degradation,
- Energy dissipation characteristics,
- System ductility, and
- Redundancy

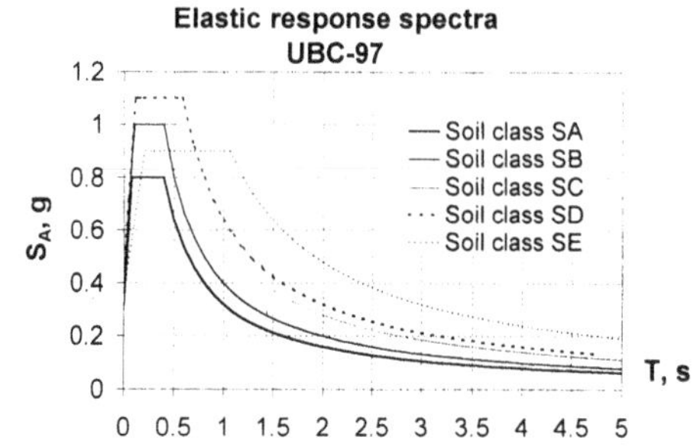

Figure 1.1.5: Elastic response spectra of UBC-97. Z=0.4

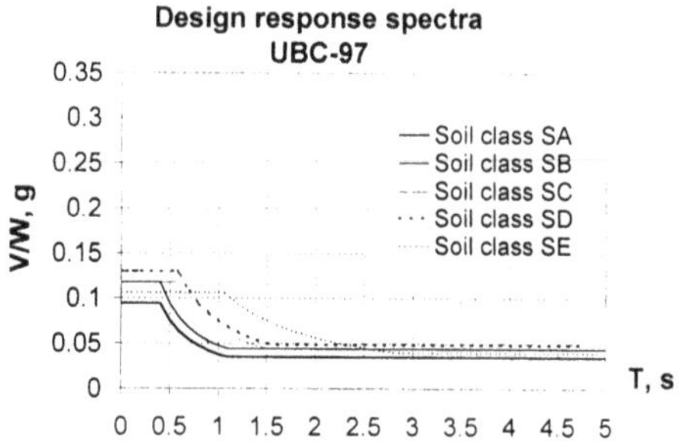

Figure 1.1.6: Design response spectra of UBC-97. Z=0.4, R=8.5

AIJ_{LSD}–90

The Japanese Standard for Limit State Design of Steel Structures is quite scarce in its provisions for determining the seismic action. For the distribution coefficient of seismic shear force, vibration characteristic coefficient and zone factor it refers to "The Building Standard Law and Enforcement Order". The expected value of the acceleration earthquake ground motion in the return period of 50 years, it is temporarily assumed that A_e=0.2g.

Three soil profiles are specified: firm, medium and soft. The elastic design response spectra for these soil profiles is presented in Figure 1.1.7. The spectrum values are the nominal ones, and are not scaled by the load factor γ_E=2.0 for the ultimate limit state.

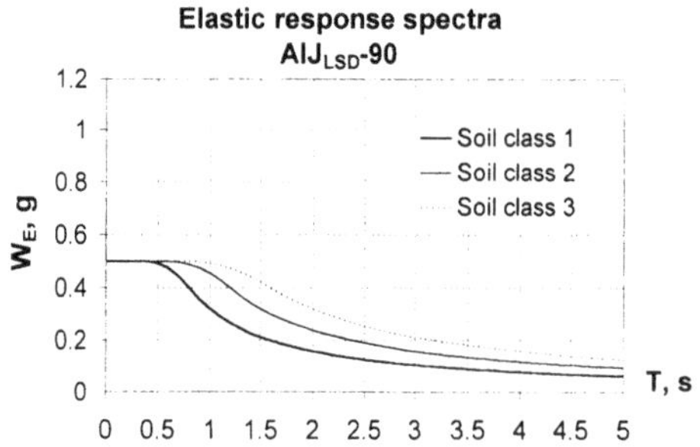

Figure 1.1.7: Elastic response spectra of AIJ_{LSD}-90. A_e=0.2 g, Z_e=1.0

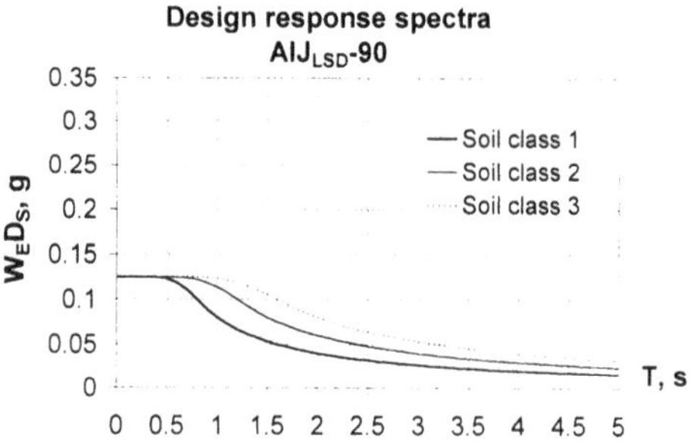

Figure 1.1.8: Design response spectra of AIJ_{LSD}-90. A_e=0.2 g, Z_e=1.0, D_S=0.25

In the ultimate limit state design against the load combination including earthquake load, the design strength of the frame (design lateral load-carrying capacity) is to be greater or equal to the required strength (required lateral load-carrying capacity), which is calculated based on the combinations of factored loads including the earthquake load stipulated in TABLE 1.1.10. For unbraced and braced frames, the required lateral load-carrying capacity of each storey of the frame may be reduced according to the structural classification and the structural characteristics factor D_S. These reduced spectra are presented in Figure 1.1.8.

No importance or risk classes for structures are specified in this code.

AIJ_L–93

Seismic loads are evaluated basically by the response spectrum method. The response spectrum of ground motion is expressed in terms of many parameters including maximum ground acceleration and velocity, associated return period conversion factors, factors depending on dynamic properties of surface ground, and natural periods and damping ratios of buildings. Intensities of ground motion are estimated in a probabilistic manner on the basis of statistical seismic data. If the response of a building is anticipated to go far beyond the elastic limit during strong ground motions, seismic loads may be reduced according to the structural characteristic factor, which is determined on the basis of hysteretic characteristics of the building.

The acceleration response spectrum determines the appropriate acceleration response value corresponding to the natural period and the damping characteristics of a structure based on the spectral characteristics of earthquake ground motion. The spectral characteristics of individual recorded earthquake ground motions reflect many physical properties depending on the seismic source, the propagation path, soil amplification and so on. Since it is impossible to take all effects into account, a limited number of parameters are chosen to represent various characteristics.

By comparing spectral characteristics of recorded motions, the following values for parameters defining spectral shape are provisionally recommended:

$$f_A = 2.5, \quad f_V = 2.0, \quad d = 0.5$$

The basic peak acceleration and the basic peak velocity are defined as values associated with a 100-year return period. They are estimated by a seismic hazard analysis based on the seismic activity around the site. The reference soil was chosen as fairly firm ground with a shear wave velocity of 400m/s to 1000m/s. The basic peak acceleration is then estimated by considering the relation between the peak velocity and peak acceleration of recorded motions on the firm soil sites as:

$$A_0 = 15 V_0$$

The approximate return period conversion factor may be expressed as (r – return period):

$$R_A = R_V = (r/100)^{0.54}$$

This formula only applies for return period between 20 and 500 years.

The soil-type modification of the acceleration response spectrum is controlled by G_A and G_V, or by T_C and one of G_A and G_V. For soft soils, detailed examination of soil amplification is recommended. The dynamic characteristics of the surface soil layer are represented by the dominant period and the amplification ratio.

Recommended values for G_A, G_V and T_C for three soil types are summarised in TABLE 1.1.9. The acceleration response spectra for three soil types are shown in Figure 1.1.9.

In many cases soil layers do not have a horizontally layered structure. Three soil types are not a sufficient representation of actual soil types, and dynamic analysis of soil amplification is recommended, particularly for soft soils and irregular soil layer structures.

TABLE 1.1.9

SOIL TYPES IN AIJ_L-93

Soil type	Description	G_A	G_V	T_C
Type I	Reference soil (Firm Ground)	1.0	1.0	0.33
Type II	Soft diluvial or Firm alluvial soil	1.2	2.0	0.56
Type III	Soft soil	1.2	3.0	0.84

The subsoil and a structure on or under the ground interact during earthquake vibrations. This phenomenon is called dynamic soil-structure interaction. There are two kinds of soil-structure interaction:

- Interaction caused by strain of the subsoil under the inertia force of the structure (inertial interaction)
- Interaction caused by the strain of the subsoil from an underground structure (kinematic interaction)

The soil-structure interaction has an influence on the evaluation of seismic design loads as follows:

- elongation of natural period,
- change of vibration mode,
- occurrence of radiation damping, and
- change of input motion amplitude.

The natural period including the dynamic soil-structure interaction (coupled natural period) can be roughly estimated in the following equation, which gives the exact value if the building is represented by a one-mass system:

$$T_1 = \frac{1}{\sqrt{1-\eta_{SR}}} T_{B1}$$

where,

T_{B1} - fundamental natural period of a structure with rigid base support

η_{SR} - sway-rocking ratio, that is, the ratio of horizontal displacement at the top of the structure caused by sway (horizontal displacement of foundation) and rocking (rotational displacement of foundation), to the total horizontal displacement at the top of structure in the fundamental vibration mode. The vibration mode of a structure also changes because of the displacement of a foundation with dynamic soil-structure interaction.

In principle, the single structural characteristic factor D_S, is determined in each direction, based on the building's inelastic characteristics. The D_S value is taken to be 1.0 in serviceability limit state design or in the first phase of structural design according to the current Building Standard Law, for a moderate level of earthquake ground motions.

The response of ordinary buildings goes far beyond the elastic limit during major earthquakes, and this is the target of ultimate limit state design or the second phase of structural design according to the current Building Standard Law. The structural characteristic factor D_S in this state is the factor for reducing the seismic forces acting on elastic building if it were to remain elastic. Here, the structural characteristic factor D_S should be conservative because the response of short-period buildings increases with the progress of plastic deformation.

The concept of the structural characteristic factor D_S for major earthquakes is based on the energy constant hypothesis between elastic and elasto-plastic systems and is available for multi-story buildings whose damage does not concentrate on certain stories and plastic deformations occur uniformly throughout all stories. However, it is difficult to evaluate quantitatively the structural characteristic factor D_S of multi-story buildings whose damage concentrates in a certain part of a building.

The reduced design response spectra, corresponding to a structural characteristic factor D_S is given in Figure 1.1.10.

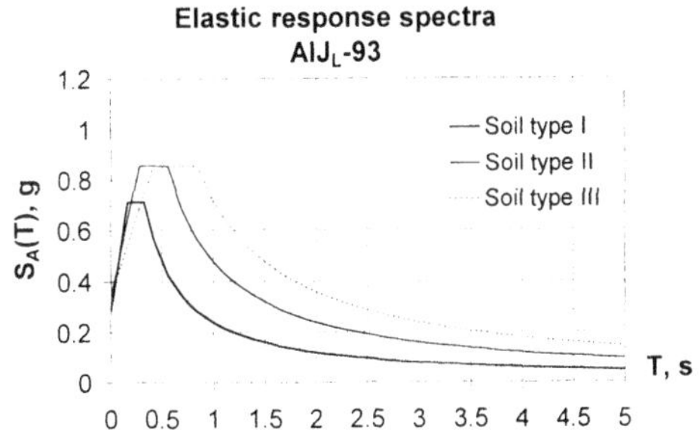

Figure 1.1.9: Elastic response spectra of AIJ_L-93. A_e=0.29 g

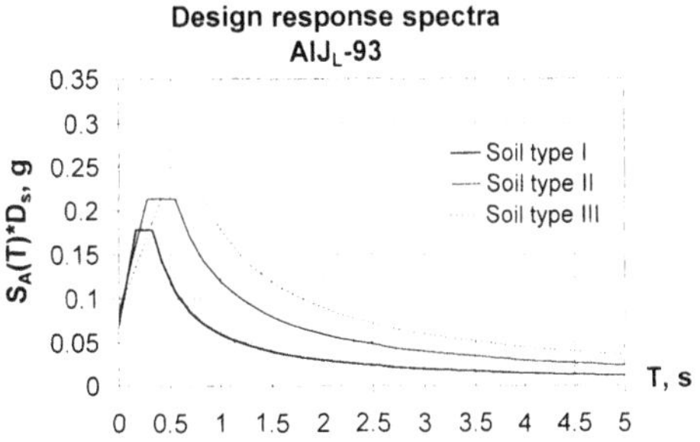

Figure 1.1.10: Design response spectra of AIJ_L-93. A_e=0.29 g, D_S=0.25

TABLE 1.1.10

EQUIVALENT SEISMIC FORCE IN EC8-94, PS 92, UBC-97, AIJ_{LSD}-90 AND AIJ_L-93.

		EC8–94	PS 92	UBC-97	AIJ_{LSD}–90	AIJ_L–93
1.	a) Design base shear force	a) $F_b = S_d(T_1) \cdot W$ T_1 – fundamental period of vibration	a) not defined explicitly	a) Total design base shear: $V = \frac{C_v \cdot I}{R \cdot T} \cdot W$ $V \le \frac{2.5 \cdot C_a \cdot I}{R} \cdot W$ $V \ge 0.11 \cdot C_a \cdot I \cdot W$	a) Not defined explicitly	a) not defined explicitly
	b) Storey shear force for simplified analysis	b) $F_i = F_b \frac{s_i \cdot W_i}{\sum s_j \cdot W_j}$ $(F_i = F_b \frac{z_i \cdot W_i}{\sum z_j \cdot W_j})$ W_i, W_j – weights of masses m_i, m_j	b) $f_r = \rho_0 \cdot m_r \cdot Z_r^\alpha \cdot \frac{\sum m_i \cdot z_i^\alpha}{\sum m_i \cdot z_i^{2\alpha}} \cdot \frac{R(T}{q}$ α = f(bracing system) [1.0, 1.5] α =1.0 for MRF $\rho_0 = 1 + 0.05\left(\frac{T}{T_C}\right)^{3/2} \ge 1.05$ for MRF	b) the concentrated force F_t at the top: $F_t = 0.07 \cdot T \cdot V \le 0.25 \cdot V$; $F_t = 0$ if $T \le 0.7$s the remaining portion distributed over the height of the structure: $F_x = \frac{(V - F_t) \cdot w_x \cdot h_x}{\sum_{i=1}^{n} w_i \cdot h_i}$	b) seismic storey shear force: $W_E = C_e \cdot A_i \cdot W_i$ $Ce = 2.5 \cdot R_t \cdot A_e \cdot Z_e$ $A_i = 1 + (\frac{1}{\sqrt{a_i}} - a_i)\frac{2T}{1+3T}$ $a_i = \frac{W_i}{W_t}$	b) seismic storey shear force: $Q_{Ei} = D_s \sqrt{\sum_{m=1}^{k} [(\sum_{j=i}^{n} w_j \beta_m u_{jm}) S_A(T_m, h_m)/g}$ Value of k may be usually be taken as 3 or less
2.	Design spectrum	$S_d(T)$ – Design spectrum for linear analysis $\alpha \cdot S \cdot [1 + \frac{T}{T_B}(\frac{\beta_0}{q} - 1)]$ for $0 \le T \le T_B$ $\alpha \cdot S \cdot \frac{\beta_0}{q}$ for $T_B \le T \le T_C$ $\alpha \cdot S \cdot \frac{\beta_0}{q}\left[\frac{T_C}{T}\right]^{kd_1} \ge 0.2 \cdot \alpha$ for $T_C \le T \le T_D$ $\alpha \cdot S \cdot \frac{\beta_0}{q}\left[\frac{T_C}{T_D}\right]^{kd_1}\left[\frac{T_D}{T}\right]^{kd_2} \ge 0.2 \cdot \alpha$ for $T_D \le T$ k_{d1}=2/3 =f(Subsoil class) k_{d2}=5/3 =f(Subsoil class) T_B – [0.10, 0.15, 0.20], s T_C – [0.40, 0.60, 0.80], s T_D – [3.0, 3.0, 3.0], s	$R(T)$ – Design response spectrum $R(T) = a_N \cdot \tau \cdot \rho \cdot R_D(T)$ $R_D(T) = R_M$ for $0 \le T \le T_C$ $R_D(T) = R_M \cdot \left[\frac{T_C}{T}\right]^{2/3}$ for $T_C \le T \le T_D$ $R_D(T) = R_M \left[\frac{T_C}{T}\right]^{2/3}\left[\frac{T_D}{T}\right]^{2/3}$ for $T \le T_D$ T_B – [0.15, 0.20, 0.30, 0.45], s T_C – [0.30, 0.40, 0.60, 0.90], s T_D – [2.67, 3.20, 3.85, 4.44], s	Not defined directly, deduced from total base shear Elastic design response spectrum (for dynamic analysis only) $\frac{C_a \cdot I}{R}$ for T=0 $2.5 \cdot \frac{C_a \cdot I}{R}$ for $T_O \le T \le T_S$ $\frac{C_v \cdot I}{R \cdot T}$ for $T_S \le T$ $T_S = C_v/2.5C_a$ $T_O = 0.2T_S$ C_a = f(soil profile type, zone) – effective peak ground acceleration C_v = f(soil profile type, zone) – velocity domain spectral response C_a and C_v can be affected by the near-source factors N_a and N_v	Normalised design response spectrum R_t : $R_t = 1$ for $T \le T_C$ $R_t = 1 - 0.2(T/T_C - 1)^2$ for $T_C \le T \le 2T_C$ $R_t = 1.6 \cdot T/T_C$ for $2T_C \le T$ T_C =[0.4, 0.6, 0.8], s	$S_A(T, h)$ – Acceleration response spectrum: $(1 + \frac{f_A - 1}{d}\frac{T}{T_c})F_h G_A R_1 A_0$, $0 < T < dT_c$ $F_h f_A G_A R_A A_0$, $dT_c < T < T_c$ $\frac{2\pi F_h f_V G_V R_1 V_0}{T}$, $T_c < T$

3.	Zone factor (α_1)	Not specified directly Part of a_g (PGA ≠ EPGA), $a_g = [0.35g, 0.25g, 0.15g]$ $\alpha_1 \cdot \alpha_0 \approx a_g$	Not specified directly, part of a_N $a_N = [0\text{-}4.5]$ = f(seismic zone, building importance class) $\alpha_1 \cdot \alpha_0 \cdot \alpha_6 \cdot \alpha_7 = a_N$	Part of Z given on a seismic zonation map and in tables $\alpha_1 \cdot \alpha_0 / g = Z$ However, Z is practically not used in the formulas, being replaced by C_a and C_v $Z = C_a$ for the soil profile type S_B (rock) $\alpha_1 \cdot \alpha_0 \cdot \alpha_3 / g = C_a$ for $T_O \le T \le T_S$	Corresponds to Z_e $Z_e = [1.0\text{-}0.7]$ $\alpha_1 = Z_e$	Not specified directly Part of A_0, given in a seismic zonation map $\alpha_1 \cdot \alpha_0 = A_0$
4.	Dynamic amplification coefficient (α_2)	Part of β_0 $\alpha_2 = \beta_0$ for $T_B \le T \le T_C$ The shape is defined by $S_d(T)$ (see 2) $\beta_0 = [2.5]$; f(Subsoil class)	Part of R_M R_M = f(soil class) $R_M = [2.5, 2.5, 2.25, 2.0]$ $\alpha_2 \cdot \alpha_3 = R_M$	Specified numerically: $\alpha_2 = 2.5$ for $T_O \le T \le T_S$ Shape defined by the base shear force formulation	Specified numerically as part of C_E: $\alpha_2 = 2.5$ for $T \le T_C$ Shape defined by R_t	Defined by f_A, f_V and the ratio $A_0/V_0 = 15$ Shape is given by $S_A(T, h)$ $\alpha_2 = f_A$ for $dT_c \le T \le T_c$ $f_A = 2.5$
5.	Soil factor (α_3)	Corresponds to $S = [1.0; 1.0; 0.9]$ 3 subsoil classes: A, B, C special studies for ground conditions not matching subsoil classes A, B, C $\alpha_3 = S$	Not specified directly Part of R_M $\alpha_2 \cdot \alpha_3 = R_M$ 4 soil types (rock, a, b, c) ⇒ 4 site classes: S_0, S_1, S_2, S_3	Not specified directly Part of C_a and C_v 6 soil profiles types: S_A, S_B, S_C, S_D, S_E, S_F special studies for soil profile type S_F required $\alpha_1 \cdot \alpha_0 \cdot \alpha_3 / g = C_a$ for $T_O \le T \le T_S$	Not present Same amplification for all soil types. Only the control period T_C is accounted for by R_t 3 soil types: (1-hard, 2-medium, 3-weak)	Defined by three parameters: G_A, G_V, and T_c, from which only two are independent $\alpha_3 = G_A$ for $dT_c \le T \le T_c$ three soil types (I, II, III) formula for evaluation of T_c when the number, thickness, depth and shear wave velocity of the soil layers are known dynamic analysis of soil amplification recommended for soft soils and irregular soil layers structures
6.	Damping factor (α_4)	Corresponds to η $\eta = 1$ for 5% viscous damping $\eta = \sqrt{7/(2+\xi)} \ge 0.7$ $\alpha_4 = \eta$	Corresponds to ρ $\rho = \left[\frac{5}{\zeta}\right]^{0.4}$; $2\% \le \zeta \le 30\%$ $\alpha_4 = \rho$ $\zeta = 2\%$ for welded steel structures $\zeta = 4\%$ for bolted steel structures	Not specified	Not specified	Damping modification factor: F_h $F_h = 1.5/(1+10h)$ for $h \ne 5\%$ $\alpha_3 = G_A$ for steel structures h=0.01 to 0.03
7.	Reduction factor (α_5)	Partially corresponds to $1/q$ q = f(material, structural system, viscous damping ≠ 5%, ductility level) $q = 5\alpha_u/\alpha_1 \approx 6$ for steel MRF q=4 for steel X CBF	Partially corresponds to $1/q$ q =f(material, structural system, possibility of efforts redistribution, plastic rotation capacity of elements, structural regularity) The behaviour factor is not part of the design response spectra.	Partially corresponds to $1/R$ R =f(material, structural system) R=8.5, 6.5, 4.5 for steel SMRF, IMRF, OMRF respectively R=5.6, 6.4 for steel SCBF, OCBF respectively R=7.0 for steel EBF	Corresponds to D_S D_S =f(material, structural classification) The behaviour factor is not part of the design response spectra. ${}_RD_S$ – unbraced frame	Corresponds to D_S D_S =f(hysteretic characteristics of the building) The behaviour factor is not part of the design response spectra D_S=1 for the in the serviceability limit state design

		$q=5\alpha_u/\alpha_1 \approx 5.5$ for steel EBF $\alpha_5=f(q, T, \eta)$ (according to the definition of the reduction factor $\alpha_5= S_e(T)/S_d(T)$) Note: a behaviour factor for vertical component q=1.0 is specified only for concrete structures	- if $T \leq T_B$ then $q' = \frac{2.5 \cdot \rho}{1-\frac{T}{T_B}\cdot\left(1-\frac{2.5\cdot\rho}{q}\right)}$ - for vertical component: $q' = \max\left(1; \frac{q}{2}\right)$ $q=5\alpha_u/\alpha_1 \approx 6$ for steel MRF q=4 for steel X CBF $q=5\alpha_u/\alpha_1 \approx 5.5$ for steel EBF $\alpha_5=f(q, T, \textit{vertical component})$	$\alpha_5 \approx R/\rho$ $\alpha_5 \approx R/\Omega_0$ for the special seismic combination	$_BD_S$ – braced frame $\alpha_5 = \frac{1}{D_S}$	Values assumed equal to those from AIJ_{LSD}-90 $\alpha_5 = \frac{1}{D_S}$
8.	Risk factor (α_6)	Part of importance factor γ_I $\alpha_6 \cdot \alpha_7 = \gamma_I$ (see also α_7)	Not present directly Part of a_N $\alpha_1 \cdot \alpha_0 \cdot \alpha_6 \cdot \alpha_7 = a_N$	Part of importance factor I $\alpha_6 \cdot \alpha_7 = I$	Not present	Not present directly Can be taken into account through the reliability index β_T when determining the load factor $_u\gamma_E$ for the ULS load combination
9.	Importance factor (α_7)	Part of importance factor γ_I $\alpha_6 \cdot \alpha_7 = \gamma_I$ 4 Importance categories of buildings: I, II, III, IV γ_I = [1.4, 1.2, 1.0, 0.8]	Not present directly Part of a_N $\alpha_1 \cdot \alpha_0 \cdot \alpha_6 \cdot \alpha_7 = a_N$ 4 Importance classes of buildings (≡ risk classes): A, B, C, D	Part of importance factor I $\alpha_6 \cdot \alpha_7 = I$ I=[1.25-1.0] 5 occupancy categories of buildings: 1, 2, 3, 4, 5	Not present	Same as α_6
10.	Ground acceleration (α_0)	Not present directly Part of α_g $\alpha_1 \cdot \alpha_0 \approx a_g$ (see also α_1)	Not present directly Part of a_N $\alpha_1 \cdot \alpha_0 \cdot \alpha_6 \cdot \alpha_7 = a_N$	Part of Z (see also α_1) $\alpha_1 \cdot \alpha_0 / g = Z$	Corresponds to A_e A_e=[0.2g] $\alpha_0 \cdot g = A_e$	Not specified directly Part of A_0, given in a seismic zonation map $\alpha_1 \cdot \alpha_0 = A_0$ (see also α_1)
11.	Return period of the design earthquake	475 years	Not specified	The ground motion shall have a 10% probability of being exceeded in 50 years ⇒ ≅ 475 years return period	50 years	100 years Return period conversion factors R_A and R_V for return periods r between 20 and 500 years: $R_A = R_V = (r/100)^{0.54}$
12.	Fundamental period of vibration	$T_1=C_tH^{3/4}$ C_t=0.085 for space steel MRF	$T_1 = 0.10 \cdot \frac{H}{\sqrt{L_x}}$ for MRF	$T = C_t \cdot (h_n)^{3/4}$ C_t=0.0853 for steel MRF	Not present	$T_1 = (0.1 \pm 0.03)N$ for steel structures

13.	Topo-graphic amplifica-tion effects	Topographic amplification coefficient assimilated to the soil factor S Considered for bidimensional topographic irregularities as: ridges and slopes with heights greater than 30 m. For slopes less than 15°, the topographical effects can be neglected. S=[1.2; 1.4]	τ - topographic coefficient [1.0-1.40] for structures near the edge of a ridge	Not present	Not present	Not present
14.	Torsional effects	Accounted for by multiplying the action effects in the individual load-resisting elements by: $\delta = 1 + 0.6 \cdot \frac{x}{L_e}$	Point of application of forces displaced in both senses by: $e_r^{'} = 0.10 \cdot L_r \cdot Z_r$	The mass at each level to be dis-placed from the calculated centre of mass in each direction a dis-tance equal to 5% of the building dimension perpendicular to the direction of the force under con-sideration.	Not present	Not present
15.	Displace-ment analysis	$\frac{d_s}{\nu} = \frac{q_d \cdot d_e \cdot \gamma_I}{\nu}$	$d_r = -\rho_0 Z_r^a \frac{\sum m_i z_i^a}{\sum m_i z_i^{2a}} \left(\frac{T}{2\alpha}\right)^2 R(T)$	$\Delta_M = 0.7 \cdot R \cdot \Delta_S$ ρ=1.0 for displacement analysis	Determined from SLS load combination	Determined from SLS load combination
16.	Vertical component of the seismic action	As defined for the horizontal seismic action, but with the ordinates reduced as follows: $S_D(T)_{vert}$=0.7 $S_D(T)_{hor}$ for $T \leq 0.15$ s $S_D(T)_{vert}$=0.5 $S_D(T)_{hor}$ for $T \geq 0.5$ s Linear interpolation for $0.15 \leq T \leq 0.5$ s	70% of the horizontal seismic action, but with modified shape of the design spectra for the spectra descending branches of $S2$-$S4$ soil classes: $R_D(T)_{vert}(S2\text{-}S4)=R_D(T)_{hor}(S1)$.	Considered by the addition of $0.5C_aID$ to the dead load effect D Zones 3 and 4 only: horizontal cantilever components shall be designed for a net upward force of $0.7 \cdot C_a \cdot I \cdot W_p$	Not present	Not present
17.	Seismic mass	gravity loads from: $\sum G_{ki} + \sum \psi_{Ei} \cdot Q_{ki}$ $\psi_{Ei} = \varphi \cdot \psi_{2i}$ ψ_{2i} = [0, 0.3, 0.6, 0.8] φ = [0.5, 0.8, 1.0]	gravity loads from: $\sum G_{ki} + \sum \phi \cdot Q_{ki}$ ϕ = [0 - 1.0]	masses from total dead load D +: 0.25 of live load in warehouses 0.48 kN/m^2 in case of partitions [0.75-1.0] of S if > 1.44 kN/m^2 total weight of permanent equipment	Not specified	Not specified
18.	Seismic load com-bination	ULS: $\sum G_{kj} + \gamma_I A_{Ed} + P_k + \sum \psi_{2i} Q_{ki}$ SLS: not defined directly	ULS: $\sum G + E + \psi_{1i} Q_{ki} + \sum_{i>1} \psi_{2i} Q_{ki}$ SLS: not defined directly	ULS: $1.2D$+$1.0E$+(f_1L+f_2S) $0.9D \pm 1.0E$ f_1 = [1.0, 0.5]; f_2 = [0.7, 0.2] special seismic load combination:	ULS: $1.1W_D$+$2.0W_E$+$0.4W_L$ $1.1W_D$+$1.7W_E$+$0.4W_S$ +$0.4W_L$ in areas with heavy snow SLS: $1.0W_D$+$0.4W_E$+$0.4W_L$	ULS: $\gamma_D D + \gamma_L L + \gamma_H H + \gamma_S S +_U \gamma_E E$ SLS: $\gamma_D D + \gamma_L L + \gamma_H H + \gamma_S S +_S \gamma_E E$

				$1.2D+f_1L+1.0E_m$ $0.9D \pm 1.0E_m$ $E=\rho E_h+E_v$ $E_m=\Omega_0 E_h$ SLS: not defined directly	$1.0W_D+0.4W_E+0.4W_S+0.4W_L$ in areas with heavy snow	Note: load factors γ are not given directly. They are to be computed based on the target reliability index.
19.	Account for P-Δ effects	Need not be considered if $\theta \le 0.10$ If $0.1 \le \theta \le 0.2$ increase the relevant seismic action by $1/(1-\theta)$ θ shall not exceed 0.3 $$\theta = \frac{P_{tot} \cdot d_r}{V_{tot} \cdot h}$$	Need not be considered if $\theta_r \le 0.10$ If $0.1 \le \theta_r \le 0.25$ increase the relevant seismic action by $1/(1-\theta)$ $$\theta_r = \frac{\delta_r \cdot P_r}{h_e \cdot F_r}$$	$P\Delta$ effects shall be included in the analysis. Need not be considered if: $$\theta = \frac{(1.2D + f_1 L + f_2 S) \cdot d_r}{V_i \cdot h} \le 0.1$$ Need not be considered in Zones 3 and 4 if $d_r/h \le 0.02/R$	Not specified	Not specified
20.	Combination of components of seismic action	SRSS technique, or: $0.3E_{Edx}+0.3E_{Edy}+1.0E_{Edz}$ $1.0E_{Edx}+0.3E_{Edy}+0.3E_{Edz}$ $0.3E_{Edx}+1.0E_{Edy}+0.3E_{Edz}$	SRSS technique, or: $\pm S_x \pm \lambda S_y \pm \mu S_z$ $\pm \lambda S_x \pm S_y \pm \mu S_z$ $\pm \lambda S_x \pm \mu S_y \pm S_z$ Generally, $\lambda=\mu=0.3$	$E=\rho E_h+E_v$ (vertical component is part of the design seismic action) $1.0E+0.3E_\perp$ for the 2 orthogonal components or, the SRSS technique	Not specified	Not specified
21.	Definition of terms	W – total weight of the building s_i, s_j – displacements of masses m_i, m_j in the fundamental mode shape z_i, z_j – heights of the masses m_i, m_j above the level of application of the seismic action T_B, T_C, T_D – response spectrum control periods α - ratio of the design ground acceleration a_g to the acceleration of gravity g a_g – effective peak ground acceleration (design ground acceleration) β_0 – spectral acceleration amplification factor for 5% viscous damping S – Soil parameter η - damping correction	ρ_0 – magnifying coefficient to take into account neglected vibration modes m_r – mass of the storey r Z_r – non-dimensional height of the storey r ($Z_r=h_r/H$) $R_D(T)$ – normalised design response spectrum T – fundamental period of vibration T_B, T_C, T_D – response spectrum control periods R_M – coefficient =f(soil class) a_N – nominal acceleration (m/s^2) ρ - damping correction coefficient q – behaviour factor τ - topographic coefficient d_r – displacement of the storey r H – Height of the structure L_x – Length in plane along x axis	T – fundamental period of vibration W – total seismic dead load h_x – height above the base to level x w_x – that portion of W located or assigned to level x Z – Seismic zone factor R – numerical coefficient representative of the inherent overstrength and global ductility capacity of the lateral-force-resisting systems I – Importance factor W_p – weight of the component D – dead load on a structural element h_n – height of the structure Δ_M – maximum inelastic response displacement,	A_i – distribution coefficient for seismic storey shear force along the height of he building W_i – accumulative weight of the building above the i-th storey W_t – total weight of the building R_t – normalised design response spectrum T_C – response spectrum control period Z_e – zone factor D_S – structural characteristics factor A_e – expected value of the acceleration earthquake ground motion g – acceleration of gravity W_D – nominal dead load W_L – nominal live load W_S – nominal snow load	Q_{Ei} – seismic shear force of the i-th story D_s – structural characteristic factor w_j – gravity load of the j-th story β_m – participation factor of the j-th vibration mode u_{jm} – m-th vibration mode of the j-th story $S_A(T, h)$ – acceleration response spectrum of single degree of freedom system having natural period T and damping ratio h T_m – natural period of the m-th vibration mode h_m – damping ratio for the m-th vibration mode k – maximum number of vibration modes taken into account in the superposition

		coefficient q – behaviour factor x – distance of the element from the centre of the building measured perpendicularly to the seismic action considered L_e – distance between the two outermost lateral load resisting elements measured as previously d_s – displacement of a point of the structural system induced by the design seismic action q_d – displacement behaviour factor, assumed equal to q d_e – displacement of the same point, as determined by a linear analysis based on the design response spectrum γ_I – importance factor ψ_{2i} – combination coefficient for quasi-permanent value of variable action i ψ_{Ei} – combination coefficient for variable action i G_{kj} – characteristic value of the permanent action j A_{Ed} – design value of the seismic action for the reference return period P_k – characteristic value of prestressing action Q_{ki} - characteristic value of variable action i P_{tot} – total gravity load at and above the storey considered d_r – design interstorey drift V – total seismic storey shear h – interstorey height E_{Edx}, E_{Edy}, E_{Edz} – action effect due to the application of the seismic action along the x, y, z axes of the structure	ϕ - coefficient of partial masses depending on the nature and duration of variable action G – dead load and long-term permanent actions E – Seismic action Q_{ki} – variable actions ψ_1, ψ_2 – load factors h_e – interstorey height δ_r – interstorey drift P_r – weight of masses at and above level r F_r – sum of seismic shear forces at and above level r S_x, S_y, S_z – deformations or forces due to horizontal and vertical components, respectively f_r –seismic shear force at level r	which is the total drift or the total storey drift that occurs when the structure is subjected to the to the design basis ground motion, including estimated elastic and innelastic contributions to the total deformation Δ_S – design level response displacement, which is the total drift or storey drift that occurs when the structure is subjected to the design seismic forces E - the earthquake load on an element of the structure resulting from the combination of the horizontal component, E_h, and the vertical component, E_v. E_h - the earthquake load due to the base shear, V. E_m - the estimated maximum earthquake force that can be developed in the structure. E_v - the load effect resulting from the vertical component of the earthquake ground motion and is equal to an addition of $0.5C_aID$ to the dead load effect, D. Ω_0 - the seismic force amplification factor that is required to account for structural overstrength. ρ - reliability/redundancy factor. S – snow load d_r – design interstorey drift V_i – total seismic storey shear h – interstorey height	W_E – nominal seismic load	n , N – number of stories g – gravity acceleration f_A – ratio of $S_A(T,\ 0.05)$ to $G_A R_A A_0$ for $dT_c \le T \le T_c$ f_V – ratio of $S_V(T,\ 0.05)$ to $G_V R_V V_0$ for $T_c \le T$ d – ratio of the lower bound period to the upper bound period of the range where $S_A(T, h)$ is constant F_h – damping modification factor A_0 – basic peak acceleration of earthquake ground motion at the reference soil V_0 – basic peak velocity of earthquake ground motion at the reference soil R_A – return period conversion factor for the peak acceleration of earthquake ground motion R_V – return period conversion factor for the peak velocity of earthquake ground motion G_A – soil type modification factor for the peak acceleration of earthquake ground motion G_V – soil type modification factor for the peak velocity of earthquake ground motion D – dead load L –live load S –snow load E –earthquake load H –earth or hydraulic pressure

Comparative Remarks

Response spectra

In a way or another, all the codes define the seismic hazard by an acceleration response spectrum, for a 5% viscous damping (value not specified in AIJ_{LSD}–90). Analytical expressions used for spectrum definition are quite different, but the shape of the curves is basically the same. However, distinction is made in some codes between the elastic response spectrum defining the seismic motion at the site, and the one used for design. Generally, the design response spectrum is modified with respect to the elastic one in order to consider the inelastic behaviour of MDOF systems designed, as compared to the elastic SDOF system used for the determination of the elastic response spectrum. Two aspects are to be underlined here. First, design spectra take into account that in the short-period range the benefits of plastic behaviour cannot be exploited, which is equivalent to a reduced value of the structural behaviour factor. Second, in the case of tall and complex buildings, the increase of vibration modes is accompanied by an increase in those modes that can produce severe local damage. As a consequence there is also a greater likelihood that high ductility demands will be concentrated in limited parts of buildings. Taking into account that in these cases there is a greater risk of damage concentration and, on the other hand, that these buildings have a greater sensivity to second-order effects, codes generally adopt more conservative criteria for longer-period structures.

EC8–94 defines two spectra: elastic response spectrum (composed of 4 branches) and the design spectrum (also 4 branches). The elastic spectrum is a uniform risk spectrum characterising seismic hazard at the site. Artificial accelerograms, if used, are to be generated from this spectrum. For the design purpose the design response spectrum is defined (also 4 branches), which incorporates the behaviour factor q. It also considers the reduced q-factor in the short period range, and amplified spectrum ordinates in the long-period range (modified parameters k_{d1}, k_{d2}). Also, a limitation on the minimum spectral acceleration to be considered in the design is given.

In PS 92 two spectra are defined: elastic response spectrum (4 branches) and design response spectrum (3 branches). The elastic spectrum is used only for complete multi-modal analysis when the structure is designed to remain in the elastic range under the design earthquake. Simplified analysis and multi-modal analysis that take account of the inelastic reserve of the structure is based on the design response spectrum having amplified values in the short-period and long-period ranges with respect to the elastic one. These modifications are intended to take account in a simplified manner of the plastic behaviour of the structure. The q factor is not part of the design spectrum, but design forces are to be reduced according to it.

UBC-97 has the elastic response spectrum defined by three branches. This is the spectrum to be used for dynamic (multi-modal) analysis. The design spectrum for simplified static procedure (2 branches) is affected by the structural behaviour factor R. It also assumes a constant acceleration spectrum plateau in the short-period range and a limitation on the minimum spectral acceleration in the long-period range.

AIJ_{LSD}–90 defines a single spectrum – the design one (3 branches). It does not contain the structural behaviour factor, neither a limitation of the spectral acceleration. The spectral acceleration in the short-period range is assumed equal to the maximum spectral amplification in the intermediate period range. Design forces may be reduced according to the structural behaviour factor once these are computed from the design spectrum.

AIJ_L–93 gives also a single spectrum composed of three branches and which is the elastic one. Design forces are to be affected by the structural behaviour factor after complete modal superposition, a procedure different from the provisions of other codes (AIJ_{LSD}–90 also adopts this approach, but consequences are irrelevant in the case of simplified static procedure covered by the code). AIJ_L–93 is the only code to adopt the same shapes for the elastic and design response spectra. Another characteristic of this code is presence in the formulation of the response spectrum of a return period conversion factor, which allows to take into account directly design earthquakes with different return periods.

The design elastic and corresponding inelastic response spectra for the maximum peak ground acceleration provided by the code and an equivalent soil class are given in Figure 1.1.11 and Figure 1.1.12. Spectra are scaled by partial load factors for the ultimate limit state earthquake combination.

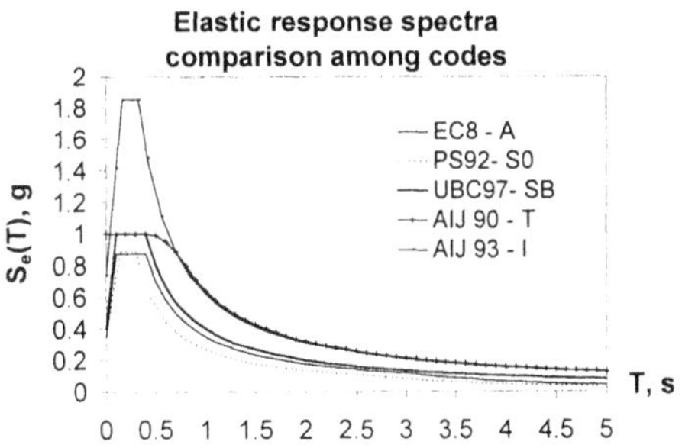

Figure 1.1.11: Elastic response spectra

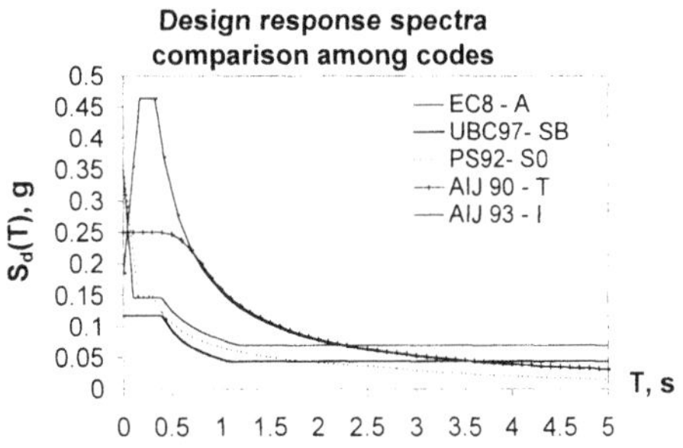

Figure 1.1.12: Inelastic design response spectra

Ground acceleration

Obviously, the seismicity of the countries to which the codes apply is different. Therefore, it is to be expected that the maximum peak ground acceleration in different national standards will not be the same. At the same time, the reference return period of design earthquake and the safety formats are different, resulting in possible different definitions of the characteristic values of seismic action. The Peak Ground Accelerations (PGA) specified in codes for the reference soil (usually rock) and for the design return period are presented in TABLE 1.1.11. In order to overcome the problem of safety formulations and different return periods, PGA was scaled by the load factors for earthquake action. It can be noted that these scaled values have similar magnitudes, except for AIJ_L–93, where the scaled PGA is almost twice that in the older Japanese code AIJ_{LSD}–90 (earthquake spectrum of the latter is virtually identical to that of Japan Building Standard Law - BSL). At the same time, in the commentary to AIJ_L–93, a value of 1.8 m/s^2 is specified when comparing the seismic coefficients of this code to that of 0.2 *g* from BSL. Due to lack of information on seismic zonation of Japan and the problems relating to the load factor calculation, this last value of 1.8m/s^2 (0.48 *g* when scaled by the load factor) was adopted for comparison among base shear forces in the considered codes.

TABLE 1.1.11

COMPARISON OF MAXIMUM VALUES OF PEAK GROUND ACCELERATION SPECIFIED IN CODES

	EC8–94	PS 92	UBC-97	AIJ_{LSD}–90	AIJ_L–93
Maximum PGA specified	0.35 *g*	≈0.36 *g* (3.5 m/s^2)*	0.4 *g*	0.2 *g*	0.29 *g* (2.8 m/s^2)
Return period	475 years	not specified	475 years	50 years	100 years
Load factor for EQ action (ULS)	1.0	1.0	1.0	2.0	2.6
Maximum PGA times load factor	0.35 *g*	0.36 *g*	0.4 *g*	0.4 *g*	0.75 *g*

* Class B buildings – "current" risk

EC8-94 introduces a relatively new concept of Effective Peak Acceleration instead of the Peak Ground Acceleration in order to characterise the design earthquake, but it is recognised there is not a unique definition and techniques for deriving the latter.

Dynamic amplification coefficient

The dynamic amplification coefficient is sometimes considered together with the soil factor, but as a rule, it is uniform among the codes, defaulting to a value of 2.5.

Soil factor

Soil characteristics and descriptions, as well as depth of the respective soil layers are used to classify the subsoil conditions in EC8–94 and PS 92. Depth of soil deposits is not so extensively used in UBC-97, while AIJ_{LSD}–90 and AIJ_{L}–93 provide only generic description of the soil types. Influence of the soil profile on the design response spectra can be associated with two aspects: shifting of the maximum amplification plateau and different amplification factors for this range. While the first aspect is uniform across the codes, i.e. for softer soils the plateau will be shifted in the higher period range, the second one is considered quite differently in the codes. EC8–94 and PS 92 provide a reduced amplification for softer soils, AIJ_{LSD}–90 keeps unchanged this coefficient, while UBC-97 and AIJ_{L}–93 provide an increase in the dynamic amplification for softer soils.

Due to the fact that codes provide different classification of subsoil conditions, the equivalent soil classes for the purpose of comparison are presented in TABLE 1.1.12.

TABLE 1.1.12

EQUIVALENT SOIL CLASSES

	EC8–94	PS 92	UBC-97	AIJ_{LSD}–90	AIJ_{L}–93
Soil class	*A*	S_0	S_B	*type 1*	*Type 1*

It is to be noted that soil amplification in UBC-97 is dependent on the level of ground shaking and increases for lower seismic zones.

Damping factor

EC8–94, PS 92 and AIJ_{L}–93 take account for damping being different from 5% by different analytical expressions, while UBC-97 and AIJ_{LSD}–90 do not provide any guidance in this aspect.

Reduction factor

All the codes allow the structures to enter in inelastic range of deformations during the design earthquake, for energy dissipation, introducing the structural behaviour factor.

The period-dependent characteristic of the structural behaviour factor is usually incorporated into the design response spectrum. Consequently, the period-independent reduction factor is given in the codes based on the material used, type of construction, force redistribution capacity and the plastic deformation capacity of elements. The structure-dependent reduction factor will depend on:

- system ductility,
- overstrength and
- redundancy.

However the definition of overstrength is commonly quite ambiguous, the last two terms being sometimes used as a unique one. In fact, the only code that defines explicitly the redundancy factor is UBC-97. EC8-94 defines the behaviour factor q as:

$$q = q_{\min} \cdot \frac{\alpha_u}{\alpha_1}$$

where:
q_{min} - is the minimum value associated to the expected performance of the structural scheme (system ductility)
α_u/α_l - the overstrength/redundancy coefficient
α_l - multiplier of the horizontal design seismic action, while keeping constant all other design actions, which corresponds to the point where the most strained cross-section reaches its plastic moment resistance
α_u - multiplier of the horizontal design seismic action, while keeping constant all other design actions, which corresponds to the point where a number of sections, sufficient for the development of the overall structural instability, reach their plastic moment resistance

Anyway, the significance of the α_u/α_l is not stated explicitly. The same procedure is adopted by PS 92, except that the α_u/α_l factor is explicitly declared as the redundancy factor.

The American code makes use of both overstrength and redundancy. The behaviour factor is given by a numerical coefficient R, which represents the inherent overstrength and global ductility capacity of the system. It should be noted that the coefficient R is an upper bound of the reduction factor. It is to be reduced in two cases. First, and namely by the redundancy/reliability factor ρ, when the structural system does not posses sufficient redundancy against lateral forces. And secondly, in the special load combination, an attempt to protect elements that may fail in brittle manner (e.g. columns, column splices) or to protect against structure overturning. In this latter case, the use of overstrength factor Ω_0 is invoked. It is to be noted here that both redundancy reliability and overstrength factors are used to amplify the earthquake load rather than to reduce R, but the effect is the same.

The Japanese code AIJ_{LSD}-90 defines a single reduction factor D_S. However, an implicit use of the overstrength/redundancy is accomplished by the adoption of plastic design for frames classified into S-I or S-II frames.

Risk/importance factor

None of the codes considered here make the difference between the risk and importance factor. PS 92 incorporates the risk/importance factor into the nominal ground acceleration to be used in design. AIJ_{LSD}-90 does not specify any classification of structures into risk/importance categories, neither do AIJ_L-93, but the latter requires "an appropriate balance between safety and economics".

Vertical component

Both EC8-94 and PS 92 specify response spectra defining vertical component of the seismic action, by modifying the corresponding horizontal spectra. PS 92 is the only one to define explicitly a reduction factor q for the vertical component, while EC8-94 defines a similar coefficient only for concrete structures. UBS-97 uses simpler formulation (dead-load proportional amplification) and, as a specific feature of this code, vertical component is mandatory, being part of the design seismic action. Japanese codes do not provide any reference to the definition of vertical component.

Second-order effects

Sensitivity to second-order effects is determined in similar ways in EC8-94, PS 92 and UBC-97. Simplified account for P-Δ effects is given by the first two design provisions. The Japanese codes do not specify any account for second -order effects

Specific features

The topographic amplification of the seismic action is covered by the European codes and is more rigorously defined by PS 92.

The American code is the only one to consider the near source-effect of the earthquake action (for design of structures located in seismic zone 4 within 10 to 15 km of active faults). The near source factors of UBC-97 are based approximately on the ratio of spectral response to that required for design of structures located at seismic zone 4 sites not near an active source. Short-period near-source factors, N_A, and long-period near-source factors, N_V, are dependent on the proximity to the active source and the fault type. The fault type is defined on the basis of the fault's maximum moment magnitude and slip rate.

UBC-97 uses a somehow arbitrary distribution of the base shear force along the height of the building, given by a concentrated force at the top of the structure, additional to the usual "triangular" distribution. It may be thought as a way to consider in a simplified manner the influence eigenmodes other than the fundamental one.

Design example

In order to overcome problems related to different formulations of the base shear force among the codes, an evaluation of the seismic forces to be used in design was performed. The structure considered is a three bay (in both horizontal directions) – five storey office building. The dimensions in plan are 18x18 m (three bays of 6 m each), while the height is 18.5 m (ground floor height – 4.5 m; the rest of floors – 3.5 m). The structural scheme is considered to be moment-resisting steel frame in both directions. The frame was considered as special moment-resisting frame in the case UBC-97 and class S-I frame in the case of Japanese codes (highest seismic performance category frames).

The dead load acting on the floors is 4.75 kN/m^2 and that for the exterior vertical surface is 1.7 kN/m^2. The live load for office buildings was taken according to the European EC1 as 3 kN/m^2. The construction site was considered to be in a high seismicity zone (see TABLE 1.1.11), and corresponding to stiff soil conditions (see TABLE 1.1.12).

The evaluation of the seismic forces performed herein is simplified as much as possible, and is intended to characterise the predesign of the structure. Only the base shear force is computed, its distribution along the height of the building being disregarded when possible. The natural period of vibration was considered as the one given by empirical formulas in the codes.

The seismic mass to be accounted for when determining seismic forces is usually determined from the dead load acting on the structure and a portion of the variable loads (live load, snow load). This approach is adopted by EC8–94, PS 92 and UBC-97. However, variable loads are accounted for only in a limited number of cases in UBC-97. Both Japanese codes do not specify any way in which the masses should be computed, therefore, only the dead load was considered.

The fundamental period of vibration determined from simplified formulas are quite scattered among the codes. EC8–94 and UBC-97 make use of the same formula for empirical computation of the first vibration mode. Much smaller values are given by the rest of the codes. It has to be noted a significant difference between the American and the other codes: UBC-97 is the only code which relies on the fundamental period of vibration determined from empirical relations for the design, while all the others rely on vibration modes and periods determined by methods of structural dynamics.

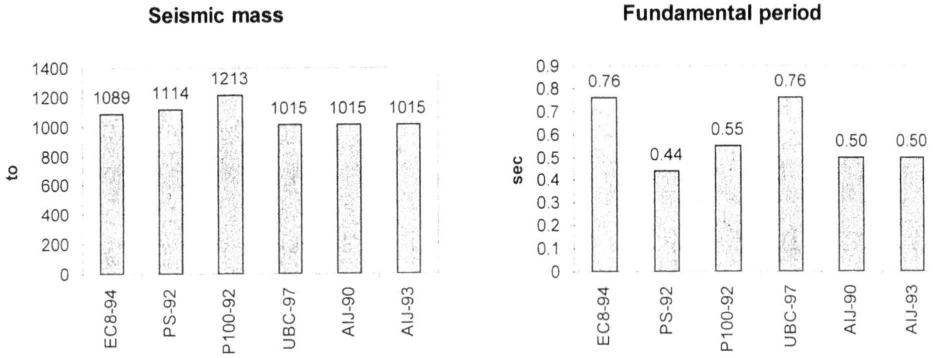

Figure 1.1.13: Seismic mass and the approximate fundamental period of vibration

It can be concluded from Figure 1.1.13 that seismic mass variation is not very important, the influence of variable loads being small. The period of vibration, instead, varies considerably among the codes, and its influence on the computed base shear force is important. Three more factors affecting the determination of base shear force are to be mentioned here: damping coefficients different from 5% are considered by PS 92 and AIJ_L–93, a magnifying coefficient to take account of the neglected vibration modes is used by PS 92, and the participation factor of the fundamental vibration mode, considered only by PS 92 and AIJ_L–93. Base shear forces as determined above are shown in Figure 1.1.14.

An important variation values of base shear force according to different codes can be noted. However, as stated earlier, these forces are intended for preliminary design only. They will change when more exact periods of vibration will be computed. The plastic analysis of the structures which is to be used in the case of Japanese codes will account for the structural overstrength/redundancy, and will somehow diminish the base shear as determined above. It is to be noted that base shears from Figure 1.1.14 will change substantially when considering soil conditions different from the stiff ones.

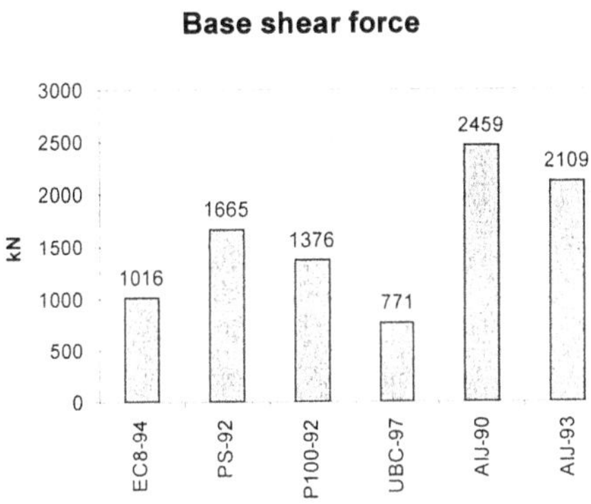

Figure 1.1.14: Comparison of base shear forces for preliminary design

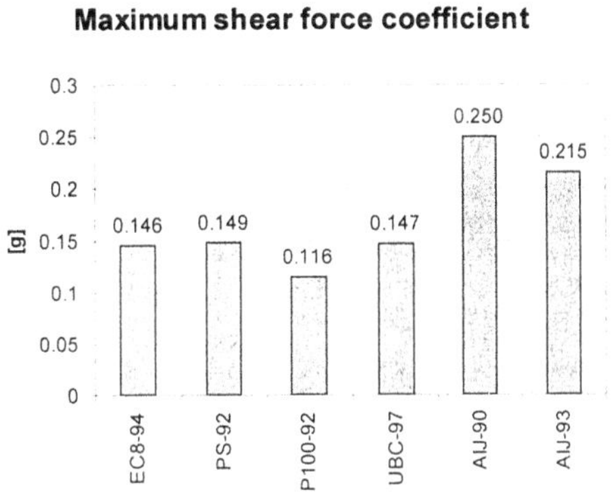

Figure 1.1.15: Maximum shear force coefficient among the codes

Yet for another comparison, the maximum base shear force coefficients are computed. They correspond to the constant spectral acceleration amplification range. This approach will remove the influence of the:

- seismic mass
- period of vibration
- damping different from 5%

As seen in Figure 1.1.15, values of base shear coefficients are much more uniform among the first three codes. The Japanese seismic provisions yield greater values, basically due to smaller reduction factors. Also, last two values are to be reduced by the structural overstrength/redundancy taken into account implicitly by the plastic analysis to be carried out according to the Japanese provisions.

SAFETY VERIFICATIONS

General

The widely accepted limit state design philosophy for seismic design considers at least two limit states associated with different levels of earthquake excitation: the service limit state for moderate earthquake shaking and the ultimate limit state for severe earthquake shaking. The limit state philosophy of modern seismic provisions is best described by the Structural Engineers Association of California (SEAOC) Recommended Lateral Force Requirements (SEAOC 1990). Structures designed in conformance with these recommendations should, in general, be able to:

- Resist minor levels of earthquake ground motion without damage
- Resist moderate levels of earthquake ground motion without structural damage, but possibly some non-structural damage.
- Resist major levels of earthquake ground motion without collapse, but likely with some structural and non-structural damage

One can achieve the first two levels of performance, called the service limit state by:

- Defining the level of moderate earthquake shaking
- Limiting stresses or internal forces in structural members
- Limiting the storey drift ratio, defined as the ratio between the interstorey drift and the storey height

The third performance level, which is often termed the ultimate limit state, can be achieved by:

- Defining the level of severe earthquake shaking
- Providing sufficient strength, ductility and deformation capacity to elements of the seismic framing system and providing a deformation-capable gravity load-resisting frame
- Limiting the maximum storey drift to ensure that the structure integrity and stability are maintained

Code provisions

EC8–94

Ultimate limit state

The safety against collapse (ultimate limit state) under the seismic design situation is considered to be ensured if conditions regarding resistance, ductility, equilibrium, foundation stability and seismic joints are met. The resistance condition requires that the following relation shall be satisfied for all structural elements (including connections), and the relevant non-structural elements:

$$E_d \leq R_d$$

where:
E_d – design value of the action effect, due to the seismic design situation, including, if necessary, second-order effects.
R_d – corresponding design resistance of the element, calculated according to the rules specific to the pertinent material and according to the mechanical models which relate to the specific type of structural system.

It shall be verified that both the structural elements and the structure as a whole possess adequate ductility taking into account the expected exploitation of ductility, which depends on the selected system and behaviour factor. Specific material-related requirements shall be satisfied, including capacity design provisions in order to obtain the hierarchy of resistance of the various structural components necessary for ensuring the untended configuration of plastic hinges and for avoiding brittle failure modes. The building structure shall be stable under the set of actions:

$$\sum G_{kj} + \gamma_I A_{Ed} + P_k + \sum \psi_{2i} Q_{ki}$$

Herein are included such effects as overturning and sliding.

Serviceability limit state

The requirement for limiting damage (serviceability limit state) is considered satisfied, if under a seismic action having a larger probability of occurrence than the design seismic action, the interstorey drifts are limited to the following values:

- $d_r/v \leq 0.004h$ (non-structural elements of brittle materials)
- $d_r/v \leq 0.006h$ (non-structural elements not interfering with the structural deformations)

where:
d_r – design interstorey drift
h – storey height
v - reduction factor to account for the lower return period of the seismic event associated with the serviceability limit state. (v=2.0 for importance category III)

The reduction factor v can also depend on the importance category of the building, and may be different for various seismic zones of a country.

PS 92

Ultimate limit state

The combinations of actions to be considered for the determination of design deformations and forces are accidental combinations for which the seismic action is affected by a coefficient γ_Q =1:

$$\sum G + E + \psi_{1i} Q_{ki} + \sum_{i>1} \psi_{2i} Q_{ki}$$

It shall be verified that under the design combinations of actions for the ultimate limit states, states of global equilibrium, resistance, or form stability, are verified by the structure, any of its substructures and the foundations. For each critical section, the following inequality shall be verified:

$$S_d \leq \frac{1}{\gamma_R} \cdot R_d\left(\frac{f_{mk}}{\gamma_{mk}}\right)$$

where, S_d represents the design action effect and R_d the corresponding design resistance determined on the basis of the characteristic resistance f_{mk} of component materials. The partial safety coefficients γ_m are applicable to resistance of materials in the seismic design situation. The coefficient γ_R represents the partial safety factor for the seismic combination, and by default is taken as equal to 1.

Serviceability limit state

It shall be verified that the non-structural elements and their connections to the structure maintain their resistance and function when the structure is subjected to the maximum displacements. The displacements of a structure are limited to the following values:

$d=H/250$ at the top of the building of height H;

$d'=h/100$ for all interstorey displacements (h – storey height)

UBC-97

Ultimate limit state

Only the elements of the designated seismic-force-resisting-system shall be used to resist design forces. The individual components shall be designed to resist the prescribed design seismic forces acting on them. The components shall also comply with the specific requirements for materials. Structures and all portions thereof shall resist the most critical effects from the following combinations of factored loads:

$$1.2D+1.0E+(f_1L+f_2S)$$
$$0.9D\pm1.0E$$

The following special load combinations for seismic design shall be used where required:

$$1.2\mathrm{D}+f_1L+1.0E_m$$
$$0.9D\pm1.0E_m$$

Serviceability limit state

Storey drifts computed using the maximum inelastic response displacement shall not exceed 0.025 times the story height for structures having a fundamental period of less than 0.7 second. For structures having a fundamental period of 0.7 second or greater, the calculated storey drift shall not exceed 0.020 times the storey height. These drift limits may be exceeded when it is demonstrated that greater drift may be tolerated by both structural and non-structural elements that can affect life safety.

AIJ_{LSD}–90

Ultimate limit state

Design condition shall be that the design strength is greater than or equal to the required strength, where the design strength is obtained by multiplying the nominal value of *ultimate* limit strength of structural element by the resistance factor; while the required strength is calculated based on the combination of factored loads derived by multiplying the nominal load by the load factor specified for the *ultimate* limit state design.

In the ultimate limit state design, the load combinations and load factors shall be given by:

$$1.1W_D+2.0W_E+0.4W_L$$
$$(1.1W_D+1.7W_E+0.4W_S+0.4W_L \text{ in areas with heavy snow})$$

Serviceability limit state

a) Design for serviceability limit strengths

Design condition shall be that the design strength is greater than or equal to the required strength, where the design strength is obtained by multiplying the nominal value of *serviceability* limit strength of structural element by the resistance factor; while the required strength is calculated based on the combination of factored loads derived by multiplying the nominal load by the load factor specified for the *serviceability* limit state design.

b) Design for serviceability limit deformations

Design condition shall be that the design value of deformation of structural element or structure (deflection, storey drift) is less than or equal to the allowable value of serviceability limit deformation, where the design value of deformation is obtained based on the combination of factored loads derived by multiplying the nominal load by load factor specified for the serviceability limit state design.
In the serviceability limit state design, the load combinations and load factors shall be given by:

$$1.0W_D+0.4W_E+0.4W_L$$
$$(1.0W_D+0.4W_E+0.4W_S+0.4W_L \text{ in areas with heavy snow})$$

Serviceability limit state design is to be based on elastic analysis. Design for serviceability limit states includes design against yielding, buckling, slipping and deflection of all structural members and storey deformations of the structure.

Storey drifts of frames shall be controlled by as appropriate in consideration of the following:

- to prevent damage to non-structural elements like exterior and partition walls
- to preserve proper function of doors and windows
- to preserve proper function of pipes, ducts, and other utilities

Unless trouble is found, the storey drift of ordinary buildings shall not be greater than 1/200 times the storey height. If analysis shows that a larger storey drift is acceptable, the limit storey drift can be enlarged, but shall not be greater than 1/120 times the storey height.

AIJ_L-93

In limit state design, structural reliability can be achieved by accounting for uncertainties of structural material strength or resistance and loading conditions. The design philosophy quantifies either reliability or failure probability with respect to two limit states: ultimate and serviceability. Design loads for the two limit states are:

Ultimate limit state

$$\gamma_D D+\gamma_L L+\gamma_H H+\gamma_S S+{}_U\gamma_E E$$

Serviceability limit state

$$\gamma_D D+\gamma_L L+\gamma_H H+\gamma_S S+{}_S\gamma_E E$$

Comparative Remarks

First of all it shall be noted that AIJ_L-93 presents generically the limit state philosophy. Partial load factors are not given directly, instead, they are to be computed based on a target reliability. However, there is insufficient data in this code to permit determination of partial load factors for design. Neither are specified any specific provisions to be satisfied by structures or structural elements (e.g. interstorey drift limits), the code being, in fact, limited to the evaluation of earthquake load (beside others) on buildings. Consequently, it will not be considered from this point on.

Three (EC8-94, PS 92 and AIJ_{LSD}-90) of the remaining four codes define two limit states to be verified for the seismic design of structures: ultimate limit state and serviceability limit state (termed deformation limit state in PS 92). UBC-97 is the only one that does not specify directly the use of the two limit states. The codes are quite homogeneous in what regard the ultimate limit state (no-collapse requirement). However, the way in which the serviceability limit state is considered is rather different.

The only code that specify and effectively use the two limit states, with corresponding severe and moderate design earthquakes is AIJ_{LSD}-90. The structure must remain in the elastic range for the serviceability limit state (structural damage not allowed), and shall satisfy certain deformation limits (limitation of non-structural damage).

EC8-94 defines the moderate earthquake for the serviceability limit state, but only non-structural damage has to be limited. PS 92 and UBC-97 go one step backward, by specifying a single design earthquake. Anyway, EC8-94, PS 92 and UBC-97 make use of a one-level design procedure, in which serviceability limit state is verified by interstorey drift limits only. The corresponding design displacements are computed parting from the ultimate limit state earthquake. It is to be noted that structural damage (yielding) under this check is not precluded directly. A review of the serviceability limit state related provisions and drift limits is presented in TABLE 1.1.13.

TABLE 1.1.13

SERVICEABILITY LIMIT STATE RELATED PROVISIONS AND DRIFT LIMITS

	EC8–94	PS 92	UBC-97	AIJ_{LSD}–90
Moderate earthquake	Yes	No	No	Yes
Definition of the serviceability limit state	Yes	Yes	No	Yes
Definition of interstorey drift limits	Non-structural	Non-structural	Structural and non-structural	Non-structural
Structural damage under moderate earthquake	Not precluded	- *	- *	Precluded
Computation of the design dis-placement	$\frac{q_d \cdot d_e \cdot \gamma_I}{\nu}$	From the design elastic spectrum	$0.7 \cdot R \cdot \Delta_S$ ρ=1.0 for displace-ment analysis	From the moderate design action
Interstorey drift limits	$h/166 - h/250$ $(0.006h - 0.004h)$	$h/100$	$h/50$ - $h/40$ (0.020h – 0.025h)	$h/200$

Note: see TABLE 1.1.10 for definition of terms

* moderate earthquake not defined

It can be seen that both interstorey drift limits, as well as the way in which the design displacements are computed vary from code to code. However, it can be noted that drift limits specified in the codes are dependent on the non-structural damage. Moreover, in the case of EC8-94 (with an average drift limit of $h/200$) and AIJ_{LSD}-90, codes which explicitly define the moderate earthquake for serviceability limit state, the drift limits are those normally assumed for non-structural deformation checks under service loads (other than earthquake ones). The drift limits imposed by PS 92 and UBC-97 can be interpreted as means to achieve the service limit state in one-level design procedure.

Further, it may be noticed that the displacement analysis is practically independent of the structural behaviour factor, conclusion valid for all the codes. The use of the behaviour factor for computation of design displacements in the case of EC8-94 and UBC-97 is merely a way to consider the elastic design response spectrum instead of the inelastic one.

Consider the drift limit for non-structural elements under service loads as being equal to *h*/200, and the above mentioned assumption of interpreting drift limits imposed by PS 92 and UBC-97 as means to achieve the service limit state in one-level design procedure. A comparison can be made in terms of the reduction factor that takes account of the lower return period of the moderate earthquake defining the serviceability limit state. The determination of the reduction factor is straightforward in the case of EC8-94 (ν=2.0) and AIJ_{LSD}-90 (ratio of partial safety factors corresponding to ultimate and serviceability limit states 2.0/0.4=5.0). By normalising code-specified drift limits to *h*/200, reduction factors of 2.0 in the case of PS 92, and 5.7 for UBC-97 ($T\geq0.7$ sec) will yield.

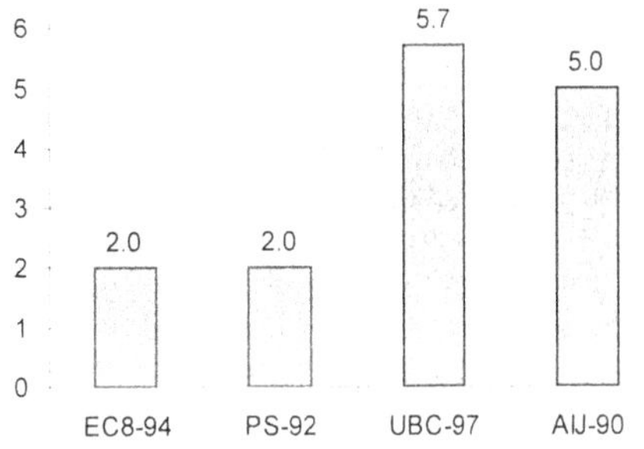

Figure 1.1.16: Serviceability earthquake reduction factor

It can be concluded from Figure 1.1.16 that moderate earthquake corresponding to the equivalent serviceability seismic action is the smallest in the case of UBC-97, while both EC8-94 and PS 92 yield the highest service seismic action, which will correspond to the most stringent deformation limitation. Finally, it must be noted that AIJ_L-93, though not considered here, provides a means for direct consideration of design earthquakes with different return periods, by mean of the R_A and R_V factors.

DESIGN CRITERIA FOR STEEL MOMENT RESISTING FRAMES

General

Generally speaking, the structural design procedure is split in several parts. Separate checks are performed at the structure, member, and cross-section levels. This simplified procedure proved adequate for design of structures in the elastic range. While the usual type of structural analysis is the linear elastic one, structure is expected to behave in the plastic range under the design earthquake. The approach based on separate checks is kept for the earthquake design of structures due to its simplicity. Anyway, much more care is needed due to design for ductility beside design for strength and deformation.

In what concerns steel structures, it has to be pointed out that the ductile nature of the structural steel material does not translate into inherently ductile structures. Although steel remains one of the most ductile of the modern structural engineering materials, special care is necessary to ensure ductile structural behaviour, even when structural stability is not a concern. The design of steel structures for ductile response requires (1) material ductility, (2) cross-section and member ductility, and (3) structural ductility (it is generally preferably to protect connections against yielding, although, in some instances, such as with semirigid connections, connection ductility may be unavoidable and substitute for member ductility). Further, a hierarchy of yielding must be imposed on a structure to ensure a desirable failure mode.

Code provisions

EC8–94

Earthquake resistant steel buildings shall be designed according to one of the following concepts:

- Dissipative structural behaviour (the capability of parts of the structure to resist earthquake actions out of their elastic range is taken into account; the behaviour factor $q > 1.0$).
- Non-dissipative structural behaviour (the actions effects are calculated on the basis of an elastic global analysis without taking into account non-linear material behaviour; $q = 1.0$).

In what regards materials, in dissipative zones, the following additional rules apply:

- In bolted connections high strength bolts in category 8.8 or 10.9 should be used in order to comply with the needs of capacity design. Bolts of category 12.9 are only allowed in shear connections.
- The value of the yield strength that cannot be exceeded by the material used in the fabrication of the structure and the tensile strength of the steel used should be specified. During the fabrication it shall be ensured that the value of the yield strength of the steel material actually used does not exceed by more than 10% the value f_y used in the design. During construction it shall be ensured that the distribution of the yield strength throughout the structure does not substantially differ from the distribution used in the design. This condition is satisfied, when the population of the ratios r_i of the actual yield strength f_{yri} to the yield strength f_{yi} considered in the design for the same component i is such, that:

$$(\max r_i - \min r_i) \leq 0.2$$

For moment-resisting frames, the behaviour factor q is determined as:

$$q = 5 \cdot \frac{\alpha_u}{\alpha_1}$$

α_l – multiplier of the horizontal design seismic action, while keeping constant all other design actions, which corresponds to the point where the most strained cross-section reaches its plastic moment resistance

α_u – multiplier of the horizontal design seismic action, while keeping constant all other design actions, which corresponds to the point where a number of sections, sufficient for the development of the overall structural instability, reach their plastic moment resistance

$$\alpha_u / \alpha_l \leq 1.6$$

When calculations are not performed in order to evaluate the multiplier α_u, approximate values of α_u/α_l may be used (α_u/α_l=1.2 for multi-storey moment resisting frames). If the building is not regular in elevation, the q-values as determined above should be reduced by 20%.

Structures with dissipative zones shall be designed such that these zones develop in those parts of the structure where yielding or local buckling or other phenomena due to hysteretic behaviour do not affect the overall stability of the structure. Structural parts of dissipative zones shall have adequate ductility and resistance. Non-dissipative parts of dissipative structures and the connections of the dissipative parts to the rest of the structure shall have sufficient overstrength to allow the development of cyclic yielding of the dissipative parts.

Sufficient local ductility of members or parts of members in compression should be ensured by restricting the width-thickness ratio b/t according to the cross sectional classes, the latter being influenced by the value of the selected behaviour factor as shown in TABLE 1.1.14.

TABLE 1.1.14

REQUIREMENTS FOR CROSS SECTIONAL CLASSES DEPENDING ON THE VALUES OF THE SELECTED Q-FACTOR

behaviour factor q	$4 < q$	$2 < q \leq 4$	$q \leq 2$
cross sectional class	class 1	class 2	class 3

Connections in dissipative zones should have sufficient overstrength to allow for yielding of the connected parts. Connections of dissipative parts made by means of butt welds or full penetration welds are deemed to satisfy the overstrength criterion. For fillet weld connections or bolted connections the following requirement should be met:

$$R_d \geq 1.20\, R_{fy}$$

where
R_d – resistance of the connection
R_{fy} – plastic resistance of the connected part

The overstrength condition for connections need not apply if the connections are designed in a manner enabling them to contribute significantly to the energy dissipation capability inherent to the chosen q-factor. The effectiveness of such connection devices and their strength under cyclic loading should be established by tests.

Moment resisting frames shall be designed so that plastic hinges form in the beams and not in the columns. The beam to column joints shall have adequate overstrength to allow the plastic hinges to be formed in the beams.

Beams should be verified as having sufficient safety against lateral or lateral-torsional buckling failure. For plastic hinges in the beams it should be verified that the full plastic moment resistance and rotation capacity is not decreased by compression and shear forces. To this end, the following inequalities should be verified at the location where the formation of hinges is expected:

$$M_{Sd}/M_{pl,Rd} \leq 1.0$$
$$N_{Sd}/N_{pl,Rd} \leq 0.15$$
$$(V_{G,Sd} + V_{M,Sd})/V_{pl,Rd} \leq 0.5$$

where:
M_{Sd}, N_{Sd} – design action effects
$M_{pl,Rd}$, $N_{pl,Rd}$, $V_{pl,Rd}$ – design resistances
$V_{G,Sd}$ – shear force due to non-seismic actions
$V_{M,Sd}$ – shear force due to the application of the resisting moments $M_{Rd,A}$ and $M_{Rd,B}$ with opposite signs at the extremities A and B of the beam

The sum of the design values for the bending moments in the adjacent cross-sections of the columns shall not be less than the sum of the resisting moments M_{Rd} of the beams connected to the columns:

$$\sum M_{Rd,b} \leq \sum M_{Rd,c}$$

At the base of the frame, the design bending moments for the connection of the columns to the foundations should be taken as:

$$M_{Sd} = M_{Sd,G} + 1.2\ M_{Sd,E}$$

where:
$M_{Sd,G}$ – bending moment due to non-seismic actions included in the combination of actions for the seismic design situation
$M_{Sd,E}$ – bending moment due to the design seismic action multiplied by the importance factor

The column shear force V_{Sd} (resulting from the structural analysis) should be limited to:

$$V_{Sd} / V_{pl,Rd} \leq 0.5$$

In framed web panels of beam/columns connections, the following assessment is permitted:

$$V_{wp,Sd} / V_{wp,Rd} \leq 1.0$$

The connections of the beams to the columns should be designed for the required degree of overstrength, taking into account the moment resistance $M_{pl,Rd}$ and the shear force ($V_{G,Sd} + V_{M,Sd}$).

PS 92

Steel structures can be designed according to one of the following concepts:

- Dissipative structural behaviour (structures are designed so as in the case of a seismic event, certain elements are subjected to plastic deformations; the location and efficiency of these dissipative zones shall be perfectly controlled; the behaviour factor $q > 1.0$).
- Non-dissipative structural behaviour (structures are designed so as to resist the design seismic action in the elastic range; $q = 1.0$).

In what regards structural steel in the dissipative and adjacent non-dissipative zones: variation of the actual yield limit and the one used in design shall not change the location of the dissipative zones. More precisely, if the maximal ratio of the actual yield limit to the design one in a dissipative zone (f_{yr}/f_y)max, is greater by more than 15% than the minimum ratio of the same type (f_{yr}/f_y)min in a non-dissipative zone, it is required to reconsider the design.

In the case of regular structures and medium irregular structures, the behaviour factor is determined as follows (for moment resisting frames):

$$q = 5\alpha_u/\alpha_l \leq 8$$

For structures designed with a behaviour factor q>1.0, the cross section of elements in dissipative zones working in compression and/or bending, shall satisfy the criteria of cross-sectional classes from TABLE 1.1.15. In order to use a behaviour coefficient q>6, all the cross-section of dissipative elements shall be of class A, and the axial design force N_{Sd} and the reduced slenderness $\bar{\lambda} = \lambda / \lambda_y$ shall satisfy the conditions:

- member in bending with double curvature: $N_{Sd}/N_{pl,Rd} \leq 0.15$ and $\bar{\lambda} \leq 1.1$
- member in bending with single curvature: $N_{Sd}/N_{pl,Rd} \leq 0.15$ and $\bar{\lambda} \leq 0.65$

where:
$N_{pl,Rd}$ – the design plastic resistance of the member to compression

The use of semirigid joints is not permitted, except for a scientific justification established by testing. Connections realised by means of butt welds or full-penetration welds do not require any weld verification. Connections realised by means of partial-penetration welds or fillet welds, as well as bolted connections, shall satisfy the following general condition:

$$S_d \le R_d/\gamma_E$$

where:
R_d – the design resistance of the connection (connecting elements as well as constitutive parts)
γ_E – the partial safety factor (γ_E = 1.0 for connections in non-dissipative zones, γ_E = 1.2 for connections adjacent to dissipative zones)
S_d – the design action effect

TABLE 1.1.15

REQUIREMENTS FOR CROSS SECTIONAL CLASSES DEPENDING ON THE VALUES OF THE SELECTED Q-FACTOR

behaviour factor q	$q \le 6$	$q \le 4$	$q \le 2$
cross sectional class	class A	class B	class C

Dissipative columns shall be of class A. Beside all the resistance checks, the following conditions shall satisfy:

- for a column in double curvature bending:

$$\frac{N_{Sd}}{N_{pl,Rd}} + 0.8\bar{\lambda} \le 1.0\text{, if } \frac{N_{Sd}}{N_{pl,Rd}} \ge 0.15$$

$$\frac{N_{Sd}}{N_{pl,Rd}} + 0.8\bar{\lambda} \le 1.6\text{, if } \frac{N_{Sd}}{N_{pl,Rd}} < 0.15$$

- for a column in single curvature bending:

$$\frac{N_{Sd}}{N_{pl,Rd}} + 1.35\bar{\lambda} \le 1.0\text{, if } \frac{N_{Sd}}{N_{pl,Rd}} \ge 0.15$$

$$\frac{N_{Sd}}{N_{pl,Rd}} + 1.35\bar{\lambda} \le 1.1\text{, if } \frac{N_{Sd}}{N_{pl,Rd}} < 0.15$$

The column shall be checked for resistance (N_{Sd} and M_{Sd}) and buckling. The design shear in dissipative columns shall be limited to 1/3 of the plastic shear resistance ($V_{Sd} \le V_{pl,Rd}/3$).

Non-dissipative columns part of dissipative structures shall be checked to resistance (N_{Sd} and M_{Sd}) and buckling, with the remark that the partial safety factor γ_E=1.2.

The design moment resistance of beams with dissipative zones is equal to the plastic moment resistance $M_{pl,Rd}$ for class A and B cross sections, and to the elastic moment resistance $M_{el,Rd}$ for class C cross sections. Beams of class A and B cross sections, when the dissipative behaviour is accomplished by flexure, shall satisfy the following where plastic hinges are to be formed:

$$M_{Sd}/M_{pl,Rd} \le 1.0\text{, with } N_{Sd}/N_{pl,Rd} \le 0.15 \text{ and } V_{Sd}/V_{pl,Rd} \le 1/3$$

If $N_{Sd}/N_{pl,Rd} > 0.15$, the beam shall be considered as an element in compression and bending.

UBC-97/ AISC-97

Seismic provisions for structural steel buildings of UBC-97 are based on the AISC-92 provisions, with some amendments. This is probably due to lack of synchronisation between the two documents. Therefore, specific issues treated here are those of the up to dated AISC-97.

AISC-97 seismic design provisions are intended for the design and construction of structural steel members and connections in the seismic force resisting systems in buildings for which the design forces have been determined on the basis of various levels of energy dissipation in the inelastic range of response. The reduction factor R and the overstrength factor Ω_0 are to be determined from TABLE 1.1.16 on the basis of the structural system.

TABLE 1.1.16

STRUCTURAL SYSTEMS

Basic structural system	Description	R	Ω_0
Moment-resisting frame system	Special moment-resisting frame (SMRF)	8.5	2.8
	Ordinary moment-resisting frame (OMRF)	4.5	2.8

The required strength of a connection or a related member shall be determined from the expected yield strength F_{ye} of the connected member, where

$$F_{ye} = R_y F_y$$

F_y is the specified minimum yield strength of the grade of steel to be used. For rolled shapes and bars, R_y shall be taken as 1.5 for ASTM A36 and 1.3 for A572 Grade 42.

Specific requirements for bolted joints comprise that all bolts shall be fully tensioned high-strength bolts. Bolted joints shall not be designed to share load in combination with welds on the same faying surface. Bolted connections for members that are part of the seismic force resisting system shall be configured such that a ductile limit-state either in the connection or in the member controls the design.

Columns, when $P_u/\phi P_n$ is greater than 0.4, shall be designed according to the following requirements: The required axial compressive and tensile strength, considered in the absence of any applied moment, shall be determined form the special load combinations. The required strengths determined in this way need not exceed either of the following:

- The maximum load transferred to the column considering $1.1R_y$ times the nominal strengths of the connecting beam or brace elements of the building.
- The limit as determined from the resistance of the foundation to the overturning uplift.

P_u – required axial strength of a column
P_n – nominal axial strength of a column
ϕ – resistance factor

Special moment resisting frames

Special moment resisting frames (SMF) are expected to withstand significant inelastic deformations when subjected to the forces resulting from the motions of the design earthquake. The design of all beam to column connections used in the seismic force resisting system shall be based upon qualifying cyclic test results that demonstrate an inelastic rotation of at least 0.03 radians. Qualifying test results shall consist of at least two cyclic tests and are permitted to be based upon one of the following requirements:

- Tests reported in research or documented tests performed for other projects that are demonstrated to reasonably match project conditions.

- Tests that are conducted specifically for the project and are representative of the project member sizes, material strengths, connection configurations, and matching connection processes.

Beam-to-column connection testing shall demonstrate a flexural strength, determined at the column face that is at least equal to the nominal plastic moment of the beam M_p at the required inelastic rotation, except as follows:

- When beam local buckling rather than beam yielding limits the flexural strength of the beam, or when connections incorporating a reduced beam section are used, the minimum flexural strength shall be $0.8M_p$ of the tested beam.
- Connections that accommodate the required rotations within the connecting elements and maintain the required strength, provided it can be demonstrated by rational analysis that any additional drift due to connection deformation can be accommodated by the building.

The required shear strength V_u of a beam-to-column connection shall be determined using the load combination $1.2D + 0.5L + 0.2S$ plus the shear resulting from the application of $1.1R_yF_yZ$ in the opposite sense on each end of the beam. Alternatively, a lesser value of V_u, is permitted if justified by rational analysis. The required shear strength need not exceed the shear resulting from the load combination $1.2D+0.5L+0.2S+\Omega_0 Q_E$.

The required shear strength R_u of the panel-zone shall be determined by applying special load combinations to the connected beam or beams in the plane of the frame at the column. R_u need not exceed the shear force determined from 0.8 times ΣR_yM_p of the beams framing to the column flanges at the connection.

Width-thickness ratios: beams shall comply with λ_p, in TABLE 1.1.17. When the ratio $\Sigma M^*_{pc}/\Sigma M^*_{pb}$ is less than or equal to 1.25, columns shall comply with λ_p in TABLE 1.1.17.

TABLE 1.1.17

LIMITING WIDTH-THICKNESS RATIOS λ_p FOR COMPRESSION ELEMENTS IN AISC-97

Description of element	Width-thickness ratio	Limiting width-thickness ratios
Flanges of I-shaped rolled beams, hybrid or welded beams and channels in flexure	b/t	$52/\sqrt{F_y}$
Webs in combined flexure and axial compression	h_c/t_w	for $P_u/\phi_b P_y \le 0.125$: $\dfrac{520}{\sqrt{F_y}}\left[1-1.54\dfrac{P_u}{\phi_b P_y}\right]$ for $P_u/\phi_b P_y > 0.125$: $\dfrac{191}{\sqrt{F_y}}\left[2.33-\dfrac{P_u}{\phi_b P_y}\right] \ge \dfrac{253}{\sqrt{F_y}}$

* F_y in ksi.

The following relationship shall be satisfied at beam-to-column connections:

$$\Sigma M^*_{pc}/\Sigma M^*_{pb} > 1.0$$

where:

ΣM^*_{pc} – the sum of moments in the column above and below the joint at the intersection of the beam and column centrelines. ΣM^*_{pc} is determined by summing the projections of the nominal flexural strengths of the column (including haunches where used) above and below the joint to the beam centreline with a reduction for the axial force in the column. It is permitted to take $\sum M^*_{pc} = \sum Z_c(F_{yc}-P_{uc}/A_g)$.

$\sum M^*_{pb}$ – the sum of the moment(s) in the beam(s) at the intersection of the beam and column centrelines. $\sum M^*_{pb}$ is determined by summing the projections of the nominal beam flexural strength(s) at the plastic hinge locations to the column centrelines. It is permitted to take $\sum M^*_{pb} = \sum(1.1R_yM_p+M_v)$, where M_v is the additional moment due to shear amplification from the location of the plastic hinge to the column centreline.

Beam-to-column connection restraint: column flanges at beam-to-column connections require lateral support only at the level of the top flanges of the beams when a column is shown to remain elastic outside of the panel-zone. When a column cannot be shown to remain elastic outside of the panel-zone, the following requirements shall apply:
a. The column flanges shall be laterally supported at the levels of both the top and bottom beam flanges.
b. Each column-flange lateral support shall be designed for a required strength that is equal to 2 percent of the nominal beam flange strength ($F_y b_f t_{bf}$).
c. Column flanges shall be laterally supported, either directly or indirectly, by means of the column web or by the flanges of perpendicular beams.

A column containing a beam-to-column connection with no lateral support transverse to the seismic frame at the connection shall be designed using the distance between adjacent lateral supports as the column height for buckling transverse to the seismic frame.

Intermediate moment resisting frames

Intermediate moment frames (IMF) are expected to withstand moderate inelastic deformations when subjected to the forces resulting from the motions of the design earthquake. IMF shall be designed so that the earthquake-induced inelastic deformations are accommodated by the yielding of members of the frame when FR moment connections are used or by yielding of connection elements when PR moment connections are used.

IMF shall conform to the requirements for SMF except for the following modifications:
The design of all beam-to-column joints and connections used in the seismic force resisting system shall be based upon qualifying cyclic test results that demonstrate an inelastic rotation of at least 0.02 radians.
Beam-to-column connection testing shall demonstrate a flexural strength, determined at the column face, that is at least equal to the nominal plastic moment of the beam M_p at the required inelastic rotation, except as follows:

- When beam local buckling rather than beam yielding limits the flexural strength of the beam, or when connections incorporating a reduced beam section are used, the minimum flexural strength shall be $0.8M_p$ of the tested beam.
- Connections that accommodate the required rotations within the connection elements and maintain the design strength are permitted, provided it can be demonstrated by rational analysis that any additional drift due to connection deformation can be accommodated by the building. Such rational analysis shall include the effects of overall frame stability including second order effects.

Both flanges of beams shall be laterally supported directly or indirectly. The unbraced length between lateral supports shall not exceed $3600r_y/F_y$. In addition, lateral supports shall be placed near concentrated forces, changes in cross-section and other locations where analysis indicates that a plastic hinge will form during inelastic deformations of the IMF.

Ordinary moment resisting frames

Ordinary moment frames (OMF) are expected to withstand limited inelastic deformations in their members and connections when subjected to the forces resulting from the motions of the design earthquake.

Beam-to-column connections shall be made with welds or high-strength bolts. Connections are permitted to be FR or PR moment connections as follows:

- FR moment connections that are part of the seismic force resisting system shall be designed for a required flexural strength M_u that is at least equal to $1.1R_yM_p$ of the beam or girder or the maximum moment that can be delivered by the system, whichever is less. Partial-joint-penetration groove welds and fillet welds shall not be used to resist tensile forces in the connections. Alternatively, the design of all beam-to-column joints and connections used in the seismic force resisting system shall be based upon qualifying cyclic test results that demonstrate an inelastic rotation of at least 0.01 radians.
- PR moment connections are permitted when the following requirements are met:
  - The nominal flexural strength of the connection shall be equal to or exceed 50 percent of M_p of the connected beam or column, whichever is less.
  - Adequate rotation capacity shall be demonstrated in the connections by cyclic testing at rotations corresponding to the design story drift.
  - The stiffness and strength of the PR moment connections shall be considered in the design, including the effect on overall frame stability.

For FR moment connections, the required shear strength V_u of a beam-to-column connection shall be determined using the load combination $1.2D + 0.5L + 0.2S$ plus the shear resulting from M_u as defined above. For PR moment connections, V_u shall be determined from the load combination above plus the shear resulting from the maximum end moment that the PR moment connections are capable of resisting.

When FR moment connections are made by means of welds of beam flanges or beam-flange connection plates directly to column flanges, continuity plates shall be provided to transmit beam flange forces to the column web or webs. Such plates shall have a minimum thickness equal to that of the beam flange or beam flange connection plate. The welded joints of the continuity plates to the column flanges shall be made with either complete-joint-penetration groove welds, two-sided partial-joint-penetration groove welds combined with reinforcing fillet welds, or two-sided fillet welds and shall provide a design strength that is at least equal to the design strength of the contact area of the plate with the column flange. The welded joints of the continuity plates to the column web shall have a design shear strength that is at least equal to the lesser of the following:

- The sum of the design strengths at the connections of the continuity plate to the column flanges.
- The design shear strength of the contact area of the plate with the column web.
- The weld design strength that develops the design shear strength of the column panel-zone.
- The actual force transmitted by the stiffener.

Continuity plates are not required if tested connections demonstrate that the intended inelastic rotation can be achieved without their use.

AIJ_{LSD}–90

Structural frames (unbraced and braced) frames are to be classified in accordance with TABLE 1.1.18. Classification of width-thickness ratio of plate elements into P-I, P-II, P-III and P-IV is based on TABLE 1.1.19.

TABLE 1.1.18

CLASSIFICATION OF STRUCTURES IN AIJ_{LSD}-90

Structural classification	Structural frame
S-I	structural frames composed of members classified into P-I and L-I, conformable to provisions for joints
S-II	structural frames composed of members classified into P-II and L-II, conformable to provisions for joints
S-III	structural frames composed of members classified into P-III and L-III, conformable to provisions for joints
S-IV	structural frames composed of members classified into P-IV and L-IV

TABLE 1.1.19

CLASSIFICATION OF WIDTH-THICKNESS RATIO OF PLATE ELEMENTS IN AIJ_{LSD}-90 (H-SHAPED CROSS-SECTIONS ONLY)

<table>
<tr><th></th><th>P-I</th><th>P-II</th><th>P-III</th><th>P-IV</th></tr>
<tr><td>Beam of H-shaped section</td><td>$\frac{(b/t_f)^2}{(20\sqrt{F_{yf}})^2}+\frac{(d/t_w)^2}{(127/\sqrt{F_{yw}})^2}\le 1$</td><td>$\frac{(b/t_f)^2}{(21\sqrt{F_{yf}})^2}+\frac{(d/t_w)^2}{(136/\sqrt{F_{yw}})^2}\le 1$</td><td>$\frac{(b/t_f)^2}{(23\sqrt{F_{yf}})^2}+\frac{(d/t_w)^2}{(148/\sqrt{F_{yw}})^2}\le 1$</td><td>$17/\sqrt{F_{yf}} < b/t_f \le 37/\sqrt{F_{yf}}$
$100/\sqrt{F_{yw}} < d/t_w \le 276/\sqrt{F_{yw}}$</td></tr>
<tr><td rowspan="2">Column of H-shaped section</td><td>$\frac{(b/t_f)^2}{(20\sqrt{F_y})^2}+\frac{(d/t_w)^2}{(93/\sqrt{F_y})^2}\le 1$</td><td>$\frac{(b/t_f)^2}{(21\sqrt{F_y})^2}+\frac{(d/t_w)^2}{(99/\sqrt{F_y})^2}\le 1$</td><td>$\frac{(b/t_f)^2}{(23\sqrt{F_y})^2}+\frac{(d/t_w)^2}{(106/\sqrt{F_y})^2}\le 1$</td><td rowspan="2">$17/\sqrt{F_y} < b/t_f \le 37/\sqrt{F_y}$
$71/\sqrt{F_y} < d/t_w \le 159/\sqrt{F_y}$</td></tr>
<tr><td colspan="3">$d/t_w \le 71/\sqrt{F_y}$</td></tr>
</table>

in which:

F_y – yield stress of plate element (t/cm^2)

b – half width of a flange of H-shaped cross-section (cm)

d – depth of a web of H-shaped cross-section (cm)

t_f – thickness of a flange (cm)

t_w – thickness of a web (cm)

Classification of the slenderness ratio of beams into L-I, L-II, L-III and L-IV is based on

TABLE 1.1.20

CLASSIFICATION OF SLENDERNESS RATIO OF BEAMS IN AIJ_{LSD}-90

Classification	L-I	L-II	L-III	L-IV
Slenderness ratio λ_b	$\le 0.7\ {}_p\lambda_b$	$\le 0.8\ {}_p\lambda_b$	$\le {}_p\lambda_b$	$> {}_p\lambda_b$

in which:

λ_b – slenderness ratio for lateral buckling of a beam $= \sqrt{M_p / M_e}$

M_p – full plastic moment (t cm)

M_e – elastic lateral buckling moment (t cm)

${}_p\lambda_b$ – slenderness ratio corresponding to the plastic limit = $0.6+3(M_2/M_1)$

M_2/M_1 – ratio of end moments, measured for unbraced length, $|M_1| \ge |M_2|$. When bending is in double curvature, M_2/M_1 is taken positive.

For columns in the frames classified into S-I or S-II (unbraced and braced frames), the following limitations on the axial force ratio and slenderness ratio, are to be satisfied:

- the maximum axial force ratio: $n_y \le 0.75$
- combination of the axial force ratio and slenderness ratio: $n_y\ {}_f\lambda_c^2 \le 0.25$

where:
n_y – axial force ratio of a column = N/N_y
N – compressive axial force of a column (t)
N_y – yield strength of a column (t)
${}_f\lambda_c$ – slenderness ratio of a column $= \sqrt{N_y / {}_f N_e}$
${}_fN_e$ – elastic buckling strength of a column in the plane of bending = $\pi^2 EI/{}_kl_c^{\,2}$ (t)
I – moment of inertia about the axis of bending (cm^4)
${}_kl_c$ – buckling length n the plane of bending (cm)

For columns in the frames classified into S-I or S-II in which plastic hinges form, the following limitations on the axial force ratio and slenderness ratio, are to be satisfied in addition to the above limitations:

- when $-0.5 < M_2/M_1 \leq 1.0$

$$n_y\, \lambda_c^2 \leq 0.10\,(1+ M_2/M_1)$$

- when $-1.0 < M_2/M_1 \leq -0.5$

$$n_y\, \lambda_c^2 \leq 0.05$$

For columns in the frames classified into S-III or S-IV, the following is to be satisfied:

$$N/{}_fN_c \leq 1.0$$

In the ultimate limit state design for unbraced and braced frames against the combination of factored loads including earthquake load, the required load-carrying capacity of each storey of the frame may be reduced according to the structural classification and the structural characteristic factor shown in TABLE 1.1.21.

$$Q_R = D_S\, \gamma_E\, W_E$$

where:
Q_R – required lateral load-carrying capacity of a storey
D_S – structural characteristic factor
γ_E – load factor for the earthquake load
W_E – nominal earthquake load

TABLE 1.1.21

STRUCTURAL CHARACTERISTICS FACTOR OF UNBRACED AND BRACED FRAMES D_S IN AIJ_{LSD}-90

Structural characteristic factor	S-I	S-II	S-III	S-IV
Unbraced frame ${}_RD_S$	0.25	0.30	0.35	0.45
Braced frame ${}_BD_S$	${}_RD_S\,(1+0.4\,\beta\,\lambda_\beta) \leq 0.5$			0.50

The maximum flexural strength of beam-end or column-end connections in the frames classified into S-I, S-II, and S-III shall satisfy:

$${}_jM_u \geq 1.3\, M_p$$

where:
${}_jM_u$ – maximum flexural strength of beam-end or column-end connections
M_p – full-plastic moment of beam in case of beam-end connection, or full-plastic moment of column in case of column-end connection

The web panels in beam-to-column connections shall be designed to satisfy:

$$_pM_y \geq \{({}_bM_y^L + {}_bM_y^R) \quad and \quad ({}_cM_y^U + {}_cM_y^L)\}$$

where:

$_pM_y$ – yield strength of the panel in a beam-to-column connection
$_bM^L_y$, $_bM^R_y$ – flexural yield strengths of the beams attached to the left and right sides of the connection panel
$_cM^U_y$, $_cM^L_y$ – flexural strengths of the columns attached to the top and bottom of the connection panel

Bolts shall not be used in the connections subjected to vibration, impact or reversal of stress. Bolts shall not be used in primary load-carrying parts in steel structures 9 m or more in height or 13 m or more in span length. However, bolts may be used in such parts when the diameter of bolt holes is not greater than the bolt diameter plus 0.2 mm and furthermore, the nuts are prevented form losing.

Comparative Remarks

Two concepts of structural design are explicitly or implicitly assumed by the codes: dissipative and non-dissipative, the former being the usual one. Therefore, most of the design provisions concern avoidance of conditions that may lead to brittle failures and adoption of appropriate design strategies to allow for stable and reliable hysteretic energy-dissipation mechanisms.

The fact that nominal and actual characteristics of structural steel used in construction often are quite different, is recognised by the codes, with the exception of the Japanese one. Increased yield strength of the structural material is safe for non-dissipative design, but not for dissipative design, as the location and ductility of dissipative zones may change due to increased yield strength of some parts. Both of the European codes require the control of this phenomenon, while the American one goes further by specifying some coefficients to determine the expected yield strength of typical US structural steels. The expected yield strength is to be used instead of the nominal one for capacity design.

EC8 and PS 92 classify the sections of the frame members in 3 ductility classes based on the width-thickness ratio of plate elements. In fact, this classification which influences the q-factor values doesn't include any provisions for members. AIJ_{LSD}-90 provides a cross-sectional classification that accounts for the interaction of web and flange. Besides, a classification of frames according to their seismic performance (ductility) is accomplished based on both cross-section and member slenderness classification, which is more correct from the theoretical point of view. AISC-97 provides as well limitations on width-thickness ratios of plate elements of cross-sections. Moreover, a direct classification of frames (special, intermediate, and ordinary MRFs) according to their seismic performance is provided. This is accomplished by different levels of connection ductility (0.03, 0.02, and 0.01 radians), and limitations on member sizing and detailing. Qualifying tests on connections are needed for their use in special MRFs.

Values of the coefficient of reduction of seismic forces (behaviour factor) reflects the ability of the structure for energy dissipation, and therefore is given in all the codes in accordance to the frame classification as described above.

Usually, an overstrength is required for beam-column connections (120% in the case of European codes, 130% in the case of Japanese code, and 110%, but based on the expected yield strength in the American code). Anyway, with the exception of AIJ_{LSD}-90, possibility to use semirigid/partially restraint connections is permitted, provided that their energy dissipation capacity is proved by qualifying tests. Panel zone is allowed to yield in the European codes, also in the American code, where the required strength of the panel zone is only 80% of the expected plastic moment capacity of the beams framing into the column.

The concept of weak beam – strong column is required by EC8-94. It is also required by AISC-97, but only for special and intermediate MRFs. At the same time, equations from this latter code are more general, accounting for such effects as haunching, weakening, and presence of axial force in the column. AIJ_{LSD}-90 do not require stronger columns, neither do PS92. However, different column checks are required by the French code, as the column is dissipative or non-dissipative.

TOWARDS FUTURE SEISMIC DESIGN CODES

The first design requirements aimed to prevent building collapse during earthquakes were issued at the beginning of the 20th century. After a major earthquake struck San Francisco in 1906, reconstruction of the devastated city proceeded with an updated building code that required consideration of a wind force of 1.44 kPa for the design of new buildings. Gustav Eiffel introduced same type of provisions in Europe. After the Great Kanto earthquake (1923), a design procedure based on a lateral force equal to 10 percent of the building's weight was implemented in Japan. The 1927 UBC introduced the first seismic design requirements in North America, partly in response to the Santa Barbara earthquake of 1925 (a single horizontal point load, equal to 7.5 percent or 10 percent of the building's total load). Seismic design codes continued to evolve and improve over time. The increase of seismic risk nowadays, mainly due to increased number of urban agglomerations in seismic zones rather than increased seismic activity, require better codes for the future.

Many structural engineers believe that the future of seismic design lies in performance-based design. It was initiated in the SEAOC's Vision 2000 report (1995) and is expected to be developed into code provisions over the next 5 to 10 years. Performance-based engineering is defined as a set of procedures for engineering buildings to have controllable and predictable performance when responding to defined levels of earthquakes, within definable levels of reliability. Present day seismic codes are for the most par life-safety oriented. Performance-based design not only considers life-safety, but higher levels of performance, such as damage control and full functionality. The increasing complexity of our society's infrastructure, extensive use of high-technology manufacturing, prevalence of hazardous materials, and large populations in vulnerable areas, necessitates going beyond mere life-safety protection levels.

A number of new devices and systems have come into use in seismic design, particularly in seismic retrofit situations. Many buildings of the future, both new and existing, will have seismic protective devices, such as dampers and other energy absorbing mechanisms.

Nonlinear analysis, up to now mostly confined to very specialised problems and situations, is expected to be more frequently used in both static (pushover) and dynamic (time-history) seismic analyses. Most seismic design procedures to date have taken force based approaches. A displacement based approach (by means of nonlinear pushover analysis) is becoming very popular, as a more effective strategy for performance based engineering. The displacement approach recognises the close relationship between displacement and seismic performance. The inter-storey drift is global response parameter, which can serve as primary criterion for both structural and non-structural design.

A tendency can be observed world-wide of unification of various seismic design codes. In Europe, the idea to develop models for an international set of codes for structural design was born in 1974 based on an agreement between several technical-scientific organisations. In a first step, the individual codes and their relevant parts are published as European Prestandards (ENV). Eurocode 8 was published in ENV version on October 1994. After a test period, its transposition into European Standard (EN) is planned.

In US, the new International Building Code (IBC) is under development. The first edition of IBC is prepared to be published in the year 2000, and will replace the three model code groups existing in the United States: the Uniform Building Code (UBC), the National Building Code (NBC), and the Southern Building Code (SBC). IBC will be the first "national" building code to be used throughout the United States.

At the end of this chapter we find very useful to list on below the conclusions of the International Workshop on Seismic Design Methodologies for the Next Generation of codes, held in Bled, Slovenia, on June 24-27, 1997:

1. Future seismic engineering practice should be based on explicit and quantifiable performance criteria, considering multiple performance and hazard levels.

2. In order to develop guidelines for performance based engineering, addition research needs to be carried out to permit the development of a generally applicable design methodology and to address unresolved issues of ground motion modelling and demand and capacity evaluation. Emphasis in research should be placed on the following issues:
   - Development of a general methodology that permits performance based design at multiple performance and hazard levels, and with due consideration given to the complete soil-foundation-structure system, all nonstructural elements, and the building contents.
   - Development of performance criteria for structural and nonstructural components and systems, with a protocol for reporting research data in a performance format.
   - Development of improved procedures for the prediction of seismic demands.
   - Development of improved procedures for the assessment of strength and deformation capacities of components and systems at all performance levels.
   - Development of improved analytical methods and of efficient but practical tools for performance evaluation.

3. The most suitable approach for seismic design to achieve the objectives of performance based engineering appears to be deformation controlled design. It is recommended that deformation controlled design be implemented in future codes, both by enhancing force-based design through verification of deformation targets and by development of direct deformation based design approaches.

4. A great need exists to improve the dissemination of knowledge that will become a prerequisite for the practice of performance based engineering. Organisations involved in relevant research should disseminate research results expediently and in a consistent manner that makes the new knowledge accessible and useable to educators, students, practising engineers, and guideline and code writers.

5. A set of model buildings representing the most common style of construction worldwide, should be selected, documented and made generally available. These buildings should be used to calibrate new techniques, compare the effectiveness of various codes, and evaluate new design procedures. The set should include buildings that have experienced damage in earthquakes and, preferably, those that are instrumented. Similarly, a set of model bridges should be selected and documented.

REFERENCES

AIJ, (1990). *Standard for Limit State Design of Steel Structures (draft)*. Architectural Institute of Japan.

AIJ, (1993). *Recommendations for Loads on Buildings*. Architectural Institute of Japan

AISC 97, (1997). *Seismic Provisions for Structural Steel Buildings*. American Institute of Steel Construction, Inc. Chicago, Illinois, USA

Bruneau M., Uang C.-M., Whittaker A. (1998). *Ductile Design of Steel Structures*, McGraw Hill

Court, A. B., Kowalsky, M. J., (1998). Performance-Based Engineering of Buildings – a Displacement Design Approach. *Structural Engineering World Wide 1998*, T109-1, Elsevier Science, San Francisco, USA

EC 8 – 94 (1994). *Eurocode 8 – Design provisions for earthquake resistance of structures*. CEN European Committee for Standardisation

Fajfar, P., Krawinkler, H., (1997). *Seismic Design Methodologies for the Next Generation of Codes*. International Workshop held in Bled, Slovenia, June 24-27. Balkema, Rotterdam.

Gallagher, R. P., Bonneville, D. R., (1998). Future of the SEAOC Blue Book. *Structural Engineering World Wide 1998*, T109-2, Elsevier Science, San Francisco, USA

Kato, B., (1995). Development and design of seismic-resistant steel structures in Japan. *Proceedings of the International Workshop on Behaviour of Steel Structures in Seismic Areas – STESSA'94*, E&FN SPON, London, UK

Kircher, C. A., (1998). New Ground Shaking Design Criteria. *Structural Engineering World Wide 1998*, T109-5, Elsevier Science, San Francisco, USA

Mazzolani, F. M., Piluso, V., (1996). *Theory and Design of Seismic Resistant Steel Frames*. E&FN SPON, London, UK

McIntosh, R. D., Pezeshk, S., (1997). Comparison of recent U.S. Seismic Codes. *Journal of Structural Engineering*, **VOL. 123 NO. 8**

PS 92 (1995). *Norme Francaise. Regles de construction parasismique PS 92*. Association Francaise de Normalisation (AFNOR)

SEAOC, (1995). Structural Engineers Association of California – Vision 2000. *Conceptual Framework for Performance Based Engineering of Buildings*, Sacramento, CA, USA

UBC 97, (1997). *Uniform Building Code, Volume 2, Structural Engineering Design Provisions*. International Conference of Building Officials, Whittier, California, USA

1.2

INFLUENCE OF THE TYPE OF SEISMIC GROUND MOTIONS

Victor Gioncu, Graziella Mateescu, Lucia Tirca, Anthimos Anastasiadis

INTRODUCTION

The occurrence of earthquakes, their consequent impact on people and the facilities they live and work in, the evaluation and interpretation of the damages caused by severe ground motions are the principal items for structural engineers designing buildings in the seismic areas. The attempts to find answer to the question "Why such a damage could happen, after a wide amount of research works?" are ethical duties for all specialists. The damage of a structure during an earthquake represents a challenge for the structural engineers to improve the design methods.

The period 1985 – 1995 has been marked by increasing in the size of losses resulted from earthquake catastrophes. The first event of this decade was the Michoacan earthquake (1985), followed by Loma Prieta (1989), Northridge (1994) and Kobe (1995) events. If the economic losses in the decades 1965 – 1975 and 1975 - 1985 were of 10 and 67 billion US $, respectively, the losses in the 1985 – 1995 decade have increased dramatically to 182 billion US $. These losses are certainly enormous, but their size pales against the possible losses of 100 to 150 billion US $ if "the Big One " occurs in San Francisco, or 800 to 1,200 billion US $, in Tokyo, if the 1923 earthquake would happen today (Berz and Smolka, 1995).

The reason for this remarkable increase in economic losses can be defined as follows:

(i) Due to the growth of the world population, the concentration in people and the industrialisation of high- risk regions is a today phenomenon. As a consequence, the occurrence of probability that a great event strikes an important city dramatically has increased. For instance, the largest earthquake ever to hit the USA was concentrated in the New Madrid seismic zone (our days Memphis City, Tennessee, built several years later) in 1812. The reports on this event were phenomenal: overflow of rivers and uprooted trees, but the damages were not so dangerous because the region was relatively inhabited. How would the situation differ today? The Memphis area is now highly populated and an earthquake of a similar magnitude could involve costs of billions of dollars (MCEER, 1998). Unfortunately, strong earthquakes could strike a great number of similar urbanised sites everywhere in the world, with enormous losses.

(ii) For long time the main purpose of the seismic design was the protection of people against serious injuries or life loss and the prevention of the buildings from collapse under the maximum intensity earthquake, while the reduction of property damages was secondary. In the last time specialists have been awarded that the only fulfil of the live protection is economically unacceptable. They realised that it is necessary to pay more attention to the reduction of the building elements damages, for the all intensity ranges of the earthquakes. Consequently, a multi-level approach has been developed (Bertero, 1996).

(iii) Each event is basically unique, offering new surprises in the vulnerability of the affected building and showing the great complexity of the phenomenon. Referring to steel structures, for a long time no serious damages during some major earthquakes were recorded, so the persuasion that steel structures are safe solutions in seismic areas was consolidated among the structural designers. But the 1985 Michoacan earthquake produced in Mexico City the first collapse of a high steel building, due to the site conditions. The 1994 Northridge and 1995 Kobe earthquakes have involved many failures in steel moment-resisting frames, especially in the beam-to-column joints, mainly due to the near-source position of the structures, where high velocity was recorded. These failures represented the start of a large activity in the research work, where the control of the ductility for severe conditions, both in members and joints plays a leading role (Gioncu, 1999).

(iv) In spite of the progress of the research works, both theoretical and experimental, there is an important gasp to remark, which could be explained through the fact that always the implementation of advanced new concepts is constrained by the professional needs, to keep the design process as simple as possible. So, in spite of the great variability of the phenomena, the design methodology based on the use of design spectra contains only few design parameters (acceleration level, earthquake and structural natural periods), that are not sufficient for a proper description of the seismic loading. In the past, due to the reduced number of records during severe earthquakes (the famous El Centro records obtained in 1940 were for long time the alone information about the time-history ground motions), the developed design methods were mainly based on hypothesis, their accuracy was not easy to be verified. During the very last years, due to the development of a large network of instrumentation all over the world, there are a large amount of records of the ground motion, for different distances from the sources and different site conditions. The analysis of this new information has emphasised the diversity of ground typologies, neglected in the current design methodology. For instance, these new information offer the possibility to consider in the design the differences in the ground motion between near-source and far-source regions, generally ignored in the code provisions. For a proper design it is imperatively required to enlarge the code provisions with the parameters imposed by the new stage of the earthquake knowledge.

(v) Due to the development of the computer science, to perform a non-linear time-history analysis of structures is no more a problem. But the real problem consists in the option of an accelerogram to represent adequately the expected earthquake at the structure site. As a result of the extensive monitoring of the areas with high seismic risk, today it is possible to dispose of large number of records. All these records have random characteristics, due to the source particularities, magnitude, travel path, soil conditions, influence of neighboring building, etc. The choice is complicated by the fact that at the same site, and as the result of the same source, the ground motions could be different in characteristics for different events. So, it is difficult to select from a large number of available records, the specific accelerogram, with similar characteristics with those to take place on the site of the analyzed structure.

(vi) Due to the great variability of the accelerogram characteristics (peak, periods, pattern, duration) able to influence the response of the structure, the response values show a very large scatter. From the randomness of these results it is difficult to emphasise the main factors influencing the structure behaviour and to separate these ones from the secondary importance factors. All the studies of the structural mechanics begin with a determinist concept to ascertain the theory. Only in the second step, when the determinist response is well known, the study is extended to the random aspects. Contradictory with this rule, the seismic analysis, using a recorded accelerogram and a time-history computer program had jumped over the determinist step, passing directly to the randomness approach. This fact complicates the discerning of the main characteristics of the structural response. All these pointed out aspects show very clear, that in the aim to explain the actual response of the structure, it is imperatively required to use artificial generated accelerograms for practical purposes. The control parameters of these accelerograms should be chosen with regard to the main characteristics of the expected earthquake on the site, neglecting the secondary factors.

The purpose of this paper is to reveal the main characteristics of different earthquake ground motions and to study their influence on the structure response.

GROUND MOTION TYPOLOGIES

Chaotic vibrations

Usually, the first key to pronounce a judgement upon the characteristics of a ground motion is the chaotic vibrations observed on a recorded chart. The motion is observed to exhibit no visible pattern or periodicity, and to establish predictive rules for these chaotic vibrations is a very difficult task. The ability of the structural engineer to select the few parameters, which really govern the phenomenon, is probably the most important quality. To obtain approximate methods with sufficient accuracy is really the essence of engineering (Massonnet, 1982).

First of all, the problem of the predictability is the study of the chaotic motion in spatially coupled non-linear systems, soil and structure. A new theory of "Chaotic Dynamics" was developed in the last period, for chaotic vibrations (Moon, 1987, Schuster, 1988), especially for fluid mechanics, mechanical and electrical systems, which could be extended to the study of the ground motions produced by earthquakes. The essence of the theory consists in the recognition that geometrical and physical non-linear dissipative systems could admit non-periodic and chaotic motions, which behave in a random way, even through determinist excitations. For some initial conditions, a periodic movement is suddenly transformed in a chaotic movement. This transformation has a sensitive dependence on the initial conditions, and even they are not stochastic or random, the determinist system could have a chaotic motion. Small differences in the initial conditions conduct to great changes of the phenomenon, and the prediction of the system behaviour becomes a very intricate problem. So, the fundamental question raises: is it possible to predict the main factors to influence the chaotic motion? Even if there are still much more unsolved than solved problems, the answer for dissipative systems is yes. There are some properties of chaotic phenomena, discovered by the new theory of chaotic dynamics, which allow the prediction of the chaotic motion. One of them is the determinist feature of the parameters of the initial conditions. So, to recognize that it is possible to study the initial conditions for different ground motion typologies in a determinist way is the major result of the "chaotic vibrations theory".

Earthquake parameters

It is very well known that it is difficult to obtain exact values of the seismic actions, because of the irregularities in the occurrence of the earthquakes, as would like to use structural engineers. Some general proper information could be obtained by the examination of the factors influencing the ground motion. The ground motion sites are influenced by the following factors (Fig. 1.2.1): source characteristics, propagation path and site soil effects.

(i) *Source characteristics* are defined by the mechanism typology, spatial description and magnitude.

The *source mechanism* may be (Chandler et al, 1992) (Fig. 1.2.2):
- interplate mechanism, produced by sudden relative movement of two adjacent tectonic plates, at their boundaries. Very large magnitude and long natural vibration periods and duration characterize interplate earthquakes;
- intraplate mechanism, associated with relative slip across geological faults within a tectonic plate. The maximum credible earthquake produced by such a mechanism is generally smaller in magnitude, natural vibration periods and duration.

The Californian, Japanese and New Zealand earthquakes are specific interplate ground motions, while the European (excepted some Romanian and Greek sources), Canadian and Australian earthquakes are specific intraplate ground motions.

The *spatial description* refers to focal depth and directionality (Fig. 1.2.3). The focal depth has a considerable influence on the earthquake behaviour, the affected area being directly related to the depth. There are following earthquakes types:
- shallow crustal earthquake, with the hypocenter situated at depth till 25 km;
- crustal earthquake, with depth from 25 to 70 km;
- intermediate earthquake, with depth from 7o to 300 km;
- profound earthquakes, for depth over 300Km.

The shallow crustal events are the most frequent for Europe and adjacent areas, over 85% being ranged till 15 km (Ambraseys and Free, 1997) (Fig. 1.2.4). The framing of earthquakes into one of these types is very important, because intermediate and profound earthquakes may produce great events in regions far from the epicenter. In exchange, the crustal earthquake effects are limited only for short distances round about the epicenter. The directionality of the wave propagation is a significant characteristic of the ground motion in this case, because the damage of the structures is generally concentrated only along the rupture direction (Fig. 1.2.3b). It is interesting to notice that the damages are the largest in the direction perpendicular to the fault.

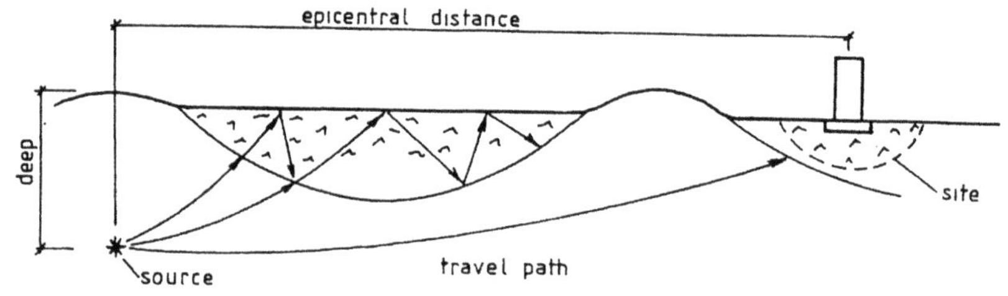

Figure 1.2.1: Factors influencing the ground motion

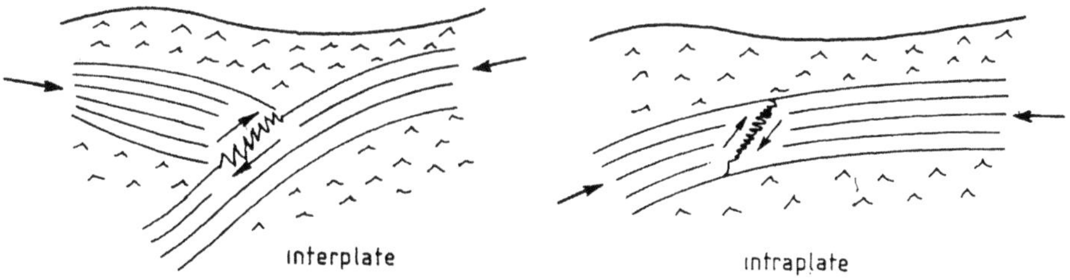

Figure 1.2.2: Mechanism types

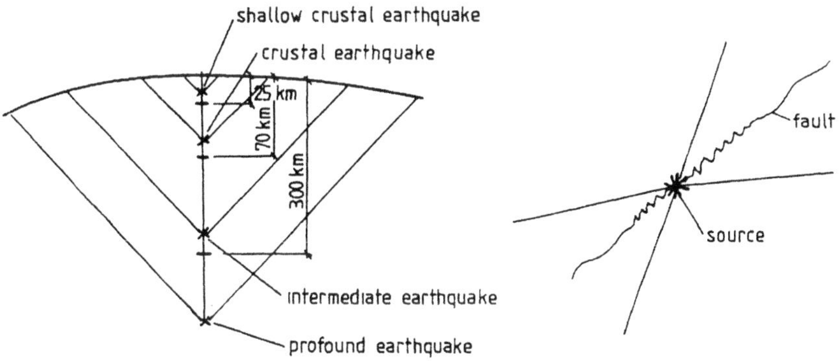

Figure 1.2.3: Influence of the source depth

The *magnitude* is determined in order to show the power of the ground motion. UBC (1997) proposed a classification of the sources in this relation, as:

- type A: faults, which are capable to produce large magnitude events (M > 7.0) and with high rate of seismic activity;
- type B: all faults other than types A and C (6.5<M<7.0);
- type C: faults, which are not capable to produce large magnitude earthquakes (M<6.5) and have a relatively low rate of seismic activity.

A classification of the earthquakes in function of the peak ground accelerations (Mazzolani and Piluso, 1996) is the following:

- low seismicity $a_g = 0.15g$
- moderate seismicity $a_g = 0.25g$
- high seismicity $a_g = 0.35g$

The more frequent ground motions for the European earthquake types have magnitudes from 4 to 5, showing low to moderate seismicity (Fig. 1.2.4).

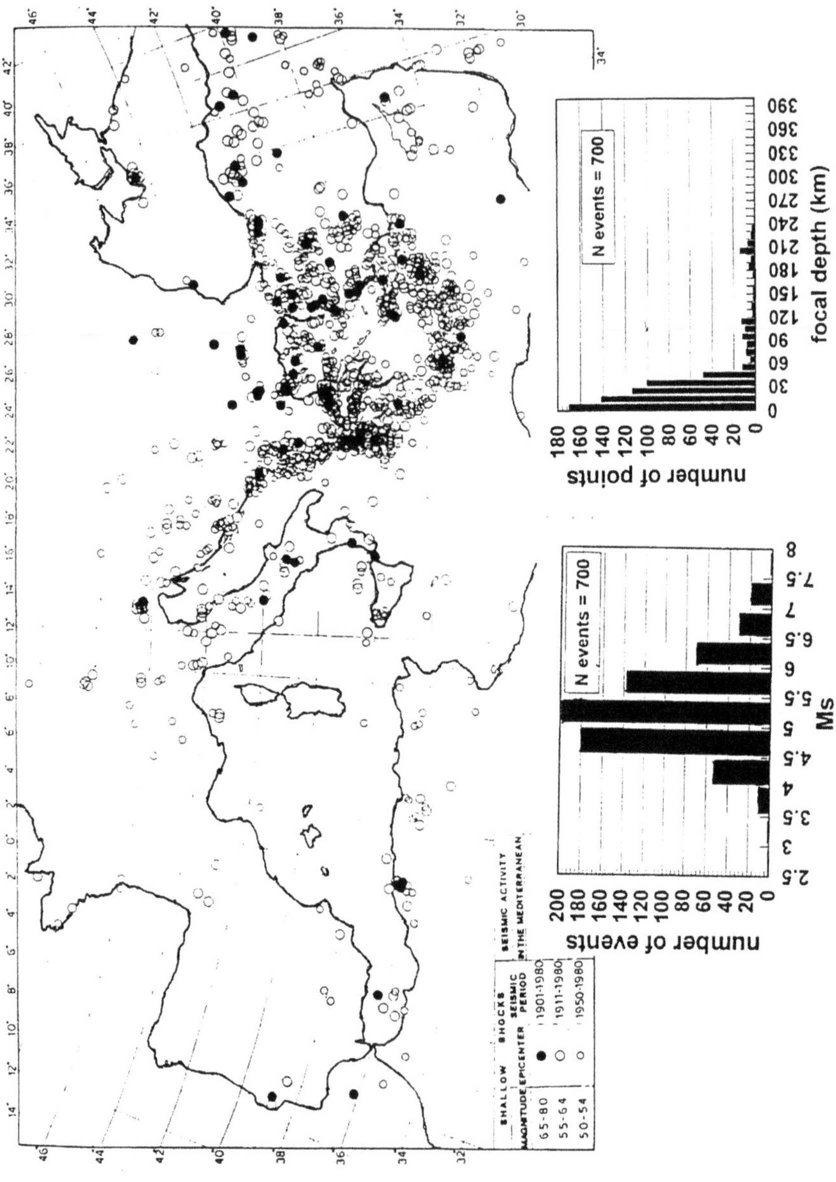

Figure 1.2.4: Mediterranean earthquakes

(ii) *Propagation path effects* are characterized by travelled soil conditions, attenuation of the ground motion and increasing of the duration.
The *travelled soil conditions* depend on the percentage of the path traveling through rock or through soft sediments and on the topographical irregularities. Results indicate accumulations of the peak acceleration in alluvium and soft rock soils.
The *attenuation of the ground motion* reveals the importance of the source-structure distance (Ambraseys and Bommer, 1991) (Fig. 1.2.5): reduced attenuation is observed for near-source sites and a significant reduction of the acceleration, in case of far-source sites. One can see that the focal depth for near-source sites has an important effect, while for far-source sites small differences were observed.
The *increasing of the duration* is influenced by the traveled soil conditions. Generally, one can observe that the increasing of the duration is proportional to the epicentral distance: the duration is generally shorter near the source and longer far from the source.

(iii) *Site soil effects* depend on local soil profile, dominant period and duration.
The *local soil profile* is characterized by multi-layers with different mechanical properties, thickness and their alternance. The classification of the soil conditions refers to rock, alluvium and soft soil sites.
The *dominant period* of the soil layers is a very important parameter, because the seismic waves may be amplified or attenuated in function of the resonance phenomenon. Generally, near the source, P-waves dominate the ground motion, far away from the source increases the importance of the S-waves. P-waves are characterized by short, while S-waves, by long periods.
The *duration* can be prolonged due to the site conditions. Rock soil has generally no effect on the increasing of the duration, but soft soil conducts to an increasing of 3 to 4 seconds.

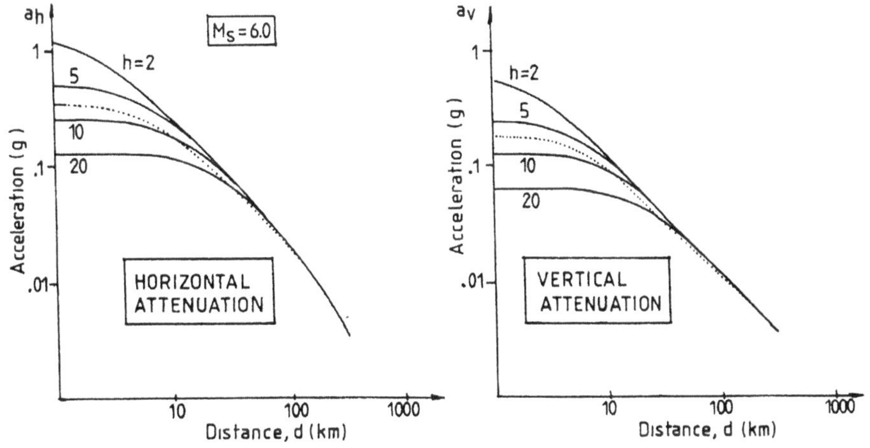

Figure 1.2.5: Attenuation of the ground motion

Near-source and far-source ground motions

During the past 20 years an ever-increasing data base of recorded earthquakes has indicated that the characteristics of the ground motion can vary significantly between recording stations, located in the same area. This is particular true for stations located near the epicentral region (Anderson and Bertero, 1987). Other important differences were marked between the characteristics of ground motions recorded in epicentral areas and at some distance from the epicenter. As a result of these findings, two main regions with different ground motions are considered (Fig. 1.2.6):

- the *near-source region* can be defined as the region within few kilometers of either the surface rupture or the projection on the ground surface of the fault rupture zone. The region is also referred as the near-field-region (Iwan, 1996);
- the *far-source region* situated at some hundred kilometers from the source.

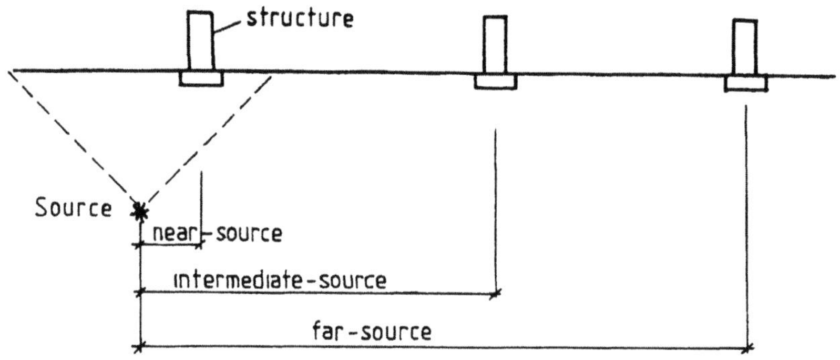

Figure 1.2.6: Earthquake types

Unfortunately, the ground motions and the design methods adopted in the majority of the codes are mainly based on records obtained from far-source fields, being incapable to describe in proper manner the earthquake action from the near-source field. Only the last version of the Uniform Building Code (1997) has introduced supplementary previsions concerning the near-source earthquakes, from the lessons learned from the last great events. The main differences between these two earthquake types are presented in Figure 1.2.7 (Gioncu and Mateescu, 1999).

Some historical earthquakes, Figure 1.2.8, recorded in epicentral areas and recent records, Figure 1.2.9, could be examined from the mentioned figures.

Characteristic	Near-source	Far-source
Main influence	Fault rupture direction 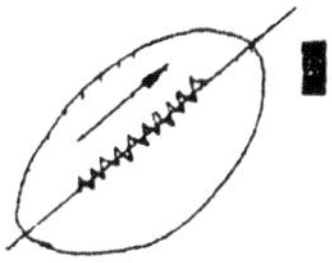	Soil stratification 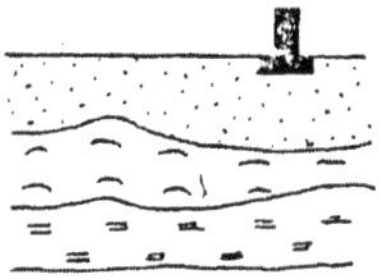
Ground motion	Impulse type 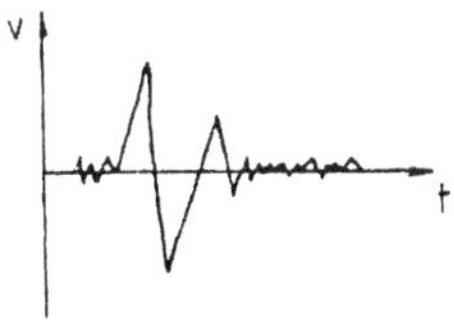	Cyclic type 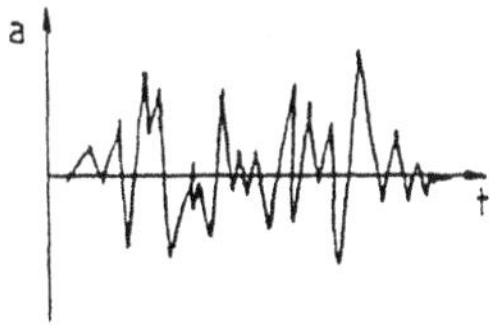
Vertical component	High values 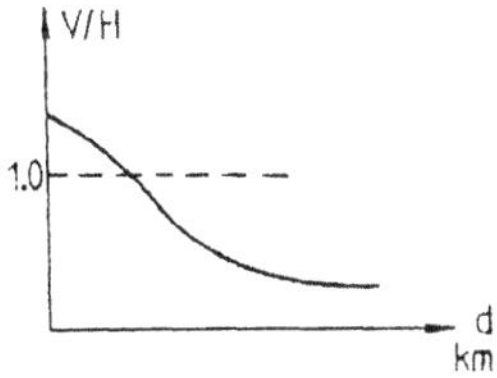	Reduced values 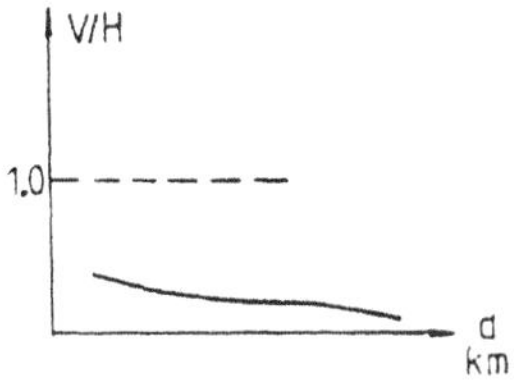
Velocity	High values 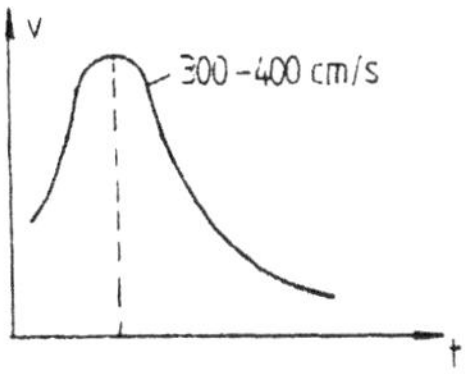	Reduced values 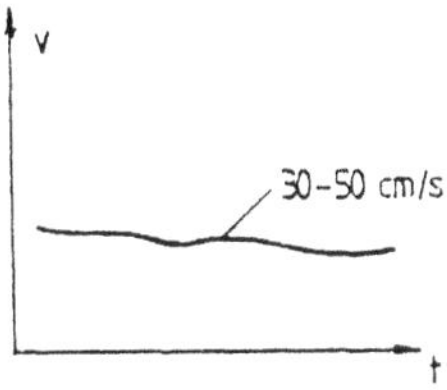

Figure 1.2.7: Near-source and far-source earthquake characteristics

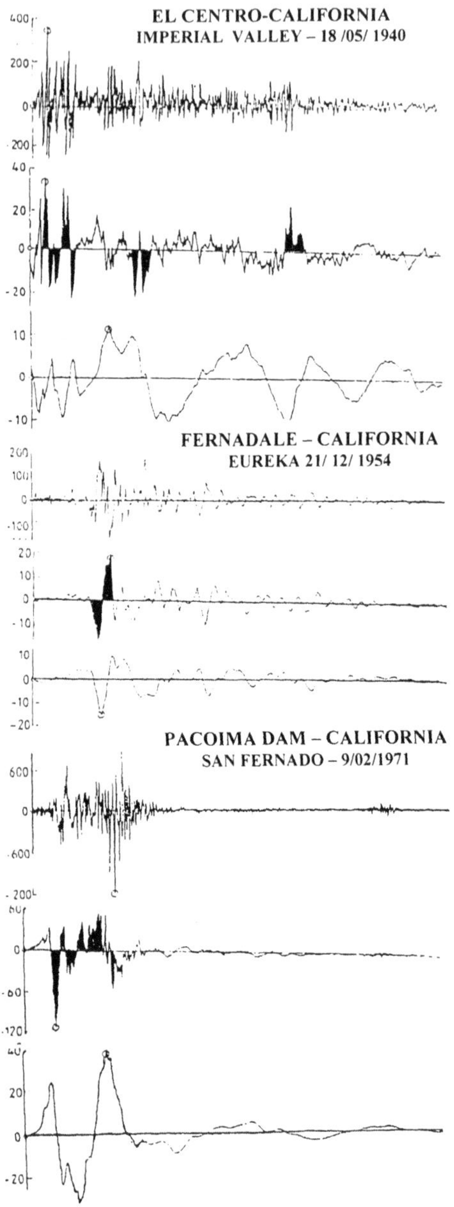

Figure 1.2.8: Historical records (data from Ifrim)

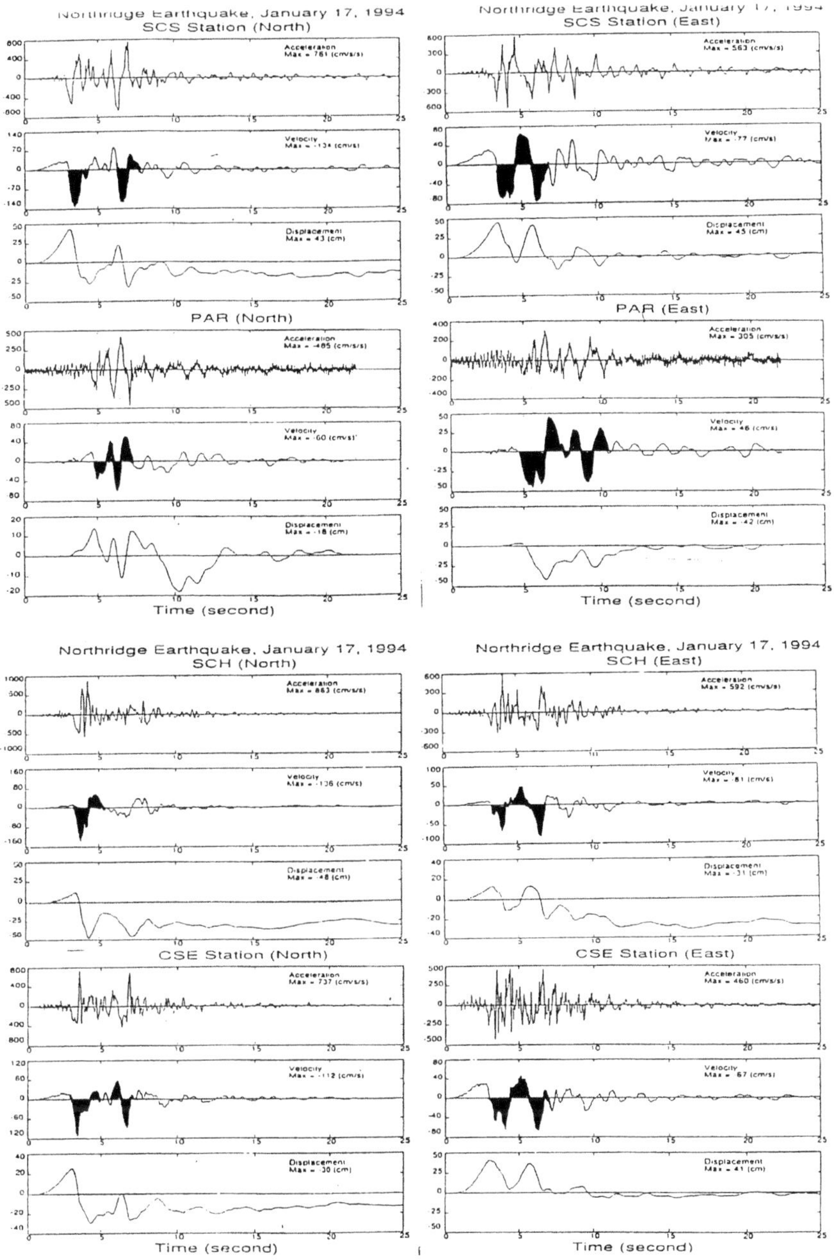

Figure 1.2.9: New records (data from Iwan)

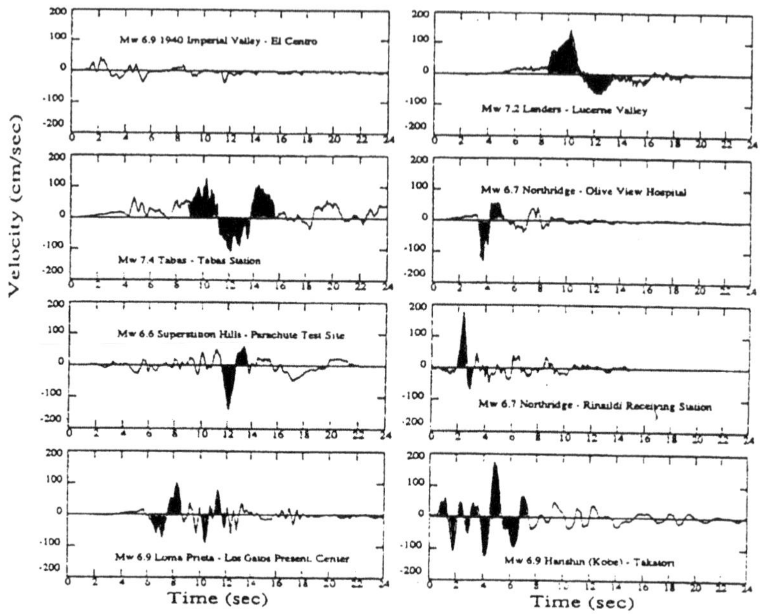

Figure 1.2.9 (continued): New records (data from Hall)

(i) The *direction of propagation* of the fault rupture has the main influence for near-source fields, the local site stratification having a minor influence. The directionality of the wave propagation was significant for Lucerne Valley, Loma Prieta and Kobe earthquakes, where the damages of the structures were concentrated along the rupture direction. In exchange, in case of far-source fields, the soil conditions for the travelling waves and the site conditions are of the first importance. The peak acceleration amplification at 400 kilometers from the source in case of the Michoacan earthquake is a very good example for this aspect. The same effect may be observed in the case of Vrancea earthquake, recorded at Bucharest, at a distance about 160 kilometers from the source, where the soil conditions have amplified the ground motion.

(ii) The *velocity pulse* is one of the main characteristics of near-source earthquakes. The ground motion could be described as acceleration, velocity or displacement pulse, or as cyclic movement. The ground motion has in the near-source fields, a distinct low-frequency pulse in acceleration and a pronounced coherent pulse in velocity and displacement. In far-source fields it can be described by cyclic movements, or in some cases, due to specific soil conditions, by acceleration pulses.

A relevant case is the Banat (Romania, 1991) earthquake. There are records in the epicenter zone, at Banloc and others at a distance of about 40 kilometers from the source, at Timisoara (Fig. 1.2.10).

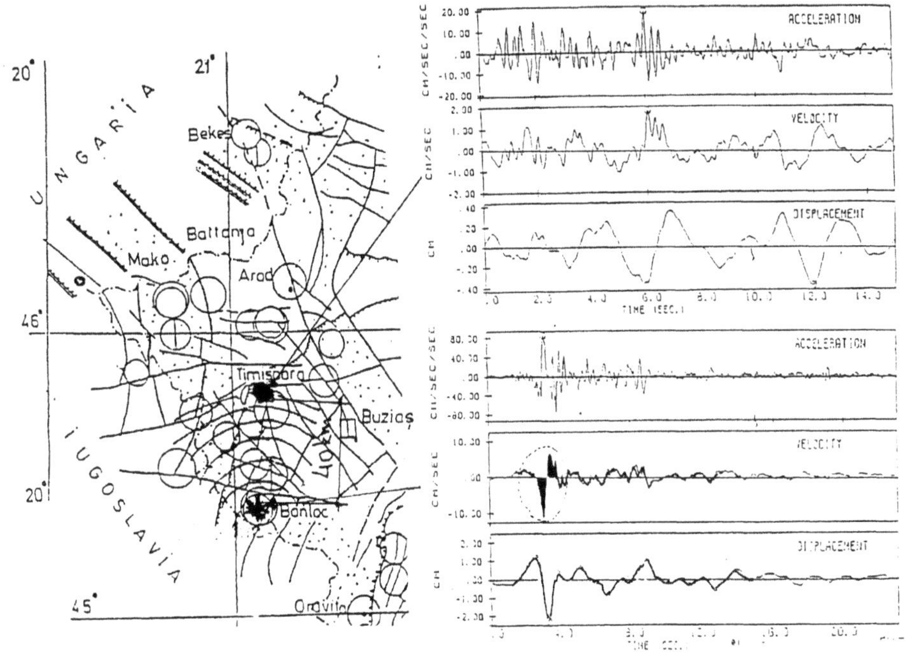

Figure 1.2.10: Banat earthquakes

All these earthquakes present a significant chaotic variation in the accelerogram records, without any possibility to notice some regularity. In exchange, if the velocity and displacement records are examined, there are some regularities to be observed. The conclusions for the near-source earthquakes refer to the following aspects:

- long duration velocity pulses are observed at all the recorded ground motions;
- peak velocity is often a better indicator of the damage potential than the peak acceleration;
- velocity pulses could be symmetrical, but in a lot of cases a well-marked asymmetry is observed;
- the ground motion could be composed by only one pulse (Banat, Lucerne Valley and some Northridge records), or two or three adjacent pulses (San Fernando, Loma Prieta, Kobe);
- some distinct pulses (El Centro, Ferndale) could compose the ground motion.

The natural period of these impulses is a very important parameter of the ground motion. Figure 1.2.11 shows the variation of the frequencies in acceleration during the Kobe earthquake. One can see a large variation of the frequency (Mohammadioun, 1997), but the pulse periods do not exceed 1 sec, in the portion of the records with maximum acceleration.

In Figure 1.2.12 the variations of the velocity periods for Kobe, Banat and Loma Prieta earthquakes are presented (Tirca and Gioncu, 1999). In each case, the beginning of the ground motion is characterized by short natural periods of 0.20 – 0.30 sec, which correspond to the base excitation. They are followed by a gradual increasing, due to the effects of local conditions. The first case is the Kobe earthquake, with a maximum period of about 0.7 sec, followed by Banat and Loma Prieta earthquakes, with maximum periods of 0.3 – 0.4 sec. A detailed examination of the periods of velocity impulses was performed for the Banat earthquake. The periods for horizontal components are of the range 0.25 – 0.80 sec, with an average value of 0.4 sec, while for vertical components, the periods are within 0.15 – 0.50 sec, with an average value of 0.22 sec.

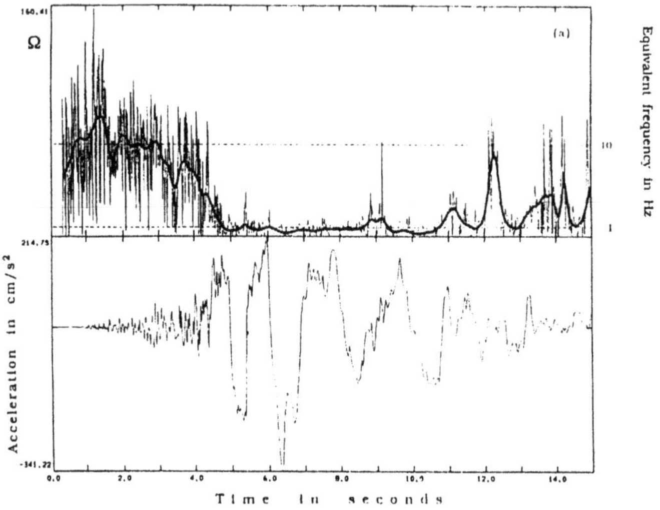

Figure 1.2.11: Natural period of velocity pulses for Kobe earthquake (after Mohammadioun)

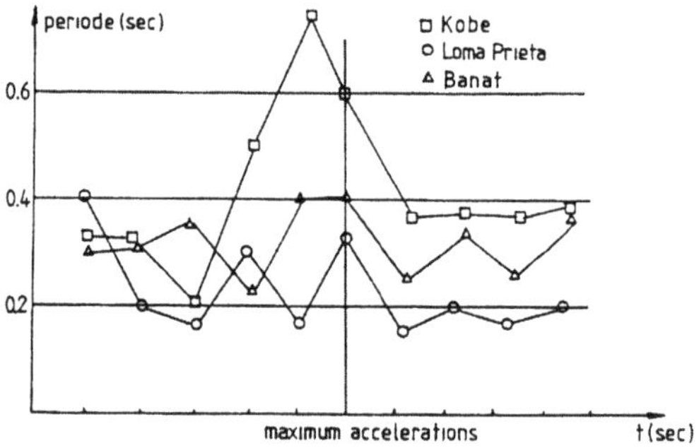

Figure 1.2.12: Variation of natural periods

As was observed from some recorded ground motions (Landers, USA 1993), the ground displacements were very important, being directionally associated with the fault-rupture process (Hall, 1995). These displacement–pulses could be ones of the most damaging earthquakes, if the structure is situated on the rupture line.

In exchange, cyclic ground motions may be noticed for far-source records, with some exceptions, bad soil conditions could generate acceleration pulses, which was the case of Bucharest (Fig.1.2.13) and Mexico City earthquakes. The distances from the epicenter were about 160 km for Bucharest and 400 km for the Mexico City earthquake.

(iii) The *vertical components* in the near-source field could be greater than the horizontal ones, due to the direct propagation of the P-waves. Figure 1.2.14 shows the ratio of vertical to horizontal spectra for the Northridge earthquake (Hudson et al, 1996). Similar aspects were related to other earthquakes (El Nashai and Papazouglu, 1997) from Table 1.2.1.

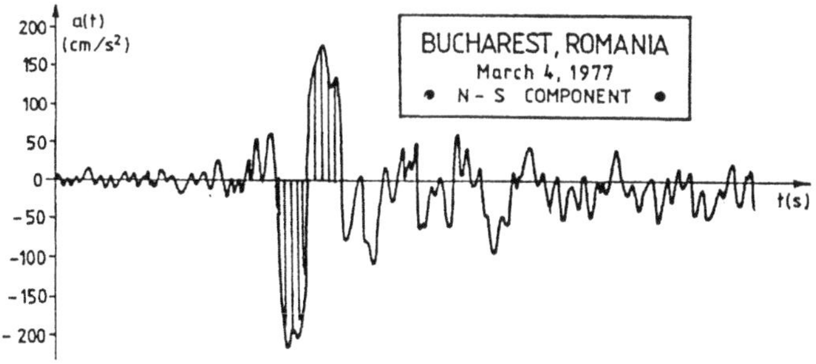

Figure 1.2.13: Bucharest (1977) earthquake

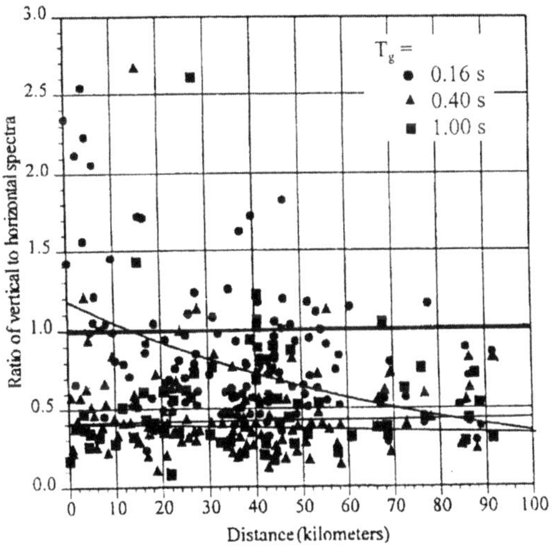

Figure 1.2.14: Vertical to horizontal component ratios for Northridge earthquake

TABLE 1.2.1
VERTICAL TO HORIZONTAL COMPONENT RATIOS
(Data from El Nashai and Papazouglu,1997)

Earthquake	Date	M	d [km]	h [km]	V/H
Nahani, Canada	85	6.79	8	6	2.15
Coyote Lake, USA	79	5.72	4	6	1.78
Gazli, USSR	76	7.10	19	10	2.28
Imperial Valley, USA	79	6.86	1	8	1.85
Imperial Valley, USA	79	6.86	1	8	3.70
Imperial Valley, USA	79	6.86	6	8	0.62
Imperial Valley, USA	79	6.86	4	8	1.17
Imperial Valley, USA	79	6.86	4	8	1.53
Imperial Valley, USA	79	6.86	5	8	1.75
Kobe, Japan	95	7.20	18	14	0.54
Kobe, Japan	95	7.20	20	14	1.96
Kobe, Japan	95	7.20	25	14	1.56
Loma Prieta, USA	89	7.17	0.1	17	0.92
Loma Prieta, USA	89	7.17	14	17	1.35
Loma Prieta, USA	89	7.17	16	17	0.91
Loma Prieta, USA	89	7.17	9	17	1.09
Loma Prieta, USA	89	7.17	7	17	1.32
Morgan Hill, USA	84	6.17	15	8	1.87
Morgan Hill, USA	84	6.17	14	8	3.94
Morgan Hill, USA	84	6.17	14	8	1.86
Morgan Hill, USA	84	6.17	17	8	1.76
Montenegru, Yug.	79	7.04	21	10	1.74
Northridge, USA	94	6.70	6	18	1.79
Northridge, USA	94	6.70	12	18	0.94
Northridge, USA	94	6.70	9	18	0.89
Whittier, USA	87	6.00	3	14	0.87
Whittier, USA	87	6.00	4	14	1.65
Whittier, USA	87	6.00	11	14	1.38

The attenuation of the ratio of vertical to horizontal components in function of the epicentral distance is presented in Figure 1.2.15 (EL Nashai and Papazouglu, 1997). One can notice a very important reduction of the ratio with the distance.

(iv) The *velocities* in the near-source field are very high. During the Northridge and Kobe earthquakes, values of 177 cm/sec were recorded at the soil level (Table 1.2.2). The spectral velocities are drown in Figure 1.2.16 showing for Northridge and Kobe earthquakes an increasing till the values of 400-500 cm/sec for short periods. The velocities of the vertical components increase dramatically near to the epicenter (Fig. 1.2.17).

So, in the case of near-source fields, the velocity is the most important parameter in the design concept, replacing the acceleration, which is the dominant parameter for the far-source field.

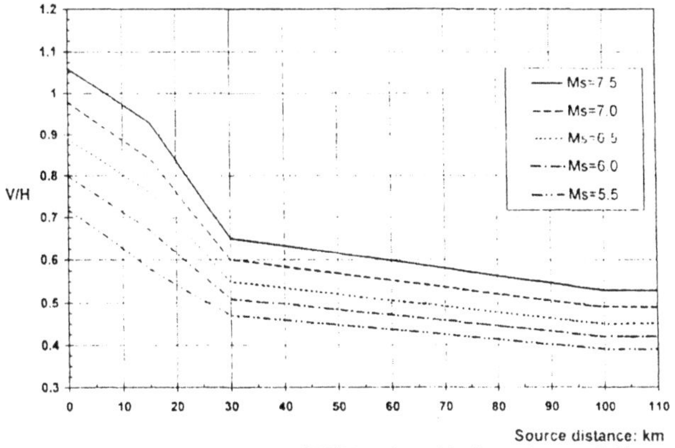

Figure 1.2.15: Attenuation of V/H ratio with the epicentral distance

TABLE 1.2.2
PEAK NEAR-SOURCE GROUND MOTIONS (Data from Hall 1995)

Earthquake	Date	M_g	Distance [km]	Acceleration [g]	Velocity [cm/s]	Displace-ment [cm]
Imperial Valley						
El Centro Array 7			1	0.65	110	41
El Centro Array 6			1	1.74	110	55
Bonds Corner	1979	6.5	4	0.81	44	15
El Centro Array 5			4	0.56	87	52
El Centro Array 8			4	0.64	53	29
Loma Prieta						
Los Gatos Presentation Center	1989	6.9	0	0.62	102	40
Lexington Dam			5	0.44	120	32
Landers						
Lucerne	1993	7.2	1	0.90	142	255
Northridge						
Rinaldi Receiving Station			0	0.85	177	50
Sylmar Converter Station			0	0.90	129	50
Los Angeles Dam			0	0.32	79	22
Sepulveda Vetrans Hospital			0	0.94	75	15
Jensen Filtration Plant	1994	6.7	0	0.85	103	38
Sylmar County Hospital			2	0.91	134	44
Van Nuys (hotel)			2	0.47	48	13
Arleta fire station			4	0.59	44	15
Newhall fire station			5	0.63	101	36
Hanshin						
Kobe (JMA)	1995	6.9	0	0.85	105	26
Kobe University			0	0.31	55	18
Takatori			0		176	

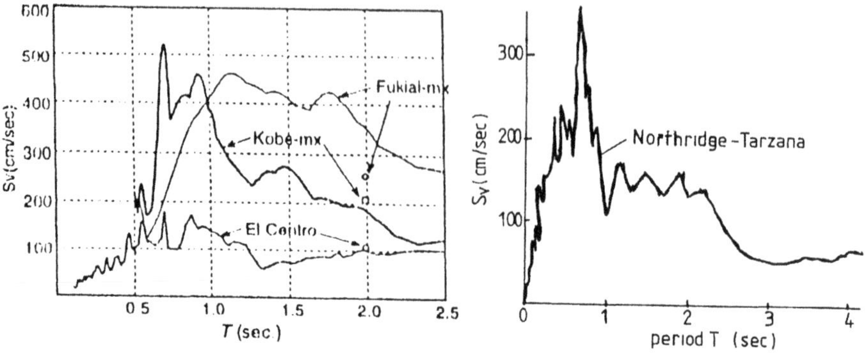

Figure 1.2.16: Velocity response spectra for Northridge and Kobe earthquakes

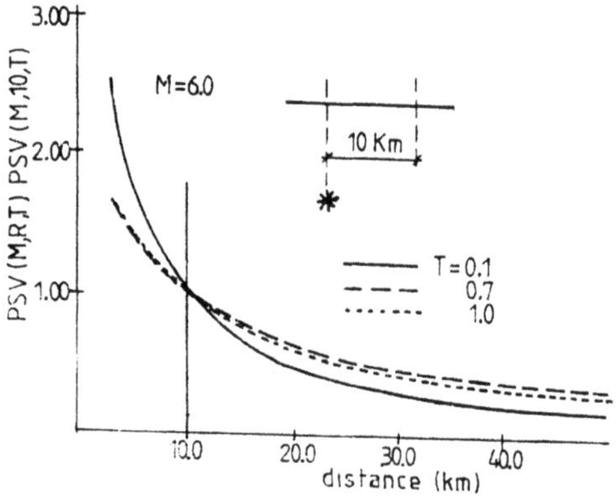

Figure 1.2.17: Velocity of vertical components near to source

The velocities of the vertical and horizontal components in case of far-source earthquakes are more reduced comparatively to the near-source areas, not exceeding 30-40 cm/sec.

THE BEHAVIOUR OF THE STRUCTURES FOR NEAR-SOURCE VERSUS FAR-SOURCE EARTHQUAKES

As a consequence of the above-mentioned differences in the ground motion, there are significant modifications in the behaviour of the structures subjected by near-source earthquakes versus the structures acted by far-source ground motions. These differences are underlined in Figure 1.2.18 (Gioncu & Mateescu 1999).

Characteristic	Near-source	Far-source
Vibration modes	Superior modes 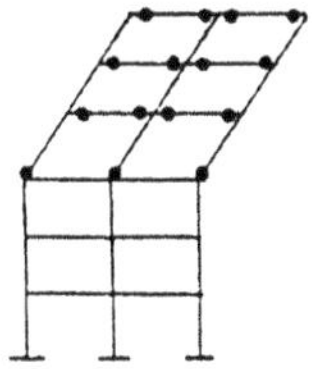	First mode 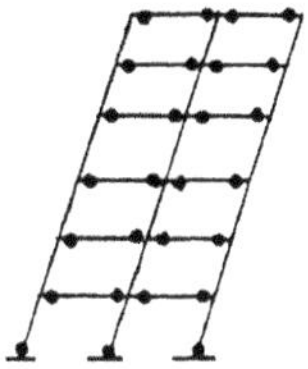
Vertical component	Increasing of axial forces 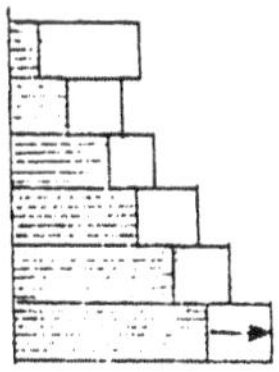	Negligible increasing 
Ductility demands	Maximum demands for upper part 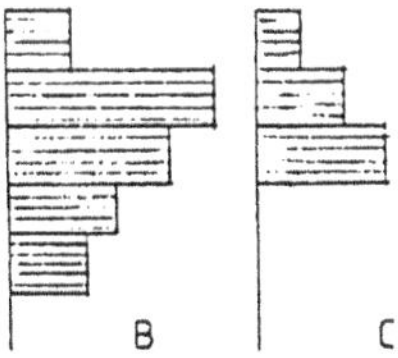	Maximum demands in the lower part 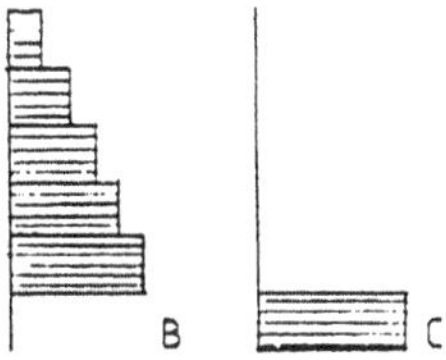
Velocity	Increasing of storey drifts 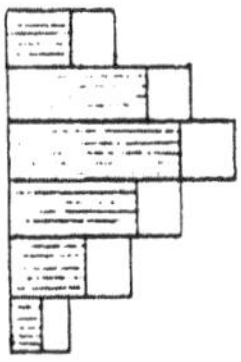	Reduced increasing of storey drifts 

Figure 1.2.18: Behaviour of structures for different ground motion types

(i) *Influence of the vibration modes*. In near-field areas due to the very short periods of the ground motion and to the pulse characteristics of the loads, the effect of higher vibration modes increases, in comparison with the case of far-field regions, where the fundamental mode is dominant. For structures subjected to pulse actions, the impact propagates through the structure as a wave, causing large localized deformations, or important interstorey drifts (Fig. 1.2.19a) (Iwan, 1995), especially for the top of the structure. In exchange, in case of far-source earthquakes, the lateral displacements of the structure show that the influence of the first vibration mode is dominant.

Thus, the classic design method based on the response of a single degree of freedom structure, characterized by the design spectrum is not adequate to describe the real behaviour of a structure situated in a near-source field. A new methodology for the design of the steel structures has to be elaborated.

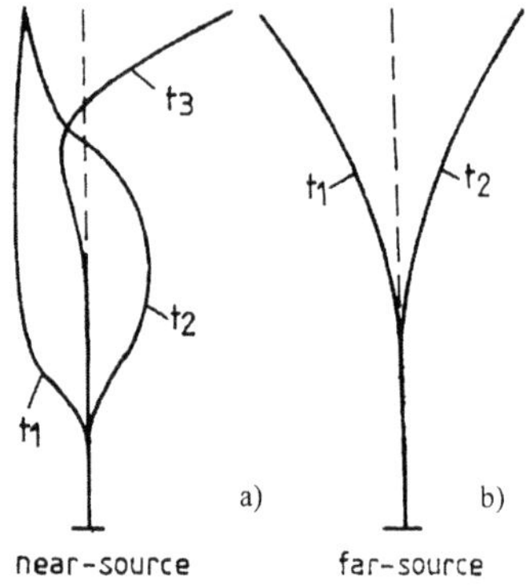

Figure 1.2.19: Influence of vibration modes

(ii) *Influence of the vertical components*. Of fundamental importance are the natural periods for the vertical vibrations of those buildings, which are more stiffer in the axial than the transverse direction, and hence possess shorter periods in vertical direction (Fig. 1.2.20a) (Papazouglu and El Nashai, 1996; Papaleontiou and Roesset, 1993). Due to the fact that the amplification range for the vertical strong-motion records is found to lie for buildings between 0.05 and 0.25, an important amplification of the vertical effects may occur.

Naeim (1998) shows that during the Northridge earthquake, the instrumented building indicated that peak vertical ground accelerations were amplified by the structures, with a factor ranging from 1.1 to 6.4. In the same time, taking into account the reduced possibilities of plastic deformation and damping under vertical displacements, the vertical behaviour can be of the first importance for the structures located in near-source fields.

The combination of vertical and horizontal components produces an increasing of the axial forces in the columns (Fig. 1.2.20b) and consequently, the increasing of second order effects. Due to this effect, additional plastic hinges may occur in the columns of the first level.

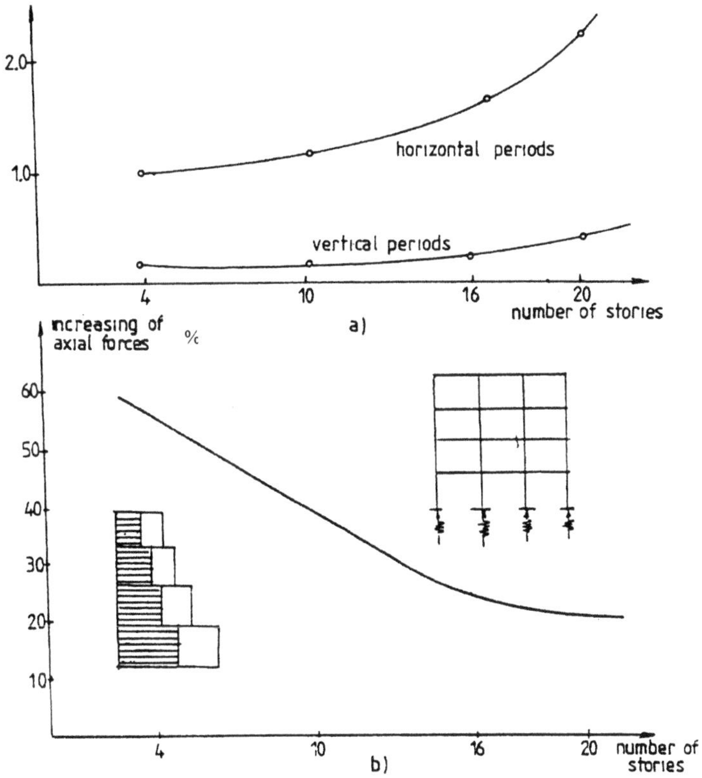

Figure 1.2.20: Influence of vertical components

(iii) *Ductility demands*. Due to the pulse characteristics of the actions, developed with great velocity, and especially due to the lack of important restoring forces, the ductility demands could be very high, so the using of the inelastic properties of the structure for the seismic energy dissipation has to be carefully examined. In the same time, the short duration of the ground motion in the near-source field is a favorable factor. A balance between the severity of the ductility demands, due to the pulse action and the effects of the short duration has to be seriously analyzed. Another important effect of the near-source earthquake is the influence of the second vibration mode, which interact with the first one, introducing irregularities in the bending moment diagram, especially in the middle high of the frame (Fig. 1.2.21a).

This situation induces a dramatic reduction of the available ductility in the middle or top parts of the frame. Because the required ductility has a maximum just in these places, the collapse of the building may occur due to insufficient ductility (Gioncu, 1999). This was a common phenomenon during the Kobe earthquake, where many buildings were damaged just in the middle storeys.

Figure 1.2.21b shows the available and required ductilities in the case of far-source earthquakes, where the two ductilities have the maximum value at the first story.

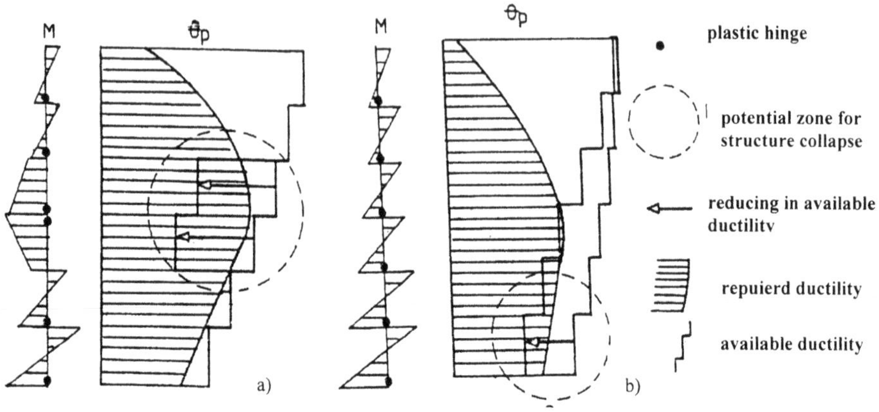

Figure 1.2.21: Ductility demands

(iv) *Velocity*. Due to the great velocity of the seismic actions, an increasing of the yield strength occurs, as an effect of the increasing of the strain rate; the increasing of the yield strength conducts to a significant reduction of the available local ductility. The ultimate strength is not so sensitive to the strain rate, consequently, the difference between the ultimate and yield strength decreases and a brittle fracture could appear. At the same time, the increasing of the velocity leads to an increasing of the ductility demands.

A fundamental problem of the seismic actions is the cumulative damage, which could involve important differences between near-source and far-source fields. The fracture occurs at the first excursion into the plastic range due to the pulse action, in first case, while in far-source fields it appears after multiple cycles as a consequence of the accumulation of plastic deformations (Fig. 1.2.22).

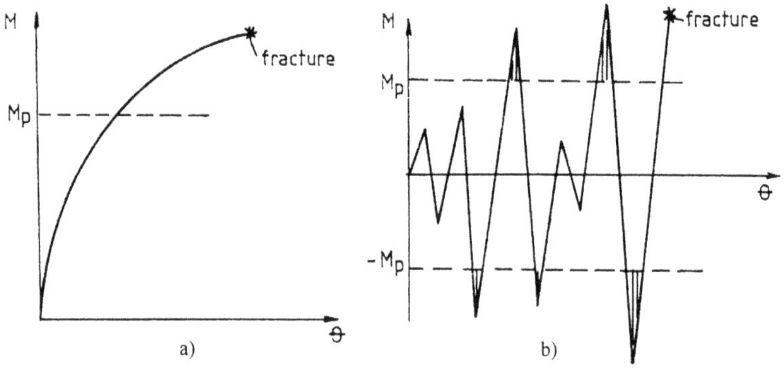

Figure 1.2.22: Fracture due to pulse or cyclic actions

ARTIFICIAL GENERATED ACCELEROGRAMS FOR PULSE ACTIONS

The seismic response of a structure with complex behaviour can be correctly calculated only through direct nonlinear dynamic time-history analysis, where the structure model is subjected to a base acceleration input. The used accelerograms have to represent adequately the characteristics of the expected earthquake at site of the structure.

As results of the extensive monitoring of the areas with high seismic risk (see USA-California and Japan) it was possible to collect a large number of records, in recent years. All these records have random characteristics. The design practice consists in the selection of the most appropriate ground motion type and to perform a time-history analysis, obtaining the randomness values, involved by the utilized acceleration-record. This procedure has some major disadvantages, as presented in the introduction. So that an alternative to the earthquake records, the use of artificial generated accelerograms, with respect to certain given restrains has been allowed by some recent codes. The control parameter should be chosen with regard to the characteristics of the expected earthquakes at the site, depending on the generic mechanism and the position of the site, in relation to the focal area.

In order to control the possibility to use artificial ground motions for the determination of the structure behaviour during a pulse earthquake, the response should be determinate for time fragments in the recorded accelerograms. Figure 1.2.23 presents the accelerogram of Bucharest 77, where the duration of the most severe 17 seconds was divided into four time intervals (denoted from a to d): three of 5 seconds each one and another of 2 seconds (Ifrim et al, 1986). The 'a' interval corresponds to the beginning, 'b', represents the major ground motion characterized by a pulse acceleration, the intervals 'c' and 'd' are the consequences of the main shock.

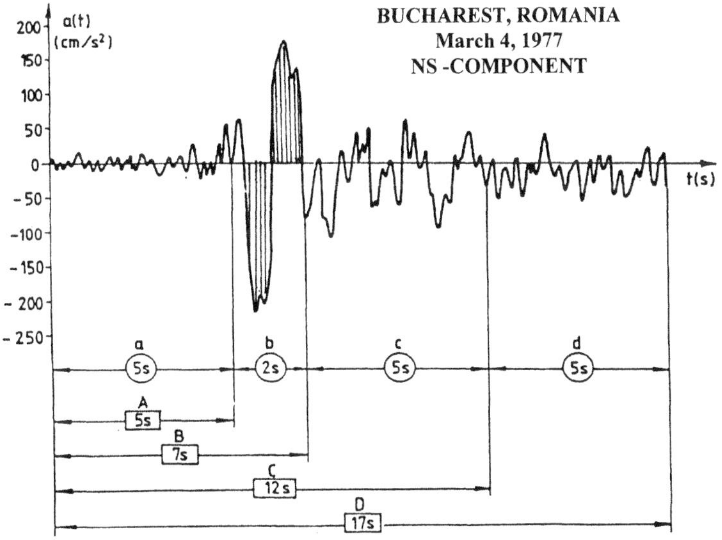

Figure 1.2.23: Bucharest 77 accelerogram (after Ifrim)

Figure 1.2.24a shows the importance of each accelerogram fragment, considering each interval as an independent earthquake. It is very clear that the response is dominated by the interval 'b', which correspond to the pulse acceleration. The superposing of the response for the different intervals is presented in Figure 1.2.24b. Curve A corresponds to the interval 'a', curve B, to 'a'+'b', curve C, to 'a'+'b'+'c' and the curve D, to the sum of the four intervals 'a' to 'd'. One can see that for periods ranged between 0.2 and 1.6 seconds, the curves B, C and D practically coincide. The simulation of the structural response using a base harmonic excitation with one and two pulses corresponding to the interval 'b' is also presented in Figure 1.2.24b.

The good correspondence of the responses for the recorded and artificial ground motion shows the potential of the using artificial accelerograms for the determination of the response of the structure.

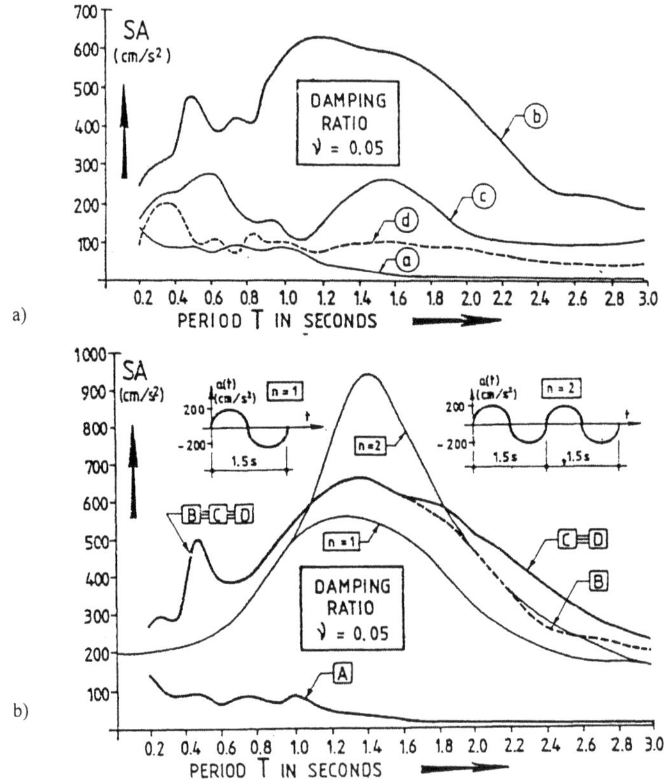

Figure 1.2.24: Sequential seismic response spectra (after Ifrim)

Considering the characteristics of the examined near-source earthquakes emphasized in the previous section, the following artificial generated ground motions, taking into account the pulse velocity were chosen (Fig. 1.2.25):

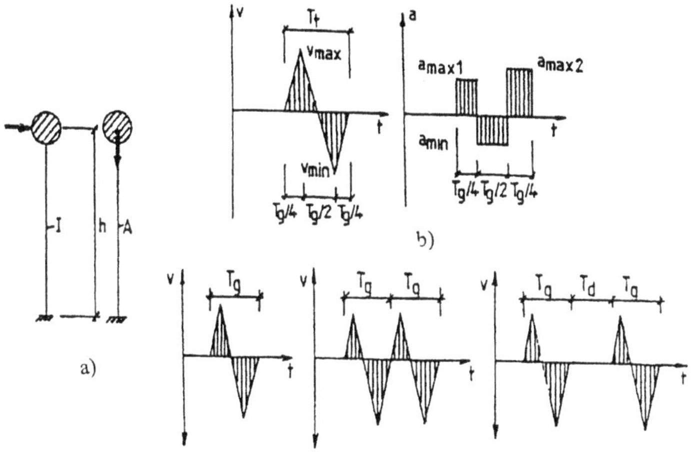

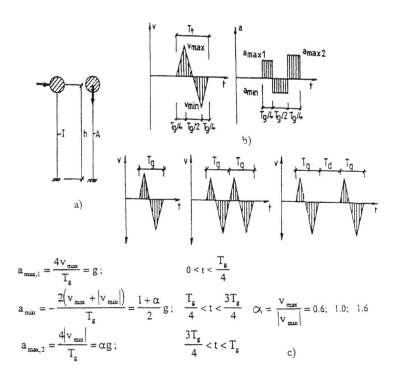

$$a_{max,1} = \frac{4v_{max}}{T_g} = g; \qquad 0 < t < \frac{T_g}{4}$$

$$a_{min} = -\frac{2(v_{max} + |v_{min}|)}{T_g} = \frac{1+\alpha}{2} g; \qquad \frac{T_g}{4} < t < \frac{3T_g}{4} \qquad \alpha = \frac{v_{max}}{|v_{min}|} = 0.6;\ 1.0;\ 1.6$$

$$a_{max,2} = \frac{4|v_{min}|}{T_g} = \alpha g; \qquad \frac{3T_g}{4} < t < T_g$$

c)

Figure1.2.25: Artificial ground motions for near-source earthquakes

a) direction of the action:
   - horizontal;
   - vertical;
b) ratio of positive and negative velocity peaks:
   0.6; 1.0 (symmetry); 1.6;
c) duration of the pulse:
   - for horizontal actions: $T_g = 0.1;\ 0.2;\ 0.3;\ 0.4;\ 0.5$ and 1.0 sec
   - for vertical actions: $T_g = 0.05;\ 0.1;\ 0.2;\ 0.3;\ 0.4$ and 0.5 sec
d) number of pulses:
   - one single pulse;
   - two adjacent pulses;
   - two distinguished distant pulses.

ANALIZED FRAMES FOR HORIZONTAL COMPONENTS

Artificial spectra for horizontal components

Figure 1.2.26a shows the spectra for horizontal motion plotted for a pulse period Tg=0.2 sec and different structure periods, as well as asymmetry ratios (Tirca et al, 2000). A comparison with the EC 8 spectrum is also presented. One can see that the EC 8 values do not cover the high amplification in the range of reduced structure periods. It is important to notice that the pulse with the highest value at the second peak gives the maximum amplification. A comparison between one single pulse, two adjacent pulses and two distinguished pulses is presented in Figure 1.2.26b.One can observe that the amplification is maximum for two adjacent pulses.

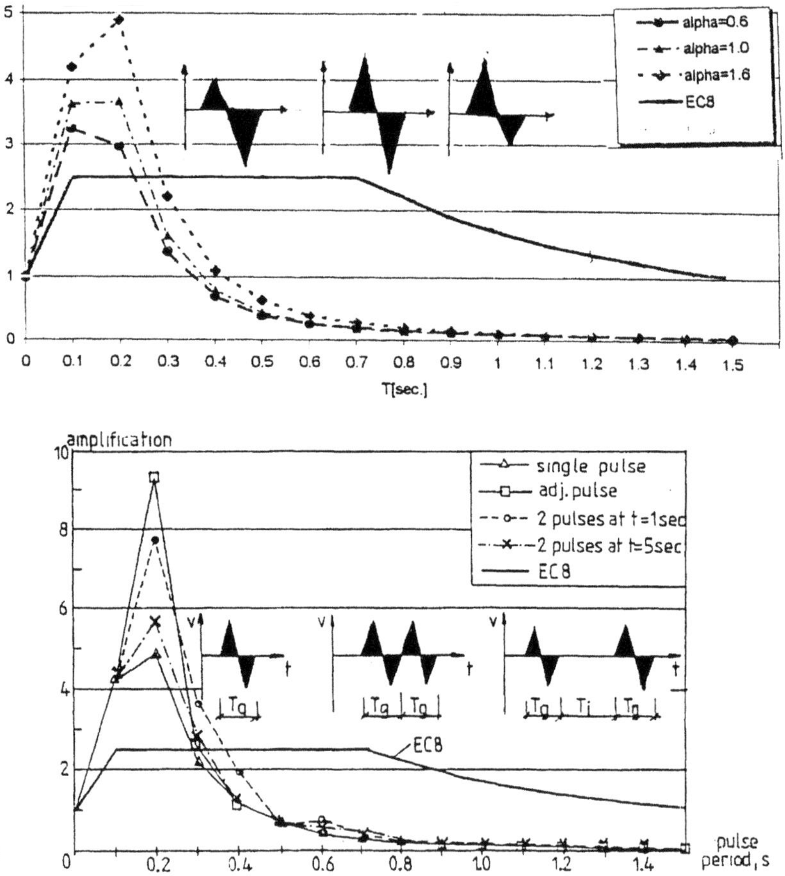

a)

b)

Figure 1.2.26: Spectra for horizontal pulses

Analysis with artificial accelerograms

The one span and six levels analyzed frames are presented in Table 1.2.3 (Tirca and Gioncu, 2000). MRF1a is an ordinary moment resisting frame with variable cross-section, MRF1b, with constant cross-section for the columns, MRF2 is a special resisting frame, where the superior modes are not considered, while MRF3 considers the second vibration mode. The collapse mechanism of MRF1 is a story one, MRF2 and MRF3 develop global collapse mechanisms. The natural period of the three frames is also presented. The used accelerograms are of artificial pulse type, with peak values of 0.35g, 0.25g and 0.15g, corresponding to high, moderate and low seismic actions. The analysis was performed using the DRAIN 2D computer program.

The formation of the plastic hinges for the MRF1 frame is presented in Figure 1.2.27a. One can see that for periods within 0.5...0,8 sec, corresponding to the frame period of the second vibration mode, the structure collapses, due to large displacements of the upper storyes. The increasing of the pulse natural period involves different mechanism types. For the first vibration mode the collapse consists of three level mechanisms. It is very interesting to notice that by MRF1b, the pattern of the collapse is drastically changed (Fig. 1.2.27b): this means that for near-source earthquakes, the column cross-sections have to be maintained constant on the frame high.

TABLE 1.2.3
CHARACTERISTICS OF THE ANALYZED FRAMES

Frame		**Columns**		**Beams**		**Periods [sec]**		
Geometry	Type	P1	P2	P1	P2	T1	T2	T3
P_2 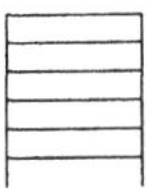 P_1	MRF1a	HE240B	HE200B	IPE360	IPE300	1.81	0.65	0.40
	MRF1b	HE240B	HE240B	IPE360	IPE300	1.80	0.64	0.39
	MRF2	HE260B	HE220B	IPE360	IPE300	1.72	0.61	0.35
	MRF3	HE450B	HE400B	IPE360	IPE300	1.35	0.40	0.18

MRF2-frame was designed to obtain a global mechanism for the first vibration mode. One can see (Fig. 1.2.28a) that for pulse periods smaller than 0.3 sec, the structure behaves elastically. For 0.4 sec, the first plastic hinge occurs at the top of the frame. With increasing of the pulse period, the formation of the plastic hinges moves down, till at T_g=0.7 sec, two plastic mechanisms appear, due to the second vibration mode. With increasing of the pulse period, the collapse mode is changing. For a period corresponding to the first vibration mode, three local plastic mechanisms occur.

Figure 1.2.29 shows the influence of the pulse period on the ductility demands. For periods near to the second mode the maximum required ductility is concentrated at the top of the structure, while for periods near to the first vibration mode, the maximum ductility demands occur at the lowest levels (Fig. 1.2.29a). The required ductilities of beams and columns resulted for the second vibration mode are more reduced than those obtained from the firs one (Fig. 1.2.29b).

MRF3-frame is sized to eliminate the story mechanism. One can see from Figure 1.2.28b that the first plastic hinges occur at the pulse period corresponding to the second vibration mode and the global mechanism is reached for a period close to the value of the first vibration mode. The maximum ductility demands are obtained at the top of the structure for all the pulse periods (Fig. 1.2.30).

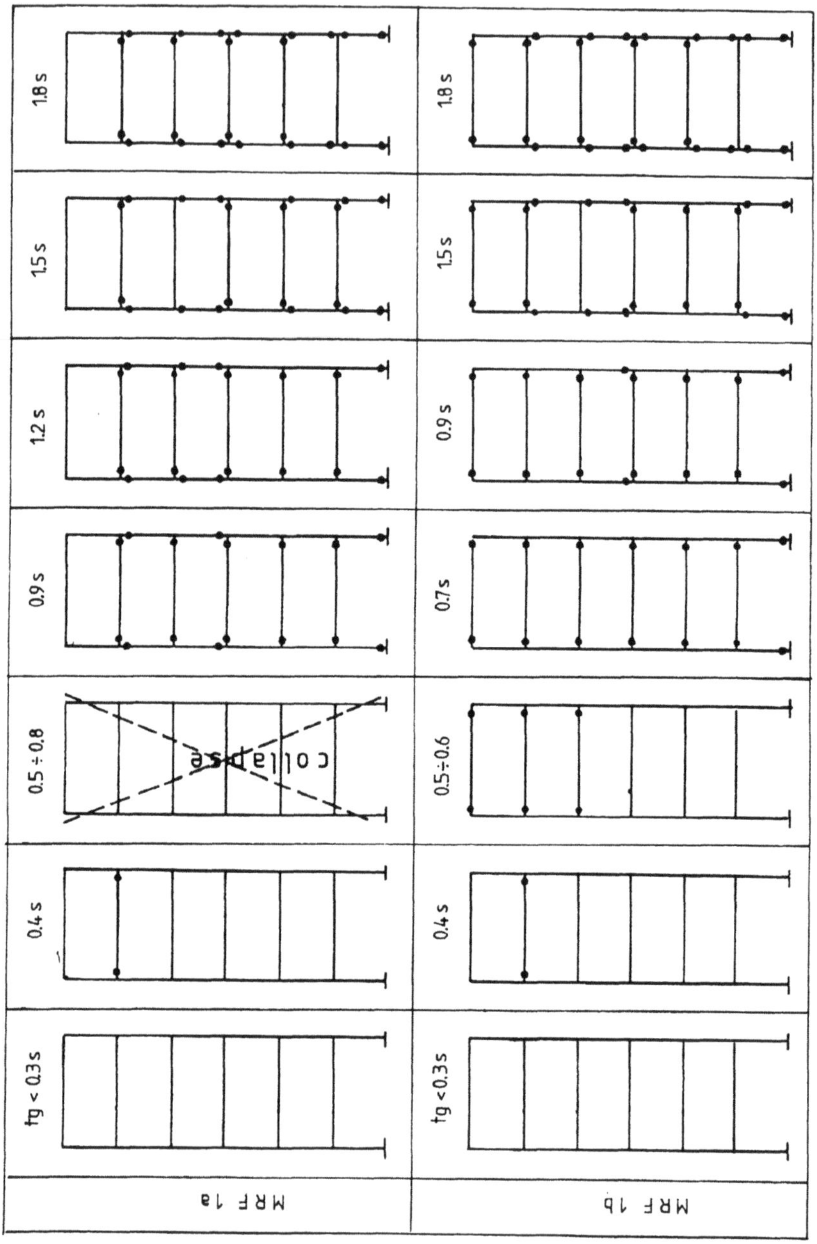

Figure 1.2.27: Collapse modes for MRF1a,b – frames

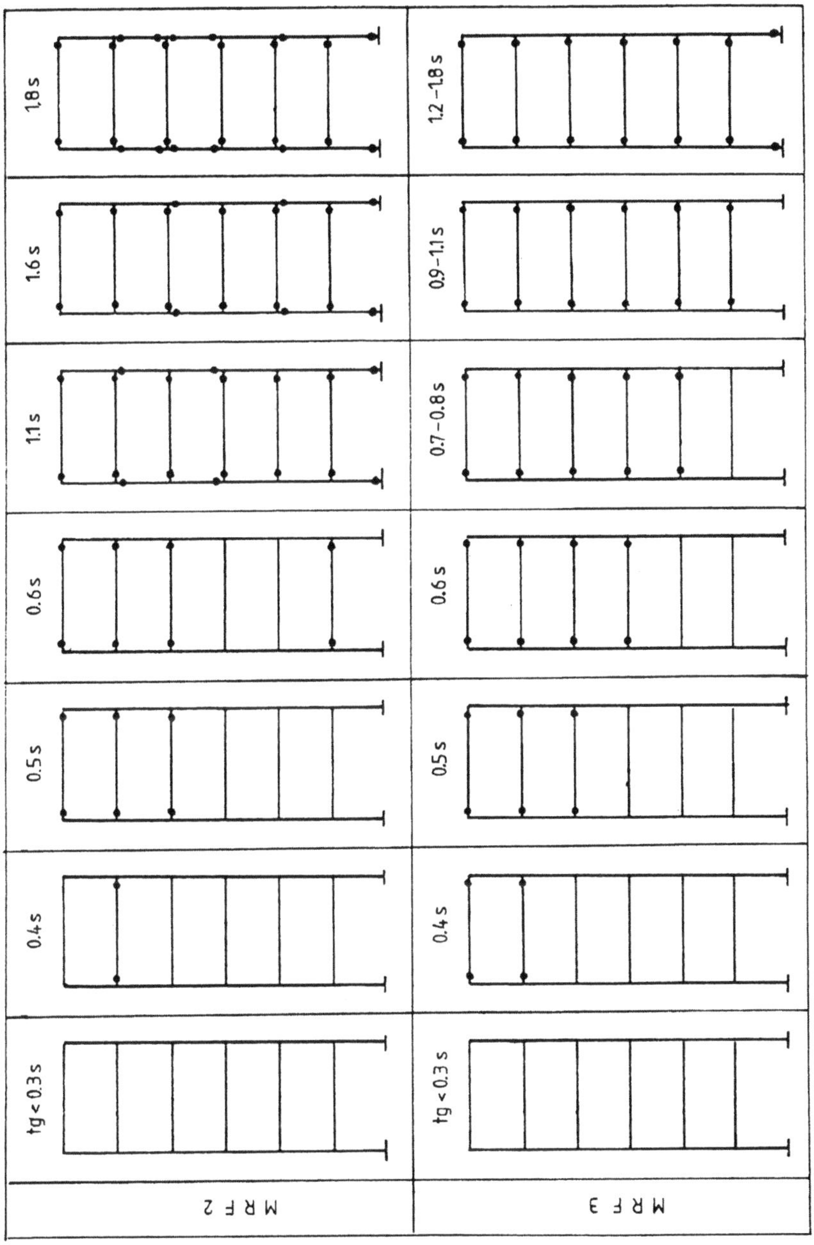

Figure 1.2.28: Collapse modes for MRF2, 3 – frames

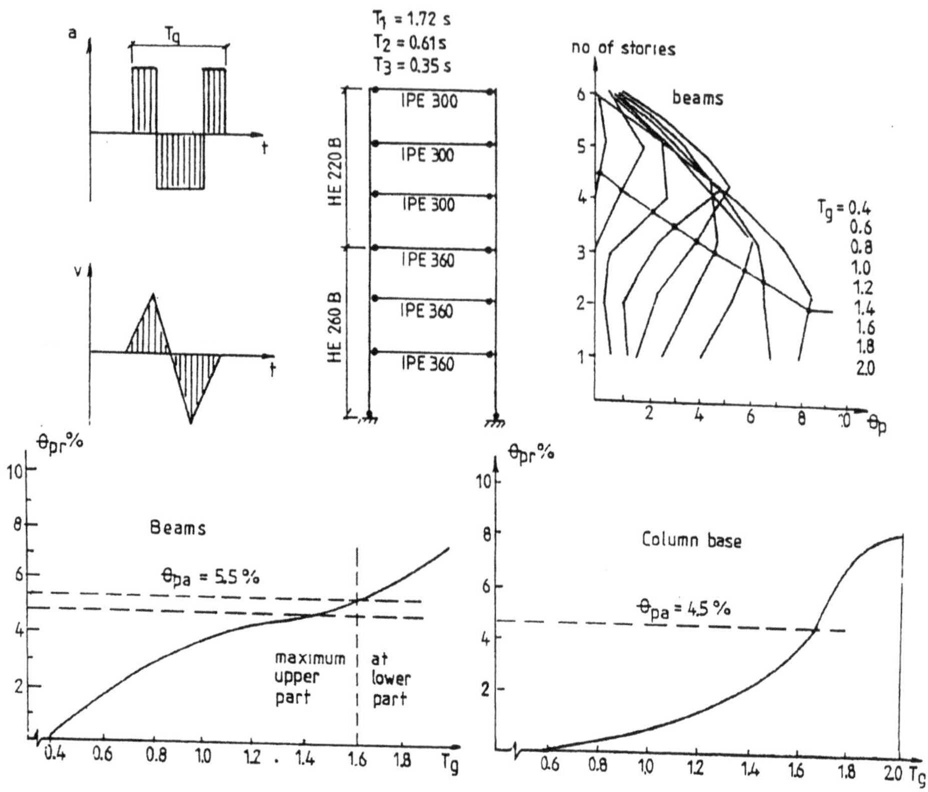

Figure 1.2.29: Ductility demands for MRF2-frame

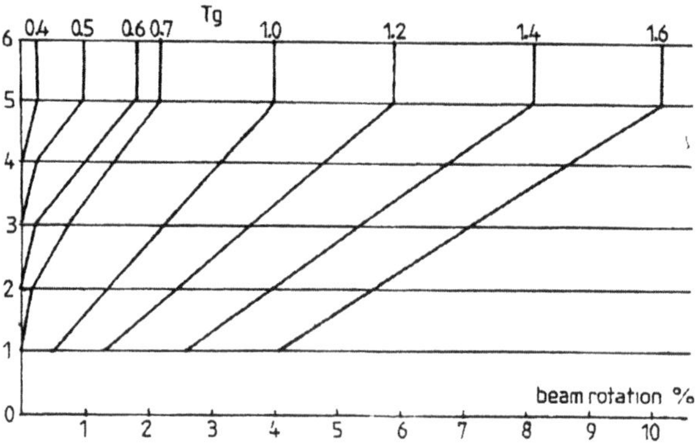

Figure 1.2.30: Ductility demands for MRF2-frame

The influence of adjacent pulses is presented in Figure 1.2.31. One can observe that the required ductility at the top of the structure increases dramatically with the pulse number.

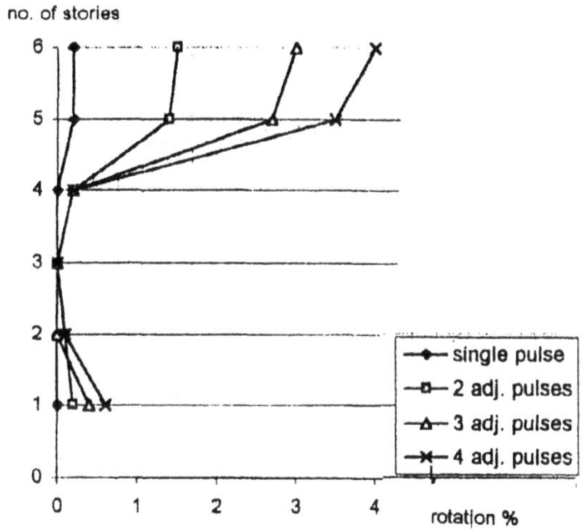

Figure 1.2.31: Influence of number of pulses

ANALYZED FRAMES FOR VERTICAL COMPONENTS

Natural periods of structures for vertical movements

The values of the natural periods for vertical movements are by far inferior to the horizontal ones due to the high axial rigidity of the columns. It is important to notice that, because of these reduced values of the natural period and the fact that a high amplification occurs in this range of periods, significant axial forces appear in the frame columns. The first natural period for multiple-level frames is determined using the Rayleigh method (Fig. 1.2.32)

The damping effects have reduced values for the vertical components, compared with the horizontal ones. A value of 2% was proposed for the viscous damping ratio. The reduced factor taking into account the effect of energy dissipation is smaller for vertical than for horizontal components; values of q=1.0...1.5 was proposed.

The vertical components can produce an important increasing of the axial forces, due to the high amplification and the reduction factors of damping and energy dissipation. This can give rise to a reduction of the rotation capacity of the columns and could extend the stability problems. All these aspects have to be introduced in the calculation of the available ductility of the columns.

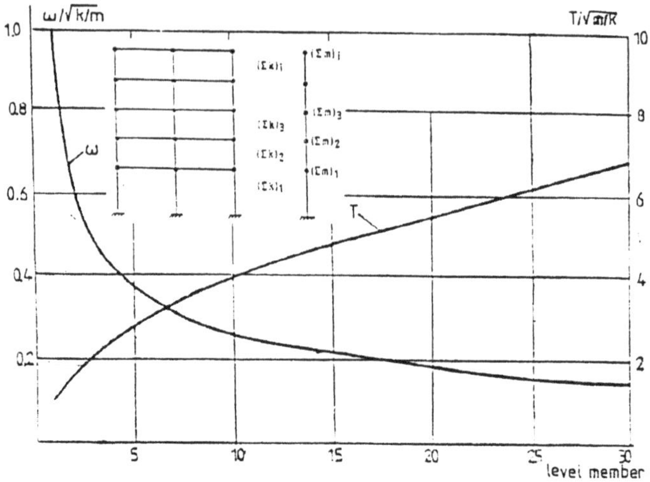

Figure 1.2.32: Natural periods of frames for vertical movements

Artificial spectra for vertical components

The spectra for vertical ground motions are presented in Figure 1.2.33 with a comparison with the EC 8 proposals. It is clear that for vertical pulses the code provisions are not sufficient, because the maximum obtained amplification corresponds exactly to the field of the vertical structure periods.

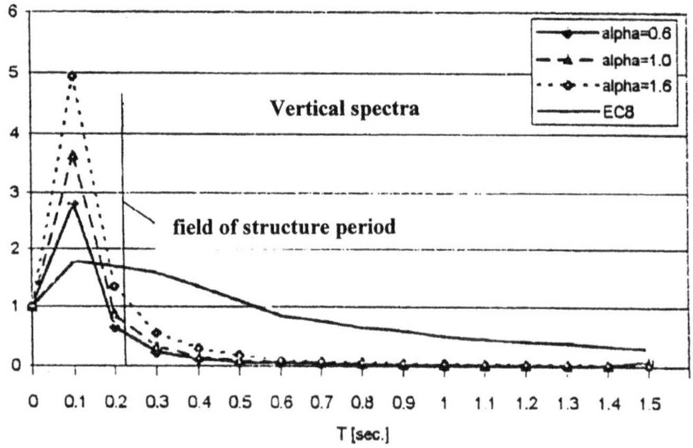

Figure 1.2.33: Spectra for vertical pulses

Analysis with artificial accelerograms

For the frame from Figure 1.2.34, a vertical, asymmetric ground motion pulse of T_g=0.15 sec was applied. The first natural period of the structure is 0.9 sec. for the horizontal, and 0.081 sec for the vertical movements, which correspond to the range of high amplification. The first plastic hinge and the collapse mode occur for acceleration values of a_p = 1353 cm/s^2 and a_c = 4709 cm/s^2, respectively. These values are very high, exceeding the recorded vertical accelerations. So, the vertical ground motion has to be considered as a factor influencing only the effects of the horizontal components. It is interesting to notice that, even if the vertical components applied to the structure are the same for each column, the formation of the plastic hinges has a marked asymmetry. This is due to the analysis imperfections, but also to the structural imperfections, characteristically for the in site frame behaviour. A very important problem to be solved in the future is the asynchronism of the vertical movement (Fig. 1.2.35), which can introduce internal forces, especially in beams and nodes.

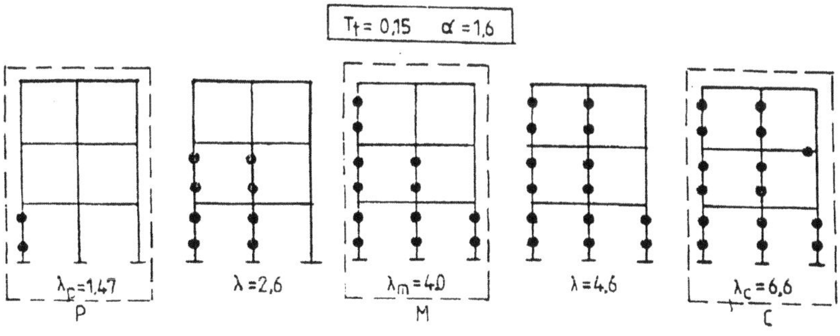

Figure 1.2.34: Analyzed frame for vertical pulse

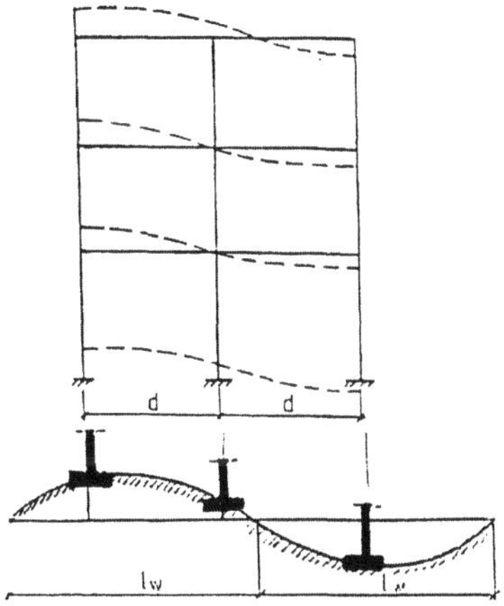

Figure 1.2.35: Asynchronism of vertical movements

Analysis with recorded accelerations

Two structures with 4 and 16 levels, respectively, were studied by Papaleontiou and Roesset (1993) for the Loma Prieta-Capitola earthquake, considering the horizontal and vertical components and their simultaneous action. The results show an important increasing of the axial forces in the bottom columns, which is greater for the structure with a less number of levels (see Fig. 1.2.20b).

CONCLUSIONS

Generated ground motions in the vicinity of the source contain large rapid pulses, which could involve a very different structural response from the response anticipated by the code provisions. In the paper, the main factors influencing the ground motion are determined for near-source and far-source regions. Important differences are remarked for these two earthquake types, the velocity pulse characteristic being the most relevant one. Aiming to study the structures subjected to velocity pulses, artificial accelerograms were generated, introducing the main parameters of the near-source ground motions.

The frame analysis for horizontal and vertical components, using artificial and recorded accelerograms has been allowed to underline the principal characteristics of the seismic structural response: the influence of the superior mode effects of the vertical components and the consequence of the very high velocities.

The main conclusion of these studies is the necessity to enlarge the code provisions by including the analysis of structures for different positions related to the potential sources.

References

Ambraseys N. N., Bonner J.J. (1991). The attenuation of ground accelerations. *Europe Earthquake Engineering and Structural Dynamics* **20,** 1179-1202.

Ambraseys N.N., Free M.V. (1997). Surface-wave magnitude calibration for European region earthquakes. *Journal of Earthquake Engineering* **1,** 1-22.

Anastasiadis A., Gioncu V., Mazzolani F.M. (1999a). New upgrading procedures to improve the ductility of steel MR-frames. *XVII Congresso CTA. Construire in acciaio: Struttura e Architettura*, 3-5 Octobre 1999, Napoli.

Anastasiadis A., Mateescu G., Gioncu V., Mazzolani F.M. (1999b). Reliability of joint systems for improving the ductility of MR-frames. "*Stability and Ductility of Steel Structures SDSS 99*"(ed. D.Dubina and M.Ivanyi), 9-11 September 1999, Timisoara. Elsevier. 259-268

Anderson J.C., Bertero V.V. (1987). Uncertainties in establishing design earthquakes. *Journal of Structural Engineering* **113:1,** 1709-1724.

Bertero V.V. (1996). The need for multi-level seismic design criteria. "*10th World Conference on Earthquake Engineering* " 23-28 June 1996, Acapulco, 25-32.

Berz G., Smolka A. (1995). Urban earthquake loss potential: Economic and insurance aspects."*10th European Engineering*"(ed.G.Duma) 28 August-2 September 1994, Vienna, Balkema, Rotterdam, **2,** 1127-1134.

Chandler A.M., Hutchinson G.L., Wilson J.L. (1992). The use of interplate derived spectra in interplate seismic regions. "*10th World Conference on Earthquake Engineering*" 9-24 July 1992, Madrid, Balkema, Rotterdam, 5823-5827.

El Nashai A.S., Papazouglu A.J. (1997). Procedure and spectra for analysis of RC structures subjected to strong earthquake loads. *Journal of Earthquake Engineering* **1:1,** 121-155.

Gioncu V. (1999): Framed structures: Ductility and seismic response. General report. "*Stability and Ductility of Steel Structures SDSS 99*"(ed. D.Dubina and M.Ivanyi), 9-11 September 1999, Timisoara.

Gioncu V., Mateescu G. (1999). Influence of type of seismic motions and vertical components. *INCO COPERNICUS RECOS PROJECT Final Report*, 1999.

Gioncu V. (2000). Effect of strain-rate on the ductility of steel members. "*Behaviour of Steel Structures in Seismic Areas. STESSA 2000*"21-24 August 2000, Montreal (manuscript).

Hall J.F. (1995a). Near-source ground motion and its effects on flexible buildings. *Earthquake Spectra* **11:4,** 569-605.

Hall J.F. (1995b). Parameter study of the response of moment-resisting steel frame buildings. *Technical Report SAC* **95-05**, 1.1-1.83.

Hudson M.B., Skyers B.N., Lew M. (1996). Vertical strong ground motion characteristics of the Northridge earthquake. "*11th World Conference on Earthquake Engineering*" 23-28 June 1996, Acapulco, CD-ROM paper 728.

Ifrim M., Macavei F., Demetriu S., Vlad I. (1986). Analysis of degradation process in structures during the earthquake. "*8th European Conference on Earthquake Engineering"Lisbon,* 65/8-72/8.

Iwan W.D. (1995). Drift demand spectra for selected Northridge sites. *Technical Report SAC* **95-05**, 2.1-2.40.

Iwan W.D. (1997). The drift demand spectrum and its application to structural design and analysis. "*11th World Conference on Earthquake Engineering*" 23-28 June 1996, Acapulco, CD-ROM paper 1116.

Mahammadioun B. (1997). Nonlinear response of soils to horizontal and vertical bedrock earthquake motion. *Journal of Earthquake Engineering* **1:1,** 93-119.

Massonnet C.E. (1982). The collapse of struts, trusses and frames: A survey of up-to-date problems. "*Collapse: The Buckling of Structures in Theory and Practice*" (eds. J.MT.Thompson and G.W.Hunt), 31 August-3 September 1982, London, Cambridge University Press, 183-208.

Mazzolani F.M., Piluso V. (1996). *Theory and Design of Seismic Resistant Steel Frames*, E&FN Spon, London, UK.

MCEER (1998): Engineeering and socio-economic impact of earthquakes: An Analysis of electricity life line disruptions in the New Madrid area. *MCEE Bulletin* – Summer 1998, 17.

Moon I.C. (1987). *Chaotic Vibrations. An Introduction for Applied Scientists and Engineers*, John Wiley & Sons, New York.

Naeim F. (1998). Research overview: Seismic response of structures. *The Structural Design of Tall Buildings* **7,** 195-215.

Papaleontiou C., Roesset J.M. (1993). Effect of vertical accelerations on the seismic response of frames. "*Structural Dynamics. EURODYN 93*"(eds. Moan et al), Balkema, Rotterdam, 19-26.

Papazoglou A.J., El Nashai A.S. (1996). Analytical and field evidence of the damaging effect of vertical earthquakes ground motion. *Earthquake Engineering and Structural Dynamics* **25,** 1109-1137.

Schuster H.G. (1998). *Deterministic Chaos. An Introduction*, VCH Verlag, Weinheim.

Tirca L., Gioncu V. (1999). Ductility demands for MRFs and LL-EBFs for different earthquake types. "*Stability and Ductility of Steel Structures*"(eds. D.Dubina and M.Ivanyi) 9-11 September 1999, Timisoara. Elsevier, 429-438.

Tirca L., Mateescu G., Gioncu V. (2000). Artificial ground motions for design of MRFs subjected to pulse type earthquakes. "*Behaviour of Steel Structures in Seismic Areas. STESSA 2000*", 21-24 August 2000, Montreal (manuscript).

Tirca L., Gioncu V. (2000). Behaviour of MRFs subjected to near-field earthquakes. "*Behaviour of Steel Structures in Seismic Ares. STESSA 2000*", 21-24 August 2000, Montreal (manuscript).

UBC (1997): Uniform Building Code. Division V, Soil Profile Types.

Chapter 2

Ductility of Members and Connections

2.1

PREDICTION OF AVAILABLE DUCTILITY BY MEANS OF LOCAL PLASTIC MECHANISM METHOD: DUCTROT COMPUTER PROGRAM

Victor Gioncu, Graziella Mateescu, Dana Petcu, Anthimos Anastasiadis

INTRODUCTION

In the design of structures for static and seismic actions, engineers have recognised the importance of the plastic design. The static analysis is accounted with the inelastic force distribution in the calculation of load effects. For the seismic analysis, the interest is intended on dissipation of the input seismic energy. The basic parameter in both approaches is the ductility, considered as the ability of the structure to undergo large plastic deformations without strength loosing.

As before the 60s the ductility notion has been used only to characterise the material behaviour, after the Baker's studies in plastic design and Housner's research work in earthquake problems, this concept has been extended at the level of the structure. Because the steel, as material, has a very good ductility, in the design practice it is generally considered that it is an excellent material for structures erected in seismic areas. But the 1985 - 1995 were very dark years for the behaviour of steel structures. In fact, the earthquakes of Mexico City (1985), Loma Prieta (1989), Northridge (1994) and Kobe (1995) have seriously compromised the idyllic image of steel as a perfect material in seismic areas. In some cases the performance of steel joints and members was very bad and large damage was produced, showing that in special conditions, the present design concepts based only on some constructional rules are not sufficient. Specialists have been award that from material properties to the structure behaviour, an important erosion of the steel qualities may occur, and the quantitative checking only for strength and rigidity, as it is required by the current code provisions, are not sufficient to assure a good behaviour of the structure, especially during strong earthquakes. In the last time, a great amount of research works were devoted for the development of a transparent methodology in which the ductility verification of the structure is quantified at the same level as the checking for strength and stiffness.

The design objective for ductility is to verify if the available ductility is greater than the required ductility. The available ductility results from the local behaviour of the nodes (joint panel, connections, members). The required ductility is obtained from the main characteristics of the ground motions and the overall structure behaviour. The present paper presents the problems of the available ductility determination, using the method of yield lines and local plastic mechanisms. This seems to be the most adequate method to determine the ductility for design proposes. For the required ductility, see the paper of Gioncu et al (2000).

BASIC PHILOSOPHY FOR AVAILABLE DUCTILITY

Definitions

Beams, columns and joints compose a structure. In the capacity design method, some critical sections are chosen to form a suitable plastic mechanism, able to dissipate an important amount of the input energy. Generally it is considered that these sections are located at the beam-ends, where plastic hinges occur during a strong earthquake. But this end of beam is tied into a node, which connects also the column. So, the local plastic mechanisms in the structure can be localised not only at the beam or column end, but also at joins, or at both, member ends and joints (Fig. 2.1.1), depending on the earthquake characteristics and the node conformation.

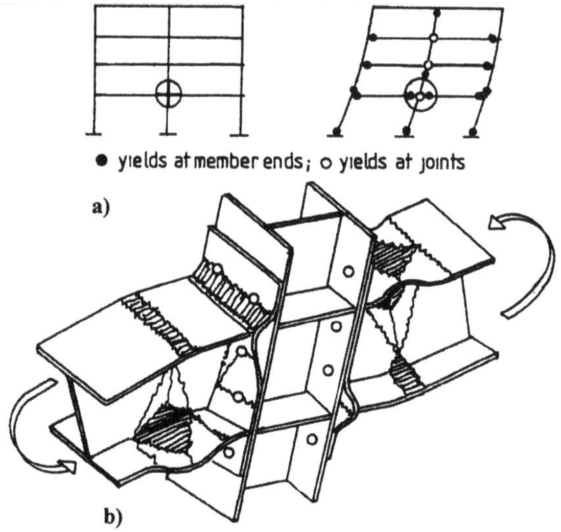

Figure 2.1.1: Local plastic mechanisms

The modern codes impose that plastic deformations occur only at the beam ends and the column bases without considering the joints, even it is well known that these could show, in some conditions, a stable behaviour. But in reality, the required conditions (the joint capacity must be 20% stronger than the adjacent members) does not assure the elastic behaviour of the joints and, as a consequence, the joint could be the weakest component of the node. So, the local ductility has to be defined at the level of the node, which is composed by members and joints.

Some definitions are necessary for the terms of node, joint, connection and panel, because of the indiscriminate everyday use in the literature (Fig. 2.1.2):
- panel zone is the portion of the web corresponding to the connection high;
- the joint is composed by panel zone and connections, the last one representing the physical components which mechanically fasten the beams and columns;
- nodal zone covers the joint and the adjacent beam and column ends, where plastic deformations may occur.

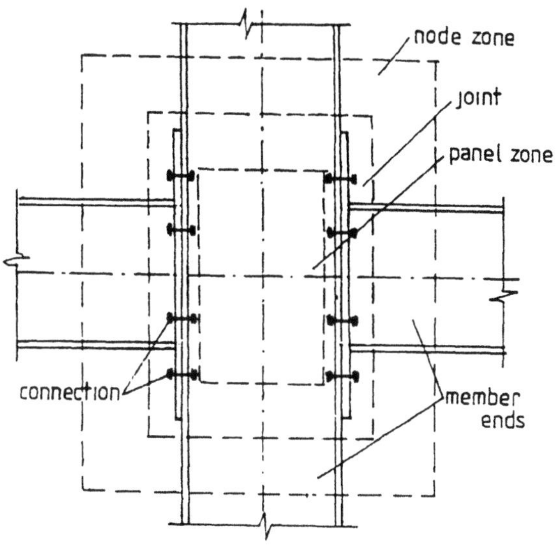

Figure 2.1.2: Panel, joint and node

Factors Influencing the Available Ductility

For an accurate determination of the available ductility, a proper methodology has to consider all the main factors which could influence the behaviour of the node. These factors are presented at the level of member and joint (Gioncu, 1997).

At the level of *members*, the local ductility depends on:

(i) *Material*. It is well known that the mechanical properties of interest for the ductility are determined from tensile tests and the result is the constitutive stress-strain-curve. This aspect refers to:
- *Steel grade*, knowing that the steel ductility decreases with the increasing of the steel quality. In the last period a tendency to use steel with high strength is noticed, so this aspect is very important for the determination of the available ductility;
- *Yield ratio*, is the ratio between yield stress and ultimate strength. The increasing of these characteristics may transform the node collapse from a ductile to a brittle one;
- *Randomness*, referring to yield stress and ultimate strength. In the resistance checking lower bound values have to be considered, and the codes ask producers to verify only these values. For the seismic behaviour, the upper bound values, which are out of control, reveal importance in the design for the ductility assessment;
- *Strain-rate*. Under seismic loads, especially near the source, where the earthquake develops very high peak velocities, the structure interior forces appear with important strain-rates, with the effect of increasing the yield stress and ultimate strength. Due to the fact that the increasing of the yield stress is more important than the ultimate strength, the yield ratio increases too, reducing the local ductility by impending the formation of plastic hinges.

(ii) *Cross-section*. The characteristic diagram for the cross-section behaviour is the moment-curvature curve, which is affected by the following parameters:

- *Cross-section type*, as double T, box, tubular, composite steel-concrete section, battened built-up section, trusses, with very different ductility response;
- *Fabrication*. The profiles can be manufactured by hot-rolling, welding or press-forming. These different fabrication types introduce different junctions between the component plates, different variations of the yield stress along the profiles and different residual stresses;
- *Wall slenderness*, which control the occurrence of the elastic and plastic buckling, framing the cross-sections in different classes: plastic, compact and semi-compact section, with different ductility properties;
- *Wall interaction*. A very important aspect is the flange-web interaction, taking into account that the flange buckling is restrained by the web, as so as the web buckling, by the flanges.

(iii) *Members*. The characteristic diagram for member behaviour is the moment-rotation curve of the plastic hinges, which depends on:
- *Strain-hardening effect*, which produces an increasing of the bending moment over the full plastic moment. The ultimate bending moment, reached at values of about (1.4...1.5)M_p can involve cracks in the tension zones;
- *Buckling of compression walls*, this phenomenon occurred in the plastic range reduces the plastic moment. The slenderness of the compression flange and the beam span play a leading role in the determination of the local ductility;
- *Moment gradient*, taking into account that the ductility for constant moment is grater than the ductility for a variable one;
- *Axial and shear forces*, the ductility being dramatically reduced in the presence of these forces. So, the level of axial and shear forces must be limited;
- *Cyclic loads*, which produce a decreasing in the local ductility, due to the cumulative damage. This effect is relevant in case of far field earthquakes, characterised by cyclic actions. The effect is more reduced in case of near field earthquakes, because of the small number of high cycles.

At the level of *joints*, the local ductility depend on following parameters:
(i) *Panel zone*. The plastic deformation of the panel zone is influenced by:
- *Panel zone type*, the panel being stiffened or unstiffened, with or without diagonal stiffeners;
- *Shear mechanism*, composed by yield lines along the buckling surface and yielding of the diagonal tensile strip;
- *Crushing mechanism*, which occurs in the compression zone of the column web, in absence of web stiffeners.

(ii) *Column flanges*. The plastic deformation of the column flanges depend on:
- *Column type*, realised as double T with free flange, or box section with simple supported flange;
- *Plastic local mechanism* of the flange, which depend on the presence of stiffening details. Yielding lines generally form the mechanism.

(iii) *Connections*. The ductility depends on:
- *Connection type*, realised as double web angles, top and bottom angles, flush end plate, extended end plate;
- *Plastic local mechanism*, composed by yield lines, fracture lines and plastic zones, taking into account of all constructional details;
- *Cyclic loading*, which can produce a fracture due to the accumulation of plastic deformations.

Classification of Joints
Joints may be classified according to their capability to restore the behavioural properties (rigidity, strength and ductility) of the connected members. With respect to the global behaviour of the connected member, two main classes are defined (EC3, 1998) (Fig. 2.1.3):

(i) *Fully restoring joints*. They are designed in such a way to have behavioural properties always equal to, or higher than those of the connected member, in terms of elastic rigidity, ultimate strength and ductility. The generalised force-displacement curve of the connection always lies above the one of the connected member. The existence of the connection may be ignored in the structural analysis.

(ii) *Partially restoring joints*. The behavioural properties of the connection do not reach those of the connected member, due to the lack of capability to restore either elastic rigidity, ultimate strength, or ductility of the connected member. The generalised force-displacement curve could fall in some part below the curve of the connected member. The structural analysis has to consider the existence of such connections.

With respect to the single behavioural property of the connected member, joints may be classified according to strength, rigidity and ductility (Fig. 2.1.4).
(i) With respect to *strength*, connections can be classified as (Fig. 2.1.4a):
- strength restoring (full strength) joints, or
- strength non restoring (partial strength) joints,

depending on whether the ultimate strength of the connected member is restored or not, referring to rigidity and ductility.

(ii) With respect to *rigidity*, joints can be classified as (Fig. 2.1.4b):
- rigidity restoring (rigid) joints, or
- rigidity non-restoring (semi-rigid) joints,

depending on whether the initial stiffness of the connected member is restored or not, referring to strength and ductility.

(iii) With respect to *ductility*, joints can be classified as (Fig. 2.1.4c):
- ductility restoring joints (ductile), or
- ductility non-restoring (semi-ductile or brittle) joints,

depending on whether the ductility of the joint is higher or lower than the ductility of the connected member, referring to strength and rigidity.

Ductile joints have ductility equal or higher than that of the connected member; elongation and rotation could be ignored in the structural analysis. Semi-ductile joints have a ductility less than the one of the connected member, but higher than its elastic limit deformation; elongation and rotation limitations have to be considered in the inelastic analysis. Brittle joints have ductility less than the elastic limit deformation of the connected member; elongation and rotation limitations have to be considered in both elastic and inelastic analysis.

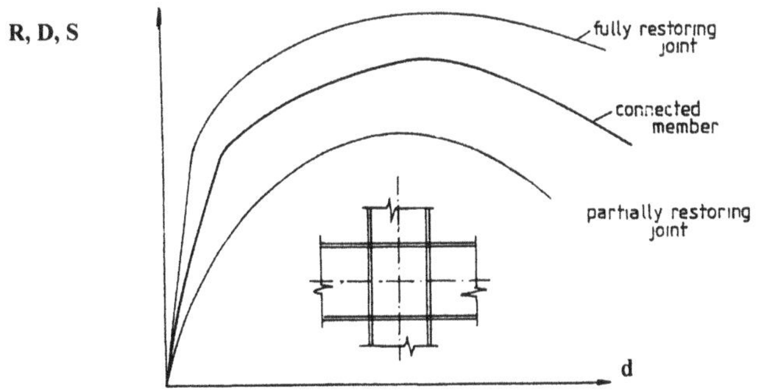

Figure 2.1.3: Fully and partially restoring joints

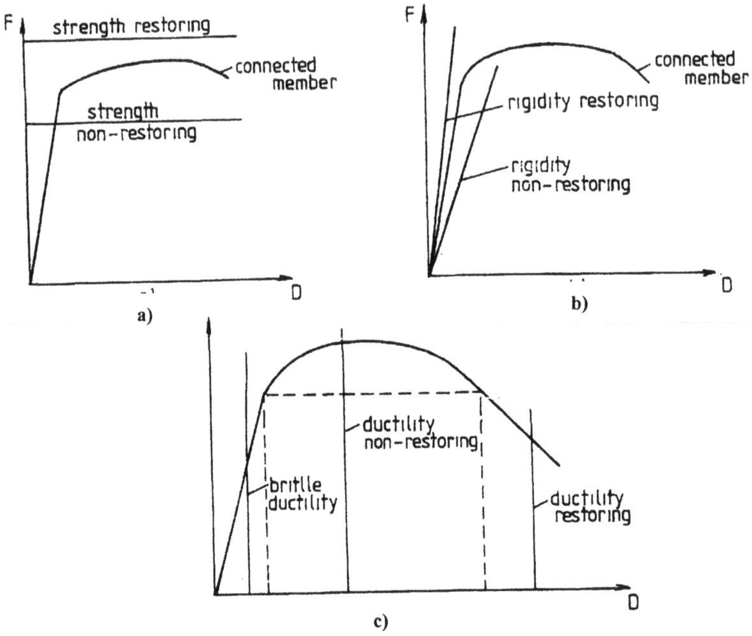

Figure 2.1.4: Joint types

Component Method for Node Available Ductility

Because nodes are composed by a lot of components, the calculation of the available ductility for this complex microstructure is a very difficult task.

A description of the behaviour of a node has to cover all the sources of deformabilities for the joint and the connected member. Due to the multitude of the influencing parameters, a macroscopic view of the complex node by subdividing it into individual basic components has provided to be most appropriate. The procedure can be expressed in three distinct steps (Jaspart et al, 1998):

- identification of the components for joint and members;
- determination of the properties of the components, including the ductility;
- assembly of the component behavioural curves to those of the node behavioural curve, from which the node ductility may be determined.

The main hypothesis of the component method states that the overall behaviour of the node is dictated by the behaviour of the weakest component (Tschemmernegg et al, 1998). From the ductility point of view, the components may be classified as (Fig. 2.1.5):
- *high ductile component*, with an almost unlimited increasing of the deformation capacity. The collapse is due to a ductile fracture, when the ultimate strains are reached;
- *limited ductile component*, when the load-deformation curve presents a moderate decreasing in capacity, after attending the maximum values. The collapse is produced by local plastic buckling, in this case;
- *reduced ductile component*, with a brittle fracture. Especially high strength bolts and some welding procedures present this behaviour type.

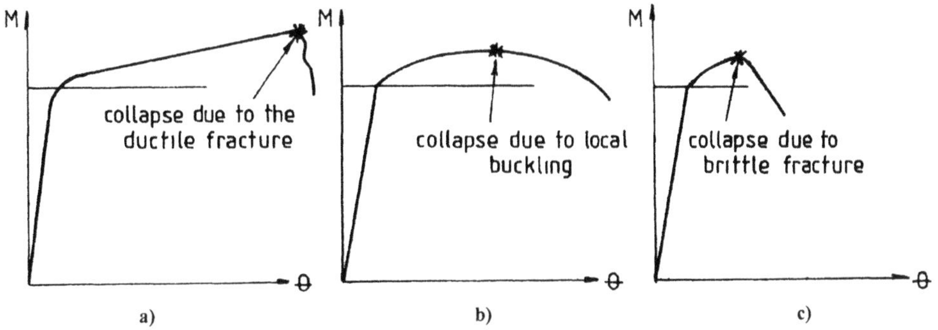

Figure 2.1.5: Ductile and brittle collapse

The assembly of the two components, for joint and member is presented in Figure 2.1.6, they being connected in series. The feature of the two components is that one is high ductile, while the other, limited or reduced ductile. One can see that the ductility of the assembly depends essentially on the weaker strength component. It is absolutely irrelevant if the stronger element is a ductile or a brittle one. A special case occurs when the ductile component provide the smallest value, but the strongest, has the carrying capacity close to the first component and an interaction between these two components may occur. It is possible that, due to some scatter in the material properties, the behaviour of the node should be determined by the strongest component with lower ductility. So, the optimistic consideration of the node behaviour as ductile does not correspond to the reality. A statistical analysis has to be considered in the estimation of the node properties.

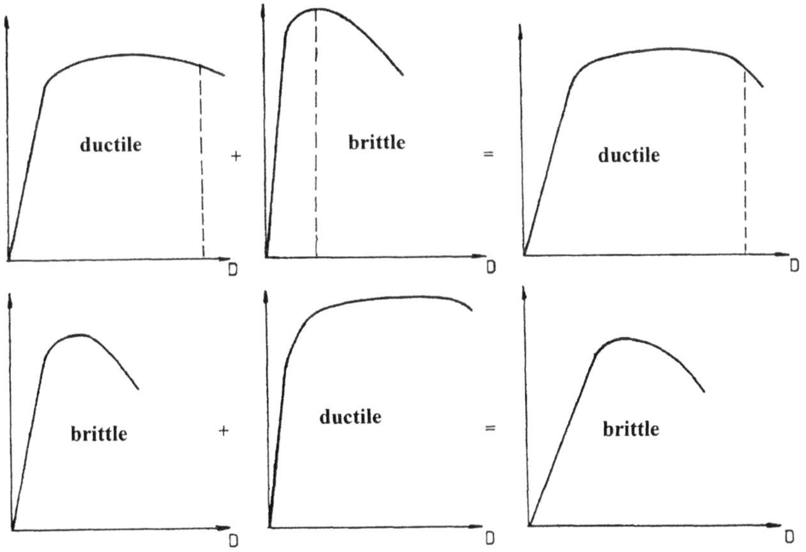

Figure 2.1.6: Component method

PLASTIC YIELD LINE METHOD

Plastic Collapse Mechanism

During experimental test on a node composes by joint and members one can observe that the plastic deformations are produced only in a limited zone, the remaining parts working in elastic field. If the plastic deformations are concentrated along some lines, a local plastic mechanism can be formed, mainly composed by a series of rigid plates separated by yield lines, along which bending can occurs. The rigid plate element is the part of the plate where forces are transmitted. The yield lines are the deformable parts, which admit plastic flows and absorb and dissipate external energy. There are the following mechanism types:
(i) *True* and *quasi-mechanism*. If the plastic mechanism is composed only of yield lines, the mechanism is called true. If it involves also plastic zones, where membrane yieldings occur, it is named quasi-mechanism (Murray, 1995). This differentiation is very important for the ductility calculation, because the quasi-mechanism involve much larger energy than the true mechanism. Generally, true mechanisms appear at thin-walled members, while the quasi-mechanism is specific for plastic and compact cross-sections.

(ii) *Mechanism* with *stationary* and/or *travelling yield lines*. There are yield lines determined by the plastic buckling pattern, being influenced by the plate configuration, the loading system and the initial imperfections. During the rotation of the plastic mechanism these yield lines remain in fixed positions, being the so-called stationary yield lines. In exchange, some other yield lines, formed after the occurrence of the buckling shape change their position, trying to reach the best mechanism pattern - these are the travelling yield lines (Kotelko, 1996). For instance, stationary yield lines characterise the plastic buckling of the compressed flanges of an I-profile, while travelling yield lines shapes the plastic deformation of the web.

(iii) *Complete* or *incomplete mechanism*. In a plastic mechanism, there are principal and secondary yield lines. The principal yield lines must be always present in the plastic mechanism pattern. In exchange, the secondary yield lines may occur only partial formed a part of the theoretical line working in elastic range. From the experimental evidences, the plastic mechanisms of the compressed flanges are complete ones, while the plastic mechanisms of the webs may be incompletely.

(iv) *Mechanism* with *yield* or *yield-fracture lines*. In a plastic mechanism there are yield lines which have larger rotations than other yield lines. During the rotation of the plastic mechanism the stresses of these lines exceed the yield stresses, working in the strain-hardening range. For important rotations, the ultimate strain may be reached and the yield lines are transformed in fracture lines, marking the ultimate rotation capacity of the plastic mechanism. This is the case of some yield lines of the buckled flanges, or the yield lines in the plates of connections.

Rigid-Plastic Analysis

The plastic collapse mechanism is composed of rigid part and local plastic lines or zones. The admissible plastic mechanism can be primarily determined considering the plate shape, edge conditions and loading systems. The work of a collapse plastic mechanism imply that the most part of energy may be absorbed in the small area of plastic lines or zones, so the elastic deformations could be neglected. The rigid-plastic analysis is based on the principle of the minimum of the total potential energy. The total energy, V, functional is defined as:

$$V = U - L_p \tag{2.1.1}$$

where U is the strain energy (internal potential energy) and L_p is the loading potential (external potential). The first variation of the energy is equal to the virtual work. The principle of the minimum of the total potential energy states that for equilibrium the first derivative vanishes. Here, only the displacement field is subjected to variation. In case of the plastic mechanism described by the displacement d_i, the principle takes the simple form for equilibrium:

$$\frac{\partial V}{\partial d_i} = 0 \tag{2.1.2}$$

The strain energy is given by the plastic mechanism work:

$$U = \sum U_{ls} + \sum U_{lt} + \sum U_z \tag{2.1.3}$$

U_{ls}, U_{lt} and U_z being the strain energies corresponding to stationary and travelling yield lines, respectively for the plastic zones. For yield lines only the rotations are involved, while for plastic zones, only the axial deformations are considered:

$$U_{ls,t} = \sum_i M_{pi}\theta_i l_i; \qquad U_z = \sum_j N_{pj}\varepsilon_j A_j \tag{2.1.4a, b}$$

where M_{pi} is the plastic moment along the ℓ line, θ_i, the rotation of this line, ℓ_i the length of the i line, N_{pj}, the axial force of the plastic zone j, ε_j, the axial deformation of this plastic zone and A_j, the area of the plastic zone j. The loading potential is given by:

$$L_p = \int_\delta P_k d\delta_k + \sum_\theta M_l d\theta_l \tag{2.1.5}$$

P_k and M_l being the external forces and moments acting on the element, δ_k and θ_l, the displacement and rotation under the external forces and moments, respectively. Taking into account of (2.1.2), results:

$$P_k = \frac{\partial U}{\partial \delta_k}; \qquad M_k = \frac{\partial U}{\partial \theta_l} \tag{2.1.6a, b}$$

A special treatment must be applied in case that the plastic mechanism contains travelling lines. Because the strain energy for these lines is a function of the geometrical parameters, χ, the pattern of the collapse mechanism could be obtained by minimisation of the equation (2.1.3):

$$U_{min} = \min_{\rightarrow\chi} U \tag{2.1.7}$$

From this condition results χ_m and the relation load- deformation could be determined.

Moment Capacity of Yield and Fracture Lines

The evaluation of the plastic moment capacity from equation (2.1.4a) is of substantial importance when the plastic mechanism approach is applied.

For *yield lines*, the full plastic moment is:

$$M_p = \frac{bt^2}{4} f_y \tag{2.1.8}$$

For an axially loaded plate, the full plastic axial capacity is:

$$N_p = btf_y \tag{2.1.9}$$

If both, bending moment and axial force act on a plate, the reduced plastic moment is (Fig.7):

$$M_{pN} = M_p \left[1 - \left(\frac{N}{N_p} \right)^2 \right] \tag{2.1.10}$$

In case of *fracture lines*, the fracture may occur when the uniform strains are reached at the section extremities. In the same way as for yielding, the ultimate moment is:

$$M_u = \int \sigma y dA = \int_e \sigma y dA + \int_{yp} \sigma y dA + \int_{sh} \sigma y dA + \int_{fp} \sigma y dA \tag{2.1.11}$$

In this relation, the integration over the cross-section is divided into four parts: elastic, yield plateau, strain hardening and failure plateau, in accordance with the idealised σ–ε curve. The term corresponding to the failure plateau has a dominant role, the other terms, as elastic and yield plateau may be neglected. So, a simple relation results for the ultimate moment:

$$M_u = \frac{bt^2}{4} f_u = \frac{1}{\rho_y} M_p \tag{2.1.12}$$

where:

$$\rho_y = \frac{f_y}{f_u} \tag{2.1.13}$$

is the yield ratio.

The ultimate axial load is:

$$N_u = btf_u = \frac{1}{\rho_y} N_p \tag{2.1.14}$$

If both bending moment and axial force act on a plate, a similar relation as (2.1.10) gives the reduced ultimate moment:

$$M_{uN} = M_u\left[1-\left(\frac{N}{N_u}\right)^2\right] = \frac{1}{\rho_y}M_p\left[1-\rho_y^2\left(\frac{N}{N_p}\right)^2\right] \tag{2.1.15}$$

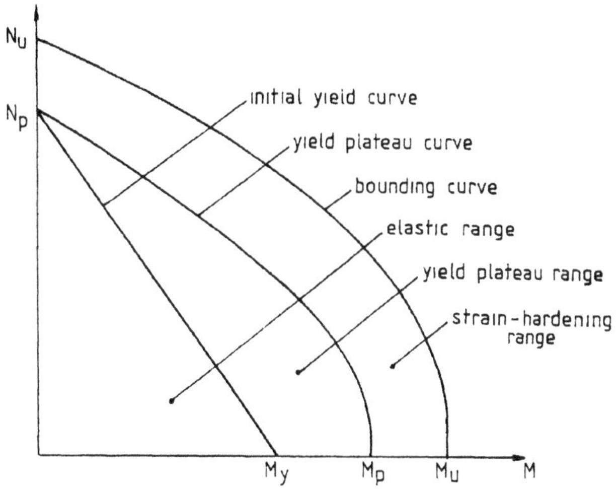

Figure 2.1.7: Bending moment-axial force interaction curves

The relation (15) is also presented in Figure 2.1.7. One can see that the hardening range is directly dependent on the yield ratio. Due to the assumed hypothesis, the collapse load is proportionally to the plastic moment.

Rotation of Fracture Lines

During strong earthquakes and also at experimental tests, cracks were noticed along yield lines, due to the attending of the ultimate strain. This can occur either in case of monotonic or in case of cyclic loading, by accumulation of plastic rotations. In the method of plastic collapse mechanism an approximation is used by the concentration of the plastic rotation along a single section (Fig. 2.1.8a).

The fracture rotation is then $\theta_f = 2\varepsilon_u = 0.22 \div 0.28$, where ε_u is the uniform strain. But it is a too severe condition, because in reality, the rotation is distributed along a plastic zone and the fracture rotation results larger than the values obtained from the above relation. Kotelko (1996) proposed a simplified relation for the fracture rotation:

$$\theta_f = 2n\varepsilon_u \tag{2.1.16}$$

where n is a coefficient determined either experimentally or by means of a minimisation procedure.

Values of $n = 3 \div 4$ correspond well to experimental results, which means that the plastic zone is composed in reality by three or four yield lines. Taking into account this aspect, the length of the plastic zone can be determined from Figure 2.1.8b as:

$$l_p = \left(\frac{M_{uN}}{M_{pN}} - 1 \right) l = \left(\frac{1}{\rho_y} - 1 \right) l \tag{2.1.17}$$

The curvature of the plastic zone is:

$$\chi = \frac{\theta}{l_p} = \frac{1}{1/\rho_y - 1} \cdot \frac{\theta}{l} \tag{2.1.18}$$

where ℓ is the distance between the two inflection points of the plastic mechanism. Taking into account that:

$$\varepsilon = \frac{t}{2} \cdot \chi = \frac{1}{2} \cdot \frac{1}{1/\rho_y - 1} \cdot \frac{t}{l} \cdot \theta \tag{2.1.19}$$

results:

$$\theta_f = 2 \left(\frac{1}{\rho_y} - 1 \right) \frac{l}{t} \cdot \varepsilon_u \tag{2.1.20}$$

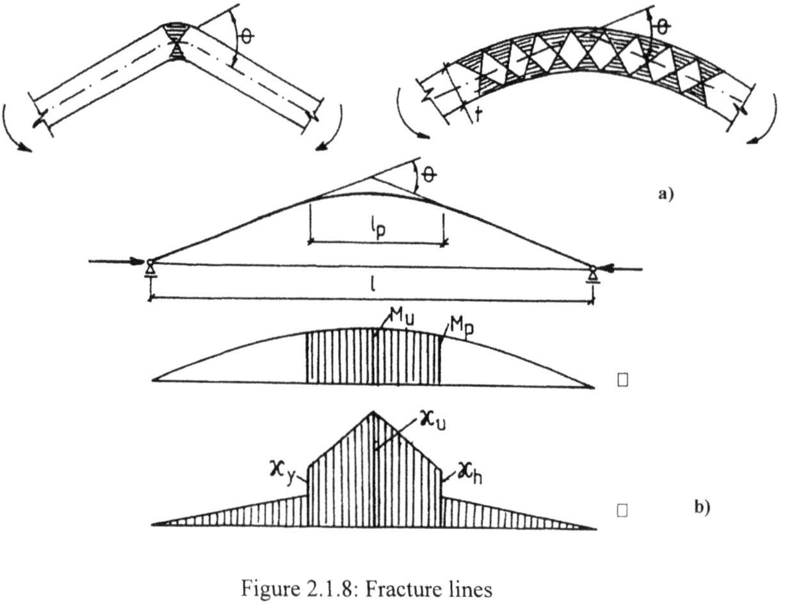

Figure 2.1.8: Fracture lines

For Fe 360 and $l/t = 6$, $\theta_f = 6.38\varepsilon_u$, which correspond very well to the values proposed by Kotelko. One can see that this ultimate ration depends on the yield ratio, the length of the plastic mechanism and the steel quality. Figure 2.1.9 shows the influence of the yield ratio on the fracture rotation. A very high reduction of the ultimate ration could be noticed, when the yield ratio increases towards 1, showing the significance of the steel quality in the prevention of a premature collapse.

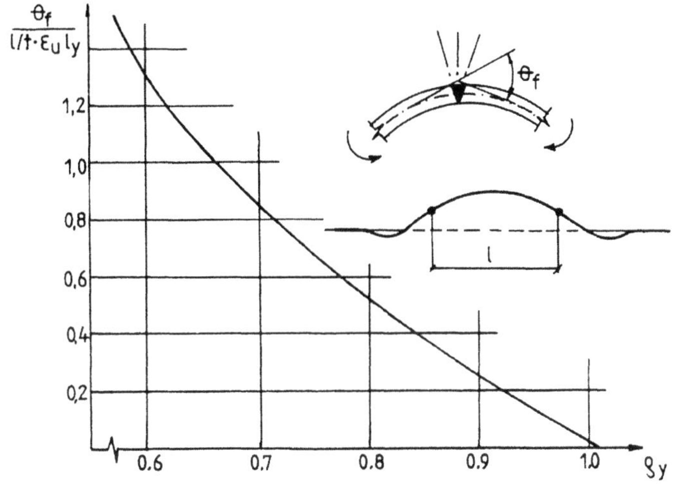

Figure 2.1.9: Influence of the yield ratio on the fracture rotation

AVAILABLE DUCTILITY FOR MEMBERS

Behaviour of an Actual Member in a Structure

The available ductility of a member has to be determined taking into account that the member belongs to a structure with a complex behaviour (Fig.2.1.10), being influenced by both gravitational and seismic actions. Due to great number of factors influencing the behaviour of the actual member, the determination of the available ductility could be a very difficult task. So, it is important to simplify the analysis by using a simple substitute member, with a very similar behaviour to the actual one. Thus, Spangemacher & Sedlacek (1992) and Gioncu & Petcu (1995, 1997) proposed the so-called standard beam. One can see that in a complex structure the inflection point divides the member in portions with positive and negative bending moments, so the actual member from a structure could be replaced by a combination of two standard beam types:
- SB 1, central concentrated load beam, for the case of members under moment gradient;
- SB 2, distributed load beam, for the case of beams with weak moment gradient. In some cases, two-concentrated load beams could replace it.

In the followings it is assumed that the seismic loads act from the left to the right side of the structure. The problem of working with standard beams is to divide the structure into single span beams, i.e. to determine the position of the inflection points. The beam is loaded by two asymmetric end-bending moments, which generally have different values, due to the fact that the two beam-ends work in different conditions:
- left bending moment: $M_A = m_A M_p$ (2.1.21a)

the increasing over the full plastic moment being the result of the interaction of the beam flange with the slab, in the compression zone;
- right bending moment: $M_B = m_B M_p$ (2.1.21b)

the increasing being produced by the strain hardening and the effect of the slab, in the tension zone.

In the following, the assumption, that after attending the maximum value, the moment remains constant, is used. The behaviour of the beam can be studied using two standard beams; the spans of these beams are $l_l = 2x_i$ and $l_r = 2(l - x_i)$, x_i being the distance to the inflection point.

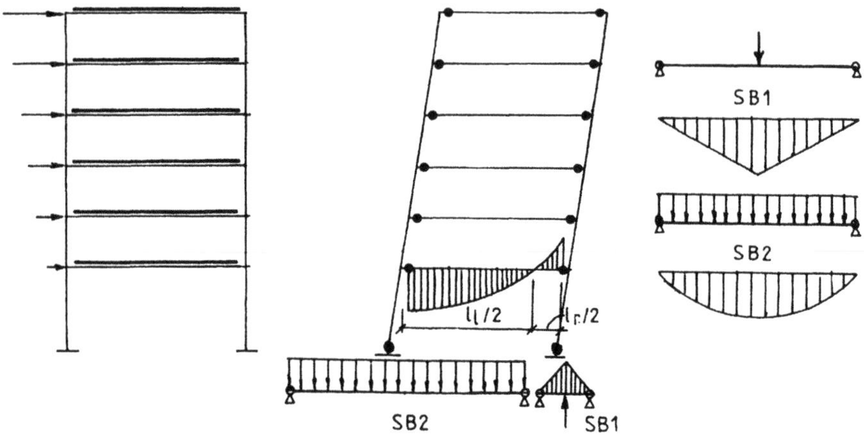

Figure 2.1.10: Definition of the standard beam

Behaviour of Actual Beams

The two beam-ends are in different conditions regarding the interaction beam-slab. For the right end the slab is in tension zone, while for the left, it is in compression. There are the following situations:

- *The interaction has to be considered* (Fig. 2.1.11a), when the constructional details assure a good co-operation between the two elements. In this case the plastic behaviour at the right end is dominated by the buckling of the lower compression flange, while the left end behaviour is controlled by crushing of the reinforced concrete slab. This situation creates great differences between the two end plastic moments, the smallest being the right end moment, a considerable movement in the inflection point being observed.

The ratio of the two plastic moments is:

$$\nu_m = \frac{M_{pl}}{M_{pr}} \tag{2.1.22}$$

where M_{pr} is the plastic moment for the right end (buckling of the lower compression flange) and M_{pl}, the moment for the left end (interaction beam-slab).

The two different cases concerning the place of the plastic hinge are given by:

- if:

$$\frac{\nu_m + 1}{4} \cdot \frac{M_B}{M_0} \geq 1 \tag{2.1.23}$$

with the bending moment of a simple supported beam:

$$M_0 = \frac{ql^2}{8} \tag{2.1.24}$$

the left plastic hinge occurs at the beam end ($M_A = M_{pl}$) and the position of the inflection point is given by:

$$x_i = \left\{1 - \frac{v_m + 1}{4}\frac{M_B}{M_0} + \left[\left(1 - \frac{v_m + 1}{4}\frac{M_B}{M_0}\right)^2 + m_A \frac{M_p}{M_0}\right]^{1/2}\right\}\frac{l}{2} \tag{2.1.25}$$

- if:

$$\frac{v_m + 1}{4}\frac{M_B}{M_0} < 1 \tag{2.1.26}$$

the left plastic hinge occurs at some distance from the beam end ($M_A \neq M_{pl}$, $M_{max} = M_{pl}$), the distance of the maximum moment may be determined by:

$$x_m = \left[1 - \left(\frac{v_m + 1}{4}\frac{M_B}{M_0}\right)^{1/2}\right] l \tag{2.1.27}$$

and the position of the inflection point will be:

$$x_i = x_m + \left(v_m \frac{M_B}{M_0}\right)^{1/2} \frac{l}{2} \tag{2.1.28}$$

The distance between the point of maximum moment and the inflection point is then:

$$\Delta_i = x_i - x_m = \left(v_m \frac{M_B}{M_0}\right)^{1/2} \frac{l}{2} \tag{2.1.29}$$

Variations of the characteristic distances are presented in Figure 2.1.11b. The A-type behaviour corresponds to the plastic hinge formed at some distance from the edge, the B-type, to the case when the plastic hinge occur at the beam edge. One can see that the interaction beam-slab has a great influence on the plastic behaviour of the beam.

• *The interaction is avoided* by some special details aiming to eliminate the incertitude on the level of beam-slab interaction and to ensure a very clear plastic behaviour. The inequality of the two end bending moments (Fig.2.1.12a) is due in this case, to the difference in the moment gradient. The moment gradient is high for the right end, and so the plastic moment increases, due to the strain hardening, $M_{pB}=M_hM_p$ and $m_B=m_h$. The left plastic hinge works with a quasi-constant moment, $M_{pA}=M_p$ and $m_A=1$. The two behaviour case are given by the following conditions:

- if:

$$\frac{1+m_b}{4}\frac{M_p}{M_0} \geq 1 \tag{2.1.30}$$

the left plastic hinge occurs at the beam end and the position of the inflection point results from:

$$x_i = \left\{1 - \frac{1+m_b}{4}\frac{M_p}{M_0} + \left[\left(1 - \frac{1+m_b}{4}\frac{M_p}{M_0}\right)^2 + \frac{M_p}{M_0}\right]^{1/2}\right\}\frac{l}{2} \tag{2.1.31}$$

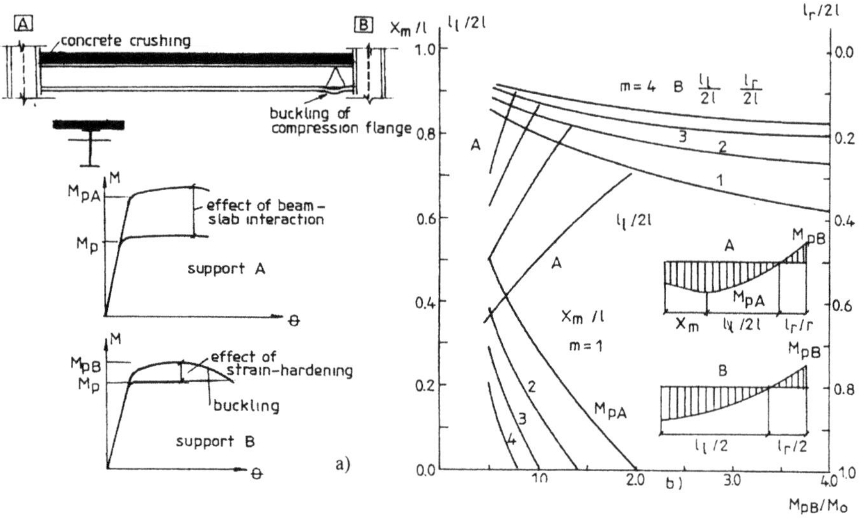

Figure 2.1.11: Standard beams for beam-slab interaction

- if:

$$\frac{1+m_b}{4}\frac{M_p}{M_0} < 1 \tag{2.1.32}$$

the left plastic hinge occurs at a certain distance from the beam end, which could be determined from:

$$x_m = \left[1 - \left(\frac{1+m_b}{4}\frac{M_p}{M_0}\right)^{1/2}\right] l \tag{2.1.33}$$

and the position of the inflection point will be:

$$x_i = x_m + \left(\frac{M_p}{M_0}\right)^{1/2}\frac{l}{2} \tag{2.1.34}$$

$$\Delta_i = x_i - x_m = \left(\frac{M_p}{M_0}\right)^{1/2}\frac{l}{2} \tag{2.1.35}$$

The distance between the characteristic points is presented in Figure 2.1.12b. It is clear that in the case when the interaction is avoided, the differences between the plastic moment values at the two ends have not a relevant influence on the plastic behaviour of the beam.

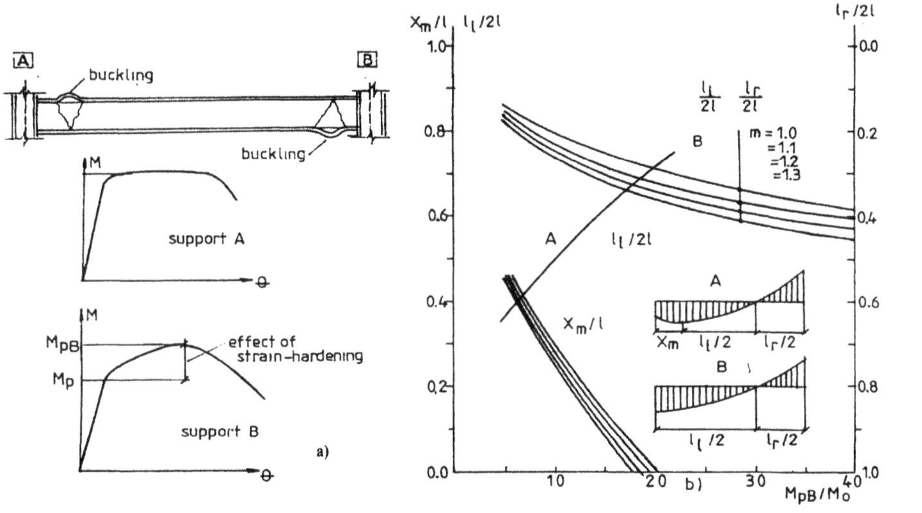

Figure 2.1.12: Standard beam for free beams

Behaviour of the Actual Beam-Column

Examining the collapse mechanism (Fig. 2.1.13) for beam-columns one can remark two different situations:
(i) *One plastic hinge* occurs at the column base, in case of the global mechanism. If M_s is the elastic bending moment at the upper end of the column, the position of the inflection point will be:

$$\frac{h_s}{h} = \frac{1}{1 \pm \frac{M_s}{M_p}} \tag{2.1.36}$$

The plus sign in the relation (36) refers to the case of a double curvature moment, while the minus corresponds to a single curvature moment. The first case corresponds to a weak column related to the beam, the second, to a strong column.

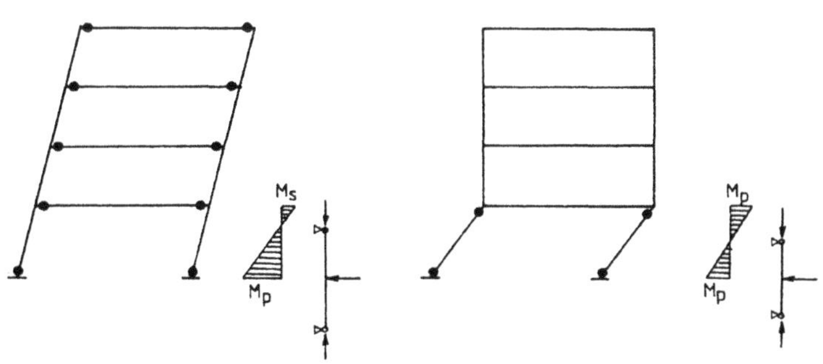

Figure 2.1.13: Standard beam for columns

(ii) *Two plastic hinges* at both column ends occur in case of a storey mechanism. The inflection point is localised at the middle of the column.

In both cases a standard beam SB1-type could be used, with the span corresponding to two times the distance between base and inflection point. The standard beam must be loaded with the axial forces corresponding to the ones determined from the structure analysis.

Behaviour of Standard Beams

The behaviour of plates in tension and compression is shown in Figure 2.1.14. The first very important aspect refers to the behaviour in the plastic range. If the strains are under control, the yielding plateau is crossed step by step. Contrary, if the stresses are controlled, a dynamic jump in strain across the yielding is produced.

Therefore, no local strain can exist between the elastic strain limit, ε_y, and the beginning of the strain-hardening, ε_h. Because during the seismic actions there is no control of the strains, it is necessary to have in mind that both tension and compression flanges work in the strain-hardening range.

The symmetry of the behaviour in tension and compression is broken due to the occurrence of the plastic buckling of the compression plate. Thus, the local buckling acts as a limitation of high stresses, while the increasing of stresses for the tension flange is not limited.

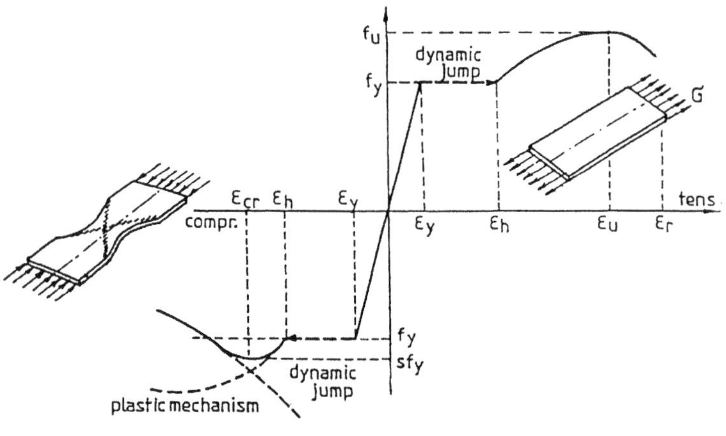

Figure 2.1.14: Behaviour of the tension and compression flanges

Figure 2.1.15a shows a three-point loaded beam where the gradient moment is the main characteristic. The moment-rotation curve is presented in Figure 2.1.15b, where the increasing of the moment over the full plastic value is due to the hardening effect. The stress distribution is shown in Figure 2.1.15c. After the elastic behaviour, yielding of the two flanges occurs and quickly a full plasticization of the cross-section appears. Due to the dynamic jump, the maximum stresses correspond to the hardening range, $\sigma_t = \sigma_c > f_y$. The symmetry in the stress distribution is broken by the buckling of the compression flange at the stress $\sigma_c = \sigma_{cr} = sf_y$, followed by decreasing of the stresses and moving down of the neutral axis. Thus, in the post-buckling range, the tension stresses remain constant $\sigma_t = sf_y$, while the compression ones decrease towards the yielding strength, f_y. Due to this fact, a reduction of the bending moment occurs after local plastic buckling.

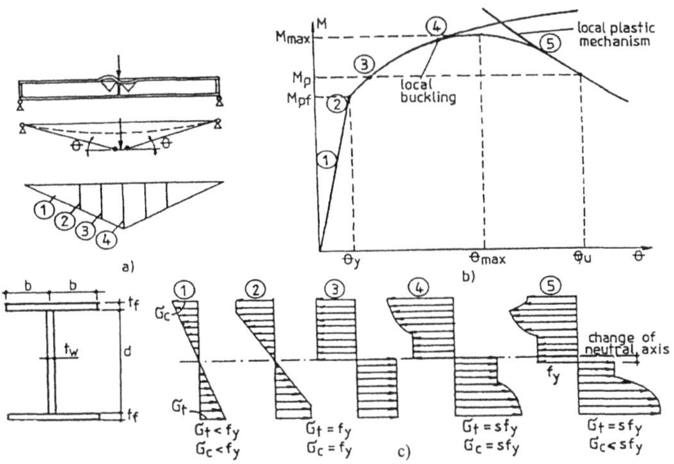

Figure 2.1.15: Stress distribution in **I**-sections

Local Plastic Mechanisms

The great majority of the experimental tests were performed on standard beams SB1, so the observations are mainly obtained from this beam type. During experimental tests one can observe that the plastic deformations are produced only in a limited zone, the remaining part of the member being in elastic field. Large rotations are concentrated in the plastic zone, which work as plastic hinges. They are amplified, if a buckling of the flanges occurs in these zones.

Examining experimental and numerical simulations of buckled shapes, presented in the papers of Gioncu et al (1989), Ivanyi (1979, 1985), Kuhlmann (1986), Spangemacher (1991), Climenhaga and Johnson (1972), Suzuki et al (1994) there are important conclusions to remark:

(i) During the plastic deformations, crumpling of the flanges and web could be observed, which form a local plastic mechanism composed by yield lines and plastic zones.
(ii) The buckling always starts by a flange crumpling, which induces the web deformations. Thus, a very important interaction between flange buckling and web deformation exists. The formation of the plastic mechanism of a flange is always complete and imposes the general shape of the plastic mechanism. The plastic mechanism for a web is sometimes partially formed and the web deformations cannot follow the large deformations, especially when high steel grade is used.
(iii) There are many forms of local plastic buckling, which depend on the geometrical proportions of the beam: in-plane buckling asymmetrical in comparison with the cross-section (Fig. 2.1.16a); out-of-plane buckling (Fig. 2.1.16b); buckling of the flange into the web, denoted as 'flange induced buckling' in EC3, which arises by very slender webs (Fig. 2.1.16c); shear plastic buckling of the web, in case of short beams (Fig. 2.1.16d).
(iv) An interaction between these local mechanisms is observed during experimental tests. In the majority of cases, the plastic buckling starts with an in-plane buckling, but due to the weakening in the lateral rigidity caused by the plastic buckling, a lateral buckling occurs. The lowering path of the moment-rotation curve is dominated by the interaction of the two buckling modes.
(v) From a theoretical point of view, two buckling shapes have to occur, therefore in practice only one of them is more developed, because of the local imperfections.

So, the member collapse is due to the formation of a general mechanism, composed by two rigid parts and two plastic hinges (Fig. 2.1.17). The plastic buckling of the compression flange and web could be replaced by a local plastic mechanism, composed by plastic zones and yield lines. The method to determine the collapse loads and the post-yielding behaviour is based on the principle of virtual work (Gioncu and Petcu, 1997). It could be written:

$$\frac{M}{M_p} = \frac{m_{nv}}{C(\chi)}\left[A(\chi) + B(\chi)\frac{1}{\theta^{1/2}}\right] \tag{2.1.37}$$

where the coefficient m_{nv} considers the influence of the axial shear forces on the plastic moment determined using the EC3 rules. The following coefficient are used in (2.1.37):

$$\chi = \frac{\beta b}{\delta d} \tag{2.1.38}$$

Is the mechanism geometrical parameter, which defines, trough β, the length of the plastic buckling wave:

$$\beta \approx 0.6\left(\frac{t_f}{t_w}\right)^{3/4}\left(\frac{d}{b}\right)^{1/4} \tag{2.1.39}$$

and through δ, the position of the rotation points of the mechanism. The coefficients A(χ), B(χ) and C are determined after some algebra as:

$$A(\chi) = \rho\left[\frac{\beta^2 b^2}{\chi^2 d^2} + \left(1 - \frac{\beta b}{\chi d}\right)^2\right]\frac{t_w}{b} + 2s\left(1 - \frac{\beta b}{\chi d}\right)\frac{t_f}{d} + 2\frac{bt_f}{d^2}\begin{Bmatrix} \beta^2/\chi, & \beta \le 1 \\ (2\beta - 1)/\chi, & \beta > 1\end{Bmatrix} \tag{2.1.40a}$$

$$B(\chi) = \rho\frac{\beta}{\chi^{1/2}}\left[\frac{(2+\chi)(1+\chi^2\alpha^2\beta^2) - \chi(2+\chi^2)}{\left[1 - 2\chi + (1-\alpha^2\beta)\chi^2\right]^{1/2}} + 2\chi\left[1 - 2\chi + (1 - \alpha^2/\beta^2)\chi^2\right]^{1/2} - \right.$$
$$\left. - \begin{Bmatrix} 1, & \beta \le 1 \\ \beta^{1/2}, & \beta > 1\end{Bmatrix}\right]\frac{t_w^2}{d^2} + 2\frac{t_f^2}{d^2}\begin{Bmatrix} (\beta+2)/\chi^{1/2}, & \beta \le 1 \\ 3(\beta/\chi)^{1/2} & \beta > 1\end{Bmatrix} \tag{2.1.40b}$$

while the coefficient C is given from the following relationship:

$$C = 2M_p\frac{1}{1 - n_p\bar{\lambda}^2} \tag{2.1.40c}$$

where the notations:

$$\rho = \frac{f_{yw}}{f_{yf}}; \qquad s = \frac{\sigma_{cr}}{f_y} \tag{2.1.41a,b}$$

are the ratios between steel grades of web and flanges and between the critical stress and yielding strength, respectively. The coefficient α from the relation (2.1.40b) is a parameter of the mechanism asymmetry, while the notations from (2.1.40c) represent:

$$n_p = \frac{N}{N_p}; \qquad \lambda = \left(\frac{N_p}{N_{cr}}\right)^{1/2} \tag{2.1.42a,b}$$

The relation (2.1.37) represent a hyperbola related to the rotation of the plastic hinge (Fig. 2.1.18).

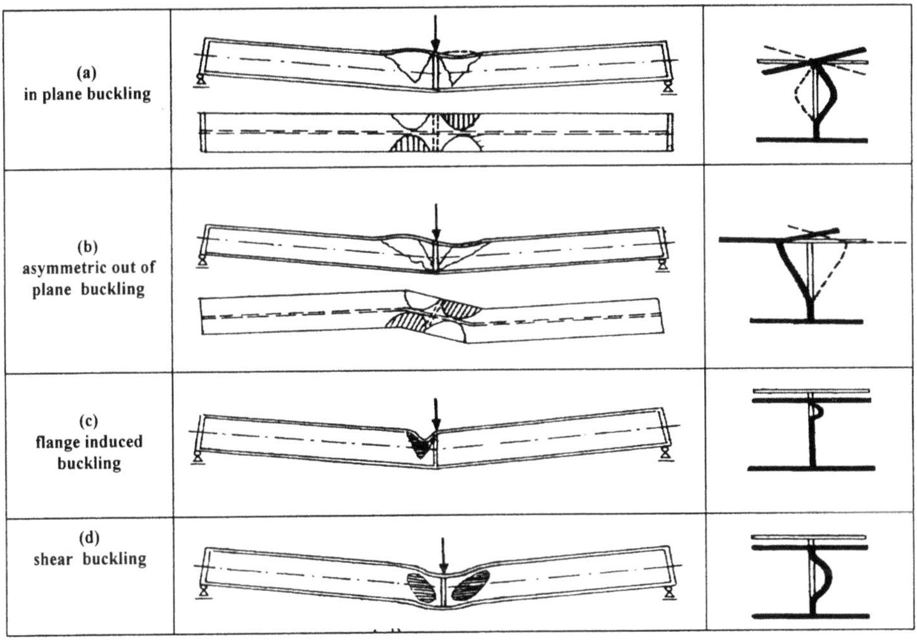

Figure 2.1.16: Plastic buckling types

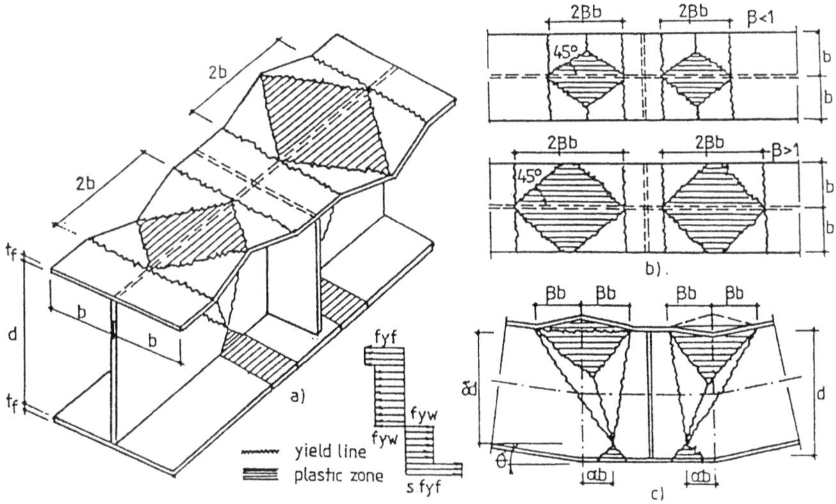

Figure 2.1.17: Plastic collapse mechanism

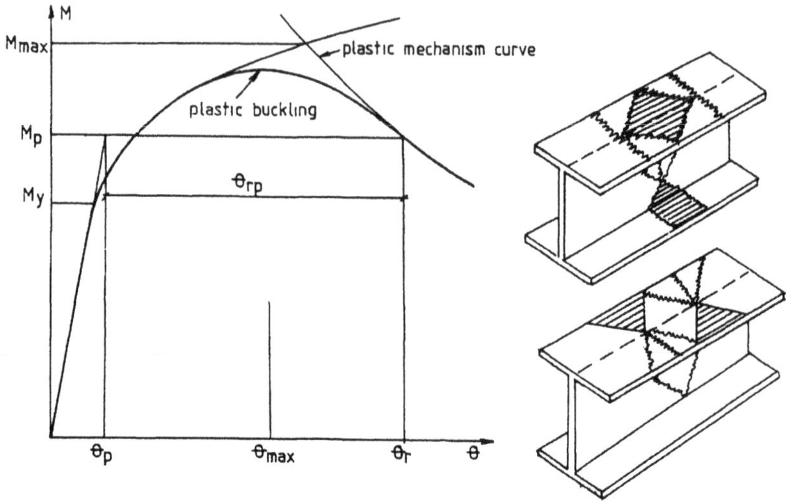

Figure 2.1.18: Moment-rotation curve

According to the cinematic theorem of plastic collapse, the minimum values of the M-θ curve are obtained by minimisation of eq. (2.1.37) with respect to the parameter χ:

$$\left(\frac{M}{M_{pnv}}\right)_{\min} = \min\frac{1}{C(\chi)}\left[A\,(\chi)+B\,(\chi)\frac{1}{\theta^{1/2}}\right] \tag{2.1.43}$$

Such a minimisation is presented in figure 2.1.19.

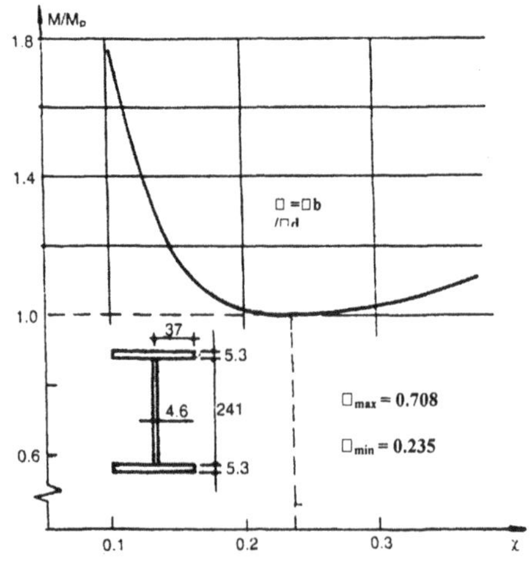

Figure 2.1.19: Minimisation of M/ M_p ratio

In the same way, the rigid-plastic curve for the out-of-plane plastic mechanism could be determined with the coefficients $A(\chi)$ and $B_{1,2}(\chi)$ from Gioncu and Petcu (1997):

$$\frac{M}{M_{pnv}} = \frac{1}{C(\chi)}\left[A(\chi) + B_1(\chi)\frac{1}{\theta^{1/2}} + B_2(\chi)\frac{1}{\theta^{1/4}}\right] \tag{2.1.44}$$

Comparatively to the in-plane mechanism, a supplementary term, which produces a more significant degradation in the post-buckling range, appears.

Definition of the Rotation Capacity

The formula to calculate the available rotation capacity is given by:

$$R_a = \frac{\theta_{rp}}{\theta_p} = \frac{\theta_r}{\theta_p} - 1 \tag{2.1.45}$$

where θ_{rp} is the ultimate plastic rotation, θ_p, the rotation corresponding to the first plastic hinge and θ_r, the total ultimate rotation (Fig. 2.1.18).

There are some proposals (see EC8, Background Document) to determine the rotation capacity for a reduced plastic moment $M_p/1.1 \approx 0.9M_p$, so the rotation capacity will be:

$$R_{a,0.9} = \frac{\theta_{r,0.9}}{\theta_p} - 1 \tag{2.1.46}$$

A comparison between numerical results using the local plastic mechanism with 85 experimental data, collected from literature, has been performed by Gioncu and Petcu (1997) finding for the experimental values a high coefficient of variation ($c_v = 0.379$). A new analysis has been performed with a statistical method, which removes the values determined with an error probability greater than 5%. Using this approach, the coefficient of variation is reduced to 0.237, and only 60 experimental results were kept. Figure 2.1.20 presents the comparison between numerical and experimental results, showing a good correspondence and giving confidence in the results obtained by using the local plastic mechanism methodology.

Rotation Capacity for Monotonic Loading

Some results obtained using the local plastic mechanism methodology are presented in Figure 2.1.21. Two different classifications are plotted there, one after EC3 (cross-section classes), the other (member classes), proposed by Mazzolani and Piluso (1993). One can see great differences between the two classifications and so it is clear that the code provisions have to consider the member classes, they being more adequate for the checking of the structure ductility than the cross-section classes. One of the main factors influencing the member ductility is the mode of fabrication. Hot-rolled profiles, widely used in the structural design, provide different ductility capacities than welded sections. In Figure 2.1.22 the influence of the web-flange junction is plotted. One can remark a relevant increasing of the plastic rotation in case of hot-rolled sections compared to the same sections, where the influence of a rigid zone is neglected (Anastasiadis and Gioncu, 1999).

A simplified coefficient of correction was proposed, in the aim to use the results without considering the junction:

$$R_r = c_r \cdot R \tag{2.1.47}$$

where:

$$c_r = \left(\frac{b}{c}\right)^2 = \left(\frac{b}{b - 0.5t_w - 0.8r}\right)^2 \tag{2.1.48}$$

with the dimensions from Figure 2.1.22.

The correlation between exact and corrected values and relation (2.1.47) confirms this simple procedure to determine the improved values of the rotation capacity of hot-rolled sections. Using these results, a classification of IPE and HEA profiles in member classes was proposed. This new classification according to the member concept bases on the criteria:

- high ductility, **H**, $R \geq 7.5$,
- medium ductility, **M**, $5 < R < 7.5$,
- low ductility, **L**, $1.5 < R \leq 4.5$.

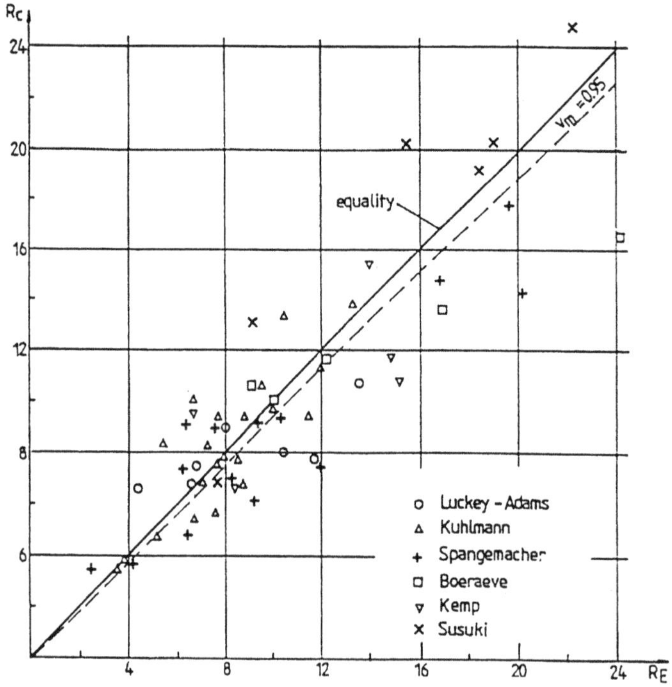

Figure 2.1.20: Comparison between experimental and theoretical results

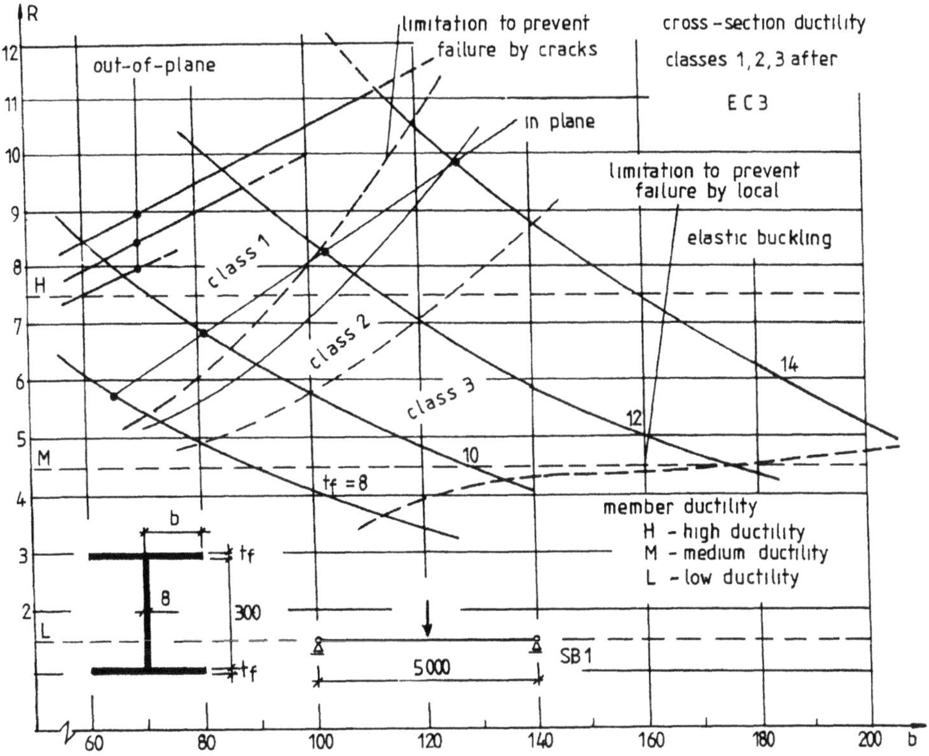

Figure 2.1.21: Influence of geometrical parameters

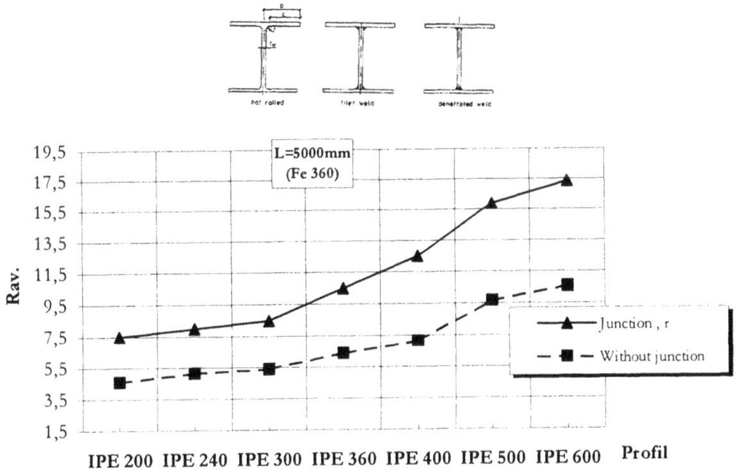

Figure 2.1.22: Influence of the web-flange junction

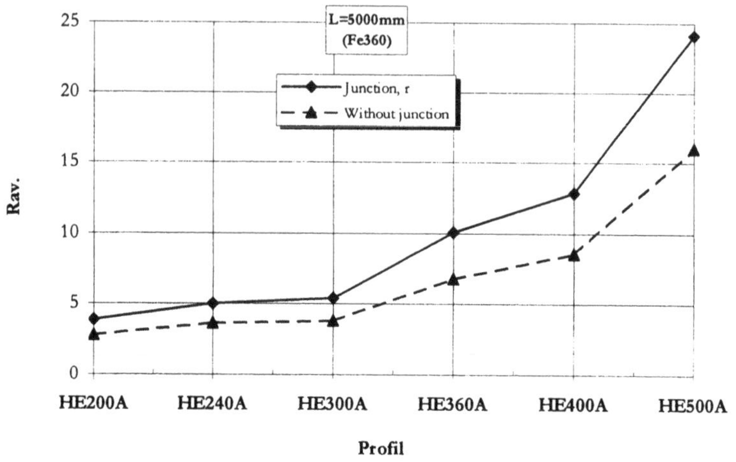

Figure 2.1.22: continued

The values presented in Table 2.1.1 show clearly that the length of the member and the member type, as well as the steel quality have a major influence on the local available ductility. From this table, the designer can select the profile, which for a given data, assures a good inelastic global behaviour of the structure.

TABLE 2.1.1
MEMBER CLASSIFICATION FOR IPE BENDING BEAMS

	L=3000			L=4000			L=5000			L=6000			L=7000		
Profil	**Fe 360**	**Fe 430**	**Fe 510**	**Fe 360**	**Fe 430**	**Fe 510**	**Fe 360**	**Fe 430**	**Fe 510**	**Fe 360**	**Fe 430**	**Fe 510**	**Fe 360**	**Fe 430**	**Fe 510**
IPE 140	H	H	H	H	H	M	M	M	M	L	L	L	L	L	L
IPE 160	H	H	H	H	H	M	M	M	M	L	L	L	L	L	L
IPE 180	H	H	H	H	H	M	H	M	M	M	M	M	L	L	L
IPE 200	H	H	H	H	H	M	H	H	M	H	M	M	M	M	M
IPE 220	H	H	H	H	H	H	H	H	M	H	H	M	M	M	M
IPE 240	H	H	H	H	H	H	H	H	M	H	H	M	H	M	M
IPE 270	H	H	H	H	H	H	H	H	M	H	H	M	H	M	M
IPE 300	H	H	H	H	H	H	H	H	M	H	H	M	H	M	M
IPE 330	H	H	H	H	H	H	H	H	M	H	H	M	H	M	M
IPE 360	H	H	H	H	H	H	H	H	H	H	H	M	H	M	M
IPE 400	H	H	H	H	H	H	H	H	H	H	H	M	H	H	M
IPE 450	H	H	H	H	H	H	H	H	H	H	H	M	H	H	M
IPE 500	H	H	H	H	H	H	H	H	H	H	H	H	H	H	M
IPE 550	H	H	H	H	H	H	H	H	H	H	H	H	H	H	H
IPE 600	H	H	H	H	H	H	H	H	H	H	H	H	H	H	H

TABLE 2.1.1 (continued)
MEMBER CLASSIFICATION FOR HE-A BENDING BEAMS

	L=3000			L=4000			L=5000			L=6000			L=7000		
Profil	**Fe 360**	**Fe 430**	**Fe 510**	**Fe 360**	**Fe 430**	**Fe 510**	**Fe 360**	**Fe 430**	**Fe 510**	**Fe 360**	**Fe 430**	**Fe 510**	**Fe 360**	**Fe 430**	**Fe 510**
HE 160A	H	H	H	M	M	L	M	L	L	L	L	L	L	L	L
HE 180A	H	H	H	M	M	L	M	L	L	L	L	L	L	L	L
HE 200A	H	H	H	M	M	L	M	L	L	L	L	L	L	L	L
HE 220A	H	H	H	M	M	M	M	M	L	L	L	L	L	L	L
HE 240A	H	H	H	H	M	M	M	M	L	M	L	L	M	L	L
HE 260A	H	H	H	H	M	M	M	M	L	M	M	L	M	L	L
HE 280A	H	H	H	H	M	M	M	M	M	M	M	L	M	L	L
HE 300A	H	H	H	H	H	M	M	M	M	M	M	L	M	L	L
HE 320A	H	H	H	H	H	M	H	H	M	H	M	M	M	M	L
HE 340A	H	H	H	H	H	H	H	H	M	H	H	M	H	M	L
HE 360A	H	H	H	H	H	H	H	H	H	H	H	M	H	M	M
HE 400A	H	H	H	H	H	H	H	H	H	H	H	H	H	H	M
HE 450A	H	H	H	H	H	H	H	H	H	H	H	H	H	H	H
HE 500A	H	H	H	H	H	H	H	H	H	H	H	H	H	H	H
HE 550A	H	H	H	H	H	H	H	H	H	H	H	H	H	H	H
HE 600A	H	H	H	H	H	H	H	H	H	H	H	H	H	H	H

For the practical design of the rotation capacity of beams and beam-columns, there are simplified design relations proposed, based on numerical tests (Gioncu and Mazzolani, 1997). The rotation capacity for beams under monotonic loads is given by:

$$R_{av.mon} = 3 \cdot 10^4 c_r \frac{t_f}{bL_{sb}} \varepsilon \left(0.8 + 0.2 \frac{f_{yw}}{f_{yf}} \right); \qquad \varepsilon = 235/f_{yf} \tag{2.1.49}$$

where:

- c_r is a coefficient taking into account the influence of the junction,
- t_f, the flange thickness,
- b, the half width of the flange,
- L_{sb}, the span of the standard beam,
- f_{yw} and f_{yf}, the yield tensile strength for web and flange

The simplified relation (2.1.49) covers the domain for practical applications. Over 150 numerical tests have been performed and a very reduced coefficient of variation (c_v = 6%) was found.

In the same manner as for beams, an approximate relationship for the beam-column rotation capacity, with respect to the effect of the axial force and moment diagrams was also proposed (Anastasiadis and Gioncu, 1999):

$$R_{av,mon} = 1481.3 c_r \left(\bar{\lambda} \frac{b}{t_f} \sqrt{f_y} \right)^{-1.33} \qquad \text{for } n_p = 0.10 \tag{2.1.50a}$$

$$R_{av,mon} = 5099.9 c_r \left(\bar{\lambda} \frac{b}{t_f} \sqrt{f_y} \right)^{-1.61} \qquad \text{for } n_p = 0.40 \tag{2.1.50b}$$

where: b/t_f is the flange slenderness and λ, a non dimensional slenderness, given by:

$$\bar{\lambda} = \left(\frac{N}{N_{cr}}\right)^{1/2} = \frac{f_{yf} \cdot W_{pf} + f_{yw} \cdot W_{pw}}{\pi^2 \dfrac{E \cdot I_c}{\mu \cdot L}} \tag{2.1.50c}$$

with: f_y, the yield tensile strength of the section and L_{sb}, the standard beam span.

Over 160 numerical tests were performed to obtain adequate statistical parameters. The results obtained with the relation (2.1.50) cover the domain of the HE 100A(B) to HE 600A(B) profiles, with a moment ratio $M_{sup}/M_{inf} = 1...0$, as well as considering the influence of the axial force, n_p.

Rotation Capacity for Seismic Loading

The rotation capacity has been studied in the previous section, exclusively under static loading. But during the seismic action, the members are affected by physical phenomena as strain-rate, cumulative damages due to local buckling, fracture, low-cycle fatigue, etc. So, some doubts arise about the performances determined under monotonic conditions, to resist seismic actions and to undergo large plastic deformations during the severe excursions in the inelastic range.

Influence of the strain rate. In case of near-source earthquakes, the velocity is very high (Gioncu et al, 2000) inducing very important strain-rates. In addition, the first important ground motion is the strongest, for these earthquakes. So, the member behaviour could be studied using impulse forces. It is well known that the strain-rate has the main influence on the increasing of the yield stress (Fig. 2.1.23a), especially for values greater than 10^{-1}/s. In the same time, the increasing of the ultimate strength is moderate, consequently, the yield ratio has the tendency to reach the value of 1, with increasing of the strain-rate (Fig. 2.1.23b). So, a reduction of the ductility occurs, especially for strain-rates grater than 10^{-1}/s, due to the increasing of the yield ratio.The influence of the yield ratio on the ductility was studied considering the fracture ductility for the rotation of a bended stub

$$\mu_{\theta fr} = \frac{\theta_{rf}}{\theta_p} = \frac{1}{2}\frac{E}{f_y}\left(\frac{1}{\rho_y} - 1\right)^2 \left(\frac{c}{t}\right)^2 \varepsilon_u^2 \tag{2.1.51}$$

with θ_{rf}, calculated from a simplified approach.

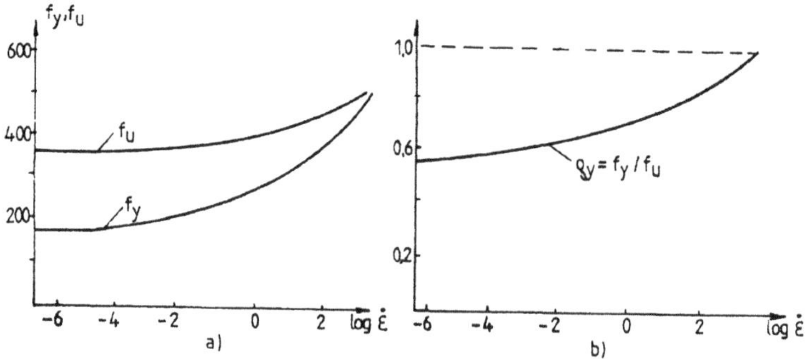

Figure 2.1.23: Influence of the strain-rate

If the values obtained from monotonic loading correspond to the plastic ductility and the values resulted from (2.1.51) represent the fracture ductility, the interaction of the two curves, determined for different yield ratios, show two distinct domains. The ductility of the domain with low yield ratio is given by the plastic deformations, while for high yield ratio, the fracture dramatically reduces the ductility (Fig. 2.1.24). The strain-rate increases the yield stress and the yield ratio. Taking into account that the randomness of the yield ratio of usual steel is about 0.60 – 0.73, and that the yield stress increases with a strain-rate of about $10^{-1}...10^{1}$ in the field of strong earthquakes, results consequently an increasing of the yield ratio, of about 0.75 - 0.95; this is the domain there plastic ductility changes to fracture ductility. Thus, it is possible that in case of near-source earthquakes, a brittle local fracture could replace the plastic ductility.

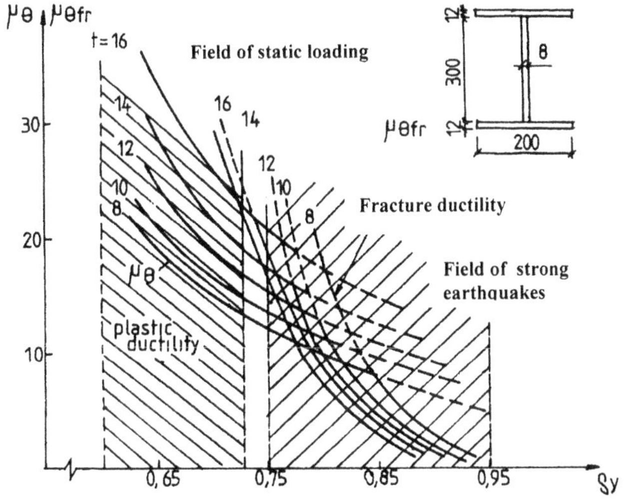

Figure 2.1.24: Plastic and fracture ductility

(ii) *Influence of the cyclic loading*. Some research works have classified the failure of the structural members during seismic actions as a low-cycle fatigue. During the earthquake, the structure resists hundreds of loading cycles, but only few cycles cause high plastic deformations. During an earthquake with short duration (fewer than 5 cycles) large plasticity occurs, while by a long duration (about 20 cycles), plastic excursions are induced. Therefore, classifying failures under repeated large deformations as belonging to the category of fatigue-failure is questionable. The low cycles with high plastic deformations cause an accumulation of these deformations along the yield lines, inducing cracks or rupture in the deformed plates, rather than a reduction of the material strength, as by fatigue under high cycle failure. Experimental results show a progressive reduction of the ductility by cyclic loading. But the effects of seismic loads are more intricate, because the movement history is a chaotic one, depending on a great number of factors. The deformations induced by the seismic motion vary from one member to another and also from one earthquake to other. Therefore, it is extremely difficult to select a particular deformation time-history to generalise the earthquake-induced deformations. Considering these uncertainties, one must adopt a caution approach to determine the member rotation capacity under seismic actions. Thus, this could be determined on the basis of monotonic loading, correcting the obtained values for cyclic loads. The parameters defining the cyclic loading are:

-the pattern of cycling loading in function of the time: constant, constant increasing or decreasing rotations (Fig. 2.1.25);

-period of the cycling loading, which may be related to the first yield rotation (ECCS Recommendations, 1986) or the rotation capacity (De Martino and Manfredi, 1994);

-collapse criteria as rotation limit, or fracture rotation.

The influence of the cycling loading is studied in the paper for three cycle types:
-constant rotation with an amplitude of $0.5\theta_{ru}$ (half of the ultimate rotation capacity), used especially in laboratory test, to determine the number of cycles until the fracture;
-continuous increasing of rotation, the first and following steps being of about $0.25\theta_{ru}$, characteristic for far field earthquakes;
- continuous decreasing of rotation, the first step being equal to the ultimate rotation, the following cycles decreasing by step with $0.25\theta_{ru}$.

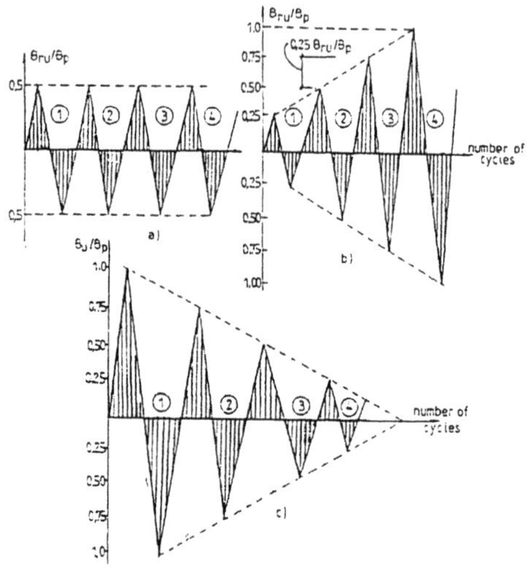

Figure 2.1.25: Cyclic loading types

The behaviour of **I**-sections has a particular feature by cyclic loading. The first semi-cycle produces buckling of the compression flange and the section rotates around a point located in, or near the opposite flange. As a result, the tension forces are very small. By the reversal semi-cycle, the compression flange buckles also. The most important observation is concerned that the opposite flange remains unchanged because of the small tension force, it being incapable to straighten the buckled flange (Fig. 2.1.26). Therefore, during the next cycle, the section works as having initial geometrical imperfections resulted from the previous cycle. In this manner, after each cycle a new rotation is superposed over the previous one. The plastic mechanism consists of two deformed shapes, each one for the corresponding flange.

Figure 2.1.27a presents the behaviour by *constant cycles*. One can see that after the third cycle a reduction in the rotation capacity, due to the cyclic loading appears. The effect of the cyclic loading could be determined from a simplified relation:

$$\left(\frac{M}{M_p}\right)_{cycl} = \alpha_1 + \alpha_2 \frac{1}{\left(\dfrac{n}{2}\dfrac{\theta_{ru}}{\theta_p}\right)^{1/2}} \leq 1 \qquad (2.1.52)$$

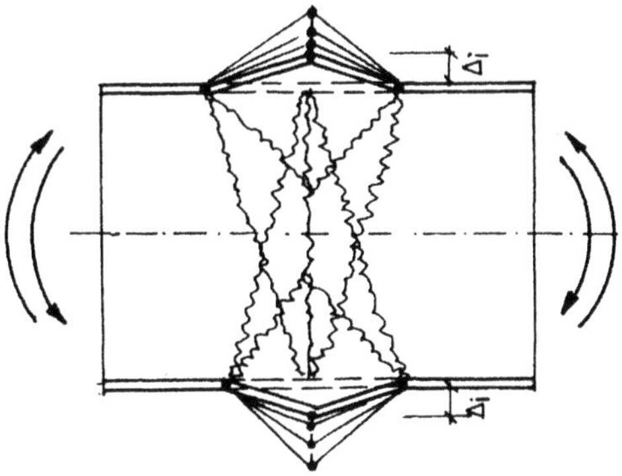

Figure 2.1.26: Plastic mechanism for cyclic loading

where the first coefficient is the term for the potential energy corresponding to the plastic zones, the second results from the yield lines. Only the last term is eroded by the cyclic loads. So, the minimum value for an infinite number of cycles will be:

$$\left(\frac{M}{M_p}\right)_{\min} = \alpha_1 \tag{2.1.53}$$

corresponding to the behaviour of plastic zones, which remain unaffected by the cyclic deformations.

The case of *increasing rotation* is presented in Figure 2.1.27b:

$$\frac{M}{M_p} = \alpha_1 + \alpha_2 \frac{1}{\left(\frac{n-1}{2}\frac{\theta_{ru}}{\theta_p}\right)^{1/2}} \leq 1 \tag{2.1.54}$$

The case of *decreasing rotation* is shown in Figure 2.1.27c. A reduction of the rotation capacity could be observed.

One can see that the rotation capacity is reduced to half in comparison with the monotonic loading, if cyclic loading acts on the member. But the use of rotation capacity defined for monotonic loading is disputable to be used for cyclic one, therefore until now there is no other adequate alternative.
Relationship (2.1.52) and (2.1.53) give the possibility to determine the number of cycles producing the flange fracture:

for constant rotation:
$$n = 2\frac{\theta_{rf}}{\theta_p}\frac{1}{\frac{\theta_{ru}}{\theta_p}} \tag{2.1.55a}$$

for increasing rotation:
$$n = 2\frac{\theta_{rf}}{\theta_p}\frac{1}{\frac{\theta_{ru}}{\theta_p}} + 1 \tag{2.1.55b}$$

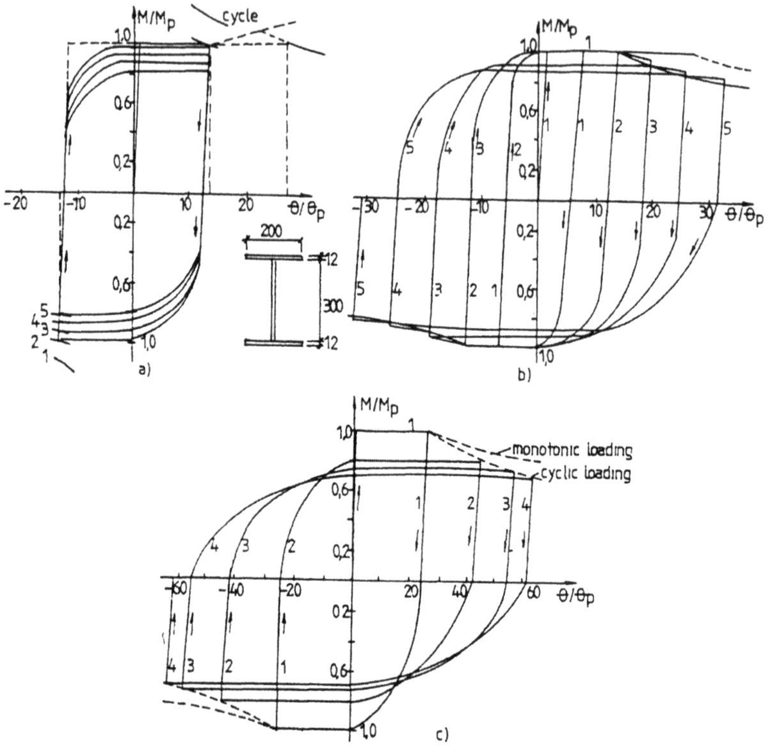

Figure 2.1.27: Moment-rotation cyclic loading

Figure 2.1.28 presents the number of cycles till the reaching of the flange fracture for the case of increasing rotation. There is a great influence of the yield ratio and the flange thickness to remark. The number of cycles decreases significantly by thick flanges, which leads to the conclusion to use rather a moderate slenderness for the member flanges.

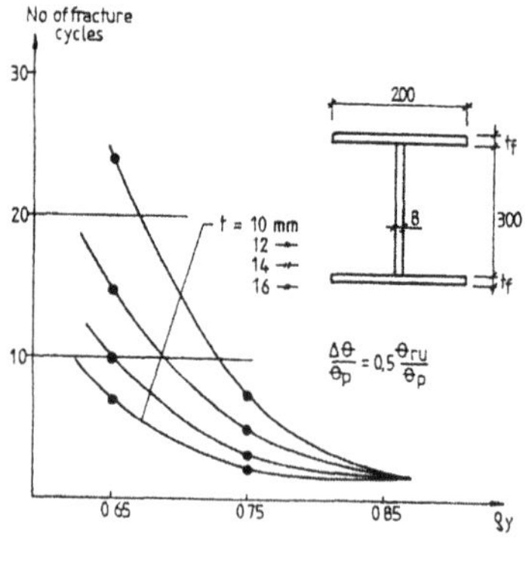

Figure 2.1.28: Number of fracture cycles

(iii) *Influence of the cyclic loading with high velocity.* The effect of high velocity cyclic loading could be introduced in the analysis by increasing the yield ratio in function of the strain-rate ε. Using the Soroushian and Choi (1987) relation for the yield and ultimate stresses, for Fe 360 steel, results:

$$\rho_{ysr} = \frac{1.46 + 0.0925 \cdot \log \dot{\varepsilon}}{1.15 + 0.0496 \cdot \log \dot{\varepsilon}} \rho_y \tag{2.1.56}$$

Figure 2.1.29 presents the number of fracture cycles in function of the strain-rate. So, the coupling of these two erosion effects, cyclic loading and strain-rate, can produce a premature fracture. Because the range of 10^{-1} to 10^{1} of the strain-rate correspond to the field of strong near-source earthquakes, one can remark an important reduction in the fracture cycles. The fracture may occur at the first, or second cycle in case of these earthquakes.

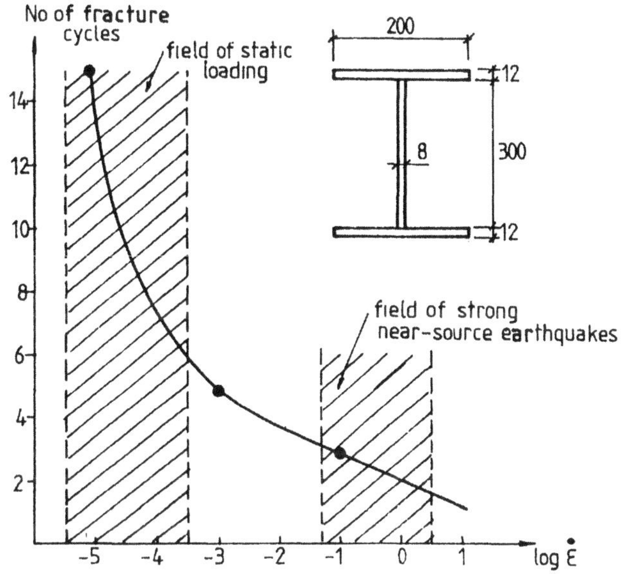

Figure 2.1.29: Cyclic loading with high strain-rate

AVAILABLE DUCTILITY OF JOINTS

Approaches Considering the Joint Behaviour

Dissipative structures are designed by allowing the plastic deformation of some zones of their members, which dissipate the earthquake input energy. Moment resisting frames (MRF) have a large number of dissipative zones, located at member end or in the joints.

Two different approaches can be taken in design of MRFs: the first is based on the location of the dissipative zones at the member ends, the second considers that the dissipative zones are located in the joints. The location of the plastic zones depends on the ratio between joint and member strength. Generally, the codes provisions assure that the joint provide sufficient overstrength to allow the yielding of the member ends, taking into account the maximum value of their yield strength. But this provision is only a theoretical one, because due to the several parameters influencing the node behaviour, it is practically impossible to prevent the plastic deformations of the joints. So it is more prudent to consider that both joints and member ends are involved in dissipation of the seismic input energy.

There are two main types of joints, welded or bolted. The common types of the welded joint, from Figure 2.1.30, are either full welded or with welded flanges and bolted web angles. The variety of bolted joints is richer in typologies; some of these types and the most used ones are presented in Figure 2.1.31: as top and seat-angles, flush end plates, or extended end plates.

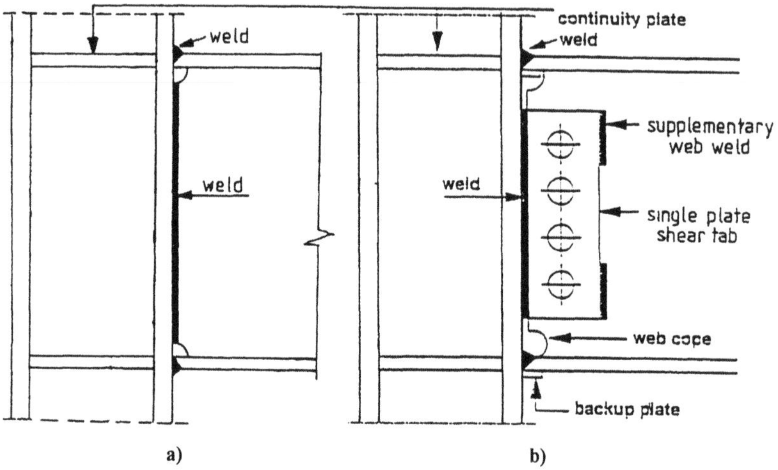

Figure 2.1.30: Welded joints

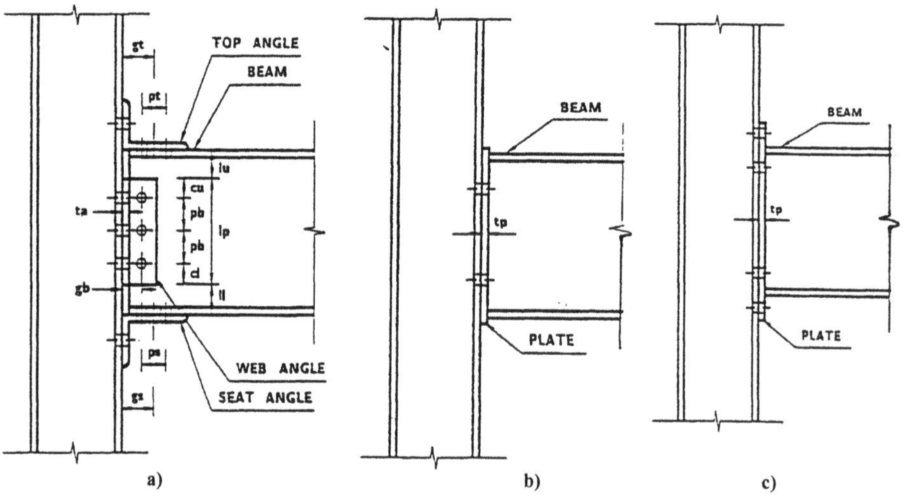

Figure 2.1.31: Bolted joints

Ductility of Joints

Some important features in the moment-rotation curve of a joint could be identified during the monotonic loading (Fig. 2.1.32). Initially, the curve is sensibly linear up to a moment M_y, with a rotation θ_y and a corresponding stiffness C_i. Non-linear effects produce a reduction of the stiffness to a value C_r, and the joint achieves the plastic moment M_p and the rotation θ_p, respectively, when plastic deformations will cause a significant increasing of the rotation. The reduced plastic stiffness of the joint is now C_p. The ultimate moment M_u and the rotation θ_u occur if local buckling or fracture of certain elements appears. Unfortunately, there are many research works, both theoretical and experimental concerning the first part of the moment-rotation curve, and very little information about the ultimate range. The post-buckling curve has a moderate slope, if the ultimate moment is reached through local buckling. But if the collapse is produced by fracture, the post-fracture behaviour shows a precipitous reduction of the joint carrying capacity.

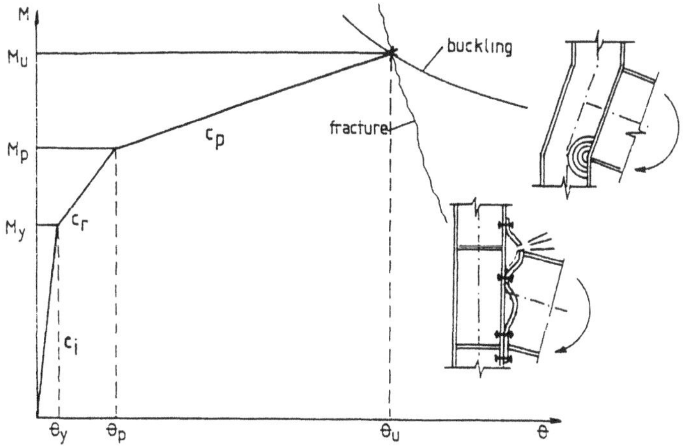

Figure 2.1.32: Moment-rotation curve for joints

The joint resistance and ductility depend on the behaviour of three distinct zones (Fig. 2.1.33): tension, compression and shear zone.

The *tension zone* deformations shall be determined taking into account the following components:
- yielding of the column web;
- deformation of the column flange;
- deformation of the connection plates;
- weld or bolt fracture.

The *compression zone* deformations have to be determined for the following components:
- crushing or buckling of the column web;
- deformation of the column flange;
- deformation of the connection plates.

The *shear zone* deformations depend on following components:
- shear buckling;
- fracture of the tensioned welds.

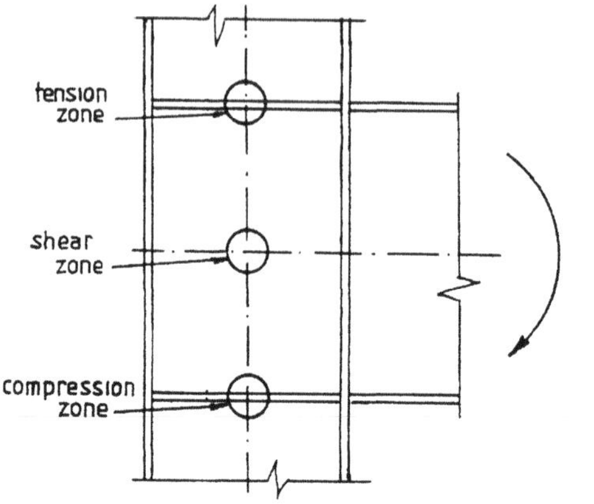

Figure 2.1.33: Zones of possible joint collapse

Collapse Plastic Mechanisms for Joints

Three major analysis techniques are available in literature concerning the design of joints:

(i) finite element method (Sherbourne and Bahari, 1994, Choi and Chung, 1996);

(ii) T-stub method (Faella et al, 1997, 1998);

(iii) Collapse plastic mechanism method (Kishi and Chen, 1987, Bernuzzi et al, 1991, Mann and Morris, 1979, Yee and Melchers, 1986, Murray, 1988, Olsen, 1997).

Between these methods certainly the best one is the use of FE, giving the most accurate results. But due to the necessity of a very dense network and the non-linear analysis for material and deformation, the computer time and the cost are very high. So, this method remains only for theoretical research work. The T-stub method is a very simple one, recommended for the design practise by EC3 (Annex J, 1997), but contains too many hypotheses, to be an enough accurate method. In these conditions, the method of collapse plastic mechanism seems to be the most adequate for design proposes. It was successfully used for the determination of the rotation capacity of the members. So, it is only necessary to extend the method to the joints. There are some important differences between the using for members or for joints, to be emphasised. The main difference is related to the failure mode, because plastic buckling determines the members failure mode, while for joints, the failure is mainly produced by fracture. For instance, the collapse plastic mechanism in Figure 2.1.34, is composed by several yield lines and two fracture lines. The differences between yield and fracture lines consist in the presence, in the first case, of bolts, which produce a reduction of the moment gradient (Fig. 2.1.35). In exchange, in the field of beam flange - end plate connection, the moment gradient is very high, so it may produce a fracture along the folded line. The yield lines along the bolts work in the field of plastic deformation (on the yield plateau), while the deformations of the fracture lines are situated in the strain-hardening range. From the theoretical point of view two fracture lines must occur along the two flange sides, but practically, observed during the experimental tests, only one is developed, the weakest one.

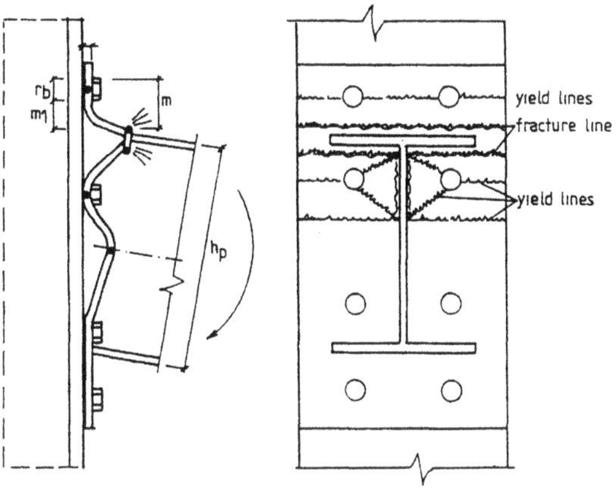

Figure 2.1.34: Collapse plastic mechanism

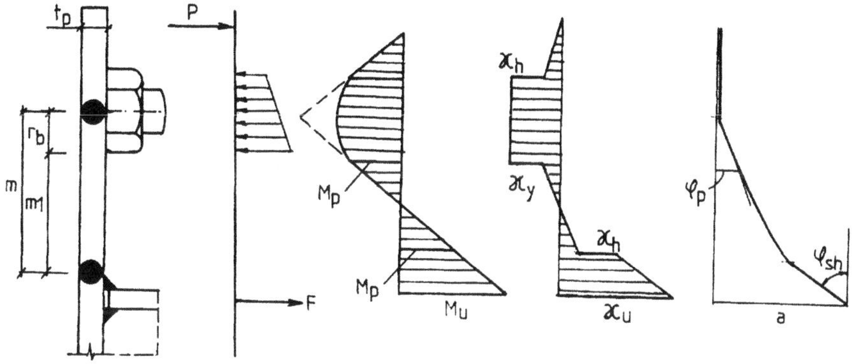

Figure 2.1.35: Yield and fracture lines behaviour

Available Ductility for Welded Joints

The collapse of welded joints may occur through:
- shear buckling of the joint panel,
- crushing of the compressed column web, or
- fracture of welding zones

For the joint panel subjected to shear, a plastic mechanism is developed considering a diagonal band tensioned in one direction and compressed in the other. The plastic mechanism consists of three yield lines formed by the buckled shape and a plastic zone (Fig. 2.1.36). The obtained curves for unreinforced and reinforced joint panels are presented in Figure 2.1.37. One can see that the decreasing in the moment-rotation curves occurs for high ductility values. So, generally, the column web in shear is not the weakest component of the joint.

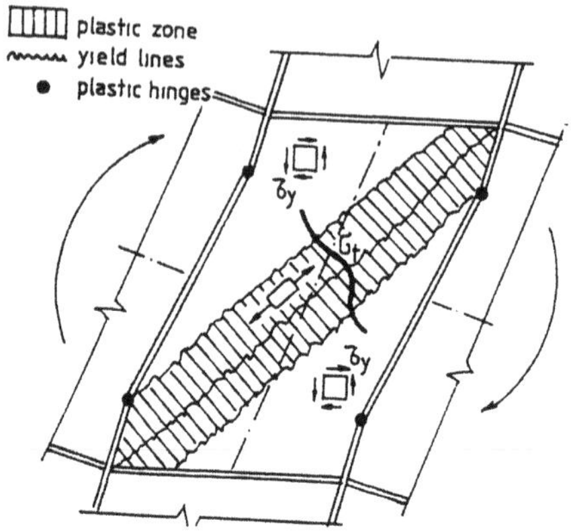

Figure 2.1.36: Collapse plastic mechanism for joint panel in shear

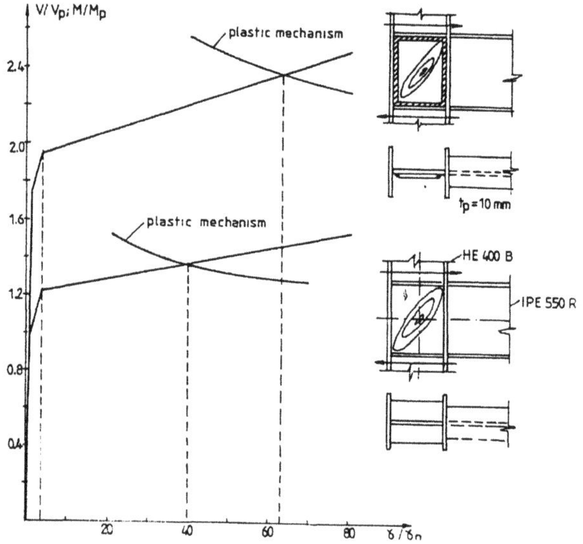

Figure 2.1.37: Ductility of panel zone in shear

The crushing of the compressed column web is studied using the collapse plastic mechanism from Figure 2.1.38. Yield lines and a plastic zone compose it. The post-crushing curve is presented in Figure 2.1.39. The ductility is also very high. Figure 2.1.40 shows the collapse types for different combinations of IPE and HEB profiles for beam and column, respectively. Due to the fact that the strength and ductility of welded joints are high, considering the shear and crushing mechanism, the failure of welded joints is produced by fracture in the weld zones.

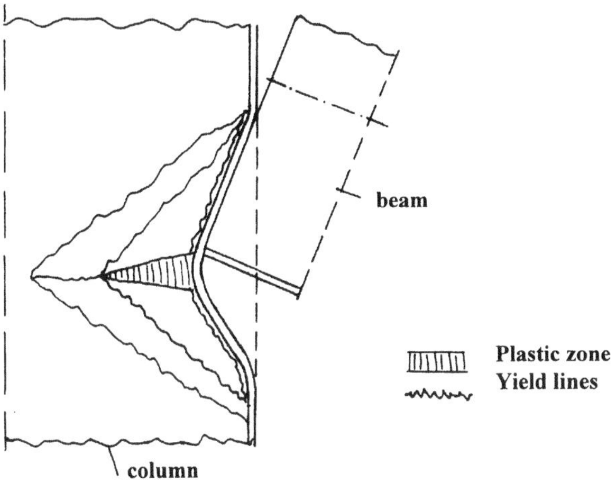

Figure 2.1.38: Collapse plastic mechanism for compressed column web

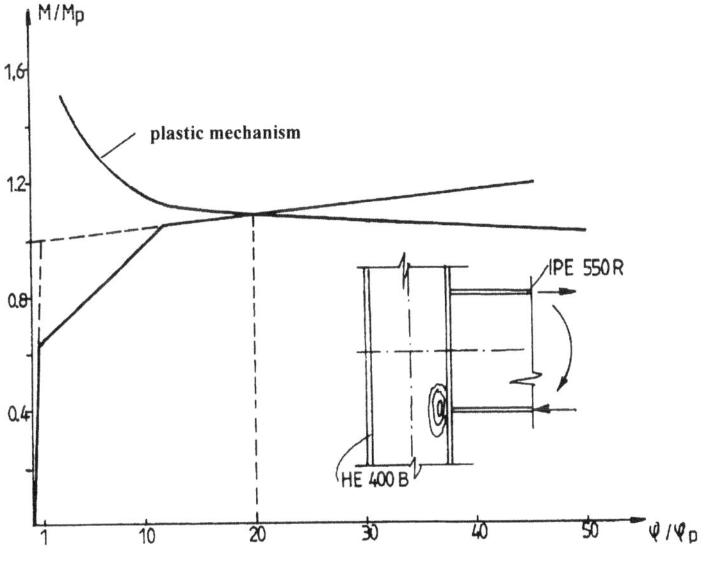

Figure 2.1.39: Ductility of crushing compressed column web

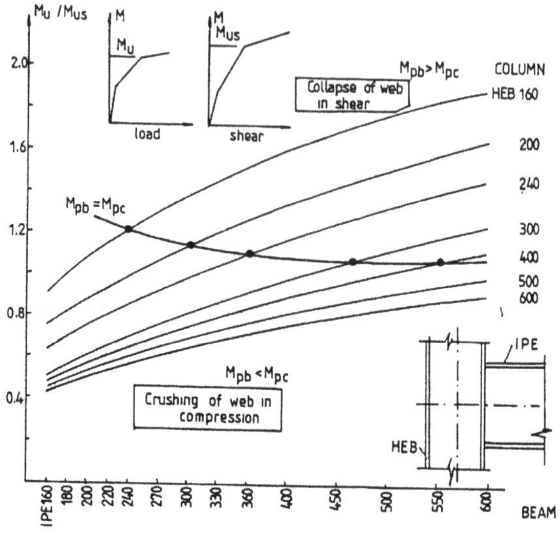

Figure 2.1.40: Collapse types

Available ductility for bolted joints

The collapse of bolted joints may occur through:
- fracture of the bolts in the tension zone;
- rupture of welds;
- yielding of column flange, end plate, column web and beam web in the tension zone;
- crushing of the compressed column web.

The collapse through yielding or crushing of the column web is not so important for bolted joints as for welded ones. In the aim to use the plastic mechanism method, the fracture of bolts or rupture of weld has to be eliminated through constructional rules.

The collapse plastic mechanism for top and seat-angles with double web-angles and extended end plate is presented in Figure 2.1.41. One can see that the mechanisms are formed by yield and fracture lines. The plastic rotations of the yield lines remain in the field of the yielding plateau, while the fracture lines exhibit rotations in the strain-hardening range. The ratio between the rotation of the fracture line of the root of beam flange-end plate and of the bolts yield line is:

$$\frac{\varphi_{sh}}{\varphi_p} = \frac{\varepsilon_h + \varepsilon_{sh}}{\varepsilon_h} \frac{1-\rho_y}{1+\rho_y} \frac{m_1}{r_b} \tag{2.1.57}$$

which is plotted in Figure 2.1.42 for Fe 360, Fe 430 and Fe 510 steel grades. Taking into account that the ratio r_b/m_1 takes values between 0.3 ...0.8, results that the rotation of the fracture line is 5...15 times greater than the rotation of the bolt lines. The Fe 360 and Fe 430 steel grades present a ductile rotation, while for Fe 510 the fracture rotation is reduced, showing a tendency towards the brittle fracture.

The ultimate rotation of the joint is given by:

$$\theta_u = \bar{\varphi}\frac{m \cdot m_1}{h_b \cdot t_p}\varepsilon_u \qquad (2.1.58)$$

where:

$$\bar{\varphi} = 2\left[2\frac{r_b}{m_1}\left(\frac{r_b}{m_1} + \frac{\rho_y}{1+\rho_y}\right)\frac{\bar{\varepsilon}_h}{\bar{\varepsilon}_h + \bar{\varepsilon}_u} + \frac{1-\rho_y}{(1+\rho_y)^2}\right] \qquad (2.1.59)$$

The geometrical parameters are presented in Figures 2.1.34 and 2.1.35, ρ_y is the yield ratio. The mechanical parameters are:

$$\bar{\varepsilon}_h = \frac{\varepsilon_h}{\varepsilon_y}; \qquad \bar{\varepsilon}_u = \frac{\varepsilon_u}{\varepsilon_y} \qquad (2.1.60a,b)$$

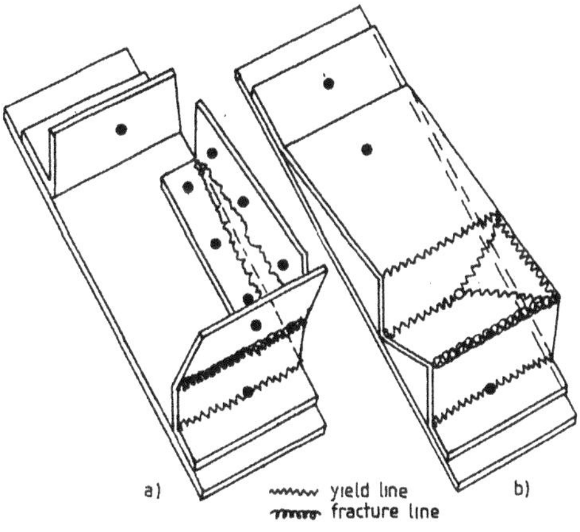

Figure 2.1.41: Collapse plastic mechanisms

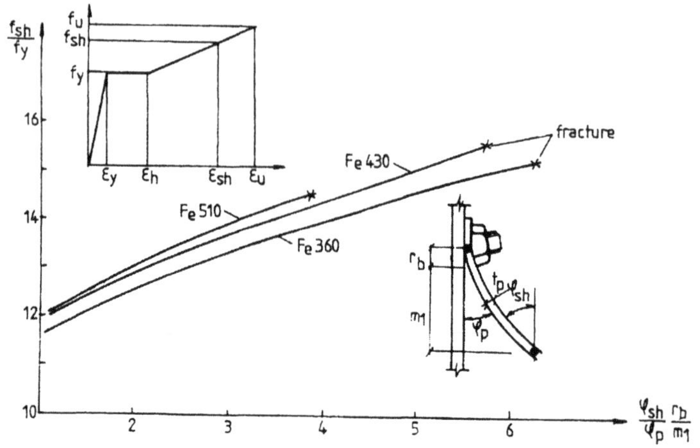

Figure 2.1.42: Rotation of fracture and yield lines

A comparison between the theoretical and experimental results is presented in Figure 2.1.43. There were collected experimental results for top and seat-angles and extended end plates. Unfortunately, the material properties and some geometrical dimensions of the specimens are not completely reported in the original works, so they have to be assumed. This fact may be an explanation of the scattering values obtained in this analysis.

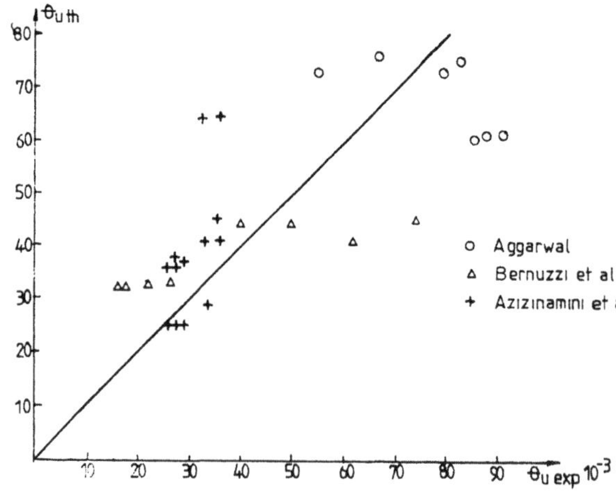

Figure 2.1.43: Correlation theoretical and experimental results

Influence of the Strain-Rate

From the relation (2.1.57) and (2.1.58) one can see that the ultimate rotation depends on the yield ratio ρ_y. The reduction of the ultimate rotation of a joint in function of the strain-rate is shown in Figure 2.1.44.

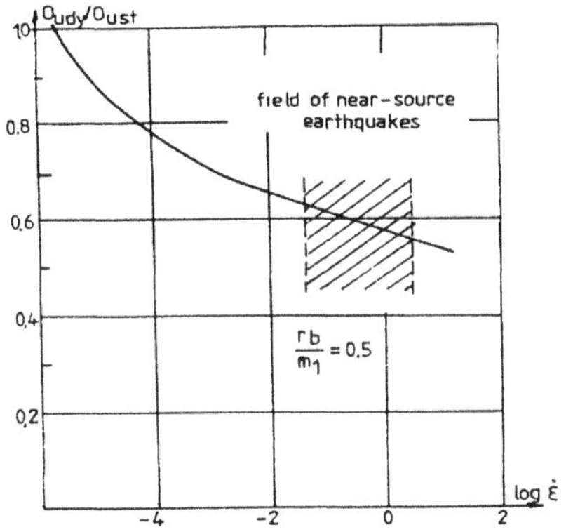

Figure 2.1.44: Influence of the strain-rate

Taking into account that the strain-rate for near-source earthquakes lies between $10^{-1}...10^{1}$, the reduction of the ultimate rotation of the joints is about 40%-50%. This can be an explanation of some joint fractures during the Northridge and Kobe earthquakes, characterised by very high-recorded velocities.

Influence of the Cyclic Loading

The inelastic rotation of the joints is significantly affected by the cyclic loading. As it is presented in Figure 2.1.45, the first semi-cycle produces a plastic displacement of the end plate, when the beam rotates around a point located in opposite position. For the unloading field, at zero force, some residual displacement occurs. The reversal semi-cycle produces a compression force, which for the maximum value reduces the residual displacement at zero. At the end of the first cycle, a residual displacement remains. The start of the second cycle is characterised by the presence of a residual displacement, so the needed force to reach a constant displacement is reduced in comparison with the first cycle. Results a continuous decreasing of the joint moment capacity and an accumulation of residual rotation, which can reach the fracture rotation after several cycles. As in the case of members, the fracture may occur at the first cycle (near-source earthquake with one important impulse), or after 2–3 cycles, if the ground motion is characterised by some adjacent impulses. In case of far-source earthquakes, characterised by cyclic loading, the fracture appears after 5–10 important cycles, producing plastic excursions in the joint elements.

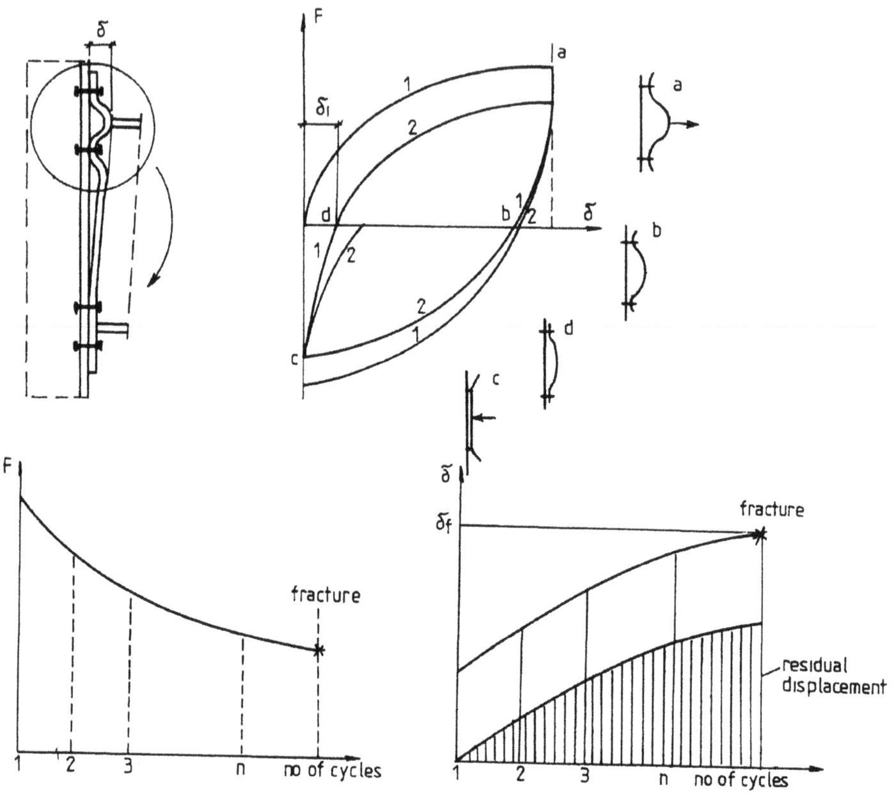

Figure 2.1.45: Influence of cyclic loading

NODE ANALYSIS

Classification of Nodes

In the previous sections the available mechanical characteristics for member and joints were determined. The following step consists in the comparison of these results:

$$
\begin{aligned}
S_{am} &> S_{aj} \quad \text{for stiffness} \\
R_{am} &> R_{aj} \quad \text{for resistance} \\
D_{am} &> D_{aj} \quad \text{for ductility}
\end{aligned}
\qquad (2.1.61a\ldots c)
$$

where the indices: 'a' refers to the available ductility, 'm' to members, 'j' to joints. In function of these inequalities the nodes may be classified in restoring or non-restoring for rigidity, strength and ductility.

The framing in restoring or non-restoring strength, for welded nodes composed by IPE and HEB profiles for the beams and columns, respectively, is presented in Figure 2.1.46 (Gioncu 1999). In Figure 2.1.47 the required end-plate thickness is plotted, for bolted nodes with external or flush end-plate joints; the beam sections are IPE profiles. Comparatively to the flush end plate, the required thickness for the extended end plate is considerably reduced (Olsen,1997).

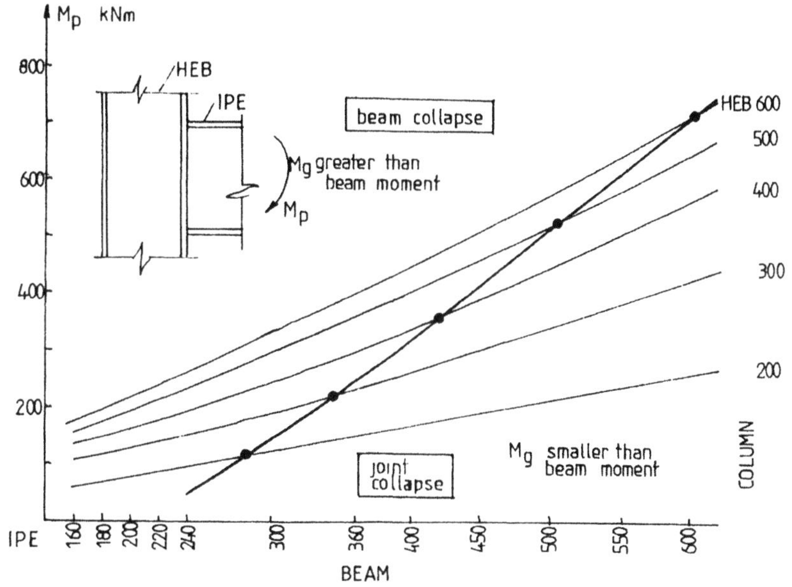

Figure 2.1.46: Welded node collapse

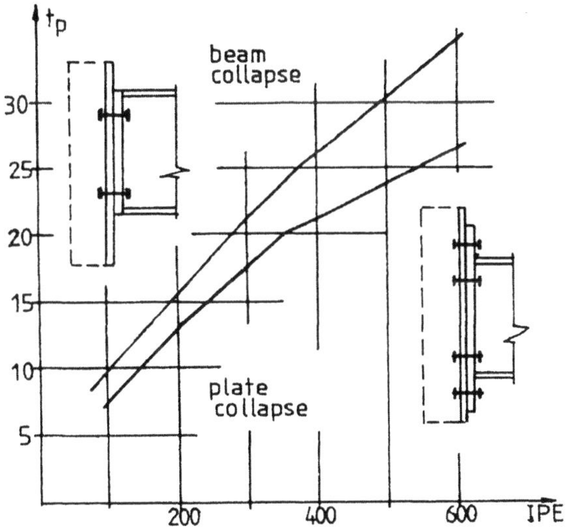

Figure 2.1.47: Bolted node collapse

New Upgrading Solutions to Improve Node Ductility

During the recent earthquakes of Northridge (1994), and Kobe (1995) many brittle fractures of welded moment resisting frames, mainly located at the beam-to-column connections have been observed. These recent events demonstrated that for an efficient earthquake design in high seismically areas the ductility control have to be assured. Brittle fractures are mainly produced by stress concentration induced by the weld technologies, or by some defects.

So, new upgrading solutions to improve the node ductility have been searched. A very promising solution is to move the plastic hinges away from the column face, by increasing or decreasing the moment capacity of the beam end. It leads to modified moment resisting frames, by using the weakening technique, like the so called 'dog-bone' solution (DB), or the strengthening techniques, like the so called 'reinforced' solution (RF) (Anastasiadis et al., 1999).

These procedures could be obtained by:
(i) cutting of beam flanges near to the beam-to-column connection, in a specific weakened zone (DB), where the formation of plastic hinges is assured due to the reduced moment capacity;
(ii) strengthening the beam flanges with welded plates, at the beam end near the column (RF), so the plastic hinge moves far from the beam-column interface, due to the increased moment capacity of the beam.

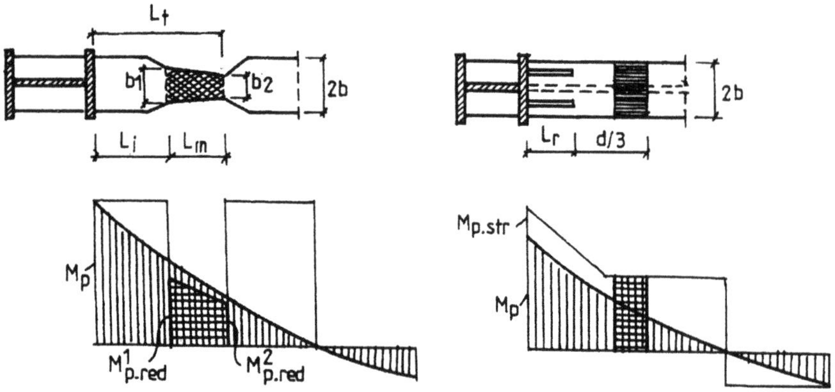

Figure 2.1.48: Weakening and strengthening solutions

Figure 2.1.49a shows the influence of the main geometrical parameters on the available plastic rotation capacity of the reduced beam section. It is observed that by reducing the beam flanges of about 45%, an increasing of the ductility, of about 40% is obtained, as compared with the unreduced beam section. The influence of the reduced beam section in relation with the column section, according to the member hierarchy criterion is plotted in Figure 2.1.50a. One can see that the weakening solution gives the possibility to reduce the column cross-section, with respect to the SC-WB (strong column – weak beam) concept.

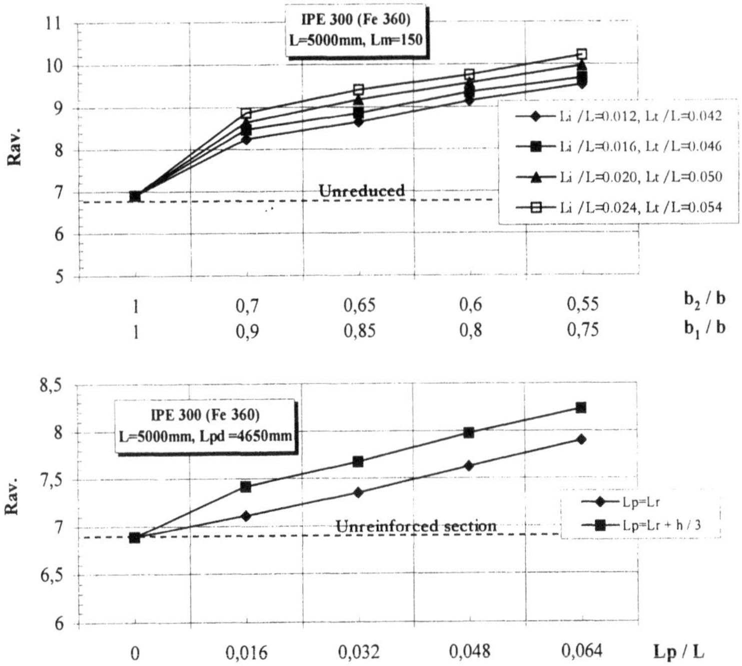

Figure 2.1.49: Reduced beam solution, reinforced beam section

The influence of the beam section reinforcing on the member hierarchy criterion is presented in Figure 2.1.50a. One can see that the extension of the reinforced zone leads to an increasing of the column section, in the same time, to an unfavourable economical impact on the structure. The influence of the ribs length and the plastic hinge position on the rotation capacity is shown in Figure 2.1.49b. The plastic rotation capacity of the reinforced beam section can be greater than the unreinforced section of about 15%.

From the comparison of the two analysed solutions one can observe the convenience of the 'dog bone' solution, because it improves the local ductility by keeping constant, or reducing the dimensions of the columns, contrary to the 'reinforced' solution.

DUCTROT MJN– COMPUTER PROGRAM

In the aim to determine the rotation capacity of a node, the DUCTROT MJN (DUCTility of ROTation for Members, Joints, Nodes) was elaborated at INCERC – Timisoara by Petcu and Gioncu (1999) on the basis of local plastic mechanism methodology.

The work with DUCTROT MJN is characterised by a very simple modification of the geometrical and mechanical parameters, if this is required during the design. The results could be presented in tables or graphics, very useful for the design practise.

The user can decide the branch to follow for the determination of:

- rotation capacity of members DUCTROT M;
- rotation capacity of joints DUCTROT J;

- rotation capacity of nodes DUCTROT N

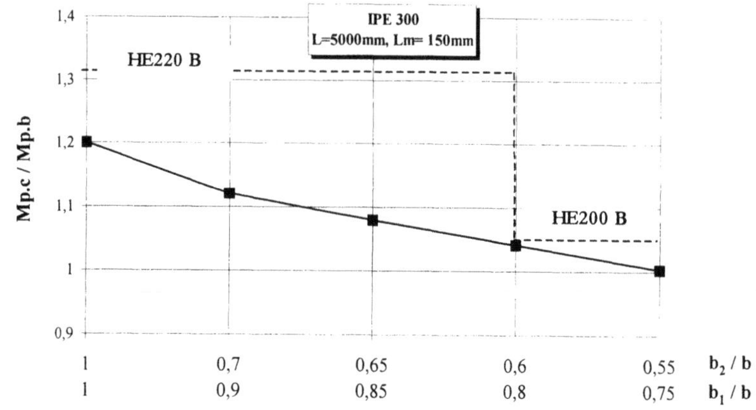

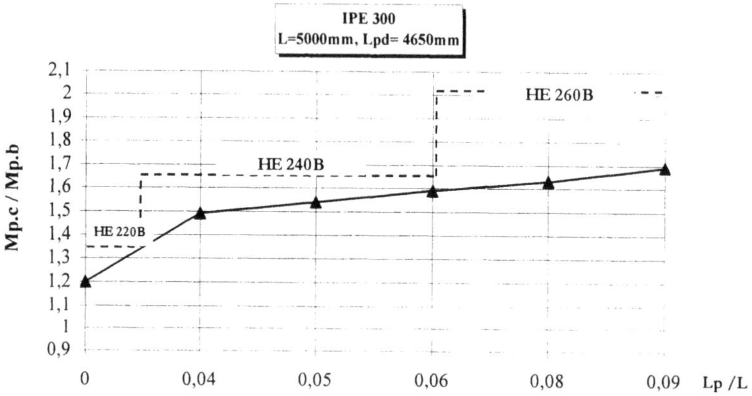

Figure 2.1.50: Influence of the "dog-bone and reinforced concept on member hierarchy criterion

Branch DUCTROT M

The branch DUCTROT M is composed by the following sequences:

- selection of the cross-section type,
- introducing of the mechanical characteristics,
- introducing of the geometrical characteristics,
- selection of the loading system and the collapse mode,
- result the mechanical characteristics of the cross-section,
- result the mechanical characteristics of the member,
- result the rotation capacity,
- rotation values for two parametric variations,
- determination of minimum rotation capacity, for different collapse modes,
- influence of the strain-rate,
- influence of the cyclic loading.

Branch DUCTROT J

The branch DUCTROT J consists of the following sequences:

- selection of the joint type,
- introducing the mechanical characteristics for profiles and bolts,
- selection of the collapse type,
- result the mechanical characteristics of the joint,
- results the rotation capacity of the joint,
- parametric study in function of two parameters,
- determination of minimum rotation capacity for different collapse modes,
- influence of the strain-rate,
- influence of the cyclic loading.

Branch DUCTROT N

The results for members and joint, obtained from the two previous branches are compared, in the aim to identify the node type. The comparison is performed from the point of view:

- rigidity,
- strength,
- ductility.

CONCLUSIONS

The use of a collapse plastic mechanism methodology allows us to determine the available ductility of members and joints with accuracy confirmed by experimental results. The research work is consolidated in a computer program DUCTROT MJN, where are the factors influencing the ductility are included.

References

Aggarwal A.K.(1994). Comparative tests on end plate beam-to column connections. *Journal of Constructional Steel Research* **Vol.30,** 151-175.

Anastasiadis A., Gioncu V. (1999). Ductility of IPE and HEA beams and beam-columns. "*Stability and Ductility of Steel Structures, SDSS 99*" (eds. D. Dubina and M. Ivanyi), 9-11 September, Timisoara, Elsevier, Amsterdam, 249-258.

Anastasiadis A., Gioncu V., Mazzolani F.M. (1999). New upgrading procedures to improve the ductility of steel MR-Frames. *XVII Congresso CTA. Construire in acciaio: Struttura e Architettura*, 3-5 Ottobre 1999, Napoli, Italy.

Azizinamini A., Bradburn J.H., Radziminski J.B. (1987). Initial stiffness of semi-rigid steel beam-to-column connections. *Journal of Constructional Steel Research* **Vol.8,** 71-80.

Bernuzzi C., Zandonini R., Zanon P. (1991). Rotational behaviour of end plate connections. *Construzioni Metalliche* **No.2,** 74-`103.

Choi C.K., Chung G.T. (1996). Refined three-dimensional finite element model for end-plate connections. *Journal of Structural Engineering* **22:11,** 1307-1318.

Climenhaga J.J., Johnson R.P. (1972). Moment-rotation curves for locally buckling beams. *Journal of Structural Division* **98:6,** 1239-1254.

ECCS TWG 1.3 (1986). Recommended testing procedure for assessing the behaviour of structural steel elements under cyclic loads. **45:86.**

EUROCODE 3. Part 1. Revised Annex J. (1997). Joint in Building Frames.

Faella C., Piluso V., Rizzano G. (1997). Prediction of bolted connections ductility. "*Behaviour of Steel Structures in Seismic Areas STESSA '97*" (eds. F.M.Mazzolani and H.Akiyama), 3-8 August, Kyoto, 10/17 Salerno, 582-591.

Faella C., Piluso V., Rizzano G. (1998). Experimental analysis of bolted connections: snug versus preloaded bolts. *Journal of structural Engineering* **124:7,** 765-774.

Gioncu V. (1997). Ductility demands. General report. "*Behaviour of Steel Structures in Seismic Areas STESSA '97*" (eds. F.M.Mazzolani and H.Akiyama), 3-8 August, Kyoto, 10/17 Salerno, 279-302

Gioncu V. (1999). Framed structures. Ductility and seismic response. General report. "*Stability and Ductility of Steel Structures, SDSS 99*" (eds. D. Dubina and M. Ivanyi), 9-11 September, Timisoara.

Gioncu V., Petcu D. (1995). Numerical investigations on the rotation capacity of beams and beam-columns. "*Stability of Steel Structures*" (ed. M.Ivanyi), 21-23 September, Budapest, **Vol.1,** 129-140.

Gioncu V., Petcu D. (1997). Available rotation capacity of wide-flange beams and beam-columns. Part 1. Theoretical Approaches. Part 2. Experimental and numerical tests. *Journal of Constructional Steel Research* **43:1-3,** 161-217, 219-244.

Gioncu V., Mazzolani F.M. (1997). Simplified approach for evaluating the rotation capacity of double T steel sections. "*Behaviour of Steel Structures in Seismic Areas STESSA '97*" (eds. F.M.Mazzolani and H.Akiyama), 3-8 August, Kyoto, 10/17 Salerno, 303-310.

Gioncu V., Mateescu G., Orasteanu S. (1989). Theoretical and experimental research regarding the ductility of welded I-sections subjected to bending. *Stability of Metal Structures*" 10-12 October, Beijing, 289-298.

Gioncu V., Mateescu G., Tirca L., Anastasiadis A. (2000). Influence of type of seismic motions. *INCO COPERNICUS RECOS* – Project. (in this volume).

Ivanyi M. (1979). Moment rotation characteristics of locally buckled beams. *Periodica Polytechnica, Civil Engineering* **23:3-4,** 217-230.

Ivanyi M. (1985). The model of interactive plastic hinge. *Periodica Polytechnica, Civil Engineering* **29:3-4,** 121-146.

Jaspart J.P., Anderson A., Steenhuis M. (1998). Derivation of the joint properties by means of component method. *"Control of Semi-Rigid Behaviour of Civil Engineering Structural Connections"*COST Conference, 17-18 September, Liège.

Kishi N., Chen W.F. (1978). Moment-rotation relations of semi-rigid connections with angles. *Journal of Structural Engineering* **116:7,** 1813-1834.

Kotelko M. (1996). Ultimate load and post failure behaviour of box-section beams under pure bending. *Engineeering Transactions* **44,** 229-251.

Kuhlmann U. (1986). Rotationskapazität biegebeanspruchter I-Profile unter Berücksichtigung des plastischen Beulens. *Technica Report, Mitteilung* **85:5.**

Mann A.P., Morris L.J. (1979). Limit design of extended end-plate connections. *Journal of Structural Engineering* **105:3,** 511-526.

Martino De A., Manfredi G. (19994). Experimental testing procedures for the analysis of cyclic behaviour of structural elements: Activity of RILEM Technical Committee 134 MJP. *"Danneggiamento Ciclico e Prove Pseudodinamiche"*(ed.E.Cosenza), 2-3 June 1994, Napoli, 1-20.

Mazzolani F.M., Piluso V. (1993). Member behavioural classes of steel beams and beam-columns. Thin-Walled Structures **19,** 337-351.

Murray J.M. (1988). Recent developments for the design of moment end-plate connections. *Journal of Constructional Steel Research* **10,** 133-162.

Murray N.M.(1995). Some effects arising from impact loading of thin-walled structures. *"Lightweight Structure in Civil Engineering"* (ed. J.B.Obrebski), 25-29 September 1995, Warsaw, **1,** 389-394.

Olsen P.C. (1997). Design of bolted endplate connections. *Journal of Structural Engineering* **43:1-3,** 119-140.

Petcu D., Gioncu V. (1999). DUCTROT MJN Computer Program Guide for users. INCERC-Timisoara.

Sherbourne A.N., Bahaari M.R. (1994). 3D simulation of end-plate bolted connections. *Journal of Structural Engineering* **120:11,** 3122-3136.

Soroushian P., Choi K.B. (1987). Steel mechanical properties at different strain rate. *Journal of Structural Engineering* **113:4,** 863-872.

Spangemacher R., Sedlacek G. (1992). Zum Nachweis ausreichender Rotationsfähigkeit von Fliessgelenken bei der Anwendung des Fliessgelenkverfahrens. *Stahlbau* **61:11,** 329-339.

Spangemacher R. (1991). Rotationsnachweis von Stahlkonstruktionen, die nach dem Traglastverfahren berechnet werden. *Dissertation,* Technische Hochschule Aachen.

Suzuki T., Ogawa T., Ikarashi K. (1994). A study on local buckling behaviour of hybrid beams. *Thin Walled Structures* **19,** 337-351.

Tschemmernegg F., Rubin D., Pavlov A. (1998). Application of the component method to composite joints. *"Control of Semi-Rigid Behaviour of Civil Engineering Structural Connections"* COST Conference, 17-18 September, Liège.

Yee Y.L., Melchers R.E. (1986). Moment-rotation curves for bolted connections. *Journal of Structural Engineering* **112:3,** 615-635.

2.2

PLASTIC DEFORMATION CAPACITY OF BOLTED T-STUBS: THEORETICAL ANALYSIS AND TESTING

Ciro Faella, Vincenzo Piluso, Gianvittorio Rizzano

INTRODUCTION

Modern codes have opened the door to advanced methods of structural analysis which account for geometrical and mechanical non-linearities (Chen and Toma, 1994). In particular, in the case of steel frames, mechanical non-linearity arises from the yielding of the member ends and/or of the beam-to-column joints. Therefore, advanced analysis methods require an accurate modelling of members and connections up to failure. This means that, in order to evaluate the ultimate load carrying capacity of steel frames, it is necessary to know the plastic deformation capacity of members and connections, i.e to model the moment-rotation curve of the system composed by the beam end and the beam-to-column joint up to the occurrence of the ultimate rotation.

Many methods for predicting the plastic rotation capacity of beams and beam-columns are available in the technical literature (Mazzolani and Piluso, 1996), but, unfortunately, regarding the evaluation of the plastic rotation supply of beam-to-column joints the gap of knowledge is still significant.

Many reviews of the available experimental data have been performed aiming at the identification of the parameters affecting the plastic rotation capacity of connections, but there has not been any attempt to face the problem from a theoretical point of view. However, these reviews have underlined the fundamental role played by the beam depth (Roeder and Foutch, 1996). In fact, there is an experimental evidence that, for any given structural detail of the beam-to-column connection, the plastic rotation supply decreases as the beam depth increases. This means that the plastic rotation supply of connections can be more efficiently investigated from the component point of view, because it is basically the ratio between the plastic deformation of the joint components at the beam flange level and the lever arm.

The weakest joint component, governing the joint flexural resistance, is generally also the main source of plastic deformation capacity. The other joint components can constitute additional sources of plastic deformation capacity whose contribution is as more significant as their strength is more close to the one of the weakest component. Therefore, a procedure to account for the contribution of the different components is necessary, provided that the plastic deformation capacity of each joint component can be properly predicted (Faella et al., 1999).

Regarding the evaluation of the plastic rotation supply of bolted beam-to-column joints, the prediction of the plastic deformation capacity of bolted T-stubs is of primary importance. In fact, there are many components of bolted connections, such as the column flange in bending, the end plate in bending and the angles in bending, which can be modelled by means of equivalent T-stubs (Faella et al., 1995, 1996a, 1996b, 1999; Jaspart et al., 1995; Yee and

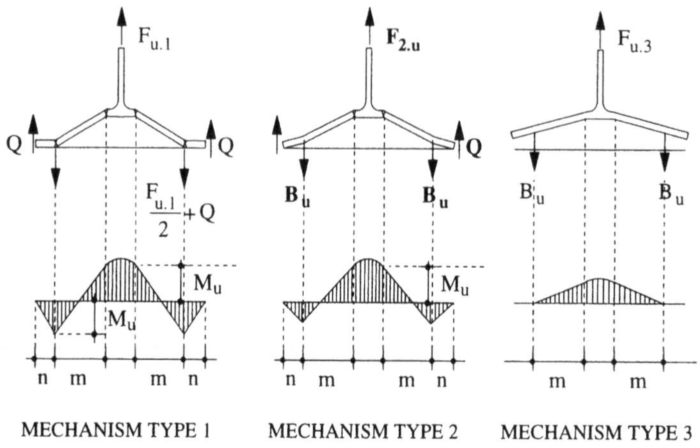

Figure 2.2.1:Collapse mechanism typologies

Melchers, 1986; Shi et al., 1996; -11]. As, in the case of bolted connections, the weakest joint component is very often one of those modelled by means of an equivalent T-stub, it is evident that the prediction of the force-displacement curve of bolted T-stubs up to failure would allow both the prediction of the ultimate plastic displacement of the weakest joint component (i.e. the main source of plastic rotation capacity) and the prediction of the contribution of some of the remaining joint components to the plastic deformation of the beam-to-column joint.

The first studies dealing with the theoretical prediction of the plastic deformation capacity of bolted T-stubs are those of Faella et al. [4,12]. The refinement of the original model [12] is herein presented [4]. In addition, the results of an experimental program devoted to the validation of the theoretical model are presented and discussed.

2 PREDICTION OF THE FORCE-DISPLACEMENT CURVE

2.1 Generality

It is well known that bolted T-stubs can fail according to three different collapse mechanisms: flange yielding (type-1 mechanism), flange yielding with bolt fracture (type-2 mechanism) and bolt fracture (type-3 mechanism) (Fig.1).

For each failure mode, the bending moment diagram along the T-stub flanges in the collapse condition can be immediately determined. Therefore, combining this diagram with the moment-curvature relationship of the T-stub flange section, the curvature diagram corresponding to the ultimate conditions can be obtained. By properly integrating the curvature diagram, the plastic rotations of the yielded zones can be computed. These plastic rotations can develop at the section corresponding at the flange-to-web connection and, eventually, at the bolt axis location.

The ultimate value of the plastic rotation of the hinges involved in the collapse mechanism is computed by assuming that failure occurs when the strain in the most extreme fibre attains the value corresponding to fracture. The constitutive law of the material is, to this scope, expressed in terms of true stress versus true strain behaviour which is modelled by means of a quadrilinear

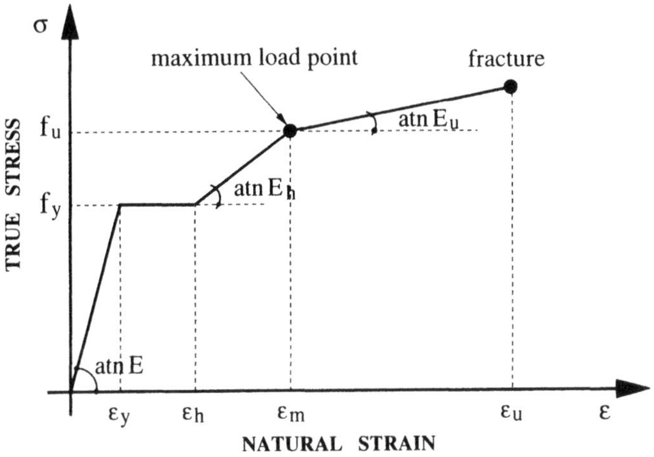

Figure 2.2.2:True stress versus true strain curve

relationship as depicted in Fig.2. The strain corresponding to the fracture of the material is evaluated according to the RILEM Recommendations [13]:

$$\varepsilon_u = \ln \frac{A_o}{A_f} \qquad (1)$$

where A_o is the initial area of the specimen and A_f is the area of the specimen in the necking zone after fracture.

In the following, only the final formulations will be presented.

As soon as the ultimate value of the plastic rotation of the hinges involved in the mechanism has been computed, the plastic displacement of the single T-element constituting the bolted T-stub can be easily evaluated from the kinematics of the mechanism (Fig.3):

- type-1 and for type-3 mechanisms:

$$\delta_p = \theta_p \, m \qquad (2)$$

- type-2 mechanism:

$$\delta_p = \theta_{p1} \, m + (\theta_{p1} - \theta_{p2}) \, n \qquad (3)$$

for $\theta_{p1} > \theta_{p2}$ and $\theta_{p2} \neq 0$, otherwise $\delta_p = \theta_{p1} \, m$.

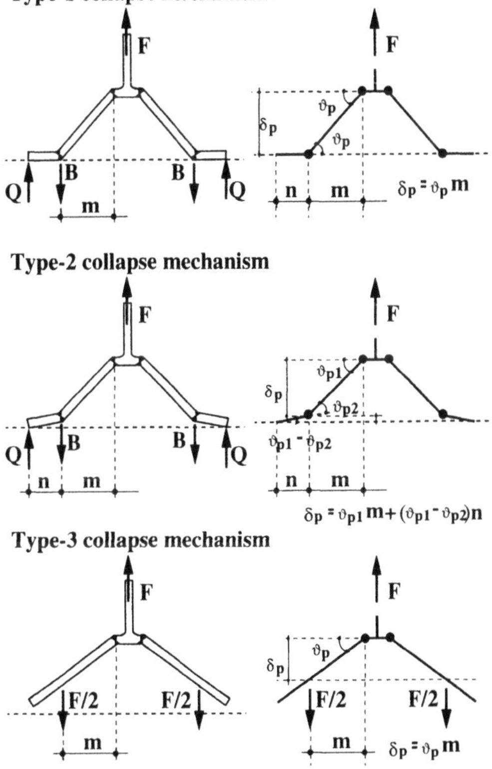

Figure 2.2.3: Kinematics of collapse mechanisms

The mathematical procedure, leading to closed form relationships, for computing the plastic rotations of the hinges involved in the collapse mechanism is detaily described in [4].

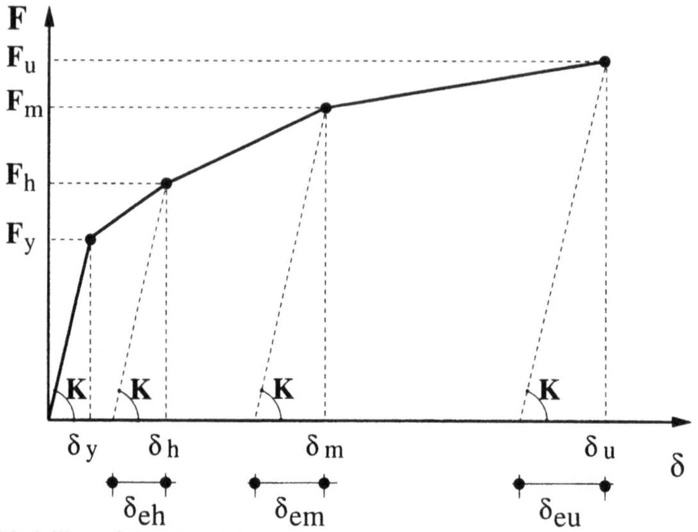

Figure 2.2.4: Modelling of T-stub axial response

Even though the primary aim of the formulations presented in [4] is the prediction of the plastic deformation supply of bolted T-stubs, they can also be used for a refined modelling of the force-displacement curve of bolted T-stubs by assuming that the point of zero moment along the T-stub flange remains unchanged during the loading process and equal to that occurring in the ultimate conditions [4]. A quadrilinear modelling of the force-displacement curve of the single T-element can be performed by identifying four characteristic points as depicted in Fig. 4. The first point (F_y , δ_y) corresponds to the attainment of the first yielding condition. The second point (F_h , δ_h) corresponds to the beginning of strain hardening, i.e. to the attainment of the strain level ε_h in the extreme fibres of the sections where the formation of plastic hinges is expected. The third point (F_m , δ_m) corresponds to the achievement of the stress f_u. Finally, the fourth point (F_u , δ_u) corresponds to the ultimate conditions, i.e. to the attainment of the strain ε_u leading to the fracture of the material.

Regarding the collapse mechanism typology, it can be identified by means of the parameter β_u, given by:

$$\beta_u = \frac{4\,M_u}{2\,B_u\,m} \tag{4}$$

where M_u is the bending moment corresponding to the attainment of the ultimate strain ε_u in the most extreme fibre of the T-stub flange and B_u is the ultimate axial resistance of the bolts.

It can be demonstrated that:

- type-1 mechanism occurs for
  $\beta_u \leq 2\,\lambda/(1+2\,\lambda)$;
- type-2 mechanism occurs for
  $2\,\lambda/(1+2\,\lambda) < \beta_u \leq 2$;
- type-3 mechanism occurs for $\beta_u > 2$

where the nondimensional parameter $\lambda = n/m$, defining the location of the prying forces, has been introduced.

2.2 Type-1 mechanism

In the case of type-1 mechanism, the load levels corresponding to the above mentioned characteristic points of the force-displacement curve can be easily computed by means of the following relationship:

$$F = \frac{(32\, n - 2\, d_w)\, M}{8\, m\, n - (m + n)\, d_w} \tag{5}$$

where d_w is the washer diameter, the nut diameter or the bolt head diameter, as appropriate. It provides:

- $F = F_y$ for $M = M_y = (b'_{eff}\, t_f^2/6)\, f_y$, where b'_{eff} is the effective width for stiffness calculation [4,14];
- $F = F_h$ for $M = M_h$, where M_h is the bending moment corresponding to the beginning of strain-hardening, i.e. to the attainment of the strain ε_h in the most extreme fibre of the T-stub flanges. It is given by:

$$M_h = \frac{1}{2}\left[3 - \left(\frac{\chi_y}{\chi_h}\right)^2\right] M_y \tag{6}$$

  where $\chi_y = 2\,\varepsilon_y/t_f$ and $\chi_h = 2\,\varepsilon_h/t_f$.;
- $F = F_m$ for $M = M_m$, where M_m is the bending moment corresponding to the occurrence of the stress f_u in the most extreme fibre of the T-stub flanges, given by:

$$M_m = \left\{\frac{1}{2}\left[3 - \left(\frac{\chi_y}{\chi_m}\right)^2\right] + \frac{1}{2}\frac{E_h}{E}\left(\frac{\chi_m - \chi_h}{\chi_y}\right)\left(1 - \frac{\chi_h}{\chi_m}\right)\left(2 + \frac{\chi_h}{\chi_m}\right)\right\} M_y \tag{7}$$

  where $\chi_m = 2\,\varepsilon_m/t_f$;
- $F = F_u$ for $M = M_u$, where M_u is the ultimate value of the T-stub flange bending moment, provided by:

$$M_u = \left\{\frac{1}{2}\left[3 - \left(\frac{\chi_y}{\chi_u}\right)^2\right] + \frac{1}{2}\frac{E_h}{E}\left(\frac{\chi_u - \chi_h}{\chi_y}\right)\left(1 - \frac{\chi_h}{\chi_u}\right)\left(2 + \frac{\chi_h}{\chi_u}\right) + \right.$$
$$\left. - \frac{1}{2}\frac{E_h - E_u}{E}\frac{\chi_u - \chi_m}{\chi_y}\left(1 - \frac{\chi_m}{\chi_u}\right)\left(2 + \frac{\chi_m}{\chi_u}\right)\right\} M_y \tag{8}$$

  where $\chi_u = \varepsilon_f/t_f$.

It is useful to underline that the ratios M_h/M_y, M_m/M_y and M_u/M_y are dependent on the material properties only.

Regarding the evaluation of the displacements corresponding to the four characteristic points of the force-displacement curve of the single T-element, they can be computed by exploiting the formulations presented in [4] which are herein summarized.

With reference to a single T-element, the displacement corresponding to the occurrence of first yielding can be computed as the ratio F_y/K, where the stiffness K is given by:

$$K = 0.5\, E \frac{b'_{eff}\, t_f^3}{m^3} \tag{9}$$

where b'_{eff} is the effective width for stiffness calculation.

Therefore, with reference to a bolted T-stub constituted by two equal T-elements, the first yielding displacement can be computed as:

$$\delta_y = 2\frac{F_y}{K} + \delta_{b.y} \tag{10}$$

where $\delta_{b.y}$ is the corresponding bolt elongation given by:

$$\delta_{b.y} = \frac{B}{E A_b/L_b} = \frac{\frac{F_y}{2} + \frac{M_y}{n}}{E A_b/L_b} \tag{11}$$

where A_b is the bolt section area and L_b is the conventional bolt length given by the sum of the thicknesses of the connected plates plus half time the sum of the bolt head thickness and of the bolt nut thickness.

The initial stiffness of the bolted T-stub can be computed according to the following relationship:

$$K_i = \frac{F_y}{\delta_y} \tag{12}$$

The displacement of a bolted T-stub, constituted by two equal T-elements, corresponding to the load level F_h, i.e. to the beginning of strain hardening, is given by:

$$\delta_h = \delta_{eh} + 2\,\delta_{ph} \tag{13}$$

where $\delta_{eh} = F_h/K_i$ is the elastic part of the displacement and δ_{ph} is the plastic displacement of the single T-element corresponding to the beginning of strain hardening, i.e. to the attainment of the curvature χ_h.

The plastic displacement of the single T-element is given by [4]:

$$\delta_{ph} = \theta_{ph}\, m = \frac{m^2}{2\,t_f}\, D(\xi_2) \tag{14}$$

where $D(\xi_2)$ is the value of the function:

$$D(\xi) = \varepsilon_y\left[2\frac{\chi_\xi}{\chi_y} - \frac{1}{\xi}\frac{M_y}{M_u}\left(3\frac{\chi_\xi}{\chi_y} + \frac{\chi_y}{\chi_\xi} - 3\right) - 1\right] \tag{15}$$

for $\chi_\xi = \chi_h$ and $\xi = \xi_2 = M_h/M_u$, where χ_ξ is the curvature value corresponding to the bending moment $\xi\, M_u$, with $\xi \leq 1$. It is useful to note that $D(\xi_2)$ is a constant depending only on the material properties.

The displacement of the bolted T-stub corresponding to the load level F_m, i.e. to the attainment of the stress f_u, is given by:

$$\delta_m = \delta_{em} + 2\,\delta_{pm} \tag{16}$$

where $\delta_{em} = F_m/K_i$ is the elastic part of the displacement and δ_{pm} is the plastic displacement of the single T-element corresponding to the attainment of f_u in the extreme fibres of the sections where plastic hinges develop, i.e. to the attainment of the curvature χ_m.

The plastic displacement of the single T-element is given by [4]:

$$\delta_{pm} = \theta_{pm}\, m = \frac{m^2}{2\,t_f}\, F(\xi_3) \tag{17}$$

where $F(\xi_3)$ is the value of the function $F(\xi)$:

$$F(\xi) = \varepsilon_y \left\{ 2\frac{\chi_\xi}{\chi_y} - \frac{M_y}{M_u}\frac{1}{\xi}\left[3\frac{\chi_\xi}{\chi_y} + \frac{\chi_y}{\chi_\xi} - 3 + \frac{E_h}{E}\frac{(\chi_\xi - \chi_h)^3}{\chi_\xi\,\chi_y^2} \right] - 1 \right\} \tag{18}$$

for $\chi_\xi = \chi_m$ and $\xi = \xi_3 = M_m/M_u$. It is useful to note that also $F(\xi_3)$ is a constant depending only on the material properties.

Finally, the ultimate displacement of the bolted T-stub corresponding to the attainment of the ultimate natural strain ε_u is given by:

$$\delta_u = \delta_{eu} + 2\,\delta_{pu} \tag{19}$$

where $\delta_{eu} = F_u/K_i$ is the elastic part of the displacement and δ_{pu} is the ultimate plastic displacement of the single T-element to be computed according to the following equation [4,12]:

$$\delta_{pu} = \theta_{pu}\, m = \frac{m^2}{2\,t_f}\,C \tag{20}$$

where C is a constant depending only on the mechanical properties of the material:

$$C = 2\left\{ \varepsilon_u - \frac{1}{2\,(M_u/M_y)}\left(3\,\bar{\varepsilon}_u + \frac{1}{\bar{\varepsilon}_u} - 3 + \frac{E_h}{E}\,C_h + \frac{E_u}{E}\,C_u \right) - \frac{1}{2} \right\}\varepsilon_y \tag{21}$$

with:

$$C_h = \frac{\bar{\varepsilon}_m^{\,3}}{\bar{\varepsilon}_u} + 3\,\bar{\varepsilon}_m\,\bar{\varepsilon}_u - 3\,\bar{\varepsilon}_m^{\,2} + 3\,\bar{\varepsilon}_h^{\,2} - 3\,\bar{\varepsilon}_h\,\bar{\varepsilon}_u - \frac{\bar{\varepsilon}_h^{\,3}}{\bar{\varepsilon}_u} \tag{22}$$

$$C_u = \bar{\varepsilon}_u^{\,2} - 3\,\bar{\varepsilon}_m\,\bar{\varepsilon}_u + 3\,\bar{\varepsilon}_m^{\,2} - \frac{\bar{\varepsilon}_m^{\,3}}{\bar{\varepsilon}_u} \tag{23}$$

where:

$$\bar{\varepsilon}_h = \frac{\varepsilon_h}{\varepsilon_y} \qquad \bar{\varepsilon}_m = \frac{\varepsilon_m}{\varepsilon_y} \qquad \bar{\varepsilon}_u = \frac{\varepsilon_u}{\varepsilon_y} \tag{24}$$

2.3 Type-2 mechanism

Regarding type-2 mechanism, it is preliminarly necessary to observe that the ultimate resistance of a bolted T-stub failing according to this mechanism typology can be expressed as:

$$F_{u.2} = \frac{2\,M_u}{m}\,(\,1 + \xi\,) \tag{25}$$

where $\xi\,M_u$ is the bending moment occurring at the bolt axis location when the bending moment at the flange-to-web connection reachs its ultimate value M_u. The parameter ξ is given by:

$$\xi = \frac{(\,2 - \beta_u\,)\,\lambda}{\beta_u\,(\,1 + \lambda\,)} \tag{26}$$

Equation (25) can be immediately derived taking into account that $F_{u.2}$ is the double of the shear force in the T-stub flange between the bolt axis and the flange-to-web connection.

By assuming that the point of zero moment remains unchanged during the loading process, the load levels corresponding to the four characteristic points of the force-displacement curve can be computed by means of the following relationship:

$$F = \frac{2\,M}{m}\,(\,1 + \xi\,) \tag{27}$$

which provides:

- $F = F_y$ for $M = M_y$;
- $F = F_h$ for $M = M_h$;
- $F = F_m$ for $M = M_m$;
- $F = F_u$ for $M = M_u$.

Regarding the T-stub displacement corresponding to first yielding, it can be still computed by means of equation (10) where, in this case, the bolt contribution is given by:

$$\delta_{b.y} = \frac{\dfrac{F_y}{2} + \dfrac{\xi M_y}{n}}{E A_b / L_b} \tag{28}$$

When the bending moment at the flange-to-web connection reachs the M_h value, the corresponding bending moment at the bolt axis is equal to ξM_h and can be expressed by means of the following nondimensional parameter:

$$\xi_2^* = \frac{\xi M_h}{M_u} = \xi \frac{M_h}{M_y} \frac{1}{M_u/M_y} \tag{29}$$

With reference to the plastic hinge located at the flange-to-web connection, at the beginning of strain hardening, the plastic rotation θ_{p1h} can be computed according to the following relationship [4]:

$$\theta_{p1h} = \frac{m}{t_f (1 + \xi)} D(\xi_2) \tag{30}$$

With reference to the plastic hinge located at the bolt axis, it can be demonstrated that $\theta_{p2h} = 0$ if $\xi_2^* \le \xi_1$ while, for $\xi_2^* > \xi_1$ ($\xi_1 = M_y/M_u$) the following relationship holds [4]:

$$\theta_{p2h} = \frac{m}{t_f} \left(\frac{\xi}{1 + \xi} + \lambda \right) D(\xi_2^*) \tag{31}$$

where $D(\xi_2^*)$ is the value of the function $D(\xi)$, Eq. (13), for $\xi = \xi_2^*$ so that χ_ξ is the curvature value corresponding to the bending moment $\xi_2^* M_u$.

As a result, the displacement corresponding to the beginning of strain hardening can be computed according to Section 2.1 where $\theta_{p1} = \theta_{p1h}$ and $\theta_{p2} = \theta_{p2h}$.

Regarding the displacement corresponding to the attainment of the stress f_u in the extreme fibre of the plastic hinge located at the flange-to-web connection, it can evaluated taking into account that the plastic rotation of the hinge located at the flange-to-web connection corresponding to the development of the bending moment M_m is given by [4]:

$$\theta_{p1m} = \frac{m}{t_f (1 + \xi)} F(\xi_3) \tag{32}$$

where $F(\xi_3)$ is given by Eq. (18) for $\xi = \xi_3 = M_m/M_u$, so that χ_ξ is the curvature value corresponding to the bending moment M_m, i.e. $\chi_\xi = \chi_m$.

When the bending moment at the flange-to-web connection reachs the value M_m, the corresponding bending moment at the bolt axis is ξM_m which can be expressed through the following nondimensional parameter:

$$\xi_3^* = \frac{\xi M_m}{M_u} = \xi \frac{M_m}{M_y} \frac{1}{M_u/M_y} \tag{33}$$

Depending on the value assumed by the above parameter, the plastic rotation of the hinge located at the bolt axis can be computed as [4]:

$$\theta_{p2m} = 0 \quad \text{for} \quad \xi_3^* < \xi_1 \tag{34}$$

$$\theta_{p2m} = \frac{m}{t_f}\left(\frac{\xi}{1+\xi} + \lambda\right) D(\xi_3^*) \qquad \text{for} \qquad \xi_1 < \xi_3^* \leq \xi_2 \tag{35}$$

$$\theta_{p2m} = \frac{m}{t_f}\left(\frac{\xi}{1+\xi} + \lambda\right) F(\xi_3^*) \qquad \text{for} \qquad \xi_2 < \xi_3^* \leq \xi_3 \tag{36}$$

Therefore, $D(\xi_3^*)$ is the value of the function (15) where $\xi = \xi_3^*$ and χ_ξ is the curvature corresponding to the bending moment $\xi_3^* \, M_u$, while $F(\xi_3^*)$ is the value of the function (18) for $\xi = \xi_3^*$, being, in this case, χ_ξ the curvature value corresponding to the bending moment $\xi_3^* \, M_u$.

As a result, the displacement corresponding to the attainment of the bending moment M_m at the flange-to-web connection can be still computed according to equation (16) where, in this case, the plastic part has to be computed according to Section 2.1 with $\theta_{p1} = \theta_{p1m}$ and $\theta_{p2} = \theta_{p2m}$.

Finally, regarding the ultimate displacement corresponding to the occurrence of the ultimate natural strain ε_u in the most extreme fibre of the plastic hinge at the flange-to-web connection, it can be still computed by means of equation (19) where the ultimate plastic displacement of the single T-element has to be computed according to Section 2.1 by assuming $\theta_{p1} = \theta_{p1u}$ and $\theta_{p2} = \theta_{p2u}$.

The ultimate plastic rotation of the hinge located at the flange-to-web connection is, in this case, given by [4]:

$$\theta_{p1u} = C \, \frac{m}{t_f\,(1+\xi)} \tag{37}$$

while, the plastic rotation of the hinge located at the bolt axis is given by:

$$\theta_{p2u} = 0 \qquad \text{for} \qquad \xi \leq \xi_1 \tag{38}$$

$$\theta_{p2u} = \frac{m}{t_f}\left(\frac{\xi}{1+\xi} + \lambda\right) D(\xi) \qquad \text{for} \qquad \xi_1 < \xi \leq \xi_2 \tag{39}$$

$$\theta_{p2u} = \frac{m}{t_f}\left(\frac{\xi}{1+\xi} + \lambda\right) F(\xi) \qquad \text{for} \qquad \xi_2 < \xi \leq \xi_3 \tag{40}$$

$$\theta_{p2u} = \frac{m}{t_f}\left(\frac{\xi}{1+\xi} + \lambda\right) G(\xi) \qquad \text{for} \qquad \xi_3 < \xi \leq 1 \tag{41}$$

where $G(\xi)$ is given by:

$$G(\xi) = \varepsilon_y \left\{ 2\frac{\chi_\xi}{\chi_y} - \frac{1}{\xi}\frac{M_y}{M_u}\left[3\frac{\chi_\xi}{\chi_y} + \frac{\chi_y}{\chi_\xi} - 3 + \frac{E_h}{E} G_h + \frac{E_u}{E} G_u \right] - 1 \right\} \tag{42}$$

with:

$$G_h = \frac{\chi_m^{\,3}}{\chi_\xi \, \chi_y^{\,2}} + 3\frac{\chi_m \, \chi_\xi}{\chi_y^{\,2}} - 3\frac{\chi_m^{\,2}}{\chi_y^{\,2}} + 3\frac{\chi_h^{\,2}}{\chi_y^{\,2}} - 3\frac{\chi_h \, \chi_\xi}{\chi_y^{\,2}} - \frac{\chi_h^{\,3}}{\chi_\xi \, \chi_y^{\,2}} \tag{43}$$

$$G_u = \frac{\chi_\xi^{\,2}}{\chi_y^{\,2}} + 3\frac{\chi_m^{\,2}}{\chi_y^{\,2}} - 3\frac{\chi_m \, \chi_\xi}{\chi_y^{\,2}} - \frac{\chi_m^{\,3}}{\chi_\xi \, \chi_y^{\,2}} \tag{44}$$

2.4 Type-3 mechanism

Regarding type-3 mechanism, four cases can be identified depending on the ratio between the bolt axial resistance and the T-stub flange flexural resistance which can be expressed through the ratio ξ given by:

$$\xi = \frac{B_u m}{M_u} = \frac{2}{\beta_u} \tag{45}$$

The first case occurs for $\xi \leq \xi_1$. In this case, the T-stub flanges are not engaged in plastic range so that $\delta_p = 0$. In this case, it can be assumed that the force-displacement curve of the single T-element linearly develops up to the achievement of the bending moment $B_u m$ at the flange-to-web connection zone.

In the second case, occurring for $\xi_1 < \xi \leq \xi_2$, the force-displacement curve of the single T-element can be assumed bilinear. The force leading to first yielding of the T-stub flanges is:

$$F_y = \frac{2 M_y}{m} \tag{46}$$

The corresponding displacement can be easily computed taking into account that in this case prying forces do not develop:

$$\delta_y = \frac{F_y}{K} \tag{47}$$

where K is given by equation (9).

The ultimate plastic displacement can be easily computed by means of equation (2) where the plastic rotation of the hinge located at the flange-to-web connection zone is given by:

$$\theta_{pu} = \frac{m}{t_f} D(\xi) \tag{48}$$

As a consequence, the ultimate displacement of the single T-element corresponding to the force $F_u = 2 B_u$ is given by:

$$\delta_u = \frac{F_u}{K} + \frac{m^2}{t_f} D(\xi) \tag{49}$$

The third case develops for $\xi_2 < \xi \leq \xi_3$. In this case the force-displacement curve of the single T-element can be assumed trilinear. The point corresponding to first yielding (F_y, δ_y) is provided by equations (47) and (46). The point corresponding to the beginning of strain hardening (F_h, δ_h) is given by:

$$F_h = \frac{2 M_h}{m} \tag{50}$$

$$\delta_h = \frac{F_h}{K} + \frac{m^2}{t_f} D(\xi_2) \tag{51}$$

which is immediately derived from Eqs. (2) and (48) taking into account that ξ_2 is the value of ξ corresponding to the occurrence of the bending moment M_h.

In addition, it is easy to recognize that the ultimate displacement corresponding to the occurrence of the force $F_u = 2 B_u$ can be computed in this case as [4]:

$$\delta_u = \frac{F_u}{K} + \frac{m^2}{t_f} F(\xi) \tag{52}$$

Finally, the last case occurs for $\xi > \xi_3$. In this case the force-displacement curve of the single T-element is quadrilinear. The point corresponding to first yielding can be still evaluated by means of equations (47) and (46) while the point corresponding to the beginning of strain hardening can be still computed through equations (51) and (50). The third characteristic point (F_m, δ_m) of the force-displacement curve of the single T-element corresponds to the development of the stress f_u in the most extreme fibres of the T-stub flange section at the flange-to-web connection zone. It can be easily recognized that this point can be identified by means of the following relationships:

$$F_m = \frac{2\,M_m}{m} \tag{53}$$

$$\delta_m = \frac{F_m}{K} + \frac{m^2}{t_f} F(\xi_3) \tag{54}$$

which is immediately obtained from Eq. (52) considering that ξ_3 is the value of ξ corresponding to the occurrence of the bending moment M_m.

Finally, the ultimate displacement corresponding to the occurrence of the force $F_u = 2\,B_u$ can be computed, in this case, as [4]:

$$\delta_u = \frac{F_u}{K} + \frac{m^2}{t_f} G(\xi) \tag{55}$$

In all cases, for any load level, the displacement of a bolted T-stub constituted by two equal T-elements bolted through the flanges is obtained by adding the bolt elongation to the sum of the axial displacements of the T-elements.

2.5 Numerical simulation

The procedure previously described for the three collapse mechanism typologies can be easily used to predict also the plastic displacement corresponding to any loading stage. To this scope it is still assumed that the point of zero moment along the T-stub flange is known and remains unchanged during the loading process [4,12].

As an example, with reference to type-1 mechanism, it can be assumed that the point of zero moment remains, during the loading process, located at the middle section between the bolt axis and the flange-to-web connection. Therefore, it can be assumed that the bending moment at the bolt axis section is equal to that at the flange-to-web connection. By denoting this bending moment with ξM_u, it means that collapse occurs for $\xi = 1$, so that the whole loading process can be simulated by varying the coefficient ξ from 0 to 1. For any given step of the loading process, i.e. for any given value of ξ, the value of the external axial force F is given by:

$$F = \frac{(32\,n - 2\,d_w)\,\xi\,M_u}{8\,m\,n - (m + n)\,d_w} \tag{56}$$

The corresponding value of the plastic displacement of the single T-element can be computed by means of the following relationship:

$$\delta_p = \theta_p\,m \tag{57}$$

where the plastic rotation θ_p can be computed with formulations similar to those described in Section 2.2.

Obviously, the overall displacement has to be computed adding to the plastic displacement the elastic one due to the T-elements and to the bolts.

The elastic part of the displacement of a single T-element can be computed as the ratio F/K where K is the stiffness of the single T-element to be evaluated according to Eq. (9). In order to compute the bolt contribution, for any given loading stage, the prying force can be computed as:

$$Q = \frac{\xi M_u}{n} \tag{58}$$

Therefore, the axial force in the bolts ($B = Q + F/2$) and the corresponding lenghtening δ_b can be computed. Finally, the overall axial displacement of the bolted T-stub is evaluated considering that it is composed by two T-elements and the bolts, so that:

$$\delta = 2\left(\frac{F}{K} + \delta_p\right) + \delta_b \tag{59}$$

A similar procedure can be applied for type-2 and type-3 mechanisms.

3 COMPARISON WITH EXPERIMENTAL EVIDENCE

In order to show the reliability of the theoretical formulations for predicting the force-displacement curve of bolted T-stubs and, in particular, for evaluating the plastic deformation capacity of the fundamental component of bolted connections, an experimental program devoted to the ultimate behaviour of bolted T-stubs has been carried out at the Material and Structure Laboratory of Salerno University. In order to cover the range of variability of the most important geometrical parameters, 10 specimens have been prepared. The specimens are constituted by the coupling of T-stub elements which have been obtained from rolled profiles of HEA and HEB series, steel grade Fe430, by cutting along the web plane. These T-stubs are connected through the flanges by means of two high strength bolts (class 8.8 or class 10.9). With reference to the notation of Fig. 5, the measured values of the geometrical properties of tested specimens are given in Table 1 where, in addition, d_h is the bolt head diameter, t_h is the thickness of the bolt head and t_n is the thickness of the bolt nut. All the specimens have been subjected, under displacement control, to a tensile axial force which is applied to the webs tightened by the jaws of the testing machine (Schenck Hydropuls S56, maximum test load 630

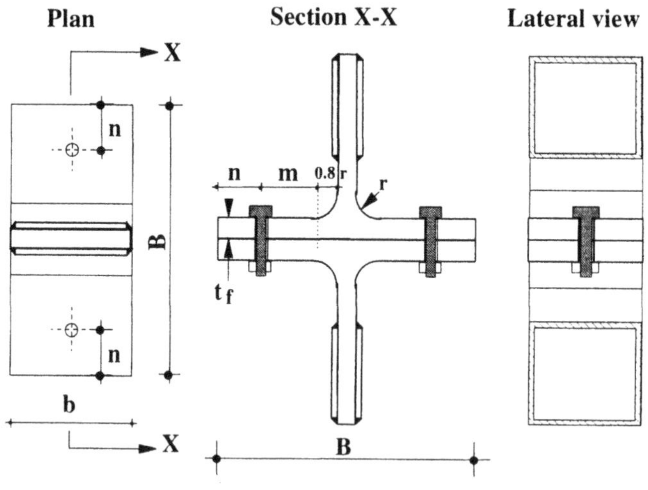

Figure 2.2.5:Tested specimens

Table 2.2.1: Geometrical properties of tested specimens

T-STUB	1	2	3	4	5	6	7	8	9	10
d_b (mm)	20.00	20.00	20.00	24.00	24.00	24.00	27.00	24.00	27.00	20.00
d_h (mm)	33.53	33.53	33.53	39.98	39.98	39.98	45.20	39.98	45.20	35.03
d_w (mm)	37.00	37.00	37.00	44.00	44.00	44.00	56.00	35.38	56.00	37.00
t_h (mm)	12.72	12.72	12.72	14.78	14.78	14.78	17.35	14.78	17.35	13.00
t_n (mm)	16.00	16.00	16.00	19.00	19.00	19.00	27.00	19.00	27.00	16.00
t_f (mm)	14.40	14.60	13.00	12.30	13.80	16.35	13.90	13.30	12.50	10.85
m (mm)	39.30	39.30	42.30	48.95	49.75	45.65	52.45	45.20	54.05	32.60
n (mm)	40.8	40.80	22.20	58.10	74.90	45.20	71.70	19.30	53.00	58.50
r (mm)	18.00	18.00	15.00	24.00	27.00	18.00	27.00	15.00	24.00	18.00
b (mm)	126.50	119.00	124.00	118.80	115.00	120.00	122.80	112.50	125.10	125.00
b'_{eff} (mm)	112.13	112.13	118.13	118.80	115.00	120.00	122.80	112.50	125.10	95.20

Table 2.2.2: Mechanical properties of tested specimens

T-STUB	1	2	3	4	5	6	7	8	9	10
f_y (N/mm^2)	291.16	264.95	273.15	299.76	317.72	280.46	307.58	269.42	300.97	293.10
f_u (N/mm^2)	517.21	501.11	504.33	543.59	546.84	527.76	543.57	482.70	552.27	514.87
E_h (N/mm^2)	3276	3171	3087	2877	3339	3360	3423	3276	3234	2163
E_u (N/mm^2)	371.11	378.32	435.95	465.24	483.15	488.11	516.05	476.80	466.65	406.32
ε_h (%)	1.358	1.135	1.171	1.285	1.362	1.202	1.318	1.055	1.290	0.600
ε_u (%)	49.18	48.12	58.70	76.77	78.43	63.83	77.88	69.40	67.75	92.50
f_{ub} (N/mm^2)	800	800	800	800	800	800	800	800	800	1000

kN, piston stroke ±125 mm). The axial displacement has been progressively increased up to the collapse of the specimens. Depending on the ratio between the bolt axial resistance and the flange flexural resistance, two types of collapse mechanism developed which correspond to type-1 and to type-2 mechanism. In the first case, the complete fracture of one of the T-stub flanges occurred at the flange-to-web connection zone. In the second case, the premature failure of one bolt prevented the complete development of the plastic reserves of the T-stub flanges.

Regarding the material properties, coupon tests have been developed for evaluating the true stress versus true strain curve of the T-stub flanges. The stress-strain curve obtained from experimental tests can be accurately represented by means of the quadrilinear model depicted in Fig. 2. The parameters of the quadrilinear representation of the stress-strain curve have been numerically obtained by imposing that the area below the quadrilinear model has to be coincident with the one below the experimental true stress-true strain curve. As a result of this energy equivalence, the material mechanical properties of the tested specimens are given in Table 2.

On the basis of the measured geometrical properties and of the measured material mechanical properties, the force-displacement curves of the tested specimens have been simulated according to the formulations presented in the previous Sections. For all the specimens, both the quadrilinear model of the F–δ curve and its complete simulation are compared with the experimental curve in Figs. 6–9. The comparison with the experimental evidence shows a very good agreement in terms of stiffness, resistance and deformation capacity between the theoretical model and the experimental results. In particular, regarding the prediction of the plastic deformation capacity, which is the primary aim of the developed theoretical model, the obtained degree of accuracy is very satisfactory. Taking into account that the prediction of the

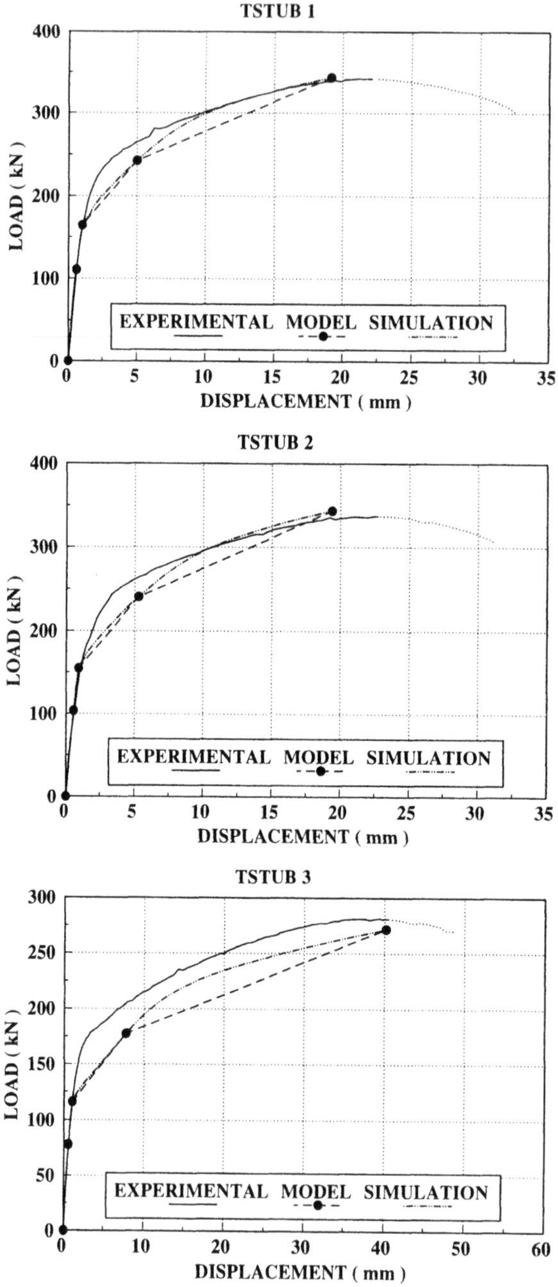

Figure 2.2.6: Comparison between the predictions obtained by means of the theoretical model and the experimental test results

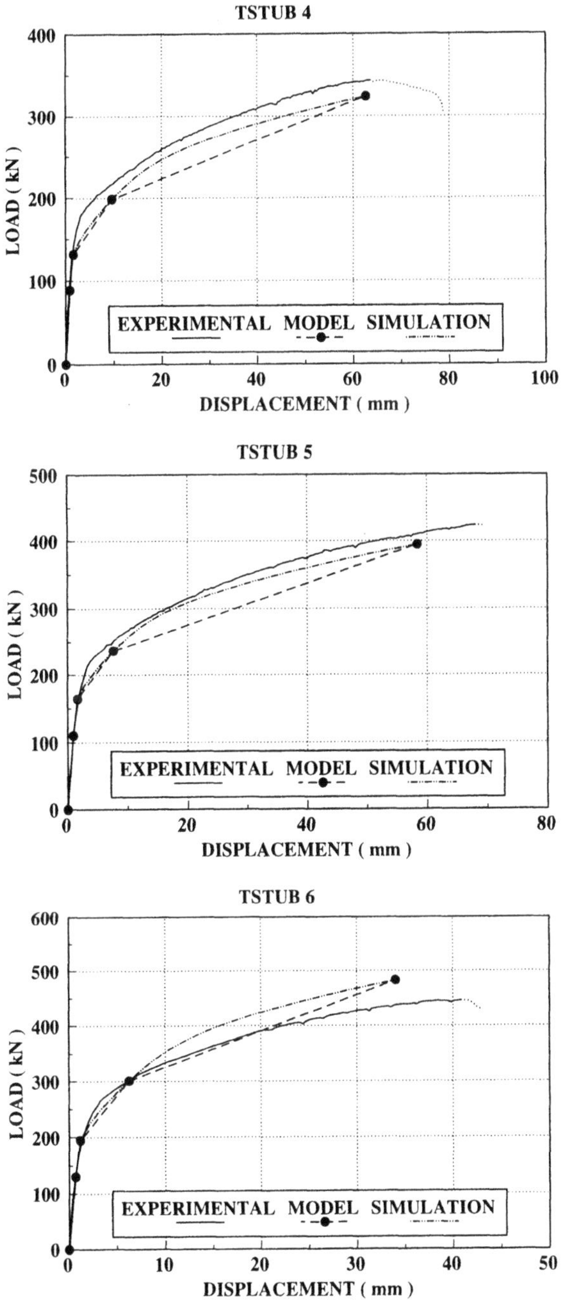

Figure 2.2.7: Comparison between the predictions obtained by means of the theoretical model and the experimental test results

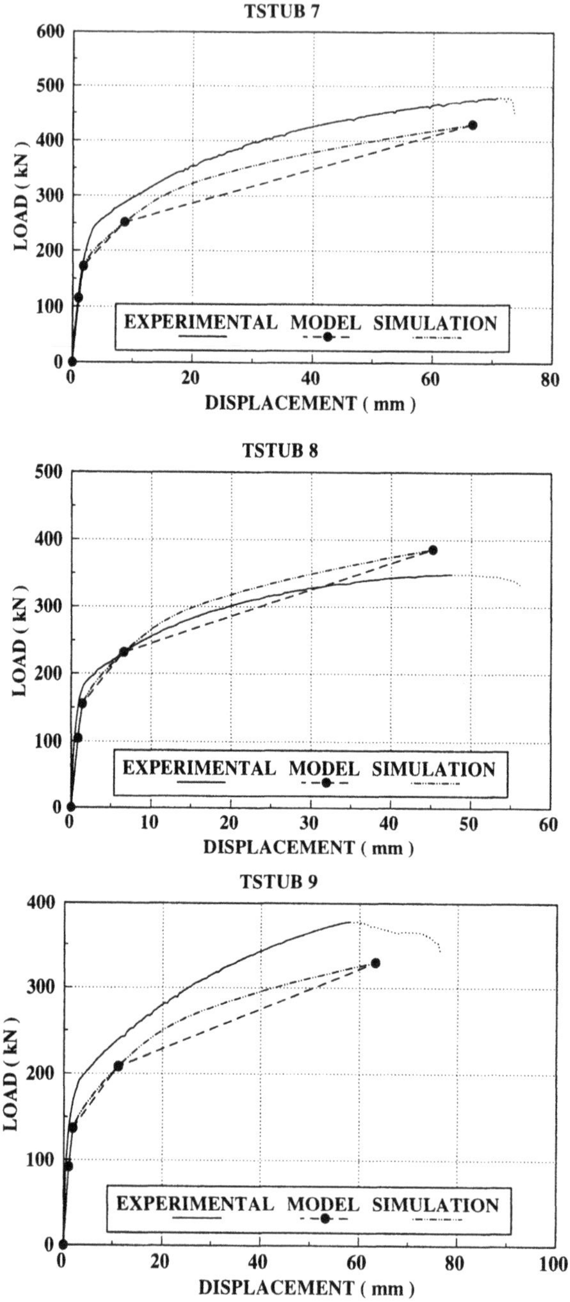

Figure 2.2.8: Comparison between the predictions obtained by means of the theoretical model and the experimental test results

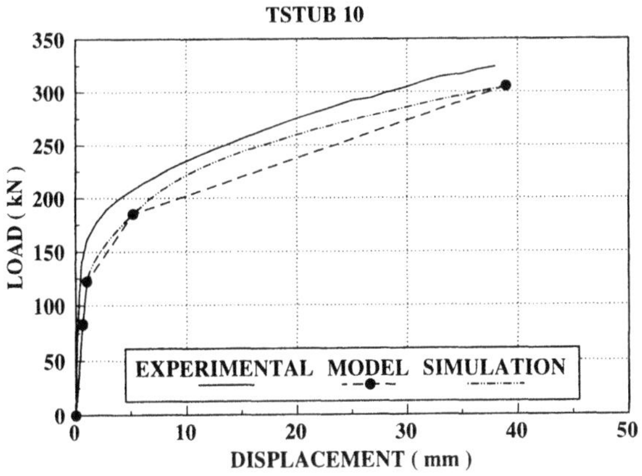

Figure 2.2.9: Comparison between the predictions obtained by means of the theoretical model and the experimental test results

plastic deformation capacity of bolted T-stubs is of primary importance to lay down the bases towards the theoretical prediction of the plastic rotation supply of bolted connections, this result is very encouraging about the possibility to solve a very complicated problem, such as the prediction of the ductility of bolted connections, by means of a theoretical approach. The importance of such result is more and more evident as soon as it is considered that, up-to-now, the plastic rotation supply of the beam-joint system has been only investigated by means of expensive experimental tests which, unavoidably, cannot cover the whole variability range of the geometrical and mechanical properties. In addition, another advantage of the developed theoretical model is its ability to underline the most important parameters governing the plastic deformation capacity of bolted T-stubs and, as a consequence, the plastic rotation supply of bolted connections.

4 CONCLUSIONS

Taking into account that the prediction of the plastic deformation capacity of bolted T-stubs is of primary importance to lay down the bases towards the theoretical prediction of the plastic rotation supply of bolted connections, the results presented in this paper are very encouraging about the possibility to solve a very complicated problem, such as the prediction of the ductility of bolted connections by means of a theoretical approach. The importance of such result is more and more evident as soon as it is considered that, up-to-now, the plastic rotation supply of the beam-joint system has been only investigated by means of expensive experimental tests which, unavoidably, cannot cover the whole variability range of the geometrical and mechanical properties of the structural details which can be adopted in designing beam-to-column connections.

5 REFERENCES

[1] W.F. Chen, S. Toma: «Advanced Analysis of Steel Frames», CRC Press, 1994

[2] F.M. Mazzolani, V. Piluso: «Theory and Design of Seismic Resistant Steel Frames», FN Spon an Imprint of Chapman and Hall, 1996.

[3] C.W. Roeder, D.A. Foutch: «Experimental Results for Seismic Resistant Steel Moment Frame Connections», Journal of Structural Engineering, ASCE, Vol.122, N.6, June, 1996.

[4] C. Faella, V. Piluso, G. Rizzano: «Structural Steel Semirigid Connections», CRC Press, 1999 (in press).

[5] J.P. Jaspart, M. Steenhuis, K. Weinand: «The stiffness model of Revised Annex J of Eurocode 3», Third International Workshop on Connection in Steel Structures, Trento, 28-31 May, 1995

[6] Y.L. Yee, R.E. Melchers: «Moment-Rotation Curves for Bolted Connections», Journal of Structural Engineering, ASCE, Vol.112, January, 1986.

[7] Y.J. Shi, S.L. Chan, Y.L. Wong: «Modeling for Moment-Rotation Characteristics for End-Plate Connections», Journal of Structural Engineering, ASCE, Vol.122, N.11, November, 1996.

[8] C. Faella, V. Piluso, G. Rizzano: «Some Proposals to Improve EC3-Annex J Approach for Predicting the Moment-Rotation Curve of Extended End Plate Connections», Costruzioni Metalliche, N.4, 1996a.

[9] C. Faella, V. Piluso, G. Rizzano: «Reliability of Eurocode 3 Procedures for Predicting Beam-to-Column Joint Behaviour», Third International Conference on Steel and Alluminium Structures, Istanbul, May, 1995

[10] C. Faella, V. Piluso, G. Rizzano: «Prediction of the Flexural Resistance of Bolted Connections with Angles», IABSE Colloquium on Semirigid Structural Connections, Istanbul, 25-27 September 1996b.

[11] J.M. Aribert: «Aspects Actuels sur le Dimensionnement Plastique des Assemblages en Construction Métallique», Construction Métallique, N.2, 1992.

[12] C. Faella, V. Piluso, G. Rizzano: «Plastic Deformation Capacity of Bolted T-Stubs», International Workshop and Seminar on Behaviour of Steel Structures in Seismic Areas, STESSA '97, 4-7 August 1997 Kyoto Japan.

[13] RILEM: RILEM Draft Recommendation - Tension Testing of Metallic Structural Materials for Determining Stress-Strain Relations under Monotonic and Uniaxial Tensile Loading, *Material and Structures*, 1990, Vol.23, pp.35-46.

[14] C. Faella, V. Piluso, G. Rizzano: «Experimental Analysis of Bolted Connections: Snug versus Preloaded Bolts», *Journal of Structural Engineering*, ASCE, Vol. 124, No.7, July 1998, pp. 765-774.

Chapter 3

Cyclic Behaviour of Beam-to-Column Bare Steel Connections

3.1 Influence of strain rate

3.2 Influence of connection typology and loading asymmetry

3.3 Influence of haunching

3.4 Influence of column size

3.1

INFLUENCE OF STRAIN RATE

Darko Beg, André Plumier, Crt Remec, Luis Sanchez

INTRODUCTION

Because of problems observed in steel beam-to-column connections after the Northridge (1994) and Kobe (1995) earthquakes, numerous tests have been made in the USA (Lu et al. 1999, Anderson et al. 1995) and in Japan (Kato et al. 1997, Akiyama et al. 1999) on welded connections, since it is the typical detail considered in these countries.

These tests have taken into account all the aspects of the problem: characteristics of base and weld material, design of connections, welding preparation and technology, strain rate effects. Many conclusions have been drawn and other research work under way will bring more, but it is already clear that the strain rate influences the connection behaviour:

- It increases yield strength f_y and tensile strength f_u, bringing higher demand on the connection.
- It increases f_y more than f_u, which generates a localisation of the plastic strains subsequently reducing the apparent ductility.
- It can propagate cracks in a brittle way.

The U.S. and Japanese studies do not include the typical European connections, since they refer to fully welded connections, involving on-site welding, while the European practice favours shop welding and on-site bolting.

The research work in this sub-chapter considers the following aspects of steel connections:

1) Test on small specimens in order to study the effect of the temperature and the strain rate on the behaviour of fillet and butt welds.

2) Test on small specimens of fillet welded connections in order to acquire information on the critical size of fillet welds and on the influence of the strain rate.

3) The influence of the fabrication processes on the mechanical properties of the weld.

4) Tests on full-scale beam-to-column connections in order to study the influence of the loading speed on the capacity to dissipate energy cyclically by plastic mechanism for various typical beam-to-column steel connections. The behaviour of some of these connection typologies was studied in a previous larger test programme under quasi-static loading condition (Plumier et al. 1998):

- Welded connection. Beam flanges welded to the column face for two different loading speeds.
- Comparison between symmetric and unsymmetric bolted connection under cyclic dynamic loading conditions.
- Symmetric bolted connection, to get information on the influence of loading speed.
- Dog-Bone connection, under cyclic dynamic loading.
- Semi-rigid partial strength connection, under quasi-static and dynamic cyclic loading.

The work within point 2 and the last three items of point 4 was done at the University of Liege, Belgium and the work within points 1, 3 and the first two items of point 4 was done at the University of Ljubljana, Slovenia.

TESTS ON SMALL SPECIMENS

Objectives

Along with full-scale tests on connections at increased strain rates it is also important to know the behaviour of the most important single components of connections. For the base metal it is well known that at higher strain rates the strength is increased and the ductility and toughness is decreased. In order to get information on the behaviour of welds at higher strain rates, a series of tension tests on specimens containing welded joints within the gauge length was executed. As additional testing parameters influencing weld properties the change in temperature and fillet weld size were chosen.

Influence of strain rate and temperature

Test specimens and material properties

Test specimens are presented in Figure 3.1.1. They are basically flat, with the initial gauge length of 80 mm and width of 40 mm. They were made in two thicknesses (10 mm and 20 mm) and with two types of welded joints placed within the gauge length: fillet welds and full penetration butt welds. Fillet welds were designed according to Eurocode 3 to resist as much as steel plates connected with these welds. At 10 mm thick specimens fillet weld throat thickness was supposed to be 6 mm, but after completition it proved to be around 8 mm.

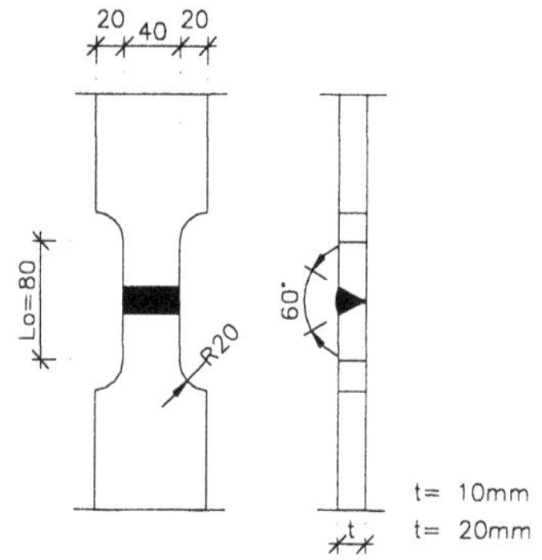

a) Test specimens with full penetration butt welds

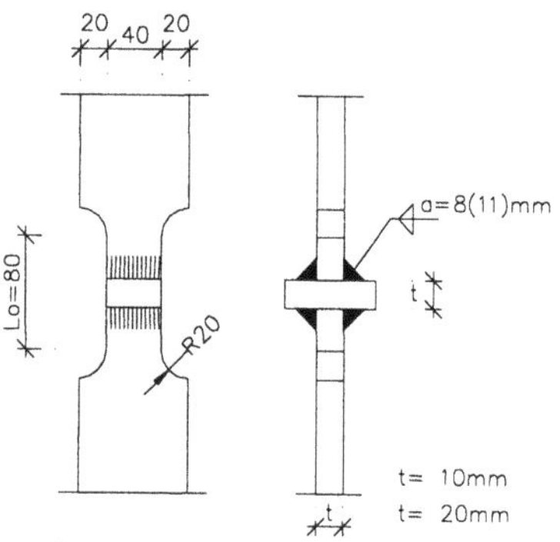

b) Test specimens with filet welds

Figure 3.1.1: Geometry of test specimens

Test specimens were arranged in four groups according to their thickness and the weld joint type (see Table 3.1.2). Each group contained eight test specimens. They were supposed to be tested at different strain rates (0.0003 s^{-1}, 0.01 s^{-1}, 0.05 s^{-1}) and temperatures (+20°C, 0°C, -20°C, -40°C). Due to some problems with the identification of the base material, one specimen from each of the first two groups was used to determine the material properties, and the tests at -40°C were omitted in those two groups.

To determine the mechanical properties of base metal two standard tensile tests and a series of Charpy V-notch tests for each plate thickness were performed. The results are gathered in Table 3.1.1 and Figure 3.1.2. From these results it is clear that the material with the thickness of 20 mm can be classified according to EN 10025 as S 235JR and the thinner material (10 mm) which is much tougher as S 275J2.

TABLE 3.1.1

MATERIAL PROPERTIES OF BASE METAL

Specimen	Thickness (mm)	f_y (MPa)	f_u (MPa)	A	Z
1.8.1	20	243	418	0.326	0.680
1.8.2	20	229	413	0.333	0.678
2.8.1	10	320	485	0.347	0.709
2.8.2	10	326	480	0.344	0.710

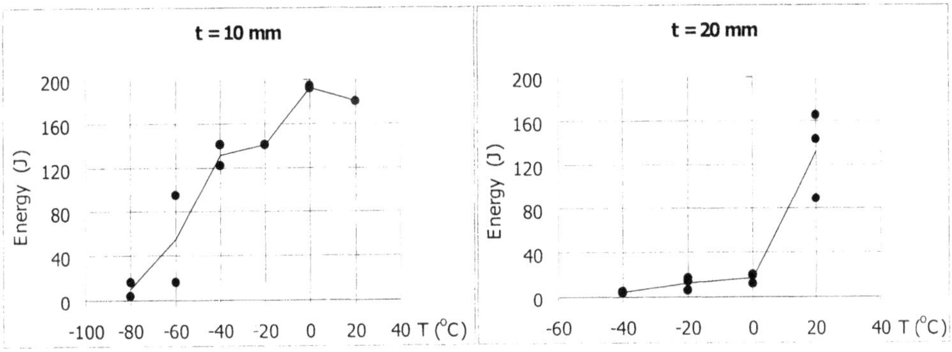

Figure 3.1.2: Charpy V-notch test results

Description of tests

All tests were run in ZWICK 700 kN machine. The displacement control was used in order to control the strain rate. The highest strain rate of 0.05 s^{-1} was chosen according to the testing machine capacity at the applied test specimen geometry (theoretically possible strain rate 0.07 s^{-1}). For the tests at lower temperatures test specimens were cooled in the solution of water and liquefied nitrogen. The specimens were wrapped in alu-foil to slow down their heating. This procedure allowed to perform only the tests at higher strain rates (0.01 and 0.05 s^{-1}), which were completed in short time. Slower tests at a strain rate of 0.003 s^{-1} were executed only at the room temperature, because the heating rate of the test specimens was certainly too high for longer testing time. During the tests the force-displacement relationship was recorded and the failure mode was observed.

Test results

The most important results concern the change in strength and ductility in relation to the change in strain rate and temperature. They were obtained from stress-strain relationship that was deduced from the measured force-displacement relationship and test specimen's geometry. These results are: yield strength (f_y), defined according to ECCS testing procedure (ECCS, 1986), ultimate tensile strength (f_u) and maximum (rupture) strain (ε_{max}). The results are shown in Table 3.1.2. Regarding the failure mode, it is interesting to note that in all cases the failure took place in the parent metal and not in the welds.

For easier analysis of the results the main parameters, such as f_y, f_u, ε_{max}, f_u/f_y, are plotted in Figure 3.1.3, as the function of testing strain rate and temperature. They are always normalised with the value of the corresponding parameter at the lowest strain rate 0.0003 s^{-1} and room temperature.

TABLE 3.1.2
TEST SPECIMEN DATA AND TEST RESULTS

Specimen	Thickness (mm)	Type of weld	Strain rate (s^{-1})	Temp. (°C)	f_y (MPa)	f_u (MPa)	ε_{max}	Failure mode
1.1	20	FW	0.05	+20	283	471	0.309	BM
1.2	20	FW	0.0003	+20	255	450	0.315	BM
1.3	20	FW	0.01	+20	291	470	0.302	BM
1.4	20	FW	0.01	-20	328	494	0.312	BM
1.5	20	FW	0.05	-20	313	507	0.327	BM
1.6	20	FW	0.01	0	298	485	0.304	BM
1.7	20	FW	0.05	0	291	493	0.324	BM
2.1	10	FW	0.05	+20	349	519	0.238	BM
2.2	10	FW	0.0003	+20	358	497	0.233	BM
2.3	10	FW	0.01	+20	359	511	0.224	BM
2.4	10	FW	0.01	-20	393	545	0.227	BM
2.5	10	FW	0.05	-20	395	552	0.231	BM
2.6	10	FW	0.01	0	367	530	0.224	BM
2.7	10	FW	0.05	0	361	534	0.223	BM
3.1	20	FPW	0.05	+20	281	458	0.337	BM
3.2	20	FPW	0.0003	+20	244	437	0.405	BM
3.3	20	FPW	0.01	+20	274	447	0.342	BM
3.4	20	FPW	0.01	-20	300	476	0.367	BM
3.5	20	FPW	0.05	-20	296	484	0.388	BM
3.6	20	FPW	0.01	0	276	461	0.392	BM
3.7	20	FPW	0.05	0	285	468	0.346	BM
3.8	20	FPW	0.05	-40	317	502	0.374	BM
4.1	10	FPW	0.05	+20	350	503	0.278	BM
4.2	10	FPW	0.0003	+20	315	483	0.322	BM
4.3	10	FPW	0.01	+20	334	496	0.286	BM
4.4	10	FPW	0.01	-20	373	525	0.301	BM
4.5	10	FPW	0.05	-20	363	531	0.302	BM
4.6	10	FPW	0.01	0	356	511	0.281	BM
4.7	10	FPW	0.05	0	352	514	0.288	BM
4.8	10	FPW	0.05	-40	396	548	0.300	BM

FW – Fillet weld, FPW – Full penetration butt weld, BM – Base metal

The following conclusions can be drawn from these diagrams:

- Both parameters describing the strength - f_y and f_u increase with decreasing temperature and increasing strain rate. The maximum increase of 24% for f_y is reached at the test specimen 4.8 (-40°C, strain rate 0.05) and the maximum increase of 15% for f_u is reached at the test specimen 3.8 (-40°C, strain rate 0.05). Both specimens, 3.8 and 4.8, were tested at the most severe conditions. The diagrams are more regular in the case of tensile strength that was measured directly. Yield strength was established indirectly according to the ECCS testing procedure. Regarding only the room temperature f_y increased by 10% and f_u by 5% at most. Regarding only the tests with the highest strain rate 0.05 s^{-1} f_y increased by 10% at -20°C and 13% at -40°C. The figures for f_u are even smaller.

- The results for the parameters describing ductility (ratio f_u/f_y and rupture deformation ε_{max}) show less order than the results for f_y and f_u. A general tendency of a moderate decrease of the ratio f_u/f_y with increasing strain rate and decreasing temperature is evident. The largest drop of 9% was observed in group 4. In the case of full penetration butt welds (test groups 3 and 4) the rupture strain ε_{max} decreases when the strain rate is increasing. The maximum drop of 17% is observed for the specimen 3.1. The relation of ε_{max} to change in temperature is very weak and exhibits even some small unexpected increase at lower temperatures.

- In the case of fillet welds there is almost no change of ε_{max} in relation to the changes of temperature and strain rate (less than 4% in relation to the test at 20°C and the slowest strain rate). This can be explained by large deformability of fillet weld joints clearly visible in the form of a gap between the plates in T joint after the tests. This deformability compensates the drop of ε_{max} at higher strain rates.

- Regarding the different material properties of the base metal and associated thickness the increase of f_y and f_u is larger at thicker and less tough material.

- Regarding the type of weld used in connection the only important difference arises from the difference in the deformability of welded joints.

- Comparing the influence of temperature and strain rate it is evident that the decrease in temperature produces more or less linear increase in f_y and f_u, while the increase of strain rate produces nonlinear response with more pronounced increase of f_y and f_u at the beginning (strain rate 0.01 s^{-1}) and more gradual increase at higher strain rates.

The lesson which can be learned from the tests made is that in simple loading conditions with increasing load (tensile test) the decrease of temperature, increase of strain rate or even the combination of both does not cause any premature failure, if the welds are designed to be as strong as the plates connected.

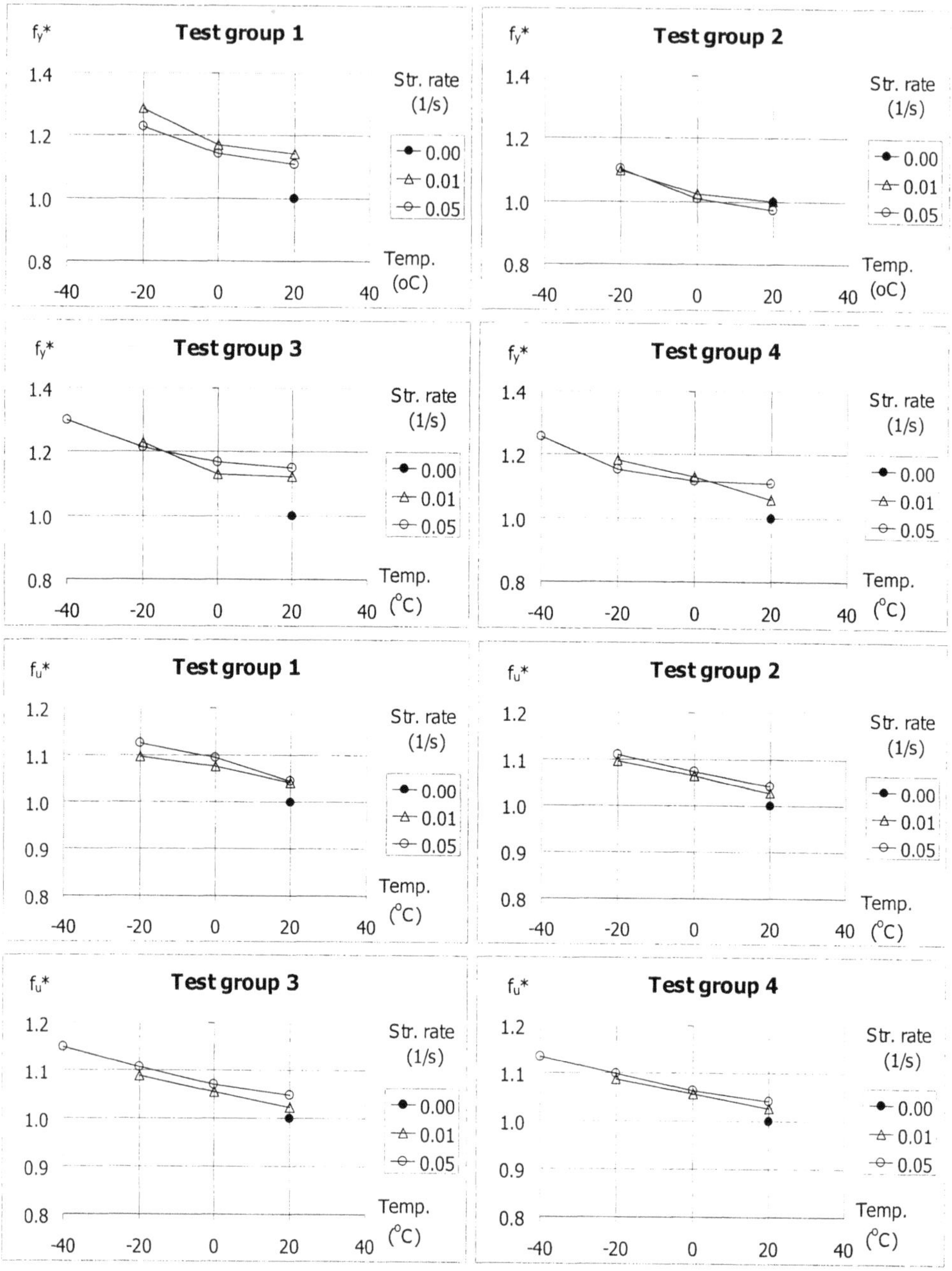

$f_y^* = f_y/f_y(T=20^oC$, str. rate$=0.0003\ s^{-1})$
$f_u^* = f_u/f_u(T=20^oC$, str. rate$=0.0003\ s^{-1})$

Figure 3.1.3: Diagrams with test results as a function of temperature and strain rate

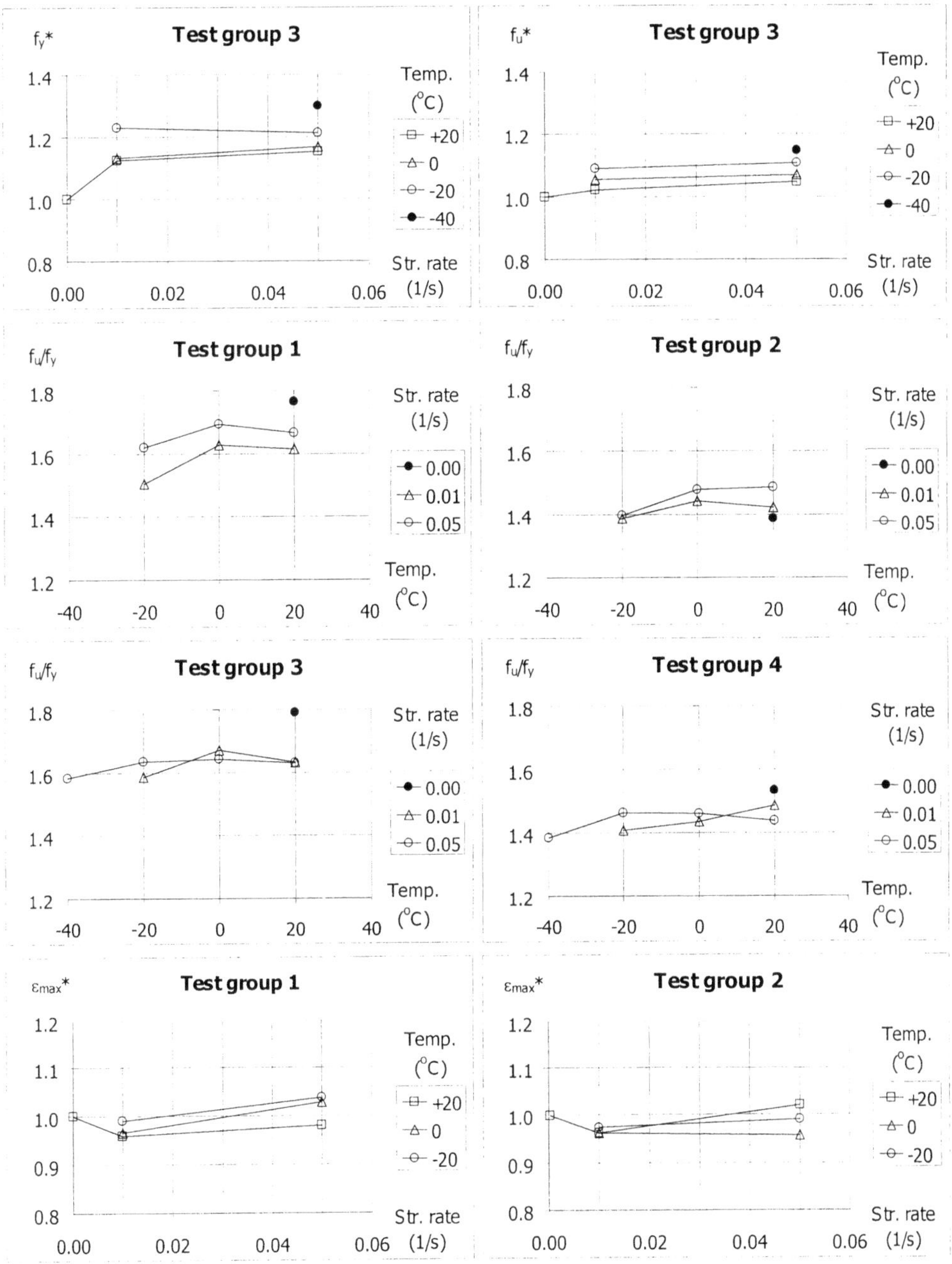

$f_y^* = f_y/f_y(T=20^\circ C, \text{str. rate}=0.0003\ s^{-1})$
$\varepsilon_{max}^* = \varepsilon_{max}/\varepsilon_{max}(T=20^\circ C, \text{str. rate}=0.0003\ s^{-1})$

Figure 3.1.3 (continued)

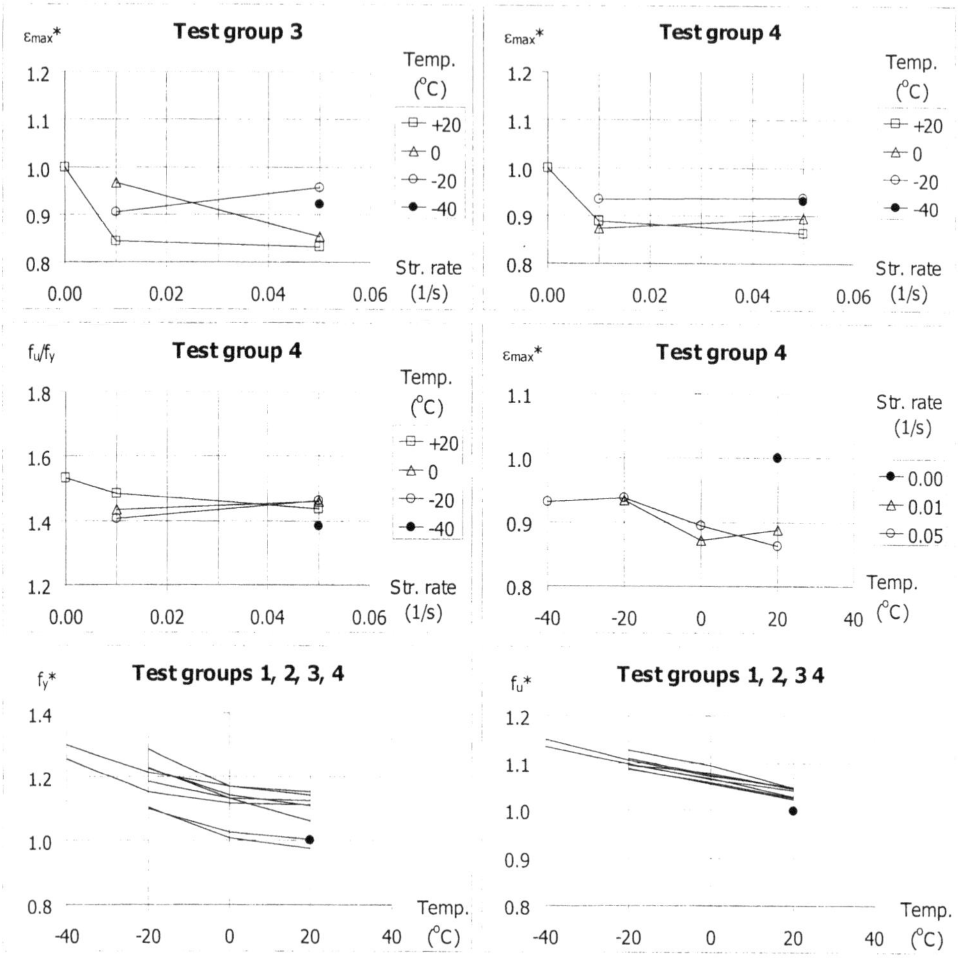

$f_y^* = f_y/f_y(T=20^{\circ}C, \text{str. rate}=0.0003\ s^{-1})$
$f_u^* = f_u/f_u(T=20^{\circ}C, \text{str. rate}=0.0003\ s^{-1})$
$\varepsilon_{max}^* = \varepsilon_{max}/\varepsilon_{max}(T=20^{\circ}C, \text{str. rate}=0.0003\ s^{-1})$

Figure 3.1.3 (continued)

Influence of strain rate and fillet weld size

Test specimens and material properties

There are four series of 5 specimens with the size of fillet welds being 6 mm, 8 mm, 10 mm and 12 mm respectively. See Figure 3.1.4.

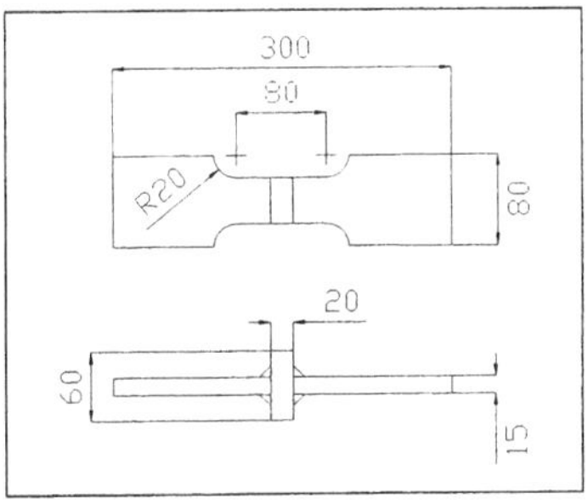

Figure 3.1.4: Small specimens

Because dynamic behaviour of material is related to the ability to support shocks, the material of the tensile specimens was tested on toughness (Charpy V-notch test with three specimens at each temperature). The cuts were made perpendicular to the weld, in the base material, with a net section of 0.8 cm^2. The results of the tests are shown in Table 3.1.3.

TABLE 3.1.3
RESULTS OF CHARPY TESTS

N° specimen	Test temperature (°C)	Absorbed Energy (J)			Resilience (J/cm^2)			Weldability Class
fw 2/1	-20	9	10	7	11.25	12.5	8.75	B
fw 2/2	0	10	12	11	12.5	15	13.75	
fw 2/3	+20	155	143	153	193.8	178.8	191.3	

Description of tests

Tests are run in a DARTEC 300 kN machine with displacement control.

In each series, a standard tension test is run in order to determine:
- f_y yield stress.
- e_y yield displacement of the 80 mm basis of reduced width.

It was intended to run the cyclic tests in each series between $\pm e_y$.

The first test run cyclically raised the problem of overall buckling of the specimen. In spite of the very compact aspect of the specimen designed especially to avoid buckling problems and in spite of the hydraulic grips fixing it to the machine, plastic buckling took place progressively with the cycles. This generates unexpected bending moments in the grips, which cannot be tolerated because they may break the grips. Because of the high potential impact of damage to the testing machine, it was decided to revise the test program on small specimens in the following way:
- The effect of strain rate is studied by means of tensile test at various strain rates.
- The critical size of fillet welds is studied in dynamic conditions, without the cyclic effect.

This option may be considered acceptable because dynamic cyclic testing is realised on the full size specimens described in the second part of the report.

Test results

From these diagrams, significant characteristics are deduced:

- f_y and ε_y, yield stress and yield displacements established according to the ECCS testing procedure.
- $\varepsilon_y = e_y / 80$ mm.
- f_u, maximum stress in a specimen.
- $\varepsilon_{ult} = e_{ult} / 80$ mm, ultimate strain in a specimen.

The results of f_y, ε_y, f_u, and ε_u are presented in Table 3.1.4, together with the failure mode observed in the tests.

TABLE 3.1.4

EXPERIMENTAL VALUES OF f_y, ε_y, f_u AND ε_u DEDUCED FROM THE TEST

Specimen	a (mm)	dε/dt (1/s)	f_y (N/mm^2)	ε_y	f_u (N/mm^2)	ε_u	Mod. Failure
Sqfw11	6	-----	253.0	0.0114	-----	-----	In the weld
Sqfw13	6	0.00031	241.2	0.0101	380.0	0.182	In the weld
Sqfw14	6	0.02500	275.0	0.0156	405.8	0.172	In the weld
Sqfw15	6	0.10000	289.2	0.0137	401.7	0.133	In the weld
Sqfw21	8	-----	236.6	0.0086	410.4	-----	Outside
Sqfw22	8	0.00031	258.3	0.0110	389.0	0.269	Outside
Sqfw23	8	0.02500	263.0	0.0108	420.4	0.301	Outside
Sqfw25	8	0.10000	296.1	0.0127	427.3	0.281	Outside
Sqfw31	10	-----	267.8	0.0109	416.6	-----	Outside
Sqfw32	10	0.00031	249.2	0.0106	386.9	0.261	Outside
Sqfw33	10	0.02500	263.7	0.0106	426.1	0.266	Outside
Sqfw35	10	0.10000	280.2	0.0116	434.0	0.264	Outside
Sqfw41	12	-----	266.1	0.0105	421.4	-----	Outside
Sqfw42	12	0.00031	266.4	0.0132	394.0	0.255	Outside
Sqfw43	12	0.02500	272.1	0.0110	424.3	0.260	Outside
Sqfw45	12	0.10000	297.8	0.0129	435.5	0.240	Outside

It is clear from the results that, independently of the strain rate applied, the critical size experimentally established for the filled welds is between 6 and 8 mm.

Other graphs and tables were produced to set forward the effect of strain rate. Table 3.1.5 indicates the ratio between the property (f_y, f_u, ε_{ult}) at high ($\dot{\varepsilon} = 0.1\,s^{-1}$) and low strain rate ($\dot{\varepsilon} = 0.0031\,s^{-1}$), where subscript 'o' indicates that the value corresponds to $\dot{\varepsilon} = 0.0031\,s^{-1}$.

TABLE 3.1.5

RATIO BETWEEN PROPERTIES AT HIGH STRAIN RATE AND LOW STRAIN RATE

	f_y/f_{yo}	$\varepsilon_y/\varepsilon_{yo}$	f_u/f_{uo}	$\varepsilon_{ult}/\varepsilon_{ult.o}$	f_{uo}/f_{yo}	f_u/f_y	Mod. Failure
Series 1	1.20	1.36	1.06	0.73	1.58	1.39	In the weld
Series 2	1.15	1.16	1.10	1.04	1.51	1.44	Outside
Series 3	1.12	1.10	1.12	1.01	1.55	1.55	Outside
Series 4	1.12	0.98	1.11	0.94	1.48	1.46	Outside

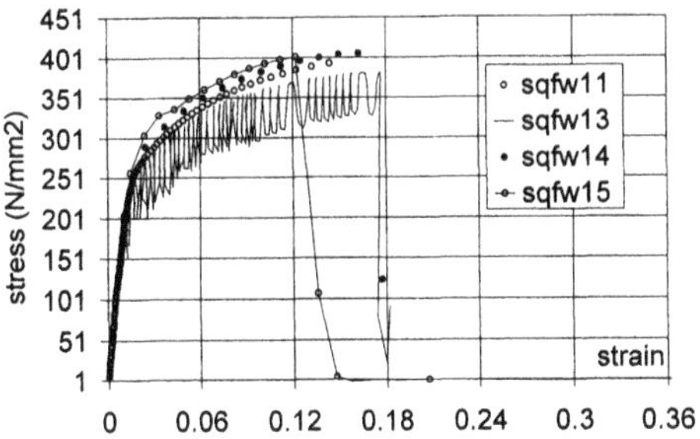

Figure 3.1.5: Series 1 (a= 6mm)

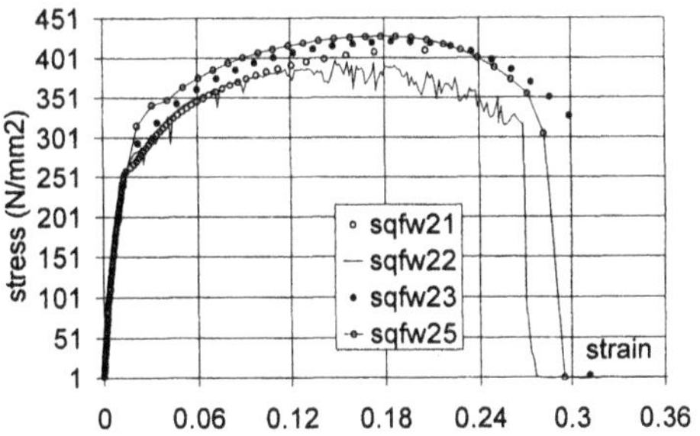

Figure 3.1.6: Series 2 (a=8 mm)

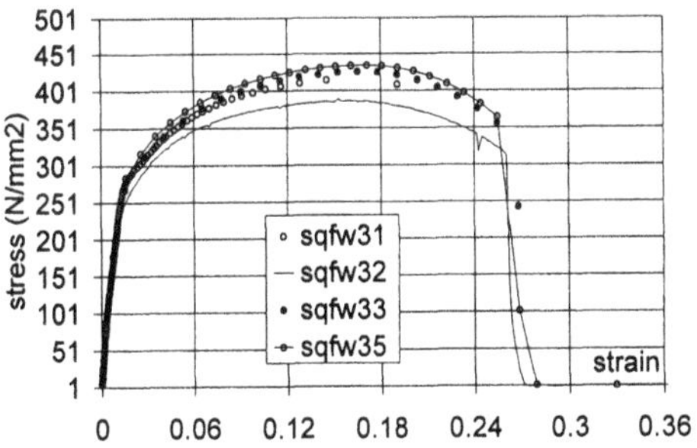

Figure 3.1.7: Series 3 (a=8 mm)

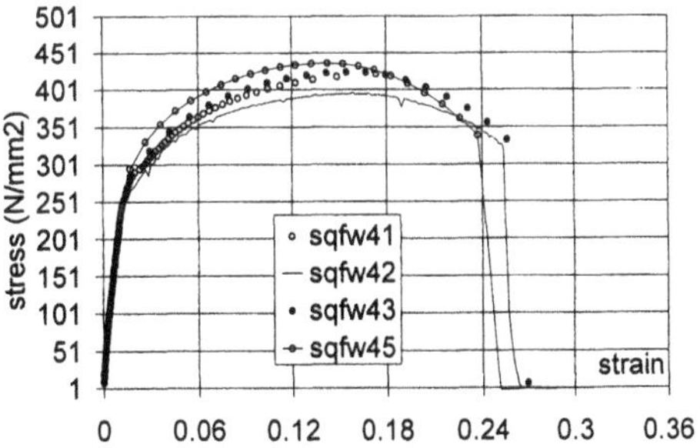

Figure 3.1.8: Series 4 (a=8 mm)

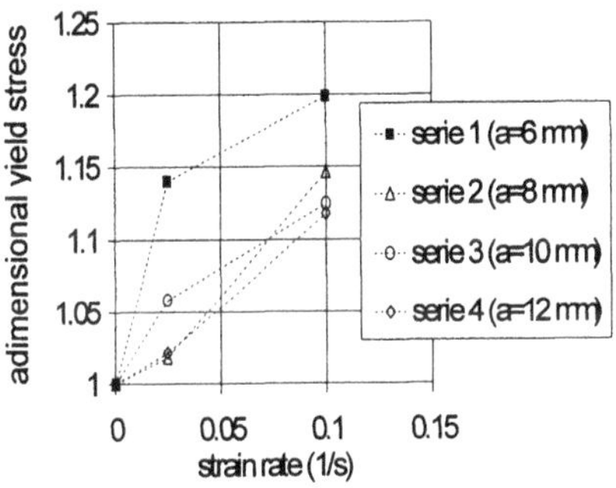

Figure 3.1.9: $f_y(\dot{\varepsilon})/f_y(\dot{\varepsilon}=0)$

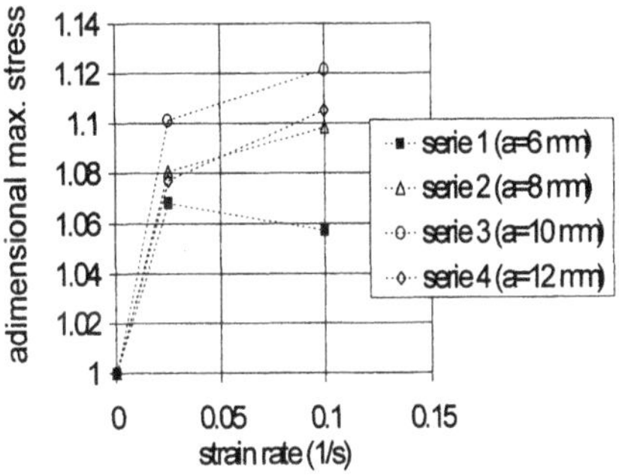

Figure 3.1.10: $f_u(\dot{\varepsilon})/f_u(\dot{\varepsilon}=0)$

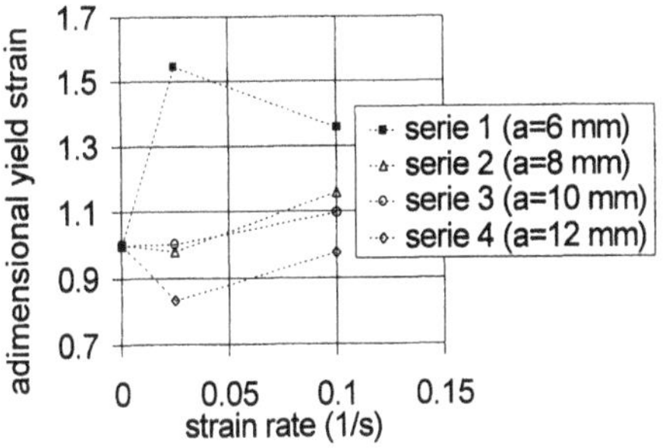

Figure 3.1.11: $\varepsilon_y(\dot{\varepsilon})/\varepsilon_y(\dot{\varepsilon}=0)$

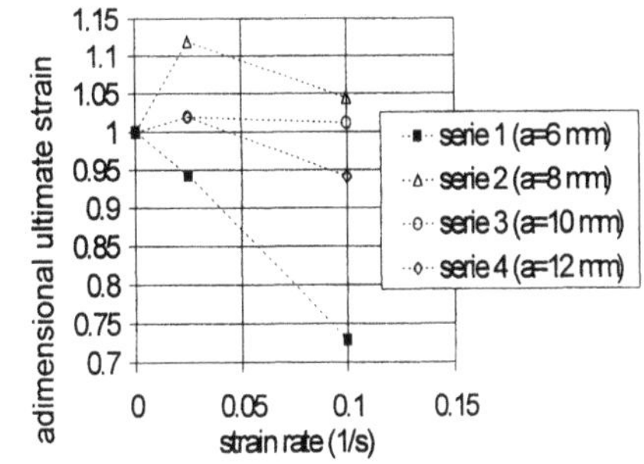

Figure 3.1.12: $\varepsilon_u(\dot{\varepsilon})/\varepsilon_u(\dot{\varepsilon}=0)$

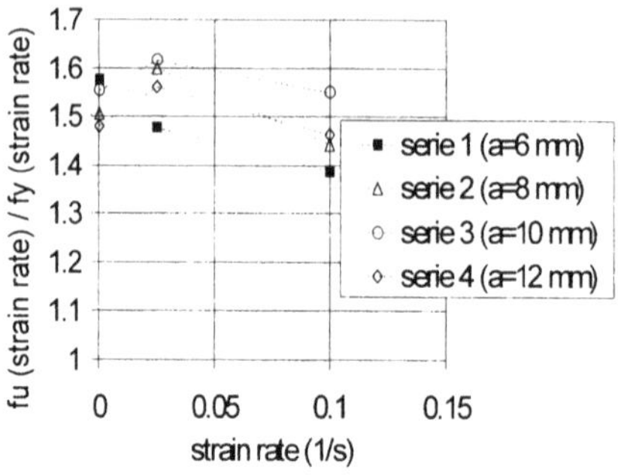

Figure 3.1.13: f_u/f_y versus strain rate

Discussion of test results

Effect of Strain Rate

It is very clear from the experimental values that strain rate values around $0.1\,s^{-1}$ change significantly the yield stress in comparison to the yield stress deduced from standard tensile test. This is true independently of the failure mode (weld or base metal).

Strain rate values around $0.1\,s^{-1}$ increase the ultimate stress, which is less sensitive to strain rate than yield stress. For the specimens with the failure in the weld, the value of the ultimate stress for the $0.1\,s^{-1}$ strain rate is lower than the ultimate stress for $0.0251\,s^{-1}$, because the rupture takes place before the maximum stress can be attained.

Tests made in the base material show that the ratio f_u/f_y decreases when the strain rate increases. In these tests, the influence of the strain rate in the weld is also studied: when the strain rate is increased to $0.0251\,s^{-1}$, the ratio f_u/f_y increases for three specimens with rupture in the base material and for $0.1\,s^{-1}$ the ratio f_u/f_y decreases for three specimens. This unexpected result is very relevant, because of the important influence of this ratio on the steel connections behaviour.

Critical size of fillet welds

Fillet welds designed to Eurocode 3 are strong enough to avoid failure in the welds. This conclusion is valid independently of the strain rate.

Indeed, for an 40 x 15 mm thick plate in Fe 360, the design resistance per weld must be: $F_{Rd} = \dfrac{40 \times 15 \times 235}{2} = 70\,500\ N$

The weld size needed is: $a = \dfrac{F_{Rd} \times R_k \times \sqrt{3} \times \beta_w}{f_u} \times \gamma_{MW}$ $\qquad a = \dfrac{70\,500 \times 0.8 \times \sqrt{3} \times 1.25}{360 \times 40} = 8.5\ mm$

Indeed, failure occurs in the weld when a = 6 mm and in the base material when a = 8 mm or more.

TESTS ON FULL-SCALE WELDED CONNECTIONS

Objectives

Although bolted beam to column connections are most popular in European steel construction practice, welded connections are also used widely. Two full scale dynamic tests on welded connections made by full penetration butt welds between a beam and a column were performed to supplement the results of the cyclic tests on welded connections, presented in Chapter 3.2. This multi-specimen testing program was prepared and executed at Istituto Superior Tecnico in Lisbon, Portugal, by L. Calado in order to establish a cumulative damage model. Additionally the influence of a column size was analysed.

Two dynamic tests were designed and executed to give the information on connection behaviour at higher strain rates. In addition the influence of fabrication processes on mechanical properties of welds was also investigated.

The influence of fabrication process on the quality of welds

The main objective of this study was to analyse the mechanical properties of welds in respect of different welding consumables and different welding processes produced and used in Slovenia. Standard tensile tests and Charpy V-notch tests were used for this purpose.

Test input parameters

- Parent metal: S 355 J2 grade was used as most demanding amongst mild structural steel grades regarding the toughness properties of welds.

- 5 combinations of welding processes and consumables:
  - Manual metal arc welding (MMA):
    - flux covered basic electrode (EN 499) - EVB 50
    - flux covered rutile electrode (EN 499) - RUTILEN 12
  - Metal arc active gas welding (MAG):
    - wire electrode (EN 449) - VAC 60
    - flux cored (basic) wire electrode (EN 758) - FILTUB
  - Flux cored wire metal arc welding without active gas shield:
    - flux cored (basic) wire electrode (EN 758) - FR-203MP

- Two types of welds
  - full penetration butt weld
  - fillet weld

Test specimens

Tensile and Charpy V-notch test specimens were cut out of tee-butt weld joints. These joints were selected to simulate beam flange to column flange or beam flange to end-plate welded joint as a part of moment connections (Figure 3.1.14). All full penetration welds were welded on two sides with back chipped groove and inspected by ultrasonic testing. No defects were observed.

Tensile test specimens were cut out only from full penetration joints. Additional piece was attached from the other side of the thicker plate to ensure sufficient length for test specimens (Figure 3.1.15).

Charpy V-notch standard test specimens were cut out at three different locations (see Figure 3.1.16).

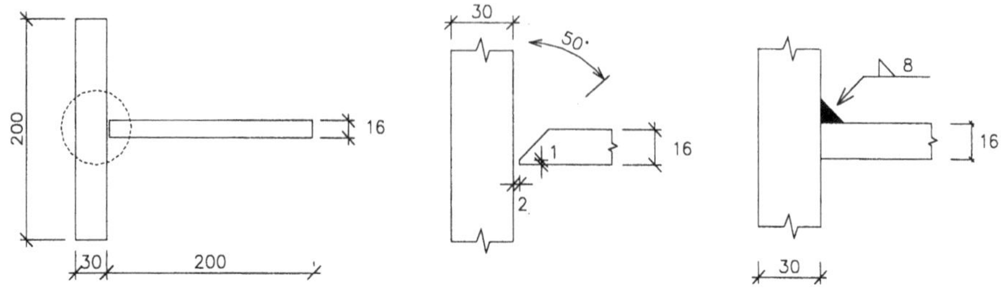

Full penetration single bevel weld

Fillet weld

Figure 3.1.14: Welded joints prepared for the fabrication of test specimens

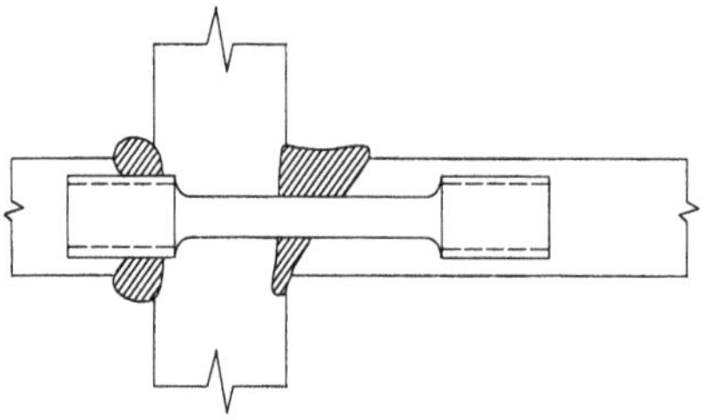

Round tensile specimens

initial diameter: $d_o = 10$ mm
initial gauge length: $l_o = 50$ mm

Figure 3.1.15: Position of tensile test specimens

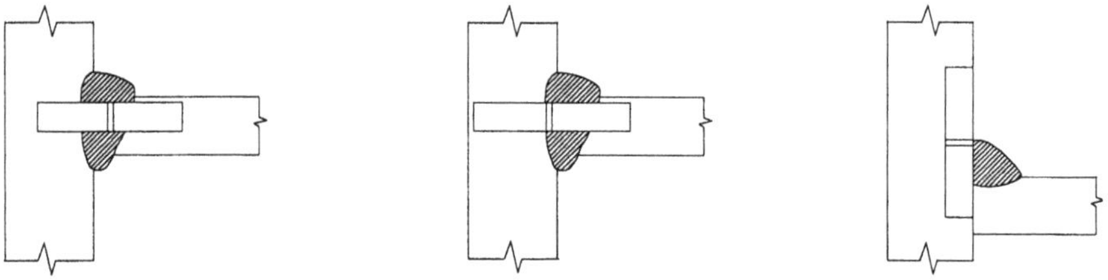

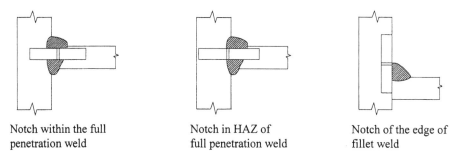

Notch within the full penetration weld

Notch in HAZ of full penetration weld

Notch of the edge of fillet weld

Figure 3.1.16: Position of Charpy V-notch test specimens

Tensile tests

Two tensile tests per every welding procedure - welding consumable combination were made and additional four tests for basic metal. The average results are shown in Table 3.1.6.

All the obtained values of yield and tensile strength meet the requirements for steel grade S 355. Except in the case of NR-203 MP welding consumable the rupture always occurred outside the weld.

TABLE 3.1.6

TENSILE TEST RESULTS

Test specimen		f_y (Mpa)	f_u (Mpa)	Failure position
Welding process	Consumable			
MMA	EVB 50	355	572	Base metal
	RUTILEN 12	383	563	Base metal
MAG	VAC 60	395	582	Base metal
	FILTUB 12B	388	577	Base metal
Without gas shield	NR – 203MP	373	582	Weld
-	Base metal	389	571	Base metal

Charpy V-notch tests

The following test parameters were taken into account:

- 5 combinations of welding process – consumable.
- 3 different notch locations (see Figure 3.1.16).
- 5 different temperatures (6 for fillet welds): -60°C, -40°C, -20°C, 0°C, +20°C (+60°C only for fillet welds).
- 3 specimens for each temperature.

The absorbed energy as a function of temperature is plotted for all three positions of a notch in Figure 3.1.17. Only average results for every temperature are plotted in these diagrams. In the case of fillet welds where the notch was located in the base metal at the edge of the weld the influence of welding process or consumable is not very important. Only rutile electrode RUTILEN 12 violated the toughness requirements for steel grade S 355 J2 (27 J of absorbed energy at -20°C). In the case of full penetration welds with notch located within the weld or in HAZ the influence of choice of consumable is more pronounced. With the notch in HAZ all test specimens fulfilled the toughness requirements and with the notch within the weld again only rutile electrode violated this condition.

Very good results in all three cases were obtained by MMA welding using basic electrode EVB 50. Comparing the results for two types of consumables used in MAG welding process, they are quite similar. Only at the notch position within the weld the results for flux cored wire FILTUB 12B are better.

For the fabrication of full-scale specimens used in dynamic tests on moment connections two types of welding were selected which are very common in Slovenia:

- Site welding (welded connections SW1, SW2 – this section): MMA welding procedure and EVB 50 flux covered basic electrode.

- Shop welding (beam to end plate welds for bolted connections UB1, UB2 and SB1 – see next section): MAG welding procedure and VAC 60 wire electrode.

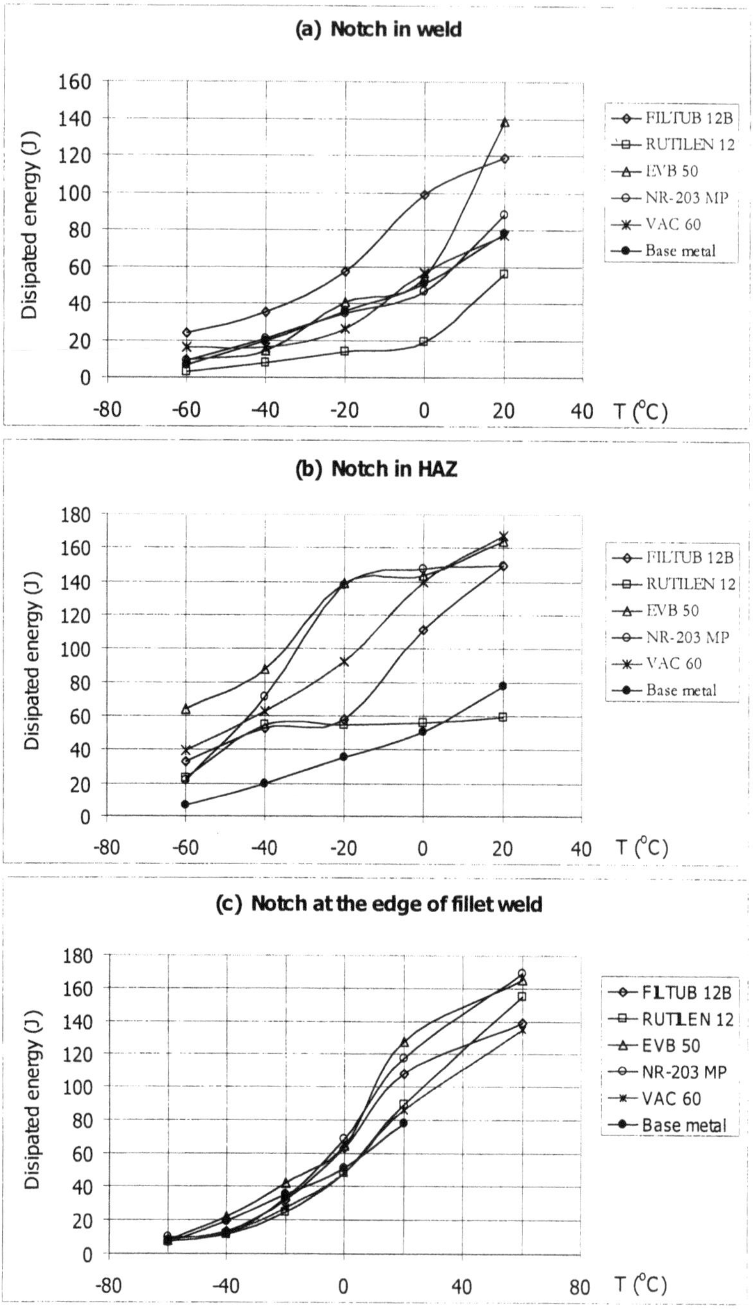

Figure 3.1.17: Charpy V-notch test results for three different notch locations

Full-scale tests on welded connections

Description of test specimens

Tee-shaped beam-column assembly was chosen to represent a part of a real frame around the connection. The beam part of the assembly was made of IPE 300 hot rolled profile and the column part of HEB 200 hot rolled profile. The specimens were produced by a steel fabricator implementing standard workmanship quality. All welds in the connection were full penetration butt welds. In the beam flanges single bevel welds were welded on two sides with back chipped groove.

This type of connections is usually made on site and therefore the MMA welding procedure was used. Round openings (scallops) in the beam webs enabled easier flange welding.

All the welds were checked by ultrasonic testing and no defects were found. The geometry of the test specimens SW1 and SW2 is shown in Figure 3.1.18 and is similar to the geometry of the test specimens from the BCC6 group with a medium size column (see Chapter 3.2). The only important difference is that at specimens SW1 and SW2 the column part was shorter because it was simply supported on both ends in comparison to built-in ends in the BCC6 group of test specimens.

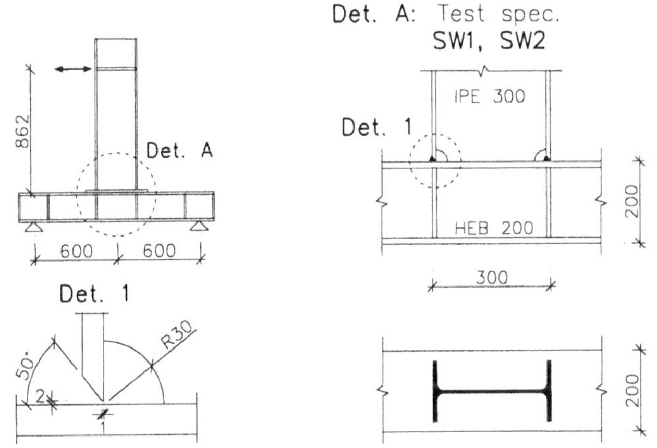

Figure 3.1.18: Test specimen geometry

Material properties of test specimens

Two tensile tests were made from the material taken from column and beam flanges and webs. The average results are shown in Table 3.1.7. The toughness of beam flange material was determined by standard Charpy V-notch tests (see Table 3.1.8). The quality of material ordered was S235JR, but the actual properties are somewhat higher and practically fulfil the requirements for S275J2 according to EN 10025.

Test set up and instrumentation

The test set-up and instrumentation are presented in Figure 3.1.19. The test frame consists of a horizontal part that supports specimens and a vertical part that supports the actuator. The force introduced into the specimens was measured with a load cell placed between the actuator and the specimen. The displacements of the beam at the level of the actuator, relative rotations between the beam and the column and shear deformations of column web panel were measured with displacement transducers. There were also some strain gauges set up on beam flanges for strain measurements.

TABLE 3.1.7

MATERIAL PROPERTIES OF BASE METAL

Specimen	f_y (MPa)	f_u (MPa)	f_y/f_u	A (%)
SW1 – beam flange	306	432	1.41	36.8
SW1 – beam web	366	467	1.28	25.2
SW1 – colum flange	305	429	1.41	32.4
SW1 – colum web	313	440	1.41	32.0
SW2 – beam flange	313	428	1.37	33.4
SW2 – beam web	317	440	1.39	33.2
SW2 – colum flange	299	431	1.44	36.8
SW2 – colum web	350	460	1.31	29.6

TABLE 3.1.8

CHARPY V-NOTCH TEST RESULTS

Specimen	Temp. (°C)	Absorbed Energy (J)		
SW1, SW2 – beam flange	+20	154	161	163
	0	135	130	135
	-20	28	38	19

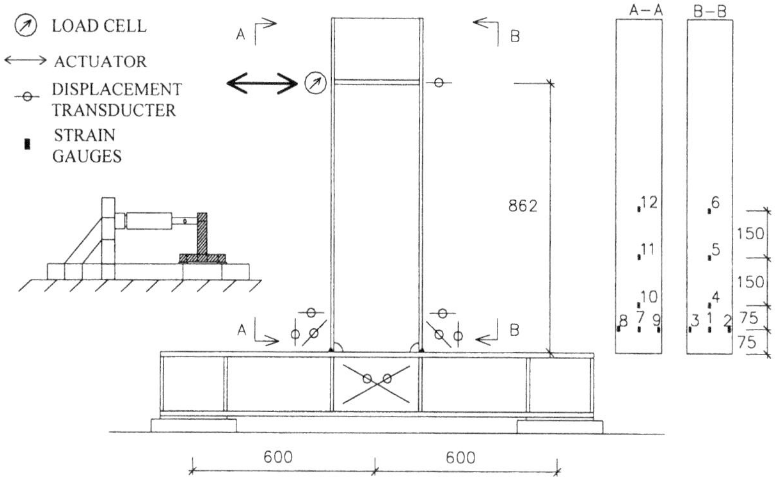

Figure 3.1.19: Test set-up and instrumentation

Displacement control following the sinusoidal pattern with constant amplitude of approximately two times yield displacement was applied during the tests. The first test was run at a frequency of 0.3 Hz and the second one at a frequency of 0.5 Hz. The dynamic displacements were introduced by a hydraulic actuator with the stroke of ± 200 mm and the capacity of ± 250 kN. For every test specimen the yield displacement was established during the first static part of the test according to ECCS testing procedure. The test input parameters are gathered in Table 3.1.9.

TABLE 3.1.9
LOADING HISTORY: DISPLACEMENT AMPLITUDE AND FREQUENCY

Test specimen	Frequency (Hz)	v_y (mm)	Displacement amplitude (mm)
SW1	0.3	9	± 18
SW2	0.5	10	± 18

Test results

From the measured data the most important results are presented in the form of the following diagrams:

- Moment-rotation diagrams for the whole connection and for the web panel. The moment is determined at the column axis (Figures 3.1.20 and 3.1.26).
- Cumulative dissipated energy in relation to the number of cycles (Figures 3.1.22 and 3.1.28).
- Energy dissipated in each cycle (E_c) normalised by the energy dissipated in the second cycle (E_2) as a function of the number of cycles (Figures 3.1.23 and 3.1.29).
- Moment amplitude (positive and negative) normalised by the moment amplitude in the second loading cycle as a function of the number of cycles (Figures 3.1.21 and 3.1.27).
- Strain rate in the first few dynamic loading cycles (Figures 3.1.24 and 3.1.30).

As a failure criterion 50% loss of strength capacity (moment amplitude) or 50% loss of energy dissipated in each cycle was selected, both with reference to the value at the second cycle.

Test SW1

The test was stopped at the 38th cycle when the strength decrease was evident. Namely, test SW1 was the first dynamic test of this kind run on the described equipment and it was stopped to prevent some possible damage of the equipment due to the sudden collapse of the test specimen. From the moment amplitude diagram (Figure 3.1.21) it can be deduced that the failure criterion is met at 39 cycles. The failure was caused by the fracture in the beam flange, which was preceded by a stable crack growth. The crack was initiated in the middle of the beam flange at the weld toe and then it spread to the flange edges (Figure 3.1.25). The first signs of local buckling in the beam flanges were observed. A similar crack appeared with some delay on the other flange. The other crack appeared also in the web starting from the scallop corner after the beam flanges lost some strength.

The hysteretic behaviour is stable with large energy dissipation (Figures 3.1.20, 3.1.22 and 3.1.23). Strain rates in the first few dynamic cycles measured in flanges at a distance of 7.5 cm from the column face are on average of the magnitude around 0.01 s (see Figure 3.1.24).

Test SW2

The results are very similar to those of test specimen SW1. Test specimen SW2 was able to withstand 35 dynamic loading cycles up to the loss of 50% of its strength capacity (Figure 3.1.27). Stable crack propagation was observed before the final flange rupture (Figure 3.1.31), and the hysteretic behaviour is also similar (Fig. 3.1.26, 3.1.28 and 3.1.29). The strain rate of 0.01/s was measured in the flange 7.5 cm from the column face. Higher strain rates with the magnitude between 0.05/s and 0.1/s (close to earthquake conditions) are expected nearer to the column face.

The test results are gathered in Table 3.1.10. During the test the temperature on the surface of the beam-to-column flange weld was measured. The increase of the temperature due to plastic deformations was 16.5°C and 7.5°C. The increase is not very large, because the largest plastic deformations occurred in the beam flanges somewhat aside from the weld.

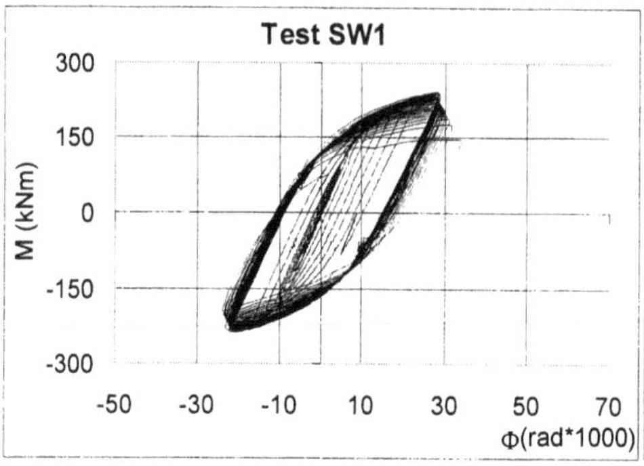

Figure 3.1.20: Moment-rotation diagram

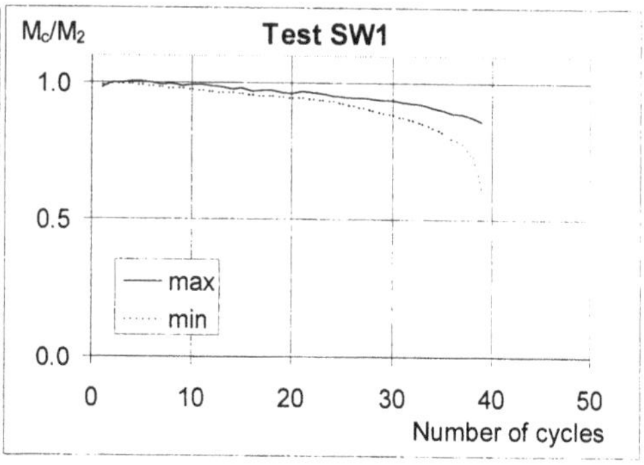

Figure 3.1.21: Moment amplitude

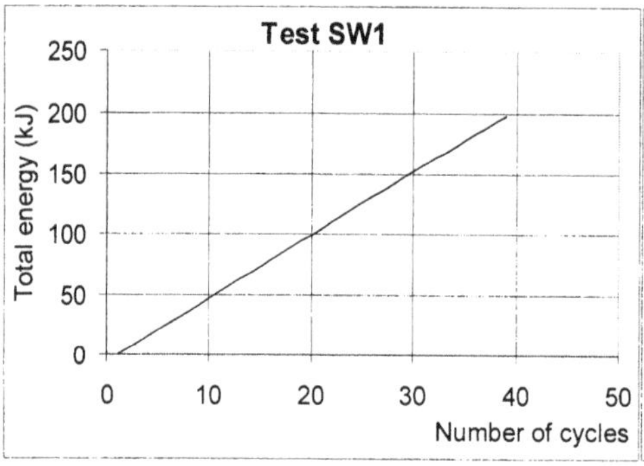

Figure 3.1.22: Cumulative dissipated energy

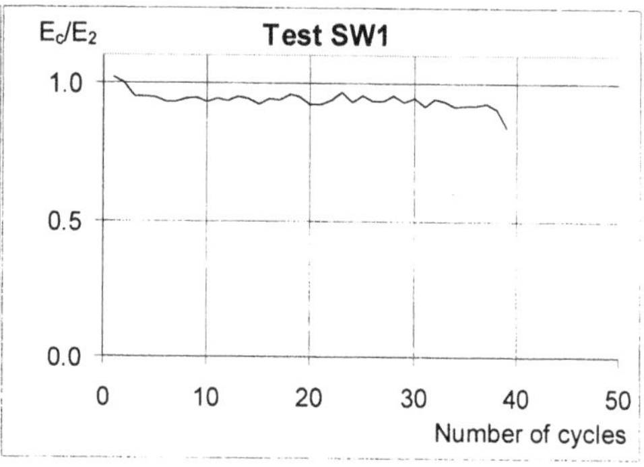

Figure 3.1.23: Energy dissipated in each cycle

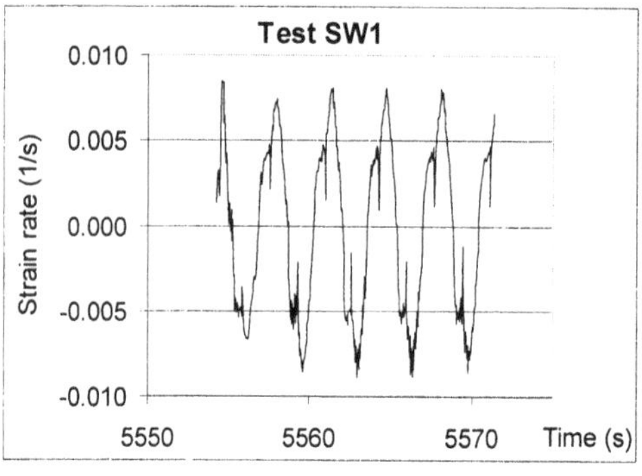

Figure 3.1.24: Strain rate at strain gauge No. 8

Figure 3.1.25: Photo of SW1 test specimen after failure (beam flange)

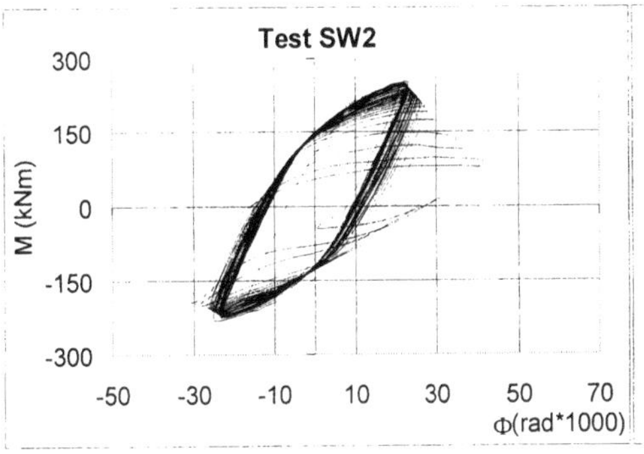

Figure 3.1.26: Moment-rotation diagram

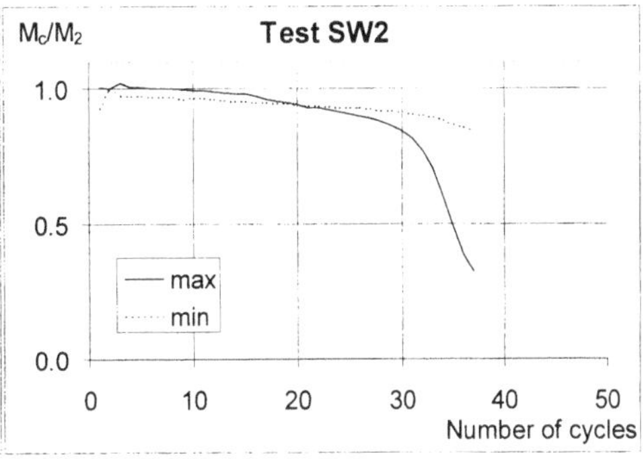

Figure 3.1.27: Moment amplitude

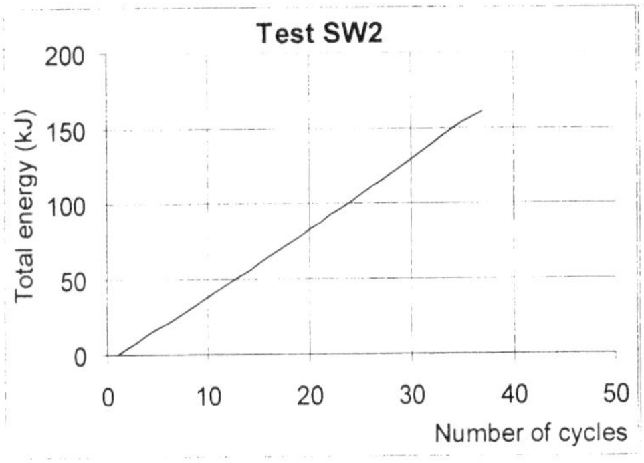

Figure 3.1.28: Cumulative dissipated energy

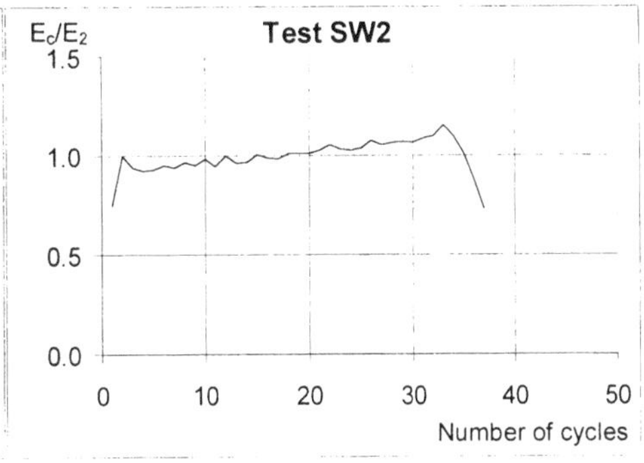

Figure 3.1.29: Energy dissipated in each cycle

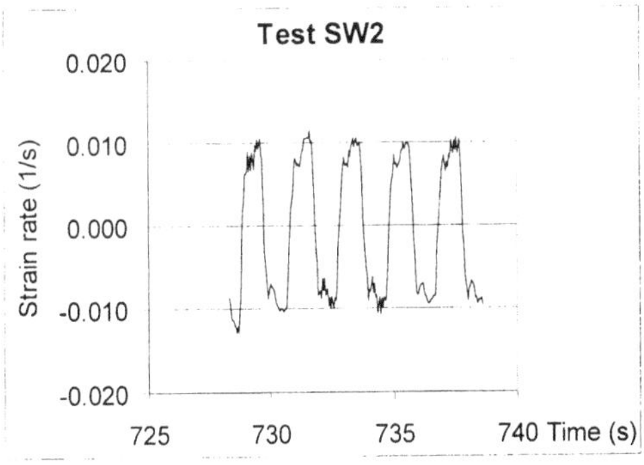

Figure 3.1.30: Strain rate at strain gauge No. 1

Figure 3.1.31: Photo of SW2 test specimen after failure (beam flange)

The main parameters needed for low cycle fatigue assessment are also presented in Table 3.1.10.

TABLE 3.1.10

RESULTS OF DYNAMIC TESTS ON WELDED CONNECTIONS

Specimen	Number of cycles	f_y (Mpa)	v_y (mm)	Δv (mm)	Δv_{pl} (mm)	$\Delta T(^oC)$	Failure mode	Failure position
SW1	39*	306	9	38	20	16.5	Stable crack prop.	Beam – weld toe
SW2	35*	313	10	35	15	7.5	Stable crack prop.	Beam – weld toe

* Strength criterion

On the basis of the cyclic tests on test specimen group BCC6 with the same geometry as test specimens SW 1 and SW2 (see Chapter 3.2) Calado (Calado et al. 1998a) defined low cycle fatigue S-N curves taking into account different definitions of the strain range S (A-Krawinkler and Zohrei; B-Ballio and Castiglioni; C-Feldman, Sedlacek, Weynard and Kuck).

The test results for SW1 and SW2 tests as well as for the symmetric bolted connection SB1 (see next section) with similar geometry and similar behaviour at dynamic tests are plotted in Figure 3.1.32 in relation to Calado S-N curves. The results lie below the S-N curves (white marks), which was expected due to negative influence of the higher strain rates applied at the tests. This influence can be assessed from the tests realised in Liege on bolted connections (see next section).

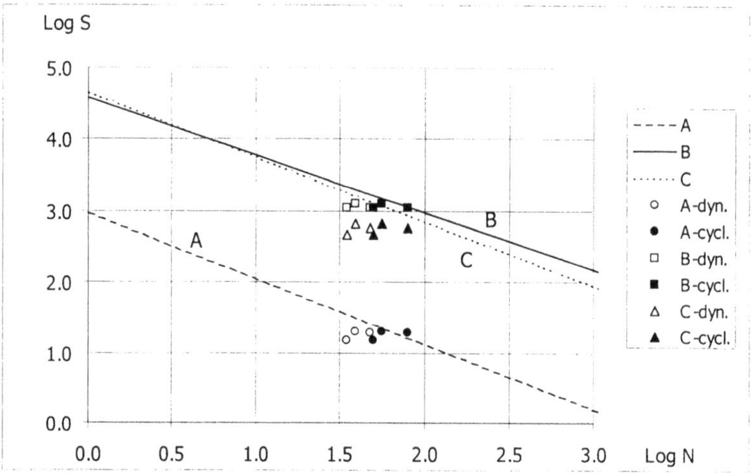

Figure 3.1.32: Low cycle fatigue assessment of the test results

At the values 1.4 (SW1, SW2) and 1.5 (SB1) of the base metal ratio f_u/f_y the increase in the number of cycles at the failure due to slow cycling is estimated to 44% and 66%, respectively (see Figure 3.1.51). With this modification the results for Krawinkler, Zohrei and particularly for Ballio-Castiglioni strain range definition agree very well (black marks).

TESTS ON FULL-SCALE BOLTED END-PLATE CONNECTIONS

Two series of tests are presented in this section, the first one in order to study the influence of the loading speed at symmetric bolted connections (tests performed at the University of Liege), and the second one to compare the behaviour of unsymmetric and symmetric bolted end-plate connectionsunder dynamic cyclic condition (tests performed at the University of Ljubljana).

Influence of loading speed at symmetric bolted connections

Typical European rigid full strength bolted connections, involving welding at the shop and bolting on site, were tested. See Figure 3.1.35 for detailed description of specimens with low steel grade (group L) and Figure 3.1.36 for the specimens with high steel grade (group H). The end plate of the beam is welded with K preparation.

Some common descriptions concerning the tests of this section and the following two sections

Tested specimens

All the basic properties of the beams and connections of the tested specimens are summarised in Table 3.1.11. Detailed description can be found in the sections corresponding to each specimen group.

TABLE 3.1.11

BRIEF DESCRIPTION OF THE TESTED SPECIMENS

Beam section	f_y beam (N/mm^2)	Connection type	Slow rate strain specimen	Fast rate strain specimen	For details see Figure
IPE 450	316	Beam with welded end plate bolted to the column. Type A1	LS1	LF1	3.1.55
IPE 450	273	Beam with welded end plate bolted to the column. Type A1	LS2	LF2	3.1.55
IPE 450	273	Beam with welded end plate bolted to the column. Type A1	LS3	LF3	3.1.55
IPE 450A	405	Beam with welded end plate bolted to the column. Type A1	HS1	HF1	3.1.56
IPE 450A	405	Beam with welded end plate bolted to the column. Type A1	-	HF2	3.1.56
IPE 450A	405	Beam welded to the column with dog-bone. Type C3	-	DOGF	3.1.75
2UPN 300	290	Semi-rigid connection by 2x4 transverses bolts in web of U. Type B	US	UF	3.1.83

Test set up

The imposed displacement is a sinusoidal function, the first with a period of 40 seconds, and the second, in order to simulate real earthquake conditions, with a period of 2.5 seconds.

The imposed displacement, v=±40mm, is two times the yield displacement v_y; this is because the actuator used has a displacement range of 100 mm, which corresponds to about 5 v_y for the tested beams and geometry of the test set up.

Doubler plates in the web of the column are adopted to avoid panel zone mechanism because the test program intends to concentrate on plastic beam mechanisms. Thus, no plastic shear mechanism in the panel zone of the column is studied in this project.

IPE 450 beams are likely to be used for spans of about 8 m. Given the shape of the bending moment diagram under the combination of static and seismic loading the point of contra-flexure is about 2 m away from the connection. This defines the axis for the load application.

The test program intends to concentrate on beam plastic phenomena. From this it is derived that the column size and fixing in the test set up can be such that they minimise the column deformation and facilitate the testing work. The overview of the test set up is given in Figure 3.1.33.

The measurements during the cyclic tests was the following (Figure 3.1.34):

- Load P applied by the actuator is measured by means of a load cell.
- Load P is measured indirectly by the strain measurement at the two flanges of the steel profile at the middle of the beam profile length.
- Displacement v under load is measured by a LVDT. It measures the relative displacement between the point of the beam where the load is applied and an external reference.
- The displacement used to pilot the test is the relative displacement of the actuator.
- Additional displacement transducers are installed in order to take into account settlements in supports.

Figure 3.1.33

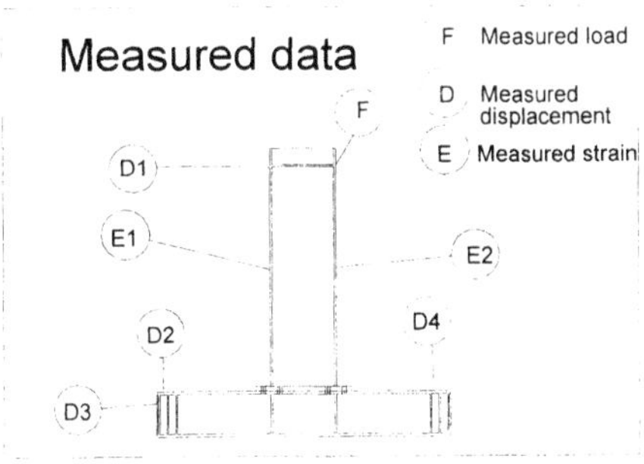

Figure 3.1.34

Tests on steel base material

Three different steel grades used in the beam profiles used in the rigid connection were put to a standard tensile test. The results are shown in Table 3.1.12.

Charpy V-notch tests were performed on beam material for each of the three following specimens (LS1, LS2, HS1) that represent three different steel grades used. The cuts were made in the lamination direction, with a net section of 0.8 cm^2. The results of the test are shown in Table 3.1.13.

TABLE 3.1.12
RESULTS OF TENSILE TEST

Specimens	f_y	f_u	f_u/f_y	A (%)
LS1, LF1	316	429	1.36	35
LS2, LF2, LS3, LF3	273	406	1.49	38
HS1, HF1, HF2, DOGF	405	500	1.23	31

TABLE 3.1.13
RESULTS OF CHARPY TEST FOR SPECIMENS LS1, LS2, HS1

N° specimen	Temperature (°C)	Absorbed Energy (J)			Resilience (J/cm^2)			Weldability Class
LS1-1	+20	154	151	176	192.5	188.8	220	C
LS1	0	25	95	35	31.25	118.8	43.75	
LS1	-20	10	9	13	12.5	11.25	16.25	
LS2-2	+20	163	143	145	203.8	178.8	181.3	D
LS2	0	155	125	140	193.8	156.3	175	
LS2	-20	154	100	140	192.5	125	175	
HS1-1	+20	160	169	155	200	211.3	193.8	D
HS1	0	160	155	145	200	193.8	181.3	
HS1	-20	135	165	156	168.8	206.3	195	

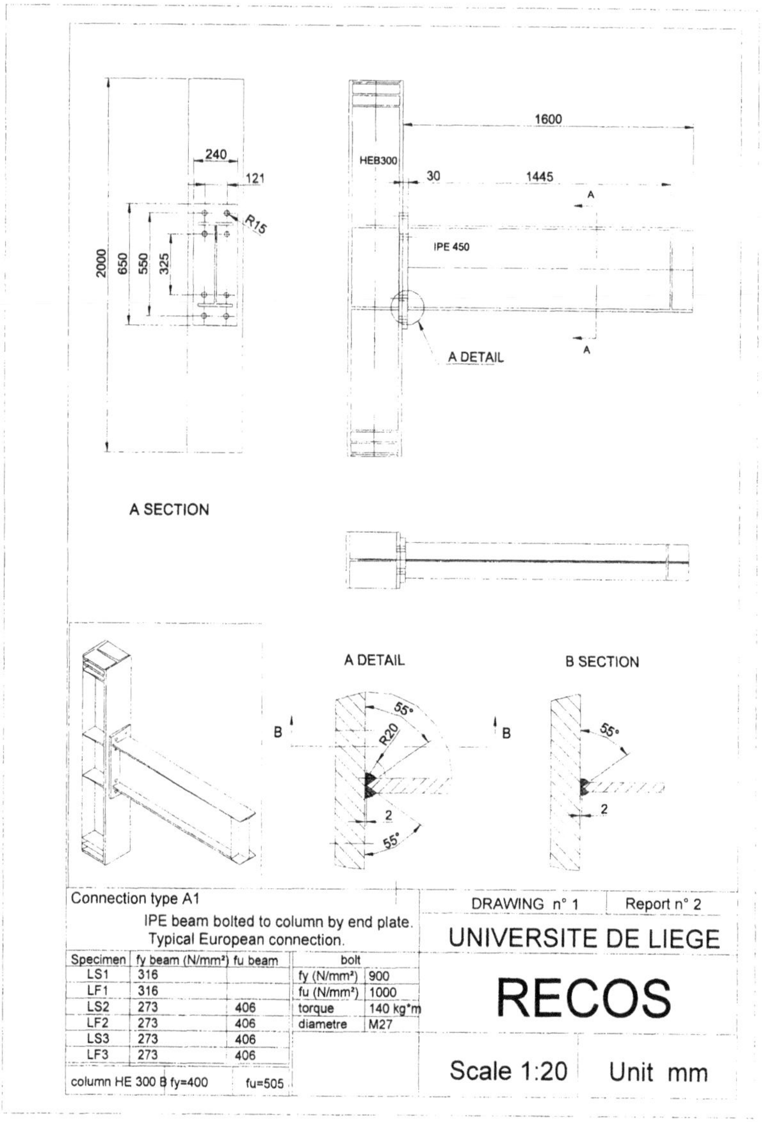

Figure 3.1.35: Specimen group L

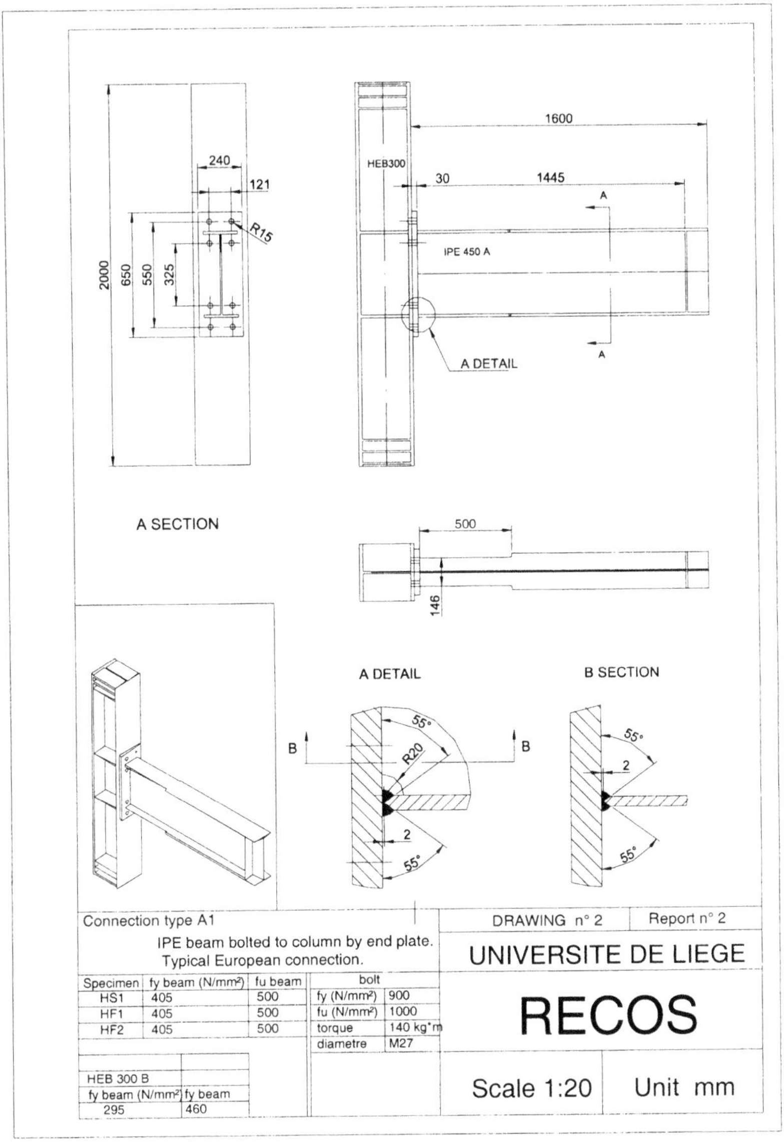

Figure 3.1.36: Specimen group H

Results of tests on specimens made with low strength steel (L group)

The results of this group are summarised in Table 3.1.14. The failure mode for the six specimens of this group is the same, with fracture in the flange close to the weld and with a stable propagation of the crack. Some local buckling of the flange beam can be observed (see Figure 3.1.37). No differences can be observed in the failure mode for slow or fast strain rate. The number of cycles withstood in average up to collapse is by 34 % lower when higher load speed is applied and the total energy dissipated up to collapse is on average lower by 37 %.

Figure 3.1.37: Specimen LS3. Failure mode

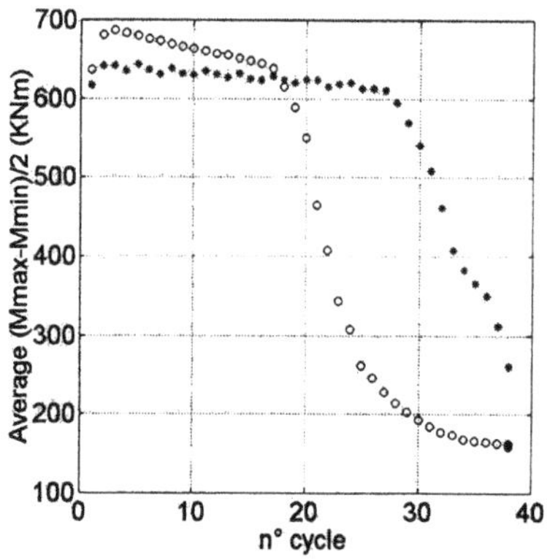

Figure 3.1.38: Group L. Average for 3 specimens tested at low (*) and high (o) loading speed

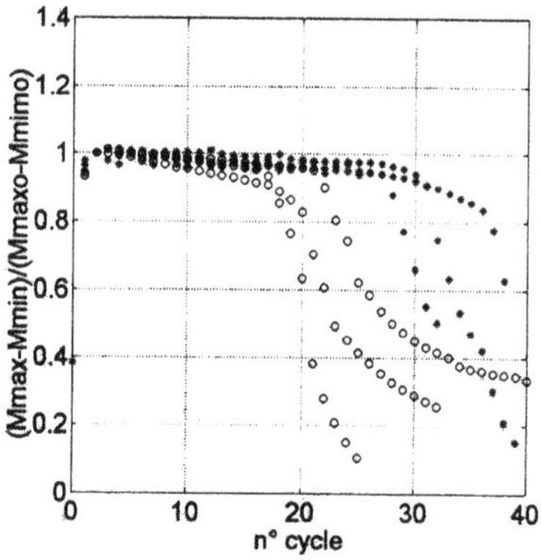

Figure 3.1.39: Group L. Results for the six specimens

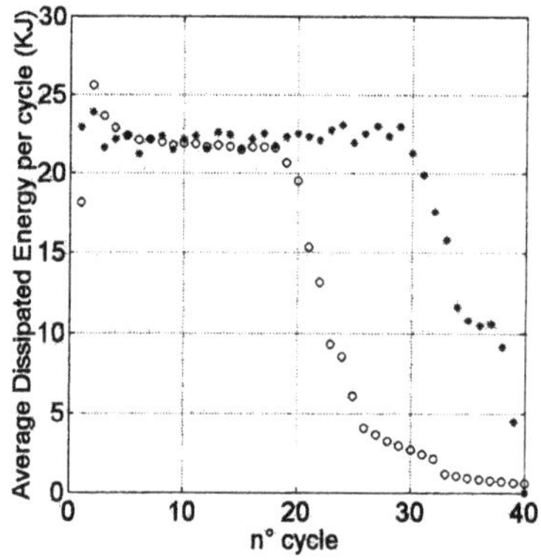

Figure 3.1.40: Group L. Average for 3 specimens tested at low (*) and high (o) loading speed

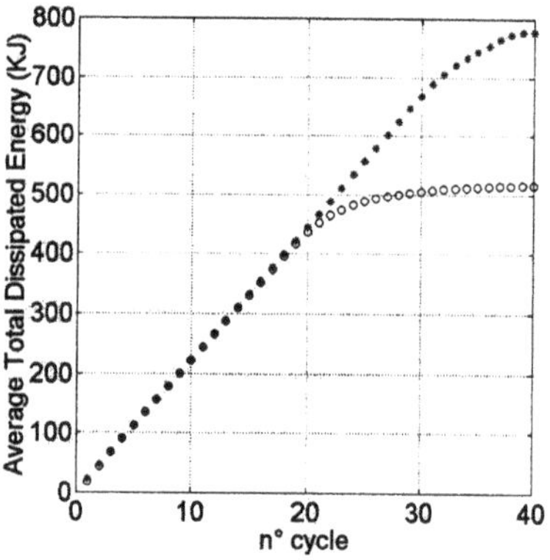

Figure 3.1.41: Group L. Average for 3 specimens tested at low (*) and high (o) loading speed

TABLE 3.1.14

RESULTS FOR THE LOW STEEL GRADE EUROPEAN RIGID CONNECTION

Specimen	N° cycles up to collapse	Mean	Cyclic dissipated energy mean (kJ)	Mean	Total dissipated energy up to collapse (kJ)	Mean
LS1	33	33.3	18.8	22.1	620.0	743.3
LS2	38		25.5		970.0	
LS3	29		22.1		640.0	
LF1	24	21.7	18.8	21.7	450.0	466.7
LF2	21		21.4		450.0	
LF3	20		25.0		500.0	

Results of tests on specimens made with high strength steel (H group)

The test results for the specimens HS1, HF1 and HF2 are shown in Table 3.1.15. In the HS1 specimen, the flange crack propagation was unstable, producing much noise and the failure was completely brittle; local buckling can not be observed. HF1 and HF2 failure mode is the propagation of the flange cracks close to the weld, but the crack propagation was more ductile than for the specimen HS1.

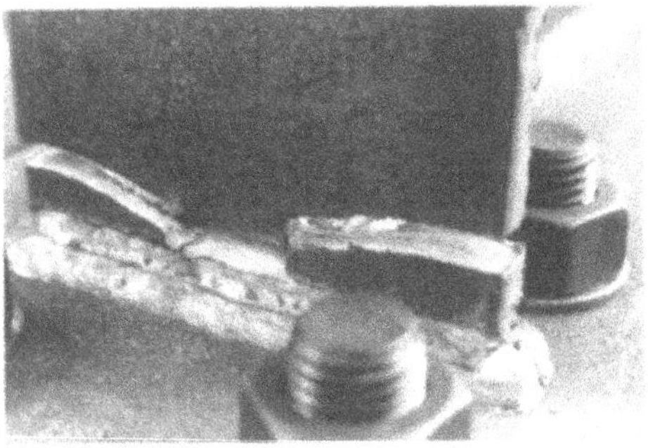

Figure 3.1.42: Specimen HS1

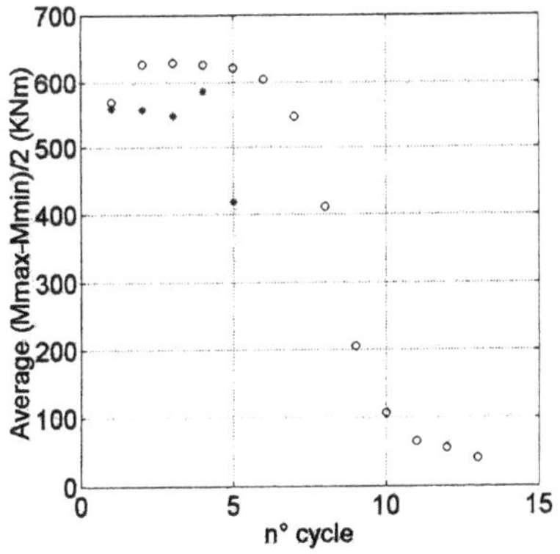

Figure 3.1.43: H Group. Average for specimens tested at high (o) and low (*) loading speed

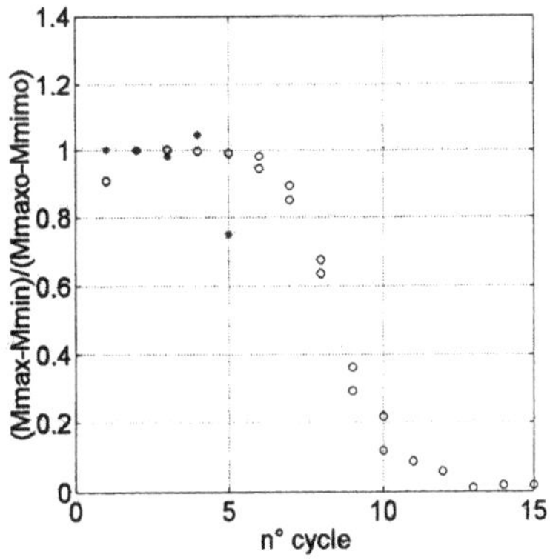

Figure 3.1.44: Group H. Results for the three specimens tested

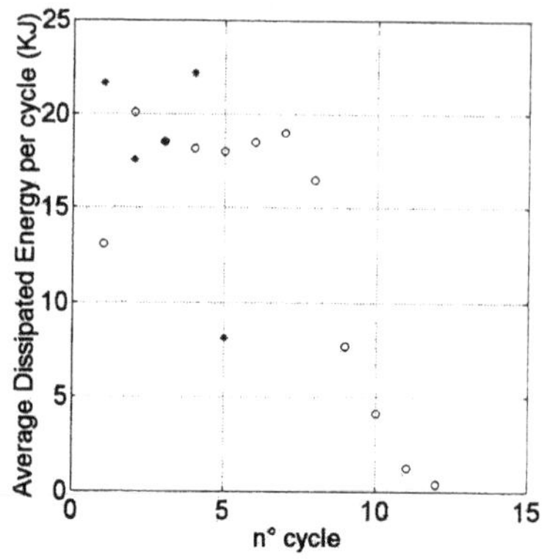

Figure 3.1.45: Group H. Average for specimens tested at high (o) and low (*) loading speed

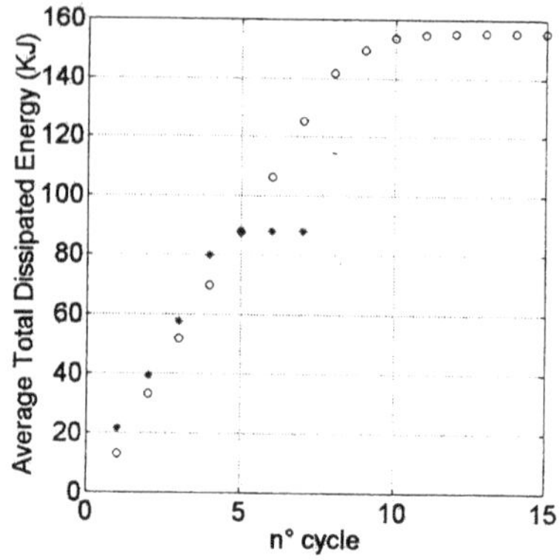

Figure 3.1.46: Group H. Average for specimens tested at high (o) and low (*) loading speed

The average number of cycles withstood up to collapse in group H (6.6 cycles) is 76% lower than in the group L (27.5). This worse behaviour was to a certain degree expected, because of the lower f_u/f_y value of the group H base steel material and because of the higher steel grade used.

The average number of cycles withstood up to collapse in H group (8 cycles) is 100% higher when the higher load speed is applied than when the slow load speed is applied (4 cycles), and the total energy dissipated up to collapse is 70 % higher. The number of specimens (one specimen tested at low speed and two tested at high speed) is too small to obtain definitive conclusions. It is necessary to perform more tests for this steel grade, at least 3 specimens for each loading speed.

TABLE 3.1.15

RESULTS FOR THE HIGH STEEL GRADE EUROPEAN RIGID CONNECTION

Specimen	N° cycles up to collapse	Mean	Cyclic dissipated Energy mean (kJ)	Mean	Total dissipated Energy up to collapse (kJ)	Mean
HS1	4	4	21	20.8	83	83
HF1	8	8.0	16.9	17.6	135.0	141.0
HF2	8		18.4		147.0	

Discussion of test results

The strain rate has increased the yield moment M_y and the ultimate moment in all the specimens tested (see Figures 3.1.47, 3.1.48 and 3.1.49). This means that an increase in loading speed increases strength demand in the welded zone.

The results show that the value of M_y of group H has been more sensitive to the strain rate than for group L. This result was not expected, because the higher the steel grade, the less sensitive M_y should be to the strain rate. Further tests are necessary for group H.

The ratio of ultimate moment to yield moment must be higher for the specimens tested at high loading speed, because yield moment is more sensitive to the strain rate than ultimate moment. Figures 3.1.48 and 3.1.49 show that, for all the specimens tested at high loading speed, this ratio is higher, as expected. This means that when the loading speed is increased, the plastic hinge length and the beam ductility are reduced.

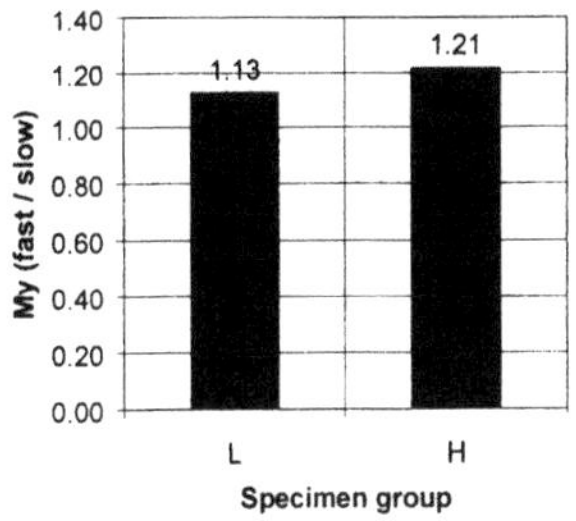

Figure 3.1.47

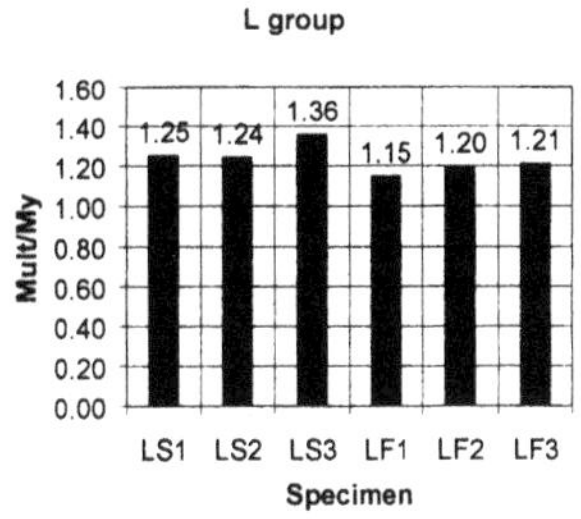

Figure 3.1.48

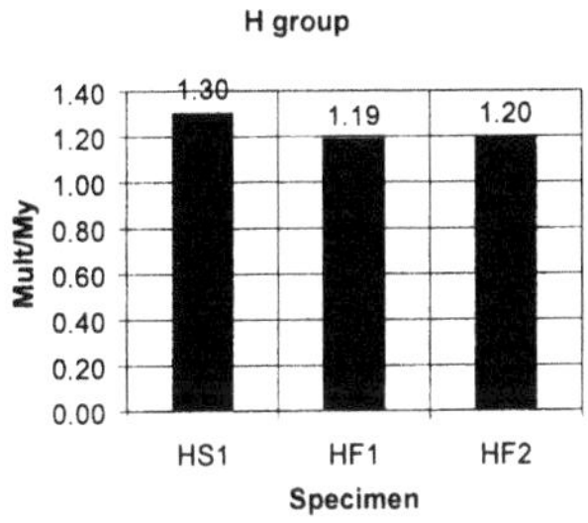

Figure 3.1.49

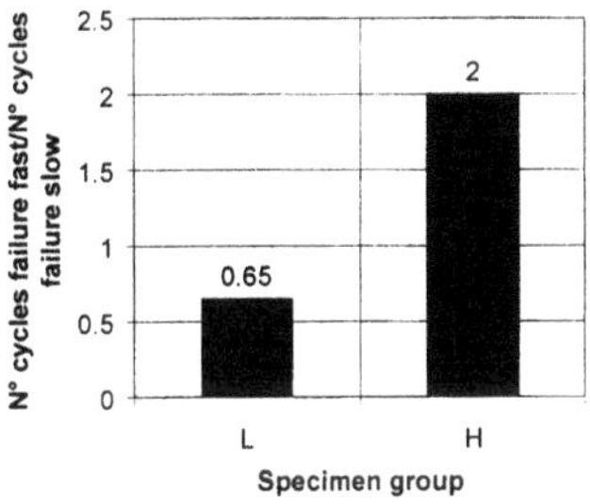

Figure 3.1.50

For group L, there is an average reduction in the number of cycles by 35 % when the load period is decreased from 40 s to 2.5 s (see Figure 3.1.50). For group H, there is an increment of 100 % (it must be remembered that in this group the only specimen tested at low speed has a very brittle failure mode, and this result needs confirmation).

Figure 3.1.51 shows that, as it was expected, the steel with f_u/f_y=1.49 is more sensitive to loading speed than the steel with f_u/f_y=1.36. The steel f_u/f_y=1.23 is the steel of group H.

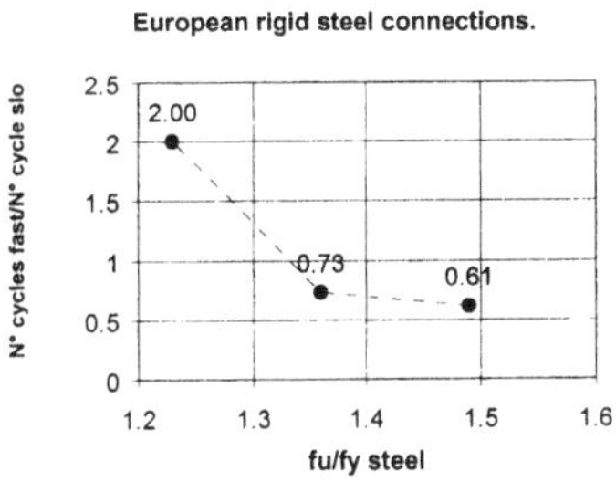

Figure 3.1.51

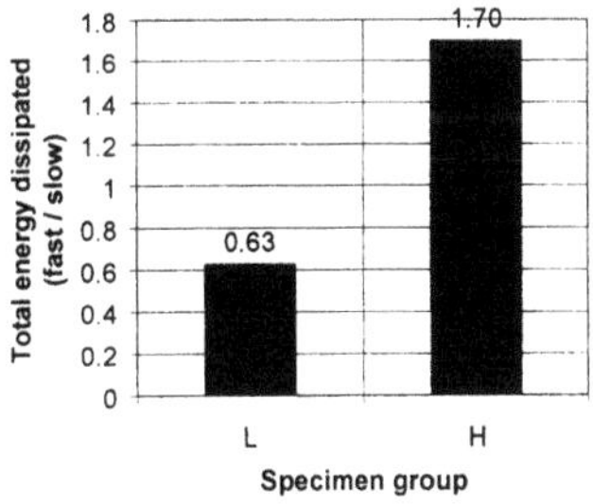

Figure 3.1.52

Figure 3.1.52 represent the ratio of the total energy dissipated for each group of specimen up to collapse. For the two groups the values of this last ratio are a little lower than the values in Figure 3.1.50; this indicates that on average higher strain rates bring a decrease in the energy dissipated per cycle.

Figure 3.1.53 shows for group H and L the same ratio of total energy as in the previous Figure, but as a function of f_u/f_y of the steel base steel material.

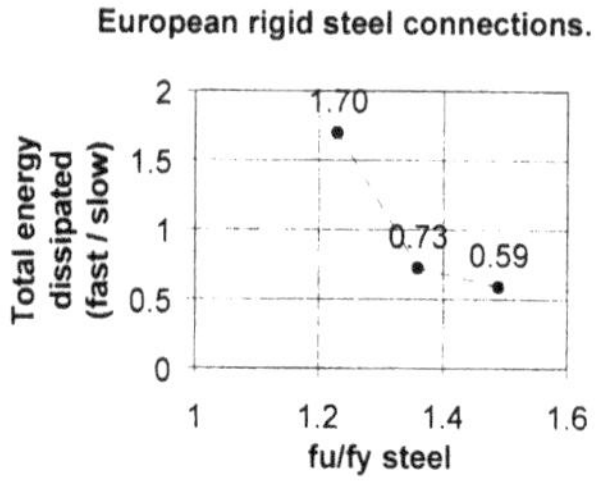

Figure 3.1.53

The results of the test were also processed in the term of S-N fatigue curves. The EC3 curve for high cycle fatigue for detail category 112 that corresponds to transverse butt welds is far from the test results. The results are relatively close to the detail category curve 50 for low steel grade.

The test results have very good agreement with the S-N curves proposed in the reference (Calado et al. 1998b) for beam-to-column welded connection, with butt weld in K grooves connecting the flange of the beam to the column flange. This proposed curve depends on the welding details and the imposed displacement (a threshold displacement is defined that governs the failure mode). Three types of failure mode are defined for welds with K preparation (B1= brittle, M = mixed and D = ductile).

B1 is a failure mode in which a crack forms in the centre of the weld between the beam flange and the end plate and propagates towards the edges. It corresponds to the failure mode of all tested specimens of group H. Some specimens, where initiation of local bucking can be observed with a ductile crack propagation (group L), could be between failure modes B1 and M.

Using 0.9 for the parameter related to the weld quality, the threshold displacement limit between brittle and mixed failure is 112 mm for group L and 124 mm for group H. As the applied displacement is 80 mm, the S-N curve to be used is the one corresponding to B1 failure. If larger displacements were applied, local buckling of the beam flange would protect the welding zone from higher strength demands, see reference (Bernuzzi et al. 1999).

From Figure 3.1.54, it can be seen that this approach is conservative for all specimens tested (including the one that collapsed in only four cycles with very brittle and fast crack propagation). From the specimen group L, it can be seen that the influence of loading speed on the connection behaviour in terms of low cycle fatigue is not negligible; it is as important as the influence of the failure mode.

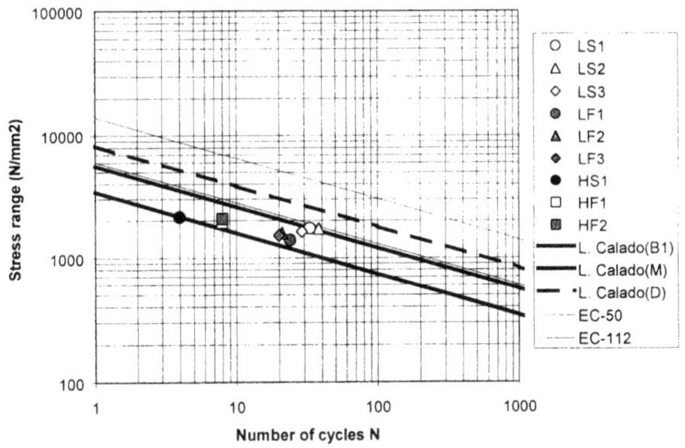

Figure 3.1.54: Fatigue strength curve

Symmetric and unsymmetric full-scale bolted end-plate connections under dynamic loading

Objectives

The unsymmetric bolted end-plate connections with a row of bolts at each side of the upper flange and one row of bolts above the bottom flange (see Figure 3.1.55) are primarily suitable to resist gravity bending moments in non-sway frames. They can be used also in low-rise sway frames in non-seismic regions or for moderate wind loading, because in both mentioned cases moments from gravity loads prevail.

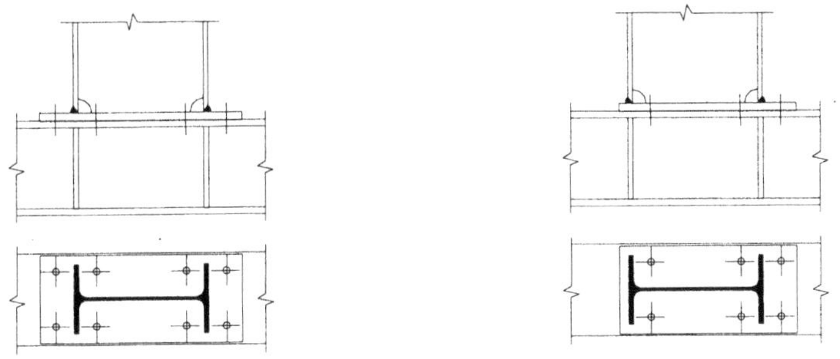

Symmetric extended end-plate bolted connection

Unsiymmetric end-plate bolted connection

Figure 3.1.55: Symmetric and unsymmetric bolted end plate connection

This type of connections was in the past sometimes used also in some seismic regions. For instance, it was adopted in Slovenia mainly from the German practice, usually overlooking the fact that Slovenia is a seismic region and Germany is not. Old Slovenian code for seismic resistance of structures, which is still in application in parallel with Eurocode 8, has for steel structures implicitly incorporated the behaviour factor 5 or even more. Such behaviour factors require fully ductile design. However, before the Northridge and Kobe earthquakes steel structures were recognised as ductile without much argument. The consequence of such an approach is the fact that in Slovenia a certain number of sway frames was designed using unsymmetric end-plate connections, the resistance and ductility characteristics of which can be questionable.

In order to get an insight into the behaviour of unsymmetric bolted end-plate connections, a small testing program was elaborated. Two typical full-scale tee beam-column assemblies containing unsymmetric bolted end-plate connection were tested under dynamic loading, simulating the earthquake conditions. For a comparison, the third dynamic test was executed on the usual symmetric extended end-plate connection.

Description of test specimens

The geometry of test specimens is shown in Figure 3.1.56. Tee-shaped beam-column assembly in all three cases consists of HEB 200 column section and IPE 300 beam section. Test specimens UB1 and UB2 with unsymmetric connection are nominally equal and preloaded with the bending moment equal to 30% of beam plastic moment to simulate the effect of gravity loading.

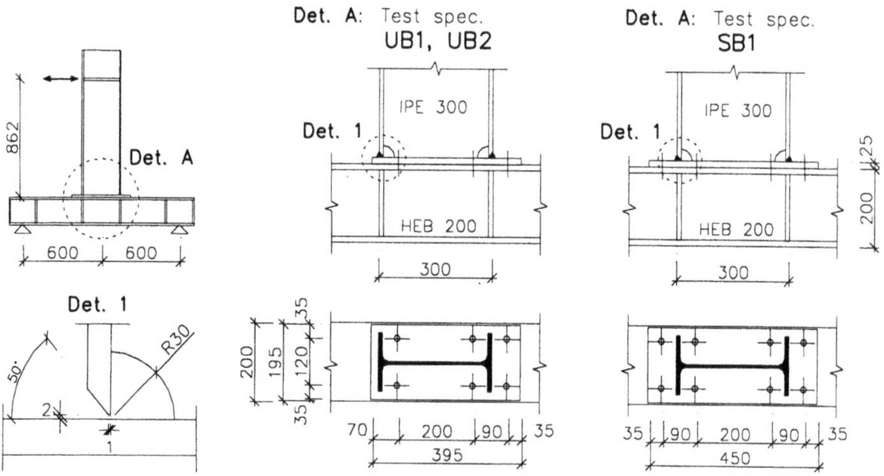

Figure 3.1.56: Test specimen geometry

The beams were welded to the end-plates by full penetration single bevel butt welds and the final root run was made from the other side. MAG welding procedure was used. M20 bolts in the connections were of grade 10.9 and half of the full preloading was applied. The connections were designed as full-strength and rigid except the column-web panel which did not confirm these requirements.

Material properties of test specimens

For each of the beam flange and web, column flange and web and end plate two tensile tests were made. The average results are presented in Table 3.1.16.

Charpy V-notch tests were made for the material taken from the beam flanges (Table 3.1.17), where the most severe strain and stress conditions are to be expected.

TABLE 3.1.16
MATERIAL PROPERTIES OF BASE METAL

Specimen	f_y (MPa)	f_u (MPa)	f_y/f_u	A (%)
UB1, UB2 – beam flange	314	429	1.37	33.8
UB1, UB2 – beam web	326	439	1.35	34.1
UB1, UB2 – colum flange	303	440	1.45	34.0
UB1, UB2 – colum web	350	463	1.32	28.6
UB1, UB2 – end-plate	360	604	1.68	25.8
SB1 – beam flange	269	403	1.50	37.1
SB1 – beam web	339	434	1.28	36.4
SB1 – colum flange	276	430	1.56	35.8
SB1 – colum web	309	438	1.42	33.4
SB1 – end-plate	297	410	1.38	36.5

TABLE 3.1.17
CHARPY V-NOTCH TEST RESULTS

Specimen	Temp. (°C)	Absorbed energy (J)		
UB1, UB2 – beam flange	+20	154	161	163
	0	135	130	135
	-20	28	38	19
SB1 – beam flange	+20	161	150	147
	0	125	149	152
	-20	128	125	137

Test set-up and instrumentation

Test set-up and instrumentation are the same as for the tests on welded connections, described in the previous section (see Figure 3.1.19). The tests were run under displacement control following the sinusoidal pattern with constant amplitude of approximately two times the yield displacement. The constant frequency of 0.5 Hz was applied. In each test the yield displacement was established in the first static part of the test according to the ECCS testing procedure. The details are shown in Table 3.1.18.

TABLE 3.1.18
LOADING HISTORY: DISPLACEMENT AMPLITUDE AND FREQUENCY

Test specimen	Frequency (Hz)	v_y (mm)	Displacement amplitude (mm)	Preloading
UB1	0.5	9	± 18	~ 0.3 M_P
UB2	0.5	10	± 16	~ 0.3 M_P
SB1	0.5	9	± 20	–

To simulate the effect of the gravity loading the static displacement causing the bending moment equal to 30% of the beam plastic moment was introduced for test specimens UB1 and UB2 prior to dynamic loading. The preloading certainly caused tension at the stronger part and favourable compression at the weaker part of the connection, as in real frame situation.

During the test the applied force was measured with a load cell. The arrangement of displacement transducers was such that it enabled measurements of relative displacements of a beam free end to the column, relative beam-column rotations and shear deformation of the column web panel. Strains in the beam flange were measured by strain gauges (see Figure 3.1.19).

Test results

For each test specimen the results are presented in the form of characteristic diagrams as in the previous section: moment-rotation diagram and diagrams with positive and negative moment amplitude, total energy dissipated and energy dissipated in each cycle plotted in relation to the testing cycles (Figures 3.1.57 to 3.1.72).

Test UB1

In the first dynamic cycle the bolts at the weaker side were loaded over the yield strength and elongated plastically. As the result the end plate separated from the column flange face in every cycle. Although the bolts at the stronger side of the connection were designed according to EUROCODE 8, allowing for the overstrength of 20%, the actual material properties of the hot rolled profiles were much higher and caused a slight exceeding in yield strength of these bolts. As a result the slip in the connection occurred in every cycle of about 1 mm in each direction. No additional damage or premature failure due to this slip and no overloading of the end plate were observed.

The specimen was able to withstand 195 cycles up to the loss of one half of its strength capacity. The final failure was caused by the fracture in the beam flange with a stable crack. The crack started in the middle of the flange at the weld toe and propagated to the edges until complete fracture of the flange (Figure 3.1.61).

The failure criterion of decreasing hysteretic energy of an individual cycle was reached at 93 cycles, which is approximately one half of the strength criteria (Figure 3.1.60). The hysteretic energy per cycle decreased rapidly in the first 30 or 40 cycles, probably due to the plastification of the bolt. After that the decreasing to the final failure was more gradual with the values somewhat below 50% of the initial value.

Hysteretic behaviour at constant amplitude cycling can be seen from the moment-rotation diagram (Figure 3.1.57). The loops are unsymmetric and strong pinching effect mainly due to bolt plastification is present. The unloading branches exhibit large stiffness similar to the initial connection stiffness.

Test UB2

The test specimen UB2 exhibited somewhat larger yield displacement of 10 mm than specimen UB1, but displacement amplitude $2v_y$ = 20 mm was too large for the two bolts at the weaker part of the connection and they failed in the first dynamic loading cycle. The bolts were replaced by new ones and the test was restarted.

The test results are very similar to those of UB1 specimen. The only real difference is that no drop of hysteretic energy per cycle was observed in the first part of the test and the decisive failure criterion was loss of strength capacity at 118 cycles (Figure 3.1.64).

Test SB1

In general the behaviour of the SB1 test specimen was very similar to that of welded specimens SW1 and SW2 described in Chapter 3.1.2. This is not surprising, because the end plate and bolts were strong enough to cause the failure to be induced only in the beam flange by low cycle fatigue. Stable crack propagation in the base metal above the weld preceded the thorough fracture of the flange (Figure 3.1.62), accompanied by local buckling in the region of the fracture. The 0.5 reduction failure criterion was reached due to the loss of strength capacity at 48 cycles (Figure 3.1.70), which is comparable to tests SW1 and SW2 on welded connections (see Chapter 3.1.2).

The test results are summarised in Table 3.1.19. At unsymmetric connections due to bolt plastification plastic strains in flanges and dissipated energy were smaller and the number of cycles at the failure was greater.

The highest strain rates derived from strain measurements on beam flanges at the location 7.5 cm from the endplate are between $0.01s^{-1}$ and $0.015s^{-1}$ for unsymmetric connections (see Figure 3.1.67) and around $0.02s^{-1}$ for symmetric connection.

Nearer to the endplate and weld where the moment reaches the maximum, the strain rate is greater, estimated to the values between $0.05s^{-1}$ and $0.10s^{-1}$, which is close to the strain rates expected in earthquakes.

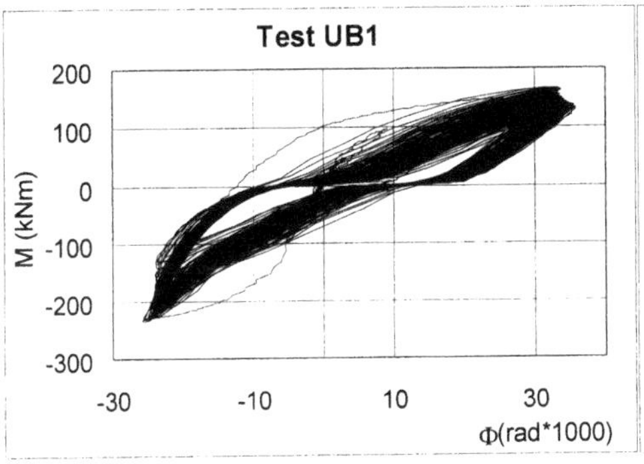

Figure 3.1.57: Moment-rotation diagram

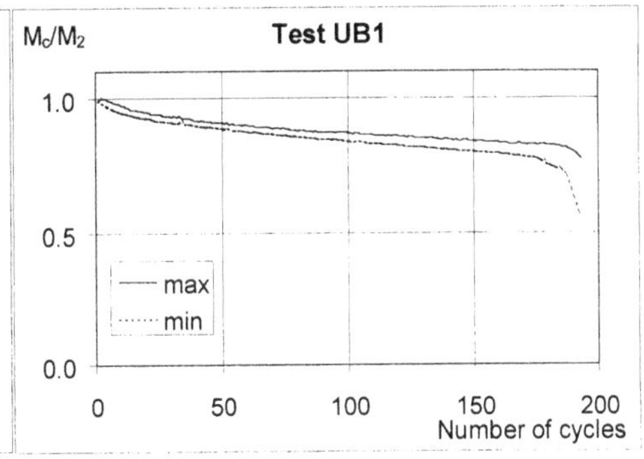

Figure 3.1.58: Moment amplitude

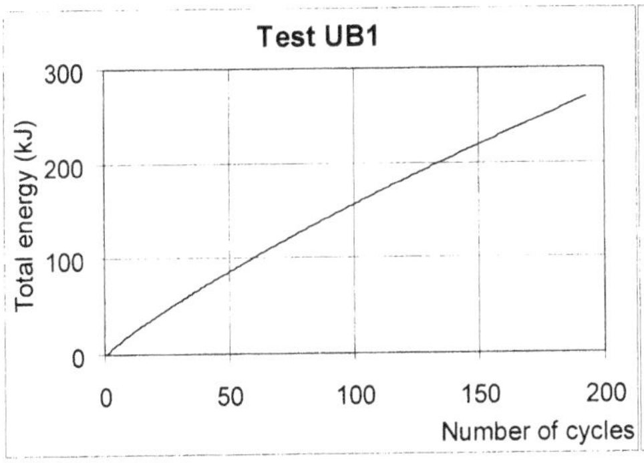

Figure 3.1.59: Cumulative dissipated energy

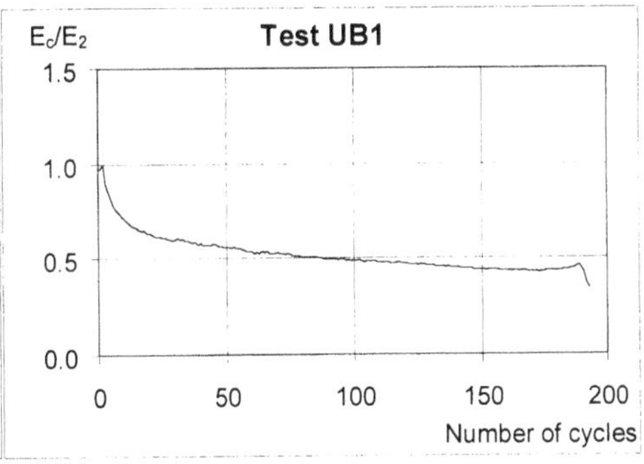

Figure 3.1.60: Energy dissipated in each cycle

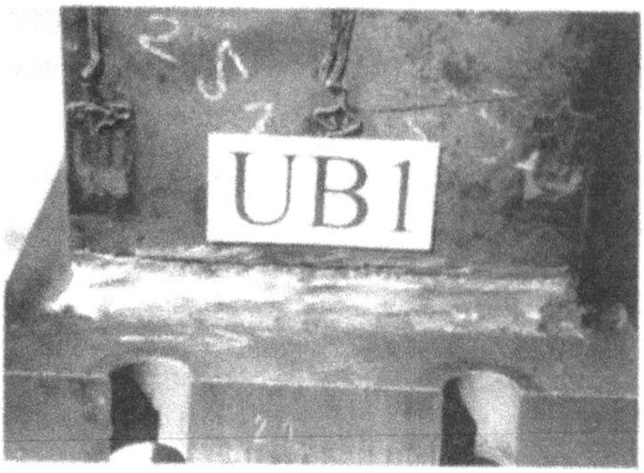

Figure 3.1.61: Photo of UB1 test specimen after failure (beam flange)

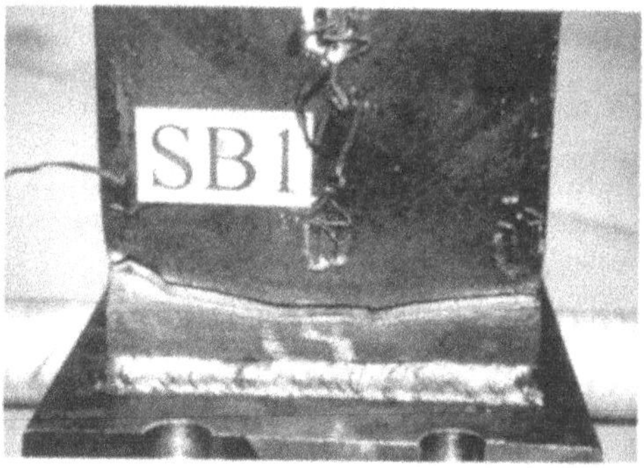

Figure 3.1.62: Photo of SB1 test specimen after failure (beam flange)

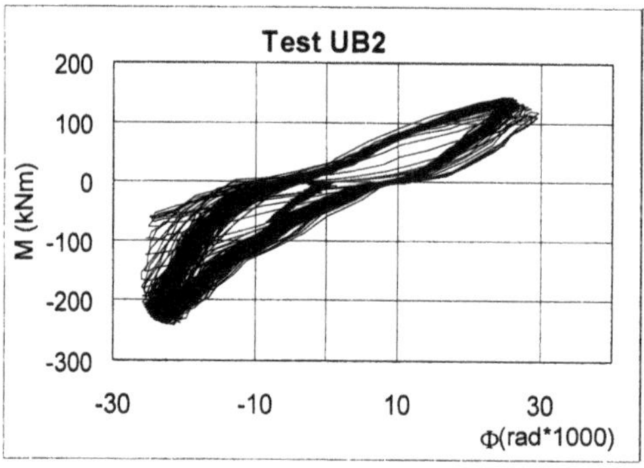

Figure 3.1.63: Moment-rotation diagram

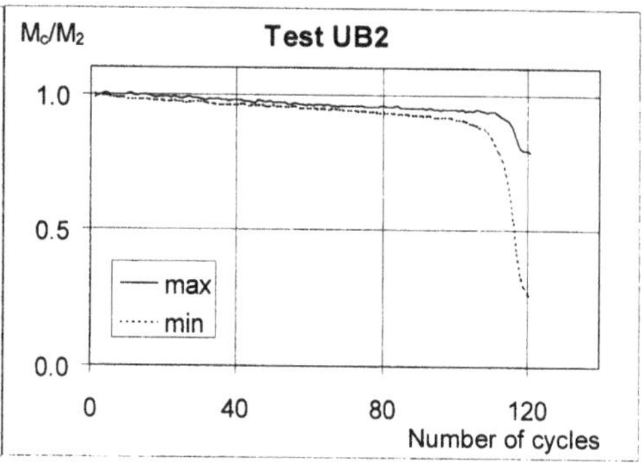

Figure 3.1.64: Moment amplitude

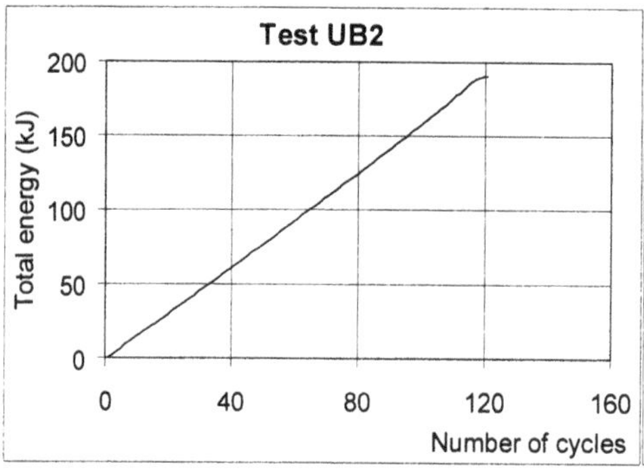

Figure 3.1.65: Cumulative dissipated energy

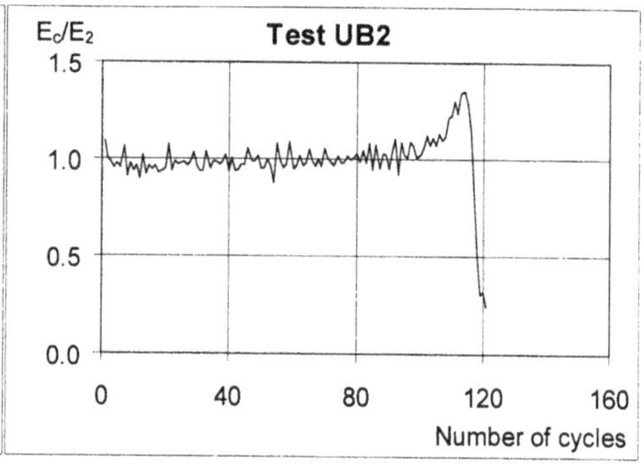

Figure 3.1.66: Energy dissipated in each cycle

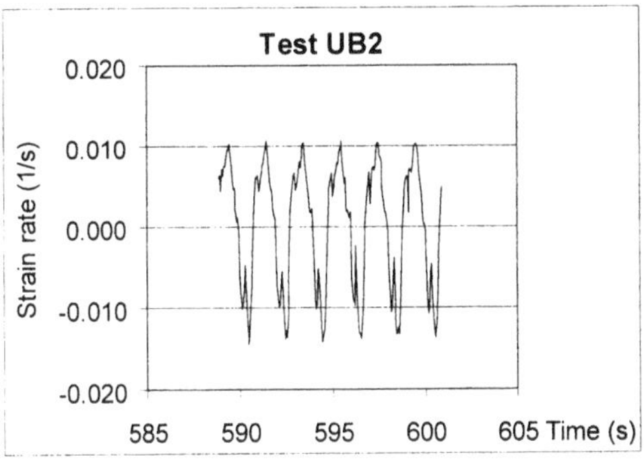

Figure 3.1.67: Strain rate at strain gauge No. 9

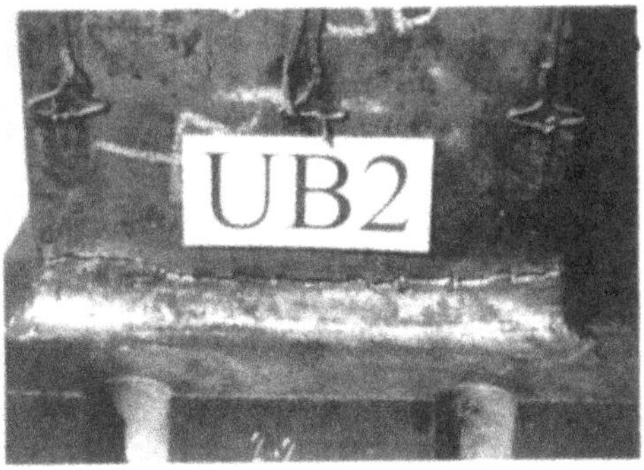

Figure 3.1.68: Photo of UB2 test specimen after failure (beam flange)

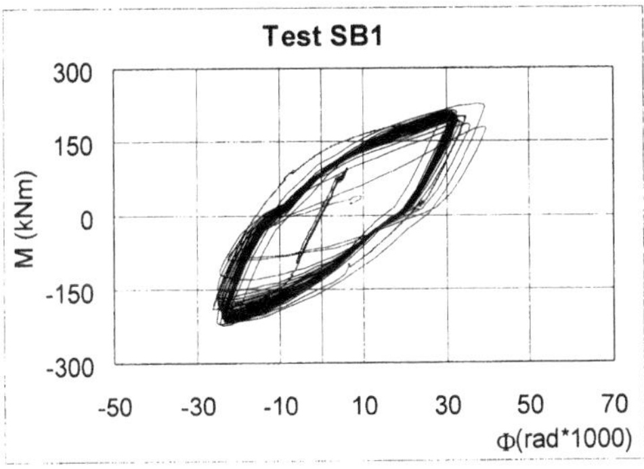

Figure 3.1.69: Moment-rotation diagram

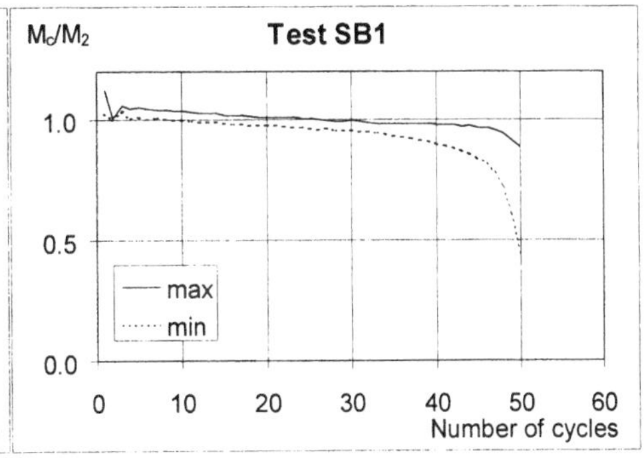

Figure 3.1.70: Moment amplitude

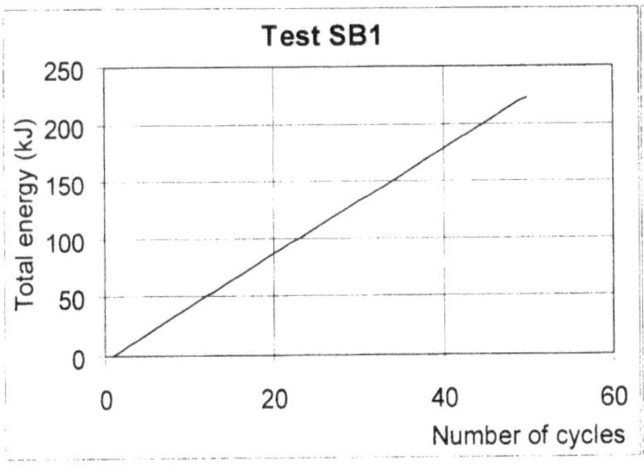

Figure 3.1.71: Cumulative dissipated energy

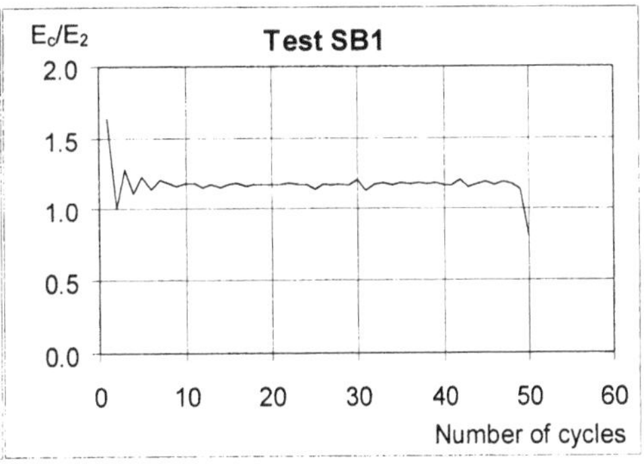

Figure 3.1.72: Energy dissipated in each cycle

TABLE 3.1.19

RESULTS OF DYNAMIC TESTS ON BOLTED END-PLATE CONNECTIONS

Specimen	Number of cycles	f_y (Mpa)	v_y (mm)	Δv (mm)	Δv_{pl} (mm)	ΔT (°C)	Failure mode	Failure position
UB1	93**	315	9	35	17	8.7	Stable crack prop.	Beam – weld toe
UB2	118*	313	10	32	12	15.5	Stable crack prop.	Beam – weld toe
SB1	48*	269	9	37.5	19.5	26.1	Stable crack prop.	Beam – base metal

** Dissipated energy criterion, * Strength criterion

During the tests the temperature on the surface of beam flange-to-endplate weld was also measured. For unsymmetric specimens the increase of temperature was only 8.7°C and 15.5°C, because the largest plastic strains developed outside the welds. At symmetric connection the temperature was then measured at the location of the fracture in the flange and the temperature increased by 26.1°C. At the room temperature this increase is not very important, but at lower temperatures it can help to achieve better toughness.

Comments on test results

The following conclusions can be drawn out from the test results:

- Unsymmetric bolted connections exhibit strong pinching effect and in each cycle they dissipate only one third of the energy dissipated at symmetric connections.

- Symmetric bolted connections (with strong enough end plates and bolts) behave as welded connections with similar connection configuration.

- The behaviour of unsymmetric bolted connections in frames subjected to earthquake loading was studied by Beg (Beg at al.1999). The results of nonlinear dynamic analysis show that in moderate earthquakes after bolt failure on the weaker side of the connection the frames behave better than it was expected.

DOG-BONE CONNECTIONS

This specimen is the typical US frame connection with the beam flanges welded to the column, but the typical flat backing bar used in the weld is replaced by a new design of bar which should have created conditions for a good first full penetration weld (see detail B in Figure 3.1.73). In the current practice, this bar would be an extruded steel wire with the designed cross section. In the present research program, the bar is machined out of a rectangular bar. Previous testing has shown that this backing bar gives bad welds and a poor behaviour of the connection in cyclic quasi static test made on beam without dog-bone. In this research a dog-bone was realized in the beam in order to prevent brittle failure in the welded connection. Only one specimen was tested.

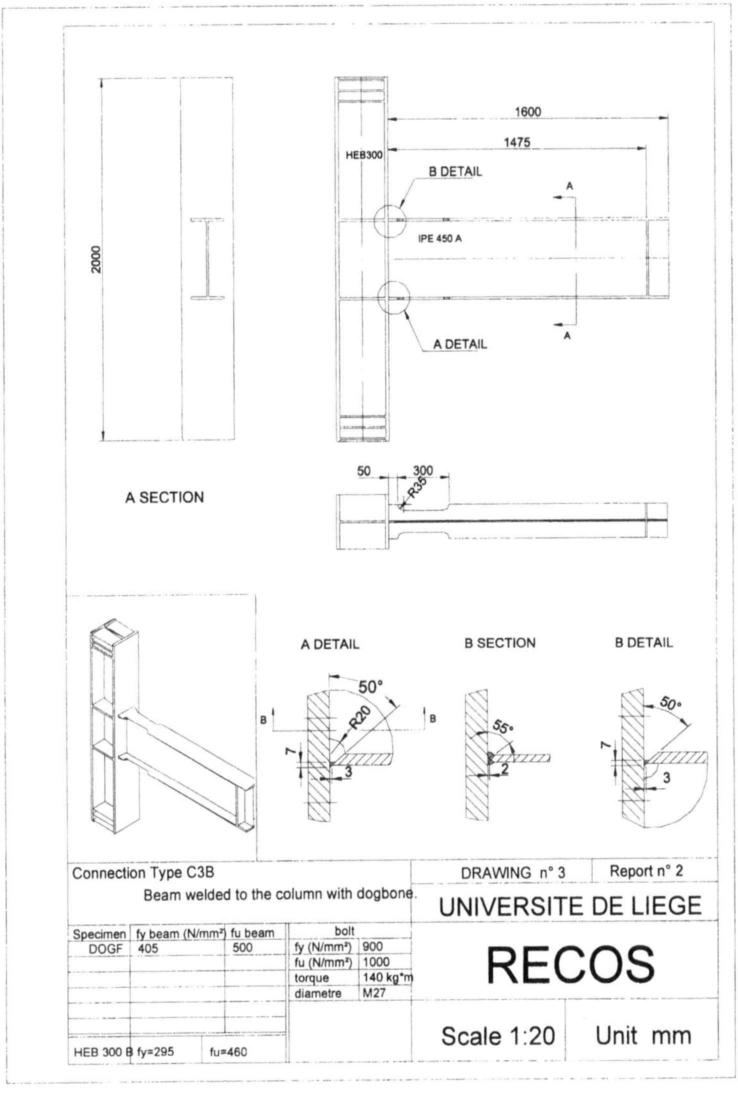

Figure 3.1.73

Specimen dimensions, steel profiles, base material properties and weld details can also be found in the previous Figure.

The imposed displacement is a sinusoidal function, with a maximum imposed displacement of v=±40mm, which is expected to be the double of the yield displacement. The period of the imposed displacement is 2.5 seconds.

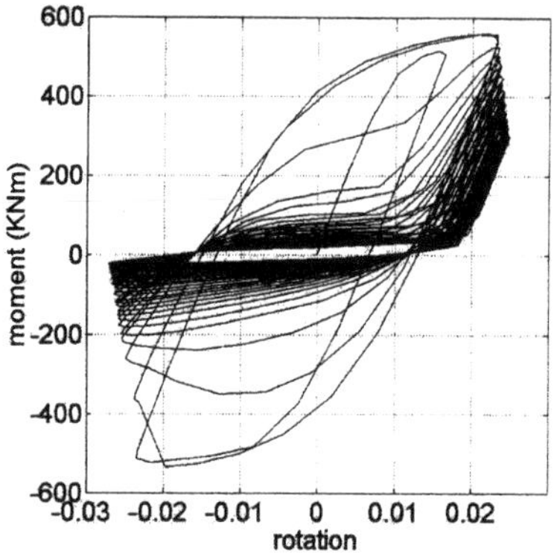

Figure 3.1.74: Specimen DOGF

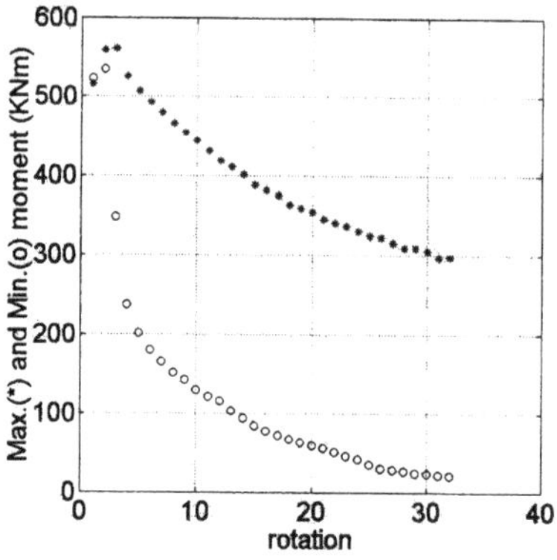

Figure 3.1.75: Specimen DOGF

It can be seen from the moment-rotation diagram (Figure 3.1.74) and from Figure 3.1.76 how the flange that works in tension in the first cycle is detached, showing that this new welding detail design did not realise a full penetration weld.

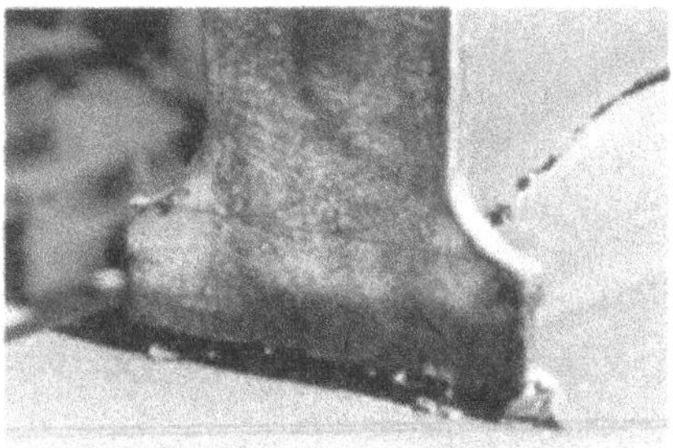

Figure 3.1.76: Failure mode specimen DOGF

The design of the Reduced Beam Section has been made in the following way.

The plastic moment realised in the cyclic quasi static test was 575 KNm, expressed at the face of the column flange. The corresponding moment at mid length of RBS, given the dimension of the test set up, is $575 \times (1475 - 200)/1475 = 479\ KNm$.

The dog-bone must be designed in such a way that M_p in the RBS zone is less than 497 KNm. With the dimensions of Figure 3.1.73: $M_p = 422$ KNm < 497 KNm .

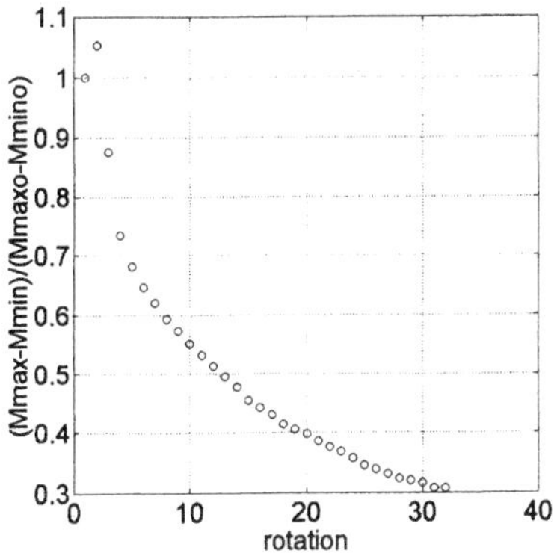

Figure 3.1.77: Specimen DOGF

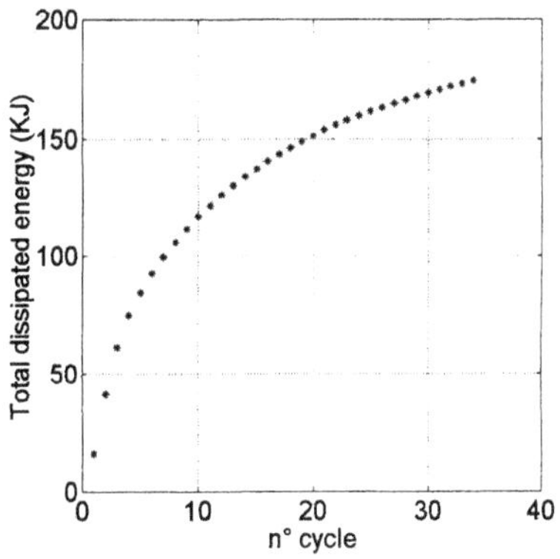

Figure 3.1.78: Specimen DOGF

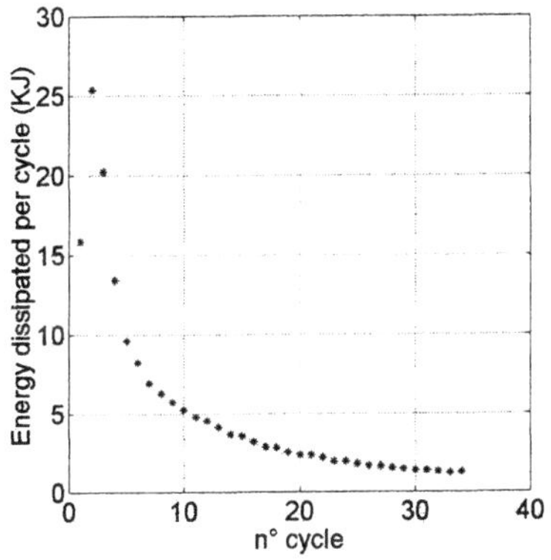

Figure 3.1.79: Specimen DOGF

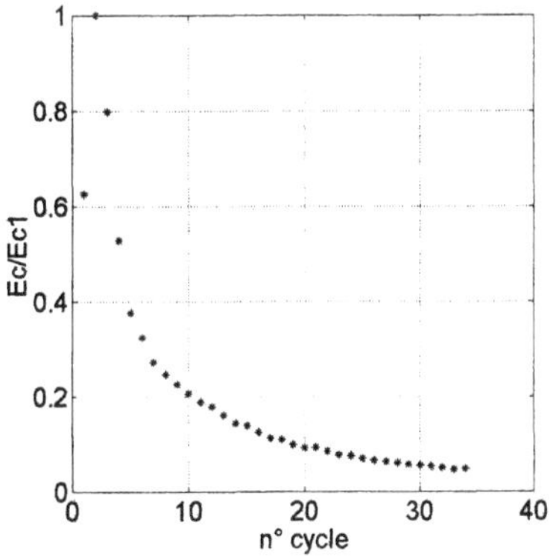

Figure 3.1.80: Specimen DOGF

The specimen was able to withstand only 4 cycles up to the loss of 50 % of its strenght capacity and the failure took place in the weld. Thus, it can be concluded that the dogbone did not protect the connection in this case. This bad result can be due to the high stress concentration in the welding detail, where localized plastic strains may exist, in spite of an average stress in the beam flange which is, by the design mentioned above, only 0.85 fy. This results support the option of SAC research project (SAC, 1997), which accepts RBS in new design, but not in repairs.

TABLE 3.1.20

RESULTS FOR SPECIMEN DOGF

Specimen	N° cycles up to collapse	Means	Cyclic dissipated Energy mean (kJ)	Means	Total dissipated Energy up to collapse (kJ)	Means
DOGF	4	-----	18.75	-----	75	-----

PARTIAL STRENGTH CONNECTIONS

Two full-scale specimens with semi-rigid partial strength beam to column connections in which bolts are perpendicular to the bent beam were tested, see Figure 3.1.81.

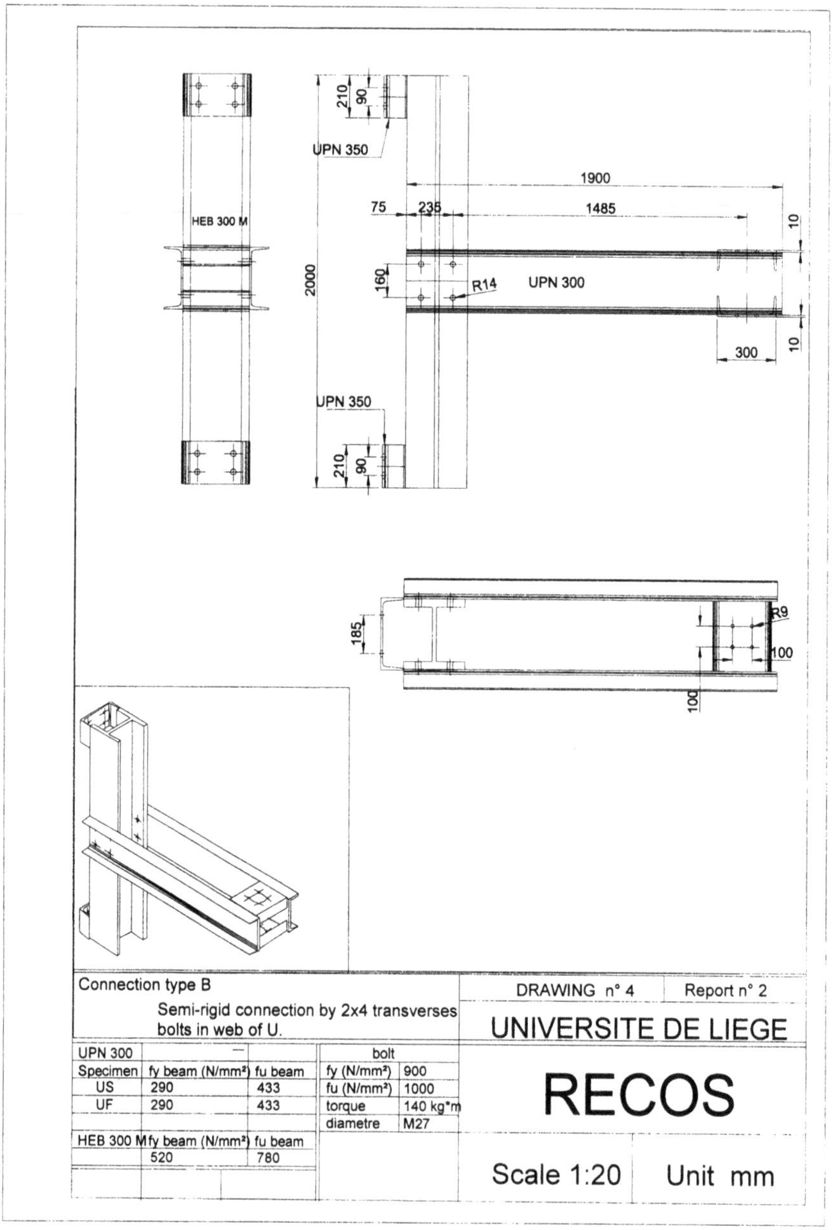

Figure 3.1.81

An interesting point of this connection is extremely low cost in the preparation and during the erection phase, as well as a good support surfaces to floor slabs or metal decking. It does not transfer the beam plastic moment to the column. It is intended to develop energy dissipation through the ovalization of the bolt holes and through friction between the web of the UPN beam and the flange of the HEM column. Both mechanisms work in the following manner:

- Friction gives a constant resistance throughout the displacement.
- Bearing resistance provides the increases in resistance that can be seen both at the left and the right side of the moment-rotation diagrams.

Specimen dimensions, steel profiles, base material properties, bolt and weld details can be found in Figure 3.1.81. The bolts are pre-stressed by means of a 1400 Nm torque.

The imposed displacements are sinusoidal functions, with a maximum imposed displacement of v=±40mm. The periods of the imposed displacement are 40 seconds for the specimen US and 2.5 seconds for the specimen UF.

Figure 3.1.82: Test set up

It can be seen from the moment-rotation diagrams that friction gives almost all the energy dissipation (see Figure 3.1.83 and 3.1.84). For the imposed displacement, the increase in resistance due to plastic deformation of the bolt holes is small and the ovalization of the bolt holes cannot be observed (see Figures 3.1.85 and 3.1.86).

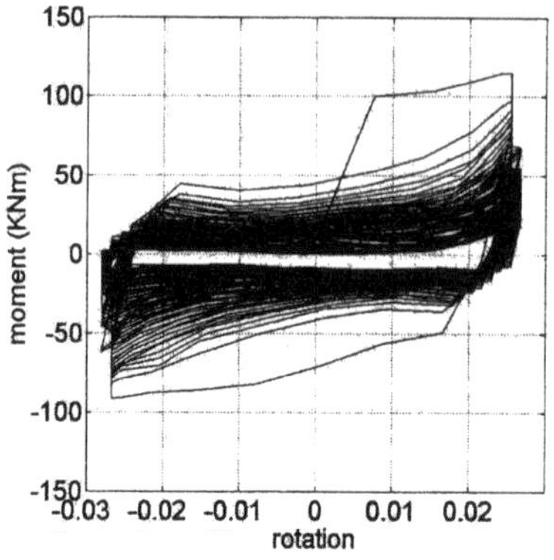

Figure 3.1.83: Specimen US

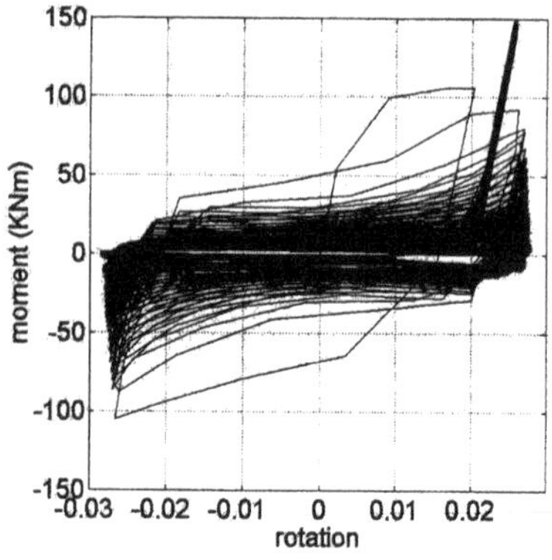

Figure 3.1.84: Specimen UF

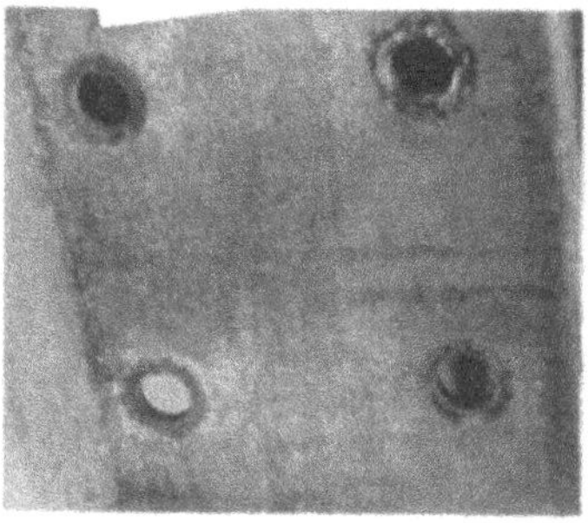

Figure 3.1.85: Specimen US

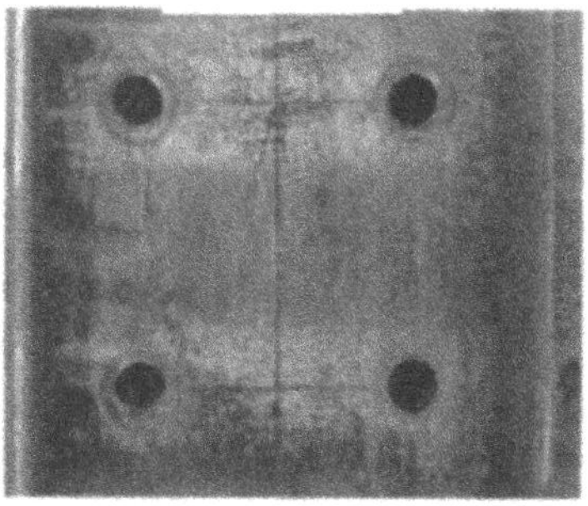

Figure 3.1.86: Specimen UF

In Figure 3.1.87 it can be seen that the reduction in maximum strength capacity with the number of cycles is faster for the specimen tested at high loading speed (note that the maximum moment has no important influence on the energy dissipation, because of the relatively small imposed displacement).

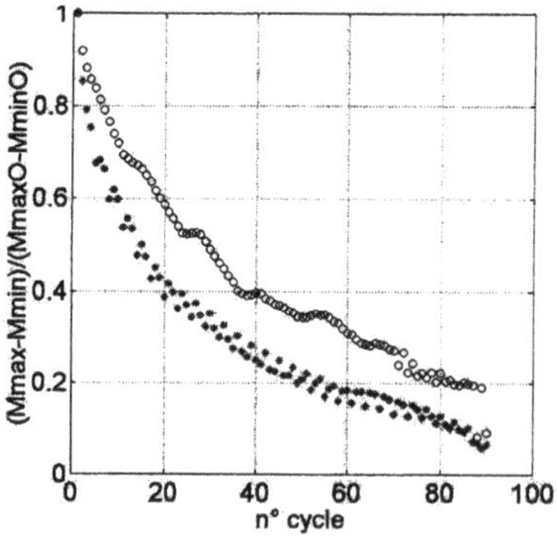

Figure 3.1.87: US (o) and UF (*)

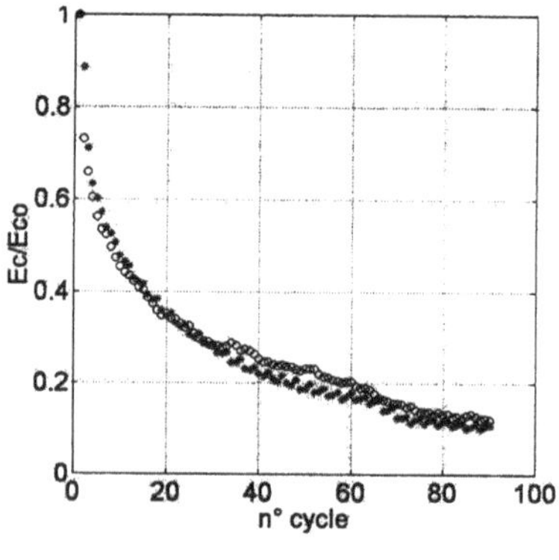

Figure 3.1.88: US (o) and UF (*)

On the contrary, no important influence of the loading speed can be observed in the reduction of dissipated energy per cycle with the number of cycles (Figure 3.1.88). For the imposed displacement the energy dissipation is basically due to friction and the loading speed does not influence the reduction of friction with the number of cycles. For the first 18 cycles, the reduction in E_c is a little higher for the specimen US; later it becomes lower.

The total energy dissipated is always greater for the specimen tested at low loading speed (Figure 3.1.89). It corresponds to the idea that an increment in velocity brings a reduction in friction (see Figure 3.1.90).

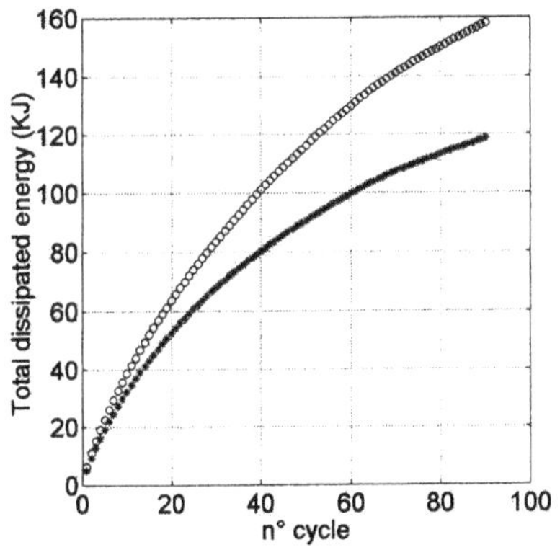

Figure 3.1.89: US (o) and UF (*)

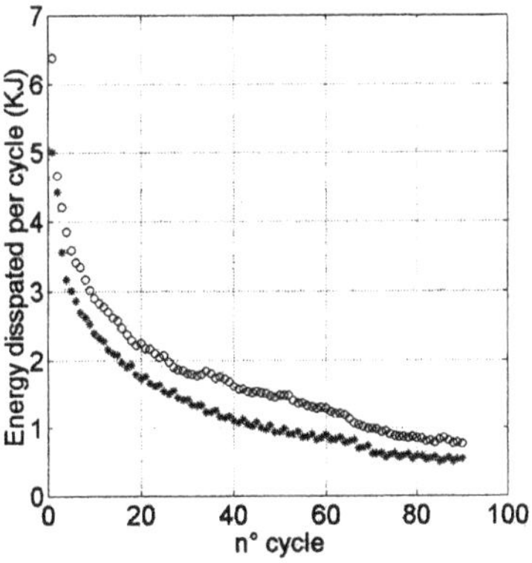

Figure 3.1.90: US (o) and UF (*)

The loading speed has a negative effect on the number of cycles sustained up to collapse. The reduction is 40%, when defining collapse as 50% reduction of strength capacity (18% for a 75% strength reduction).

TABLE 3.1.21

RESULTS FOR SEMI-RIGID CONNECTION SPECIMENS

Specimen	Collapse (50% of max. strength reduction)			Collapse (75% of max. strength reduction)		
	N° cycles up to collapse	Average dissipated energy per cycle (kJ)	Total dissipated energy up to collapse	N° cycles up to collapse	Average dissipated energy per cycle (kJ)	Total dissipated energy up to collapse
US	25	2.96	74	49	2.37	116
UF	15	2.80	42	40	2.00	80

Table 3.1.21 shows that the dissipated energy average per cycle is lower for the high speed loading (2.8 kJ for each cycle), friction decreases by 7 % in average during the loss of 50 % of strength capacity and by 16 % in average during the loss of 75 % of strength capacity.

The total energy dissipated up to collapse is 43 % and 31 % lower when the loading speed is increased, respectively, when defining collapse as 50% and 75% reduction of strength capacity.

MAIN CONCLUSIONS

On the influence of loading speed

Tests on small specimens

From test on small specimens with fillet weld connection it can be concluded that critical weld sizes prescribed in EC3 are strong enough, independently of the strain rate.

With full penetration butt welds, the failure always occurred in the base metal, which means that properly fabricated butt welds of the size tested are not more critical at higher strain rates than the base metal.

For the base steel material, in spite of the dispersed results, it has been shown in the literature how yield strength is more sensitive to strain rate than ultimate strength, and thus the ratio of ultimate to yield strength decreases with the strain rate. This clear conclusion for the base material is not so evident from our tests, where not only the base material plays a role, but also the weld metal and the heat affected zone.

Ultimate strength and yield strength increase at increased strain rates, however for their ratio, which also has a relevant influence on the beam ductility, the results are rather scattered and it is not possible to draw any specific conclusion.

Tests on beam-to-column steel connections

For the specimens tested there is no evidence that the loading speed could affect the failure mode of the steel beam-to-column connection. Nevertheless more tests are necessary on rigid connections made of higher steel grades and for larger beam depth.

The number of cycles that the specimens withstood up to collapse decreases with the loading speed. This conclusion contrasts from the Japanese studies made after the Kobe earthquake on beam-to-column steel connections (for two connection typologies: flange welded to the column and bolted web, and fully welded), where it was concluded that loading speed does not affect the number of cycles withstood up to collapse (Kato et al, 1997).

Loading speed decreases the ratio of ultimate to yield moment, reducing apparent ductility, causing plastic localisation, and thus for the same imposed displacement increasing strain demands.

Unsymmetric bolted connections are suitable only for loading conditions which cause very moderate plastic excursions. At the displacement amplitude of $2v_y$ and favourable preloading of 30% of the beam plastic moment the carrying capacity of bolts at the weaker side of the connection was almost exhausted. Hysteretic behaviour exhibits strong pinching effects mainly due to bolt plastification at the weaker side of the connection. Dissipated energy per cycle is only one third of the energy dissipated by the corresponding symmetric bolted connections. Symmetric welded and bolted connections were also subjected to small plastic deformations (displacement amplitude $2v_y$) in contrast to the tests of Calado (see Chapter 3.2) on the same connection configuration, where large displacement amplitudes were imposed. Regarding low cycle fatigue, our results lie on the S-N curve obtained by Calado from large amplitude tests.

For the semi-rigid connection, the loading speed reduces the energy dissipated per cycle. The reduction in the number of cycles, although evident, could be related to the definition of collapse used which is based on strength reduction (one half of strength capacity). However strength does not play an important role in energy dissipation at the imposed displacement. Further tests are necessary for higher displacements.

On the influence of temperature

Tests on small specimens

The results show that the decrease in temperature produces almost linear increase in yield and ultimate strength. This increase is very similar for all strain rates and for yield strength it amounts to around 10% as the temperature decreases from 20°C to -20°C.

The real implication of the increase of strength and decrease of ductility due to strain rate and temperature changes can be expected in earthquake conditions. Although at the first glance the uncertainty of earthquake loading is much higher than 10 or 20% increase of strength under severe loading conditions (Bruneau et al. 1998), at stronger earthquakes plastic hinges are formed at beam ends near beam-column connections, irrespective of the exact magnitude of the earthquake loading. Columns and connections that are normally protected by overstrength design, can be damaged or can even collapse due to additional strain rate or temperature induced increase of beam strength.

REFERENCES

Akiyama, H. (1999). Evaluation of Fractural Mode of Failure in Steel Structures Following Kobe Lessons, *General Reports and Keynote Lectures of SDSS '99 Colloquium, Timisoara.*

Anderson, C.A., Johnston, R.G., Partridge, J.E. (1995). *Post Earthquake Studies of a Damaged Low Rise Office Building.* University of Southern California, Department of Civil Engineering, Report No. CE 95-07, Los Angeles.

Beg, D., Remec, C., Skuber, P. (1999). Behaviour of Unsymmetric Bolted Connections Subjected to Dynamic Loading, *Proc. of SDSS '99 Colloquium Timisvara*, 191-198.

Bernuzzi, C., Castiglioni, C.A. & Vajna de Pava, S.(1999) Behaviour of beam-to-column joints in Moment-Resisting Steel Frames, *Proc. of 6th International Colloquium on Stability and Ductility of Steel Structures, Timisoara*, 199-210.

Bruneau, M., Uang, C.M., Whittaker, A. (1998). *Ductile Design of Steel Structures.* McGraw-Hill.

Calado, C., Bernuzzi, C., Castiglioni, C.A. (1998a). Structural Steel Components under Low-cycle Fatigue: Design Associated by Testing, *Proc. of Structural Engineers World Congress, San Francisco*, Paper T196-4.

Calado, L., C.A. Castiglioni, P. Barbaglia & C. Bernuzzi (1998b) Seismic design criteria based on cumulative damage concepts, *Proc. of 11th European Conference on Earthquake Engineering, Paris*, 604.

ECCS (1986). *Recommended Testing Procedure for Assessing the Behaviour of Structural Steel Elements under Cyclic Loads.* European Convention for Constructional Steelwork, Publication No. 45, Brussels.

Kato, B. (Editor) (1997). *Kobe Earthquake Damage to Steel Moment Connections and Suggested Improvement.* Japanese Society of Steel Construction, Technical Report No. 39, Tokyo.

Lu, L.W., Riches, J.M., Mao, C. (1999). Critical Issues in Achieving Ductile Behaviour of Welded Moment Connections, *General Reports and Keynote Lectures of SDSS '99 Colloquium, Timisoara.*

Plumier, A., Agatino, M.R., Castellani, A., Castiglioni, C.A. & Chesi, C.(1998) Resistance of steel connections to low-cycle fatigue, *Proc. of 11th European Conference on Earthquake Engineering, Paris*, 297.

SAC Joint Venture (1997), *Evaluation, Repair, Modification and Design of Welded Steel Moment Frame Structures*, Federal Emergency Management Agency, FEMA-267 Interim Guidelines.

3.2

INFLUENCE OF CONNECTION TYPOLOGY AND LOADING ASYMMETRY

Dan Dubina, Daniel Grecea, Adrian Ciutina, Aurel Stratan

INTRODUCTION

Steel Moment Resisting Frames are more and more used nowadays, because they confer clear spans and architectural freedom. The design of such structures, on the other hand, is difficult to accomplish due to large flexibility under lateral loads. Moment Resisting Frames dissipate the large amount of energy induced by the earthquake in plastic hinges that can be formed in beams, columns or beam-to-column joints. Some recent strong earthquakes (Kobe, 17 Jan. 1995 - 7.2 Mg. on Richter scale, Northridge, California, 17 Jan. 1994 - 6.7 Mg. on Richter scale) have shown that in the case of Moment Resisting Frames the most vulnerable points are the beam-to-column joints, and the fact that the understanding of their behaviour during the earthquakes is still scarce. During the last years, many investigations were made on the subject, but no final conclusions have been drawn.

Till recent days, the usual beam-to-column joints have been classified either as rigid or nominally pinned. The European Code EUROCODE 3 introduces the concept of semi-rigidity for joints (in terms of stiffness and strength), and offers by Annex J computation formulae for their characteristics (by component method), but this is only for monotonic loading.

Starting from here, appears the imperative need of testing the beam-to-column joints to cyclic loading in order to better understand the phenomena produced in case of a strong ground motion and by this to assure a safer design both for structures and for their inhabitants.

The present work describes investigations on beam-to-column joints, carried out at the laboratory of steel structures at the Civil Engineering Faculty of Timisoara. First are described the joints that have been tested, their characteristics, the loading system and procedure. Further are given the theoretical characteristics of the joints, computed according to EUROCODE 3. The results of the tests present the experimental characteristics of the tested joints and the behavioural curves. Finally, a comparison between the theoretical (by EUROCODE 3) and the experimental characteristics is made, as well as the resulting conclusions.

SPECIMENS AND TESTING SET-UP DESCRIPTION

Three typologies of beam-to-column joints have been tested from a total of 12 specimens. All the joints are double sided. For all the specimens the design steel grade was S235 (f_y=235 N/mm^2, f_u=360 N/mm^2), beams being IPE 360 and columns HEB 300 – see Figure 3.2.1.

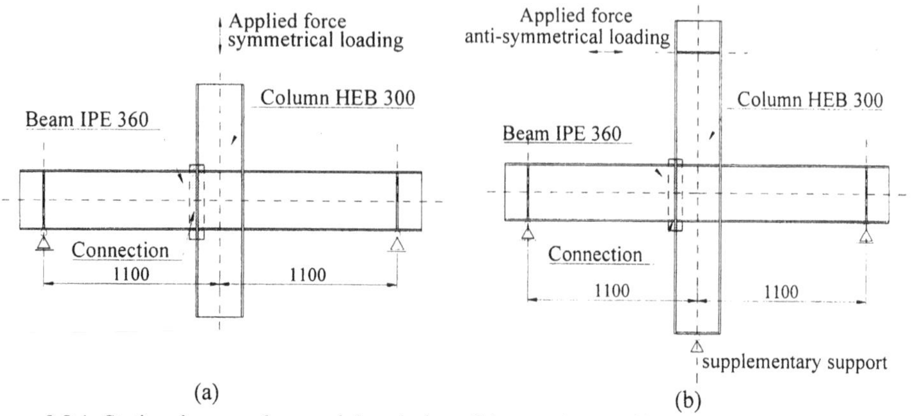

Figure 3.2.1: Static scheme and general description of the specimens: (a) symmetrical loading and (b) anti-symmetrical loading

The mechanical characteristics are given in TABLE 3.2.1, while the measured geometrical characteristics are given in TABLE 3.2.2. Mechanical characteristics have been determined on samples extracted from the joint components: beam web and flange, column web and flange, end plate, web plate, cover plate and stiffeners. The tensile test was conducted in accordance with SR EN 10002-1 (1990) code referring to tensile tests of metallic samples. There can be observed the great values of yielding and ultimate strengths for the case of beam and column web and flanges, while for the end plate were obtained the smallest values, the yielding strength being near the limit of S235 steel. It resulted very clear, that the steel used for beams and columns seems to be S275 rather that S235. Although the strengths for the case of end-plate samples are lower, their ductility is also lower, the greater material ductility being obtained for web plates. The geometrical characteristics have been determined by direct measurement of the joint components.

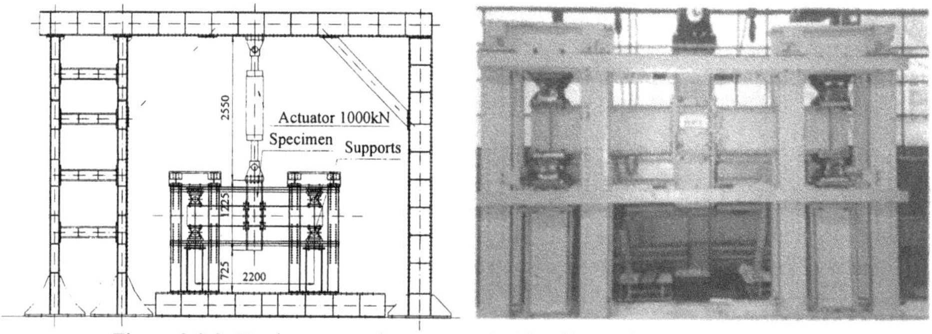

Figure 3.2.2: Testing set-up for symmetrical loading: scheme and real set-up

Two types of loading were applied: symmetrical and anti-symmetrical (see Figure 3.2.2 and Figure 3.2.4) and three connection typologies were tested (see Figure 3.2.3):

Type 1 - Figure 3.2.3 a:

- 2 symmetric cruciform extended end plate bolted connections (prestressed 10.9 M20 bolts) – specimens XS-EP1 and XS-EP2

- 2 anti-symmetric cruciform extended end plate bolted connections (prestressed 10.9 M20 bolts) specimens - XU-EP1 and XU-EP2

Type 2 - Figure 3.2.3 b:

- 2 symmetric cruciform welded connections (full-penetration welds) specimens - XS-W1 and XS-W2
- 2 anti-symmetric cruciform welded connections (full-penetration welds) specimens - XU-W1 and XU-W2

Type 3 - Figure 3.2.3 c:

- 2 symmetric cruciform connections with welded cover plates (full-penetration welds) and welded web plate (bolted for erection) specimens - XS-CWP1 and XS-CWP2
- 2 anti-symmetric cruciform connections with welded cover plates (full-penetration welds) and welded web plate (bolted for erection) specimens - XU-CWP1 and XU-CWP2

TABLE 3.2.1

MECHANICAL CHARACTERISTICS RESULTED FROM TENSILE TESTS.

Sample type	Sample	f_{yf} N/mm^2	$f_{y\,mean}$ N/mm^2	f_{uf} N/mm^2	$f_{uf\,mean}$ N/mm^2	A%	A% $_{mean}$
BEAM IPE 360							
Beam flange	C1/1	321.7	**329.8**	456.6	**463.2**	38.3	**42.2**
	C1/2	335.0		462.7		45.5	
	C1/3	332.8		470.3		42.7	
Beam web	C2/1	356.3	**348.4**	465.5	**464.0**	48.8	**51.8**
	C2/2	340.7		470.9		52.0	
	C2/3	348.3		455.6		54.5	
COLUMN HEB 300							
Column flange	C3/1	312.4	**313.0**	444.8	**449.8**	44.7	**45.2**
	C3/2	311.3		451.6		45.1	
	C3/3	315.3		453.0		45.8	
Column web	C4/1	338.6	**341.8**	461.5	**464.4**	44.5	**44.5**
	C4/2	340.0		466.7		45.3	
	C4/3	346.7		465.0		43.7	
COVER PLATE AND STIFFENERS							
Cover plate & Stiffeners	C5/1	275.7	**273.2**	455.9	**459.1**	47.4	**46.0**
	C5/2	272.0		460.0		44.3	
	C5/3	272.0		461.3		46.4	
END PLATE							
End plate	C6/1	261.9	**248.3**	415.0	**416.0**	43.0	**42.6**
	C6/2	243.0		415.7		42.8	
	C6/3	239.9		417.3		42.1	
WEB PLATE							
Web plate	C7/1	315.3	**313.6**	434.7	**436.7**	51.4	**52.2**
	C7/2	312.2		436.7		52.2	
	C7/3	313.3		438.8		53.0	

A% represents the elongation (in per cents) between two marks on the sample.

TABLE 3.2.2
GEOMETRICAL CHARACTERISTICS MEASURED FROM ELEMENTS.

Specimen	Element type	Measured Values [mm]					
		t_f	t_w	b_f	h	r	t_{plate}
XS-EP 1	Beam IPE 360	12.1	8.2	175.5	363.0	18.0	-
	Column	18.4	11.6	297.0	303.0	27.0	-
	Stiffener	-	-	-	-	-	15.0
	End-plate	-	-	-	-	-	20.2
XS-EP 2	Beam IPE 360	12.2	8.4	171.3	360.8	18.2	-
	Column	18.2	11.6	299.0	301.0	27.0	-
	Stiffener	-	-	-	-	-	15.1
	End-plate	-	-	-	-	-	20.2
XS-W 1	Beam IPE 360	11.6	7.9	172.2	361.7	18.0	-
	Column	19.0	11.4	297.3	302.3	27.0	-
	Stiffener	-	-	-	-	-	15.0
XS-W 2	Beam IPE 360	11.6	8.0	172.1	360.8	18.0	-
	Column	18.4	11.4	298.0	303.2	27.0	-
	Stiffener	-	-	-	-	-	15.0
XS-CWP 1	Beam IPE 360	12.4	8.2	172.8	364.2	18.0	-
	Column	19.6	11.3	299.9	302.4	27.0	-
	Stiffener	-	-	-	-	-	15.3
	Cover plate	-	-	-	-	-	14.8
	Web plate	-	-	-	-	-	10.9
XS-CWP 2	Beam IPE 360	12.4	8.3	170.8	363.3	18.0	-
	Column	18.4	11.3	299.7	303.0	27.0	-
	Stiffener	-	-	-	-	-	15.1
	Cover plate	-	-	-	-	-	14.7
	Web plate	-	-	-	-	-	10.9
XU-EP 1	Beam IPE 360	12.5	7.8	171.9	361.0	18.0	-
	Column	18.3	11.7	300.0	301.6	27.0	-
	Stiffener	-	-	-	-	-	15.1
	End-plate	-	-	-	-	-	19.9
XU-EP 2	Beam IPE 360	12.2	8.0	171.4	361.8	18.0	-
	Column	18.6	11.5	297.8	302.3	27.0	-
	Stiffener	-	-	-	-	-	15.0
	End-plate	-	-	-	-	-	20.3
XU-W 1	Beam IPE 360	12.5	8.0	171.6	360.9	18.0	-
	Column	18.7	11.6	297.7	303.4	27.0	-
	Stiffener	-	-	-	-	-	15.0
XU-W 2	Beam IPE 360	11.6	8.0	172.1	361.7	18.0	-
	Column	18.5	11.7	300.2	303.5	27.0	-
	Stiffener	-	-	-	-	-	15.0
XU-CWP 1	Beam IPE 360	12.1	8.0	171.0	362.8	18.0	-
	Column	18.7	12.0	298.0	302.6	27.0	-
	Stiffener	-	-	-	-	-	14.9
	Cover plate	-	-	-	-	-	14.9
	Web plate	-	-	-	-	-	10.9
XS-CWP 2	Beam IPE 360	12.3	8.2	172.2	363.5	18.0	-
	Column	18.8	11.7	300.1	303.6	27.0	-
	Stiffener	-	-	-	-	-	15.1
	Cover plate	-	-	-	-	-	14.9
	Web plate	-	-	-	-	-	10.9

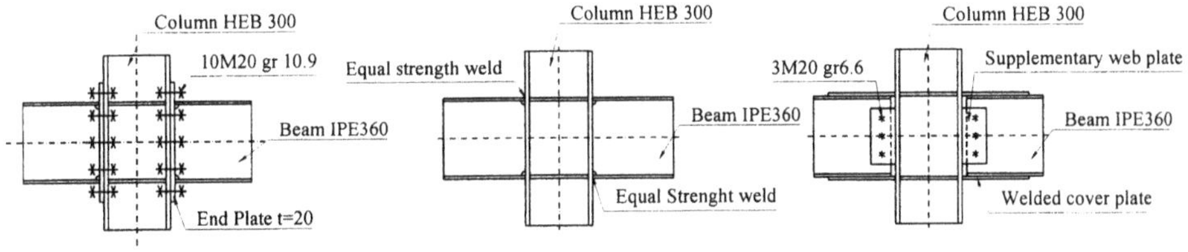

(a) (b) (c)

Figure 3.2.3: Connection configurations: (a) bolted, (b) welded and (c) with cover welded plate

The assemblages were realised by welding. The manufacturer was requested to provide welds of class I between the elements with the weld resistance at least equal to the resistance of the parent material (full resistant). A detail of the edge preparation prior welding is given in Figure 3.2.5.

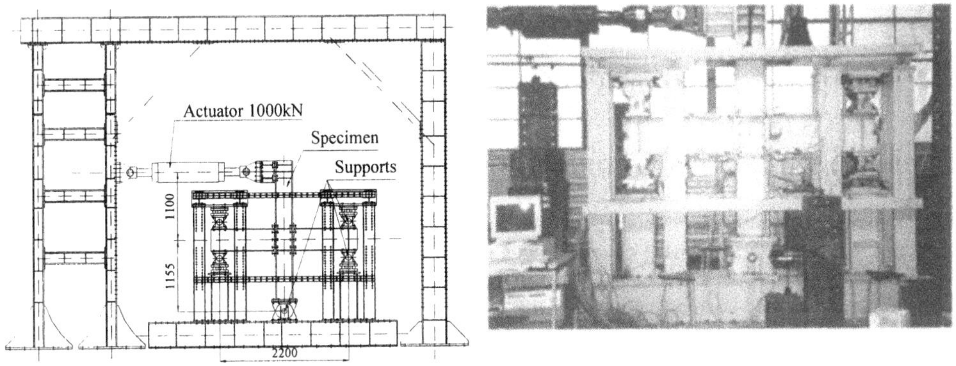

Figure 3.2.4: Testing set-up for anti-symmetrical loading: scheme and real set-up

For the case of bolted connection, the bolts have been prestressed to a moment of 620 Nm as according to Romanian Standard for prestressed bolts.

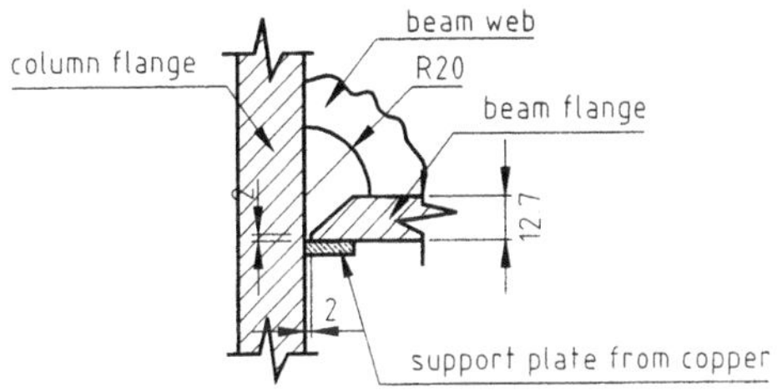

Figure 3.2.5: Detail of edge preparation

At a closer visual inspection on the welds of the joints, there was observed that not all the welds on the joints have been fully penetrated. As a conclusion, there are serious doubts that the welding procedure was made according to the designer provisions. To see the influence of the welding, the root welds of the joint XS-W1 that presented visible weld unconformities to design procedure were re-welded in the laboratory.

The cyclic tests were performed in the Laboratory of Steel Structures from the Civil Engineering Faculty of Timisoara, by means of an actuator that introduces the cyclic loading in the specimens. The apparatus permits the cyclic loading, maximum force of the actuator being +/-1000 kN and the maximum stroke of +/-200 mm. Figure 3.2.2 shows the testing set-up for the symmetric loading, while the Figure 3.2.4 shows the set-up for anti-symmetrical loading.

The joint specimens are simply supported, system that permits rotation and translation of the beam central axis. For the anti-symmetrical tests, a supplementary support at the column base was introduced, in order to allow the column bottom end to rotate but not to translate. Both systems were built such that they simulate a sub-structure from a sway-frame. A guiding system was also made, in order to allow the specimen to move only in the vertical plane and not have lateral displacements.

The loading system is composed basically from the actuator (+/-1000 kN) that permits static and dynamic loading, pressured by a hydro-mechanical unit, and guided by the computer software named HYDROMAX. The entire system (a second actuator of +/-500kN also exists), including the software, were delivered by the QUIRY Company within the European projects COPERNICUS "RECOS" and TEMPUS 011297.

The data acquisition system permits the acquisition of the displacement transducers, inclinometers, strain gauges etc, through a computer guided data logger.

COMPUTED PROPERTIES ACCORDING TO ANNEX J OF EUROCODE 3

The joints (three different joint configurations and two load types) have been designed according to Eurocode 3, Annex J. The main characteristics of the joints (computed by the measured mechanic characteristics from TABLE 3.2.1) are listed in TABLE 3.2.3, and the joints' classification, according to the above mentioned code is given in TABLE 3.2.4.

TABLE 3.2.3
JOINTS' CHARACTERISTICS COMPUTED BY ANNEX J OF EC3

SPECIMEN	$S_{j,ini}$ [kNm/rad]	$M_{j,Rd}$ [kNm]	$M_{pl.beam}$ [kNm]	$M_{j,Rd}/M_{pl.beam}$
XS-EP1	142932.2	262.7	335.9	0.78
XS-EP2	140886.8	261.3	338.2	0.77
XS-W1	□	319.4	323.9	0.99
XS-W2	□	320.2	324.7	0.99
XS-CWP1	□	468.1	346.0	1.35
XS-CWP2	□	464.0	343.0	1.35
XU-EP1	43727.2	169.2	339.2	0.50
XU-EP2	43718.2	169.1	335.4	0.50
XU-W1	68792.1	163.6	340.5	0.48
XU-W2	69062.1	164.1	325.5	0.50
XU-CWP1	75597.2	178.6	335.0	0.53
XU-CWP2	74963.1	177.4	341.8	0.52

$S_{j,ini}$ represents the initial computed joint stiffness, while $M_{j,Rd}$ is the computed plastic moment of the joint. $M_{pl.beam}$ represents the plastic moment of the beam. The slight differences in $M_{pl.beam}$ values are due to the differences in the measured geometrical properties of the beam components.

The obtained results show a great difference between the moments and rigidities for symmetrical loaded specimens and the anti-symmetrical ones, generally speaking the $M_{j,Rd}$ for anti-symmetrical specimens being 0.4-0.65 from the symmetrical ones. On the other hand, while for the XS series the

$M_{j,Rd}$ are increasing from XS-EP to XS-CWP specimens, for the XU series the moments are nearly the same. The change in the loading type (symmetrical – anti-symmetrical) makes the welded (W) and the cover plated (CWP) joints to shift from rigid and full-strength to semi-rigid and partial-strength. This fact can be explained by the change of the weakest component from beam flange and web in compression (symmetrical loading) to web panel in shear (anti-symmetrical loading).

TABLE 3.2.4

JOINTS' STIFFNESS AND RESISTANCE CLASSIFICATION ACCORDING TO EC3

SPECIMEN	EC3 Stiffness Classification	EC3 Resistance Classification	Weakest Component
XS-EP	Semi-rigid	Partial-resistant	End-plate in bending
XS-W	Rigid	Equal-resistant	Beam fl. & web in compr.
XS-CWP	Rigid	Full-resistant	Beam fl. & web in compr.
XU-EP	Semi-rigid	Partial-resistant	End-plate in bending
XU-W	Semi-rigid	Partial-resistant	Web panel in shear
XU-CWP	Semi-rigid	Partial-resistant	Web panel in shear

DATA PROCESSING

Measuring System

The actuator force induced in the column top end is delivered to the acquisition system and used for computing the moments in the connections. The actuator displacement is also delivered to the acquisition system, but only for informative purposes, not being used properly for computing the rotations of the joints and connections. The main instrumentation used for data acquisition is shown in Figure 3.2.6 for symmetrical loading and Figure 3.2.7 for anti-symmetrical loading.

The main displacements that are measured for symmetrical loaded specimens are the displacements under the column base (transducers 1 and 2), being used for direct determination of connection plastic rotation. There are also measured the panel zone deformations and the beam inclinations. It is to be mentioned that the panel zone is not subjected to shear deformations in the case of symmetrical loading. Consequently, displacement transducers 5 and 6 did not indicate any deformations.

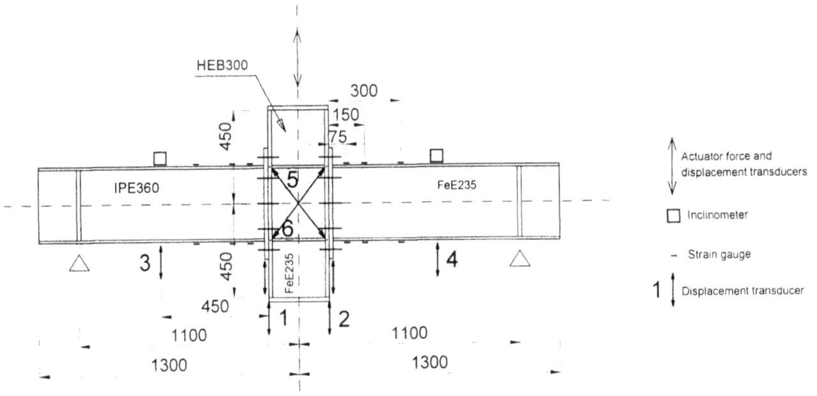

Figure 3.2.6: Instrumentation for symmetrical loading

In the case of anti-symmetrical loading, the actuator force is oriented horizontally, and the entire joint rotate with respect to the bottom articulation. The most important deformations are those of the column panel zone, measured by the transducers 1 and 2, and those of the connection, measured by the displacement transducers 3-6.

For each specimens more transducers have been used in order to determine different partial rotations of sub-components. When conducting first tests, additional displacement transducers have been installed for measuring support displacements. These proved to be negligible and have been removed for the remaining tests. On the contrary, for the anti-symmetrical loaded joints, displacements of the bottom pinned support (transducer 10 – see Figure 3.2.7) were significant and this transducer was kept.

Data Processing for Symmetrical Loading

Since the plastic rotation was mainly expected to occur in the connections, the moment computed is at the column face. Starting from the actuator force P, the theoretic span between supports L and the height of the column h_c, the bending moment can be found by the formula:

$$M=P(L-h_c)/4 \qquad (3.2.1)$$

For determining the global rotation of the joint, the transducers used are 1 and 2, their measured displacement being denoted by δ_1 and δ_2. The global rotation of the joint is given by:

$$\phi_G = \frac{1}{l}\left(\frac{\delta_1+\delta_2}{2} - \frac{Pl^3}{6EI_b} - \frac{Pl}{2Gh_b t_{wb}}\right) \qquad (3.2.2)$$

where:

- l – the free span of the beam, between the beam support and column flange.
- E – modulus of elasticity
- G – shear modulus
- I_b – moment of inertia of the beam
- P – actuator force
- $h_b t_{wb}$ – the shear area

Practically, the term: $\frac{Pl^3}{6EI_b}+\frac{Pl}{2Gh_b t_{wb}}$ accounts for the elastic rotation of the beam on the length l, being considered clumped and simply supported at the ends.

Data Processing for Anti-symmetrical Loading

Geometry and basic instrumentation for unsymmetrical loading is presented in Figure 3.2.7.
Rotations and moments are considered at the column face for all configurations. The moment is determined as follows:

$$M = \frac{H}{L}\cdot P \cdot L_b \qquad (3.2.3)$$

where: M – bending moment at the column face

- H – column height
- L – horizontal distance between beam supports
- P – actuator force
- L_b – clear length of the beam

The behaviour of anti-symmetrically-loaded joints is more complex than for the symmetrically loaded ones. Shear forces in the column web panel result from the combined action of forces in the tension and compression zone of the joint and the shear forces resulting from the moment distribution in the column. Due to these shear forces, additional deformations occur in the panel zone, which are not really rotations. But they lead to a change in angle between the axis of the column and the axis of the connected beam, as shown in Figure 3.2.8.

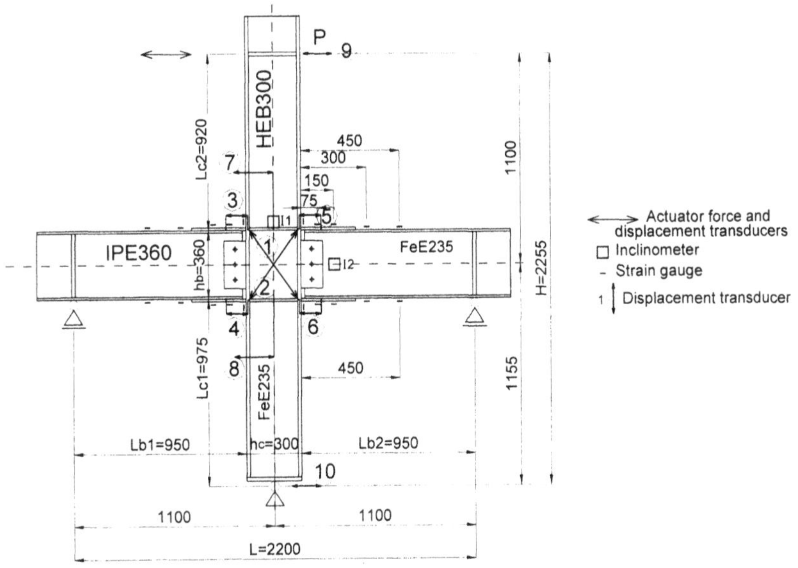

Figure 3.2.7: Instrumentation for anti-symmetrical loading

Rotations that may be defined in the case of anti-symmetrically-loaded joints are:
(1) Panel zone rotation γ has two components, (γ_1 and γ_2). Each of these components is difficult to estimate, and therefore are considered equal for further calculations ($\gamma_1 = \gamma_2 = \gamma/2$). Overall panel zone rotation angle is determined from displacement transducers 1 and 2:

$$\gamma = \frac{\sqrt{a^2 + b^2} \cdot (\delta_1 - \delta_2)}{2 \cdot a \cdot b}; \qquad (a = h_c - t_{fc};\ b = h_b - t_{fb}) \qquad (3.2.4)$$

where: δ_1 and δ_2 are displacements measured by transducers 1 and 2 respectively
a and b are dimensions of the panel zone (see Figure 3.2.8a)
h_c – column depth
t_{fc} – thickness of the column flange
h_b – beam depth
t_{fb} – thickness of the beam flange

(2) Connection rotation ϕ_c is determined from transducers 3 and 4 for the left side, and from transducers 5 and 6 for the right side (see Figure 3.2.8b):

$$\phi_c^{left} = \frac{\delta_3 - \delta_4}{b} \qquad \phi_c^{right} = \frac{\delta_6 - \delta_5}{b} \qquad (3.2.5)$$

The rotation determined in this way comprises also rotations in the portion of the beam adjacent to the column or end plate. The positive rotation is considered to be clockwise.

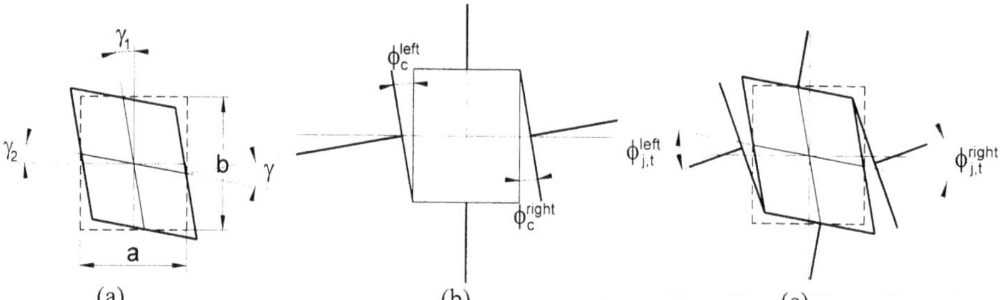

Figure 3.2.8: Definition of rotations at the boundary of panel zone

(3) Total joint rotation $\phi_{j,t}$ (see Figure 3.2.8c) is the rotation between beam and column at the panel zone boundary, and is given by:

$$\phi_{j,t} = \gamma + \phi_c \tag{3.2.6}$$

It can be determined for the left and right sides, but the average value was considered for simplicity.

(4) Elastic joint rotation $\phi_{j,el}$ is the global rotation of the column from the elastic deformation of the specimen under the applied force (see Figure 3.2.9). It is determined as:

$$\phi_{j,el} = \frac{\delta_{top,el}}{H} \tag{3.2.7}$$

$\delta_{top,el}$ is the displacement of the top of the column under the force P (see Figure 3.2.9a), considering the panel zone as infinitely rigid and is given by the following relation (see Figure 3.2.7 for definition of geometry):

$$\delta_{top,el} = \frac{1}{3}\frac{P}{EI_b}\left(\frac{H}{L}\right)^2\left(L_{b1}^3 + L_{b2}^3\right) + \frac{1}{3}\frac{P}{EI_c}\left(L_{c1}^3 + L_{c2}^3\right) + \frac{P}{GA_{fb}}\left(\frac{H}{L}\right)^2\left(L_{b1} + L_{b2}\right) + \frac{P}{GA_{fc}}\left(L_{c1} + L_{c2}\right) \tag{3.2.8}$$

where:

I_c – moment of inertia of the column
A_{fb} – beam shear area
A_{fc} –column shear area

Determination of $\delta_{top,el}$ from the above equation is slightly modified for XU-CWP joints in order to take into account for the increased moment of inertia where cover plates are present.

(5) Global joint rotation $\phi_{j,g}$ (see Figure 3.2.9b) is determined from the displacement of the column top, cleared of the bottom support displacements, from which the elastic rotation is subtracted:

$$\delta_{top} = \delta_9 - \delta_{10} \tag{3.2.9}$$

$$\phi_{j,g} = \frac{\delta_{top}}{H} - \phi_{j,el} \tag{3.2.10}$$

(6) In order to establish a correlation between the global and the total joint rotations, the global joint rotation from the panel zone deformation $\phi_{j,\gamma}$ need be introduced (see Figure 3.2.9c):

$$\phi_{j,\gamma} = \frac{\gamma}{2}\frac{1}{H}\left[L_{c1} + L_{c2} - h_b + \frac{H}{L}\left(L_{b1} + L_{b2} - h_c\right)\right] \qquad (3.2.11)$$

A relation can be established between the global joint rotation ($\phi_{j,g}$) and rotations at the boundary of the panel zone (γ, through $\phi_{j,\gamma}$, and ϕ_c):

$$\phi_{j,g} = \phi_{j,\gamma} + \phi_c^{average} \qquad (3.2.12)$$

It should be mentioned here that global joint rotation $\phi_{j,g}$ would be more appropriate when performing a global structural investigation, but it depends on the dimensions of the specimens. More general information is provided by the total joint rotation $\phi_{j,t}$.

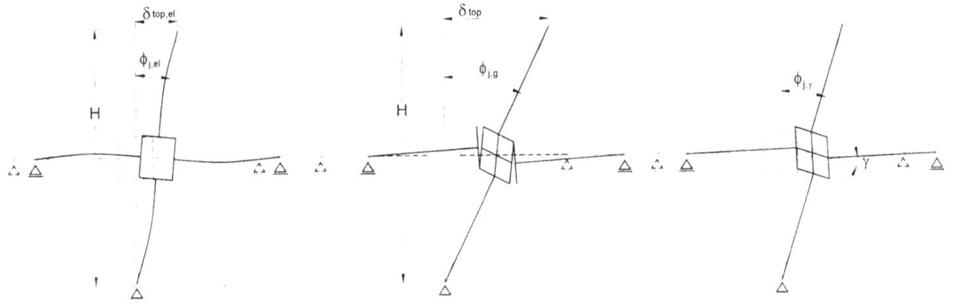

Figure 3.2.9: Definition of global joint rotations

LOADING HISTORY

The loading history was made according to the ECCS Recommendations simplified procedure, (see Figure 3.2.10), in which were performed three cycles for each even multiplier of the displacement e_y, which represents the characteristic conventional yielding displacement of the joint. It was assumed as the displacement at the column end (for both symmetrical and anti-symmetrical). Prior the plastic cycles, the simplified ECCS procedure was used in order to find the displacement e_y and the corresponding force F_y, so as to ensure that at least four levels of displacement have been performed before the conventional yielding displacement.

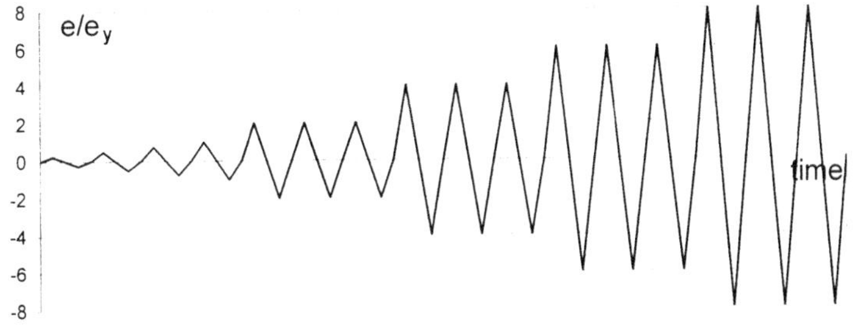

Figure 3.2.10: Load history - recommended ECCS procedure

The end of the experiment was considered when, the final force load applied to the joint was at the most half of the maximum load applied to that joint. In some cases, due to premature failure of joints (XS-W1), or due to unexpected events during the test (one support felt during the XU-W2 testing), the experiment was stopped earlier. The applied loading speed was quasi-static. The total duration of a cycle was 8 minutes, the loading speed depending on the amplitude of displacement imposed. This slow rate of loading was imposed actually by the data acquisition system rate of recording.

TESTING RESULTS

The loading history applied to the specimens adopted the ECCS simplified procedure. Some elastic cycles were applied so as to ensure that at least four levels of displacement were reached before the yield displacement. When determining the yield displacement, some progressively increasing plastic cycles were applied, which are not shown in the moment-rotation graphs, for reasons of clarity. These preliminary plastic excursions are shown in the energy graphs, where all plastic cycles are presented (up to the 50% drop of moment capacity).

Symmetrical Joints

TABLE 3.2.5 gives the loading history for the XS joint series, in terms of number of plastic cycles at each displacement level.

TABLE 3.2.5

LOADING HISTORY OF SYMMETRICAL SPECIMENS

Plastic range	Number of plastic cycles					
	XS-EP1	XS-EP2	XS-W1	XS-W2	XS-CWP1	XS-CWP2
$\pm e_y - \pm 2e_y$	4	4	5	3	3	**
$\pm 2e_y$	3	3	3	3	3	**
$\pm 4e_y$	4	3	3	4	3	**
$\pm 6e_y$	-	6	3	-	3	**
$\pm 8e_y$	-	-	2	-	9	**
TOTAL	11	16	16	10	21	**

** failure of the loading column end-plate

XS-EP specimens

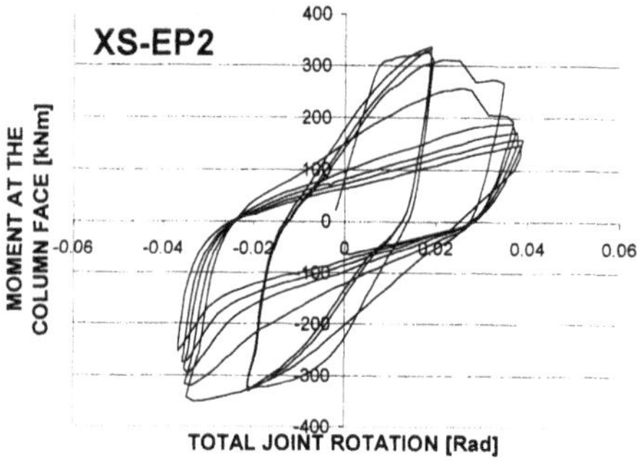

(a)

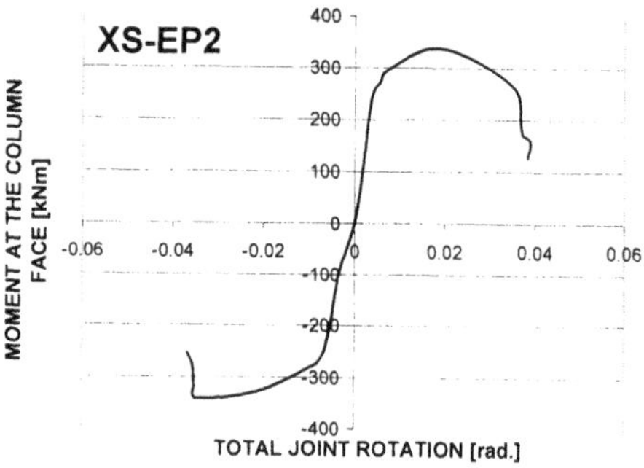

(b)

Figure 3.2.11: Cyclic Moment-Rotation curve (a) and the envelope curve (b) for specimen XS-EP2

The cyclic moment-rotation curve is shown in Figure 3.2.11a, where the rotation represents the global rotation of the joint.

During the symmetrical loading, the two connections began working in the plastic range but as the load increased, only one connection plastified, the other remaining, from that moment on, within the elastic range. This is due to the detailing of the connection: detailing of welds, slight differences in geometrical characteristics or mechanical ones. It resulted a difference in behaviour between the left and the right connections, the plastified connection dissipating the most plastic energy and developing a greater plastic rotation. This can be better emphasised in Figure 3.2.12 for specimen XS-EP2: (a) for the left connection (plastified) and (b) for the right connection. The Figure 3.2.14 shows also the two connections at the end of the experiment.

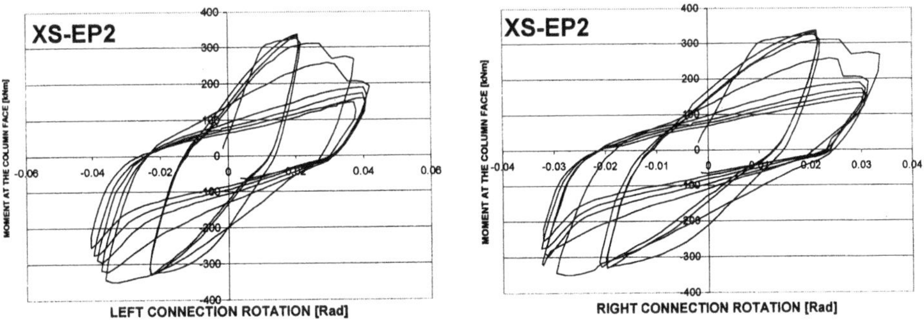

(a) (b)

Figure 3.2.12: Behavioural moment-rotation curve for left and right connections, specimen XS-EP2

The XS-EP specimens are bolted connections with extended end-plate. Before the collapse, the plastic energy was dissipated through the plastic cracking of the end-plate or beam flanges, plastic deformation of the end-plate (visible) or from the local buckling of the beam flanges, a fact that indicates that the connection components are very closely designed. In Figure 3.2.13 is shown the plastic energy dissipated during cycles (a) and the total plastic energy dissipated (b), computed as the hysteretic area from the moment-rotation curve. It can be observed that during the cycles of the same amplitude (3x2e_y, 3x04e_y, etc) the dissipated energy decreases, fact that indicates plastic degradation of the joint. But the rate of energy decreasing is as greater as the multiplier of the e_y is bigger. This confirm the low cycles fatigue approaches.

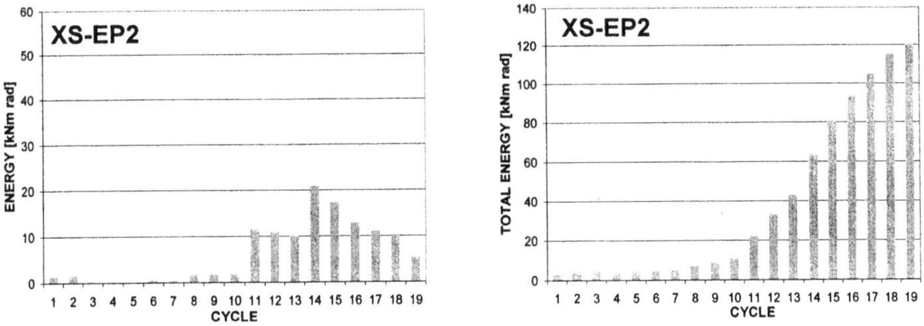

(a) (b)

Figure 3.2.13: Energy dissipated / cycle (a) and total energy dissipated (b) for specimen XS-EP2

The collapse of the specimens was due to the cracking of the end-plate and of the beam, near to the column. It is to be underlined that the cracks were initiated from the weld of beam-to-end-plate or from the heat affected zone located near the welds. Can also be observed a slippage due to the bolts, which permit a larger rotation of the joint. During the test of XS-EP2, a bolt situated in the second row failed. This is not unusual for bolted connections with extended end-plate, the bolts situated in the first row inside the beam being the most stressed ones in tension, the tensioned flange of the beam acting on them as on a T-stub. Generally, these kind of connections have good rotation capacity and energy dissipation.

Another important characteristic of the joint behaviour that shows better the ultimate rotation, and the rotation capacity of the joint is the envelope curve moment - rotation, which joins the peaks of the pairs (moment-rotation). In Figure 3.2.11b is given the envelope curve moment-rotation for the specimen XS-EP 2.

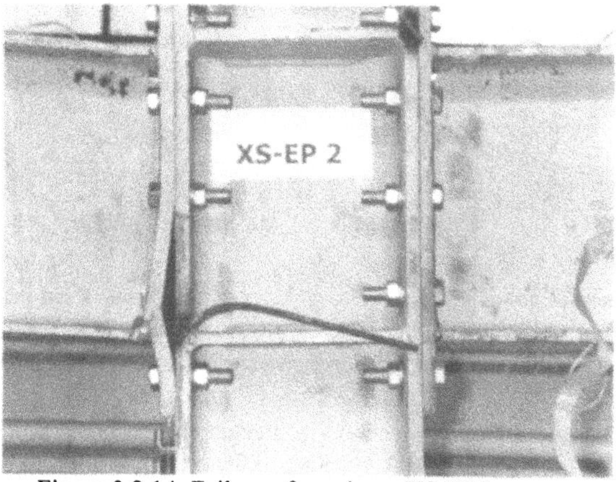

Figure 3.2.14: Failure of specimen XS-EP2

Figure 3.2.14 shows the failure of the joint, with the end-plate plastic deformation and cracking. It can also be seen the place from where the bolt has collapsed. The right connection is almost intact.

XS-W specimens

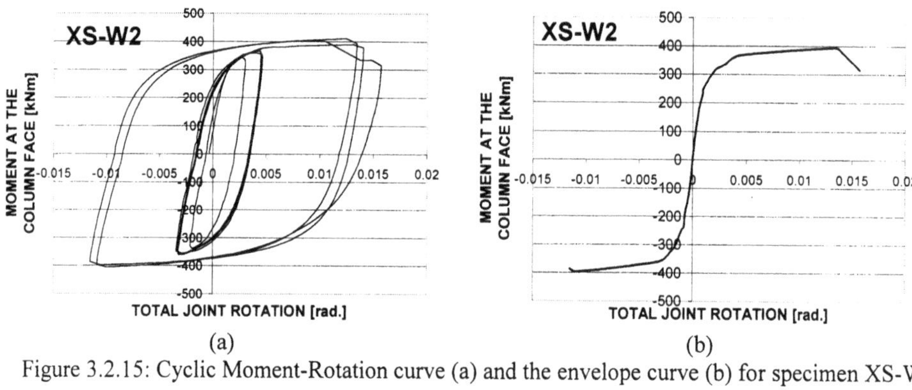

(a) (b)

Figure 3.2.15: Cyclic Moment-Rotation curve (a) and the envelope curve (b) for specimen XS-W2

The specimens have the beam directly welded on the column flange. As to their behaviour, they resisted to a greater bending moment compared to the end-plate connection specimens (XS-EP), but the collapse was brittle and sudden. The behavioural curve moment-rotation of the joint is shown in Figure 3.2.15a, where the moment is computed at the column face, while the rotation represents the global rotation of the joint. During the first cycles, the cycles were asymmetrical, the force actuator being unable to act at the necessary force in negative forces (due to mechanical reasons). This effect was present during the entire testing of the specimen XS-W1 and can be observed in the first plastic cycles on the behavioural curve moment-rotation for specimen XS-W2 (see Figure 3.2.15a).

As it can be observed they present a reduced plastic rotation capacity, fact that affects also the dissipated energy both on cycle and total (see Figure 3.2.16). Although the moment at the column face is significantly greater that the one obtained in the case of XS-EP specimens, the total dissipated energy remains only about half of the one in the case of end-plated joints.

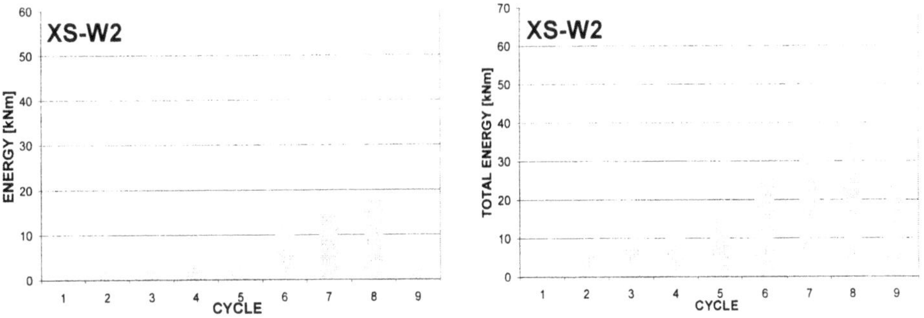

(a) (b)

Figure 3.2.16: Energy dissipated / cycle (a) and total energy dissipated (b) for specimen XS-W2

The envelope moment-rotation curve shows a very small softening branch, also due to premature and brittle failure (see Figure 3.2.15b). It can be observed also the asymmetry in positive and negative ultimate rotations, the joint showing a greater rotation capacity in the positive loading. There have been important differences between the specimens XS-W1 and XS-W2 in terms of their behaviours, as follows:

- the specimen XS-W1 resisted to a larger numbers of cycles
- the same specimen resisted to a greater maximum moment and showed a decreased stiffness
- the total energy dissipated of specimen W2 was considerable bigger (it resisted to a large number of cycles) than that of specimen W1 even if the energy dissipated per cycle was about the same (see Table 3)

These differences between the two specimens can be explained by the only difference that existed between them: the welding details. For the specimen XS-W2 the welds have been the ones provided by the manufacturer, while for the specimen XS-W1, the welds have been adjusted in the testing laboratory, by re-welding the welding root. The welding procedure seems to be the key-point for this type of joints, but not only.

For the first cycles, a visible beam flange buckling was observed, but later the plastification concentrates in the beam flanges and web cracks. The collapse of the specimens was sudden due to brittle rupture of the beam flanges and web (see Figure 3.2.17). At the specimen XS-W1, this rupture was total, the beam being out of contact with the column. For these specimens also, the cracks initiated from the welds or from the heat-affected zone.

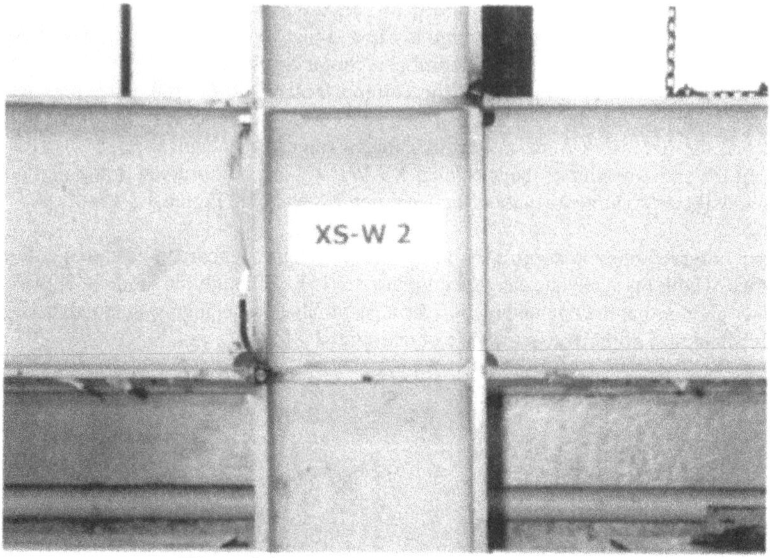

Figure 3.2.17: Failure of specimen XS-W2

XS-CWP specimens

By the detail characteristics of this type of connections, these specimens are rigid from the point of view of both resistance and strength, the cover welded plate assuring an additional strength for the connected beam. That is why, at the symmetrical loading, the connections did not work in plastic range almost at all. The behavioural curve shown in Figure 3.2.18a represents the moment at the column face in ordinate, but the theoretical rotation of the joint in abscissa. At the first plastic cycles, there was also an asymmetry of loading, the actuator attaining its maximum capacity in compression.

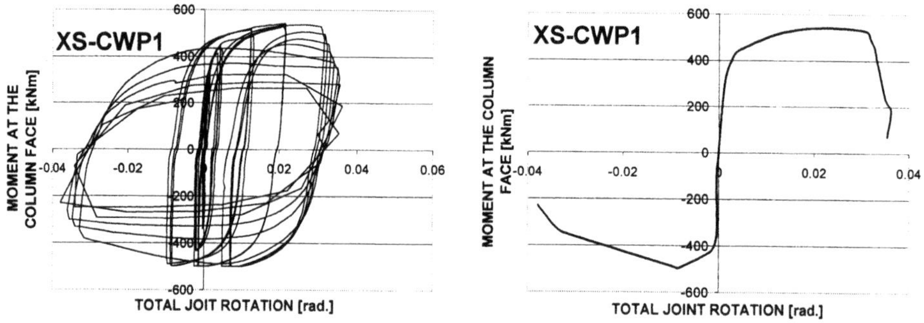

(a) (b)

Figure 3.2.18: Cyclic Moment-Rotation curve (a) and the envelope curve (b) for specimen XS-CWP1

This is only theoretically true, because the plastic articulation formed in the beam, at the end of the cover plate. Figure 3.2.19 illustrates the real beam rotation, in terms of the same moment values. It should be said that the connection plastic rotation is very small, the connection having behaved practically only within elastic range.

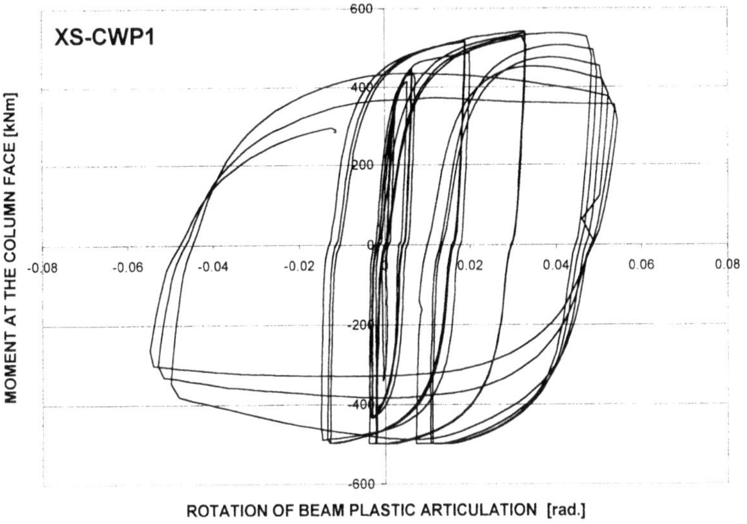

Figure 3.2.19: Rotation of beam plastic articulation for specimen XS-CWP1

The great values in the plastic rotation show the ductility of the beam element rather than the joint behaviour. The energy dissipated in the cyclic behaviour of the specimen XS-CWP1 is shown in Figure 3.2.20a, while the total dissipated energy is shown in Figure 3.2.20b. Both values of cyclic energy and total energy are greater than the ones obtained for specimens XS-EP and XS-W, but this case presents the plastic energy dissipated in the beam. The same phenomenon is present also for this specimen, that within a series of loading of the same amplitude (e.g. $6e_y$), the energy dissipated decreases, and also the decreasing amplitude is greater as the multiplier of the elastic displacement is greater.

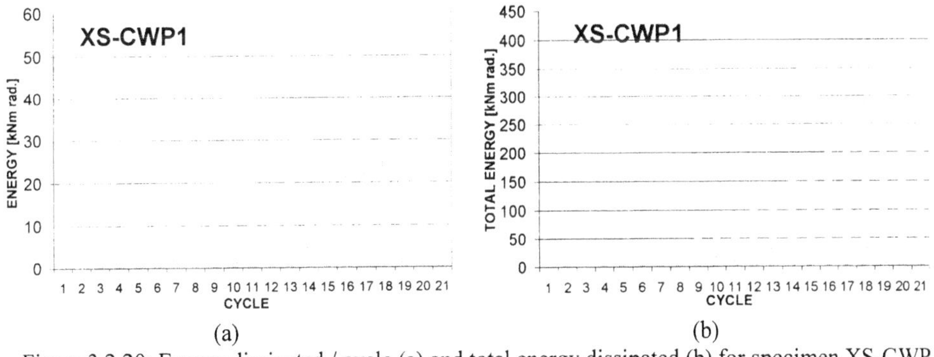

Figure 3.2.20: Energy dissipated / cycle (a) and total energy dissipated (b) for specimen XS-CWP1

For this specimen, the envelope curve (presented in Figure 3.2.18b) shows a behaviour that is closed to an ideal moment-rotation curvature, with the mention that in the negative range the cut envelope curve obtained was affected by the impossibility of the actuator to work in the negative range for cycles of $6e_y$.

Figure 3.2.21 shows the beam plastic hinge formed in the beam web and flanges, in the beam web the plastic articulation following two shear lines formed during the loading. The flanges buckled each at the time the flange was in compression. It can be observed that the beam-to-column connections have no plastic deformations. Unfortunately, during the testing of the specimen XS-CWP2 the column loading end-plate failed by weld rupture, so the specimen was lost.

Figure 3.2.21: Failure of specimen XS-CWP1

Anti-Symmetrical Joints

Loading history for anti-symmetrical joints is presented in TABLE 3.2.6.

TABLE 3.2.6

LOADING HISTORY OF ANTI-SYMMETRICAL SPECIMENS

Plastic range	Number of plastic cycles					
	XU-EP1	XU-EP2	XU-W1	XU-W2	XU-CWP1	XU-CWP2
$\pm e_y - \pm 2e_y$	7	-	4	-	4	1
$\pm 2e_y$	3	3	3	3	3	3
$\pm 4e_y$	3	3	3	3	3	3
$\pm 6e_y$	3	3	3	3	3	3
$\pm 8e_y$	18	28	17	14	3	6
$\pm 10e_y$	-	-	-	-	34	15
TOTAL	34	37	30	23	50	31

XU-EP specimens

The first signs of inelastic deformations were observed in the panel zone, where paint started to blister already at the $\pm e_y$ cycles. Plastic deformations in the panel zone increased progressively with the number of cycles. Deformations of end plate were observed starting with cycles of $\pm 2e_y$, a gap being formed between the end plate and column flange in the tension zone. First cracks in the welds between the beam bottom flange and end plate appeared at the $\pm 6e_y$, respectively $\pm 4e_y$ for the XU-EP1 and XU-EP2 specimens. Limited buckling of the beam flanges was also observed. Deformation of the end plate was also given by the loosening of bolts, which decreased much the stiffness of the connection. Cracking of welds has shown at the top flange only at $\pm 8e_y$ displacement levels. After a number of plastic excursions at $\pm 8e_y$, complete rupture of the extended part of the end plate appeared. Starting from this point, the extended end plate began to act as a flush end plate (see Figure 3.2.23a).

Panel zone showed stable hysteresis loops over the entire loading history, with an important strain hardening. It was the main source of deformation and resistance up to the point when important degradation of moment and stiffness occurred due to rupture of the end plate. The inelastic demand on the panel zone started to decrease at this point, leading to its "relaxation" (see Figure 3.2.22a). On the other hand, extended end plate connection showed a continuous degradation of both stiffness and moment over the loading history (see Figure 3.2.22b).

In the case of XU-EP2 specimen, cracking of the beam flange in the Heat Affected Zone (HAZ), at the root of the weld access hole occurred. Specimen failure culminated with complete fracture of the beam web and top flange at the right connection (see Figure 3.2.23b) and rupture of two bolts below the tensioned flange for the left connection. An important drop in moment capacity accompanied it.

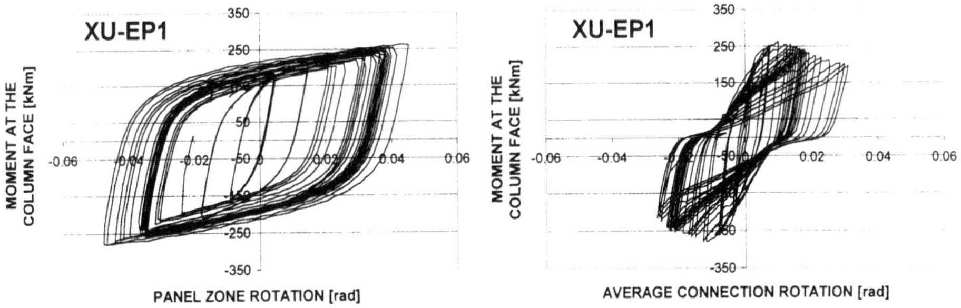

a) b)

Figure 3.2.22: Moment-Rotation relationships for specimen XU-EP1

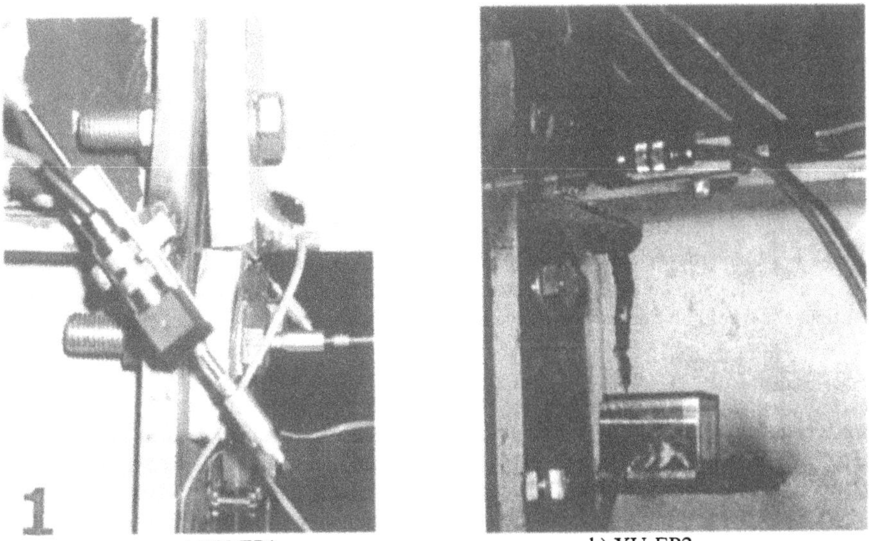

a) XU-EP1 b) XU-EP2

Figure 3.2.23: Rupture of end plate (a) and failure of beam near the beam to end plate connection (b)

Dissipated energy in a cycle and the cumulated one for XU-EP1 specimen is shown in Figure 3.2.24. While the dissipated energy is quite constant in the groups of three cycles up to $\pm 6e_y$, it begins to degrade when displacement levels of $\pm 8e_y$ are reached, mainly due to rupture of the extended end plate.

The cyclic behaviour of the joint in terms of total joint rotation and its envelope are shown in Figure 3.2.25. Maximum moment reached during testing was –280 kNm for XU-EP1 and 260.3 kNm for XU-EP2. The maximum total joint rotations attained before the 50% capacity reduction are of 0.060 and 0.062, respectively.

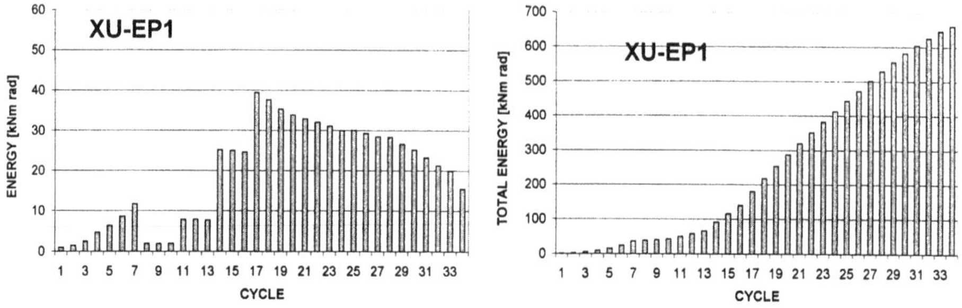

a) Dissipated energy per cycle b) Cumulated dissipation of energy per cycle

Figure 3.2.24: Energy dissipation in the specimen XU-EP1

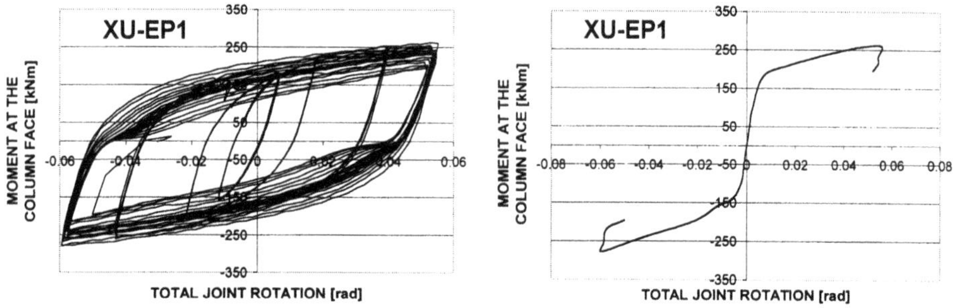

a) Cyclic behaviour b) Envelope of hysteresis loops

Figure 3.2.25: Total joint rotation versus moment relationships for XU-EP1 specimen

XU-W specimens

Panel zone was again the weakest component. It showed important deformations (blistering of paint) at deformations levels exceeding $\pm e_y$. First cracks appeared at $\pm 6e_y$ in the root of the welds between beam bottom flange and column flange. Top flange welds cracked only at the first $\pm 8e_y$ cycles.

In the case of XU-W1 specimen, starting with the 9th cycle at $\pm 8e_y$ cracks in beam bottom flanges propagated progressively into the column flange, which was ruptured on the beam flange width. Column flanges were pulled out, together with the column web tearing (see Figure 3.2.27a). This phenomenon occurred first for the right beam, and continued in a smaller extent on the left side. During successive cycles top flanges ruptured at the root of weld access hole (see Figure 3.2.27b). At the end of the 16th cycle of $\pm 8e_y$ top beam flanges were completely ruptured, and bottom ones caused extensive pull out of the column flange and tearing of the web. Minor buckling of beam flanges occurred during the test.

The XU-W2 specimen showed a similar behaviour, but it was stopped in the 14th $\pm 8e_y$ cycle due to problems at one of the supports. Beam bottom flange welds cracked extensively at the end of the 11th $\pm 8e_y$ cycle. During last cycles, cracks were observed at the welds between transversal stiffeners and column flange, showing initiation of column flange pullout.

As can be observed from Figure 3.2.26a, shear of the panel zone brought the main contribution to the energy dissipation in the joint. Only minor plastic deformations occurred at the beam end (see Figure 3.2.26b), mainly during the last cycles, when effective rupture of the beam top flanges and column pullout at the bottom beam flanges occurred. The term connection rotation is used to denote rotation at the beam end for compatibility among different types of specimen. During the last cycles, due to important degradation of the beam end, demand in inelastic deformations of the panel zone is reduced.

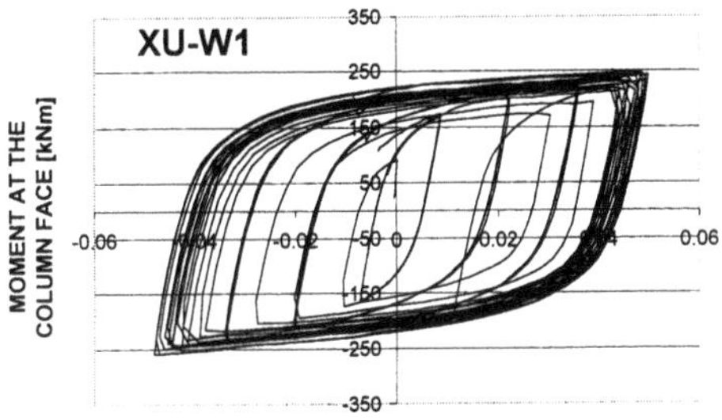

a)

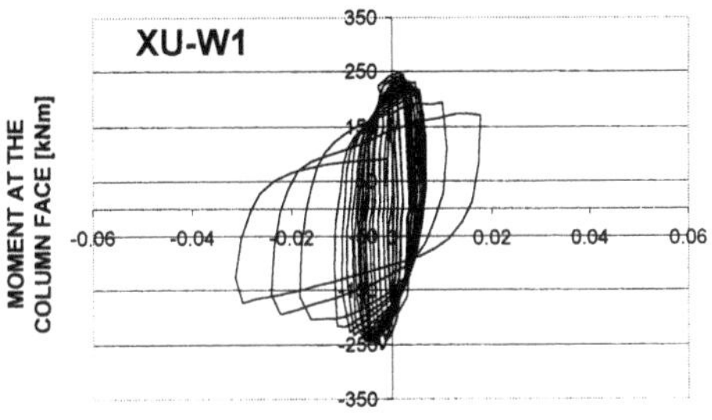

b)

Figure 3.2.26: Moment-Rotation relationships for specimen XU-W1

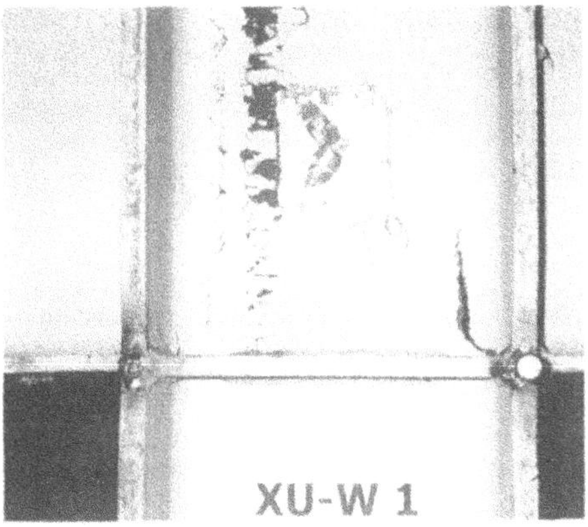

a) XU-W1

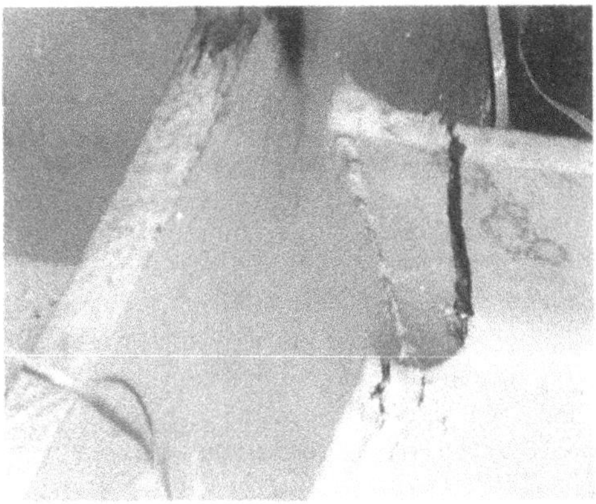

b) XU-W1

Figure 3.2.27: Rupture of column web (a) and failure of beam top flange (b)

The energy dissipated per cycle and the cumulated dissipation of energy are shown in Figure 3.2.28. Energy begins to degrade starting with the 4th cycle of $\pm 8e_y$ in the case of XU-W1 specimen, and in the 6th cycle of $\pm 8e_y$ in the case of the XU-W2 one. During anterior loading history, energy dissipation has a stable character.

Cyclic behaviour of the XU-W1 specimen in terms of total joint rotation and its envelope is shown in Figure 3.2.29. The maximum moment attained during the loading history is -258.6 kNm for XU-W1 specimen and -252.6 kNm for the XU-W2 one. Maximum rotations in the joint were of 0.052 radians for both specimens.

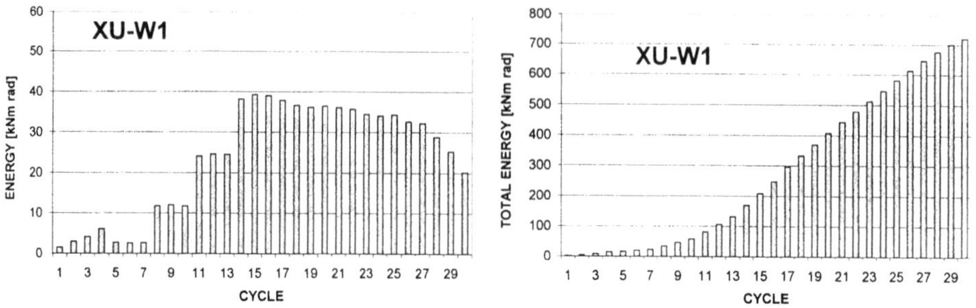

a) Dissipated energy per cycle b) Cumulated dissipation of energy per cycle

Figure 3.2.28: Energy dissipation in the specimen XU-W1

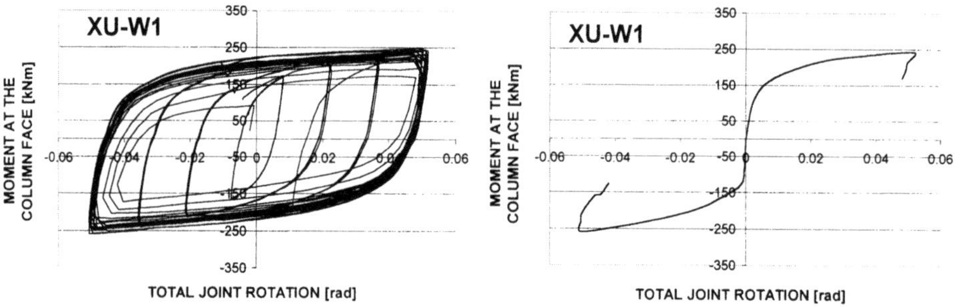

a) Cyclic behaviour b) Envelope of hysteresis loops

Figure 3.2.29: Total joint rotation versus moment relationships for XU-W1 specimen

XU-CWP specimens

Paint in the panel zone began to blister at deformation levels of $\pm e_y$. Through the loading history up to $\pm 10e_y$ panel zone continued to show increasing distortion, without any visible cracks. At the 14th cycle of $\pm 10e_y$ first cracks appeared in the welds at the lower part of bottom cover plates for the XU-CWP1 specimen. Similar cracks appeared at the XU-CWP2 specimen already at the 2nd cycle of $\pm 8e_y$. This is explained by the fact that root of the welds between the cover plate and beam flange were rewelded at the XU-CWP1 specimen.

In the 22nd cycle at $\pm 10e_y$ column web in the panel zone was slightly buckled at the XU-CWP1 specimen. A crack appeared at the left inferior part of the column web in the 27th cycle of $\pm 10e_y$. In the following cycles the fissure spread along the whole bottom edge of the bottom of the panel zone, and then also on the two lateral sides. A significant drop in load capacity was noticed. At the end of the test column web in the panel zone was completely torn on three sides, being deformed out of its plane (see Figure 3.2.30a).

A sudden and deep crack occurred in the 4th cycle of $\pm 10e_y$ at the root of the weld between the right bottom cover plate and column flange for the XU-CWP2 specimen. Three cyclès later a similar sudden crack accentuated at the left connection. In the 11th cycle of $\pm 10e_y$ a crack was formed between the bottom transversal stiffener and the column flange. One cycle later the two cracks on the interior and exterior of the column flange increased suddenly and formed a single one. At the end of the test, the right column flange was ruptured apart, with the crack extending into the column web (see Figure 3.2.30b).

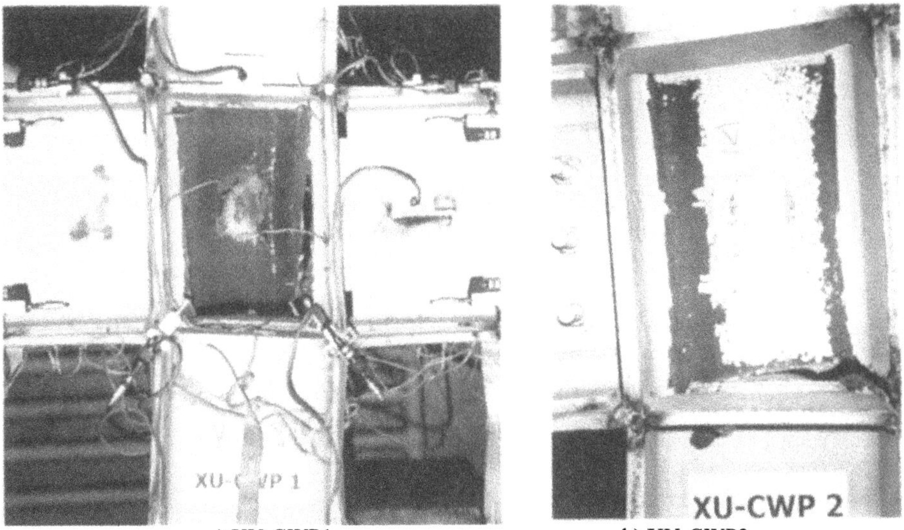

a) XU-CWP1 b) XU-CWP2

Figure 3.2.30: Failure of column web (a) and crack through the column flange and web (b)

The energy dissipated per cycle and the cumulated dissipation of energy are shown in Figure 3.2.31. The XU-CWP1 specimen showed a very stable energy dissipation capacity up to the 30th cycle of $\pm 10e_y$. For the other one, it begins to degrade already at the 4th cycle of $\pm 10e_y$, but the degradation is not so steep.

Cyclic behaviour of the XU-CWP1 specimen in terms of total joint rotation and its envelope is shown in Figure 3.2.32. The maximum moment attained during the loading history is -299.1 kNm for XU-CWP1 specimen and 301.5 kNm for the XU-CWP2 one. Maximum rotations in the joint were of 0.064 and 0.060 radians, respectively.

It has to be underlined that for this type of joint and loading, plastic deformations concentrated in the panel zone. Plastic hinge did not form in the beam. Practically, the total joint rotation (see Figure 3.2.32a) was the same with the rotation given by the panel zone.

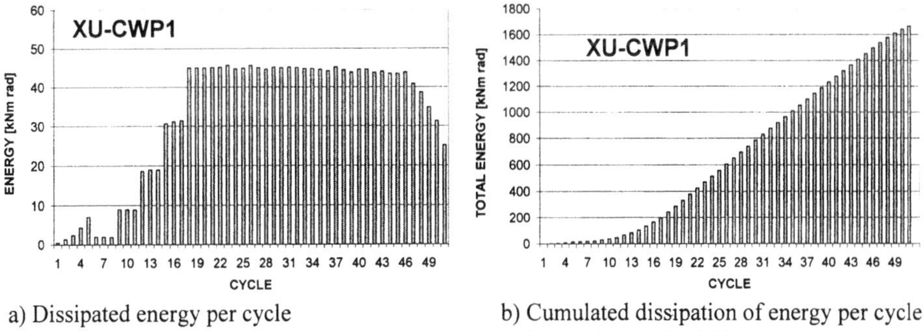

a) Dissipated energy per cycle b) Cumulated dissipation of energy per cycle

Figure 3.2.31: Energy dissipation in the specimen XU-CWP1

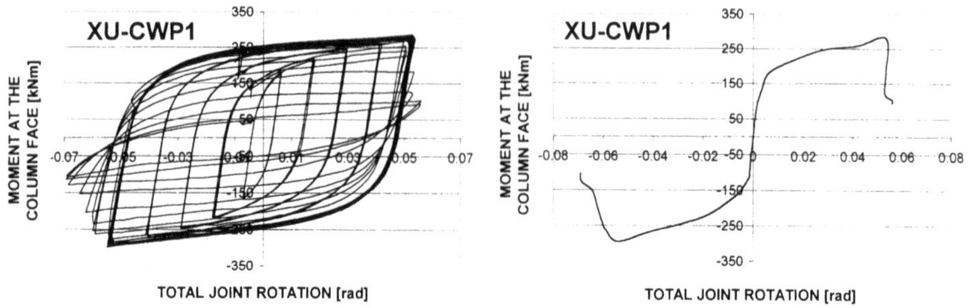

a) Cyclic behaviour b) Envelope of hysteresis loops

Figure 3.2.32: Total joint rotation versus moment relationships for XU-CWP1 specimen

General Results from the Tests

TABLE 3.2.7 comprises the main parameters monitored during the tests:

P_y – the force corresponding to joint yielding

δ_y,– the yielding displacement corresponding to P_y

M_{max} – the maximul bending moment obtained at the column face for the test

ϕ_u – the ultimate (maximum) rotation of the joint

Monitored parameters are also the maximum energy dissipated in a cycle and the total energy dissipated during a test. The number of plastic cycles to failure is considered an important parameter describing the plastic performances of the joint.

TABLE 3.2.7

MAIN CHARACTERISTICS OF CYCLIC TESTS

SPECIMEN	P_y [kN]	δ_y [mm]	M_{max} [kNm]	ϕ_u^+ [rad]	ϕ_u^- [rad]	Max. En/cycl. [kNm rad]	Total En. [kNm rad]	Nr. of pl. cycles
XS-EP 1	569.12	10	334.17	0.031	0.033	20.83	76.74	11
XS-EP 2	522.38	6.5	337.94	0.039	0.037	20.84	120.15	16
XS-W 1	642.40	4.36	441.94	0.028	0.010	19.71	125.20	16
XS-W 2	677.88	4.85	412.12	0.017	0.013	18.05	64.69	10
XS-CWP 1	678.40	5.60	542.01	0.036	0.038	45.06	390.00	21
XS-CWP 2	Accidental failure of the column loading end plate							
XU-EP 1	170.0	15.00	280.0	0.055	0.060	39.4	661.50	34
XU-EP 2	170.0	15.00	260.3	0.057	0.062	42.1	924.60	37
XU-W 1	177.0	12.80	258.6	0.052	0.051	39.3	721.00	30
XU-W 2	177.0	12.80	252.6	0.052	0.050	38.0	611.70	23
XU-CWP 1	178.3	11.00	299.1	0.054	0.064	45.7	1666.80	50
XU-CWP 2	178.3	11.00	301.5	0.060	0.060	52.0	1051.20	30

It can be observed that generally, there are not big differences between the two specimens of the same type. Anyway, a few comments should be added. First, the XS-W1 specimen was deliberately different from the XS-W2 specimen, by re-welding of the weld roots of the former specimen (as explained earlier). Secondly, although the results are similar in terms of maximum values of rotations and moments resisted by the joints, the failure mechanism and the number of plastic cycles are sometimes substantially different. This is particularly true for the XU-CWP joints.

Due to the different statical schemes for the two types of loading, the bending moment in the node for anti-symmetrical scheme is obtained by half the force needed for the symmetrical one. But this is not the cause of the drastic drop in the yield force P_y from the XS to XU series. A change in the loading type caused also a change of the joint resistive components. Test results are in accordance with the expected joint behaviour as it can be seen in TABLE 3.2.4.

In what concerns the XS series, joint resistance and rigidity are expected to increase in the range EP-W-CWP. Results in terms of maximum moment attained during testing confirm this trend. It should be noted that the maximum moment for all joints is computed at the column face. The behaviour of the three types of joints is quite different:

- XS-EP joints showed a good ductility mainly due to the end plate in bending and partially due to local buckling of beam flanges. Anyway, failure was achieved not only by rupture of the end plate, but also by rupture of the beam flange at its connection to the end plate (in the weld or in the heat affected zone). Mean maximum rotation attained (0.035 radians) shows a good plastic behaviour of XS-EP joints. The failure was a ductile one.
- Ductility of XS-W joints was mainly affected by the brittle and sudden failure of the beam to column connection. The mean maximum rotation (0.017 radians) proves this fact. This type of joint is especially affected by the quality of welds at the beam to column connection.
- The objective of XS-CWP joints, reinforced at the beam to column connection, was accomplished, the plastic zone shifting from the column face connection into the beam. Therefore, its ductility is practically given by the beam ductility. The equivalent rotation at the column face (0.037 radians) is smaller than the real one in the plastic hinge.

The dissipated energy, both the maximum and the cumulated one have comparable values for XS-EP and XS-W specimens, and are substantially greater for XS-CWP specimen. The number of hysteresis loops to failure is again much greater for the latter case.

In the case of XU series, the maximum moment attained is expected to have close values, taking into consideration that the main resisting component is the web panel in shear. This was proved to be true, with the specification that the XU-W joint showed the smallest maximum moments and the XU-CWP the biggest values. Their ductility is comparable (0.051-0.059 radians mean values), all the joints proving good ductility. Behaviour of the three types of joints is in general similar, being governed by the shear behaviour of the column web panel, but there are also some particularities:

- Beside shear in the panel zone, behaviour of XU-EP joints was influenced by the end plate in bending. At the first plastic cycles the ductility demand was distributed between the two components, participating together to the plastic excursions. While panel zone had stable hysteresis loops, the behaviour of the end plate was characterised by significant degradation (due to loosening of bolts and rupture of the extended part of the end plate).
- In the case of XU-W joints, plasticity was spread between the panel zone and in a smaller extent the beam flanges. These joints have been the least ductile, failure occurring by brittle fracture of beam flanges and pullout of the column flange.
- The web panel governed exclusively the behaviour of XU-CWP joints. Failure was due to ductile degradation of the panel zone, which was finally torn apart. A concern should be expressed here about this type of joints, as the second specimen failed by complete rupture of the column flange.

The maximum dissipated energy is higher for the XU-CWP joints. The cumulated energy is considerably higher for the same type of joints. This is partially caused by the increased number of plastic cycles.

In what concerns the differences between the XS and XU series, change of loading type led to important differences between the two series. Generally, a drop in maximum moment is observed for the anti-symmetrical loading. Anyway, this drop is different among the connection types as follows: 15% for the end plate joints and about 40% in the case of welded and cover plated joints. This fact is explained by close resistance of the extended end plate and the web panel, both components being involved in the plastic mechanism. While for the other two cases the web panel was the main participating component. Joint rotations are considerably higher for XU series. Improved ductility in the case of XU joints is given by good rotation capacity and stable hysteresis loops of the web panel in shear. Anti-symmetrical joints have generally increased energy dissipation capacity with respect to the symmetrical ones. This fact is given by the increase of both maximum energy dissipated per cycle and number of cycles (case of EP and W joints), or only increased number of cycles (case of CWP joints).

COMPARISON BETWEEN THE EC3 AND EXPERIMENTAL RESULTS

TABLE 3.2.8 comprises the results of the experimental tests compared to that of EC 3 – Annex J, in terms of joint bending moments, rotational stiffness and ultimate rotation attained. It should be noted that for this comparison, the joint characteristics are computed with the measured strengths and dimensions of the joint components.

TABLE 3.2.8

COMPARISON BETWEEN COMPUTED AND EXPERIMENTAL JOINT CHARACTERISTICS

SPECIMEN	$M_{j,Rd}^{(exp)}$ [kNm]	$M_{j,Rd}^{(th)}$ [kNm]	$S_{j,ini}^{(exp)}$ [kNm/rad]	$S_{j,ini}^{(th)}$ [kNm/rad]	$\phi_y^{(exp)}$ [rad]	$\phi_u^{(exp)}$ [rad]
XS-EP 1	252.12	262.7	69978	142932.2	0.0036	0.031
XS-EP 2	277.05	261.3	63985	140886.8	0.0043	0.039
XS-W 1	343.14	319.4	220473	□	0.0016	0.028
XS-W 2	281.58	320.2	291077	□	0.001	0.017
XS-CWP 1	382.83	468.1	261697	□	0.0014	0.036
XS-CWP 2	**	464.0	**	□	**	**
XU-EP 1	144.4	169.2	44017	43727.2	0.0033	0.060
XU-EP 2	132.2	169.1	46713	43718.2	0.0028	0.062
XU-W 1	145.6	163.6	59429	68792.1	0.0026	0.052
XU-W 2	123.3	164.1	47794	69062.1	0.0026	0.052
XU-CWP 1	148.3	178.6	62939	75597.2	0.0024	0.064
XU-CWP 2	145.0	177.4	55411	74963.1	0.0026	0.060

The yielding bending moment $M_{Rd}^{(exp)}$ is computed according to the ECCS procedure, as in Figure 3.2.33, resulting at the intersection of the $S_{j,ini}$ line and the tangent to the envelope curve $S_{j,ini}/10$ line. The intersection point corresponds to the pair ($M_{Rd}^{(exp)}$,ϕ_y).

Comparing the experimental and computed values of joint moment capacity, it can be observed that generally, close values are obtained for the XS series. An exception is the XS-CWP joint, which showed considerably lower experimental value. In the case of XU series, all experimental values are lower than the ones computed by Annex J of EC3. The difference between the computed and measured yielding bending moment could be explained by several causes:

- Annex J of EC3 does not consider cyclic loading neither strain hardening
- On the other hand, the procedure applied for determining the experimental yielding moments is a conventional one and is greatly influenced by the initial stiffness of the joint

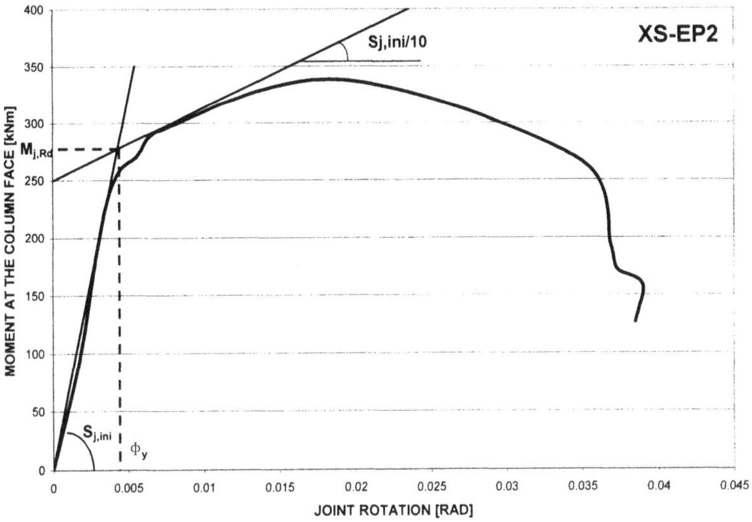

Figure 3.2.33: Definition of $M_{Rd}^{(exp)}$ and ϕ_y.

In what concerns the initial stiffness of the joints, numerical and experimental results agree fairly well for the XU series, while significant differences are noticeed for XS series. Anyway, stiffness is much lower for the anti-symmetrical joints both from experimental and computed stiffness values. This fact is again given by the deformability of the panel zone.

CONCLUDING REMARKS

As it was expected, the loading type (symmetrical or anti-symmetrical) significantly affects the response parameters of beam-to-column joints. The main component that brings the difference is the panel zone in shear. The most important consequences on the cyclic behaviour of beam-column joints are the reduced moment capacity and, (in general) increased ductility with more stable hysteresis loops in the case of anti-symmetrical loading.

These tests were conducted under limiting cases of load asymmetry. The two loading types affect significantly joint properties in terms of initial stiffness, moment and rotation capacities. Therefore, when modelling the joint for structural analysis, different characteristics should be used for gravitational and lateral loading.

Investigation of the different joint typologies revealed the importance of detailing of the connection and the welding procedure, as well as its quality. Defective welding was responsible for such phenomena as crack initiation and early cracks in welds or heat-affected zone.

Bolted end-plate joints showed an increased rotation capacity and more ductile behaviour with respect to welded joints. Extended end plate connections should be designed so as to prevent brittle failure by bolt rupture. Loosening of bolts during cycle reversals has lead to stiffness degradation. Another aspect characteristic to anti-symmetrical bolted joints is the distribution of ductility demands between the end-plate (connection) and the panel zone.

Generally, failure was brittle for welded joints and ductile for the other ones in the case of symmetrical loading. The ductile behaviour was due to connection (bolted joint) and due to shifting of the plastic hinge away from the column face in the case of CWP joint. Participation of panel zone to the plastic mechanism significantly increased the ductility of the anti-symmetrically-loaded joints. Anyway, welded joints failed in a brittle manner in this case, too.

Generally, the joint with cover and web plates showed a good behaviour. Anyway, care should be taken when designing such joints due to potential problems caused by increased moment at the column face.

REFFERENCES

Ciutina A., Stratan A. (1999). Cyclic tests on beam to column connections. *Second international conference of PhD students*, Miskolc, Hungary

Eurocode 3 Part 1.1 (1992). *General rules and rules for buildings.* CEN, Brussels, Belgium

ECCS, (1986). *Recommended Testing Procedures for Assessing the Behaviour of Structural Elements under Cyclic Loads*, European Convention for Constructional Steelworks, Technical Committee 1, TWG 1.3 – Seismic Design, No.45

Grecea D. (1999). Caracterisasion du comportamnt sismique des ossatures metalliques – Utiliation d'assemblages a resistance partielle. *PhD Thesis.* INSA Rennes, France

SAC Joint Venture (1995). Connection Test Summaries. *Report No. SAC-96-02*, Sacramento, California, USA

SR EN 10002-1(1990). *Metallic materials tensile testing.* CEN, Brussels, Belgium

Suita K., Nakashima M., Morisako K. (1998). Tests of welded beam-column subassemblies. *Journal of structural engineering.* November, 1236-1252

3.3

INFLUENCE OF HAUNCHING

Peter Sotirov, Nikolay Rangelov, Ognian Ganchev, Tzvetan Georgiev, Jordan Milev, Zdravko Petkov

INTRODUCTION

In the last 100 years steel structures have became one of the most popular types of structures used for buildings and engineering construction works. Several generations of structural engineers have been working on the development of design methods and fabrication technologies in the field of steel structures. Nowadays structural steel is a widespread construction material. The industry has created a set of structural shapes that most effectively accommodate common needs of buildings and civil works. All that results in a variety of different structural systems able to cover every need of modern society. Frame structures are one of the most used structural systems because of their architectural versatility and technological simplicity. They are often adopted for industrial buildings and for public houses as well.

The task for optimal design of frame steel structures is sometimes more art than science. Structural engineers are supposed to find the best balance between strength and stiffness demand on one side and bearing capacity supply on the other. Logically the sizes of frame elements are dominantly governed by the pattern of the bending moment diagram. The last is typical especially for one story large span industrial frames. Quite often industrial frames are designed with constant-height columns and rafters cross-sections because of technological benefits and variable depth of the transition part between rafter and column. This part of the frame is defined as haunch, having its primary role to cover the increasing bending moment demand near to the column.

Moment resisting steel frames (MRSF) are applicable not only for one-storey buildings but often they form the main skeleton of multi-storey ones. As an addition to the architectural versatility and technological simplicity typical for frames, their advantages of good ductile performance have to be emphasized, which is of vital importance for construction in regions with high seismicity. Current seismic codes assign the largest behaviour factor for MRSF and therefore the lowest seismic design forces. The distances between columns in multi-storey MRSF are usually smaller compared with the typical spans of industrial portal frames. In those cases the application of haunches helps the designer not only to develop more economical solution but also to establish additional benefits concerning seismic behavior. Haunches may be classified as type 1 with tapered web and constant flange width and type 2 with variable width of the flanges and constant web depth (Figure 3.3.1). A combination between both the types above is also possible, however it is too complicated for practical realization.

Several advantages and disadvantages of these main types of haunching might be pointed out. Which variant would be more appropriate depends on the particular case and technical requirements for the designed structure. Haunches type 1 are more effective because the increase of the web depth leads to a significant increase of the moment of inertia and bending capacity. An increase of the panel zone height is another result, which reduces the panel zone yielding. In that case a greater beam plastic rotation is necessary in order to achieve equivalent seismic energy dissipation. Haunching type 1 however sometimes may require removal of architectural finishes such as column cladding or false ceiling. By the application of constant height haunches the majority of the previously pointed disadvantages are avoided, but the effect of strengthening and stiffening is lower. Moreover, the beam flange width is limited by the dimensions of the column section, and on the other hand it cannot be too large because of undesirable effects of premature local buckling.

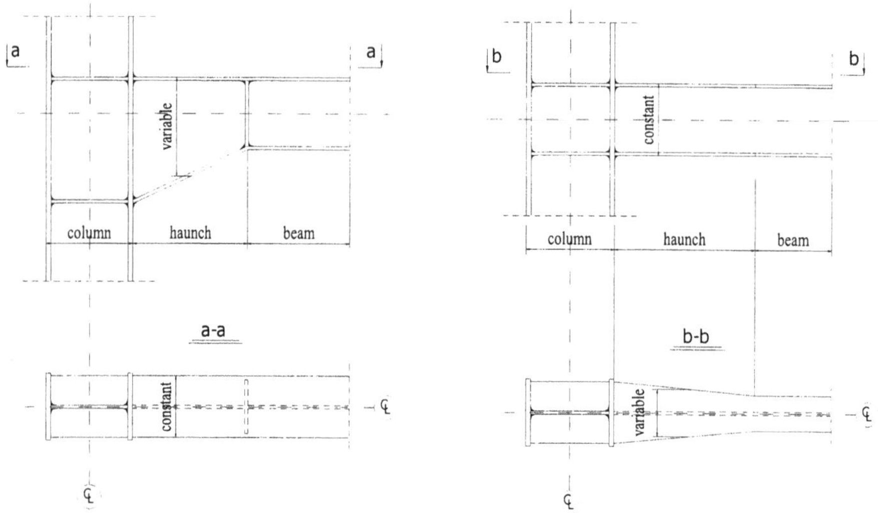

Figure 3.3.1: Types of haunching

A lot of factors like technology restrictions for fabrication, available steel profiles on the market, design traditions etc. reflect on the variety of used haunches. Haunches are applicable both in the case of built-up sections or hot rolled profiles, with bolted or welded connections.

Application of haunches in MRSF is often combined with the column-tree configuration (Figure 3.3.2), which is quite popular in Japan, in the former USSR, Bulgaria and some other countries. The stub-beams are welded to the columns by the industrial manufacturer and the remaining beam segments are field-connected to the stub-beams. Following this manner of design, fabrication and assembly allows all welds of columns, stub-beams and stiffeners to be shop executed with superior welding processes under high quality control.

The growth of engineering knowledge about seismic performance of steel frames and the conclusions drawn after recent severe earthquakes reflect on the development of beam-to-column connection design. The experience of Northridge (1994) and Kobe (1995) earthquakes unfortunately proved that no type of beam-to-column joint could be classified as 'prequalified' one. The influence of haunching to the seismic behaviour of steel moment resisting frames is of high importance. Proper application and detailing of haunched joints could be a powerful tool for ductile design of steel frames.

In this sub-chapter, two different haunched joint typologies are proposed, and the results of cyclic testing of six full-scale beam-to-column assembly specimens are summarized. The experimental study was carried out in the Steel Structures Research Laboratory of the Faculty of Civil Engineering, UACEG, Sofia.

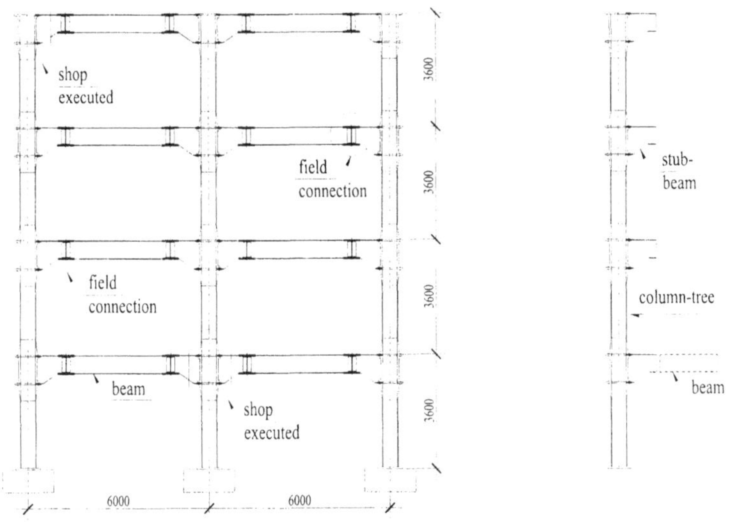

Figure 3.3.2: Column-tree configuration

RECENT STRATEGIES FOR IMPROVEMENT OF BEAM-TO-COLUMN JOINTS

The good seismic behaviour of the moment resisting steel frames and their ability to dissipate input seismic energy and thus to sustain strong ground motions depend mostly on the ductile behaviour of the beam-to-column joints. The column-tree configuration widely used in Japan (Figure 3.3.3*a*) and the so-called 'prequalified' joints in the USA (Figure 3.3.3*b*) have been considered to possess stable hysteretic behaviour. However, after Northridge (1994) and Kobe (1995) earthquakes, damages were found in the majority of beam-to-column connections, thus demonstrating that the popular details do not provide the expected ductile behaviour. Various brittle failure modes and restricted rotation capacity were observed (Bruneau *et al.*, 1998). The lessons from the recent earthquakes came to prove the need to develop new beam-to-column joints.

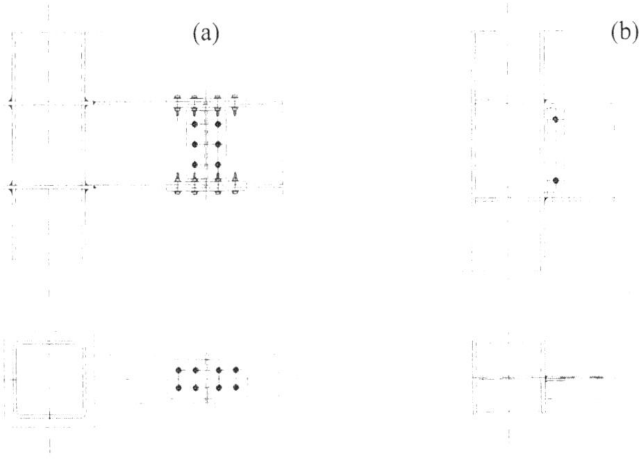

Figure 3.3.3: Most popular MRSF details used in Japan (a) and USA (b)

Two key strategies for improvement of the joint behaviour have been developed: (a) strengthening the connection and (b) weakening the beam that frame into the connection. Both of them aim at relocating the plastic hinge away from the column face as illustrated in Figure 3.3.4, thus avoiding the problem of poor ductile behaviour and potential fragility of the welds.

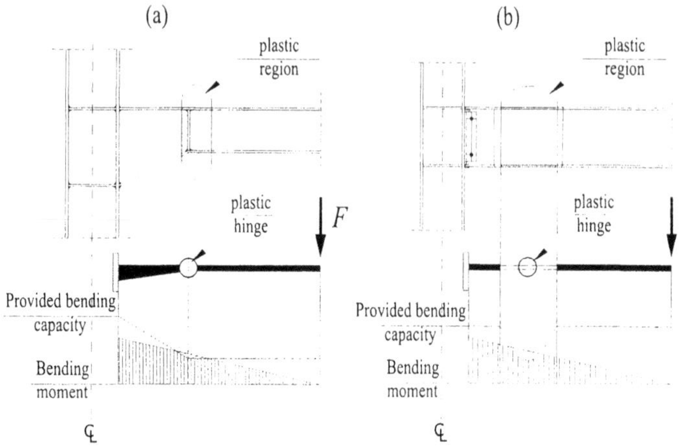

Figure 3.3.4: Principal ideas of the two strategies for joint behaviour improvement: (a) strengthened connection and (b) weakened beam

Within the frame of the *strengthening approach*, a connection stronger than the beam can be obtained by means of haunches (top and bottom, or bottom haunches only), cover plates, upstanding ribs, side plates *etc.* For more details, the reader is referred to Bruneau *et al.* (1998). Strengthening strategy for frame joints is often combined with column-tree configuration.

The basic concept of the *weakening approach* (beam-strength-reduction strategy) is that, in order to guarantee the development of plastic deformations in a particular beam section, the latter is designed with intentionally reduced strength. Consequently, plastic moment capacity is firstly achieved at that section and the location of the plastic hinge is controlled through the design. The most popular detail of this type is the so-called "dog bone" originally proposed and tested by Plumier (1990). Besides the classical "dog bone" profile, several variants of flange shapes have been suggested to achieve reduced beam section. Engelhardt *et al.* (1996) investigated the case of circular cuts (Figure 3.3.5*b*). Chen et al. (1996) proposed to taper flanges according to a linear profile so as to approximately follow the varying bending moment diagram (Figure 3.3.5*c*). The same idea was also suggested by Iwankiw and Carter (1996).

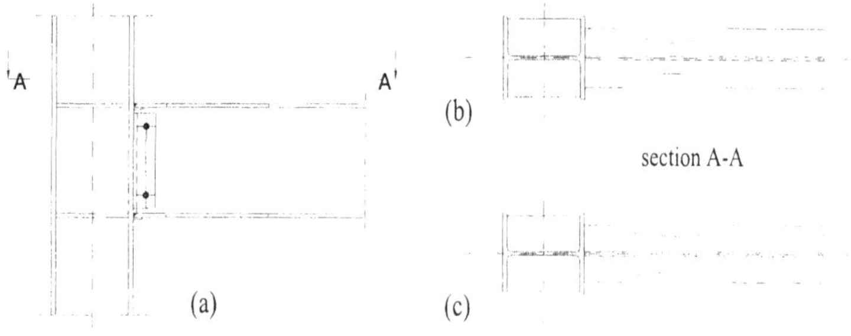

Figure 3.3.5: Reduced-beam-section designs

The "dog bone" philosophy is practically more suitable for the case of hot-rolled beams, where the reduction of beam strength is realized simply by cutting out part of the flanges. In the case of built-up sections the same result may be achieved either by shaving the ordinary fabricated beam flanges or by design of special flange profiles.

The various design solutions proposed in the literature possess some merits and disadvantages, and it seems that no detail may be assumed to be perfect. Therefore new ideas are still worth exploring. For example, recently Popov et al. (1996) suggested a combination between the two strategies, namely a cover plates connection with circular "dog bone" cuts in the beam flanges.

PROPOSED BEAM-TO-COLUMN JOINTS

In the study reported here, an improvement of the hysteretic behaviour of the beam-to-column connections in moment-resisting frames is pursued by using tapered beam flanges in two variants. Both of them are suitable for the case of built-up beams which have the advantage to be proportioned more flexibly to attain a better balance between internal forces and resistance, and the column-tree configuration with fully shop-welded rigid beam-to-column connections and field-welded erection splices in the beams. It is worth noting that high ductility can be achieved by several types of connections, however fully welded rigid connections are still reckoned to possess the highest dissipation capacity (Mazzolani & Piluso, 1996).

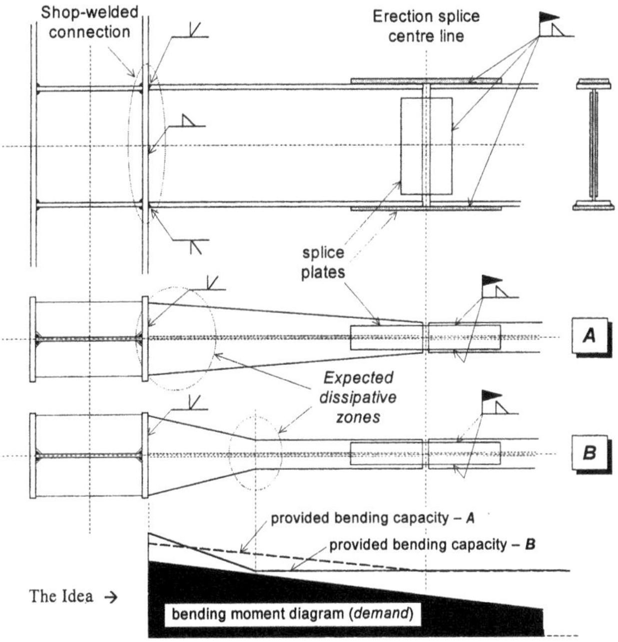

Figure 3.3.6: Proposed beam-to-column joints

The two types of joints are shown in Figure 3.3.6. In type A the flanges are tapered to follow approximately the varying bending moment diagram. Thus, on one hand the solution is more economical, and, on the other hand, under an earthquake load, the yielded zones in flanges are expected to spread over the entire portion that follows the moment diagram. As a result, the energy dissipation capacity is believed to be higher than in the conventional non-tapered flanges due to the larger volume of plasticity. Nevertheless, the expected dissipative zone is mostly near the column face.

Detail B is of the 'new generation' joints. The shifting of the plastic hinge from the column flange is achieved by a specific flange profile, following the idea illustrated in Figure 3.3.6. Thus both the above-mentioned strategies are incorporated in one detail. On one hand, the connection is strengthened without additional cover plates and ribs, keeping a constant beam depth, and, on the other hand, the beam is weakened at a predetermined location to provoke the development of a plastic hinge. This solution seems more economical than those mentioned in the previous sub-section and reserves the main advantage of possessing predictable behaviour.

The design of such a joint is based on the following procedure. Firstly, the location of the dissipative zone is selected so that the notional plastic hinge will affect neither the connection nor the beam splice. Based on experimental observations, it is suggested to choose the distance to the column face about one beam depth. The beam section is sized for the design internal forces at that point. Then the usual capacity design procedure is applied to the cross-section at the connection, the connection itself and the column, including the panel zone.

Obviously, the idea of the considered joints can be transferred to the case of column-tree configuration with hot-rolled columns and beams, where only the stub-beams must be built-up.

EXPERIMENTAL STUDY

Specimens, test setup and instrumentation

The test beam-to-column assemblies are full-scale, extracted from an especially designed two-bay four-storey regular moment-resisting frame. The specimens were fabricated by a Bulgarian industrial manufacturer. The column sections were produced by automatic welding under flux, whereas the beam sections, small in size for the automatic welding equipment, were built-up by semi-automatic welding under CO_2 shield. The full penetration welds between beam and column flanges were also made under gas shield. The erection splices were hand-welded with covered electrodes in the laboratory, thus simulating the field conditions. The steel grade used corresponds to S235 (EN10025).

A total of 6 specimens were fabricated in two series: 3 identical ones (labelled A-1, A-2 and A-3) of type A, and 3 identical specimens (B-1, B-2 and B-3) of type B. The layout, cross-sections and main dimensions are shown in Figure 3.3.7.

The test setup was designed to accommodate specimens in a horizontal position as shown in the figure. The load was applied to the cantilever beam end by a hydraulic actuator, through a clevis bolted to the beam end plate. The actuator had a displacement range of ±200mm and a capacity of ±200kN. No axial load was applied to the column. To prevent out of plane motion of the beam, lateral bracing was provided near the beam end and near the beam splice.

A total of 14 inductive displacement transducers were installed to measure the global response of the specimens: beam end displacement, joint rotation, panel zone shear deformation, and possible movement of the supports (Figure 3.3.7). Since the displacement history was used to command the actuator, the actual beam end displacement was somewhat smaller than that assigned to the actuator. Therefore the displacement measured by *IT1*, δ_1, was regarded as the true imposed displacement. Additionally it was 'cleared' of the support movements. A *Keithley DAC-02* card was used to send the displacement command signal to the actuator servovalve. The applied load was measured by a load cell attached to the actuator.

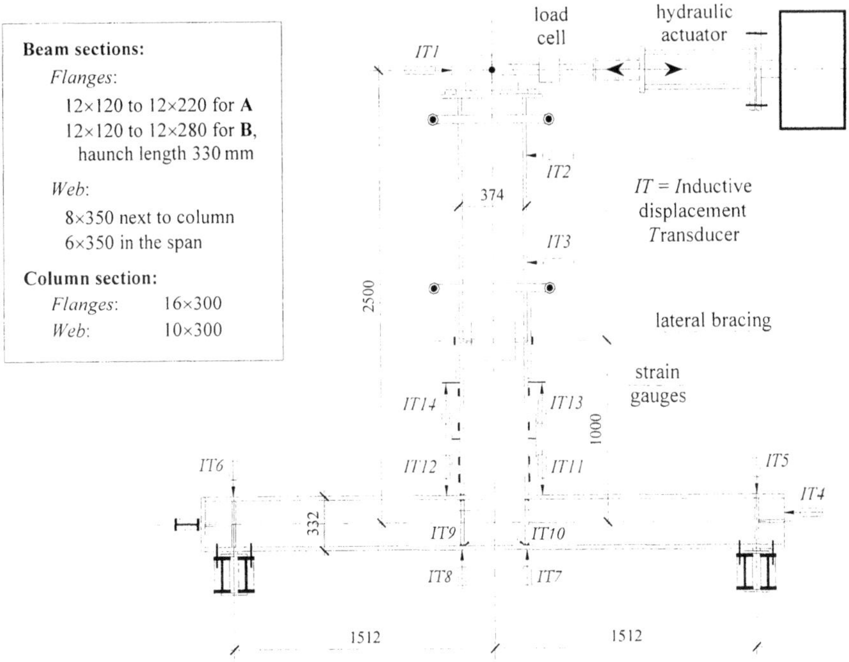

Figure 3.3.7: Test specimens, experimental setup and inductive displacement transducers

Strain gauges were used to investigate the local response of the beam flanges including the splice plates. The gauges had a guaranteed range of at least 0.5% strain, however, more than 1% strains were successfully read during the tests and valuable information was gained on the local response and the strain distributions. Figure 3.3.8 shows the arrangement of the strain gauges and some additional notation.

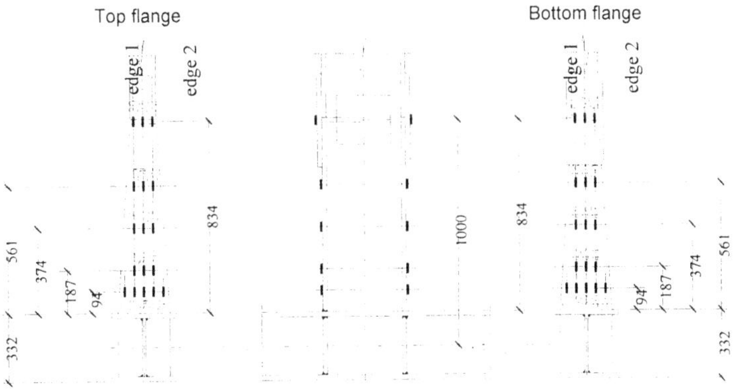

Figure 3.3.8: Strain gauges arrangement

All the instruments were connected to a PC-based data acquisition and control system with software especially developed by the Laboratory staff.

Loading history

The testing programme was based on the recommendations of ECCS (1986). The specimens were tested under displacement control, following a loading history consisting of stepwise increasing deformation cycles. Initially, four cycles in elastic range were applied with amplitudes of $\pm 0.25v_y$, $\pm 0.50v_y$, $\pm 0.75v_y$, and $\pm v_y$, where v_y was the expected first-yield displacement of the beam end. Eventually, correction of v_y was made. Then the testing continued in the plastic range with three full cycles at each amplitude level $\pm 2v_y$, $\pm 3v_y$, and so on, instead of the recommended by ECCS (1986) $\pm(2+2n)v_y$. The latter was only applied to specimen A-3. For specimens A-3 and B-3 an initial displacement corresponding to approximately 30% of the plastic bending capacity was applied to simulate the effect of the gravity loading. A typical loading history and the corresponding actuator force response for specimen A-2 are shown in Figure 3.3.9.

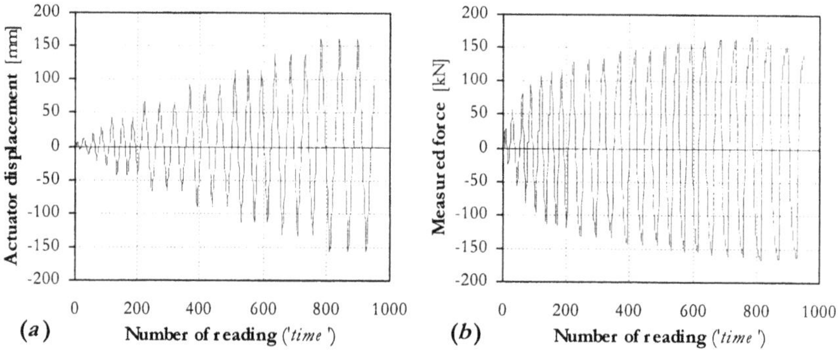

Figure 3.3.9: Typical loading history (*a*) and force response (*b*)

Data processing

The most important parameters that characterize the specimen behaviour are the applied load *F*, beam end displacement, joint rotation, panel zone shear deformation, beam deflection, plastic rotation, and dissipated energy. The total displacement of the beam end, δ_{total}, is caused by the deformations of the beam itself, column and panel zone. The joint rotates on a certain angle, θ_{joint}, and, due to shear deformations γ, the panel zone changes its initial configuration. Thus the total displacement of the beam end, δ_{total}, can be separated into three components: displacement due to deflection of the beam itself, δ_{beam}, displacement caused by rigid joint rotation, δ_{joint}, and a contribution from the panel zone shear deformation, δ_γ. This is illustrated in Figure 3.3.10.

The above-mentioned parameters are obtained on the basis of the measured displacements, excluding the support movements read by *IT4*, *IT5* and *IT6*. For example, the total beam end displacement, δ_{total}, is obtained as

$$\delta_{total} = \delta_1 - \delta_4 - (\delta_6 - \delta_5) \times (L_b / L_c), \tag{3.3.1}$$

where L_b and L_c are the axial length of the beam and column, respectively (Figure 3.3.10), and δ_i is the displacement read by the ith transducer *ITi* (see Figure 3.3.7). Accordingly, the joint rotation and the panel zone shear deformation are determined as

$$\theta_{joint} = (\delta_7 - \delta_8) / d_{7\text{-}8} - (\delta_6 - \delta_5) / L_c, \tag{3.3.2}$$

$$\gamma = (\delta_{10} - \delta_9) \times d/_{2ab}, \tag{3.3.3}$$

$d_{7\text{-}8}$ being the distance between *IT7* and *IT8*, and *d*, *a* and *b* being the diagonal and the two dimensions of the panel.

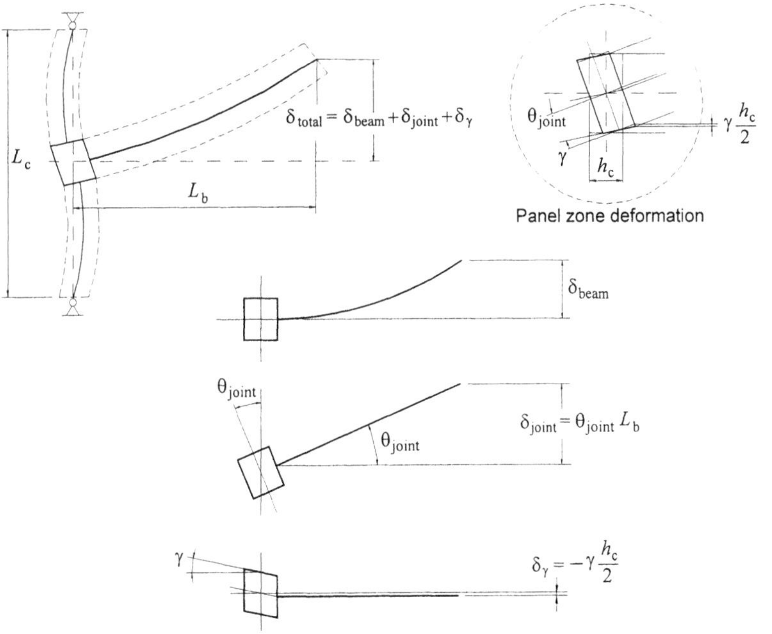

Figure 3.3.10: Contributions in total beam end displacement

As illustrated in the figure, the contribution of the column (joint rotation) and the panel zone to the beam end displacement is

$$\delta_{column\,\&\,PZ} = \delta_{joint} + \delta_\gamma = \theta_{joint} \times L_b - \gamma \times h_c / 2 \tag{3.3.4}$$

and the displacement due to beam deflection only can correspondingly be obtained as

$$\delta_{beam} = \delta_{total} - \delta_{column\,\&\,PZ}. \tag{3.3.5}$$

The total rotation, θ_{total}, is calculated as

$$\theta_{total} = \delta_{total} / L_{ref}, \tag{3.3.6}$$

where L_{ref} is the "reference" length of the beam, measured to the column face for type A specimens and to the theoretical location of the plastic hinge for type B specimens, see also Figure 3.3.6. Correspondingly, the beam contribution is obtained as

$$\theta_{beam} = \delta_{beam} / L_{ref}. \tag{3.3.7}$$

The plastic portions of the above rotations are extracted as

$$\theta_{total.pl} = \theta_{total} - M/K_{total}, \tag{3.3.8}$$

$$\theta_{beam.pl} = \theta_{beam} - M/K_{beam}, \tag{3.3.9}$$

where $M = F \times L_{ref}$, and K_{total} and K_{beam} are the elastic stiffnesses determined from the «$M - \theta_{total}$» and «$M - \theta_{beam}$» curves, respectively. For this reason, for each specimen a representative portion of the relationship was chosen and fit in a least-squares sense to determine the "elastic" slope.

The dissipated energy can be determined as the plastic work based on the hysteretic area of the experimental loops. In this study, using the relationships «$M - \theta_{total}$» and «$M - \theta_{beam}$», the total dissipated energy and the beam contribution were calculated as functions of the cumulative beam end displacement. Additionally, the local response was monitored by the strain gauges, and strain distributions in the beam flanges at various displacement cycles were plotted.

Experimental behaviour

The cyclic response of the joints is illustrated in Figure 3.3.11 in terms of force–displacement hysteretic loops for each specimen. The total displacement of the beam end, plotted in the left charts, is divided into displacement due to beam deflection and displacement resulting from column and panel zone deformation as described above. In this format one can estimate the contribution of the two notional components to the total response.

Since the good seismic behaviour of the beam-to-column joints is owing to the energy dissipated primarily through plastic deformations, the hysteretic behaviour is better represented by the moment versus plastic rotation relationships. These are plotted in Figure 3.3.12, where the beam contribution is also distinguished on the background of the total response.

As seen in the figures, initially (at cycles $\pm v_y$, $\pm 2v_y$ and $\pm 3v_y$) the first tested specimen A-1 exhibited a stable hysteretic behaviour. Plastic deformations appeared first in the panel zone in shear, and the contribution of the beam was low. The review of the local response of the flanges shows that yielding has also spread over the entire portion between the column and splice, tending to concentrate near the column face. Unfortunately, specimen A-1 failed by lateral-torsional buckling during the first $\pm 4v_y$ cycle because of the improper bracing of this first specimen.

Braced properly, A-2 showed very stable cyclic behaviour and sustained up to the three $\pm 7v_y$ cycles. Similarly to A-1, yielding was first noticed in the panel zone and developed very extensively. Thus, plastic rotations were mostly owing to panel zone shear deformations. As a result, the beam contribution to the energy dissipation rested within 1/3 of to the total energy dissipation capacity, as will be illustrated later. Nevertheless, a very extensive plasticity spread over the entire flanges, almost uniformly distributed. During the second $\pm 6v_y$ cycle the bottom flange buckled locally. However, this was not accompanied by strength degradation. No visible cracks or fractures were observed there. An interesting effect resulted from the local buckling. Despite the symmetrical total response due to the symmetrical imposed displacements, a lack of symmetry is seen in the beam and column & panel zone responses (Figure 3.3.11*a*) which balance each other. The reason is the variable beam stiffness in positive and negative direction resulting from the flange local buckling. Finally, when trying to reach a level of $+8v_y$, the specimen failed due to fracture of the top splice plate. The latter was overloaded because of some lack of overstrength and different section areas of the two splice plates. The review of the local response shows that prior to failure, large plastic strains had developed in the splice plate which, together with the stress concentration near the welds, led to low-cycle fatigue.

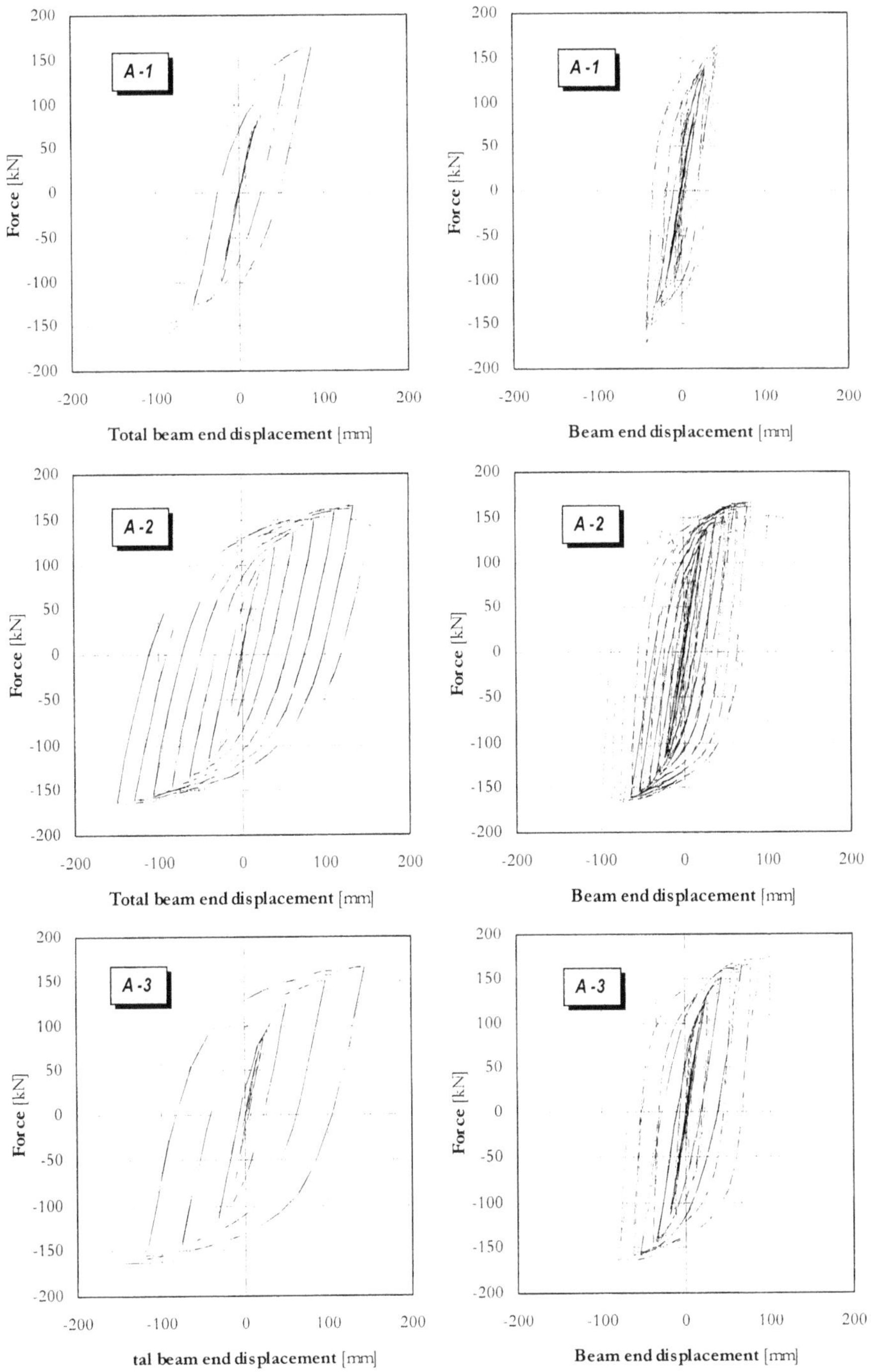

Force versus beam end displacement: total (left charts) and separated (right charts)
ibution (——) and column & panel zone contribution (– –), type A specimens

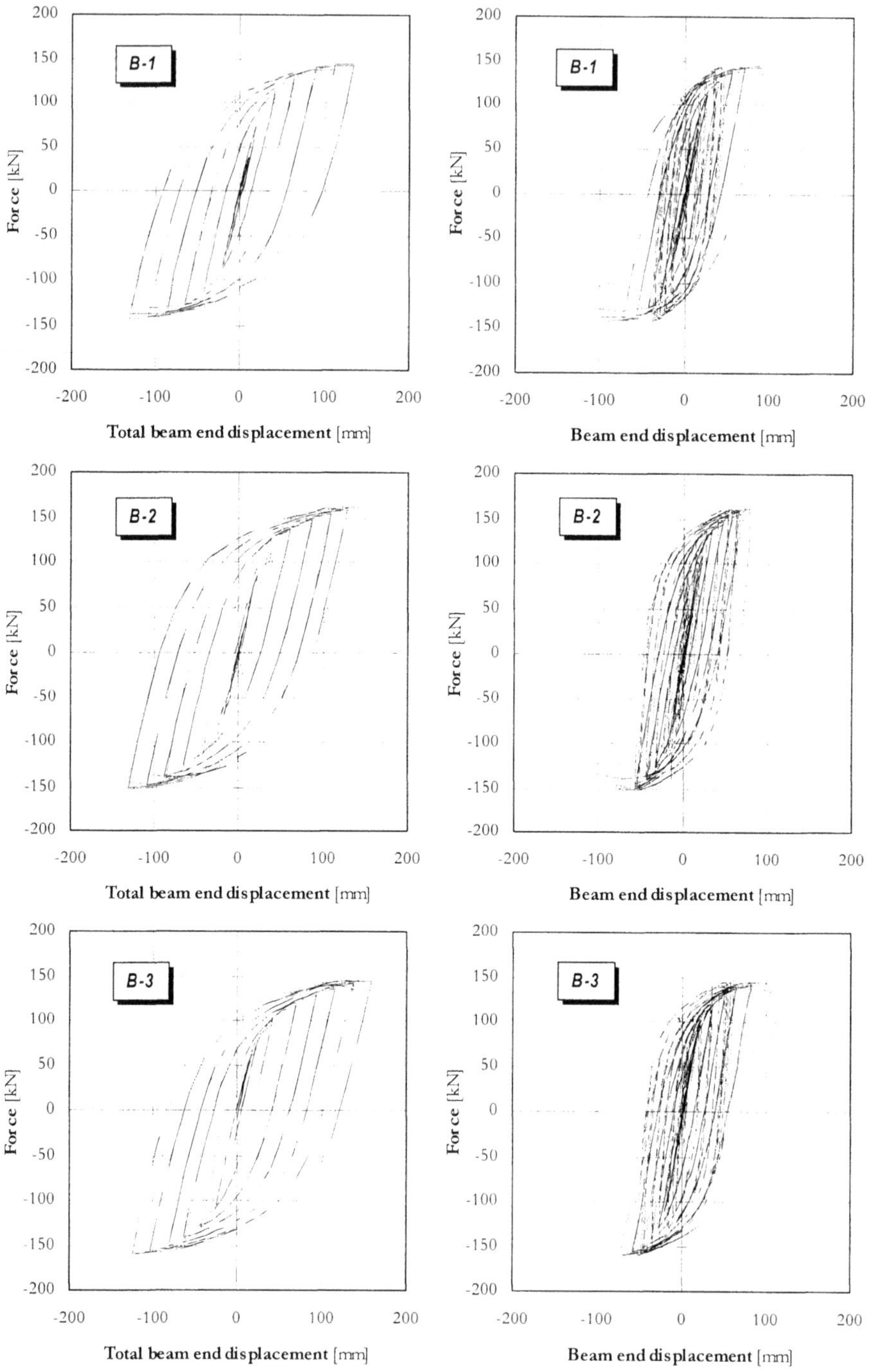

Figure 3.3.11(b): Force versus beam end displacement: total (left charts) and separated (right charts) beam contribution (——) and column & panel zone contribution (– –), type B specimens

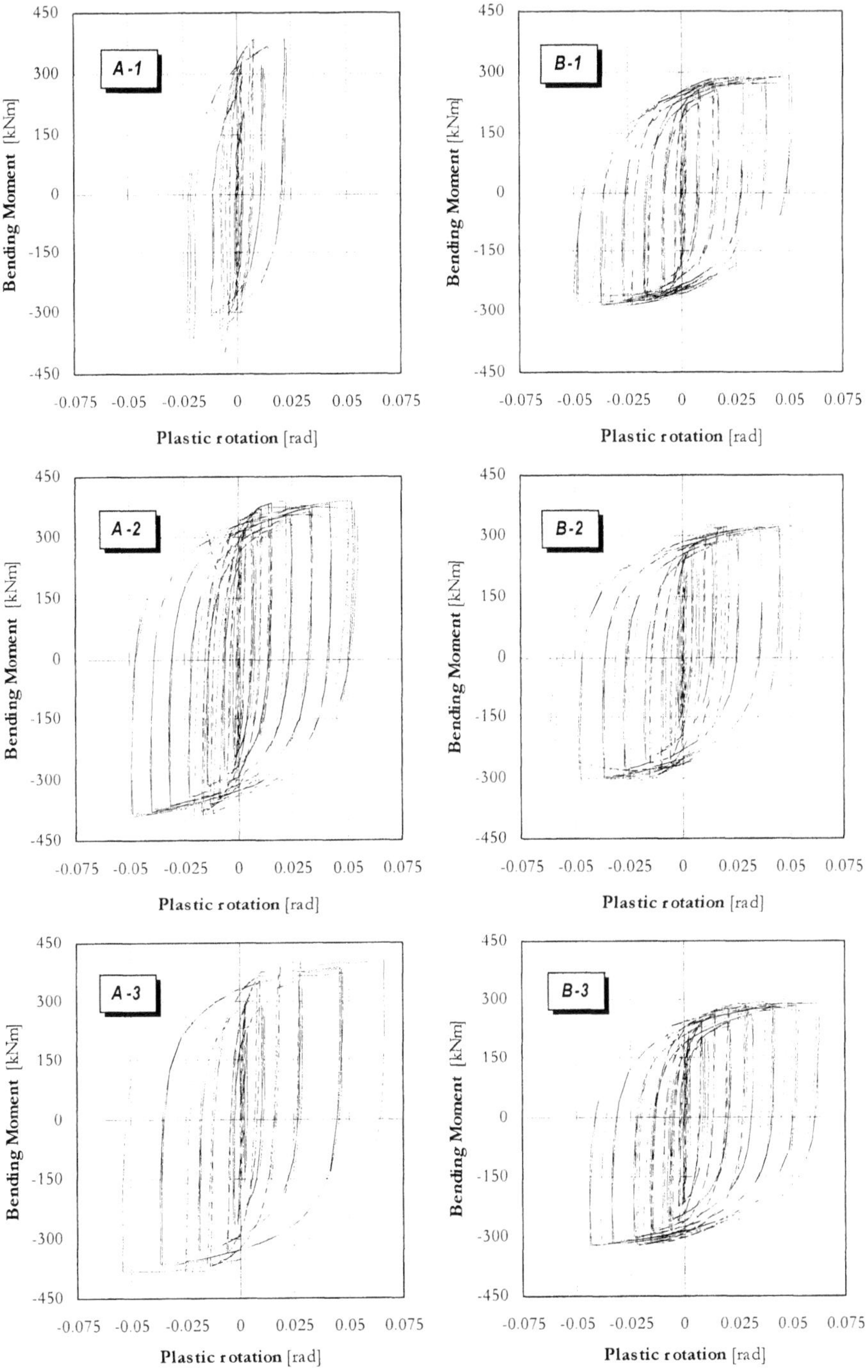

Figure 3.3.12: Bending moment versus total (——) and beam (– –) plastic rotation at the "reference" cross-section

When testing A-3, an initial displacement corresponding to approximately 30% of the plastic bending capacity was applied to simulate the effect of gravity loading. Therefore the response relationships (Figures 3.3.11*a* and 3.3.12) show some lack of symmetry which effect however cannot be regarded as significant. A larger increment of $2v_y$ was used for the consecutive cycles, thus following clearly the ECCS procedure. The comparison between the response relationships of A-2 and A-3 clearly shows a negligible influence of the different loading histories — a conclusion that evidently cannot be generalized. The specimen A-3 sustained up to the first $\pm 8v_y$ cycle. Generally, the observed behaviour was similar to that of A-2. Yielding started in the panel zone and developed very extensively both there and in the flanges. The beam contribution to the plastic rotations and to the energy dissipation was also within 35%. No local buckling took place in this specimen. The failure occurred suddenly during the second half of the first $\pm 8v_y$ cycle, due to low-cycle fatigue fracture of the top splice plate as in A-2. However, no other cracks or any other sign of fracture appeared elsewhere. Thus the good behaviour of the shop-welded full penetration welds was clearly proven.

The relationships plotted in Figures 3.3.11*b* and 3.3.12 show an excellent cyclic hysteretic response of specimen B-1. The contribution of the plastic zone in the beam was about 60% both for the rotation and for the energy dissipation. This was due to the fact that the specimen behaved as predicted and yielding started almost simultaneously in the panel zone and at the location of the designed dissipative zone. The readings from the strain gauges demonstrate a concentration of plasticity at the predetermined location. A classical plastic hinge developed in this specimen at the desired place, accompanied by local buckling. This test was not led to final fracture. Nevertheless, no cracks were observed anywhere in the specimen.

When testing specimen B-2, during the second $\pm 2v_y$ cycle, the longitudinal support device failed and the test was interrupted. After the repair, a full $\pm v_y$ "warming up" loading cycle was performed, prior to continuing with the testing protocol. The specimen behaved perfectly up to the first $\pm 7v_y$ cycle as illustrated in the figures. Because of the earlier flange local buckling, the same balancing effect as described for A-2 can be noticed between beam and panel zone responses. The relative contribution of the beam is apparently higher compared with type A specimens. During the $\pm 5v_y$, and especially $\pm 6v_y$ cycles a strong flange–web interactive local buckling developed. Finally, during the second $\pm 7v_y$ cycle, the resulting secondary bending stresses at the crest of the buckles led to a low-cycle fatigue rupture of the buckled flange, followed by tearing of the web. No other cracks were detected either in the web-to-flange fillet welds in that region, or elsewhere.

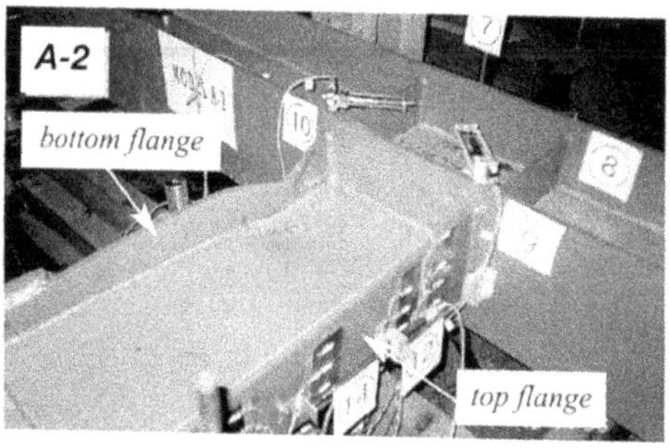

(*a*)

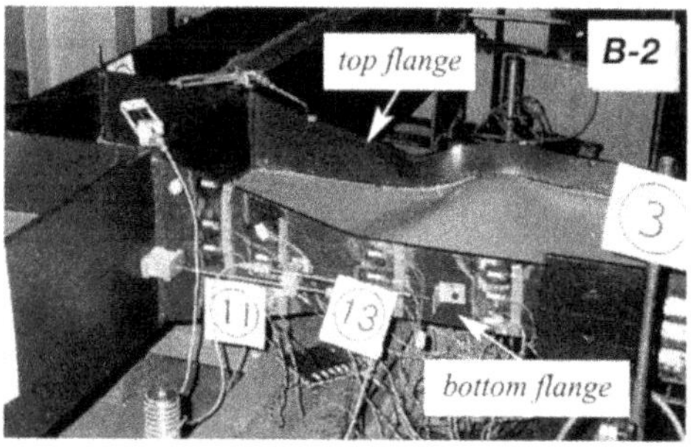

(*b*)

Figure 3.3.13: Typical local buckling patterns in type A and type B specimens

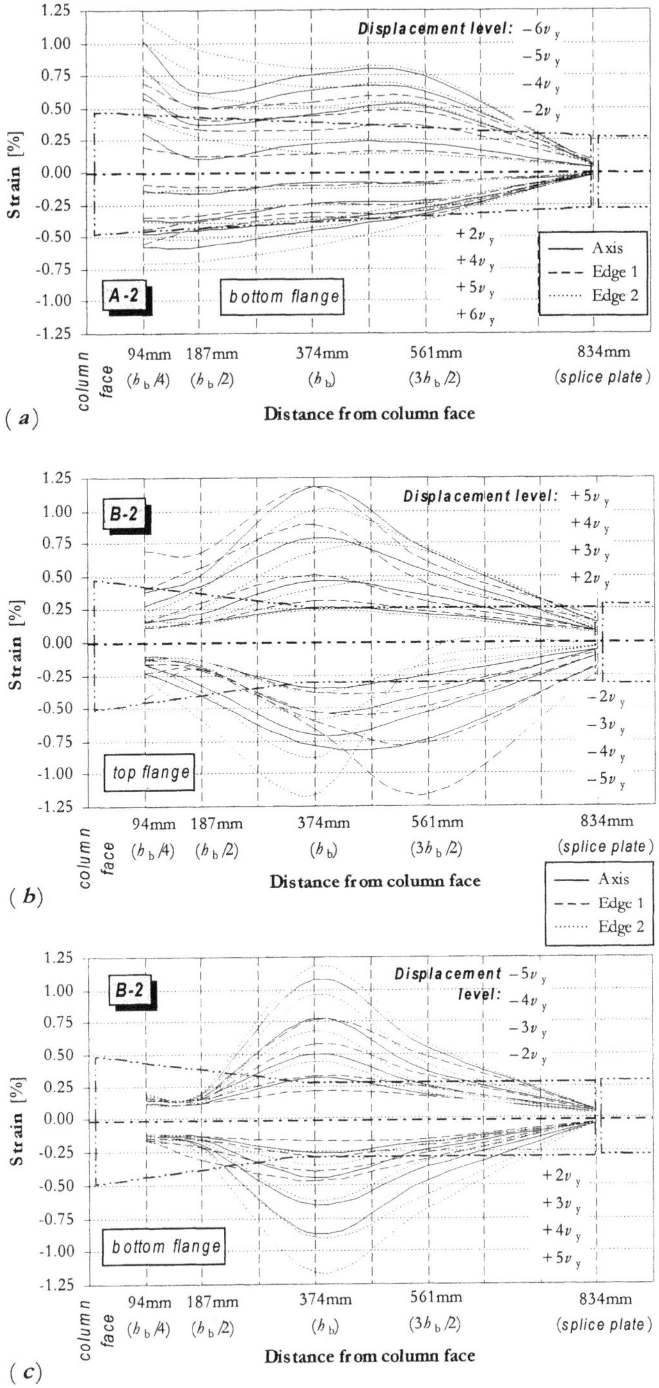

Figure 3.3.14: Typical strain distributions in the flanges at different displacement levels

Similarly to A-3, specimen B-3 was tested after imposing an initial displacement, which was chosen the same as in A-3. Despite that, the specimen sustained the three full $\pm 7v_y$ cycles, and failed when trying to start the next $+8v_y$ cycle. As can be seen from the response curves (Figures 3.3.11*b* and 3.3.12) there is some effect of asymmetry due to the initial loading. Nevertheless, the behaviour was very similar to that of specimens B-1 and B-2. Contrary to B-2, in this specimen local buckling appeared in both the flanges, however the buckling pattern was not so localized. Therefore the secondary bending stresses must have been smaller, and probably this could explain the fact that B-3 sustained some more cycles than B-2. The failure mode was the same as in B-2 — a low-cycle fatigue rupture of the buckled flange. It is worth repeating that, in spite of the large bends of the flange plate, no cracks were noticed in the web-to-flange fillet welds in that region.

Typical plastic hinges formed by local buckling in both specimen types are shown in Figure 3.3.13. In type A specimen (A-2) the plastic hinge has developed near the column face and is not well pronounced. In contrast, in type B a classical plastic hinge is well seen exactly at the designed location, accompanied by very strong flange and web local buckling.

The idea of the two joint types is best illustrated by the strain distributions. Typical strain distributions in the flanges of both specimen types are shown in Figure 3.3.14 where the same specimens A-2 and B-2, as in Figure 3.3.13, are selected.

In type A (Figure 3.3.14*a*), it is well seen that yielding has spread over the entire portion between the column and splice, thus illustrating the idea of more volume in plasticity (and therefore more dissipated energy). However, the plastic deformations apparently tend to concentrate near the column face. Note the lack of symmetry in the strain distribution in positive and negative (moving towards the bottom flange) actuator displacement which is in close relation with the local buckling pattern (compare with Figure 3.3.13). One can also observe a "perturbation" of the "strain flow" in the vicinity of the column flange. The same effect is also seen in A-1 and A-3, and is probably due to the post-welding residual stresses/strains around the full penetration welds.

The strain distributions shown in Figure 3.3.14*b* and *c* definitely prove the design idea of type B joints. A concentration of plasticity at the designed location is very clearly demonstrated. Since the bottom flange of B-2 did not undergo local buckling, the strain distribution shown in Figure 3.3.14*c* is as "perfect" as predicted. In the top flange (Figure 3.3.14*b*), the strain distributions for the two edges are shifted in correspondence with the flange torsional buckling (Figure 3.3.13*b*), which evidently does not change the general picture of strains.

A very important parameter characterizing the structural behaviour of a joint under seismic loading is its *energy dissipation capacity*. Herein, it is determined as the plastic work based on the hysteretic area of the experimental loops shown in Figure 3.3.12, and is plotted in Figure 3.3.15 versus the relevant cumulative beam end displacement.

It is seen from the figure that the contribution of the plastic zones in the beams of type A specimens is only about 30%. The review of the local response, however, shows that the flanges have yielded within the whole region between the column and the splice. Therefore, it is believed that the energy dissipation capacity would be higher than in the case of a constant cross-section, where the yielded zone would be more localized and therefore the plastic volume would be smaller. The higher gradient of the curves representing A-3 is due to lack of local buckling and probably to the different loading history.

For type B, the contribution of the beam plastic zones is approximately 50% and only for B-1 it is about 60%. Note the very close agreement of the curves for the total dissipated energy, demonstrating the similar behaviour of the three specimens.

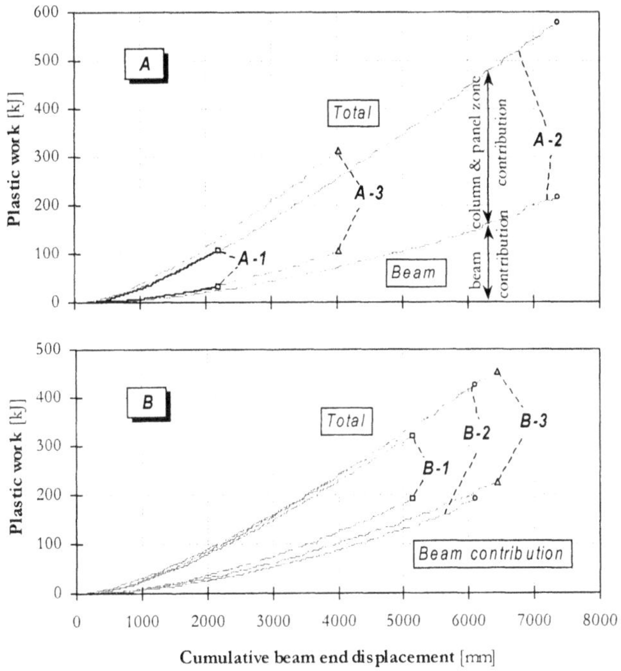

Figure 3.3.15: Dissipated energy

Comments and discussion

Behaviour.— In type A specimens, plastic deformations started to develop first in the panel zone and later in the beam flanges near the column. Satisfactory stable hysteretic behaviour was anyway observed, due however to a great extent to the panel zone contribution. All type B specimens exhibited a very stable cyclic hysteretic response, and behaved exactly as predicted. Plastic zones developed at the designed location, accompanied by flange local buckling in the three specimens. Both in A-3 and B-3, the effect of initial gravity loading was negligible and vanished well within the plastic range.

Plastic rotational capacity.— Excluding A-1 because of the improper bracing, all joints exhibited adequate ductility. In both types A and B, total plastic rotations of 0.05 rad were reached without a brittle fracture, which is well above the demand of 0.03 rad adopted recently (Bruneau *et al.*, 1998). However, due to the different contribution of the panel zone, beam-only plastic rotations reached in type A were 0.02–0.025 rad, whereas in type B they were up to 0.03–0.04 rad. It means therefore that in case of type A joints, the ductility demand to the panel zone will be higher to provide the same rotational capacity.

Energy dissipation capacity.— To compare the dissipation capacity of the different specimens, the non-dimensional dissipated energy is defined as

$$E^* = \frac{E}{F_{\mathrm{pl}}\, v_{\mathrm{tot}}}, \tag{3.3.10}$$

where E is the dissipated energy (plastic hysteretic work), F_{pl} is the force at the beam end corresponding to the theoretical plastic resistance moment at the 'reference' cross-section, and v_{tot} is the total cumulative beam end displacement. Thus, referring the dissipated energy to that of a relevant perfectly plastic system, it is believed that not only the different bending resistance of the two types of specimens, but also some variations in the loading history may be accounted for.

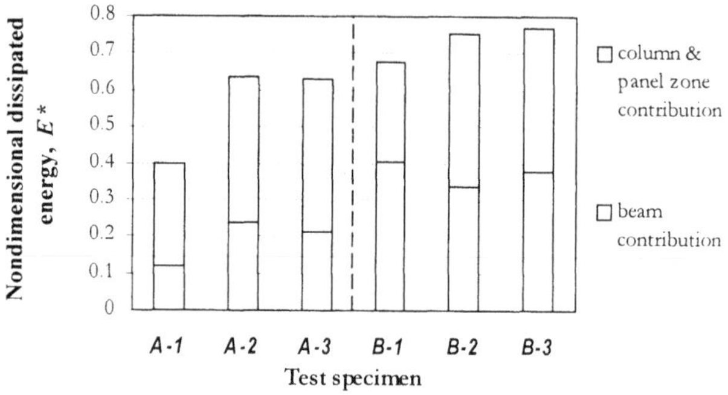

Figure 3.3.16: Relative dissipation capacity

The comparative results are shown in Figure 3.3.16. In this format, firstly, the different contributions of the two components – beams and panel zones – are well seen. Secondly, the relative dissipation capacity of type B specimens is on average some 20% higher. Moreover, the portion of the energy dissipated only by the beam plastic zones in specimens B is almost twice higher, which undoubtedly is an additional advantage.

ANALYTICAL MODELLING

To study the non-linear frame response by DRAIN-2DX, the test data were used to model the panel zone and beam plastic hinge behaviour. The test specimens were modelled by linear elastic beam elements (column and beam), connected with zero-length bilinear elastic-plastic rotational springs with strain hardening (Figure 3.3.17). The variable elastic stiffness of the beam within the haunches was described by analytical expressions.

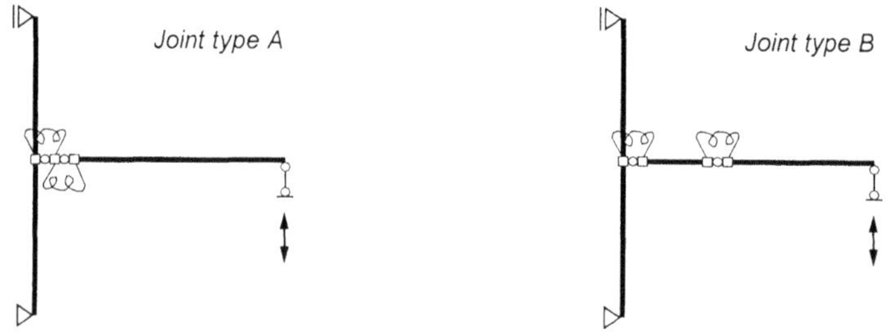

Figure 3.3.17: Models for analytical study

In the models, the elastic deformations in the region of the plastic hinges are neglected. Therefore, the strain hardening slopes are obtained from the experimental hysteretic curves «$M-\theta_{total\ pl}$» and «$M-\theta_{beam.pl}$» (Figure 3.3.12). In order to derive the characteristics of the spring representing the column and panel zone response, which generally is more difficult to be extracted separately, the following procedure is applied as illustrated in Figure 3.3.18.

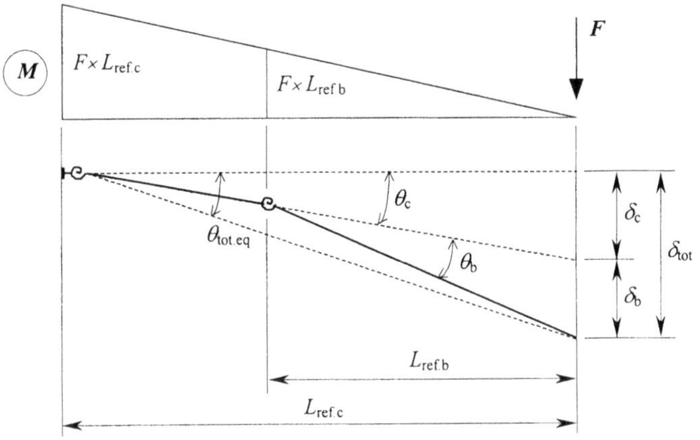

Figure 3.3.18: Idealized beam tip deflection in plastic range

The total deflection of the beam end is due to the rotations of the springs, therefore, for small deflections one can obtain:

$$\delta_{tot} = \delta_b + \delta_c = \theta_b \times L_{ref.b} + \theta_c \times L_{ref.c}. \tag{3.3.11}$$

After introducing the stiffness coefficients $K_{\theta.c.pl}$ and $K_{\theta.b.pl}$ of the two rotational springs, representing the column and panel zone and the beam, respectively, and $K_{\theta.total}$ for the total response, the beam end displacements can be written as:

$$\delta_b = \frac{FL_{ref.b}^2}{K_{\theta.b.pl}}, \tag{3.3.12}$$

$$\delta_c = \frac{FL_{ref.c}^2}{K_{\theta.c.pl}}, \tag{3.3.13}$$

$$\delta_{tot} = \frac{FL_{ref.c}^2}{K_{\theta.total}}. \tag{3.3.14}$$

By substituting the above three equations Eqn (3.3.11) can be rewritten in the form:

$$\frac{1}{K_{\theta.total}} = \frac{1}{\alpha^2 K_{\theta.b.pl}} + \frac{1}{K_{\theta.c.pl}}, \tag{3.3.15}$$

where $\alpha = L_{ref.c} / L_{ref.b}$.

The stiffness coefficients $K_{\theta.total}$ and $K_{\theta.b.pl}$ can be easily derived from the corresponding experimental hysteretic curves. Then the stiffness of the column and joint rotation $K_{\theta.c.pl}$ can be obtained from Eqn (3.3.15).

Using the above procedure, the models of specimen types A and B are studied by DRAIN-2DX program. The same displacement–controlled loading history as in the tests is applied. The analytical and the experimental results for specimens A2 and B2 are compared in Figures 3.3.19 and 3.3.20.

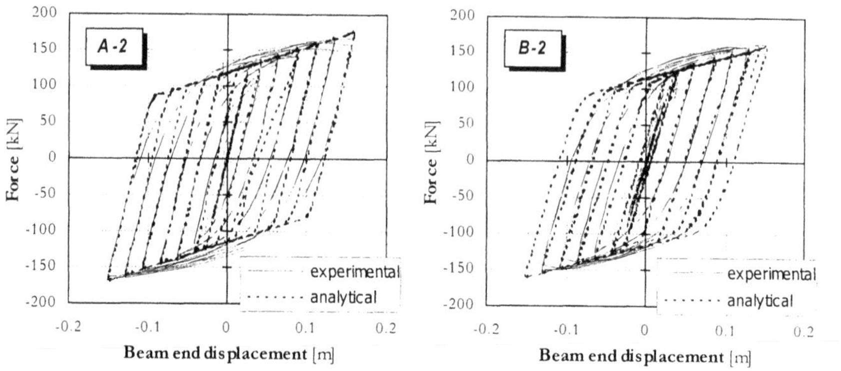

Figure 3.3.19: Force versus beam end displacement hysteretic loops — comparison between experimental and analytical results

As can be seen in Figure 3.3.19, the analytical hysteretic loops are close to the experimental ones. The higher non-coincidence can be observed in the corners, which is due to the simple analytical model with bilinear characteristics of the springs. It is possible to achieve better agreement, if several connected in parallel rotational springs are used. However, determination of their characteristics would be more complicated.

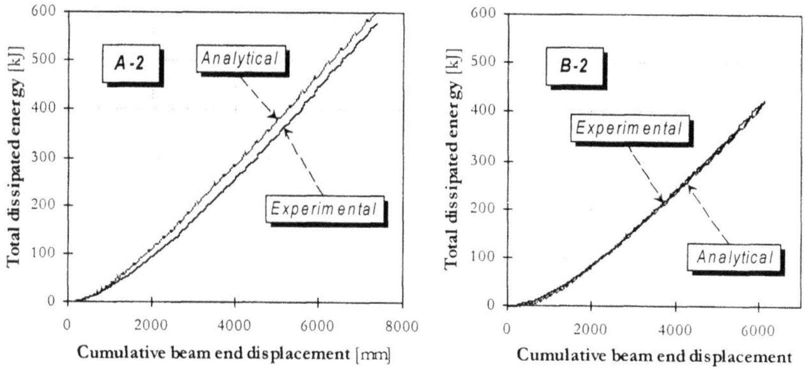

Figure 3.3.20: Dissipated energy versus cumulative beam end displacement — comparison between experimental and analytical results

Figure 3.3.20 illustrates that even the model with simple rotational springs represents very well the energy dissipation capacity of the specimens. Especially for specimen B2, the analytical and experimental curves practically coincide. There is some discrepancy for specimen A2, however it is even smaller than the discrepancy between the relationships representing the three different specimens of type A (see Figure 3.3.15).

CONCLUSIONS

Not only the extensive research carried out, but also recent earthquakes in Japan and California demonstrated a lack of deformation capacity of the connections typically used in practice and thus generated an urgent need for improved detailing. Though new strategies have been developed aiming at effectively moving the plastic hinge away from the column face, it still seems that no detail can be assumed perfect and new solutions are still worth exploring.

Haunching has been find to be an efficient tool for solving the problem of poor ductile behaviour and potential fragility of welded beam-to-column connections. In this sub-chapter, a type of haunching through variable-breadth flanges is used to develop two improved joint solutions. In joint type A, the flanges are tapered to follow approximately the varying bending moment diagram. The goal is not only to save material, but to spread yielding over the entire portion that follows the moment diagram and thus to dissipate more energy than in the case of conventional non-tapered flanges where the plastic zone is more localized. Detail B belongs to the 'new generation' joints, in which a shift of the plastic hinge from the column flange is achieved by a specific flange profile, see Figure 3.3.6. Both variants are considerably suitable for the case of column-tree configuration.

An experimental study is reported on six specimens — 3 of type A and 3 of type B — following the ECCS (1986) Recommendations. Both specimen types exhibited stable hysteretic behaviour with adequate ductility. The plastic rotation capacity of type A was however largely due to the flexible panel zone. The energy dissipation capacity of both types was high, type B being some 20% better. Though the two variants were quantitatively comparable, the most significant advantage of type B lies in its predictable behaviour. The lack of a structural 'fuse' in type A leads to some variable behaviour and imposes higher demands to the panel zone.

It should be concluded that both joint typologies considered exhibited satisfactory cyclic behaviour and high dissipation capacity. However, type B, having behaved exactly as predicted and designed, possesses some definite advantages, therefore is suggested for the design practice.

REFERENCES

Bruneau, M., Uang, C.-M. and Whittaker, A. (1998). *Ductile Design of Steel Structures*, McGraw-Hill, New York.

Chen, S.-J., Yeh, C.H. and Chu, J.M. (1996). Ductile steel beam-to-column connections for seismic resistance. *Journal of Structural Engineering*, ASCE, **122**(11), 1292-1299.

ECCS (1986). *Recommended Testing Procedures for Assessing the Behaviour of Structural Elements under Cyclic Loads*. European Convention for Constructional Steelwork, Publication □ 45, Brussels.

Engelhardt, M.D., Winneberger, T., Zekany, A.J. and Potyraj, T.J. (1996). The Dogbone Connection: Part II. *Modern Steel Construction* **36**(8), 46-55.

Iwankiw, R.N. and Carter, C.J. (1996). The Dogbone: A New Idea to Chew On. *Modern Steel Construction* **36**(4), 18-23.

Mazzolani, F.M. and Piluso, V. (1996). *Theory and Design of Seismic Resistant Steel Frames*, E & FN Spon, London.

Plumier, A. (1990). *New Idea for Safe Structures in Seismic Zones*, University of Liège, IABSE Symposium, Brussels.

Popov, E.P., Blondet, M. and Stepanov, L. (1996). *Application of "Dog Bones" for Improvement of Seismic Behavior of Steel Connections*, Report No.UCB/EERC 96/05, University of California, Berkeley, USA.

3.4

INFLUENCE OF COLUMN SIZE

Luis Calado

INTRODUCTION

In this sections the results obtained from the experimental tests performed in the Laboratory of Structures and Strength of Material of Instituto Superior Técnico, Lisbon on two alternative connection solutions, namely fully welded connections (WW) and top and seat with web angle (TSW), designed for the same beam-to-column joints are presented and discussed.

Four experimental tests have been executed on six (three welded and three bolted) different series of specimens, for a total of 24 tests. The test program was planned with the aim of assessing the comparative behaviour of bolted and welded connections, and for defining the effect of the column size and of the panel zone design on the behaviour of the two types of connection, varying the applied loading history.

Two series of full-scale specimens have been designed and tested, namely a fully welded specimen series (BCC5, BCC6 and BCC8) and a top and seat with web angle specimen series (BCC9, BCC7 and BCC10). The specimens of the two series are T-shaped beam-column subassemblages, consisting approximately of a 1000 mm long beam and of 1800 mm long column.

The material used for the columns, beams, and angles is steel S235 JR. In each series the cross section of the beam is the same (IPE300), while the column cross section has been varied, being respectively HE160B for the BCC5 (WW) and BCC9 (TSW) specimens, HE200B for the BCC6 (WW) and BCC7 (TSW) specimens, and HE240B for the BCC8 (WW) and BCC10 (TSW) specimens. In both series, the continuity of the connection through the column has been ensured by horizontal 12 mm thick plate stiffeners, fillet welded to the column web and flanges.

Testing lay-out are illustrated in Figure 3.4.1 and consists of a foundation block (a), a supporting girder (b), a reaction wall (c), a power jack (d), a lateral frame (e) (excluded from the figure), to prevent out-off plane displacements, and a purposely designed device for displacement de-amplification (f), to measure horizontal displacements (v). Forces (*F*) were applied by means of an actuator acting at the top of the beam, at a distance *h* to the external upper flange of the column.

Several electrical displacement transducers (LVDT) were used to monitor displacements of the column supporting plates, vertical and horizontal displacements of the beam, deformation of the panel zone and beam-to-column relative rotations. Different arrangements of LVDT's positioning were adopted for the different classes of specimens.

A typical instrumentation set-up is shown in Figure 3.4.2. An automatic testing technique was developed to allow computerised control of the power jackscrew, of the displacement and of all the transducers used to monitor the specimens during the testing process.

As far as loading history is concerned, in order to follow the cyclic behaviour of the specimen up to the complete failure with increasing amplitude deformations, the so-called Complete Testing Procedure provided by ECCS Recommendations (1986) has been firstly applied. In addition, since the testing program is mainly focused at the determination of beam-to-column joint response under arbitrary loading histories, as suggested by ATC Guidelines (1992), three different inelastic constant amplitude deformation levels have been also considered.

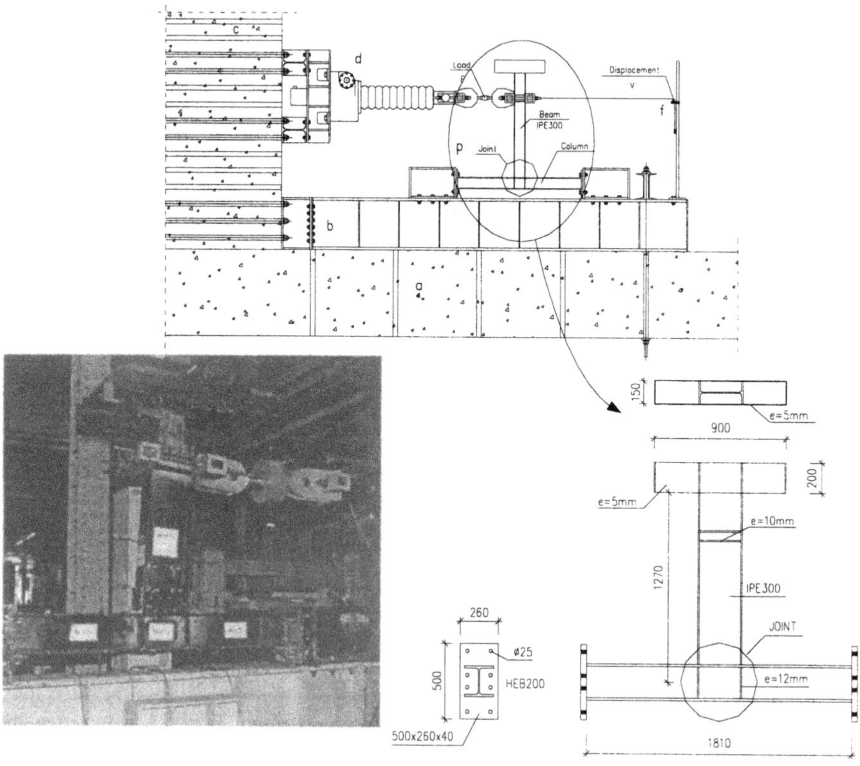

Figure 3.4.1: Test set-up.

This allows damage phenomena due to cumulated plastic excursions to be properly evaluated. In particular, displacement amplitude (v_{max}) equal to ± 37.5 mm, ± 50.0 mm and ± 75.0 mm have been considered.

In conclusion, the following cyclic sequence has been executed: (a) for all the specimens, four cycles in elastic range, with a semi-amplitude equal to: ¼, ½, ¾, 1 v_y, where v_y is the elastic displacement which has been conventionally and preventively assumed equal to 10 mm; then: (b) in case of constant amplitude tests, repeated cycles at v_{max}, until the complete failure of the specimen; (c) in case of ECCS procedure, three cycles at each amplitude $i \cdot v_y$ (with i=2,3,4,...) up to the complete failure of the specimen.

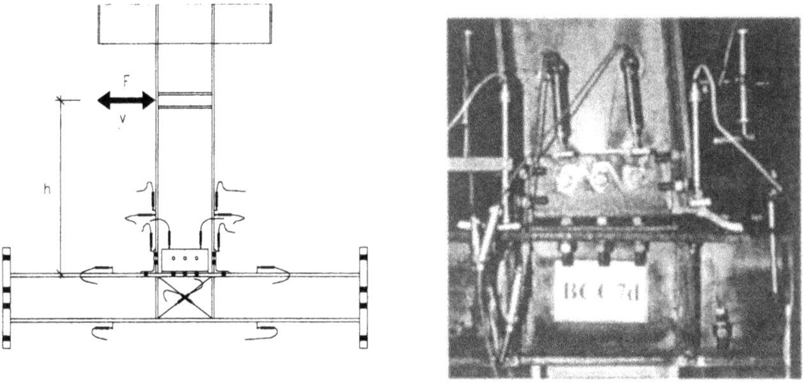

Figure 3.4.2: Typical instrumentation set-up.

TESTS ON FULLY WELDED CONNECTIONS

The specimen geometry of the fully welded connections is shown in Figure 3.4.3. In these specimens the beam flange have been connected to the column flange by means of complete joint penetration grove welds, while fillet welds have been applied to both sides of the beam web and the column flange.

In the welding process, carried out through a technique commonly adopted in the European practice, special care has been paid in order to minimise porosity and defects, thus reducing the potentials of crack formation.

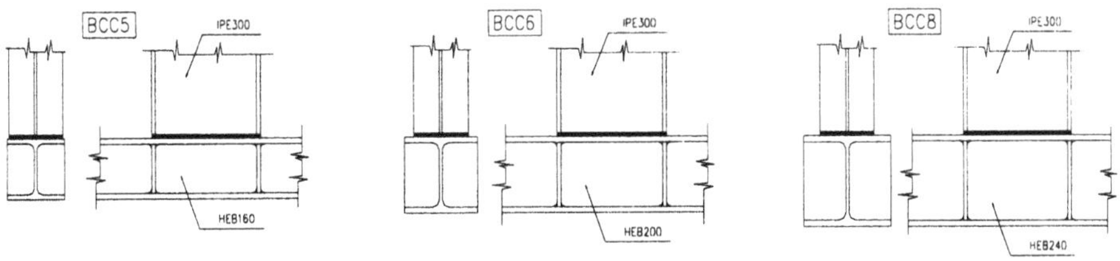

Figure 3.4.3: Specimen geometry and detail of the fully welded connections

In Table 3.4.1 the mean values of the coupon test performed for each section type and for both web and flange elements are presented.

TABLE 3.4.1
MATERIAL CHARACTERISTICS

MEMBER SECTION	Section element	f_y (MPa)	ε_y (%)	f_u (MPa)	ε_u (%)
HEB 160	*web*	395.56	0.56	490.08	22.35
	flange	323.13	0.43	460.22	27.71
HEB 200	*web*	401.62	0.49	489.35	22.30
	flange	312.56	0.38	434.91	36.75
HEB 240	*web*	309.00	0.55	469.40	28.40
	flange	300.50	0.51	457.40	34.00
IPE 300 (HEB 160)	*web*	305.54	0.35	412.60	32.65
	flange	274.78	0.38	404.55	36.05
IPE 300 (HEB 200)	*web*	304.62	0.40	418.03	34.45
	flange	278.62	0.45	398.76	37.10
IPE 300 (HEB 240)	*web*	299.50	0.56	450.55	37.70
	flange	292.00	0.44	445.15	35.80

In Table 3.4.2, the adopted nomenclature is reported for each analysed specimen. In particular, size of the column, distance *h* relative to the point of application of loading actuator and displacement history type is reported.

TABLE 3.4.2
ADOPTED NOMENCLATURE FOR TESTED SPECIMENS

DISPLACEMENT HISTORY	**COLUMN SECTION** (Actuator position *h* [mm])		
	HEB 160	**HEB 200**	**HEB 240**
v_{max} =± 37.5 mm	BCC5D (*h*=862)	BCC6D (*h*=862)	BCC8B (*h*=760)
v_{max} =± 50.0 mm	BCC5A (*h*=862)	BCC6A (*h*=862)	BCC8A (*h*=765)
v_{max} =± 75.0 mm	BCC5B (*h*=862)	BCC6B (*h*=862)	BCC8C (*h*=760)
ECCS	BCC5C (*h*=862)	BCC6C (*h*=862)	BCC8D (*h*=763)

Test results have been represented in terms of force – displacement *(F-v)* curves, being *F* the applied force and *v* the displacement at the level of the applied force, Calado et al (1999c), Mele et al (1997, 1999a), Pucinotti (1998).

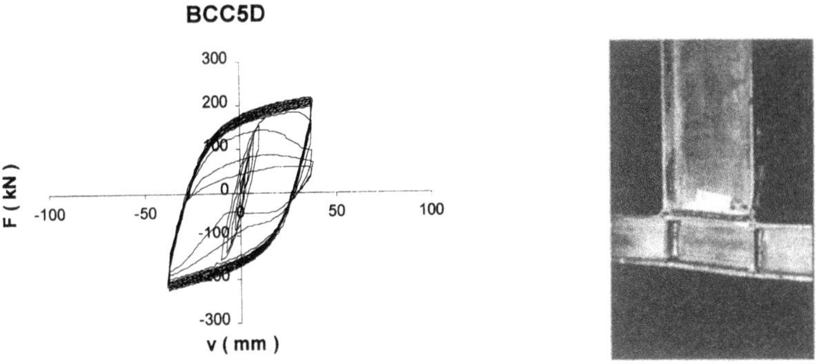

Figure 3.4.4: Histeresis loops and failure mode of BCC5D.

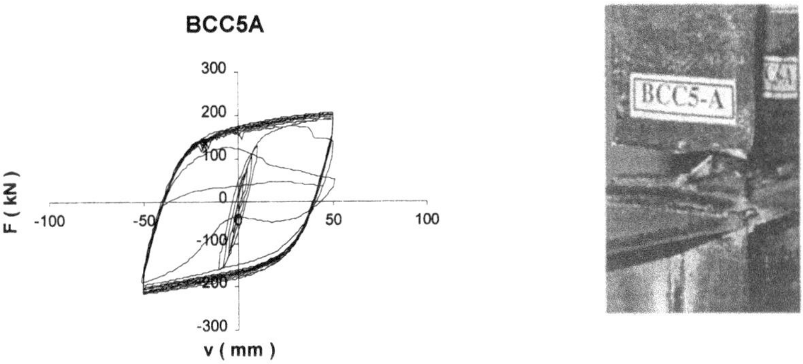

Figure 3.4.5: Histeresis loops and failure mode of BCC5A.

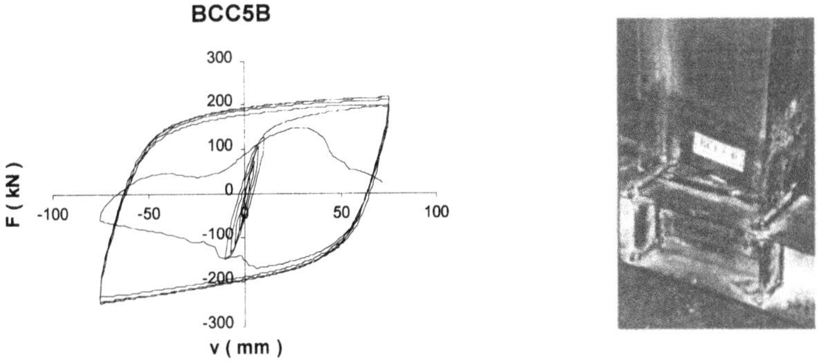

Figure 3.4.6: Histeresis loops and failure mode of BCC5B.

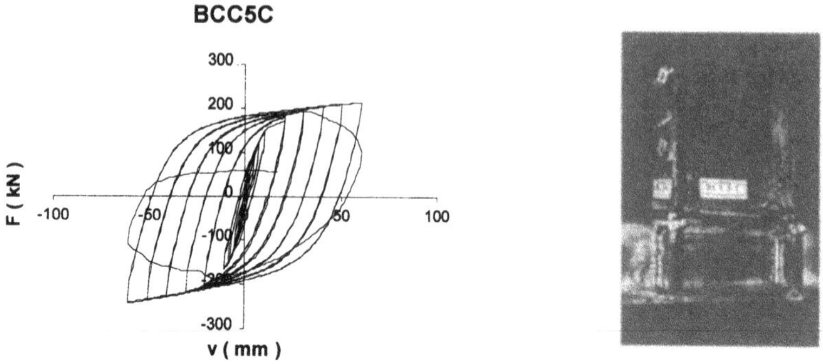

Figure 3.4.7: Histeresis loops and failure mode of BCC5C.

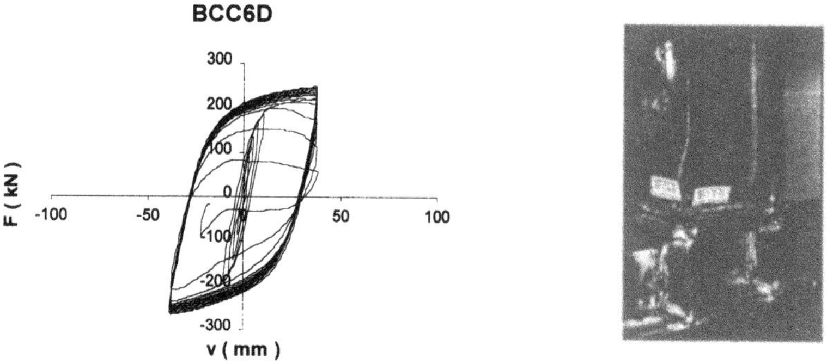

Figure 3.4.8: Histeresis loops and failure mode of BCC6D.

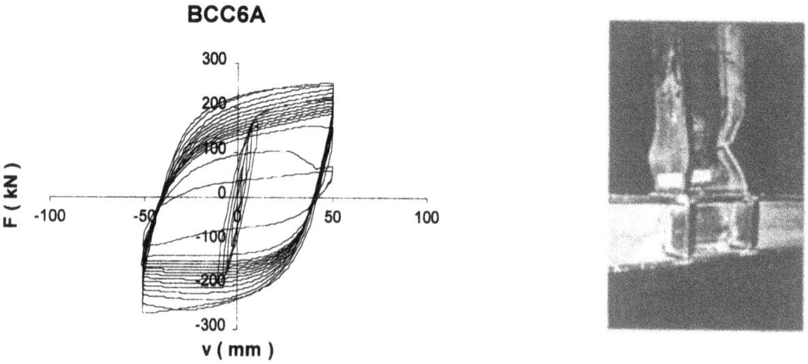

Figure 3.4.9: Histeresis loops and failure mode of BCC6A.

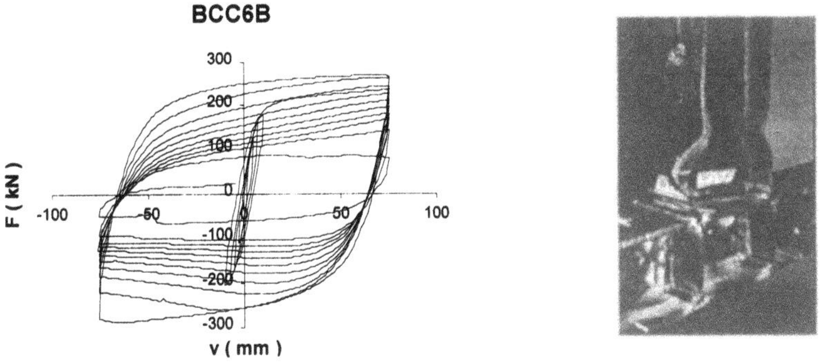

Figure 3.4.10: Histeresis loops and failure mode of BCC6B.

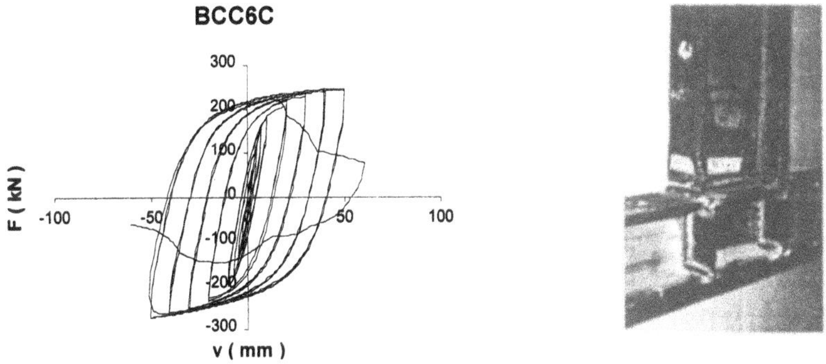

Figure 3.4.11: Histeresis loops and failure mode of BCC6C.

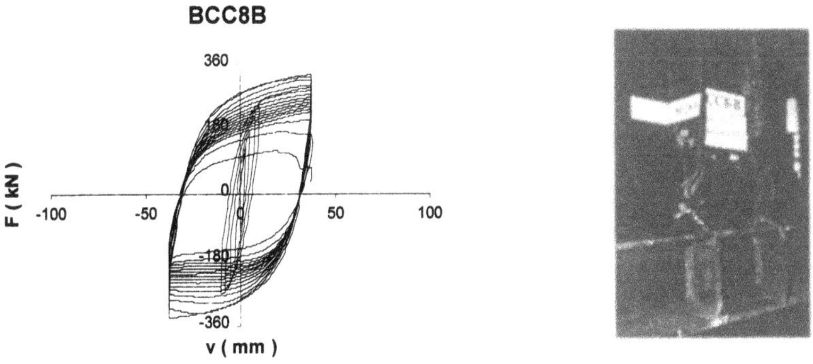

Figure 3.4.12: Histeresis loops and failure mode of BCC8B.

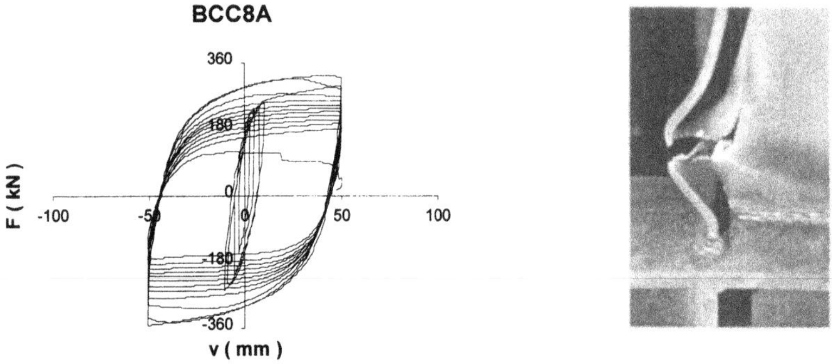

Figure 3.4.13: Histeresis loops and failure mode of BCC8A.

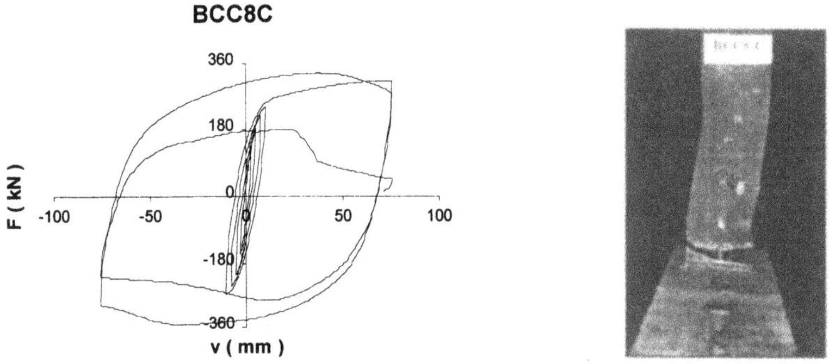

Figure 3.4.14: Histeresis loops and failure mode of BCC8C.

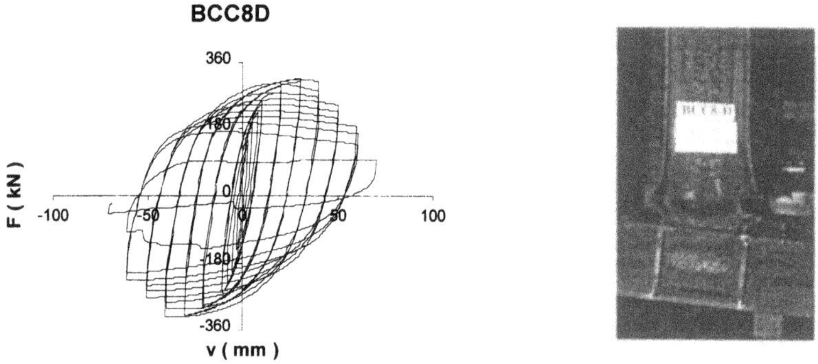

Figure 3.4.15: Histeresis loops and failure mode of BCC8D.

As can be derived from the curves reported in Figure 3.4.4 to 3.4.7 the cyclic behaviour of the specimen BCC5 is characterised by a great regularity and stability of the hysteresis loops up to failure with no deterioration of stiffness and strength properties. In the very last cycle the specimen has collapsed with a sudden and sharp reduction of strength, due to fracture initiated in the beam flange and propagated also in the web. During the tests, significant distortion of the joint panel zone has been observed, while no remarkable plastic deformation in the beam occurred.

Throughout the test program, two different kinds of cyclic behaviour have been observed for the BCC6 specimens, Figures 3.4.8 to 3.4.11. In some cases the behaviour of the specimens is close to the behaviour observed for the BCC5 type, with almost no deterioration of the mechanical properties up to the last cycle, during which the collapse occurred. For the other tests a gradual reduction of the peak moment at increasing number of cycles is evident. In these cases, starting from the very first plastic cycles, local buckling of the beam flanges occurred, and a well-defined plastic hinge has formed in the beam. The contribution of the panel zone deformation has not been as significant as in the BCC5 specimen type.

The hysteresis loops obtained from the tests on the BCC8 specimens, Figures 3.4.12 to 3.4.15 show a gradual reduction of the peak moment starting from the second cycle, where the maximum value of the applied moment has been usually registered. This deterioration of the flexural strength of the connection is related to occurrence and spreading of local buckling in the beam flanges and web. A well-defined plastic hinge in the beam has formed in all the tested specimens. In the specimens BCC8 the panel zone deformation has not been remarkable, and the plastic deformation took place mainly in the beam.

The BCC5 specimens, even though able to experience high deformation levels, have shown sudden failure modes in all the cyclic tests, with hysteresis loops practically overlaid and no degradation of the flexural strength up to the very last cycle, where a sharp decay of the load carrying capacity occurred due to fracture, generally developed in the proximity of the weld, Calado & Mele (1999, 2000).

On the contrary the BCC8 specimens have exhibited a typical ductile behaviour, with formation of a well-defined plastic hinge in the beam starting from the first plastic cycles, and a slow decrease of the peak moment at increasing number of cycles up to the collapse. These results confirm major findings reported in El-Tawil et al. (1999), where it is emphasised that although weak-panel design can substantially reduce beam plastic rotation demands and effectively contribute to global connection ductility, the stress state arising at high levels of applied rotation increases the potentials for brittle fracture collapse mode of the connection.

The BCC6 specimens displayed a behaviour sometimes closer to the BCC5 ones (tests BCC6C and BCC6D), sometimes to the BCC8 ones (tests BCC6A and BCC6B), depending on the applied loading sequence. Also with regard to the final collapse of the specimens, in the former cases it involved fracture in the beam starting at or close to the weld location, while in the latter cases it was due to the cracking in the buckled zones of the beam flanges.

In order to assess the influence of the load history on the hysteretic response of fully welded connections, per each specimen, the results obtained have been re-elaborated in terms of accumulated energy until failure and average of the accumulated energy, Figures 3.4.16.

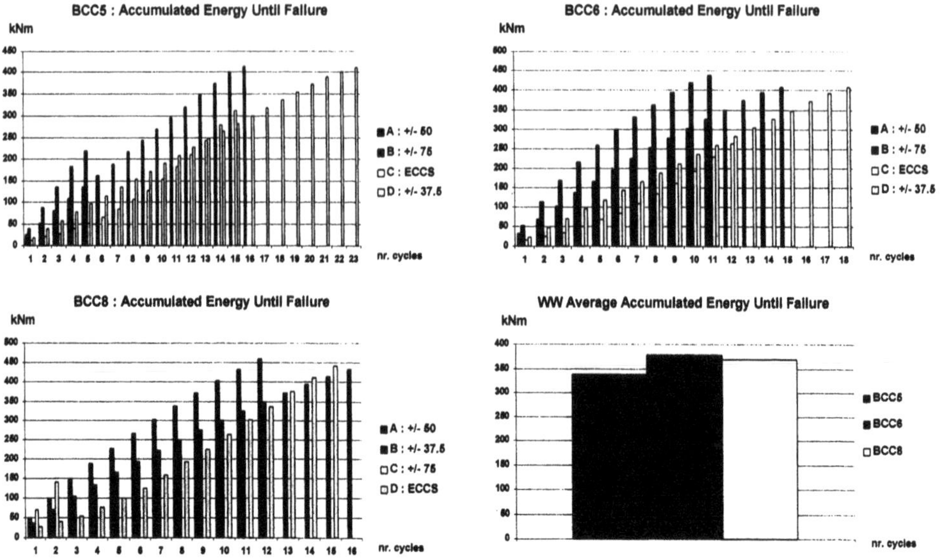

Figure 3.4.16: Accumulated energy and average of the accumulated energy of WW connections.

As expected, the accumulated energy depends strongly on the loading history. However, in terms of the average of the accumulated energy, it seems to be approximately the same for all specimens experimentally tested.

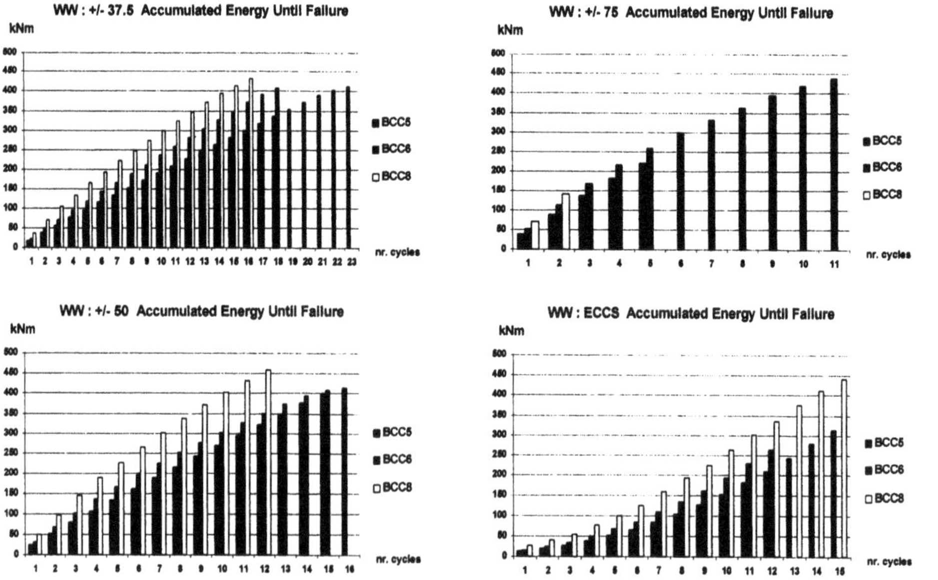

Figure 3.4.17: Accumulated energy of WW specimens.

The influence of the column size can also be analysed from the diagrams of the accumulated energy, Figure 3.4.17, and the elastic rigidity, Figure 3.4.18, ECCS (1986).

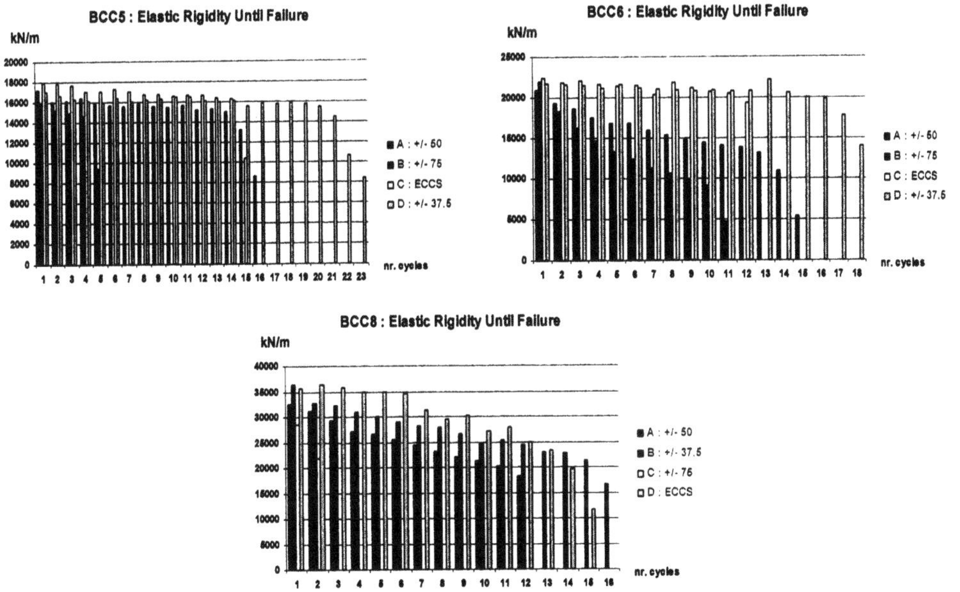

Figure 3.4.18: Elastic rigidity per cycle of WW specimens.

From these figures it can be observed that the behaviour of WW connections is affected by the column size. However the influence on the behaviour of the connection, in terms of the flexural strengths of the beam (M_{pb}), column (M_{pc}) and panel zone ($M_{p,pz}$), is better understood when computed on the basis of the nominal and actual yield stress, Table 3.4.3.

TABLE 3.4.3

MOMENT CAPACITIES OF THE WW SPECIMENS

SPECIMEN		M_{pb} (kNm)	M_{pc} (kNm)	$M_{p,pz}$ (kNm)
BCC5	*Nominal values*	147.6	83.2	91.1
	Actual values	173.0	114.0	149.8
BCC6	*Nominal values*	147.6	151.1	132.4
	Actual values	175.0	201.0	220.7
BCC8	*Nominal values*	147.6	247.5	182.9
	Actual values	183.0	316.0	239.7

From the simple comparison among the nominal plastic moments reported in Table 3.4.3, it can be observed that in the three WW specimens the weakest component of the joint configuration is respectively: the column for the BCC5 specimen, the panel zone for the BCC6 specimen, the beam for the BCC8 specimen. On the basis of the above calculations, and confirmed by the experimental tests, the specimens exhibited inelastic deformations mainly in the column panel zone (BCC5), in the beam and in the panel zone (BCC6) and mainly in the beam (BCC8), showing that the column size plays a fundamental rule in this type of connection.

TESTS ON TOP AND SEAT WITH WEB ANGLE CONNECTIONS

The specimen geometry of the top and seat with web angle connections is presented in Figure 3.4.19. In these specimens 120x120x10 angles have been adopted to connect the beam flange and web to the flange of the column. Two rows of bolts are placed on each leg of the flange angles, while on the legs of the web angles there is only one row of two bolts. The bolts are M16 grade 8.8 (yield stress f_{yb}=640 MPa, ultimate stress f_{ub}=800 MPa, A_s=157 mm^2), preloaded according to the EC3 provisions, i.e. at $F_{P,CD}$= 0.7 f_{ub} A_s = 87.9 kN.

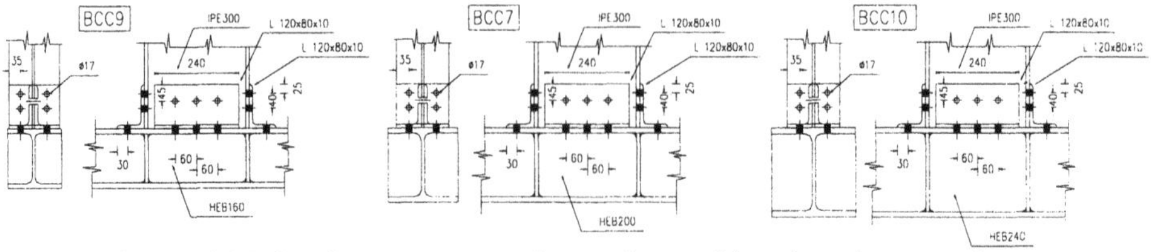

Figure 3.4.19: Specimen geometry of top and seat with web angle connections.

In Table 3.4.4 the mean values of the coupon test performed for each section type, for both web and flange elements and angles are presented.

TABLE 3.4.4

MATERIAL CHARACTERISTICS

MEMBER SECTION	Section element	f_y (MPa)	ε_y (%)	f_u (MPa)	ε_u (%)
HEB 160	*web*	348.63	0.53	490.31	31.5
	flange	303.43	0.84	453.12	44.5
HEB 200	*web*	302.31	0.55	434.39	38.2
	flange	282.29	0.75	433.27	43.3
HEB 240	*web*	290.28	0.52	447.91	39.6
	flange	304.74	0.81	454.4	39.1
IPE 300	*web*	315.67	0.66	451.26	41.1
	flange	304.71	0.68	452.63	42.7
L 120x10		252.23	0.52	420.14	44.5

In Table 3.4.5, the adopted nomenclature and the distance *h* relative to the loading actuator and displacement history type are reported. As performed for WW connections, test results have been represented in terms of force – displacement *(F-v)* curves, Calado et al (1999a, 1999b).

TABLE 3.4.5

ADOPTED NOMENCLATURE FOR TESTED SPECIMENS

DISPLACEMENT HISTORY	COLUMN SECTION (Actuator position *h* [mm])		
	HEB 160	HEB 200	HEB 240
v_{max} =± 37.5 mm	BCC9C (*h*=793)	BCC7D (*h*=780)	BCC10B (*h*=765)
v_{max} =± 50.0 mm	BCC9A (*h*=792)	BCC7A (*h*=780)	BCC10A (*h*=765)
v_{max} =± 75.0 mm	BCC9B (*h*=790)	BCC7B (*h*=780)	BCC10D (*h*=765)
ECCS	BCC9D (*h*=793)	BCC7C (*h*=770)	BCC10C (*h*=760)

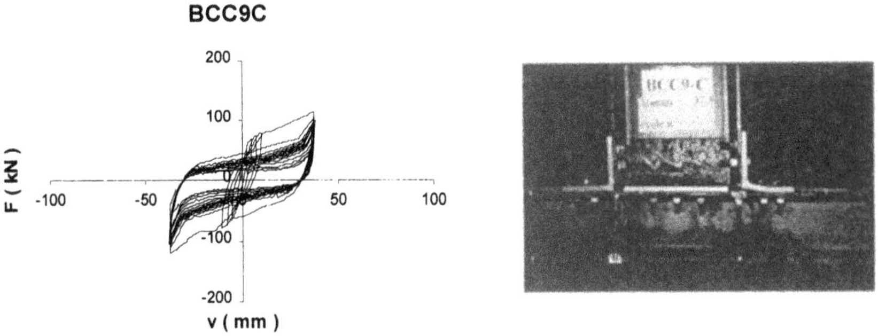

Figure 3.4.20: Histeresis loops and failure mode of BCC9C.

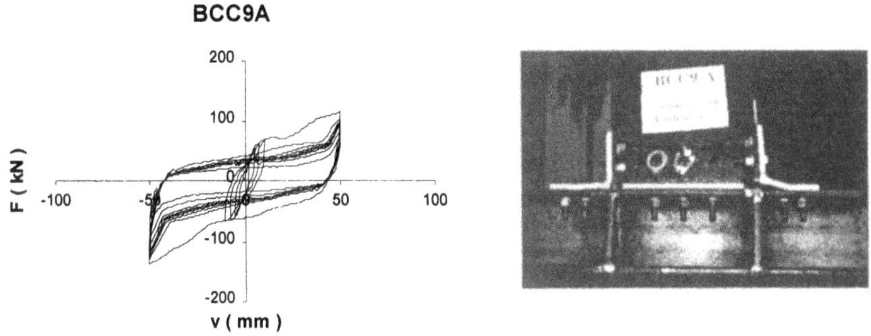

Figure 3.4.21: Histeresis loops and failure mode of BCC9A.

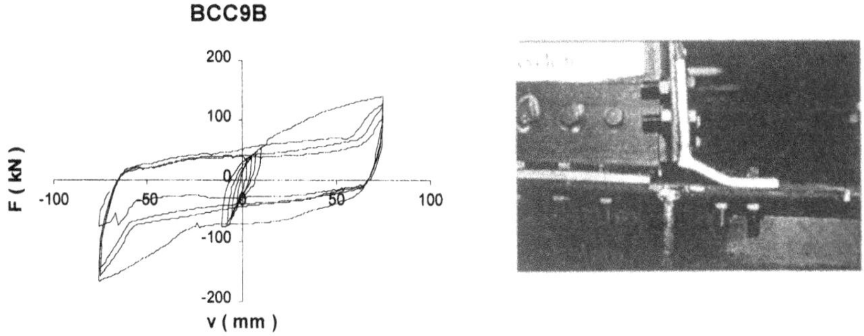

Figure 3.4.22: Histeresis loops and failure mode of BCC9B.

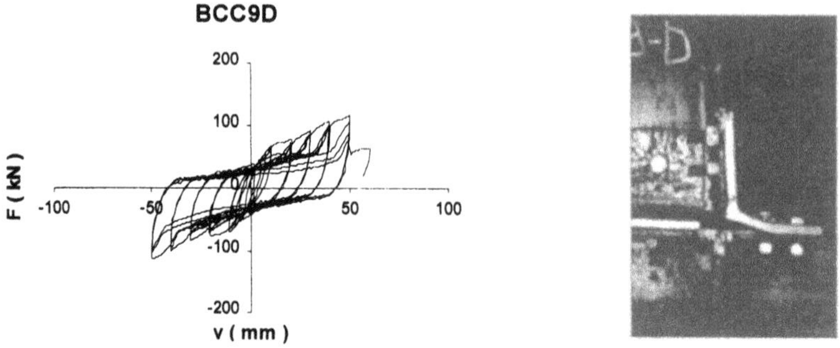

Figure 3.4.23: Histeresis loops and failure mode of BCC9D.

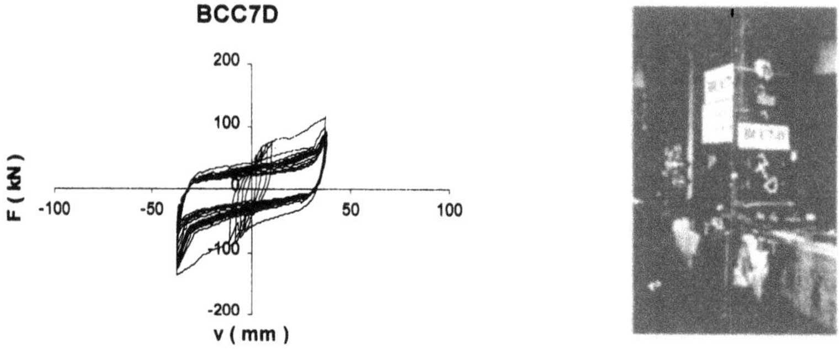

Figure 3.4.24: Histeresis loops and failure mode of BCC7D.

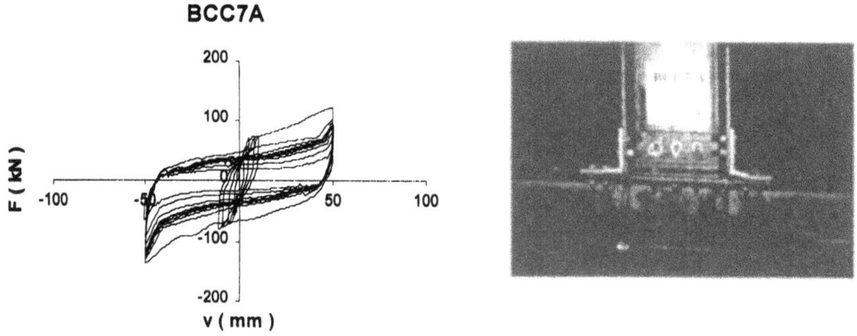

Figure 3.4.25: Histeresis loops and failure mode of BCC7A.

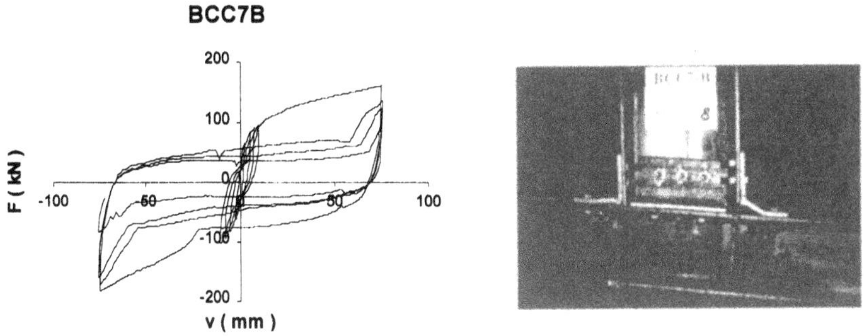

Figure 3.4.26: Histeresis loops and failure mode of BCC7B.

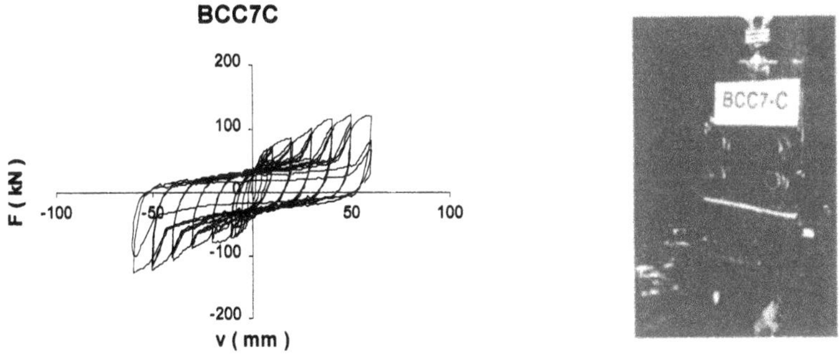

Figure 3.4.27: Histeresis loops and failure mode of BCC7C.

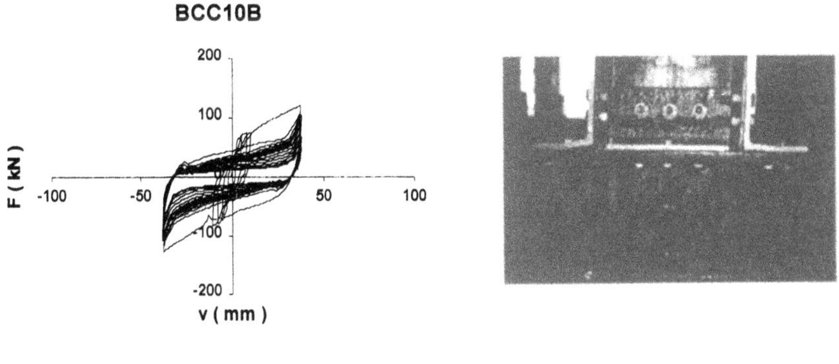

Figure 3.4.28: Histeresis loops and failure mode of BCC10B.

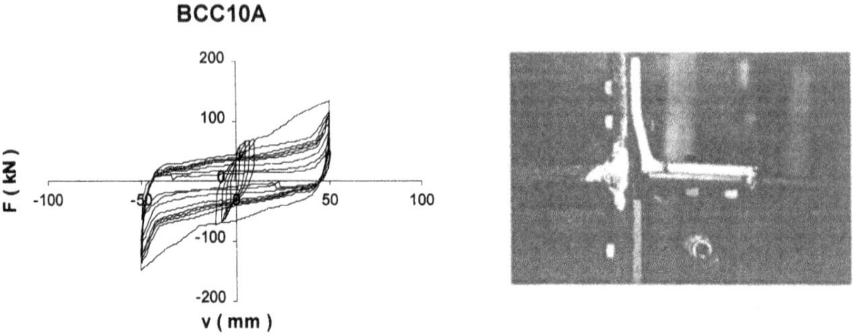

Figure 3.4.29: Histeresis loops and failure mode of BCC10A.

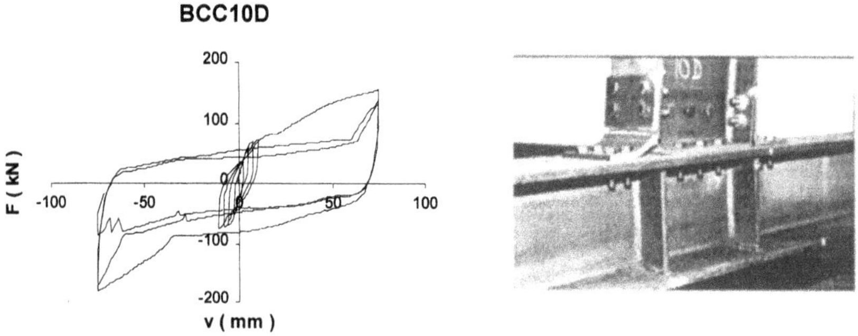

Figure 3.4.30: Histeresis loops and failure mode of BCC10D.

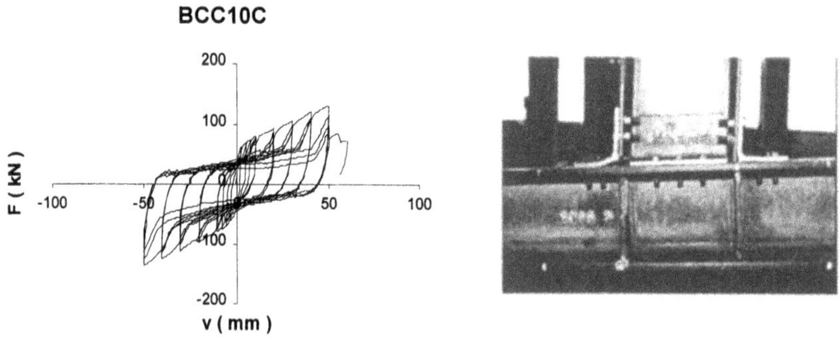

Figure 3.4.31: Histeresis loops and failure mode of BCC10C.

As can be derived from the curves reported in Figures 3.4.20 to 3.4.31, the shape of hysteresis loops of the three TSW specimens is very similar. The cyclic behaviour, the phenomena observed during the tests and the collapse modes are the same for the three specimen series, showing that the column size doesn't have and important rule for this type of connections.

The cyclic behaviour of the top and seat with web angle connections is characterised by bolt slippage and yielding and spreading of plastic deformation in the top and bottom angles, cyclically subjected to tension. Plastic ovalization of the bolt holes have also been observed mainly in the leg of the angle adjacent to column flange. The experimental curves, typical of this type of connection, shows pinched hysteresis loops, with a large slip plateau (very low slope of the experimental curve) and subsequent sudden stiffening. In fact when the specimen position is at the origin of the displacements, due to the concomitant effects of bolt slippage, hole ovalization and the plastic deformation of the angle legs adjacent to the column flange, the beam is completely separated from the column (gap open).

At large applied displacements, which impose large rotations to the connection, the contact of the compression angle and the beam web to the column flange (gap closure) give rise to sudden stiffening of the connection, which is evident in the experimental curves. No significant rotation of the column and distortion of the panel zone have been observed throughout the experimental tests carried out on the three specimens. At each step on the test, slight deterioration of the joint resistance in the three applied cycles can be observed in the experimental curves, mainly due to yielding and spreading of plastic deformation in the top and bottom angles, cyclically subjected to tension. In all the tests carried out on the three specimen series, the collapse of the connection occurred due to fracture in the leg angle located on the beam flange, immediately after the fillet.

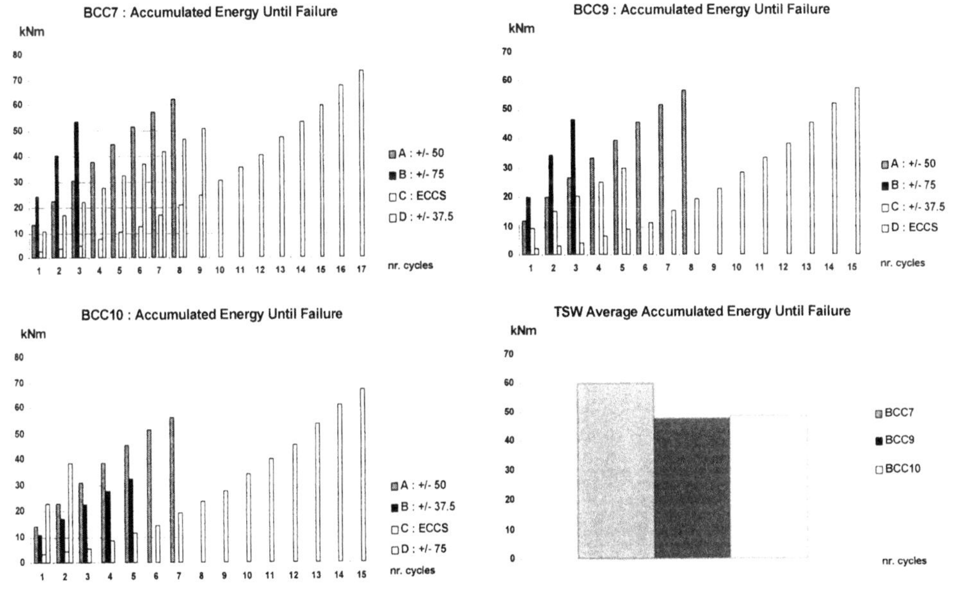

Figure 3.4.32: Accumulated energy and average of the accumulated energy of TSW connections.

Negligible scatters can be observed in the moment capacity of the three connection series, as it is expected, since the inelastic behaviour of the connection is governed by the angle. Also the maximum values of global rotation experienced by the specimens is the same for the BCC9 and BCC10 series, and slightly larger for the BCC7 one. The influence of the load history on the hysteretic response of TSW connections has been analysed in terms of accumulated energy until failure and average of the accumulated energy, Figure 3.4.32.

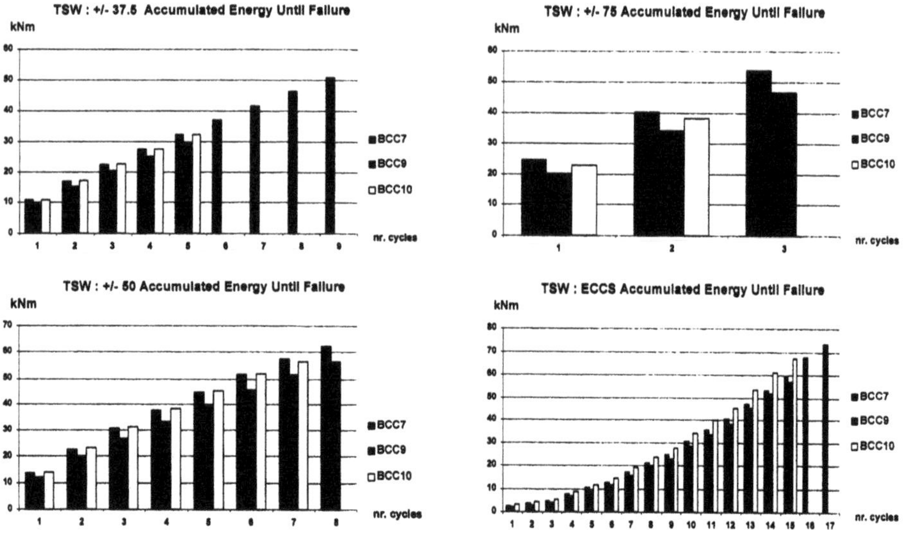

Figure 3.4.33: Accumulated energy of TSW specimens.

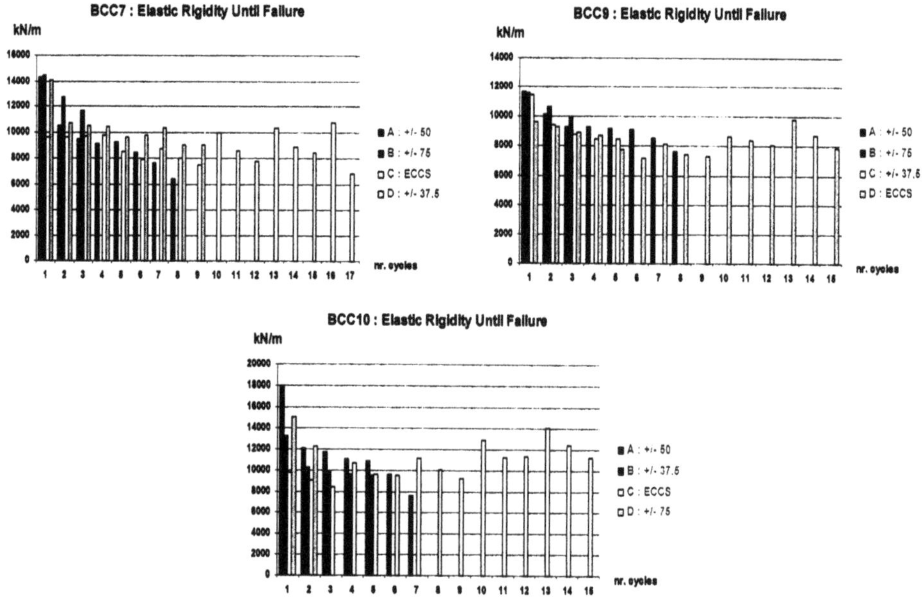

Figure 3.4.34: Elastic rigidity per cycle of TSW specimens.

As expected, the accumulated energy depends strongly on the loading history. However, in terms of the average accumulated energy, it seems to be approximately the same for all specimens experimentally tested.

The influence of the column size can also be analysed from the diagrams of the accumulated energy, Figure 3.4.33, and the elastic rigidity, Figure 3.4.34.

From these figures it can be observed that the behaviour of TSW connections is not affected by the column size.

This influence on the behaviour of the connection, in terms of the flexural strengths of the beam (M_{pb}), column (M_{pc}), bolt slippage (M_{slip}) and angle yielding ($M_{y,angle}$) have been computed on the basis of the nominal and actual yield stress and are reported in Table 3.4.6.

TABLE 3.4.6.
MOMENT CAPACITIES OF THE TSW SPECIMENS

SPECIMEN		M_{slip} (kNm)	$M_{y,angle}$ (kNm)	M_{pb} (kNm)	M_{pc} (kNm)
BCC9	*Nominal values*	32 - 47.5	23.3	147.6	83.2
	Actual values		28.1	198.2	123.4
BCC7	*Nominal values*	32 - 47.5	23.3	147.6	151.1
	Actual values		28.1	198.2	194.4
BCC10	*Nominal values*	32 - 47.5	23.3	147.6	247.5
	Actual values		28.1	198.2	305.7

From the experimental tests it was observed that the two major phenomena of the behaviour of the top and seat with web angle connections were the slippage of bolts and the yielding of the tension angle.

From the comparison between the bending moments corresponding to bolt slippage and angle yielding, it derives that the specimens are "slip critical" connections, since slippage of top and seat angle bolts occurs at a load level higher than the one corresponding to yielding of the tension angle.

FULLY WELDED VS TOP AND SEAT WITH WEB ANGLE CONNECTIONS

By comparing the two series of experimental tests it must be noticed that the three WW specimens show significant differences in the initial stiffness, maximum strength and deformation capacity, thus confirming the strong effect of the column cross section size. On the contrary, the three TSW specimens present quite close experimental responses.

This difference between the behaviour of WW and TSW specimens is mainly related to the design of the specimens, since in the TSW connections the weakest component is the same in the three specimens (the angle in tension), thus the beam, column and panel zone strength ratios does not affect the response of the specimens.

Slight scatters can be observed in the initial stiffness, due to the different column and panel zone deformability, but the non-linear portion of the curve and the maximum bending moment are very similar.

As already evidenced, the behaviour of the WW connections is affected by the column dimensions since the three combinations of beam and column framing in the joint give rise to panel zone strength values respectively: smaller than, approximately equal to and larger than the plastic moment of the beam, for the BCC5, BCC6 and BCC8 specimens. These observations are confirmed by analysing Figures 3.4.35 and 3.4.36 where the envelopes of force – displacement curve for WW and TSW are presented.

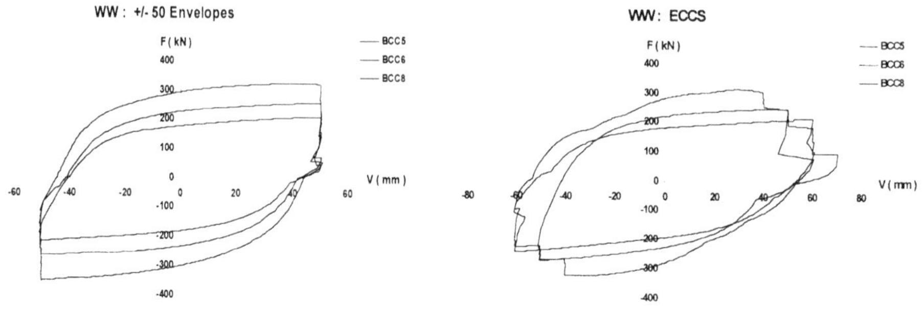

Figure 3.4.35: Envelops of the force – displacement curves for WW connections.

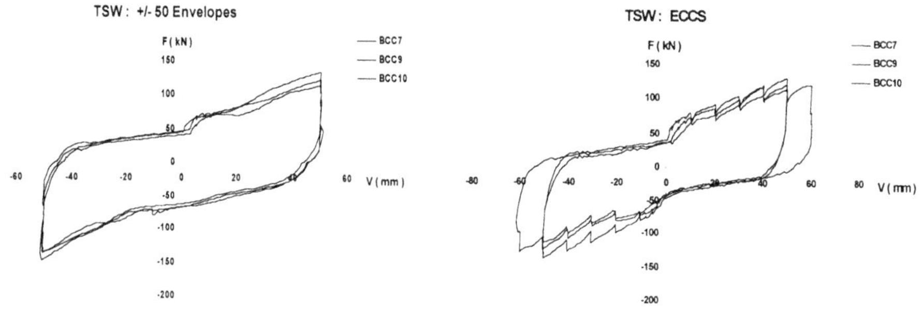

Figure 3.4.36: Envelops of the force – displacement curves for TSW connections.

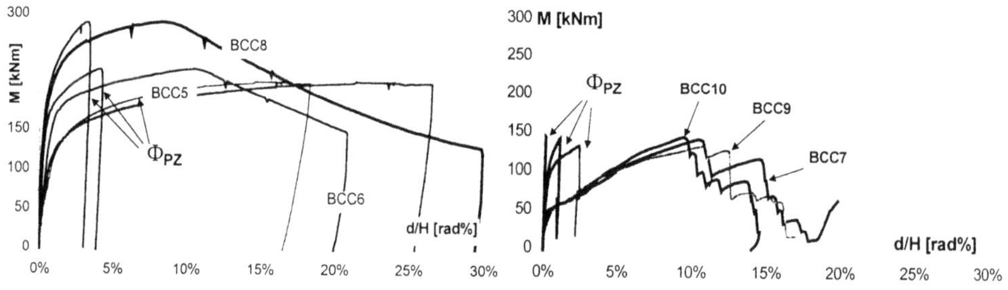

Figure 3.4.37: Monotonic experimental curves for WW and TSW specimens.
(adapted from Calado & Mele (2000))

This is also confirmed by monotonic tests performed by Calado & Mele (2000), Figure 3.4.37, in which it was noticed that while in the WW specimens the panel zone distortion Φ_{PZ} significantly contributes to the specimen global rotation d/H (at the maximum value of the bending moment registered in the relevant test), though at different extent in the three specimens, a completely different order of magnitude of this contribution is registered for the TSW connections. For the TSW specimens, the rotations due to bolt slippage Φ_{slip} computed on the basis of the LVDTs measured displacements constitutes a major contribution to the total rotation d/H.

Figure 3.4.38 shows that both the TSW and the WW specimens can be classified, according to the EC3, as semirigid joints, even though the TSW specimens are very close to the boundary of the flexible joint behaviour, while all the WW connections are able to sustain bending moment larger than the beam plastic moment capacity. Thus, even though WW specimens can be classified as full strength connections, they experience the maximum bending moment at large deformation levels.

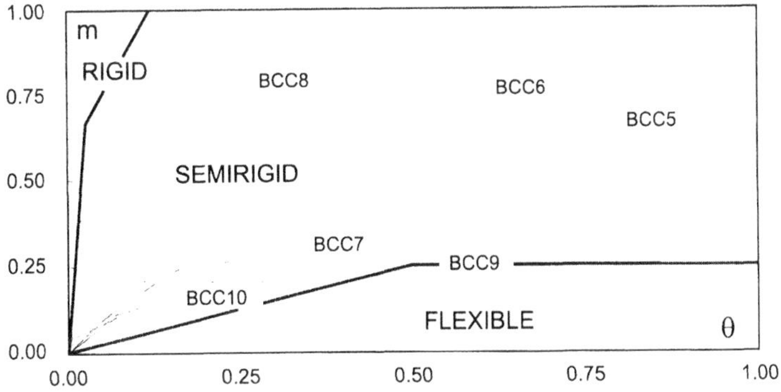

Figure 3.4.38: Classification of the specimens according to EC3.

The EC3 classification system for joints has been recently criticised for several reasons. One of the main drawbacks of the approach is that considers the stiffness and the strength criteria separately, thus possibly leading to some inconsistencies when both the serviceability and the ultimate state are considered; concerning this point in Nethercot et Al., (1998) an unified classification system in which connection stiffness and strength are considered simultaneously, is proposed. Furthermore in Hasan et Al., (1998) the weakness of the idea of adopting the stiffness of the beam as the only parameter for defining the initial stiffness of the connection is discussed and the need of defining a non-linear classification system, with no reliance on the beam stiffness is emphasised. However the definition of a unique approach, fully exhaustive, is not a simple task, Mele et al (1999b).

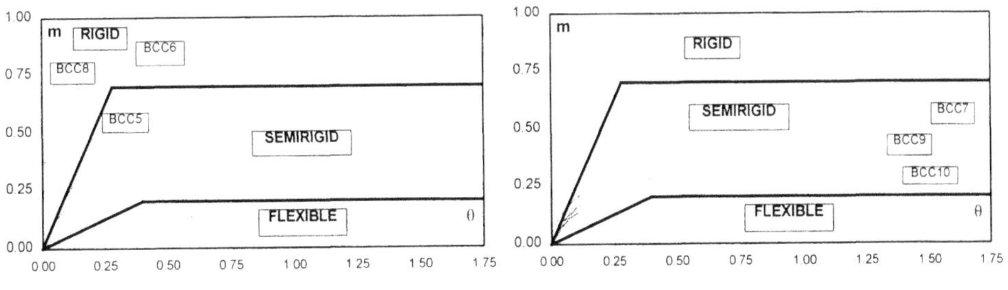

Figure 3.4.39: Classification of WW and TSW specimens according to Bjorhovde et al (1998).

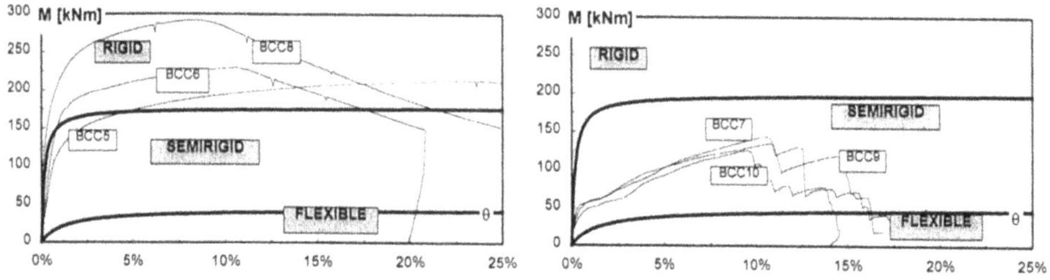

Figure 3.4.40: Classification of WW and TSW specimens according to Hassan et al (1998).

In order to show the differences arising from the application of different methods, in Figures 3.4.39 and 3.4.40 the approach proposed by Bjorhovde et Al., (1990) is quite close to the EC3 one, and the new non-linear system proposed by Hasan et Al., (1998) are applied for classifying the TSW and WW large scale connections described in this paper. From figure 3.4.39 it derives that the TSW connections can be defined as semirigid according to both the two proposal as well as to the EC3. On the contrary for the WW connections (Figure 3.4.40) the two systems provides results quite different from the EC3 ones, according which the three connections are all defined as semirigid (Figure 3.4.38). Both systems classify the BCC8 specimen as fully rigid, while the BCC6 and BCC5 specimen behave as semirigid joint at serviceability limit state and rigid at the ultimate limit state.

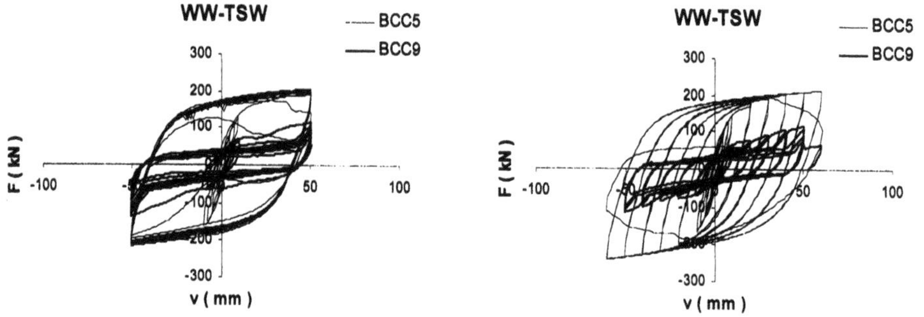

Figure 3.4.41: Comparison between force – displacement curves of WW and TSW connections.

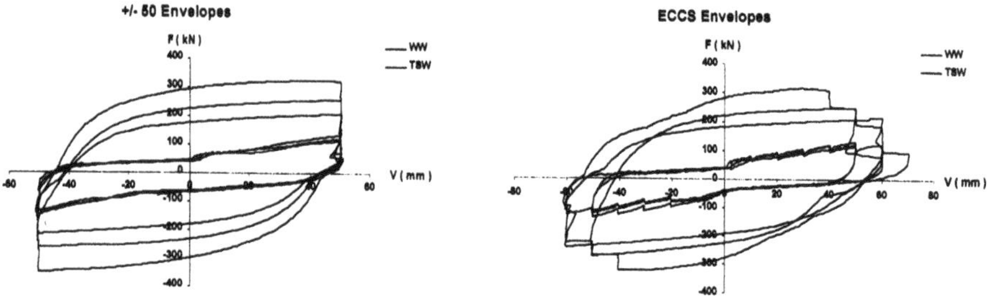

Figure 3.4.42: Comparison between envelopes of WW and TSW connections.

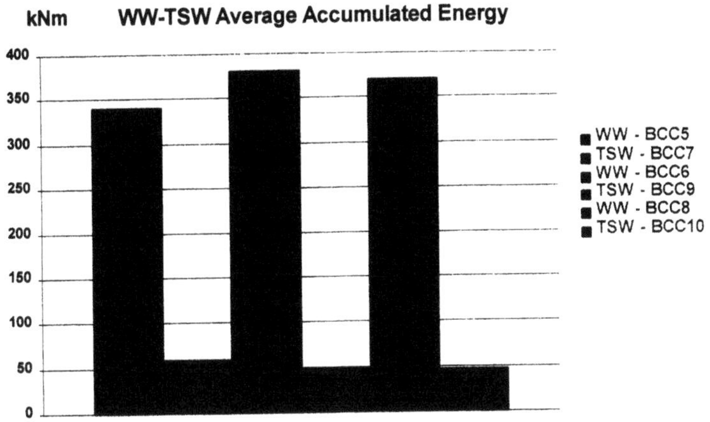

Figure 3.4.43: Average of the accumulated energy for WW and TSW connections.

Re-elaboration of tests performed by D'Avino (1998), Cappuccio (1998) and Calado & Mele (2000) have also demonstrated that, for the same displacement history, WW specimens exhibit greater maximum strength (Figure 3.4.41 and 3.4.42) and accumulated energy than TSW connections, Figure 3.4.43.

As a conclusion, and based on this experimental research it is pointed out that the column size does not affect the behaviour of the top and seat with web angle connections, which is mainly related to the tension angle geometry and strength properties. The column size has on the contrary demonstrated to affect at large extent all the response parameters (stiffness, strength and deformation capacity) of welded connections.

REFERENCES

ATC 24 – Guidelines for Cyclic Seismic Testing of Components of Steel Structures, (1992), *Applied Technology Council.*

Bjorhovde, R., Colson, A. & Brozzetti, J., (1990), Classification system for beam-to-column connections. *Journal of Structural Engineering, ASCE*, Vol. 116

Calado, L., De Mattes, G., Landolfo, R. & Mazzolani, F. M., (1999a), Cyclic Behaviour of Steel Beam-to-Column Connections: Interpretation of Experimental Results, *SDSS'99 – Stability and Ductility of Steel Structures*, Timisoara, Romania, pp 211-220

Calado, L., Landolfo, R. & De Matteis, G., (1999b), Fracture Resistance Design of Bolted Joints, *II Encontro Nacional de Construção Metálica e Mista*, Coimbra, Portugal, pp 577-588.

Calado, L., Mele, E. & De Luca, A., (1999c), Experimental Investigation on the Cyclic Behaviour of Welded Beam-to-Column Connections. *Proc. 2nd European Conference on Steel Structures*, Praha, Czech Republic, Paper No.215.

Calado, L. & Mele, E., (1999), Experimental Research Program on Steel Beam-to-Column Connections (phase-1), *Report ICIST DT n° 1/99*, Instituto Superior Técnico, Lisbon, Portugal.

Calado, L. & Mele, E, (2000), Experimental Behaviour of Steel Beam-to-Column Joints: Fully Welded vs Bolted Connections, *Proc. XII World Conference on Earthquake Engineering*, Auckland, New Zealand.

Cappuccio, R, (1998), Il Comportamento Ciclico dei Collegamenti Saldati Trave-Colonna Nelle Strutture Sismoresistente di Acciaio, *Graduation in Civil Engineering Thesis*, University of Naples, Italy (in italian).

D'Avino, M, (1998), Il Comportamento Ciclico dei Collegamenti Bullonati Trave-Colonna Nelle Strutture Sismoresistente di Acciaio, *Graduation in Civil Engineering Thesis*, University of Naples, Italy (in italian).

ECCS (1986). Seismic Design. Recommended Testing Procedure for Assessing the Behaviour of Structural Steel Elements under Cyclic Loads. Technical Committee 1 - Structural Safety and Loading, TWG1.3 – *Rep. No.45.*

El-Tawil, S., Vidarsson, E., Mikesell, T., and Kunnath, S. K. (1999). Inelastic Behaviour and Design of Steel Panel Zones. *Journal of Structural Engineering, ASCE*, Vol.125, No.2. pp 183-193.

Hasan, R., Kishi, N. & Chen, W. F.,(1998), A New Nonlinear Connection Classification System. *Journal of Constructional Steel Research.*, Vol. 47, pp 119-140.

Mele, E., Calado, L., Pucinotti, R. (1997). Indagine Sperimentale sul Comportamento Ciclico di Alcuni Collegamenti in Acciaio. *Proc. 8th ANIDIS*, Taormina, Italy, pp 1031-1040.

Mele, E., Calado, L., De Luca, A. (1999a), Experimental Behaviour of Beam-to-Column Welded Connections: Effect of the Panel Zone Design. *Proc. XVII C.T.A.*, Napoli, Italy, pp149-158.

Mele, E., Calado, L. & Di Sarno, L, (1999b), Monotonic Behaviour of beam-to-Column Connections: Tests / Modelling / Eurocode 3, *Proc. XVII C.T.A.*, Napoli, Italy, pp159-168.

Nethercot, D. A., Li, T. Q. & Ahmed, B., (1998), Unified Classification System for Beam-to-Column Connections. *Journal of Constructional Steel Research,* Vol. 45, pp 39-45.

Pucinotti, R., (1998), I Collegamenti nelle Strutture in Acciaio: Analisi Teoriche e Sperimentali, *PhD in Strutural Engineering*, University of Reggio Calabria, Italy (in italian)

Chapter 4

Cyclic Behaviour of Beam-to-Column Composite Connections

4.1

PRESENT SITUATION

Jean Marie Aribert and Alain Lachal

GENERAL

Steel-concrete composite construction appears very competitive today in Europe. Its advantages are well-known, as the clear increase in stiffness and resistance of the steel elements (beams and columns) due to the concrete participation, the improved fire resistance specially when the steel elements are partially or fully encased by concrete, the ease and rapidity of erection leading to low construction costs, etc. In particular, beam-to-column joints in composite building frames can be designed in hogging bending as composite elements for little increase in work on site. The key feature is the provision of continuous slab reinforcement to act in tension across the joint. But two aspects should be underlined about this topic: on the one hand, many types of joint can be imagined; on the other hand, design codes do not give efficient rules at the moment for composite joints. So, Eurocode 4-part 1.1 (1992) is content with describing what is meant by a composite joint, leaving the designers to adapt the detailed rules given in Annex J of Eurocode 3, part 1-1 (1992) for the steel components. Nevertheless, the Composite Sub-Group of Working Group 2 of the COST-C1 Project has published more recently detailed provisions and a model code for the design of composite joints subject to static or quasi-static loading (COST-C1, 1999); some parts of this document may be useful to interpret better the present chapter though devoted to the cyclic behaviour of joints.
The dynamic and seismic performances of steel-concrete composite frames are well admitted generally, but often designers are in front of many questions for which no convincing answers or minimal guidance exist. So, in Annex D of Eurocode 8 Part 1-3 (1995) where are adopted the same behaviour factors q for composite structures as for steel structures apparently because of lack of knowledge, odd concepts are advanced without real scientific background as the one recommending the use of partial shear connection or the other satisfying the overstrength principle of composite joints by the contribution of the steel parts alone.

The main objective of the present chapter is to provide experimental information on the risk of degradation of the shear connection and on the global behaviour (expressed by moment-rotation curves) of composite joints subject to cyclic repeated loads. The testing program deals with joints currently used in buildings and associated with solid slabs or composite ones (where concrete is cast into a profiled steel sheeting), the shear connection being ensured by welded headed studs or nailed cold formed angles. To restrict a little the field of investigation, the joint of the steelwork parts is systematically a bolted flush end plate (such a joint is often employed for buildings in Europe, noting that an extended end plate is not advantageous due to restraining into the concrete slab).

Before the joint tests properly so-called, a preliminary investigation is developed from several Push–Out and Push-Pull tests considering four types of shear connection. Interpreting the test results , it would appear necessary to adapt modified characteristics of the shear connectors under cyclic repeated loads and to control their fatigue resistance in most cases. But introducing of such a local characterization into a modelling of global behaviour of joints would be an utopian procedure in comparison with the actual state of knowledge of the behaviour of composite joints in monotonic loading. Nevertheless, an approximate determination of the moment-rotation skeleton curve (enveloping the cycles) is justified at the end of the chapter including a rotation boundary due to premature rupture of bolts by low-cycle fatigue. So, that opens a possible way towards practical design.

First of all, a general survey is made of the relevant literature to this chapter.

REFERENCES

COST C1 / European Commission. (1999). *Composite steel-concrete joints in frames for buildings : design provisions,* Brussels.

Eurocode 3, Part 1-1 ENV 1993-1-1. (1992). *Design of composite steel structures – general rules and rules for buildings*, CEN – Brussels.

Eurocode 4, Part 1-1 ENV 1994-1-1. (1992). *Design of composite steel and concrete structures – general rules and rules for buildings,* CEN – Brussels.

Eurocode 8, Part 1-3 ENV 1998-1-3. (1995). *Design provisions for earthquake resistance of structures - Annexe D : specific rules for steel-concrete composite buildings*, CEN – Brussels.

4.2

RESEARCH WORKS FROM LITERATURE

Jean Marie Aribert and Alain Lachal

WORKS CONCERNING SHEAR CONNECTORS

The majority of Push-Pull tests carried-out during the fifteen last years concerns the high-cycle fatigue of stud shear connectors for bridge applications; for example Oehler and Coughlan (1986), Naithani et al. (1988), Oehler (1991), Gattesco and Giuriana (1996), Taplin and Grundy (1997).
In low-cycle fatigue with high slip amplitude, the seismic response of stud shear connectors for buildings was studied by Hawkins and Mitchell (1984) and by Astaneh et al (1993).
Bursi and Ballerini (1997) analysed the low-cycle behaviour of welded stud connectors under variable and constant sequential-phased displacement histories. Eleven Pull-Push specimens were tested. A comparison between the load-slip curve provided by two specimens tested in a monotonic process and the skeleton curves of the other specimens tested under cyclic loading, led them to observe a reduction both of strength and ductility due to the detrimental effects of reversed displacements. The mean shear resistance predicted by Eurocode 4 Part 1-1 appeared unsafe if directly applied to seismic design. The authors concluded in favour of a need of additional studies in order to expand the experimental data base in this field.

Pacurar, Litan and Petran (1995) tested 4 Push-Pull specimens with studs of 20 mm diameter: two tests under monotonic loading and two tests under repeated cycles of pulsating loading (50 000 cycles between 110 kN and 160 kN) followed by a monotonic loading up to failure. Comparisons between monotonic and cyclic results led the authors to similar conclusions as those above-mentioned, namely a reduction both of strength and ductility due to cyclic pulsating loading.
In conclusion, very few Push-Pull tests were conducted in low-cycle fatigue process,.and essentially stud shear connectors were studied.

WORKS CONCERNING COMPOSITE BEAM-TO-COLUMN JOINTS

Lee and Lu (1989) tested three composite joints under cyclic loading process, namely two major axis joints (one exterior and the other interior) and one minor exterior joint. The effect of the composite slab and the panel zone on the joint behaviour were investigated more particularly. Theoretical predictions of the sub-assembly behaviour were developed and compared with the test results in term of rotational stiffness, moment resistance, ductility and dissipated energy.

Ermopoulos et al.(1995) developed an analytical simple model for composite joints under monotonic or cyclic loading, which can be incorporated with a structural analysis program. The strength and the deformability of the joint and the panel zone were taken into account separately by non linear springs. One advantage of the model was to simulate the joint behaviour using a reduced number of parameters.

An experimental investigation involved two cyclic tests on a cruciform joints with bolted end plates, studs and composite beams and columns. Satisfactory behaviour was observed, except when large column deformations were reached leading to a drop of rigidity. The authors proposed to restrict the flange deformability by stiffening the column web.

Bursi and Ballerini (1997) carried out quasi-static cyclic and pseudo-dynamic tests with controlled displacement on composite beams with full and partial shear connections at the University of Trento. They concluded in a satisfactory ductility factor (about 4), a shortcoming Eurocode 4 prediction of the maximum resistance of the composite beam when subject to cyclic loads, and an acceptable behaviour of partial shear connection compared to the one of full shear connection. Two-dimensional finite element analyses based on smeared-crack formulations including discrete stud connectors with a sophisticated inelastic concrete model available in the ABAQUS code were also performed.

Bursi, Gramola and Zandonini (1997) tested a few welded beam-to-column joints made of steel columns and beams with a concrete deck and designed to behave as rigid and full strength. Reinforced plates were adopted to reduce the stresses in the welds. The neutral axis was kept in the steel section by adopting a minimum amount of reinforcement.

Hajjar and Leon (1996) tested composite beam-to-column joints to explain the mode of failure initiated close to the bottom flange of steel beams during the Northridge earthquake on 1994. It was observed that the strains near the bottom flange were significantly larger than those near the top flange explaining the predominance of bottom flange failure. This experimental observation appeared a consequence of the composite floor action.

Leon (1997) appraised the seismic capabilities of semi-rigid joints in the world-wide context of codes and summarized several important characteristics to design partially restrained composite connections (PR-CC). He evaluated their effect on the dynamic performance of frames (flexibility effect, loss of resistance, required rotation, change in period, dynamic stability). A typical semi-rigid joint (with web angles) was recommended while giving detailed rules for design.

Matsuo and al. (1997) analysed the shear yielding and the maximum resistance of the panel zone of a composite beam-to-steel H column joint. Experimental results of 13 specimens were given. The panel moment - shear deformation relationship was formulated considering the Bauschinger effect with an isotropic and kinematic hardening rule; also the theoretical resistance was evaluated using limit plastic analysis.

Zandonini, Bernuzzi and Bursi (1997) gave a large state-of-the-art review on rigid and semi-rigid composite joints under seismic actions.
Regarding joints to be used in semi-rigid frames, in addition to the above-mentioned paper of Leon, the authors underlined the study of Wang et al. (1996) who carried out tests on composite beam-to-column joints with bolted flush end plates (it was suggested to strengthen the column web panel zone in order to reduce severe distortions when unbalanced beam moments are applied to the joints). Also were mentioned Ebato and Morita (1995) who analysed the hysteretic behaviour of a joint consisting in a composite beam with the bare steel joint welded to a hollow section column without concrete and without stiffeners; the semi-rigid joint behaviour was satisfactory. And Rosa et al. (1994) who proposed a new simple connection detail to join the bottom of the beam web to the column, providing resistance moment values of about 0.8 and 0.5 times the hogging and sagging plastic resistance moments of the beam, respectively.

Considering the specific aspect of the slab participation in the moment transfer between beams and columns, a large experimental investigation was carried out at the ELSA laboratory on full-scale composite frames, by Pinto and Calvi (1996) In their tests, a full degree of shear connection was adopted systematically to design the composite beams. More recently, Plumier, Doneux and Bouwkamp (1998) analysed the validity of existing models of force transfer from the slab to the joint using comparison with a few tests. Even if the exact mechanism of force transfer is not yet fully clear, a code formulation could be expected in the next future about conditions on the transverse rebars ensuring a high ductility in bending without crushing of the concrete slab nor local buckling of the steel part.

In conclusion, many works exist in the literature which plead in favour of the efficiency composite beam-to-column joints under seismic actions. However, most of the test results are global and qualitative, without referring to significant criteria required in seismic design. The fact that usual composite joints are generally semi-rigid and partial strength, hence dissipative under seismic actions, may explain the difficulty to establish analytical or theoretical formulations of far-reaching.

REFERENCES

Astaneh-Asl A., McMullin K.M. Fenves G.L. and Fukuzava E. (1993). Innovative semi-rigid steel frames for control of the seismic response of buildings. *Report UCB/EERC.*

Bursi O. and Ballerini M. (1997). Low-Cycle Behaviour and analysis of Steel-Concrete Composite Substructures. *Proc. Int. Conf. Innsbruck*, 615-620.

Bursi O.S., Gramola G. and Zandonini R. (1997). Quasi-static cyclic and pseudodynamic tests on composite substructures with softening behaviour. *5th Int. Col. On Stability and Ductility of Steel Structures*, Nagoya, Japan.

Ebato K., Morita K. and Sugiyama T. (1995). Finite element analysis of semi-rigid composite joint. *Proc. Fourth Pacific Struct. Steel Conf.*, Singapore, 9-16.

Ermopoulos J. Ch., Vayas I., Petrovites N.E., Soffianopoulos D.S. and Spanos Chr. (1995). Cyclic behaviour of composite beam-to-column bolted joints. *Steel Structures - Eurosteel '95*, Kounadis, Balkena, Rotterdam.

Gattesco N. and Giuriana E. (1996). Experimental study on stud shear connectors subjected to cyclic loading. *Journal of Constructional Steel Research* **38-1**, 1-21.

Hajjar F.J. and Leon R.T. (1996). Effect of floor on the performance of SMR connections. *Proc. Elev World Conf. On Earth. Eng.* **656**. Elsevier.

Hawkins N.M. and Mitchell D. (1984). Seismic response of composite shear connections. *J. of Structural. Eng.* **110-9**, 2120-2136.

Lee Seung-Joon and Lu Le-Wu. (1989). Cyclic tests of full-scale composite joint subassemblages. Member, ASCE. *Journal of Structural Engineering* **115:8**.

Leon R.T. (1997). Seismic design of composite semi-continuous frames. *Proc. Int. Conf. Innsbruck,* 657-662

Matsuo A., Nakamura Y., Salib R.W and Matsui Y. (1997). Behaviour of the Composite Beam-to-Steel H Column Connection. *Proc. Int. Conf. Innsbruck*, 882-883.

Naithani K.C., Gupta V.K. and Gadh A.D. (1988). Behaviour of shear connectors under dynamic loads. *Materials and Structures* **21:125**, 359-363.

Oehler D.J. and Coughlan C.G. (1986). The shear stiffness of stud shear connections in composite beams. *Journal of Constructional Steel Research* **6**, 273-284.

Oehler D.J. (1991). Deterioration in strength of stud connectors in composite bridge beams. *J. of Structural. Eng.* **116:12**, 3417-3431.

Pacurar V., Litan M., Petran I. (1995). On the elastic connectors behaviour under cyclic loading. *Steel Structures - Eurosteel '95*, Kounadis, Balkena, Rotterdam.

Pinto A.V. and Calvi G.M. (1996). Ongoing and future research in support of Eurocode 8. *Proc. of the 11th World Conference on Earthquake Engineering*, Paper n° 2055.

Plumier A., Doneux C. Bouwkamp J.G. and Plumier C. (1998). Slab design in connection zone of composite frames. *Proc. of the 11th European Conference on Earthquake Engineering*, Balkema, Rotterdam.

Rosa A.M, Lu L.W. and Viscomi B.V. (1994). Monotonic and cyclic behaviour of semi-rigid composite connections with ATLSS connectors. *Proc. Fifth U.S. Nat. Conf. On Earth. Eng.*, Chicago, 881-890.

Taplin G. and Grundy P. (1997). Incremental slip of stud shear connector under repeated loading in Composite Construction. *Proc. Int. Conf. Innsbruck*, 145-150.

Wang J.Y., Wong Y.L. and Chan S.L.Y. (1996). Experimental study of semi-rigid composite end-plate connections under cyclic loading. *Proc. Int. Conf. On Advances in Steel Structures '96*, Hong Kong, 489-494.

Zandonini R., Bernuzzi C. and Bursi O.S. (1997). Steel and steel-concrete composite joints subjected to seismic actions - General report. *Proceedings of STESSA '97*, Kyoto, Japan, (Ed. F. Mazzolani and H. Akiyama), 511-529.

4.3

CYCLIC BEHAVIOUR OF SHEAR CONNECTORS

Jean Marie Aribert and Alain Lachal

EXPERIMENTAL PROGRAM

The experimental program reported in this sub-chapter involves 30 tests of both Push-Out type (with a regular increasing displacement up to failure) and Push-Pull type (with fully reversal cyclic loading). The tested specimens may be classified in four groups presented in figure 4.3.1.

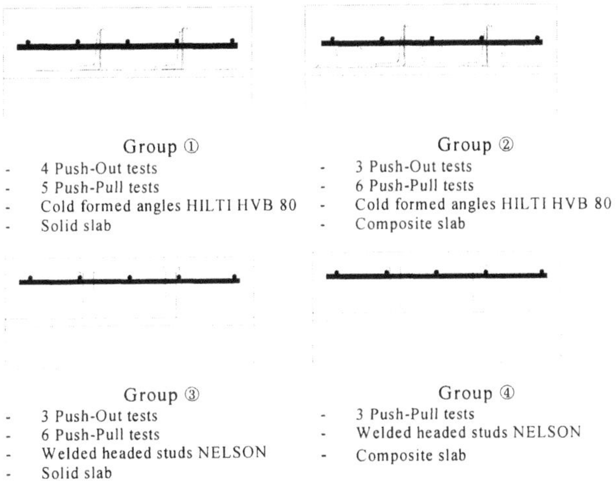

Figure 4.3.1 : Types of tested shear connectors

All the specimens were fabricated according to the recommendations given in Eurocode 4-1-1 (Chapter 10) for Push-Out tests, using one HEB 200 steel section and two concrete slabs each having *120 x 400 mm* cross-section and including a single reinforcement layer with *4ϕ10 mm* longitudinal rebars of *400 mm* length and *5ϕ8 mm* transverse rebars of *350 mm* length. It should be noted that a single reinforcement layer per slab was adopted to simulate a similar arrangement as the beam-to-column joints tested in sub-chapter 4.4.

EXPERIMENTAL ARRANGEMENT

A detail arrangement which was discovered improving the cyclic procedure for Push-Pull tests consists of introducing a small gap of about 0.5 mm between the upper and lower rigid supports (figure 4.3.2). Effectively it was observed that permitting a free transversal displacement of the two slabs at each reversal load led to avoid any significant rotation of them (therefore any significant detachment from the steel section).

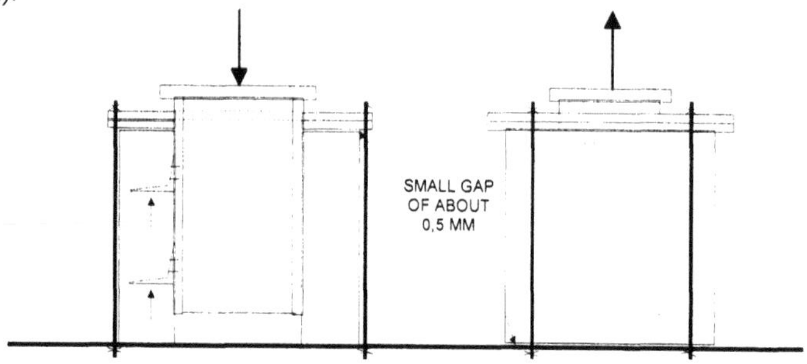

Figure 4.3.2 : Experimental arrangement

Several locations concerning the potentiometric transducers to measure the slip displacements between the solid or composite slab and the steel flange were compared (figure 4.3.3), exactly:

- at the upper end of the slab,
- at the level of connector rows,
- at the middle of the two connector rows.

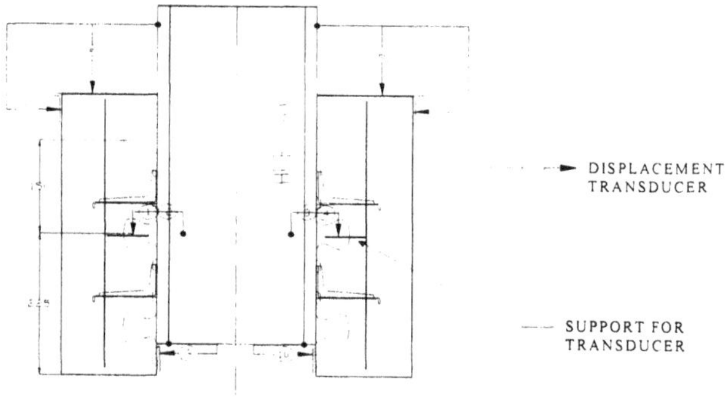

Figure 4.3.3: Measuring arrangement

In fact, differences observed between these different arrangements remain small.
Nevertheless it was noted that the most reliable slip evaluation was obtained at the middle of the two connector rows, not just at the interface between the steel flange and the concrete slab, but between the steel flange and the reinforcement layer in the slab. Finally, the third type of measure was selected, needing to plan at the time of the specimens fabrication a thin plate support going through the concrete slab and the profiled sheeting up to the reinforcement layer. In fact, four transducers were used to balance possible unsymmetrical effects during the loading of the specimen and to secure the servo-control of the actuator operating system directly from the slip measures.
Transversal displacements between the slabs and the steel section were also measured at the upper and lower parts of the specimen (figure 4.3.3) to make sure of negligible detachment.

LOADING PROCEDURE AND FAILURE CRITERION

Three loading procedures were used, as illustrated by figures 4.3.4.

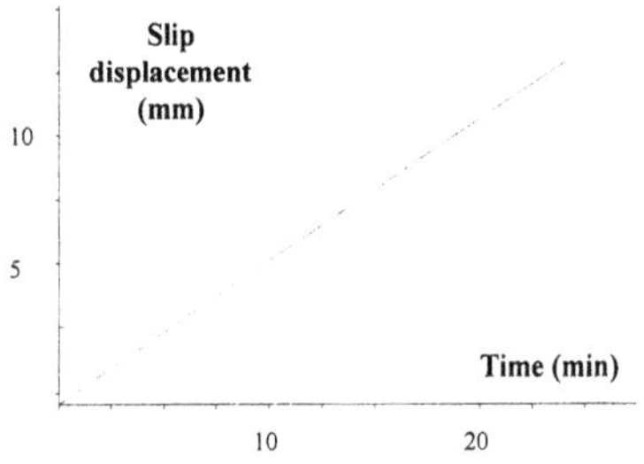

Firstly, a monotonic loading procedure with a constant slip rate chosen so that failure does not occur before 15 min. (as recommended in the French NAD of Eurocode 4-1-1).

Figure 4.3.4a : Monotonic loading

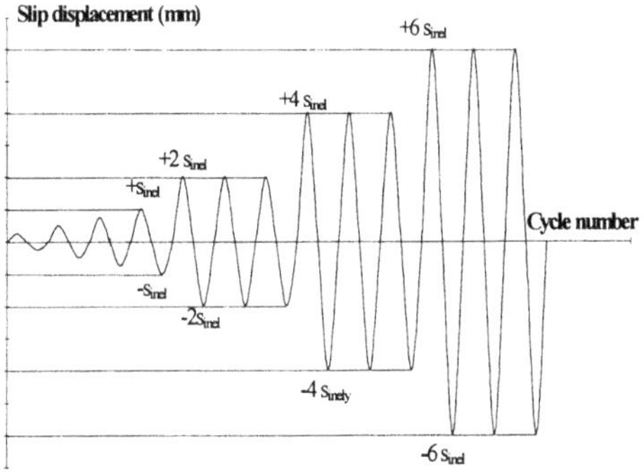

Secondly, a cyclic loading procedure according to ECCS Recommendations (1985) where the displacement history is defined on the basis of a conventional inelastic slip s_{inel} (see its determination farther on).

The cyclic loading should be controlled to produce the following increases of slip displacement :

- *4* cycles, successively for the ranges : $\pm s_{inel}/4$, $\pm s_{inel}/2$, $\pm 3s_{inel}/4$, $\pm s_{inel}$
- followed up to failure by series of 3 cycles each with a range: $\pm 2n\ s_{inel}$ where $n = 1,2,3,...$

Figure 4.3.4b : Cyclic loading according to ECCS procedure

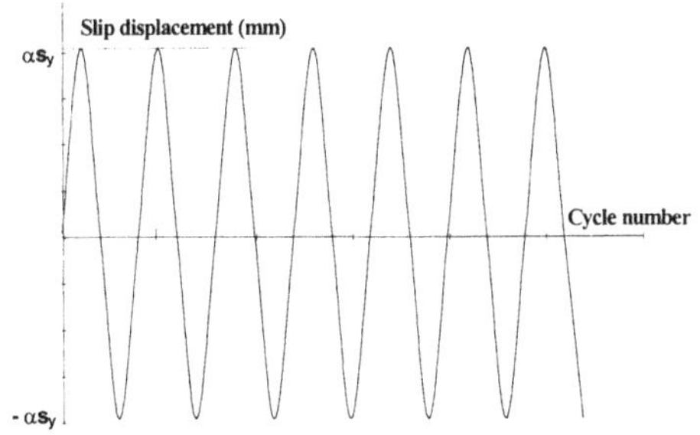

Thirdly, a cyclic loading procedure with a slip range, $\pm \alpha . s_y$ where α is a constant, to investigate directly the low-cycle fatigue resistance.

Figure 4.3.4c : Low-cycle fatigue procedure

Two types of failure were observed :

- a first type of rather brittle failure by sudden rupture of shear connector,
- a second type of failure, rather ductile by progressive lose of resistance for which the authors propose to adopt $0.2P_y$ as practical failure criterion where P_y is the elastic limit resistance of the shear connector (defined hereafter).

MONOTONIC RESULTS, INTERPRETATION AND DETERMINATION OF MAIN CHARACTERISTICS

Monotonic curves obtained from Push-Out tests are presented in figures 4.3.5a, 4.3.5b and 4.3.5c respectively for Groups ①, ② and ③. For all the groups it is observed a large scattering of the curves. More precisely, the deviation from the mean exceeds 10% limit recommended in Chapter 10 of Eurocode 4-1-1 (1992) when 3 tests are considered only, which seems to demonstrate the real difficulty to fabricate identical specimens in a same group, specially when a single reinforcement layer is set up into the slab. The average curves deduced from the monotonic results are presented in figures 4.3.6a, 4.3.6b and 4.3.6c.

As a preliminary, the cyclic procedures need to determine experimental elastic limit characteristics P_y^- *and* P_y^+ as well as associated slip values s_y^- *and* s_y^+ based on the monotonic curves of Push-Out and Pull-Out tests respectively. Taking into account the symmetrical behaviour perfectly observed for Push and Pull actions, only one couple of elastic characteristics P_y and s_y is given here.

As reminder, the classical ECCS definition consists of defining values P_y and s_y at the intersection between the initial tangent of slope k to the monotonic curve and the particular tangent having a slope equal to k/10; but such a definition cannot be adopted here considering the poor values obtained for P_y (examples of application of this definition are presented in figures 4.3.6a, 4.3.6b and 4.3.6c). It is the reason why another P_y definition corresponding to a slip displacement twice the elastic one is preferable to the previous one (the geometrical construction illustrating the second definition is drawn in figures 4.3.6a, 4.3.6b and 4.3.6c). In addition, it should be noted the difficulty to draw practically the initial tangent to the monotonic curve, so that the present authors are incited to propose a similar definition to the one given in Revised Annex J of Eurocode 3-1-1 (CEN, 1997); then, the initial stiffness is defined as the slope of the secant line joining the origin and the point on the monotonic curve located at ordinate $2/3\ P_y$; this geometrical construction requiring a short iterative procedure.

Obviously, the so-obtained values P_y are lower than the characteristic resistances P_R which should be considered when designing a shear connection at ultimate limit state, as specified in Eurocode 4-1-1. Here it seems reasonable to adopt $P_R = 0.85\ P_{max}$ where P_{max} is the maximum mean shear resistance of the connectors. Elastic limit characteristics P_y and s_y, as well the shear resistance P_R and the slip capacity s_u defined as the slip for which the shear force falls below P_R on the monotonic decreasing branch (figures 4.3.6a, 4.3.6b and 4.3.6c) are collected in table 4.3.1.

TABLE 4.3.1
CHARACTERISTICS OF REFERENCE

TYPE OF SHEAR CONNECTOR	P_y	s_y	P_{max}	P_R	s_u
Group ① :Cold Formed Angles with Solid Slab	29.5 kN	0.6 mm	39 kN	33.2	12.9 mm
Group ② :Cold Formed Angles with Composite Slab	19 kN	0.12 mm	27.8 kN	23.6	8.1 mm
Group ③ :Welded Headed Studs with Solid Slab	72.5kN	0.3 mm	103.4 kN	87.9	13.4 mm
Group ④ :Welded Headed Studs with Composite Slab	58 kN	0.3 mm	—	—	—

The P_R values are not very different from the characteristic values given by Eurocode 4-1-1 for welded headed studs (*82.5 kN* according to clause 6.3.2) or specified by the manufacturer for cold formed angles (namely *22 kN*). All the slip capacities s_u are higher than the required *6 mm* to have the right to consider these connector types as ductile (see clause 6.1.2 in EC4-1-1).

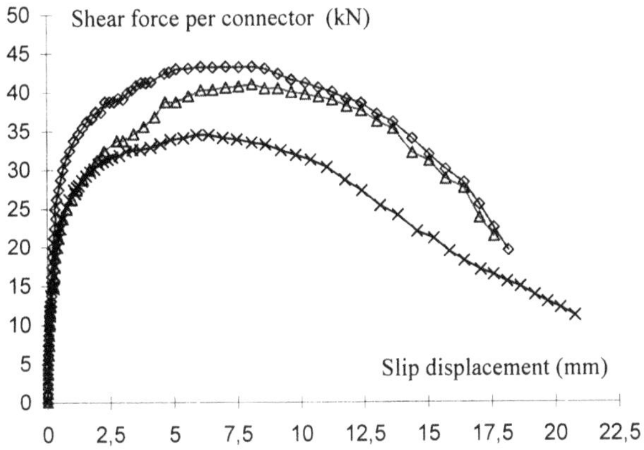

a : Group ①
Cold formed angle with solid slab

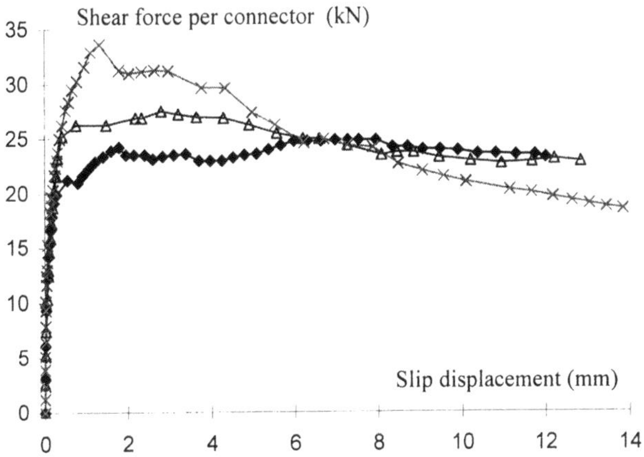

b : Group ②
Cold formed angle with composite slab

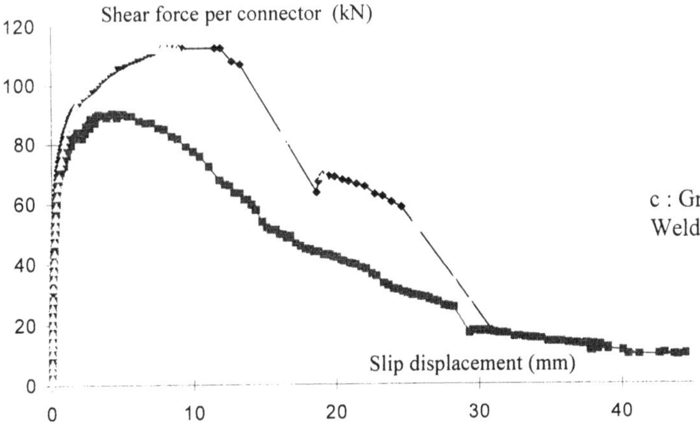

c : Group ③
Welded headed stud with solid slab

Figure 4.3.5 : Monotonic curves

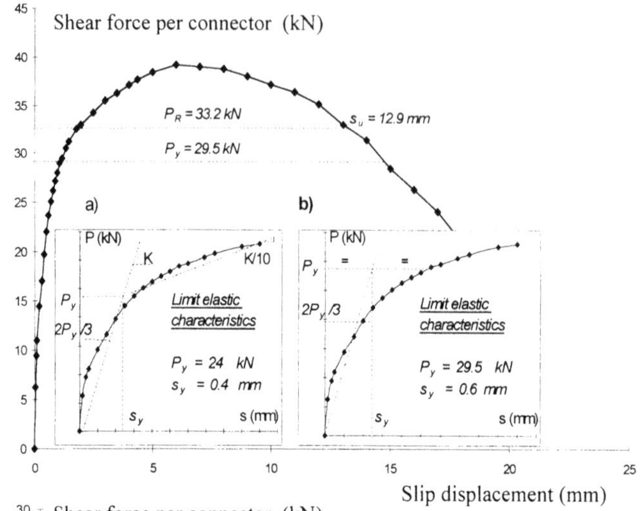

a- Group ① : Cold formed angle with solid slab

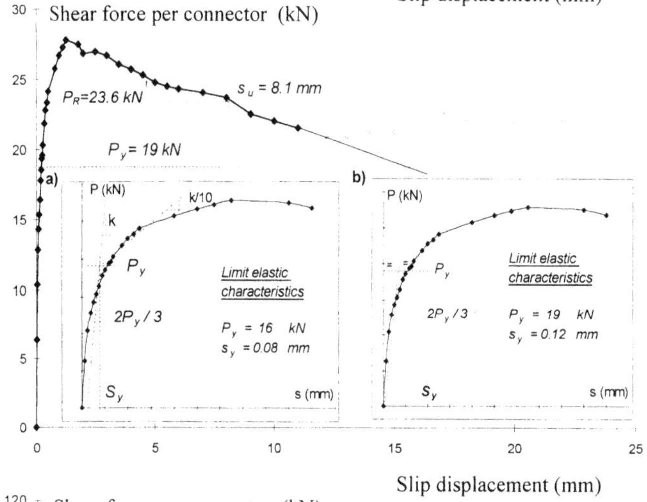

b- Group ② : Cold formed angle with composite slab

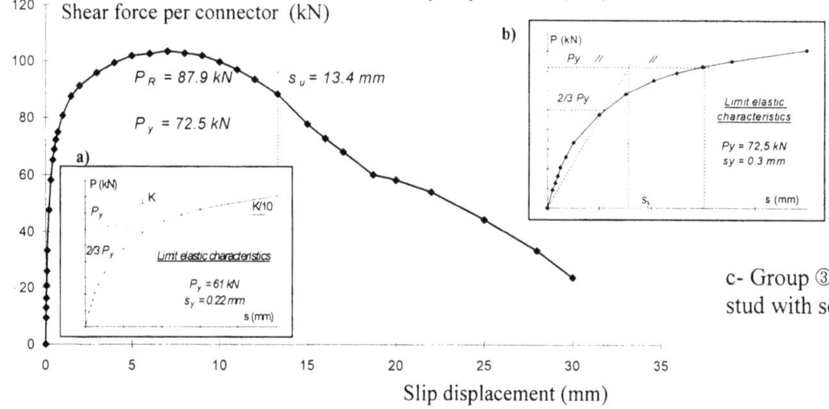

c- Group ③ : Welded headed stud with solid slab

Figure 4.3.6 : Average monotonic curves. Determination of the main characteristics

RESULTS OF CYCLIC TESTS ACCORDING TO ECCS PROCEDURE

For each Group ①, ② and ③, one curve of ECCS cyclic type is presented in figure 4.3.7. It appears that the responses of the specimens subject to full reversed displacements express a reduction both of maximum strength and ductility due to yielding and fatigue of the shear connectors and damage of the concrete. Observations can be made also comparing the half-cycles located in the push part (lower part « - » in figures 4.3.7) and the half-cycles in the pull part (upper part « + »in the same figures). For the tests of Groups ① and ③ (figures 4.3.7a and 4.3.7c with a solid slab), the strength and ductility are much larger in the push part (which corresponds to the direction of the first loading action) than the ones in the pull part. This result is very likely a consequence of the concrete crushing near the connector from the start of loading and at each increase of displacement beginning a new series of cycles. For the tests of Group ② (figure 4.3.7b with a composite slab), the result is opposite: peak strengths in the pull part are higher than in the push part, though a push action was applied at the start of loading. This odd result may be explained by some concrete confinement due to favourable behaviour of the profile steel sheeting in reversal loading; but such an explanation needs to be confirmed by more thorough analyses.

Different failure modes were observed as follows:

- Group ① : a ductile failure obtained on the 20th cycle;
- Group ② : a ductile failure obtained on the 30th cycle;
- Group ③ : a rupture by shear of the stud cross-section located just above the weld on the 20th cycle.

To propose a pragmatic interpretation of ECCS cyclic curves, it is possible to determine, from the peak strengths, the skeleton curve in each group of tests and to use it as an equivalent monotonic curve. So, the skeleton curves are plotted in figure 4.3.8 with comparatively the corresponding real monotonic curves (given above in figures 4.3.6). When the slip range exceeds a value of about ±1 mm or ±2 mm as the case may be, a large decreasing of the skeleton curve occurs with approximate linear variation up to failure. From the skeleton curves (figure 4.3.8), shear resistances expressed as a fraction of P_y have been determined for two slip amplitudes: ±2 mm and ±6 mm. These values are given in table 4.3.2.

Reminding that these two amplitudes are generally adopted to characterize non ductile and ductile shear connectors respectively, and considering results in table 4.3.2, connector types of Groups ① and ③, though very ductile under monotonic loading, can be considered only as non ductile under cyclic loading provided that a reduction of their static resistance is adopted. Here the reduced resistance goes from $0.75\ P_y$ to $0.95\ P_y$.

TABLE 4.3.2

SKELETON CURVE INTERPRETATION

Group of tests	Shear resistance measured for the slip displacement : + 2 mm	- 2 mm	Shear resistance measured for the slip displacement : + 6 mm	- 6 mm
①	$0.75\ P_y$	$-0.9\ P_y$	$0.3\ P_y$	$-0.4\ Py$
②	$> P_y$	$< -P_y$	$0.8\ P_y$	$-0.8\ P_y$
③	$0{,}85\ P_y$	$-0{,}94\ P_y$	Failure before ± 6 mm	

At the present stage of investigation, it may be concluded that shear connectors defined as ductile (according to Eurocode 4-1-1) under monotonic action should be considered as non ductile under cyclic repeated actions, and therefore under seismic actions (the single existing curve for cold formed angles with composite slab in Group ② does not allow to contradict such a point of view).

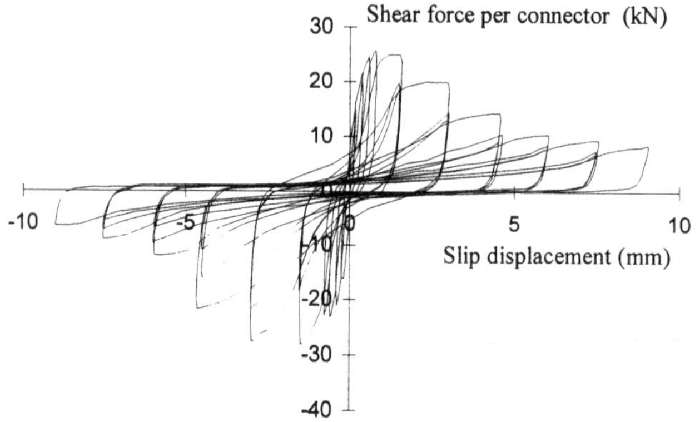

a : Group ①
Cold formed angle with solid slab

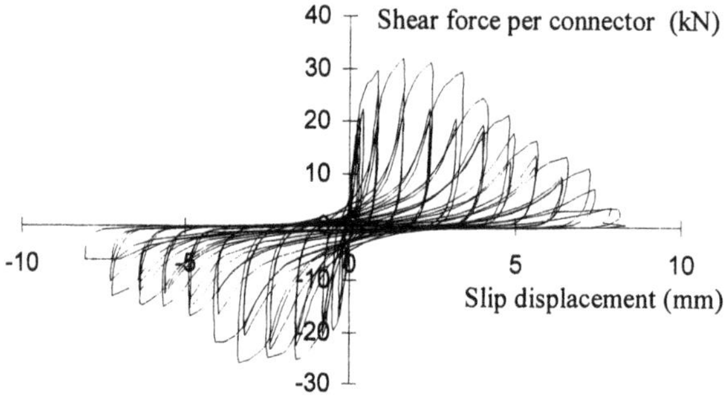

b : Group ②
Cold formed angle with composite slab

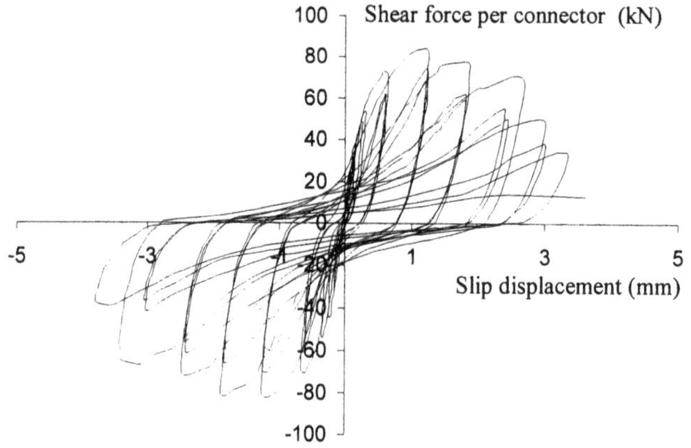

c : Group ③
Welded headed stud with solid slab

Figure 4.3.7 : Cyclic tests (ECCS type of procedure)

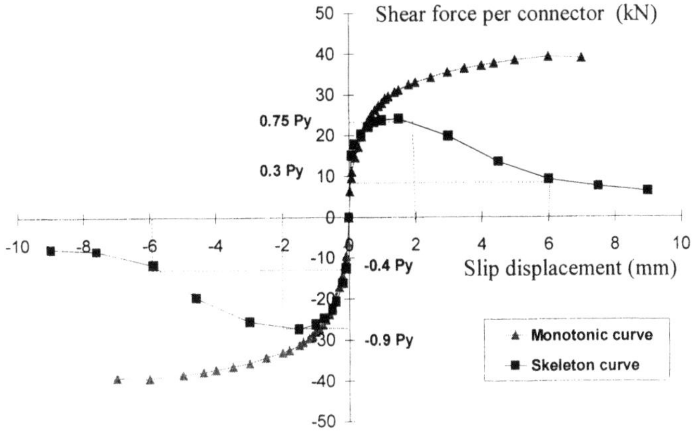

a : Group ①
Cold formed angle with solid slab

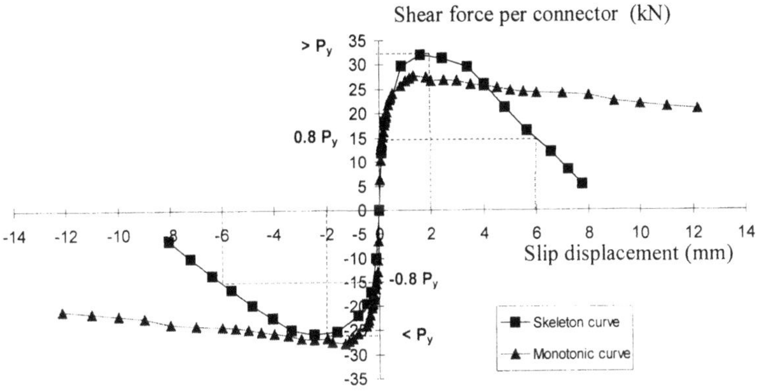

b : Group ②
Cold formed angle with composite slab

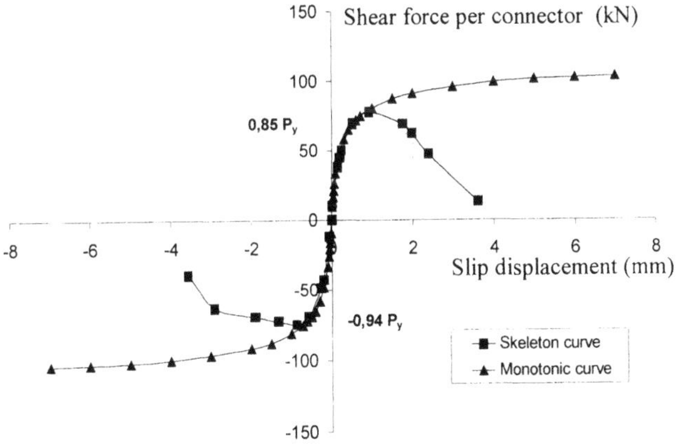

c : Group ③
Welded headed stud with solid slab

Figure 4.3.8 : Skeleton curves

RESULTS AND INTERPRETATION OF LOW-CYCLE FATIGUE TESTS

Results of four tests selected in each group are presented in figure 4.3.9. A reduction of the shear resistance appears as previously and depend on both the slip range Δs_i and the number of cycles N.

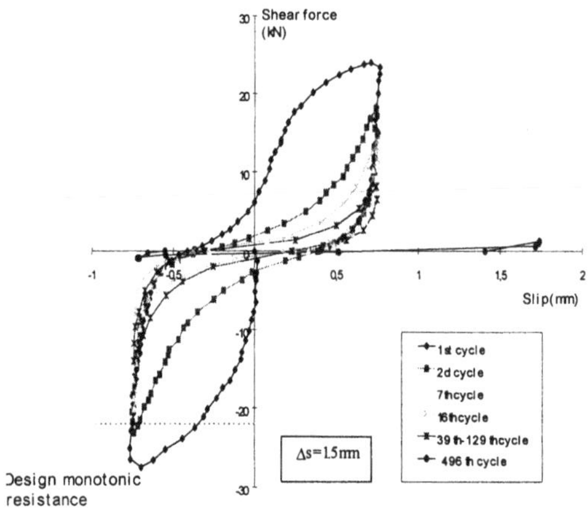

a) Cold Formed angle with solid slab
Group ①

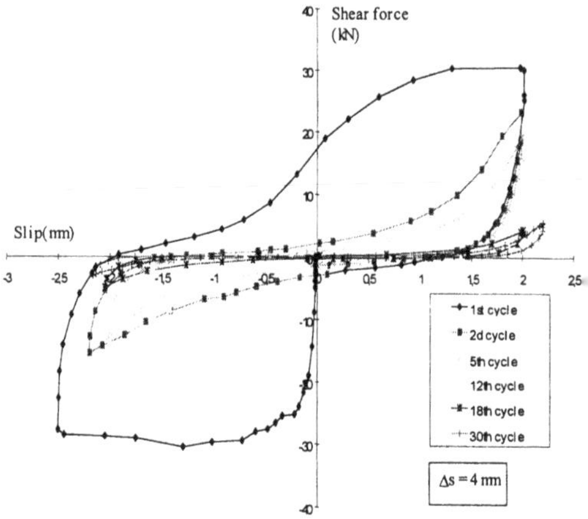

b) Cold Formed angle with composite slab - Group ②

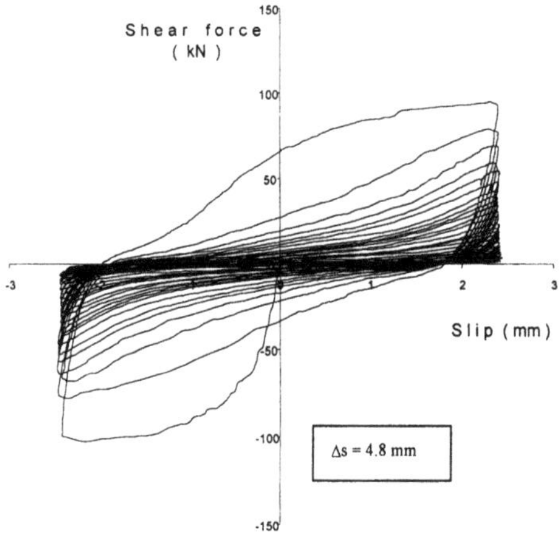

c) Welded headed stud with solid slab
Group ③

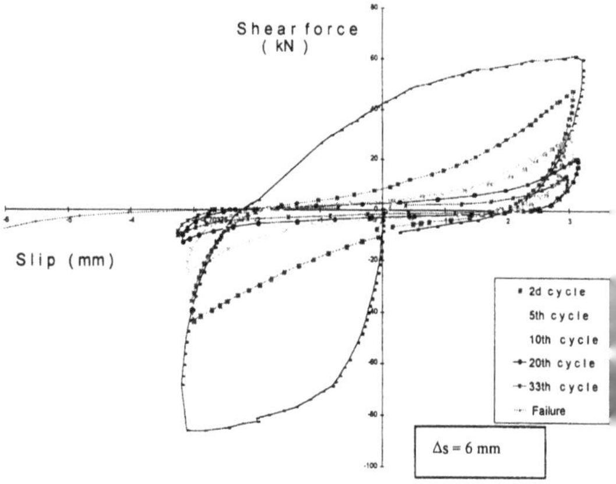

d) Welded headed stud with composite slab
Group ④

Figure 4.3.9 : Low-cycle Fatigue tests

A damage model has been introduced in order to formulate analytically all the low-cycle fatigue results. This model is based on the usual fatigue rule for design of steel structures, via the variation of the shear stress range $\Delta\tau$ versus the total number of cycles *N* (Wöhler diagram). In the case of high-cycle fatigue, the mathematical expression has the general form :

$$N\,\Delta\tau^{m} = K \qquad (4.3.1)$$

where exponent *m* and term *K* have constant values, *K* depending on the fatigue strength category of the considered detail. According to Ballio and Castiglioni (1994), also to Bernuzzi, Calado et al (1997) for steel beam-to-column joints under cyclic reversal loading, the above relationship may be generalized for low cycle fatigue in the case of quasi elastic-perfectly plastic behaviour by evaluating $\Delta\tau$ as follows:

$$\Delta\tau = \frac{\Delta s}{s_y}\,\tau_y \qquad (4.3.2)$$

where: τ_y is the shear stress of the connector corresponding to its elastic limit shear resistance P_y,
s_y is the corresponding limit slip of the shear connector, and
Δs the constant slip range applied to the shear connector.

Therefore, relationship (4.3.1) becomes : $N\left[\frac{\Delta s}{s_y}\,\tau_y\right]^m = K$ (4.3.3)

When cycles of different amplitudes Δs_i are applied to the specimen, the Miner's rule on linear damage cumulation may be assumed so that the damage index is given by :

$$D = \frac{1}{K}\sum_{i=1}^{\ell} N_i\left[\frac{\Delta s_i}{s_y}\,\tau_y\right]^m \qquad (4.3.4)$$

where: N_i is the number of cycles with same range Δs_i and
ℓ is the number of different ranges Δs_i to be considered.
As reminder, value *D=1* corresponds theoretically to the rupture of the detail.

It should be noted that the fatigue tests can be interpreted directly by means of relationship (4.3.3) whereas the ECCS type tests need to use relationship (4.3.4).

As for the calculation of τ_y, the following may be used :

$$\tau_y = \frac{P_y}{A_{sc}}$$

where A_{sc} is the shear cross-section of one connector and P_y is the elastic limit shear resistance of the connector (see table 4.3.1).

For welded studs : $A_{sc} = \frac{\pi d^2}{4}$

where *d* is the shank diameter of the studs. For the cold formed angles HVB 80, the shear cross-section is equal to $A_{sc} = 81{,}4\ mm^2$ and is located *22 mm* above the connector foot, as specified by the manufacturer (Hilti Company). The nominal yield strength of the angles is *295 N/mm²*.

To investigate the validity of the damage model presented above, the low-cycle fatigue results are plotted in figures 4.3.10 with log-log co-ordinates for Groups ①, ③ and ④. The equation obtained from a linear regression in each group is mentioned in figures 4.3.10 with the associated coefficient of correlation R^2. An index located near each point in figures 4.3.10 indicates the cyclic loading type of the corresponding test: (F) for low-cycle fatigue and (ECCS) for ECCS loading procedure.

For connectors of Groups ① and ③ with a solid slab, in comparison with current fatigue strength curves for steel structures (chapter 9 of Eurocode 3-1-1), a more important damage due to concrete weakness appears (the slope constant m of the fatigue strength curves is about *2.2*, therefore clearly lower than *5*, and the detail category corresponding to the constant *log K*, is about: $\Delta\tau_c = 20\ N/mm^2$).

For connectors of Group ④ (welded headed stud with a composite slab in figure 4.3.10c), the slope constant is slightly greater (about *2.8*) and the detail category is about *60 N/mm²*.

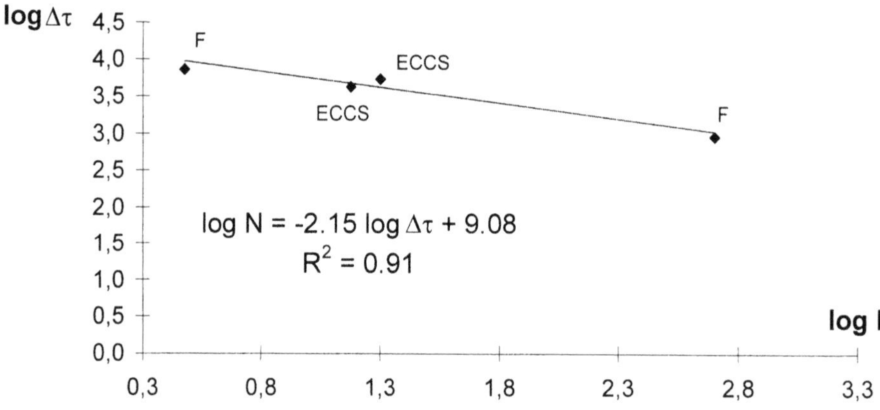

Figure 4.3.10a : Low cycle fatigue curve
Group ① : cold formed angle with solid slab

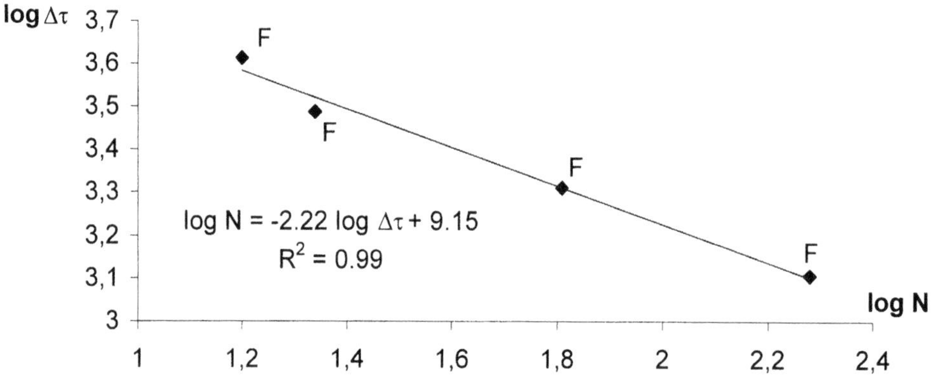

Figure 4.3.10b : Low cycle fatigue curve
Group ③: welded headed stud with solid slab

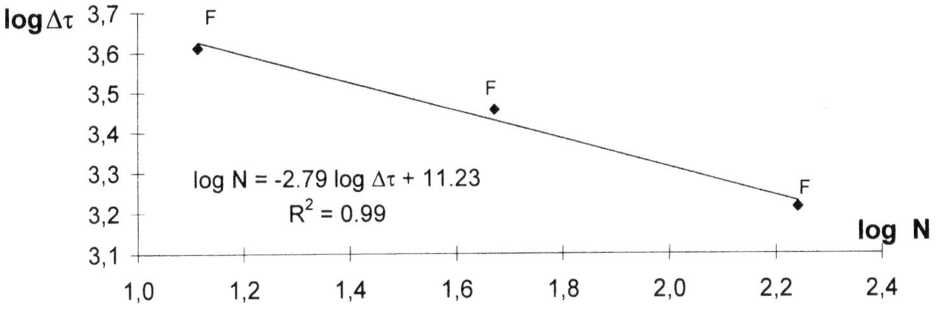

Figure 4.3.10c: Low cycle fatigue curve
Group ④: welded headed stud with composite slab

Contrary to Groups ①, ③.and ④ the low-cycle fatigue results of Group ② are too scattered to be fitted by a linear curve, as observed in figure 4.3.11. This is probably a consequence of the greater deformability of this type of connection where damage of the concrete encased in the transverse ribs tends to prevail against the resistance of the shear connectors.

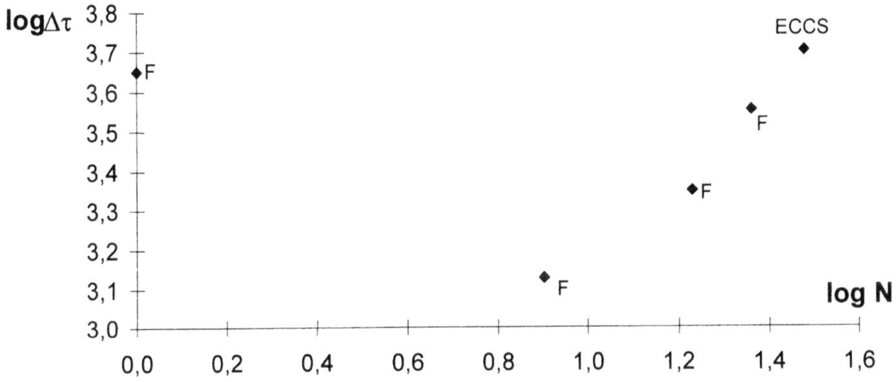

Figure 4.3.11: Low cycle fatigue results
Group ②: Cold formed angle with composite slab

Regarding the Push-Pull test results on the whole, a severe weakness under fatigue actions should be taken into account if large slip ranges beyond the elastic domain were accepted (for example, Δs varies from 1.5 mm to 6 mm in figures 4.3.9). Damage models similar to (4.3.3) and (4.3.4) could be more developed to control the fatigue resistance of shear connectors. However, it would be premature to conclude at the present stage without testing composite beam-to-column joints to know the real cyclic behaviour and resistance of the whole shear connection adjacent to the joint.

REFERENCES

Ballio G. and Castiglioni C.A. (1994). Seismic behaviour of steel sections. *Journal of Constructional Steel Research* **29**, 21-24.

Bernuzzi C, Calado L. and Castiglioni C.A. (1997). Behaviour of steel beam-to-column joint under cyclic reversal loading: an experimental study. *5th Int. Col. On Stability and Ductility of Steel Structures*, Nagoya, Japan.

ECCS (1985). *Recommended testing procedure for assessing the behaviour of structural steel element under cyclic loads,* Committee TWG 1.3.

Eurocode 4, Part 1-1 ENV 1994-1-1. (1992). *Design of composite steel and concrete structures – general rules and rules for buildings,* CEN – Brussels.

Revised Annex J of ENV 1993-1-1 (January 1997). *Steel joints in building frames,* CEN-Brussels.

4.4

BEHAVIOUR OF FULL SCALE JOINTS

Jean Marie Aribert and Alain Lachal

EXPERIMENTAL PROGRAM

TABLE 4.4.1a
EXPERIMENTAL PROGRAM – GROUP 1

BARE STEEL and COMPOSITE BEAM-TO-COLUMN BOLTED JOINTS
On major axis with a cruciform arrangement and symmetrical loading

Common elements are :

- **HEB 200 COLUMN – steel grade S 235**
- **IPE 360 BEAM – steel grade S 235**
- **End plate connected in the steelwork part (450×200×15)**
- **HR 10.9 bolts - Φ18 mm (tightened for 1/3 nominal preload))**

Stiffeners on the column web located at the beam flanges level

TEST REFERENCE	TYPE OF LOADING	SHEAR CONNECTORS Type	SHEAR CONNECTORS Number (per half span)	PROFILED SHEETING	SLAB	REINFORCEMENT Longitudinal (Transversal)
J1	monotonic (sagging bending)	Test of reference without slab				
J2	monotonic (hogging bending)	Test of reference without slab				
J3	monotonic test of reference (hogging bending)	Cold-formed angles HILTI HVB 80 (80×60×24.3)	N=20	COFRASTRA 40	1000×120	10ϕ10 rebars (ϕ8 each 10 cm)
J4	cyclic (ECCS procedure)	Cold-formed angles HILTI HVB 80 (80×60×24.3)	N=20	COFRASTRA 40	1000×120	10ϕ10 rebars (ϕ8 each 10 cm)
J5	cyclic (ECCS procedure)	Cold-formed angles HILTI HVB 80 (80×60×24.3)	N=10	COFRASTRA 40	1000×120	10ϕ10 rebars (ϕ8 each 10 cm)
J6	cyclic (ECCS procedure)	Cold-formed angles HILTI HVB 80 (80×60×24.3)	N=30	COFRASTRA 40	1000×120	10ϕ10 rebars (ϕ8 each 10 cm)

As reminder, the general objective of this sub-chapter is to investigate the risk of degradation of the slab and the shear connection in composite joints under cyclic repeated loads.
The choice of cruciform beam-to-column type of joint with symmetrical loading was adopted intentionally to have not the distortion effect of the steel column web panel to be included as a supplementary component in the joint rotation. 10 tests were planed and presented in two groups in tables 4.4.1a and 4.4.1b
- The first group (table 4.4.1a; tests J1 to J6) involves 6 tests with the same steelwork part. The bare steel joints J1 and J2 and the composite joint J3 are three monotonic tests of reference whose data are useful to define the ECCS loading procedure applied to the composite joints J4 to J6. The main investigated parameter in this group is the degree of shear connection.
- The second group (table 4.4.1b; tests J7 to J10) involves 4 tests with the same steelwork part again but of higher beam and column depths and with a solid slab and stud connectors. Steel joints J7 and J8 and composite joint J9 are monotonic tests of reference useful to the cyclic test. The investigated parameters may be considered as the beam and column depths and the types of slab and shear connectors in comparison with the tests of the first group.

TABLE 4.4.1b
EXPERIMENTAL PROGRAM – GROUP 2

BARE STEEL and COMPOSITE BEAM-TO-COLUMN BOLTED JOINTS
On major axis with a cruciform arrangement and symmetrical loading

Common elements are :

- **HEB 240 COLUMN – steel grade S 235**
- **IPE 450 BEAM – steel grade S 235**
- **End plate connected in the steelwork part (580×240×20)**
- **HR 10.9 bolts - Φ22 mm (tightened for 1/3 nominal preload))**
- **Stiffeners on the column web located at the level of the beam flanges**

TEST REFER-ENCE	TYPE OF LOADING	SHEAR CONNECTORS Type	SHEAR CONNECTORS Number (per half span)	PROFILED SHEETING	SLAB	REINFOR-CEMENT Longitudinal (Transversal)
J7	monotonic (sagging bending)	Test of reference without slab				
J8	monotonic (hogging bending)	Test of reference without slab				
J9	monotonic test of reference (hogging bending)	Welded headed studs ϕ19mm h=100mm	N = 10	solid slab	1000×120	10ϕ14 rebars (ϕ8 each 10 cm)
J10	cyclic (ECCS procedure)	Welded headed studs ϕ19mm h=100mm	N = 10	solid slab	1000×120	10ϕ14 rebars (ϕ8 each 10 cm)

Complementary data about the mechanical properties of materials concern:

(a) The concrete material of normal weight with the mean cylinder compressive strength (measured at age 28 days) as follows:

Tests of Group 1 (table 4.4.1a) : $f_{cm} = 24.3\ N/mm^2$
Tests of Group 2 (table 4.4.1b) : $f_{cm} = 27.0\ N/mm^2$

(b) The reinforcement composed of ribbed bars with high bond action in accordance with the French Norm A 35 NF 016, with steel grade S 500 – Category 3 (ensuring high ductility, at least elongation $\varepsilon_{us} = 5\%$ at maximum strength). Coupons taken from different rebars and tested on a Dartec machine gave for each group the following measured mean values:

Group 1: $f_{ys} = 568\ N/mm^2$; $\varepsilon_{us} = 6\%$ (for $\Phi = 10\ mm$)
Group 2: $f_{ys} = 568\ N/mm^2$; $\varepsilon_{us} = 7\%$ (for $\Phi = 14\ mm$).

(c) The structural steel: Tension tests on coupons taken from steel beam flanges gave the following mean values for the yield strength f_y and the ultimate tensile strength f_u :

Group 1: $f_y = 282\ N/mm^2$ and $f_u = 404\ N/mm^2$
Group 2: $f_y = 347\ N/mm^2$ and $f_u = 463\ N/mm^2$

(d) The welds made by the manual electric arc method were full penetration butt welds using basic welding consumables in accordance with the French Norm A 81 NF 309 (with guaranteed mechanical properties $Re = 520\ N/mm^2$, $Rm = 596\ N/mm^2$ and $A\% = 26$).

Considering the plastic resistance of the rebars in tension, a degree of shear connection N/N_f may be evaluated now for each composite joint similarly to the concept of Eurocode 4.1.1:

$$N/N_f = \frac{N \times P_R}{A_s \times f_{ys}} \tag{4.4.1}$$

where A_s is the cross-sectional area of longitudinal reinforcement.
Numerically:

Test J5: $N/N_f = \dfrac{10 \times 23600}{785.4 \times 568} = 0.53$; Partial shear connection.

Test J3 and J4: $N/N_f = \dfrac{20 \times 23600}{785.4 \times 568} = 1.06$; Full shear connection.

Test J6: $N/N_f = \dfrac{30 \times 23600}{785.4 \times 568} = 1.59$; More than full shear connection.

Test J9 and J10: $N/N_f = \dfrac{10 \times 87900}{1539.4 \times 568} = 1.01$; Full shear connection.

In addition, it should be noted that the following arrangements of shear connectors were adopted:

- 2 angles per row for J5, J3 and J4;
- 3 angles per row for J6 in the zone adjacent to the joint and covering 4 rows;
- 1 stud per row for J9 and J10 but 2 studs per row in the zone adjacent to the joint and covering 2 rows.

EXPERIMENTAL ARRANGEMENT AND INSTRUMENTATION

Experimental arrangement for Groups 1 and 2 is shown in figures 4.4.1a and 4.4.1b. A vertical load P was applied to the upper end cross-section of the column by means of an hydraulic jack. Each beam was supported by a double roller bearing, located at 1.73 m from the column axis. For tests J3 to J6, the slab was set under the steel beam (figure 4.4.1a) whereas for tests J9 and J10 the slab was set in normal position (figure 4.4.1b).
In this sub-chapter, all the joint moment values will be defined at the load-introduction cross-section (i.e. the interface between the end plate and the column flange). To deduce the corresponding moment values at the joint centre, the previous moment values should be multiplied simply by the following factors:

$m_1 = 1.061$ for Group 1
$m_2 = 1.075$ for Group 2

Here the choice of the load-introduction cross-section may be justified by the fact that there is no distortion at all of the column web panel and the components contributing to the joint resistance are only those of the bolted end plate, the slab and the shear connection. In addition, any comparison between the joint and the adjacent composite cross-section with regard to flexural stiffness and moment resistance is more significant (see farther on).

The rotation of the joint was mainly obtained by means of two inclinometers attached in the middle of the plates welded directly to the upper and lower beam flanges transversally to the beam axis; so, the rotation measure was not affected by the shear distortion of the beam web (Figure 4.4.2). In fact, a mean rotation was determined, namely:

$$\Phi = \frac{\Phi_1 + \Phi_2}{2} \tag{4.4.2}$$

where Φ_1 and Φ_2 are the rotation (in opposite sense) given by the inclinometers on each side of the joint.

It should be noted that the above rotation measure appeared more accurate, specially under cyclic loading, than the ones which could be deduced from the displacement in the vertical direction through the centroid of the column (transducers 3 and 4 in figure 4.4.2) and from displacements in horizontal direction given by transducers attached to the upper and lower edge flange of the beams (for example, transducers 13, 14, 15, 16, 33 and 34).
Other displacements were measured in significant zones of the joint (figure 4.4.2) with a view of future modelling developments:

- strain on the external face of the slab (transducers 11 and 12);
- end plate displacements in front of each bolt row (transducers 7, 8, 9, 10, etc);
- end plate displacements between the bolt rows (transducers 5 and 6).

In addition, displacements between the reinforcement layer of the slab and the steel flange were measured to evaluate the slip distribution, as illustrated in figure 4.4.3.

LOADING HISTORIES

For the monotonic tests, increases of rotation were applied continuously by means of a servo controlled actuator up to failure of the joint.
For the cyclic tests, the ECCS procedure was followed using two conventional elastic rotations ϕ_y^+ and ϕ_y^- deduced from the monotonic tests, in sagging and hogging bending respectively (see hereafter).

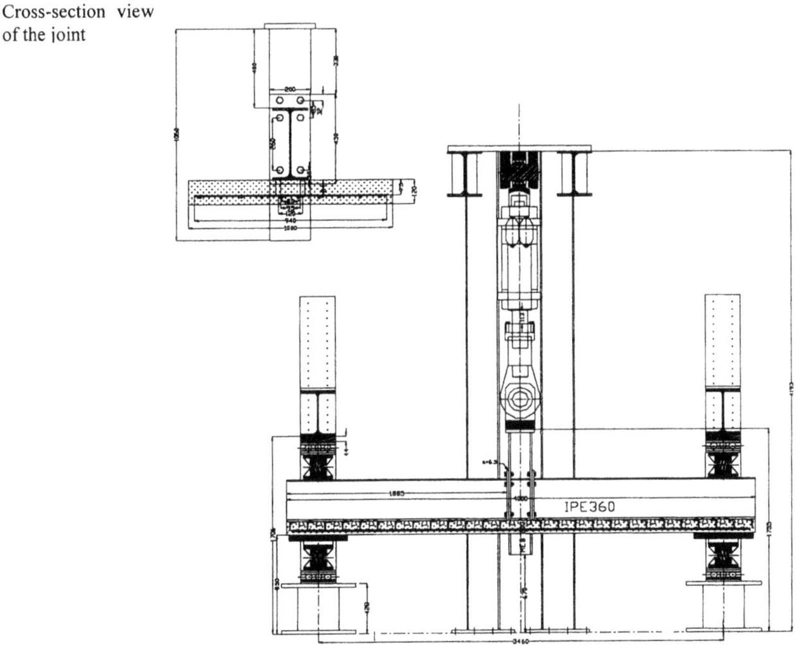

a) Group 1

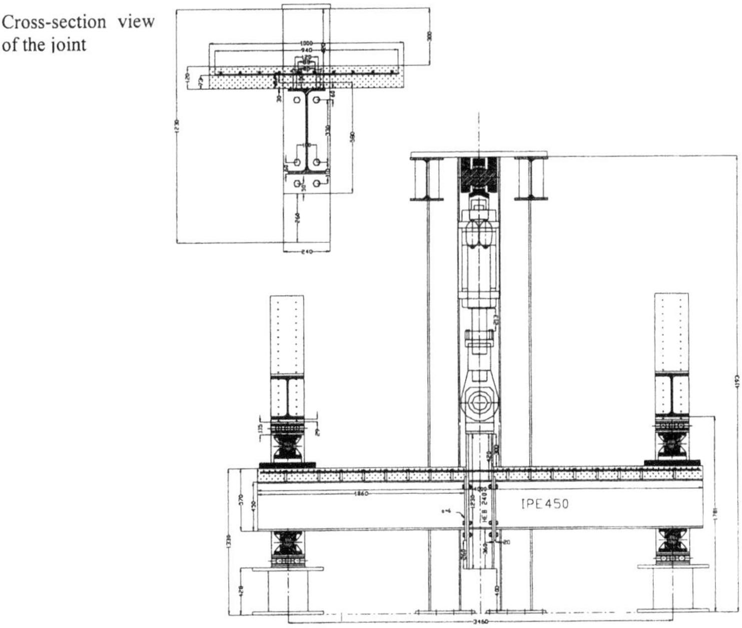

b) Group 2

Figure 4.4.1 : Experimental arrangement

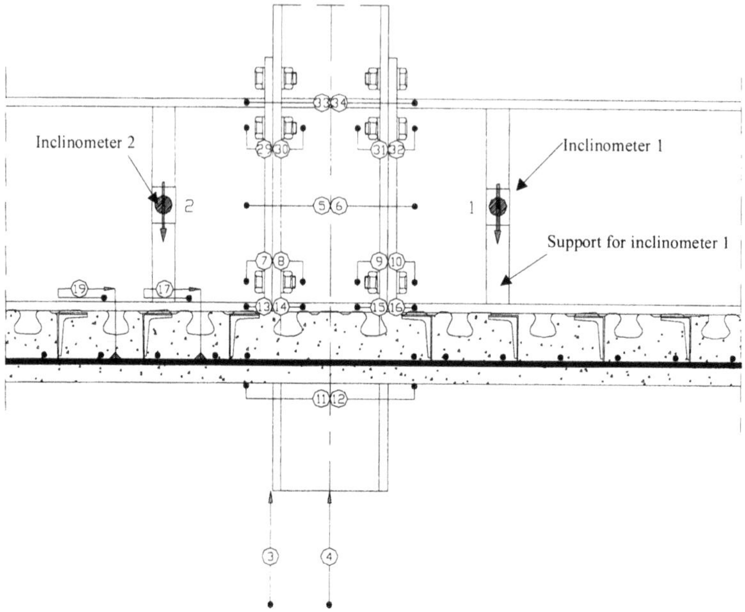

Figure 4.4.2: Displacement measuring arrangement (test J4)

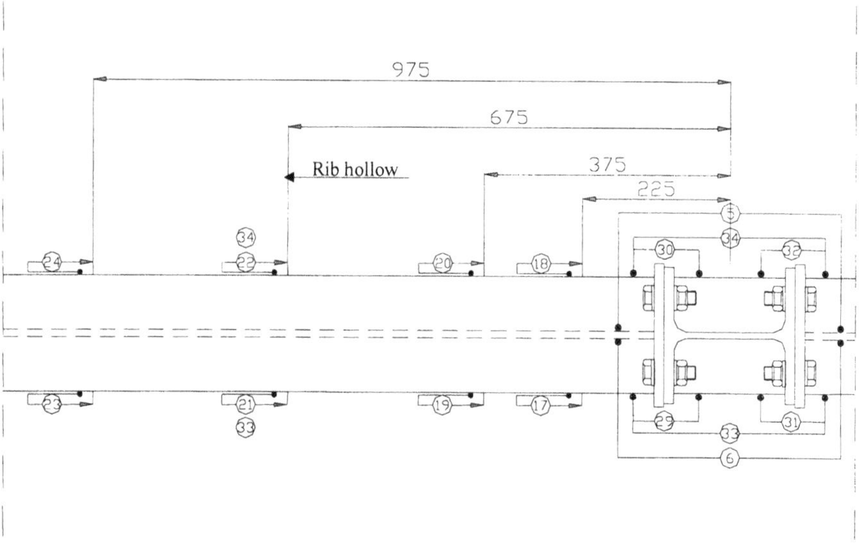

Figure 4.4.3: Slip measurement along the half-span of beam (test J4)

RESULTS OF MONOTONIC TESTS AND INTERPRETATION

For each group, four moment-rotation curves based on three monotonic tests are collected in the same figure.
Figure 4.4.4a concerns the monotonic moment-rotation curves relevant to Group 1. Curve J1 (test in sagging bending) and curve J2 (test in hogging bending) represent the behaviour of bare steel joints identical with the steelwork part of composite specimens J3, J4, J5 and J6. The composite specimen J3 was tested in monotonic loading under hogging bending.
Monotonic curves for Group 2 are presented (as for Group 1) in figure 4.4.4b where specimens J7 and J8 tested in sagging and hogging bendings respectively represent the contributions of the steelwork part of composite tests J9 and J10, specimen J9 being tested in monotonic loading under hogging bending (as specimen J3).
It should be noted that the maximum rotation values were close to *30 mrd* for all the monotonic tests except for test J1 where the maximum rotation was limited to *22 mrd* (due to premature rupture of bolts).
Same types of failure were observed between the corresponding tests for Groups 1 and 2; i.e.:

Tests J1 and J7: rupture in tension of the internal bolts close to the steel flange which is opposite to the slab in the composite tests;

Tests J2 and J8: rupture in tension of the internal bolts close to the steel flange which is connected to the slab;

Tests J3 and J9: rupture of longitudinal rebars accompanied simultaneously by rupture of bolts in tension close to the slab.

Yielding and large deflection of end plates and steel column flanges occurred at all the failure stages.

One objective of the monotonic tests was to provide each group with complete monotonic moment-rotation curves of reference to determine the conventional characteristics $(M_y^{\pm}, \Phi_y^{\pm})$ necessary for the definition of the ECCS procedure applied to the cyclic tests. For hogging bending, this determination uses directly the monotonic curves of composite tests J3 and J9. But for sagging bending the monotonic curves of tests J1 and J7 should be replaced by modified curves J1* and J7* which take account of the slab action. Evaluating the transfer of the centre of compression of the joint from the middle of the compression steel flange to the middle of the slab depth in compression, the present authors propose, for a given rotation Φ, to multiply the initial bending moment (corresponding to tests J1 and J7) by the factor:

$$m_2 = \frac{h_b - \frac{t_{fb}}{2} + \left(h_t - \frac{x}{2}\right)}{h_b - t_{fb}} \tag{4.4.3}$$

where: h_t is the total depth of the slab and
x is the depth of the stress block for the concrete.

From test J1, the maximum tension force of the steel flange is about *676 kN* when $\Phi = 20$ *mrd*; hence, for tests J4 and J6 (with full shear connection), calculation gives $x = 32.7$ *mm* and $m_2 = 1.32$.
For test J5 (with partial shear connection $N/N_f = 0.53$), $m_2 = 1.17$ may be determined by linear interpolation.
From test J7, the maximum tension force of the steel flange is about *1105 kN*; hence, for test J10 (with full shear connection), $x = 48.1$ *mm* and $m_2 = 1.24$ are obtained..
Finally, it may be retained for all the relevant tests the only practical value: $m_2 = 1.25$.

Contrary to the modified procedure used for the monotonic Push-Out tests, the usual ECCS definition (using the intersection between the initial tangent and the tangent with a slope equal to the initial

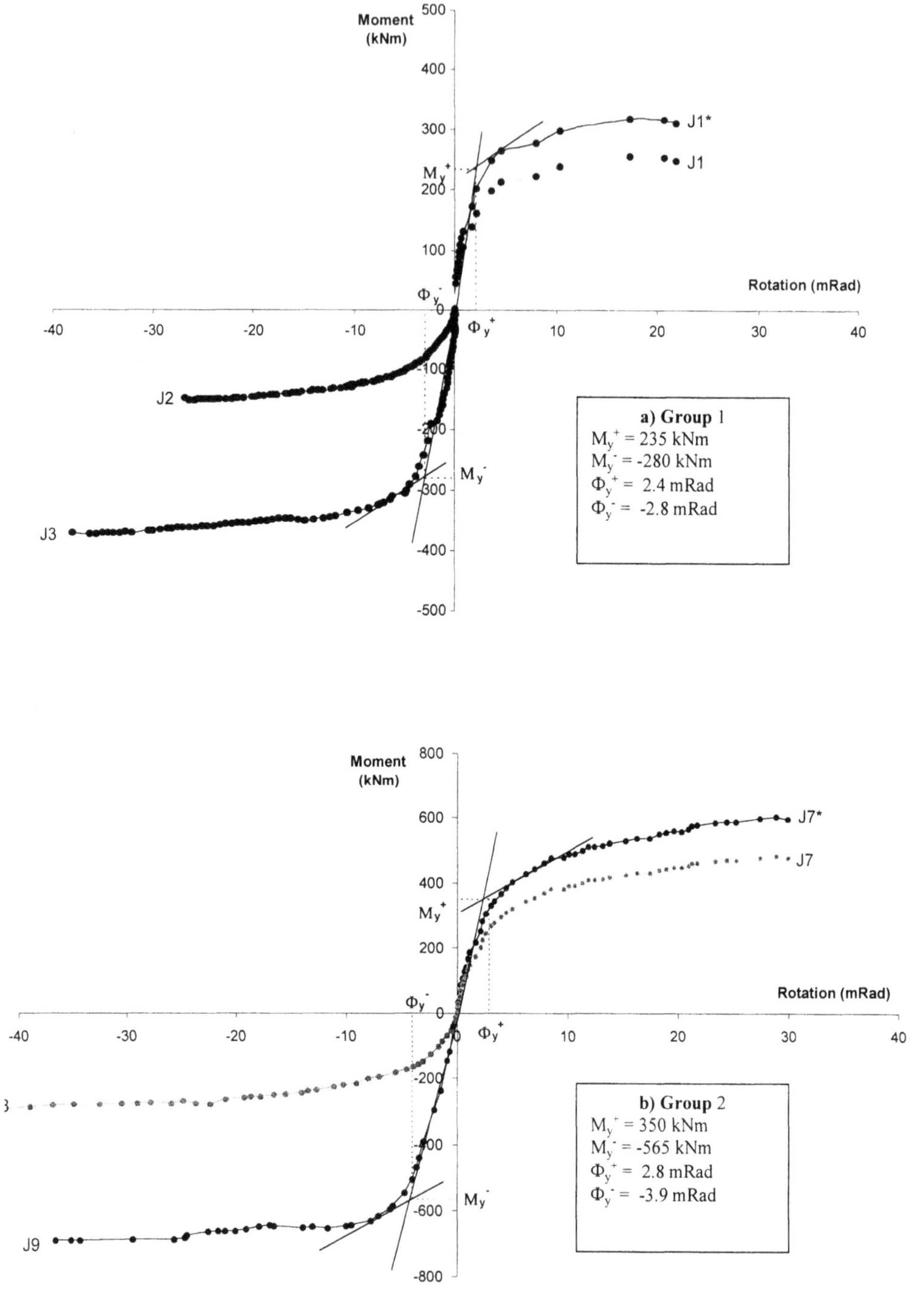

Figure 4.4.4: Reference moment-rotation curves in monotonic loading

tangent divided by 10) appeared well adapted to determine the elastic limit characteristics $M_y^{\pm}$ and $\Phi_y^{\pm}$ (keeping nevertheless the same secant definition as above for the initial tangent). Geometrical constructions and characteristic values obtained for each group are given in figures 4.4.4a and 4.4.4b.

CLASSIFICATION OF THE TESTED JOINTS BY MOMENT RESISTANCE AND ROTATIONAL STIFFNESS FROM A STATIC VIEW-POINT

In seismic design, joint classification by moment resistance (overstrength or partial strength) is more important than classification by rotational stiffness; in fact, the influence of the rotational stiffness on the q factor for structures has not been clearly demonstrated up today, Aribert and Grecea (see section 8.3 of the present book).
Criteria of classification issued from Eurocode 3 are applicable to Eurocode 4.1.1 (clause 4.10.5) for joints tested under monotonic loading. Although this classification is really correct only with theoretical values of moment resistance and rotational stiffness calculated in accordance with the Eurocode models, the authors consider that it is significant enough to compare the experimental values $M_y^{\pm}$ and $M_y^{\pm}/\Phi_y^{\pm}$ with moment resistances and rotational stiffnesses of the adjacent composite beams (calculated with measured mean values of the material properties).

Group 1:

1°/ In hogging bending

- Plastic resistance moment of the composite cross-section

$$M_{pl,b}^{-} = 385\ kN.m$$

hence:
$$\frac{M_y^{-}}{M_{pl,b}^{-}} = \frac{280}{385} = 0.73$$

→ The joint is partial strength.

- Just for information, the classification as rigid joint needs to satisfy the following condition:

$$S_j^{-} \rangle \frac{8E_a I_2}{L_b}$$

justified here because of the cruciform arrangement of testing (i.e. without sway displacement).
The rotational stiffness of the joint is:

$$S_j^{-} = \frac{M_y^{-}}{\Phi_y^{-}} = \frac{280 \times 10^6}{2.8 \times 10^{-3}} = 100 \times 10^9\ \text{N.mm}.$$

Adopting the second moment of area calculated neglecting concrete in tension:

$$I_2 = 215 \times 10^6\ mm^4$$

and the span

$$L_b = \frac{b}{0.2} = \frac{1000}{0.2} = 5000\ mm$$

where b is the effective width of the slab here equal to the slab width, the following value is obtained:

$$\frac{8E_a I_2}{L_b} = \frac{8 \times 210\,000 \times 215 \times 10^6}{5\,000} = 72 \times 10^9\ N.mm.$$

Hence, the joint is rigid in hogging bending.

2°/ In sagging bending

- $M^+_{pl,b} = 515\ kN.m$; $\dfrac{M^+_y}{M^+_{pl,b}} = 0.45$

⟶ The joint is clearly partial strength in sagging bending.

- Just for information:

$$S^+_j = \frac{M^+_y}{\Phi^+_y} = \frac{235 \times 10^6}{2.4 \times 10^{-3}} = 98 \times 10^9\ N.mm$$

Second moment of area taking into account the concrete in compression:

$$I_1 = 488 \times 10^6\ mm^4$$

$$\frac{8E_a I_1}{L_b} = \frac{8 \times 210\,000 \times 488 \times 10^6}{5\,000} = 164 \times 10^9\ N.mm \ \rangle\ S^+_j$$

⟶ The joint would be semi-rigid in sagging bending

Group 2

1°/ In hogging bending

$$M^-_{pl,b} = 815\ kN.m; \quad \frac{M^-_y}{M^-_{pl,b}} = 0.69\ ;$$

$$S^-_j = 145 \times 10^9\ N.mm\ ;$$

$$\frac{8E_a I_2}{L_b} = \frac{8 \times 210\,000 \times 470 \times 10^6}{5\,000} = 158 \times 10^9\ N.mm$$

⟶ The joint is partial strength and could be almost rigid.

2°/ In sagging bending

$$M^+_{pl,b} = 1025\ kN.m; \quad \frac{M^+_y}{M^+_{pl,b}} = 0.34\ ;$$

$$S^+_j = 125 \times 10^9\ N.mm\ ;$$

$$\frac{8E_a I_1}{L_b} = \frac{8 \times 210\,000 \times 898 \times 10^6}{5\,000} = 302 \times 10^9\ N.mm$$

⟶ The joint is clearly partial-strength and semi-rigid.

From a seismic point of view, it is underlined that specimens for Groups 1 and 2 are partial-strength in hogging and sagging bendings; consequently they should be considered as dissipative elements in seismic design.

LOADING ECCS PROCEDURE AND CYCLIC MOMENT-ROTATION CURVES

The moment-rotation curves obtained for the cyclic tests J5, J4, J6 and J10 are presented in figures 4.4.5, 4.4.6, 4.4.7 and 4.4.8.

For all the tests, hysteresis loops exhibit a slight unstable behaviour. As confirmed farther on by ECCS parameters of interpretation, the degradation of strength, stiffness and energy absorption capacity of the composite joints is due not only to concrete damage but also to permanent deformations in tension of the steelwork part in front of the bolt rows.

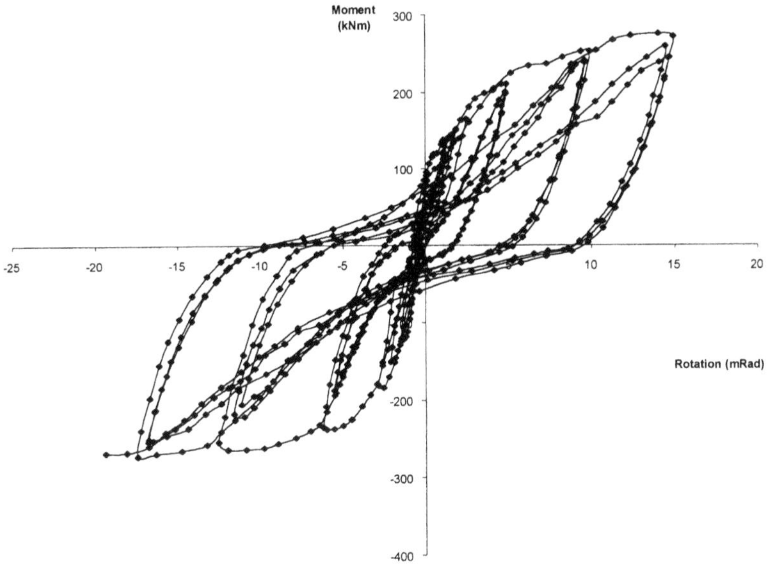

Figure 4.4.5: Moment-rotation curve in cyclic loading
Group 1 : Test J5 (N=10 angles per half span)

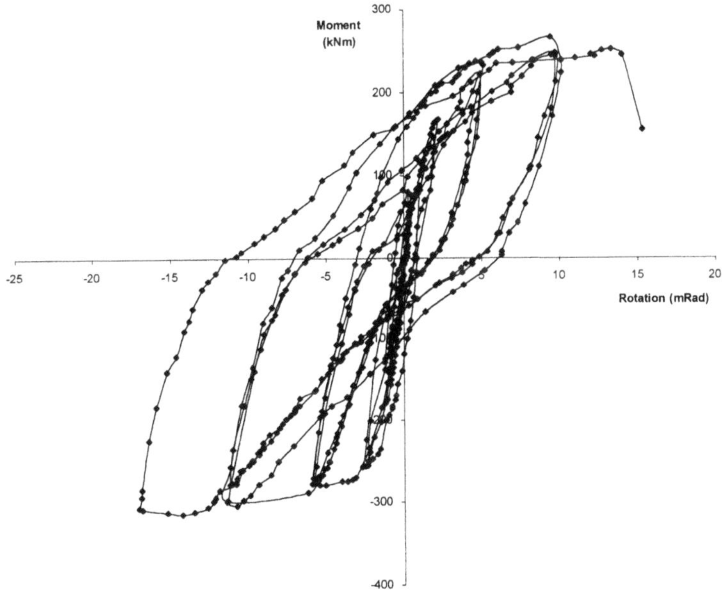

Figure 4.4.6: Moment-rotation curve in cyclic loading
Group 1 : Test J4 (N=20 angles per half span)

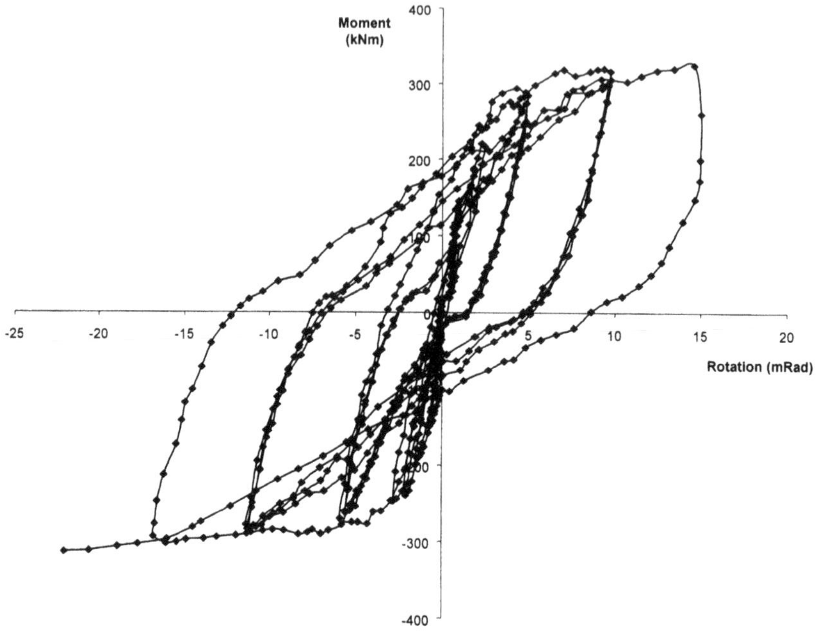

Figure 4.4.7 : Moment-rotation curve in cyclic loading
Group 1 : Test J6 (N=30 angles per half span)

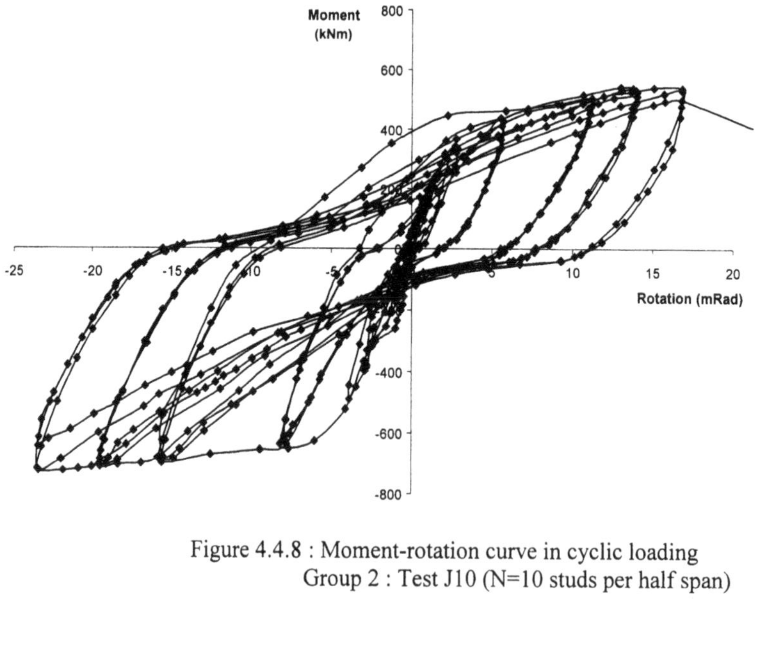

Figure 4.4.8 : Moment-rotation curve in cyclic loading
Group 2 : Test J10 (N=10 studs per half span)

For all specimens failure occurred in sagging bending by fracture in the zone of the steelwork part located on the opposite side to the slab.

- For test J4: by rupture of both interior bolts and end plate;
- For test J5: by cracking in the welds connecting the beam flange to the end plate;
- For test J6: by rupture of interior bolts;
- For test J10: by rupture of external bolts.

In all cases yielding and large deflection of the end plate appeared.

INTERPRETATION BY MEANS OF NON-DIMENSIONAL ECCS PARAMETERS

To make easier and more accurate the interpretation of cyclic tests, the following non-dimensional parameters have been calculated according to the definitions given by Mazzolani and Piluso (1992)

- The strength degradation ratio:

$$\varepsilon_i^{\pm} = \frac{M_i^{\pm}}{M_y^{\pm}} \tag{4.4.4}$$

where $M_i^{\pm}$ is the peak moment of the ith half-cycle

- The stiffness degradation ratio:

$$\xi_i^{\pm} = \frac{\tan\alpha_i^{\pm}}{\tan\alpha_y^{\pm}} \tag{4.4.5}$$

which corresponds to the ratio between the secant stiffness for a given half cycle (defined between the zero ordinate and $M_i^{\pm}$ peak value) and the initial elastic stiffness.

- The absorbed energy ratio:

$$\eta_i^{\pm} = \frac{A_i^{\pm}}{A_y^{\pm}} \tag{4.4.6}$$

between the energy absorbed in a real half-cycle corresponding to perfect elasto-plastic behaviour with the same displacement amplitude $A_y^{\pm}$.

All these non-dimensional ratio parameters have been expressed as a function of the partial ductility ratio:

$$\mu_i^{\pm} = \frac{\Phi_i^{\pm}}{\Phi_y^{\pm}} \tag{4.4.7}$$

where $\Phi_i^{\pm}$ is the absolute value of the rotation in the ith half-cycle.

Selecting the $\varepsilon_i^{\pm}, \xi_i^{\pm}$ *and* $\eta_i^{\pm}$ values at the first half-cycle of each series for a given range of rotation, the variations of $\varepsilon_i^{\pm}(\mu_i^{\pm})$, $\xi_i^{\pm}(\mu_i^{\pm})$ and $\eta_i^{\pm}(\mu_i^{\pm})$ (whose curves have been smoothed appropriately) are shown for all the cyclic tests in figures 4.4.9a, 4.4.10a and 4.4.11a for positive rotations and in figures 4.4.9b, 4.4.10b and 4.4.11b for negative rotations.

Some comments about these variations of non-dimensional parameters may be done:

- Group1

Test J4 (composite slab fully shear connected by cold formed angles) shows some difficulties to reach the elastic resistance limit in hogging bending (figure 4.4.9a). The stiffness ratio is decreasing, particularly in sagging bending (figure 4.4.10b), which may be explained by the premature deformation of bolts and end plate. However, a small decreasing is observed for the absorbed energy ratio (from *0.6* to *0.4* for η^- and *0.8* to *0.7* for η^+ in figures 4.4.11a and 4.4.11b).

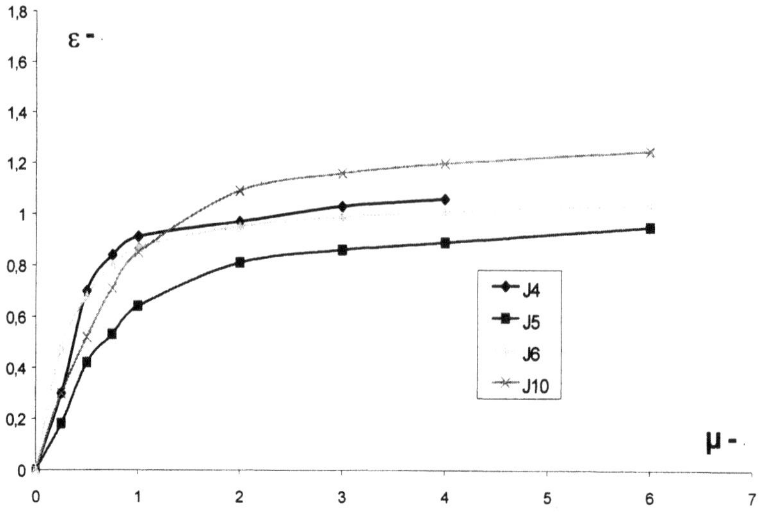

a/ Hogging bending

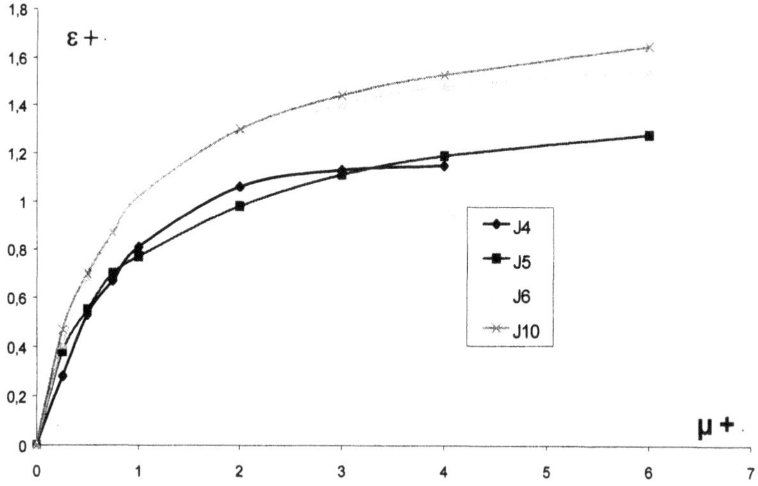

b/ Sagging bending

Figure 4.4.9 : Resistance ratio versus partial ductility

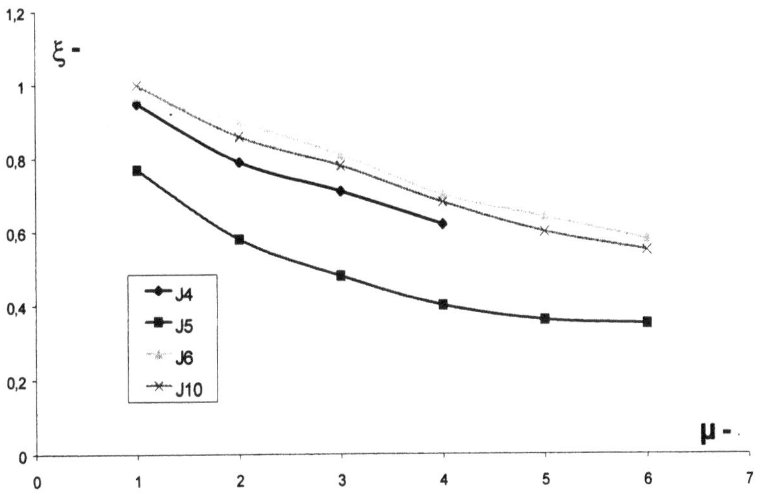

a/ Hogging bending

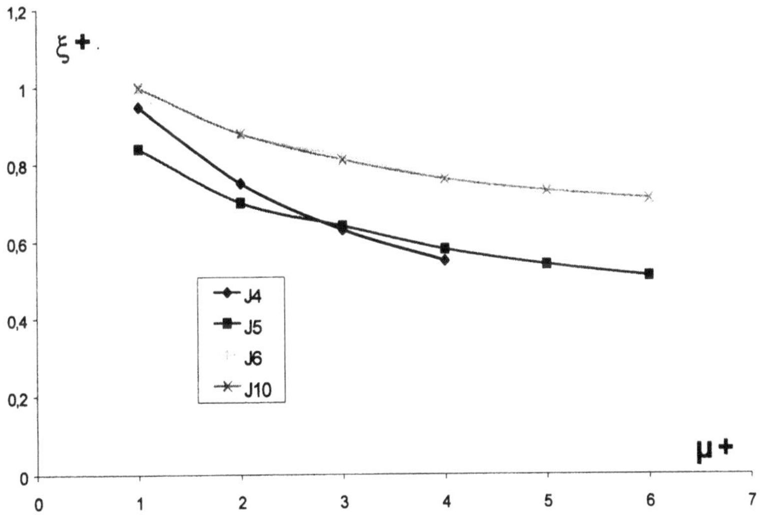

b/ Sagging bending

Figure 4.4.10 : Stiffness ratio versus partial ductility

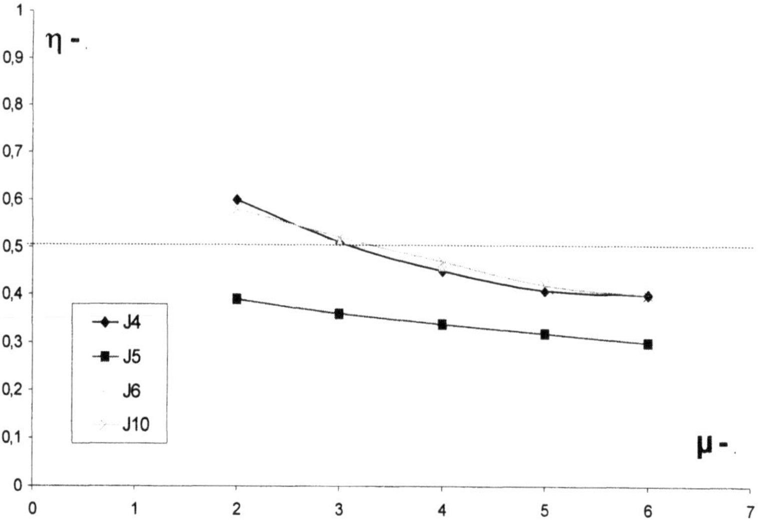

a/ Hogging bending

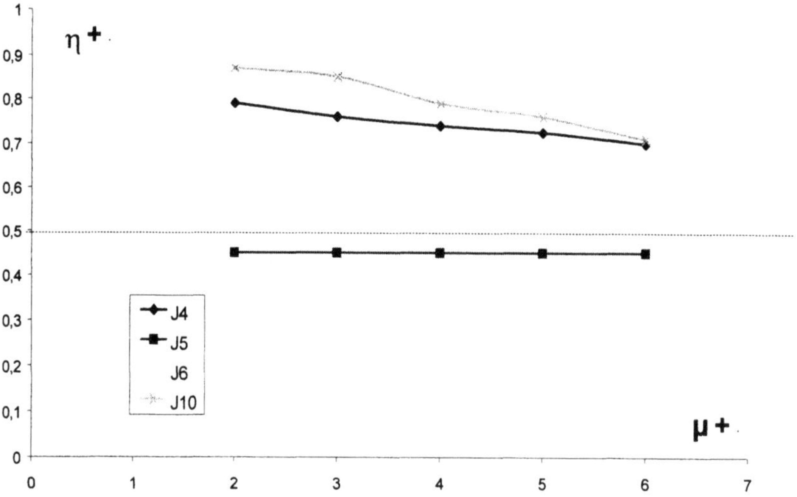

b/ Sagging bending

Figure 4.4.11 : Absorbed energy ratio versus partial ductility

Test J5 (composite slab partially shear connected by cold formed angles) does not reach the elastic resistance limit in hogging bending (figure 4.4.9a). The decreasing of the stiffness ratio is more pronounced than for other tests (figures 4.4.10a and 4.4.10b) and the absorbed energy ratio is clearly 30% lower than test J4 with a full shear connection (figures 4.4.11a and 4.4.11b).
Test J6 (composite slab more than fully connected by cold formed angles) gives similar performances than test J4 in hogging bending but a clear increase of the non-dimensional parameters in sagging bending. Apparently, the use of 3 angles nailed in a narrow width of 120 mm in the middle of first ribs close to the joint (necessary to ensure a more than full shear connection) seems to lead to a loss of shear connection efficiency probably due to a greater damage of concrete in the ribs, specially under cyclic loading. The use of a group of shear connectors concentrated on a short width in a rib should be more investigated with a view of seismic design.
In general, one may be surprised at the poor performances of test J4 and J6. A possible explanation may be a reduction of force transfer between the reinforcement and the shear connectors due to the high position of the reinforcement in the slab; the reinforcement lies 8 cm above the steel beam flange, just at the upper end of the cold formed angles. More investigation should be also needed to clarify the effect of the reinforcement position with regard to detailed seismic rules.

• Group 2
Test J10 (solid slab fully shear connected by welded headed studs) expresses the best performance of all 4 tested joints. Its curves $\varepsilon^{\pm}(\mu^{\pm})$ go clearly beyond the elastic resistance limit in hogging and sagging bendings (figures 4.4.9a and 4.4.9b). In figures 4.4.10a and 4.4.10b, we observe a normal decrease of the stiffness ratio and in figures 4.4.11a and 4.4.11b a small decrease of the absorbed energy ratio very close to test J6. It should be pointed out that the reinforcement layer of specimen J10 was located 2 cm below the heads of the stud connectors and consequently more efficient.

Taking into account the previous interpretation by means of ECCS parameters, it appears that the use of partial shear connection in composite joints under cyclic loading should not be advised. In addition, the cyclic performances of the composite joints under cyclic loading cannot be satisfactory if the performances of the steelwork part are not well controlled by an appropriate design of end-plate/column flange thicknesses, of the bolt diameter, etc.

SKELETON CURVES AND MONOTONIC CURVES

Skeleton curves obtained from the peak hogging and sagging bending moments are drawn in figure 4.4.12a for tests J4, J5 and J6 of Group 1 and in figure 4.4.12b for test J10 of Group 2 (it is mentioned again that the relevant peak moments are those of the first half-cycle of each series). Compared to the corresponding monotonic M-Φ curves, the skeleton curves of Group 1 are less favourable, specially under hogging bending and for partial shear connection (test J5) as it could be foreseen. On the other hand, monotonic and skeleton curves of Group 2 appear in good agreement. Another important remark concerns the limited rotation capacity of the skeleton curves, about *15 mrd* in sagging bending; as reminder, the physical reason of this limitation is the rupture of the steelwork part. Obviously, the rotation in hogging bending has to be stopped because of the type of cyclic loading procedure itself.

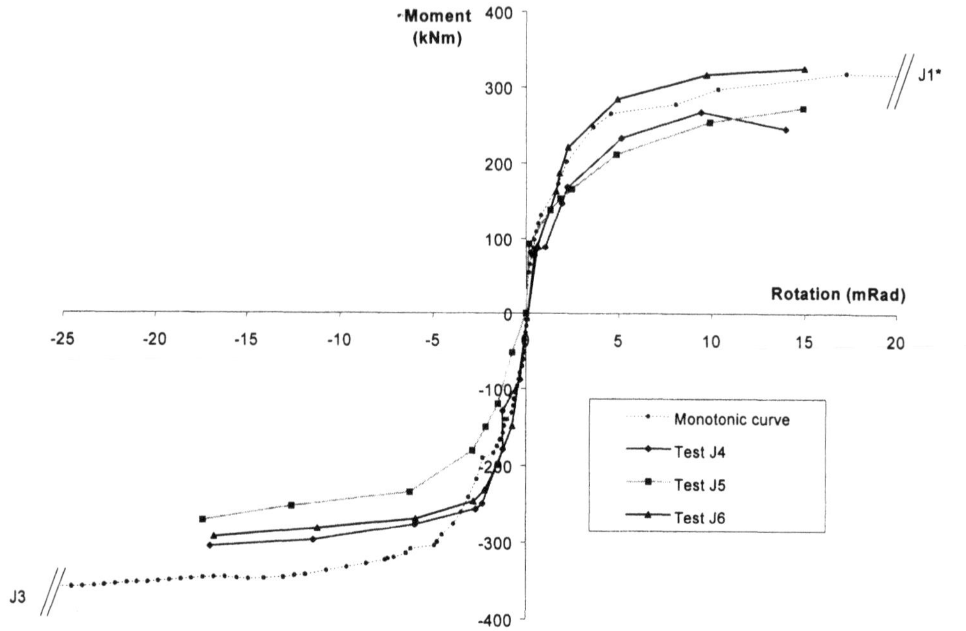

a) Group 1

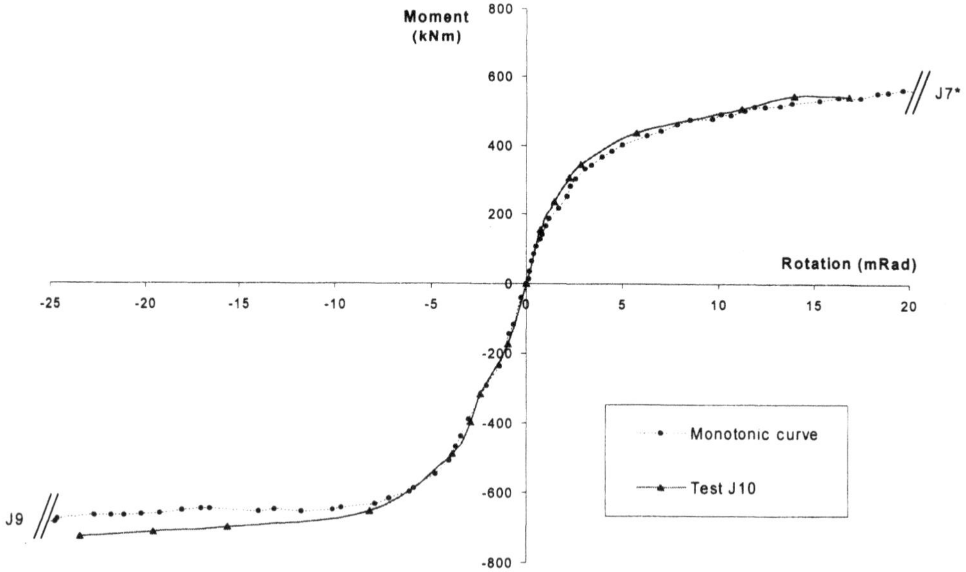

b) Group 2

Figure 4.4.12 : Skeleton curves

SLIP DISTRIBUTIONS

Slip distributions measured along the beam for the hogging bending moment peaks are represented in figures 4.4.13a,b,c,d dealing with different cycles of rotation ranges ($\Phi_y^{\pm}, 2n\Phi_y^{\pm},...$) and for tests J5, J4, J6 and J10 respectively. In fact, it is difficult to obtain accurate measures of slip at the interface between the slab and the steel flange, specially under cyclic loading where the more important degradation of the slab and possible effects of residual deformations at each reversal loading may affect the measures. So, the experimental curves given in figures 4.4.13a,b,c,d should be considered rather qualitative than quantitative.

Outside a zone of 400 mm length from the column axis, the slip tends to become uniform along the beam allowing to adopt a reference value of slip for all the tests.

For test J4, the slip does not exceed *2 mm* at the first cycle of range $(6\,\Phi_y^{+},\ 6\,\Phi_y^{-})$, just before failure..

For test J5, the maximum slip before failure is *3.5 mm*, hence more important as it could be foreseen. Taking into account the Push-Pull test results, it is not excluded that a beginning of failure, hidden by the premature failure of bolts, was set in motion in the shear connection.

For test J6, the slip does not exceed *0.6 mm*, which is in accordance with the more than full shear connection used for this test (to be compared to *2 mm* slip for test J4).

For test J10, the maximum slip lies under *1.2 mm*.

These qualitative results tend to demonstrate that a full shear connection ($N/N_f = 1.06$ for test J4 and *1.01* for test J10) remains subject to limited slips lying within the critical range defined by the Push-Pull tests (see figure 4.3.8). This much is certain that the use of more than full shear connection leads clearly to the safe side (test J6).

SIMPLIFIED THEORETICAL APPROACH

The simplified approach which is proposed hereafter allows to evaluate suitably the skeleton curve of moment-rotation cycles when assuming that:

- the shear connection is full (or more than full);
- the moment-rotation monotonic curves of the bare steel joint in sagging and hogging bendings are established previously either by appropriate tests (like J1 and J2 for Group 1 and J7 and J8 for Group 2) or by sophisticated numerical modelling, Aribert, Lachal and Dinga (1999).

The above experimental interpretation has demonstrated clearly that the skeleton curve is practically superposed on:

- the monotonic curve of the bare steel joint in sagging bending provided that this curve is transformed by anamorphosis using the multiplier m_2 (see the above determination of curves J1* and J7*);
- the monotonic curve of the composite joint in hogging bending (like J3 for Group 1 and J9 for Group2) for which one of the authors has given a specific theoretical method consisting in an approximation by a tri-linear curve ABC, Aribert (1996 and 1999), Anderson, Aribert and Kronenberger (1998).

In addition, a great care should be taken in controlling the resistance in low-cycle fatigue of details of the bare steel joint, specially the fatigue resistance of the tension bolts. Usual approach based on the Eurocode 3-1-1 formulation (Chapter 9) appears well appropriate and lead to cut the relevant curves at maximum boundary rotations apparently severe as it was observed in tests J4 and J10.

Figure 4.4.14 illustrates the validity of the theoretical approach comparing the so-calculated and experimental curves. The detailed background of all the calculations is given hereafter for whatever purpose it may serve.

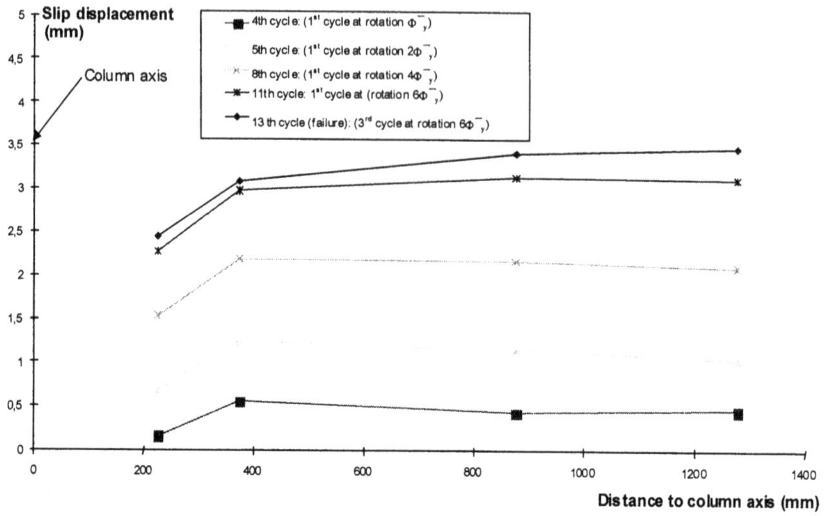

a) Test J5 ($N/N_f = 0.53$)

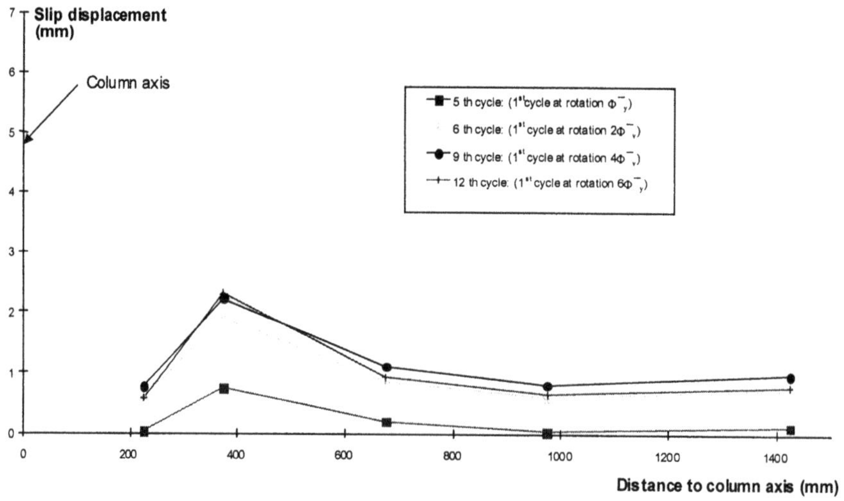

b) Test J4 ($N/N_f = 1.06$)

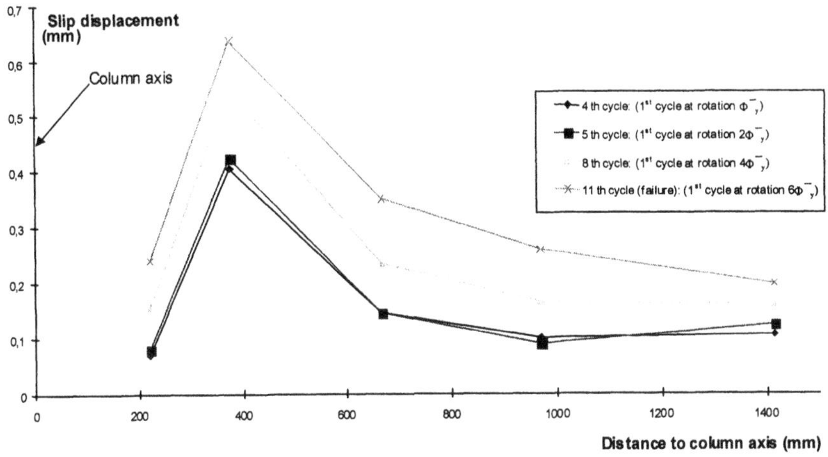

c) Test J6 ($N/N_f = 1.59$)

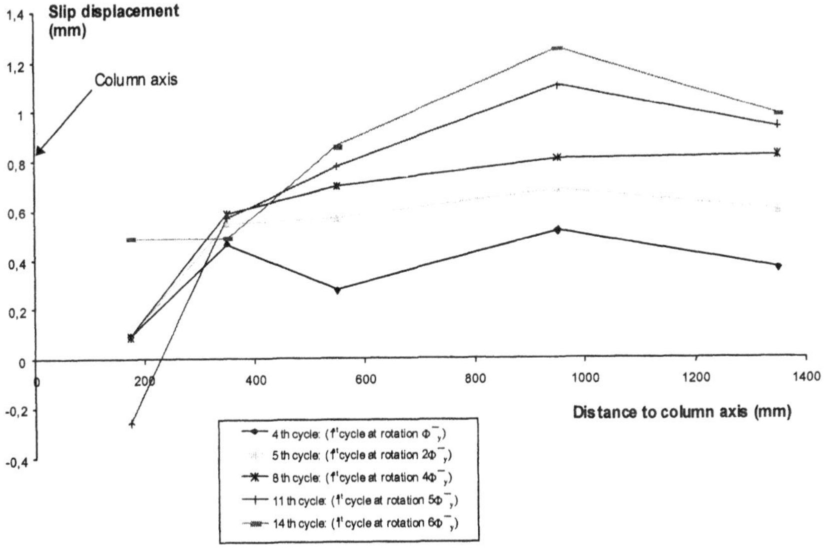

d) Test J10 ($N/N_f = 1.01$)

Figure 4.4.13 : Slip distributions when the slab is tension

(i) Evaluation of the moment-rotation curve of monotonic test J9 ($N/N_f = 1.01$)

• Data:

Distance between the longitudinal reinforcing bars in tension and the mid-steel flange in compression:

$$h_s = 535.3\ mm\ ;$$

Distance between the longitudinal reinforcing bars in tension and the centroid of the beam's steel section :

$$d_s = 316\ mm\ ;$$

Flexural stiffness of the beam's steel section :

$$E_a I_a = 7.085 \times 10^{13}\ N.mm^2\ ;$$

Cross-sectional area of the longitudinal reinforcement :

$$A_s = 1539.4\ mm^2\ ;$$

Number of shear connectors $N = 10$ distributed along the length $\ell = 1609\ mm$ in hogging bending;
Stiffness of one shear connector:

$$k_{sc} = P_y/s_y = 72.5/0.30 = 241.7\ kN/mm$$

Resistance of one shear connector:

$$P_R = 87.9\ kN$$

Distance between the external face of the column flange and the first row of shear connectors:

$$h_c/2 + a = 243/2 + 128.5 = 250\ mm.$$

• Determination of the approximate tri-linear curve ABC.

Point A:

Translational stiffness of the reinforcement:

$$K_s = \frac{E_s A_s}{h_c/2 + a} = 1169.9\ kN/mm$$

Translational stiffness of the shear connection:

$$\alpha = \frac{E_a I_a}{d_s^2 E_s A_s} = 2.43$$

$$\beta = \left[(1+\alpha) N k_{sc} \ell d_s^2 / (E_a I_a)\right]^{0.5} = 4.335$$

$$K_{sc} = N k_{sc} \Big/ \left(\beta - \frac{\beta - 1}{1+\alpha} \times \frac{h_s}{d_s} \right) = 899.2\ kN/mm$$

$$F_s^{(A)} = K_{sc} s_y = 269.8\ kN$$

Corresponding rotation of the joint:

$$\Phi^{(A)} = \frac{1}{h_s} \left(\frac{F_s^{(A)}}{K_s} + s_y \right) = \underline{0.99 \times 10^{-3}\ rd}$$

Moment applied to the joint:

$$M^{(A)} = M_a(\Phi^{(A)}) + F_s^{(A)} h_s$$

where $M_a(\Phi^{(A)})$ is the resistance moment of the bare steel joint. The moment-rotation curve of test J8 gives: $M_A(\Phi^{(A)}) = 71.2\ kN.m$

Hence: $M^{(A)} = 71.2 + 269.8 \times 0.5353 = \underline{215.6\ kN.m}$

Point B:

$$F_s^{(B)} = F_{ys} = A_s f_{ys} = 874.4\ kN$$

$$s^{(B)} = s^{(A)} \times 2F_s^{(B)} / F_s^{(A)} = 1.94\ mm$$

$$\Phi^{(B)} = \frac{1}{h_s}\left(\frac{F_s^{(B)}}{K_s} + s^{(B)}\right) = \underline{5.02 \times 10^{-3}\, rd}$$

Test J8 ⟶

$$M_a(\Phi^{(B)}) = \underline{175.8\ kN.m}$$

$$M^{(B)} = M_a(\Phi^{(B)}) + F_s^{(B)} h_s = \underline{643.9\ kN.m}$$

Point C:

Ultimate elongation of the reinforcement:

$$\Delta_{su} = (h_c / 2 + a)\varepsilon_{su} = 250 \times 0.07 = 17.5\ mm$$

Hence:

$$\Phi^{(C)} = \frac{1}{h_s}\left(\Delta_{su} + s^{(B)}\right) = \underline{36.3 \times 10^{-3}\, rd}$$

$$M^{(C)} = M_a(\Phi^{(C)}) + F_s^{(B)} h_s = 280.9 + 874.4 \times 0.5353$$

$$\underline{M^{(C)} = 749\ kN.m}$$

Note: Perfect agreement with the experimental rupture of the rebars.

(ii) Evaluation of the moment rotation curve of monotonic test J3 ($N/N_f = 1.06$)

Same calculation procedure, but using the following data:

$h_s = 443.8\ mm;\ d_s = 270\ mm;\ E_a I_a = 3.364 \times 10^{13}\ N.mm^2;\ A_s = 785.4\ mm^2;\ N = 20;\ \ell = 1630\ mm;$

$k_{sc} = P_y / s_y = 19 / 0.12 = 158.3\ kN/mm;\ P_R = 23.6 kN;\ h_c / 2 + a = 100 + 70 = 170\ mm.$

Point A:

$$K_s = 877.8\ kN/mm$$

$$\alpha = 3.09;\ \beta = 6.763$$

$$K_{sc} = 711.9\ kN/mm$$

$$F_s^{(A)} = 85.4\ kN$$

$$\Phi^{(A)} = \frac{1}{443.8}\left(\frac{85.4}{877.8} + 0.12\right) = \underline{0.49 \times 10^{-3}\, rd}$$

Test J2 ⟶

$$M_a(\Phi^{(A)}) = 29.9\ kN.m$$

$$M^{(A)} = 29.9 + 85.4 \times 0.4438 = \underline{67.8\ kN.m}$$

Point B:

$$F_s^{(B)} = A_s f_{ys} = 446.1\ kN$$

$$s^{(B)} = 0.12 \times 2 \times 446.1 / 85.4 = 1.25\ mm$$

$$\Phi^{(B)} = \frac{1}{443.8}\left(\frac{446.1}{877.8} + 1.25\right) = \underline{3.96 \times 10^{-3}\, rd}$$

Test J2 ⟶

$$M_a(\Phi^{(B)}) = 92.0\ kN.m$$

$$M^{(B)} = 92.0 + 446.1 \times 0.4438 = \underline{290.0\ kN.m}$$

Point C:

Ultimate elongation:

$$\Delta_{su} = 170 \times 0.06 = 10.2\ mm$$

$$\Phi^{(C)} = \frac{1}{443.8}(10.2 + 1.25) = \underline{25.8 \times 10^{-3}\, rd}$$

Test J2 ⟶

$$M_a(\Phi^{(C)}) = 150.8\ kN.m$$

$$M^{(C)} = 150.8 + 446.1 \times 0.4438 = \underline{348.8\ kN.m}$$

Note: Perfect agreement with the experimental rupture of the rebars.

(iii) Cutting of curves J7* and J1* due to low-cycle fatigue of external bolts

Test J7*

Determination of the elastic limit force in the tension steel flange:

$$F_{yf} = \left(M_y^+ / 1.25\right) / \left(h_b - t_{fb}\right) = 350 \times 10^6 \ / \ 1.25 \ / (450 - 13.5) = 641\,500\ N$$

Corresponding axial force in one external bolt

$$B_y = F_{yf} / 4 = 160\,400\ N$$

Tension stress in the bolt of nominal diameter $\Phi 22$:

$$\sigma_y = B_y / A_r = 160\,400/303 = 530\ N/mm^2$$

Basing on Chapter 9 in Eurocode 3-1-1 (1992), use of detail category 36 is specified for ordinary bolts and preloaded bolts in tension (see Table 9-8-1).

So, the damage index calculated after 3 cycles for the controlled rotation $5\Phi_y^+$ is equal to:

$$D = \frac{1}{K}\left[\left(1 \times 1^3 + 3 \times 2^3 + 3 \times 4^3 + 3 \times 5^3\right) 530^3\right]$$

where: $K = 95.477 \times 10^9$

Let be: $D = 0.92$

Hence, there is no rupture in principle. But, if a supplementary cycle for $6\Phi_y^+$ is performed, the damage index becomes:

$$D = \frac{1}{K}\left[\left(1 \times 1^3 + 3 \times 2^3 + 3 \times 4^3 + 3 \times 5^3 + 6^3\right) 530^3\right] = 1.26$$

Now,there is risk of bolt rupture from a point of view of design code, which is in acceptable agreement with the experimental result (in fact, the rupture was observed after the second $6\Phi_y^+$ cycle). Consequently, curve J7* can be adopted as skeleton curve in sagging bending provided that it is cut at the boundary rotation:

$$6\Phi_y^+ = 6 \times 2.8 \times 10^{-3} = \underline{16.8 \times 10^{-3}\, rd}$$

That leads also to a boundary rotation in hogging bending for the skeleton curve J9:

$$6\Phi_y^+ = 6 \times (-3.9 \times 10^{-3}) = \underline{-23.4 \times 10^{-3}\, rd}$$

Test J1*

$$F_{yf} = 235 \times 10^6 \ / \ 1.25 \ / (360 - 12.5) = 541\,000\ N$$

$$B_y = F_{yf} / 4 = 135\,250\ N$$

Tension stress in the bolt of nominal diameter $\Phi 18$:

$$\sigma_y = B_y / A_r = 135\,250 / 192 = 704\ N/mm^2$$

The damage index calculated after 3 cycles for $4\Phi_y^+$ is equal to:

$$D = \frac{1}{K}\left[\left(1 \times 1^3 + 3 \times 2^3 + 3 \times 4^3\right) 704^3\right] = 0.79$$

If a supplementary cycle for $6\Phi_y^+$ is performed, the damage index becomes:

$$D = \frac{1}{K}\left[\left(1 \times 1^3 + 3 \times 2^3 + 3 \times 4^3 + 6^3\right) 704^3\right] = 1.58$$

In principle, there is a bolt rupture as confirmed by test J4.

Consequently, curve J1* can be adopted as skeleton curve in sagging bending provided that it is cut at the boundary rotation:

$$4\Phi_y^+ = 4 \times 2.4 \times 10^{-3} = \underline{9.6 \times 10^{-3}\, rd}$$

Then, the associated boundary rotation in hogging bending is:

$$4\Phi_y^- = 4 \times (-2.8 \times 10^3) = \underline{-11.2\ mrd}$$

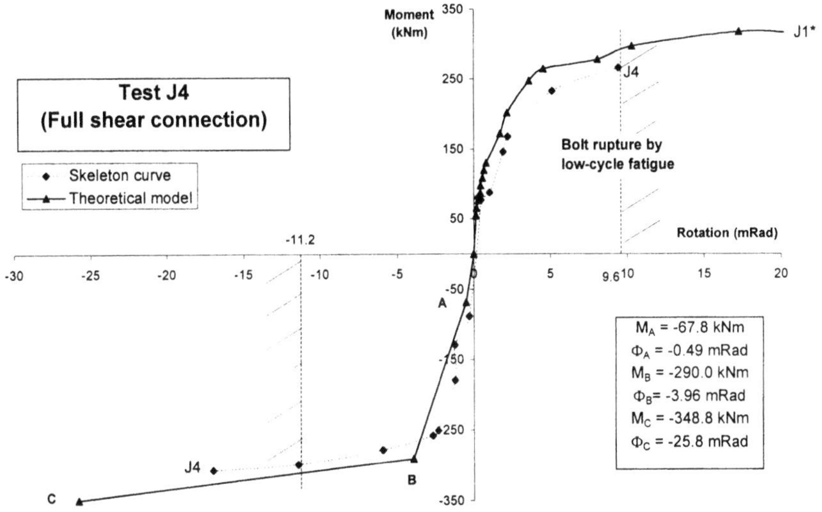

a) Group 1

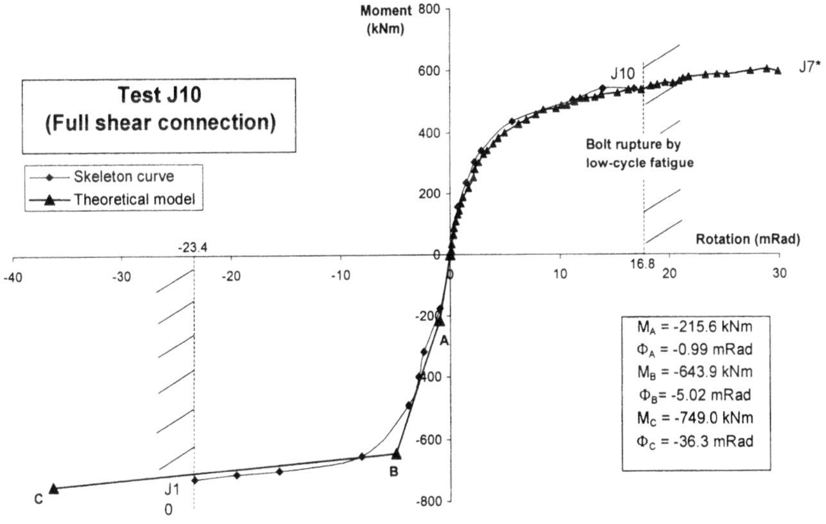

b) Group 2

Figure 4.4.14 : Comparison of the skeleton curves with the simplified theoretical model

REFERENCES

Anderson D., Aribert J.M. and Kronenberger H.J. (1998). Rotation capacity of composite joints. *COST-C1. International Conference – Session 2.* Liège, Belgium.

Aribert J.M. (1996). Influence of slip of the shear connection on composite joint behaviour. *Connections in Steel Structures* III, 11-22. Pergamon, Trento.

Aribert J.M. (1999). Theoretical solutions relating to partial shear connection of steel-concrete composite beams and joints. *Steel and Composite Structures International Conf. – Session 3,* Delft, 7/1-7/16.

Aribert J.M., Lachal A. and Dinga O. (1999). Modélisation du comportement d'assemblages métalliques semi-rigides de type poutre-poteau boulonnés par platine d'extrémité. *Revue Construction Métallique* **1**, 25-46.

ECCS (1985). *Recommended testing procedure for assessing the behaviour of structural steel element under cyclic loads,* Committee TWG 1.3.

Eurocode 4, Part 1-1 ENV 1994-1-1. (1992). *Design of composite steel and concrete structures – general rules and rules for buildings,* CEN – Brussels.

Mazzolani F.M. and Piluso V. (1996). *Theory and Design of Seismic Resistant Steel Frames*, E & FN Spon.

4.5

CONCLUSIONS

Jean Marie Aribert and Alain Lachal

The main contributions of the present investigation may be summarized as follows:

• The Push-Pull tests demonstrate clearly that shear connectors defined as ductile under monotonic loading should be classified as non ductile under cyclic repeated loads. Moreover, the requirement of ± 2 mm of slip capacity may lead to reduce the static shear resistance. These results are deduced from a specific method of interpretation proposed by the authors.

• The resistance of the shear connectors in low-cycle fatigue can be formulated analytically including damage cumulation (relationships (4.3.3) and (4.3.4)), as confirmed by three types of tested shear connection. Nevertheless, the effect of the steel profiled sheeting needs to be still clarified.

• For the cyclic behaviour of the tested composite beam-to-column bolted joints which were partial-strength joints and therefore dissipative elements, a distinction may be done for the shear connection based on the usual static definition of N/N_f degree.

• The use of partial shear connection in composite joints under cyclic loading and seismic actions should not be advised, though no failure of this component was observed here. Nevertheless, the ECCS parameters of interpretation (dealing with the degradation in strength, stiffness and absorbed energy) and the slip distributions directly measured at the steel-concrete interface tend to confirm the above recommendation.

• The rotation capacity under cyclic loading may be strongly reduced by the premature rupture of the steelwork part in sagging bending, specially the bolts (possibly accompanied by the rupture of end plate or welds). A theoretical determination of this rupture may be based on the usual fatigue formulation of Eurocode 3, as proved by a few numerical calculations at the chapter end. Consequently, the need should be underlined to well design the connection details joining the bottom of the steel beam to the column flange; in future researches, other systems than a simple bolted end plate have to be imagined to improve the rotation capacity.

• Other parameters seem affect significantly the cyclic behaviour of the composite joints, in particular the high position of the reinforcement with respect to the shear connectors and the concentration of shear connectors in the ribs of composite slabs.

• The skeleton curves enveloping the peaks in hogging and sagging bending moments may be useful to the seismic design of composite joints when considering an equivalent static approach. In this way, a simplified theoretical method is presented briefly at the chapter end and illustrated with numerical calculations showing a good agreement with the experimental curves when the shear connection is full.

Chapter 5

Re-elaboration of Experimental Results

5

RE-ELABORATION OF EXPERIMENTAL RESULTS

Luis Calado

INTRODUCTION

In the recent earthquake events of 1994 Northridge (Los Angeles) and 1995 Hyogo-ken Nanbu (Kobe) a significant population of steel structures suffered extended damage. With reference to steel buildings that did not collapse, although exhibiting significant damage, local failures of steel elements as well as of their beam-to-column joints took place, although they did not result in severe overall deformations, remaining hidden behind undamaged architectural panels, Bruneau et al. (1998).

However, these disasters underlined the need of an efficient design approach for steel structures based on the selection of the energy dissipation mechanisms, which permits to combine the stable hysteretic response of steel members and/or joints with the possibility of controlling simply, but reliably, the behavioural parameters of the relevant parts of the skeleton frame, Calado et al (1997).

Despite the several studies carried out in the past, or currently in progress, on the behaviour of subassemblages as well as of steel components under cyclic reversal loading, it has to be pointed out that the possibility to use this "aseismic" design approach implies an exhaustive understanding of the low-cycle fatigue strength of steel members and joints.

This chapter presents a re-elaboration, based on the S-N approach, of the main results of the experimental tests performed in all the laboratories involved in the RECOS project. Failure criteria for the assessment of the number of cycles to failure *(N)* and definitions (proposed by various authors) of the stress (strain) range *(S)* are briefly described. Variable amplitude loading is discussed and the Miner's rules is presented together with the rainflow counting.

A statistical analysis of the *S-N* lines based on a given probability of failure with reference to suitable levels of safety and reliability of the structural elements is analysed. This methodology together with the Ballio & Castiglioni (1995) model was applied to all test data under cyclic loading.

FATIGUE BEHAVIOUR MODELS

The success of an accurate prediction of fatigue failures (Peeker, 1997) is the selection of a good fatigue failure prediction function, which can be defined as the function that relates a parameter *(S)*, representative of the imposed cyclic actions to the fatigue endurance *(N)* of the structural detail.

Most common approaches to the fatigue behaviour modelling can be classified into three categories, depending on the fatigue failure prediction function adopted:

- the *S-N* line approach, which assumes *S* to be the nominal stress range $\Delta\sigma_0$;
- the local strain approach, which considers the local non-linear strain range $\Delta\varepsilon$;
- the fracture mechanic approach, which adopts the stress intensity factor range.

The S-N line approach

The fatigue failure prediction function, used by the *S-N* curve approach, can be expressed by the following equation:

$$N.S^m = K \tag{5.1}$$

where *N* is the number of cycles to failure at the constant stress (strain) range *S*. The non-dimensional constant *m* and the dimensional parameter *K* depend on both the typology and the mechanical properties of the considered steel component. In the *Log-Log* domain Eqn. (5.1) can be re-written as:

$$Log(N) = Log(K) - mLog(S) \tag{5.2}$$

Eqn. (5.2) represents a straight line with a slope equal to *-1/m* called fatigue resistance line, which identify the safe and unsafe regions (Fig. 5.1).

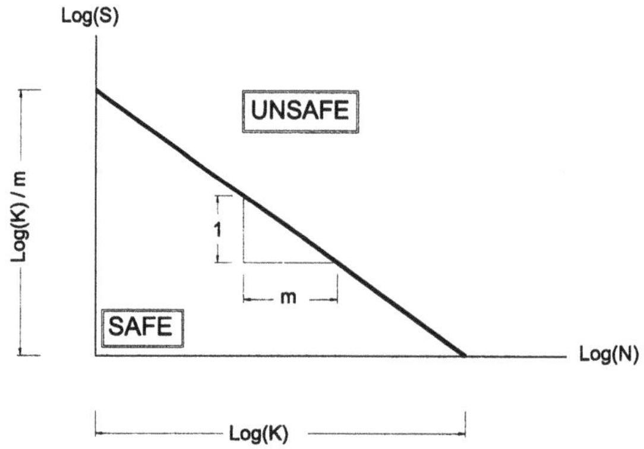

Figure 5.1: Fatigue resistance line in the *Log (S)-Log (N)* scale.

Once identified the pertinent *S-N* line, for an undamaged component, if the point of co-ordinates *Log* (n_{tot}), *Log (S)* representative of a generic loading event (where n_{tot} is the total number of cycles sustained by the component in the loading history), falls in the safe region, failure is not to be expected for the component under that particular loading event.

Because of its simplicity, the *S-N* curve approach has been introduced into many fatigue design codes. The fatigue resistance curves adopted in design standards, are built using a statistical analysis of constant amplitude fatigue tests data.

If variable amplitude loads are used, the direct assessment of the fatigue resistance is not possible and reference should be made to a cycle counting (Matsuishi & Endo, 1968) method and to a suitable damage accumulation rule. Usually the linear damage accumulation rule proposed by Miner (1945) is adopted for calculation of an effective value, S_{eq}, which is adopted instead of S as argument in the fatigue failure prediction function.

The main advantage of the *S-N* curve approach is its simplicity. As this approach is able to interpret correctly the phase of stable crack propagation, it is commonly adopted in civil engineering for the assessment of the fatigue strength of welded details where, due to the presence of fabrication imperfections, the crack initiation phase is practically absent.

The local strain approach

This fatigue model considers S to be the strain range, $\Delta\varepsilon$, which can be expressed as the sum of the elastic strain amplitude $(\Delta\varepsilon_{el})$ and the plastic strain amplitude $(\Delta\varepsilon_{pl})$ i.e. as:

$$\Delta\varepsilon = \Delta\varepsilon_{el} + \Delta\varepsilon_{pl} \tag{5.3}$$

The elastic and plastic contributions of the strain range can be expressed as:

$$\frac{\Delta\varepsilon_{el}}{2} = \left(\frac{\sigma_f - \sigma_m}{E}\right).(2.N)^b \tag{5.4}$$

and

$$\frac{\Delta\varepsilon_{pl}}{2} = \varepsilon_f.(2.N)^c \tag{5.5}$$

It should be remarked that the coefficient *2* before the term N was introduced in order to keep into account loading histories with cycles of variable amplitude. In this case, reference should be made to half-cycles, between two subsequent reversal points. Eqn. (5.5) is known as the Coffin-Manson equation (Coffin, 1954; Mason, 1954). In addition to the mean stress value σ_m, the strain relationship (Eqn. (5.3)) contains five material dependent constants: E, σ_f, b, ε_f, c. The determination of these constants is standardised (ANSI/ASTM E466-76, 1977; Landgraf et al, 1969) and their values are available in different handbooks for various materials.

Although the high precision in the fatigue failure prediction, usually the local strain approach can only be used for the assessment of the crack initiation period. Calculations of $\Delta\varepsilon$ at the crack initiator are relatively simple and if $\Delta\varepsilon$ is known, Eqn. (5.3) allows the valuation of the crack initiation period. On the other hand, calculation of $\Delta\varepsilon$ at the fatigue crack tip is more complex. Due to crack growth, the cyclic strain field at the crack tip changes at each cycle, and consequently calculation of the stable crack growth period using Eqn. (5.3) is impossible.

The fracture mechanic approach

This approach, based on the fracture mechanic, adopts the concept of the stress intensity factor, allowing to analyse the behaviour of a great variety of fatigue cracks. However, as it is based on crack propagation laws, this approach can be adopted only for the prediction of the stable crack growth period; furthermore, it requires the determination of the stress intensity factor, which might be a relatively complex procedure.

METHODS OF RE-ELABORATION OF TESTS DATA

It is commonly accepted that a modern methodology to assess the low-cycle fatigue endurance of civil engineering structures should adopt parameters related to the global structural behaviour, such as displacements, rotations, bending moments, etc. Taking this into account, the *S-N* curve approach may be adopted, considering as *S*, parameters related with the global structural ductility (e.g., interstorey drifts or joint rotations).

In order to apply Eqn. (5.1), both parameters *S* and the number of cycles to failure *(N)* should be defined, by a consistent re-elaboration of test data in accordance with the basic assumption of the selected damage model. The number of cycles to failure *N* can be identified on the basis of the failure criterion, while *S* can be defined with reference to the definitions given in the literature by various authors.

Definition of the strain range (S)

Some of the most relevant proposals available in literature for the definition of the strain range *S* are shortly presented.

Krawinkler & Zohrei proposal

The concept to connect the parameter *(S)* of the fatigue failure prediction functions with the global structural ductility was originally proposed by Krawinkler & Zohrei (1983). They proposed a relationship between the fatigue endurance and the plastic portion of the generalised displacement component, which can be expressed as:

$$N\left(\Delta\delta_{pl}\right)^{m} = K^{-1} \tag{5.6}$$

where $\Delta\delta_{pl}$ represents the plastic portion of the deformation range.

Ballio & Castiglioni proposal

Ballio & Castiglioni (1995) proposed a unified approach for the design of steel structures under low and/or high-cycle fatigue, which is based on global displacement parameters instead of local deformation parameters. The fundamental hypothesis is the validity of the following equation:

$$\frac{\Delta\varepsilon}{\varepsilon_y} = \frac{\Delta v}{v_y} = \frac{\Delta\phi}{\phi_y} \tag{5.7}$$

where ε represents the strain, v the displacement, ϕ the rotation (or the curvature), Δ the range of variation in a cycle and the subscript y identifies the yielding of the material (ε_y) as well as the conventional yielding with reference to the generalised displacement (v_y, ϕ_y) assumed as the test control parameter.

For an ideal elastic material, the relationship between the strains ε and the load F (or bending moment M) causing the displacement v (or rotation ϕ) can be written as:

$$E.\varepsilon = \sigma(F) \tag{5.8}$$

and, at yield:

$$E.\varepsilon_y = \sigma(F_y) \tag{5.9}$$

With reference to the previous Eqns. (5.7)-(5.9), the following parameters can be defined:

$$\Delta\sigma^* = E.\Delta\varepsilon = \frac{E.\Delta v}{v_y}\varepsilon_y = \frac{\Delta v}{v_y}\sigma(F_y) = \frac{\Delta\phi}{\phi_y}\sigma(F_y) \tag{5.10}$$

The term $\Delta\sigma^*$ represents the effective stress range, associated with the real strain range $\Delta\varepsilon$ in an ideal member made of an indefinitely linear elastic material. In the case of high-cycle fatigue (i.e. under cycles in the elastic range), $\Delta\sigma^*$ coincides with the actual stress range $\Delta\sigma$. In general, $\Delta\sigma^*$ can be expressed in terms of the generalised displacement component δ as follows:

$$\Delta\sigma^* = \frac{\Delta\delta}{\delta_y}\sigma(F_y) \tag{5.11}$$

Using the *S-N* curves and assuming the effective stress range $\Delta\sigma^*$ as *S*, Eqn. (5.1) can be re-written as:

$$N\left(\frac{\Delta\delta}{\delta_y}\sigma(F_y)\right)^m = K \tag{5.12}$$

Feldmann, Sedlacek, Weynand & Kuck proposal

On the basis of an extensive finite element analysis, Feldmann (1994), Sedlacek et al (1995) and Kuck (1994) investigated the behaviour of beam-to-column joints under constant amplitude cyclic loading. It appeared that the relationship between the plastic strain, ε_{pl}, at the hot spot (strain at the relevant place where first crack occurs) and the plastic rotation (ϕ_{pl}) is linear, i.e.:

$$\frac{\varepsilon_{2,pl}}{\varepsilon_{1,pl}} = \frac{\phi_{2,pl}}{\phi_{1,pl}} \tag{5.13}$$

where subscripts *1* and *2* refer to two different loading steps.

This linear relationship, when plotted in a *Log-Log* scale with the number of cycles on the vertical axis and the plastic rotations on the horizontal ones, plots as a "Wöhler" line, the slope of which *(m)* is the exponent of the Manson-Coffin equation:

$$Log(\phi_{pl}) = Log(\phi_{0,pl}) - \frac{1}{m}Log(N) \tag{5.14}$$

where $\phi_{0,pl}$ represents the maximum theoretical plastic rotation under static loading. In general, Eqn. (5.14) can be rewritten in terms of the plastic portion δ_{pl} of the generalised displacement component δ as:

$$Log(\delta_{pl}) = Log(\delta_{0,pl}) - \frac{1}{m}Log(N) \tag{5.15}$$

In accordance with the procedure proposed by Ballio & Castiglioni (1995) and with reference to Eqn. (5.13), it is possible to define and effective plastic stress range $\Delta\sigma_{pl}*$ which can be expressed in terms of the generalised displacement component δ_{pl} as follows:

$$\Delta\sigma_{pl}* = \frac{\Delta\delta_{pl}}{\delta_y}\sigma(F_y) \tag{5.16}$$

Considering the usual *S-N* curves and assuming *S* as the effective stress range *($\Delta\sigma_{pl}*$)*, Eqn. (5.1) can be re-written as:

$$N\left(\frac{\Delta\delta_{pl}}{\delta_y}\sigma(F_y)\right)^m = K \tag{5.17}$$

Bernuzzi, Castiglioni & Calado proposal

Based on the re-elaboration of an extensive experimental data (Castiglioni & Calado, 1996) of beam-to-column joints as well as of beams and beam-columns, they proposed to use the total interstorey drift (i.e., the total displacement range) Δv as parameter *S*, and consequently Eqn. (5.1) can be re-written, in the form:

$$N.\Delta\delta^m = K \tag{5.18}$$

Definition of the fatigue endurance (N)

For the prediction of the low-cycle fatigue endurance of steel structural components, it seems convenient to adopt a failure criterion based on parameters (i.e. stiffness, strength or dissipated energy) associated with the response of the component. Two failure criteria, which have been proposed (Castiglioni & Calado, 1996), are reviewed in the following.

Calado, Azevedo & Castiglioni criterion ($N_{\alpha=0.5}$)

These authors (Calado & Castiglioni, 1996; Calado & Azevedo, 1989) developed a failure criterion of general validity for structural steel components under both constant and variable amplitude loading histories that can be written in the following form:

$$\frac{\eta_f}{\eta_0} \leq \alpha \tag{5.19}$$

In this equation η_f represents the ratio between the absorbed energy of the considered component at the last cycle before collapse and the energy that might be absorbed in the same cycle if it had an elastic-perfectly plastic behaviour while η_0 represents the same ratio but with reference to the first cycle in plastic range. The value of α, which depends on several factors (such as type of the joint and the steel grade of the component) should be determined by fitting the experimental results. As it is particularly interesting to identify α *a priori*, in order to define a unified failure criterion, a value of $\alpha=0.5$ is recommended for a satisfactory and conservative appraisal of the fatigue life. In the case of variable amplitude loading histories the same criterion remains valid but should be applied on semi-cycles, which can be defined in plastic range as the part of the hysteresis loop under positive or negative loads or as two subsequent load reversal points.

Bernuzzi, Calado & Castiglioni criterion ($N_{\Delta Wr}$)

In this criterion, valid only for constant amplitude loading, is assumed that the drop of the hysteretic energy dissipation is the main parameter for the definition of the low-cycle fatigue endurance. In particular, focusing attention only on the cycles performed at the displacement range Δv_i, the relative energy drop, ΔW_r, can be defined as:

$$\Delta W_r = \left(\frac{W_{init} - W_i}{W_{init}} \right) \tag{5.20}$$

where W_{init} represents the dissipated energy in the first cycle at the assumed displacement range while W_i is the energy associated with the ith cycle at the same displacement range. Failure is assumed to occur when the energy drop is evident, i.e. the generic point of co-ordinates n_k, ΔW_{rk}, is not on the straight line.

Calado, Castiglioni & Bernuzzi criterion (Δv_{TH})

A re-analysis of the test data on beam-to-column connections (Calado et al, 1998) seems to indicate that, with reference to the force-displacement relationship, if the ductility range is lower than a threshold value (Δv_{Th}), corresponding to the displacement related to the maximum strength in a monotonic test, a brittle failure mode is to be expected, while if Δv is greater than Δv_{Th} a ductile failure mode is attained. For cycles with Δv in the range of Δv_{Th} a mixed failure mode was observed. Despite the small number of joint types analysed in this study, this seems to be a general consideration, independently on the size or the type of components (Agatino et al, 1997, Barbaglia et al, 1998).

Hence, a simplified criterion to predict the type of failure expected in a steel component for a given ductility range can be proposed based on the previous considerations. It is assumed that the main parameters governing the failure mode are: the ductility range $\Delta v/v_y$; the beam web slenderness ratio $\lambda_w = d/t_w$, of the depth of the profile (d) to the web thickness (t_w); the beam flange slenderness ratio $\lambda_f = c/t_f$, of the half width of the flange (c) to its thickness (t_f); the weld quality and the severity of the detail, globally accounted for, by the introduction of the numerical coefficient ξ.

For strong column weak beam frame structures the following equation can be proposed for the assessment of the threshold value of the ductility range cycle amplitude (Δv_{Th}) associated with transition from one failure mode to the other:

$$\Delta v_{Th} = \frac{\gamma . v_y}{\xi . \lambda_f . \lambda_w} \tag{5.21}$$

In Eqn. 5.21 ξ is a parameter related to the weld quality as well as to the severity of the detail, that might range from $\xi=1$ for good quality (or not) welds, to $\xi=0.5$ for poor quality welds.

For the non-dimensional coefficient γ, based on the available database, a value of 2000 ± 15% is proposed, independently on the component under consideration. A value of $\gamma=1700$ gives a threshold value such that lower ductility ranges are generally associated with a “brittle” failure mode, while a value of $\gamma=2300$ gives a threshold value such that larger ductility ranges are expected to be associated with a "ductile" failure mode. The range of variation ± 15% around the value γ=2000 accounts for possible "mixed" failure modes.

Since ξ is related to the weld quality as well as to the typology of the welded connection, (i.e. full penetration, simple bevel, double bevel, presence or absence of the backing bar, etc.) it was assumed, for the connections under consideration, to appraise it from the test results. For each test, the experimental value of Δv_{Th}, corresponding to the displacement related to the maximum strength in a monotonic test, was derived. As previously noticed, in the test where Δv is lower than Δv_{Th}, the maximum strength was observed at reversal points and these values were not included for the evaluation of Δv_{Th}.

In this research it is proposed that the value of α_f (Calado, Azevedo & Castiglioni criterion) should be related with $\Delta v/\Delta v_{Th}$ in order to take into account the failure modes. The value of Δv_{Th} has been appraised for all the available tests using particular attention on the value of ξ that is chosen depending on the different failure mode of the specimen and on the weld quality.

In Figure 5.2, the value of α_u (i.e. the actual value of the ratio E/E_0, experimentally determined at complete failure of the specimen) is plotted versus, $\Delta v/\Delta v_{Th}$. A conservative and satisfactory appraisal of the fatigue life can be obtained assuming in the failure criterion a value of α_f determined by fitting the experimental data so that all the performed tests plot below the $\alpha_f(\Delta v/\Delta v_{Th})$ line.

$$\begin{aligned} &\alpha_f=1-0.235*(\Delta v/\Delta v_{Th}) && \Delta v/\Delta v_{Th}<0.85 \\ &\alpha_f=1.65-(\Delta v/\Delta v_{Th}) && 0.85<\Delta v/\Delta v_{Th}<1.15 \\ &\alpha_f=0.5 && \Delta v/\Delta v_{Th}>1.15 \end{aligned} \tag{5.22}$$

In Figure 5.2 it can be seen that, in the range of $0.85<\Delta v/\Delta v_{Th}<1.15$, where a mixed failure mode is to be expected a strong dependence of α_f on the cycle amplitude Δv has been accounted for.

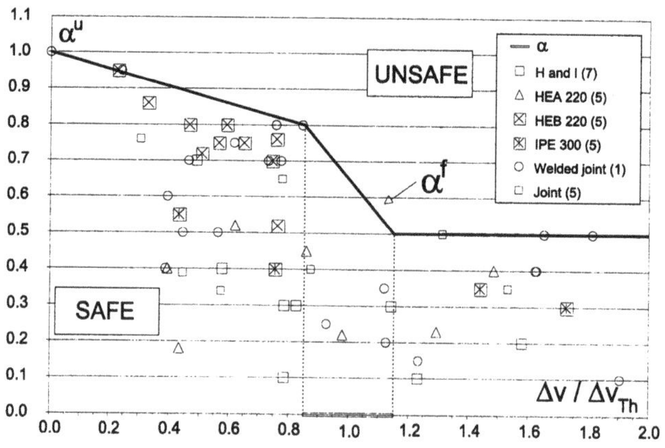

Figure 5.2: Failure criterion based on Δv of threshold.
(adapted from Calado et al (1998))

For values of $\Delta v/\Delta v_{Th}$ greater than 1.15, a ductile failure mode is to be expected, and a constant value of α_f=0.5 is assumed, in agreement with the Calado and Castiglioni proposal. For $\Delta v/\Delta v_{Th}$ lower than 0.85 a brittle failure mode is to be expected, and α_f was assumed linearly decreasing when increasing the cycle amplitude Δv.

Definition of m

The parameter *m* identifies the slope of the line interpreting, in a *Log-Log* scale, the relationship between number of cycles to failure *(N)* and the stress (strain) range *(S)*.

Focusing the attention on *m*, some discrepancies can be found in the literature among the proposal of various authors. In particular it can be noticed that Ballio & Castiglioni suggested to adopt a value of *m=3*; on the contrary, Krawinkler & Zohrei proposed an exponent *m=2*. The Feldmann approach tried to re-write the Coffin-Mason equation in terms of global deformation parameters, proposing a value *m=2* in agreement with Krawinkler & Zohrei. Both Feldmann and Krawinkler adopted the plastic range of the assumed parameters, while Ballio adopted the total cyclic excursion (i.e. elastic and plastic).

Hence, it seems that the value of the exponent *m* depends on the definition of *S*. In order to assess the exponent *m*, the data obtained in tests with cycles of constant amplitude should be plotted in terms of number of cycles to failure *(N)* and stress (strain) range *(S)* in a *Log-Log* diagram and fitted by a straight line. The slope *m* of the best fitting line can be considered the exponent to be adopted in Eqn. (5.1).

VARIABLE AMPLITUDE LOADING

Up to this point the discussion of the models for the re-elaboration of experimental data has dealt with constant amplitude loading. However, earthquake loading and most of service loading histories have variable amplitude and can be quite complex as in the case of earthquake events. Several methods have been developed to deal with variable amplitude loading using the baseline data generated from constant amplitude tests. However, the Miner's rule is mc own.

Miner's rule

The linear damage rule was firstly proposed by gren in 1942 and was further developed by Miner in 1945. Today this method is commonly known iner's rule.

The basis for the linear cumulative damage ı consists of converting random cycles into an equivalent number of constant amplitude cycles. :e, damage fractions due to each individual cycle are summed until failure occurs.

In this method, the damage fraction D_i for the i le is defined as the life used up to the ith event. Failure is assumed to occur when these damage f...ıns sum up to or exceed 1,

$$D_T = \sum_{i=1}^{n_{fail}} D_i = \sum_{i=1}^{n_{fail}} \left(\frac{1}{N_{fi}} \right) \geq 1 \tag{5.23}$$

where $D_i = (1/N_{fi})$ is the damage fraction for the ith cycle and (N_{fi}) is the fatigue life at the stress σ_i.

Often, in a sequence of variable amplitude loading, there is a repeated number of cycles of the same amplitude. Under these circumstances, it is convenient to sum the cycles at the same amplitude. Consider a situation of variable amplitude loading as illustrated in Figure 5.3. A certain stress σ_1 is applied for a number of cycles N_1, where the number of cycles to failure is N_{f1}. The fraction of the life used is then N_1/N_{f1}. Now let consider another stress amplitude σ_2 in which N_2 cycles were imposed and in which correspond N_{f2} to attend the failure. The Miner's rule states that fatigue failure is expected when such life fractions sum to unit, that is, when 100% of the life is exhausted:

$$D_T = \frac{N_1}{N_{f1}} + \frac{N_2}{N_{f2}} + \frac{N_3}{N_{f3}} + \ldots = \sum_{i=1}^{n_{int}} \left(\frac{N_i}{N_{fi}} \right) \geq 1 \tag{5.24}$$

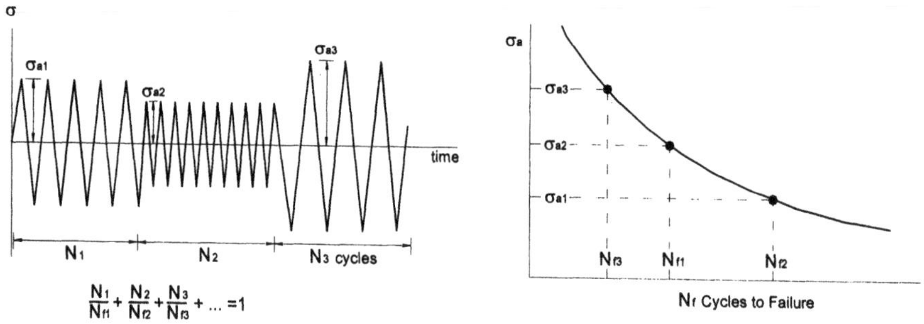

Figure 5.3: Use of the Miner's rule for the life prediction for variable amplitude loading. (adapted from Norman E. Dowling, 1993)

Considerable test data have been generated in an attempt to verify Miner's rule. In most cases these tests use a two-step story. This involves testing at an initial stress level σ_1 for a certain number of cycles. The amplitude is then changed to a second level, σ_2, until failure occurs. If $\sigma_1 > \sigma_2$ it is called a high-low test, and if $\sigma_1 < \sigma_2$, a low-high test. The results of Miner's original tests showed that the cycle ratio corresponding to failure ranged from 0.61 to 1.45 (Sines & Waisman, 1959). Other researchers have shown variations as large as 0.18 to 23.0. Most results tend to fall between 0.5 and 2.0. In most cases the average value is closed to Miner's proposed value of 1. There is a general trend that for high-low tests the values are less than 1, and for low-high tests the values are greater than 1. In other words, Miner's rule is nonconservative for high-low tests. One problem with two-level step tests is that they do not relate to many service load histories. Most load histories do not follow any step arrangement and instead are made up of random distribution of loads of various amplitudes. Tests using random histories with several stress levels show very good correlation with Miner's rule.

The difficulty in applying this method to non completely reversal amplitude loading or random loading arise from the fact that the loading histories in seismic events or loading such as wind or traffic often do not have well-defined cycles. To overcome the irregularities of real load histories, several cyclic counting methods have been developed. One of the most widely used methods is the well-known "rainflow counting method".

Rainflow counting

The original rainflow method of cycle counting derived its name from an analogy used by Matsuishi and Endo (1968) in their early work on this subject. Since that time "rainflow counting" has become a generic term that describes any cycle counting method, which attempts to identify closed hysteris loops in the stress-strain response of material subjected to cyclic loading.

The first step in implementing this procedure is to draw the strain-time history so that the time axis is oriented vertically, with increasing time downward. One could now image that strain history forms a number of "pagoda roofs". Cycles are then defined by the manner in which rain is allowed to "drip" of "fall" down the roofs. A number of rules are imposed on the dripping rain so as to identify closed hysteresis loops. The rules specifying the manner in which rain falls are as follows:

1) To eliminate the counting of half cycles, the strain-time history is drawn so as to begin and end at the strain value of greatest magnitude.
2) A flow of rain is begun at each strain reversal in the history and is allowed to continue to flow unless:
   (a) The rain began at a local maximum point (peak) and falls opposite a local maximum point greater than that from which it came.
   (b) The rain began at a local minimum point (valley) and falls opposite a local minimum point greater (in magnitude) than that from which it came.
   (c) It encounters a previous rainflow.

The foregoing procedure can be clarified through the use of an example. Figure 5.4 shows a stress history and the resulting flow of rain.

A	-45
B	25
C	-14
D	49
E	12
F	37
G	-28
H	-10
I	-45

counts	range	mean
1	39	5.5
1	18	-19
1	25	24.5
1	94	2

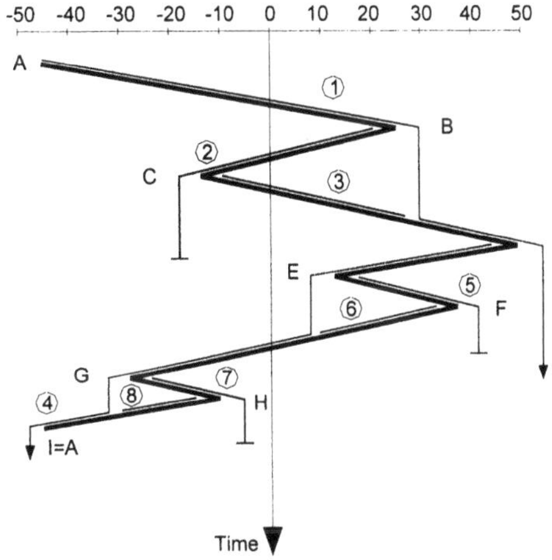

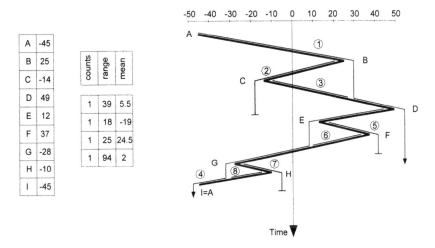

Figure 5.4: Example of rainflow cycle counting.
(adapted from Bannantine et al, 1990)

The following discussion describes in detail the manner in which each rainflow path was determined.

- Rain flows from point A over points B and D and continues to the end of the history since none of the conditions for stopping rainflow are satisfied.
- Rain flows from point B over C and stops opposite point D, since both B and D are local maximums and the magnitude of D is greater than B (rule 2a).
- Rain flows from point C and must stop upon meeting the rain flow from point A (rule 2c).
- Rain flows from point D over points E and G and continues to the end of the history since none of

the conditions for stopping rainflow are satisfied.

- Rain flows from point E over point F and stops opposite point G, since both E and G are local minimum and the magnitude of G is greater than E (rule 2b)
- Rain flows from point F and must stop upon meeting the flow from point D (rule 2c).
- Rain flows from point G over point H and stops opposite point A, since both G and A are local minimum's and the magnitude of A is greater than G (rule 2b)
- Rain flows from point H and must stop upon meeting the rainflow from point D (rule 2c).

Having completed the above, we are now able to combine events to form completed cycles. In Figure 5.4 is also shown a table with results of this example.

Equivalent strain range (S_{eq})

In case of variable amplitude loading, it appears convenient to make reference to an equivalent strain range, S_{eq}, in order to use Eqn (5.1). Using the Miner's rule the equivalent strain range, S_{eq}, can be defined as follows:

$$S_{eq} = \left(\frac{\sum_{i=1}^{L} N_i \cdot S_i^m}{N_{TOT}} \right)^{\frac{1}{m}} \tag{5.25}$$

where N_{TOT} is the total number of cycles in the loading history, N_i is the number of cycles at the same range S_i, and L is the number of different ranges S_i identified in the loading history, for instance with the rainflow counting. The parameter S_i can have one of the definitions previously described and the parameter m should be in accordance with the definition used for S_i as previously mentioned.

DEFINITION OF THE DESIGN S-N LINES

According to the limit state design method, the parameter governing the design should be defined on the basis of a statistical analysis, allowing to make reference to a value associated with a given probability of failure P_f (or of survival, $1- P_f$). Such a probability should be defined with reference to suitable levels of safety and reliability of the structure and its components. Hence, the *S-N* lines should be defined with reference to a given probability of failure P_f. In this re-elaboration, the procedure proposed in the JWG XIII-XV - Fatigue Recommendations (1994) of the International Welding Institute was used and is briefly summarised herein.

Scope of the procedure is the definition of the value $Log(K_k)$ such that the line with a slope $(-1/m)$, and intersecting the x-axis at $Log(K)$, is associated with a probability of 5% that test data plot below it. In the following, reference is made to the set of values $Log(K)$ representing the intersections with the N axis of the *S-N* lines, each one having a constant slope $(-1/m)$, and passing for every single test data point.

The basic formula to determine the design value of the random variable $X=Log(K)$ can be assumed as:

$$X_d = \mu - \phi^{-1}(\alpha) \cdot \sigma \tag{5.26}$$

where X_d is the design value, μ is the mean value of X, σ is the standard deviation of X, ϕ is the normal law of probability and α is the probability of survival.

It is possible to show that μ can be considered distributed according to the t-Student model and the standard deviation σ^2 follows the chi-square law (χ^2). As σ and μ are unknown values, their estimated values from tests, μ e σ respectively, should be associated with β, confidence level. By assuming

$$\mu = \frac{\sum_{i=1}^{n} x_i}{n} \tag{5.27}$$

as the estimated mean value of X_i, and

$$\bar{\sigma} = \sqrt{\frac{n\sum (x_i)^2 - (\sum x_i)^2}{n(n-1)}} \tag{5.28}$$

as the estimated standard deviation, the design value of X can be expressed as:

$$X_d = \bar{\mu} - \bar{\sigma} \cdot \left[\frac{t(\beta, n-1)}{\sqrt{n}} + \phi^{-1}(\alpha) \cdot \sqrt{\frac{n-1}{\chi^2(\frac{1-\beta}{2}, n-1)}} \right] \tag{5.29}$$

According to Eurocode 3 (1994) the confidence range should be *75%* of *95%* probability of survival on the *Log (N)* axis. Hence, α e β should be assumed respectively equal to *0.95* and *0.75*.

Naming γ as:

$$\gamma = \left[\frac{t(\beta, n-1)}{\sqrt{n}} + \phi^{-1}(\alpha) \cdot \sqrt{\frac{n-1}{\chi^2(\frac{1-\beta}{2}, n-1)}} \right] \tag{5.30}$$

and in the case of limited amount of data the value of γ can be obtained directly from Table 5.1 on the basis of the number data *(n)*.

In the same table, assuming a number of degrees of freedom equal to *n-1*, the value of function t for $\alpha=0.75$ and function χ for $\beta=0.125$ are reported together with the value of function ϕ in correspondence of $P_f = 0.95$. Eqn. (5.30) can hence be re-written as:

$$X_d = \mu - \sigma \cdot \gamma \tag{5.31}$$

The test data were analysed following the proposed procedure in order to evaluate K_d (i.e. the design value of the constant K) which defines the design low-cycle fatigue of the considered component.

TABLE 5.1
COEFFICIENTS FOR STATISTICAL EVALUATION

n	$t_{(0.75,n-1)}$	$\chi^2_{(0.125,n-1)}$	$\phi^{-1}_{(0.95)}$	γ
2	2.4142	0.0247	1.6449	12.1632
3	1.6036	0.2471	1.6449	5.4271
4	1.4226	0.6924	1.6449	4.1353
5	1.3444	1.2188	1.6449	3.5811
6	1.3009	1.8082	1.6449	3.2663
7	1.2733	2.4411	1.6449	3.0600
8	1.2543	3.1063	1.6449	2.9127
9	1.2403	3.7965	1.6449	2.8011
10	1.2297	4.5070	1.6449	2.7132
11	1.2213	5.2341	1.6449	2.6418
12	1.2145	5.9754	1.6449	2.5823
13	1.2089	6.7288	1.6449	2.5319
14	1.2041	7.4929	1.6449	2.4884
15	1.2001	8.2661	1.6449	2.4505

RE-ELABORATION OF EXPERIMENTAL DATA

For the assessment of the validity and accuracy, the methodology proposed by Ballio & Castiglioni (1995) for the definition of *(S)* was applied to all the tests performed in the Laboratory of Structures and Strength of Material of Instituto Superior Técnico, Lisbon.

In this re-elaboration the fatigue endurance *(N)* was assessed according the Calado, Azevedo & Castiglioni ($N_{\alpha=0.5}$) proposal.

WW – fully welded connections

For each test the relevant parameters of each specimen (Δv, v_y, f_y) are presented in Table 5.2 as well as the total number of imposed cycles *(N)* and the fatigue endurance, i.e. the number of cycles of the conventional failure N_f. The value of S_{eq} and the slop *m* of the *S-N* lines obtained with the best fitting are presented in the Table 5.3.

From Table 5.3 and Figure 5.5 it can be observed that the slop of the *S-N* line when assessed with the Ballio & Castiglioni method vary from 1.28 and 3.37. The discrepancy of the slop values can be imputed to the small number of tests for each typology and the vicinity of the results. According to Eurocode 3 (1994) to assess a fatigue line the number of data points to be considered in the analysis must be not lower that 10. On the other hand the vicinity of the results is not the most suitable for assess of the slop of the line. In effect, the numbers of cycles to failure vary from 2 to 18.

If possible, it would be preferable to have data points with very different number of cycles to failure; for instance results with N_f around ten, others around hundred and others around thousands. With this set of values, which can still be considered as low-cycle fatigue, the slop of the fatigue line will be assessed with a higher accuracy. Re-elaboration of other experimental tests performed by Ballio et al

(1997) and Calado et al (1999) on beams, beam-to-column connections and cruciform welded joint has demonstrated that the slop of the *S-N* line is approximately equal 3.

TABLE 5.2
RELEVANTE PARAMETERS OF EXPERIMENTAL TESTS
WW – FULLY WELDED CONNECTIONS
(BCC5, BCC6 AND BCC8)

Type of column	Load case	Δv (mm)	v_y (mm)	f_y (Mpa)	N	N_f
WW	**A**	100.0	8.78	274.8	20	16
HEB160	**B**	150.0	8.54	274.8	9	5
BCC5	**C**	ECCS	8.73	274.8	24	15
	D	75.0	8.68	274.8	27	22
WW	**A**	100.0	8.33	278.6	19	14
HEB200	**B**	150.0	8.26	278.6	15	8
BCC6	**C**	ECCS	8.22	278.6	21	12
	D	75.0	8.41	278.6	22	18
WW	**A**	100.0	6.87	292.0	16	12
HEB240	**B**	75.0	6.94	292.0	20	16
BCC8	**C**	150.0	6.85	292.0	6	2
	D	ECCS	6.74	292.0	21	15

TABLE 5.3
RELEVANTE PARAMETERS FOR THE ASSESSMENT OF THE S-N LINES
WW – FULLY WELDED CONNECTIONS
(BCC5, BCC6 AND BCC8)

Type of column	Load case	Nf	S_{eq} (Mpa)	m tests	Log K tests
WW	**A**	16	3104		
HEB160	**B**	5	4564	2.36	9.37
BCC5	**C**	15	2741		
	D	22	2369		
WW	**A**	14	3306		
HEB200	**B**	8	4903	1.28	5.58
BCC6	**C**	12	2529		
	D	18	2477		
WW	**A**	12	4185		
HEB240	**B**	16	3132	3.37	13.18
BCC8	**C**	2	5347		
	D	15	3765		

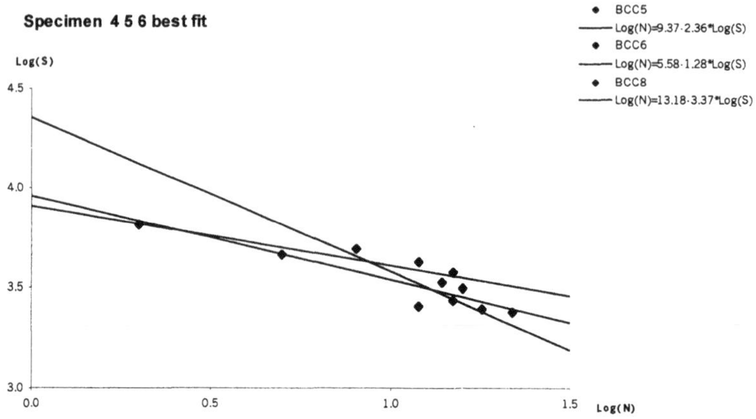

Figure 5.5: Evaluation of the *S-N* line for WW connections.

The aim of this re-elaboration is to obtain fatigue lines with an acceptable level of probability that the performance of the connection is satisfactory during its entire design life. According to the Eurocode 3 (1994) when test data are used to determine the *S-N* lines a 75% confidence interval of 95% probability of survival for Log N must be used, taking into account the standard deviation and the sample size. The JWG XIII-XV - Fatigue Recommendations (1994) of the International Welding Institute was used in this re-elaboration and the following values for the design of K_d were obtained and are presented Table 5.4.

TABLE 5.4
RELEVANTE PARAMETERS FOR THE ASSESSMENT OF THE DESIGN S-N LINES
WW – FULLY WELDED CONNECTIONS
(BCC5, BCC6 AND BCC8)

Type of column	Load case	Nf	S_{eq} (Mpa)	m	Log K (m=3)	Log K_d statistic
WW	**A**	16	3104	3	9.27	
HEB160	**B**	5	4564	3	10.28	6.15
BCC5	**C**	15	2741	3	9.14	
	D	22	2369	3	8.78	
WW	**A**	14	3306	3	9.41	
HEB200	**B**	8	4903	3	10.17	6.68
BCC6	**C**	12	2529	3	9.13	
	D	18	2477	3	8.93	
WW	**A**	12	4185	3	9.79	
HEB240	**B**	16	3132	3	9.28	5.84
BCC8	**C**	2	5347	3	11.13	
	D	15	3765	3	9.55	

In Figure 5.6 the design S-N lines for fully welded connections tested are presented.

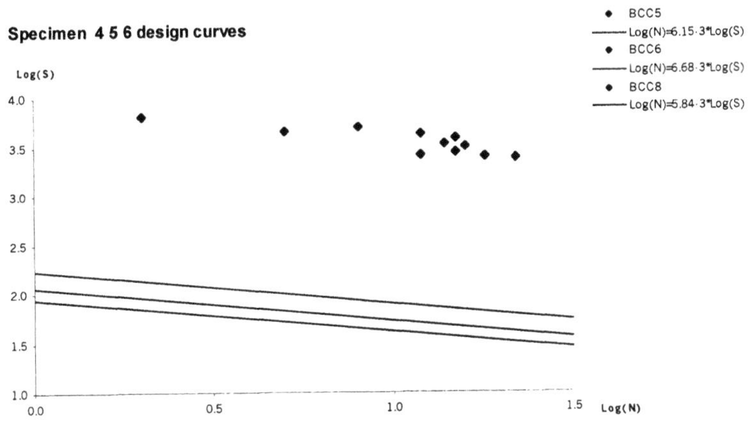

Figure 5.6: Design *S-N* lines for WW connections.

It can be observed that the *S-N* lines for these three types of connections are not so different. Some of this difference can be attributed to the column size and to the failure modes observed during the experimental tests.

TSW – top and seat with web angle connections

For each test of top and seat with web angle connections the relevant parameters of each specimen are presented in Tables 5.5 and 5.6, while in Figure 5.7 is presented the best-fitting of the test data.

TABLE 5.5
RELEVANTE PARAMETERS OF EXPERIMENTAL TESTS
TSW - TOP AND SEAT WITH WEB ANGLE CONNECTIONS
(BCC7, BCC9 AND BCC10)

Type of column	Load case	Δv (mm)	v_y (mm)	f_y (Mpa)	N	N_f
TSW	**A**	100.0	5.07	252.2	12	5
HEB160	**B**	150.0	5.22	252.2	7	3
(BCC9)	**C**	75.0	5.26	252.2	20	6
	D	ECCS	5.31	252.2	18	12
TSW	**A**	100.0	4.12	252.2	12	6
HEB200	**B**	150.0	4.10	252.2	7	2
(BCC7)	**C**	ECCS	4.06	252.2	20	13
	D	75.0	3.95	252.2	17	6
TSW	**A**	100.0	3.80	252.2	12	5
HEB240	**B**	75.0	3.72	252.2	21	4
(BCC10)	**C**	ECCS	3.65	252.2	18	12
	D	150.0	3.72	252.2	6	2

TABLE 5.6
RELEVANTE PARAMETERS FOR THE ASSESSMENT OF THE S-N LINES
TSW - TOP AND SEAT WITH WEB ANGLE CONNECTIONS
(BCC7, BCC9 AND BCC10)

Type of column	Load case	Nf	S_{eq} (Mpa)	m Tests	Log K tests
TSW	**A**	5	4632		
HEB160	**B**	3	6523	2.26	9.01
BCC9	**C**	6	3464		
	D	12	3533		
TSW	**A**	6	5867		
HEB200	**B**	2	7717	3.90	15.37
BCC7	**C**	13	4884		
	D	6	4568		
TSW	**A**	5	6192		
HEB240	**B**	4	4780	4.04	15.93
BCC10	**C**	12	5147		
	D	2	8527		

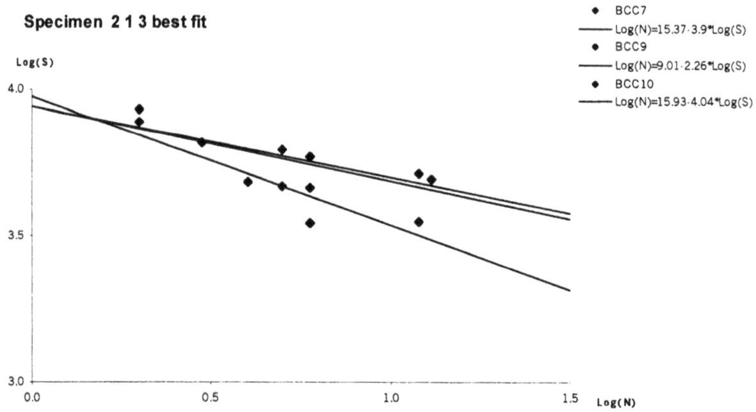

Figure 5.7: Evaluation of the *S-N* line for all TSW connections.

Similar conclusions as presented for fully welded connections can also be applied to top and seat with web angle connections. In Table 5.7 the relevant parameters for the assessment of the design *S-N* lines are presented while in Figure 5.8 the fatigue curve is shown.

For the fully welded connections tested, the influence of the column size seems to be much more important than in the case of top and seat with web angle connections. While for TSW the failure mode was always the same, failure of the angles, the plastic hinge in the beam was influenced by the size of the column, for the case of fully welded connections. This can be seen in the design fatigue lines presented in Figures 5.6 and 5.8. While for TSW the lines are near, showing a small influence of the column size, for fully welded connections, on the contrary, larger difference between the S-N lines can be observed in these case, evidencing a higher influence of the column size, as observed during the experimental tests.

TABLE 5.7
RELEVANTE PARAMETERS FOR THE ASSESSMENT OF THE DESIGN S-N LINES
TSW - TOP AND SEAT WITH WEB ANGLE CONNECTIONS
(BCC7, BCC9 AND BCC10)

Type of column	Load case	NF	$S_{eq.}$ (Map)	m	Log K (m=3)	Log K_d statistic
TSW	**A**	5	4632	3	10.30	
HEB160	**B**	3	6523	3	10.97	7.10
BCC9	**C**	6	3464	3	9.84	
	D	12	3533	3	9.57	
TSW	**A**	6	5867	3	10.53	
HEB200	**B**	2	7717	3	11.36	7.43
BCC7	**C**	13	4884	3	9.95	
	D	6	4568	3	10.20	
TSW	**A**	5	6192	3	10.68	
HEB240	**B**	4	4780	3	10.44	7.62
BCC10	**C**	12	5147	3	10.06	
	D	2	8527	3	11.49	

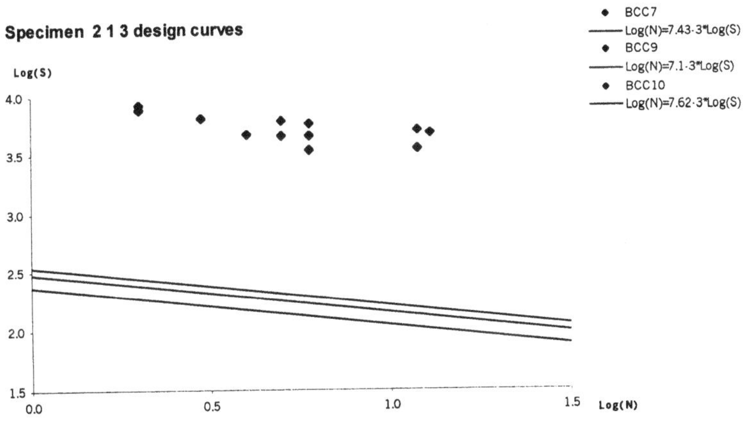

Figure 5.8: Design S-N lines for TSW connections.

COMPARISON OF ALL TESTS RESULTS WITH EC3

The experimental tests performed in most of the laboratories involved in RECOS project were not performed with the scope to assess fatigue resistance lines. For that reason it is quite difficult to re-elaborate the experimental data, because in same cases the number of data for a certain typology doesn't allow to assess a best fitting of the results, and in other cases, there are not cyclic loading histories with constant amplitude.

However, to make some comparison between the experimental data available in the RECOS project and the fatigue strength curves proposed in Eurocode 3 (1994), all the data was re-elaborated and included into those curves, which were cutted in order to include only the low-cycle fatigue range, Figures 5.9 – 5.14.

According to the Eurocode 3 (1994) the fatigue strength is defined for normal stresses by a series of $Log\ \Delta\sigma_r - log\ N$ curves, each applying to a typical detail category. Each detail category is designated by a number which represents, in N/mm^2, the reference value of the fatigue strength at 2 million cycles.

The values used are rounded values, corresponding to the detail categories given in Eurocode 3 (1994). In the zone which cycles are less than $5x10^7$ the slope of the line is 3, agreeing the Ballio & Castiglioni (1995) proposal with the value considered in the Eurocode 3 (1994).

RECOS - Lisbon

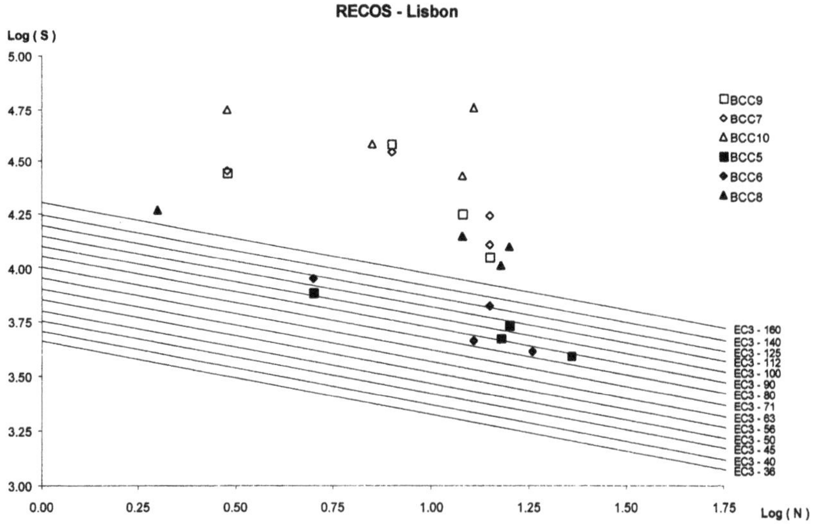

Figure 5.9: Fatigue strength curves for normal stress ranges.

RECOS – Sofia

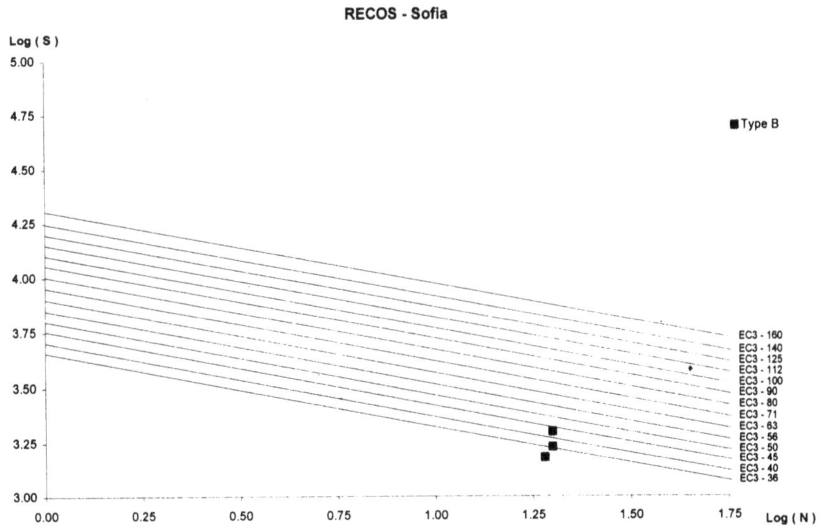

Figure 5.10: Fatigue strength curves for normal stress ranges.

RECOS – Liege

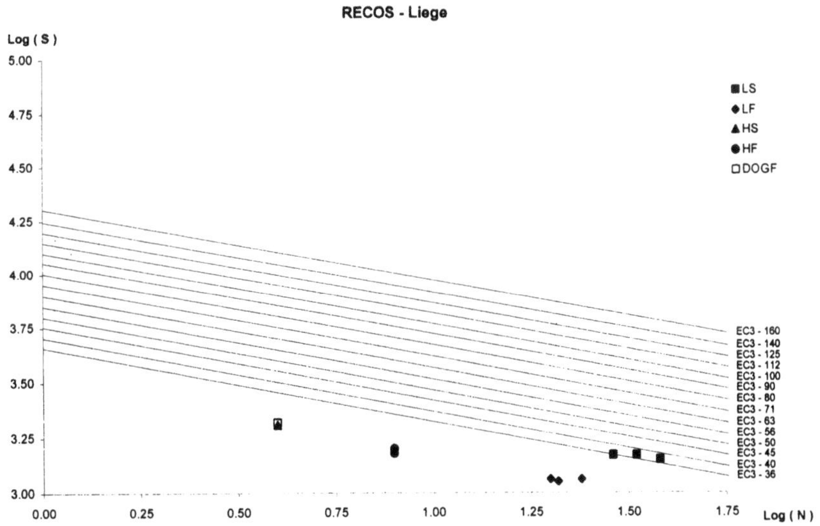

Figure 5.11: Fatigue strength curves for normal stress ranges.

RECOS - Ljubljana

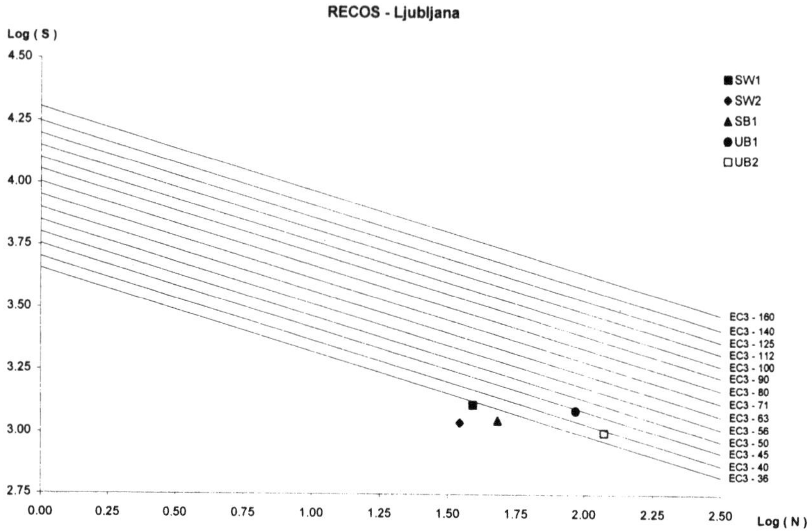

Figure 5.12: Fatigue strength curves for normal stress ranges.

RECOS – Timisoara

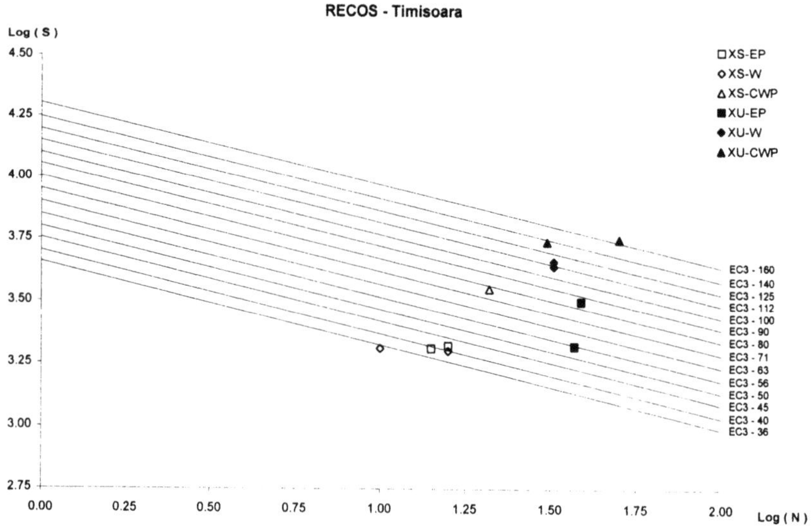

Figure 5.13: Fatigue strength curves for normal stress ranges.

RECOS – Rennes

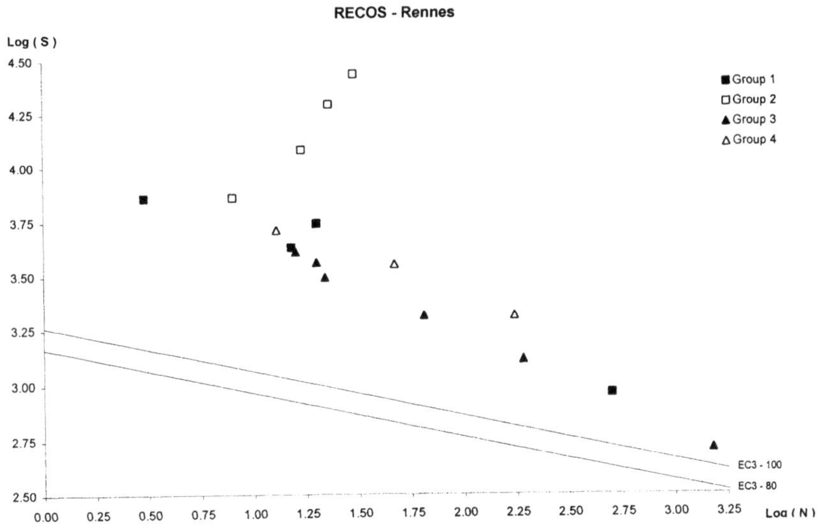

Figure 5.14: Fatigue strength curves for shear stress ranges.

In the cases of the experimental data available in Rennes, the texts were performed on bean-to-column composite connections under shear loading. For that reason the fatigue strength curves for shear stress ranges proposed in Eurocode 3 (1994) were used.

Figures 5.9 to 5.14 show a large scattering of the experimental data. The results available, are insufficient to propose design S-N lines for the typologies of connections experimentally tested.

CONCLUSIONS

In this chapter a method for the re-elaboration of test data of connections under cyclic loading based on the S-N line approach was presented. Possible definitions of the fatigue endurance *(N)* and procedures for the assessment of the stress (strain) range *(S)* were discussed, including its influence on the slope *m*.

The methodology proposed by Ballio & Castiglioni (1995) was applied to experimental data on beam-to-column connections experimentally tested on the Laboratory of Structures and Strength of Material of Instituto Superior Técnico, Lisbon, in order to analyse its applicability and accuracy.

The re-elaboration of the experimental data allows to conclude that the assessment of the slope *m* of the *S-N* lines is very dependent on the size of the simple and the vicinity of the results in terms of Log N. According to the Eurocode 3 (1994) to assess a fatigue line the number of data points to be considered in the analysis must be not lower that 10. If possible, it will be preferable to have data points with very different number of cycles to failure; for instance results with N_f around ten, others around hundred and others around thousands. With this set of values, which can still be considered as low-cycle fatigue, the slop of the fatigue line will be assessed with a higher accuracy and re-elaboration of other experimental tests performed by Ballio et al (1997) on beams and cruciform welded joint has demonstrated that the slop of the *S-N* line is approximately equal 3.

REFERENCES

Agatino, M. R., Bernuzzi, C., Castiglioni, C. A., & Calado, L., (1997), Ductility and strength of structural steel joints under low-cycle fatigue, *Proc. of the XVI Congresso C.T.A.*, Ancona, Italy, pp 9-19.

ANSI/ASTM E466-76, (1977), Standard Recommended Practice for Constant Axial Fatigue Tests on Metallic Materials, *Annual Book of ASTM Standards*, Part 10, American Society for Testing and Materials (ASTM), Philadelphia, pp. 536-540.

Ballio, G. & Castiglioni C.A., (1995), A Unified Approach for the Design of Steel Structures under Low and High Cycle Fatigue, *Journal of Constructional Steel Research*, vol. 34, pp. 75-101.

Ballio, G., Calado, L. & Castiglioni, C. A., (1997), Low Cycle Fatigue Behaviour of Structural Steel Members and Connections, *Fatigue & Fracture of Engineering Materials & Structures*, vol. 20, n° 8, pp. 1129-1146.

Bannantine, J., Comer, J. & Handrock, J., (1990), Fundamentals of Metal Fatigue Analysis, *Prentice-Hall.*

Barbaglia P., (1998), *Low-cycle fatigue behaviour of welded steel beam-to-column connections,* Laurea Thesis, Structural Engineering Dept., Politecnico di Milano, Italy.

Bruneau, M., Uang, C.M. & Whittaker, A., (1998), *Ductile Design of Steel Structures*, McGraw-Hill.

Calado L. & Azevedo J., (1989), A Model for Predicting Failure of Structural Steel Elements, *Journal of Constructional Steel Research*, vol. 14, pp 41-64.

Calado, L. & Castiglioni, C.A., (1996), Steel Beam-to-Column Connections Under Low-Cycle Fatigue Experimental and Numerical Research, *Proc. of XI World Conference on Earthquake Engineering*, Acapulco, Mexico.

Calado, L., Castiglioni, C. A. & Bernuzzi, C., (1997), Cyclic Behaviour of Structural Steel Elements: Method for Re-elaboration of Test Data, I Encontro Nacional de Construção Metálica e Mista, Porto, Portugal, pp 633-659.

Calado, L., Castiglioni, C.A., Barbaglia, P. & Bernuzzi, C., (1998), Procedure for the Assessment of Low-Cycle Fatigue Resistance for Steel Connections, *Proc. of the International Conference on Control of the Semi-Rigid Behaviour of Civil Engineering Structural Connections*, Liege, Belgium, pp 435-444.

Calado, L. Castiglioni, C. A., Mele, E. & Ferreira, C., (1999), Damage Accumulation Design of Steel Beam-to-Column Connections, II Encontro Nacional de Construção Metálica e Mista, Coimbra, Portugal, pp551-563.

Castiglioni, C.A. & Calado, L.,(1996), Seismic Damage Assessment and Failure Criteria for Steel Members and Connections, *International Conference on Advances in Steel Structures*, Hong Kong, pp 1021-1026.

Coffin, L.F., (1954), A Study on the Effect of Cyclic Thermal Stresses on a Ductile Metals, Transaction of the *American Society of Mechanical Engineers*, ASME, vol. 76, pp. 931-950.

Commission of the European Communities, (1994), Eurocode 3 - Design of Steel Structures - Part 1: General Rules and Rules for Buildings.

Dowling, N., (1993), Mechanical Behaviour of Materials – Engineering Methods for Deformation, Fracture and Fatigue, *Prentice-Hall International Editions*.

Feldman, M., (1994), Zur Rotationskapazitat vin I-profilen Statisch und Dynamisch Belasteter Trager, (in German), PhD Thesis, *Institute of Steel Construction, RWTH* Aachen, Germany.

International Welding Institute, IIW, JWG XIII-XV, 1994, Fatigue Recommendations, Doc. XIII-1539-94/XV-845-94, September.

Krawinkler, H. & Zohrei, M., (1983), Cumulative Damage Model in Steel Structures Subjected to Earthquake Ground Motion, *Computers & Structures*, vol. 16, N° 1-4, pp. 531-541.

Kuck, V. J., (1994), The Application of the Dynamic Plastic Hinge Method for the Assessment of Limit States of Steel Structures Subjected to Seismic Loading, (in German), Ph.D Thesis, *Institute of Steel Construction, RWTH* Aachen, Germany.

Landgraf, R.W., Morrow, J.D. & Endo, T., (1969), Determination of Cyclic Stress-Strain Curve, Journal of Materials, vol. 4, N° 1, *American Society for Testing and Materials (ASTM)*, Philadelphia, pp. 176-188.

Manson, S.S., (1954), Behaviour of Materials under Conditions of Thermal Stress, *National Advisory Commision on Aeronautics*: Report 1170, NACA, Cleveland, Lewis Flight Propolsion Laboratory.

Matsuishi, M. & Endo, T., (1968), Fatigue of Metals Subjected to Varying Stress, *Proceedings of the Japan Society of Mechanical Engineering*, Fukuoka, Japan, March.

Miner, M.A., (1945), Cumulative Damage in Fatigue, Trans. ASME, *Journal of Applied Mechanics*, vol. 67, pp. A159-A164.

Palmgren, A., (1924), Durability of Ball Bearing, *ZVDI*, vol. 68, n° 14, pp. 339-341 (in German)

Peeker, E., (1997), Extended Numerical Modelling of Fatigue Behaviour, Ph.D Thesis presented at the Départment de Génie Civil, *Ecole Polytecnique Federale de Lausanne*, Suisse.

Sedlacek, G., Feldmann, M. & Weynand K., (1995), Safety Consideration of Annex J of Eurocode 3, *Proc. of 3rd International Workshop on Connections in Steel Structures*, Trento, Italy, pp. 453-462.

Sines, G. & Waisman, J. L., (1959), *Metal Fatigue*, McGraw-Hill, New York.

Chapter 6

Evaluation of Global Seismic Performance

6.1 Ductility demand for semi-rigid joint frames

6.2 Interaction between local and global properties

6.1

DUCTILITY DEMAND FOR SEMI-RIGID JOINT FRAMES

Dan Dubina, Adrian Ciutina, Aurel Stratan and Florea Dinu

INTRODUCTION

The use of semi-rigid joints in frames subjected to seismic loads is a matter of controversy. None of the existing design codes include provisions for their use and, in zones characterised by high seismicity, the use of rigid full-strength joints is mandatory.

However, Astaneh, 1992 and Mazzolani & Piluso, 1996 provided some attempts of codification of semirigidity for seismic steel structures, and the evidence of the favourable effect of semi-rigid joints in such structures is continuously increasing (Astaneh et al., 1992).
According to the new generation of European seismic codes, e.g. the ECCS Manual, 1994 and Eurocode 8, a distinction is made between dissipative and non-dissipative structures. Non-dissipative structures must resist to the most severe seismic motion within the elastic range, whereas the dissipative structures are designed by allowing yielding to occur in predefined zones. During an earthquake, these zones must dissipate energy by means of hysteretic ductile behaviour in the plastic range. The formation of appropriate dissipative mechanism is related to the structural topology and to the correct sizing of its members. The ECCS Manual proposes a method to design the multi-storey steel frames so as to behave as "global dissipative structures", e.g. the collapse occurs by a global plastic mechanism. Connections in the dissipative zones must have enough over-strength as to allow for yielding of the connected parts. Both for welded and bolted joints, the resistant moment of the beam-to-column connection has to be 1.2 times the resistance of the connected members (e.g. the beams). However, there is evidence that partial strength semi-rigid joints possess sufficient plastic rotation and dissipate a large amount of energy when subjected to cyclic and dynamic loads. On the other hand, the analysis of the behaviour of multi-storey steel structures, including the performance of their connections, during the earthquakes of Northridge (California, January 17, 1994) and Hyogoken-Nanbu (Kobe, January 17, 1995) clearly shows that very stiff and full strength beam-to-column joints are not always a "panacea" for such structures.

Even more experimental and numerical studies are needed to encourage the use of partial strength semi-rigid joints within the multi-storey steel frames for buildings in seismic zones. It seems, nevertheless, that they can be used as dissipative parts of structures and to control the global performance of a structure.

In order to resist a seismic action, a structure should possess sufficient global ductility. This means that the dissipative zones within the structure should possess high levels of local ductility in order to develop adequate plastic hinges. The behavioural q-factor is a measure of the global ductility, characterising the capacities of a structure to withstand in the post-elastic range. Usually, the ductility levels for a member (beam, column), or a joint, are expressed in terms of the ultimate plastic rotation or in terms of rotation capacity.

Due to their higher flexibility, semi-rigid steel frames are prone to increases in inter-storey drifts. The inter-storey drift condition is related to the serviceability limit state which corresponds to minor, frequent earthquakes. The design objective, when building serviceability is checked, is that the building, including both structural and non-structural components, should suffer no damage, and the discomfort of the inhabitants should be minimal. The first requirement, which leads to the avoidance of damage, is satisfied by ensuring that the structure behaviour during the earthquake should remain in the elastic range. In order to fulfil the second requirement - both non-structural components damage and discomfort of inhabitants is avoided- it is necessary to provide sufficient stiffness to prevent significant deformations.

The ultimate limit state of a building in seismic circumstances could be regarded either as damage limit state or collapse limit state. The damage limit state allows some minor damages to non-structural components due to large local deformation in certain zones. The collapse limit state pertains to very infrequent severe earthquake ground motions in which both structural and non-structural damages are expected, but safety of the inhabitants must be guaranteed. Furthermore, the structure must be able to absorb and dissipate large amounts of energy.

According to the above mentioned assumptions, the following seismic performance levels of a building can be introduced (SEAOC Vision 2000, 1995):
Fully operational or serviceable: facility continues in operation with negligible damage;
Operational and Functional: facility continues in operation with minor damage and minor disruption in non-essential services;
Life safety: life safety is substantially protected, damage is moderate to extensive;
Near Collapse or Impending Collapse: life safety is at risk, damage is severe and structural collapse is prevented.

The authors' opinion (but there are others of the same opinion too, see Elnashai et al. 1996 and 1988) is that, based on seismic design performance criteria, semi-rigid steel frames could satisfy the required strength and stiffness conditions for "Operational and Functional" or "Life safety" levels. For the moment, due to the inconsistency of the methods for the evaluation of the q-factor, the design of such frames should be based on the plastic rotation demand of the beam-to-column joints and the inter-storey drift criteria.

The American code AISC 97 has already introduced ductility criteria for design of Moment Resisting Frames. Distinction is made between Special Moment Resisting Frames, Intermediate Moment Resisting Frames, and Ordinary Moment Resisting Frames, with particular requirements for detailing and ductility of joints for each of them. Therefore, the requirement for ultimate plastic rotation for joints is 0.03 rad. for Special Moment Resisting Frames, 0.02 rad for Intermediate Moment Resisting Frames, and 0.01 rad for Ordinary Moment Resisting Frames, the maximum plastic rotation being sustained by experimental evidence.

In the present study, two series of frames (6 storeys – 3 bays and 3 storeys – 3 bays) with different joint configurations are studied in different seismic circumstances in order to investigate the influence of semi-rigidity and strength of the joints on the global performances of the structures.

DUCTILITY CRITERIA

A structure is expected to develop inelastic deformations during severe earthquakes. However, the expected inelastic deformations should be limited within acceptable range to prevent the collapse of the structure. Therefore, proper design of seismic resistant structures requires that the ductility capacity of such structures exceed the ductility demand imposed by the design-imposed loads.

The conclusions drawn from the analysis of Northridge (Mahin S. A., 1998) and Kobe (Nakashima et al, 1998, Watanabe et al., 1998) earthquakes show that in the case of steel building frames, the ductility conditions have to be satisfied as in terms of resistance, stability and serviceability conditions.

The design condition for ductility check is:

$$\mu_r < \mu_a \tag{6.1.1}$$

where: μ_a - the alvailable ductility
μ_r - the required ductility

The problem is to make a correct evaluation of both parameters μ_a and μ_r.

The global ductility of Moment Resisting Frames is defined as the capacity of a structure to undergo deformations in the plastic range without substantial reduction in strength and stiffness. The global ductility depends on the local ductility of the elements, beam-to-column joints, materials and sections. The more ductile its components, the more ductile the structure.
The global ductility of the structure can be expressed in terms of horizontal displacements and q factor.

Global ductility expressed in terms of displacement

The global ductility in terms of displacement can be computed analytically or evaluated experimentally and is related to the maximum top displacement or the maximum inter-storey drift displacement (Figure 6.1.1), and is simply computed by:

$$\mu_\delta = \frac{\delta_u}{\delta_y} \tag{6.1.2}$$

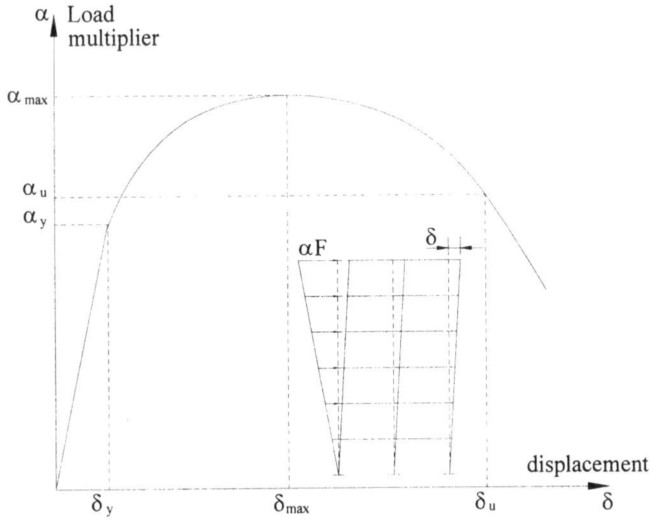

Figure 6.1.1 Idealised α-δ curve for a moment-resisting frame subjected to push-over forces

In the inelastic push-over analysis α_y and δ_y correspond to the appearance of first plastic hinge, while α_{max} and δ_{max} are the corresponding values before the collapse mechanism. For ultimate α_u and δ_u values can be used different definitions based on tests, inspections on earthquake affected structures, theoretical studies etc.

The method can also be applied in the case of dynamic time-history analysis, but a very precise accounting of the top displacement or inter-storey drift displacement should be made in order to find their maximum values during the analysis. In this case, the value of δ_y can be regarded as the displacement (inter-storey drift) for which the first plastic hinge appears.

Behavioural factor q

Figure 6.1.2. shows the definition of the q factor according to Eurocode 8.

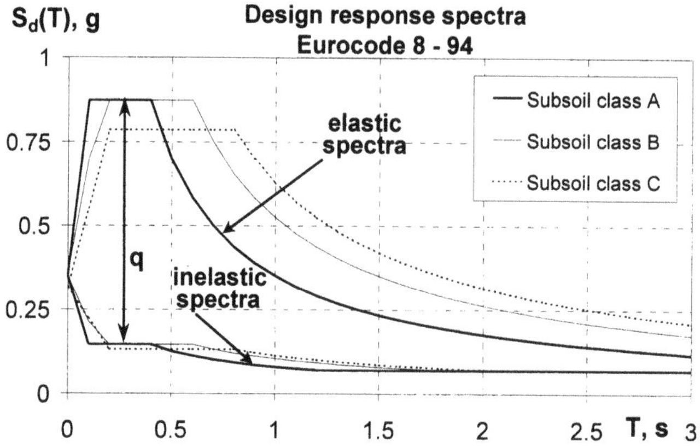

Figure 6.1.2: Definition of the design q factor (q=6 for MRF)

For practical design purposes, q factor can be evaluated by the formula:

$$q = \frac{a_u}{a_y} \tag{6.1.3}$$

where: a_u is the acceleration amplitude corresponding to the failure criterion of the structure;
a_y is the acceleration amplitude corresponding to the appearance of the first plastic hinge

Factors that influence the global ductility of a Moment Resisting Frame

On the following, a synthetic review of the factors that influence the global ductility is presented.

- *Material Ductility*

Material ductility depends on the steel quality, the yielding plateau, the hysteretic behaviour etc. All these properties can be defined by the characteristic stress-strain curve, for monotonic or cyclic loading. The material ductility is expressed in terms of tensile strains:

$$\mu_\varepsilon = \frac{\varepsilon_u}{\varepsilon_y} \tag{6.1.4}$$

where: ε_u is the ultimate steel strain
ε_y is the yielding steel strain

- *Cross-Sectional Ductility*

Cross-sectional ductility or curvature ductility takes into account the geometrical characteristics of the cross section being expressed by means of the relative curvature between two cross-sections spaced by unity:

$$\mu_\varepsilon = \frac{\chi_u}{\chi_y} \tag{6.1.5}$$

where: χ_u is the ultimate curvature
χ_y is the curvature at which begins the plastic deformation

In terms of cross-sectional elements and their slenderness ratios, EUROCODE 3 defines four cross-sectional classes: ductile, compact, semi-compact and slender.

- *Member Ductility*

The moment-rotation curve governs the behaviour of member. It also characterises the behaviour of the beam-to-column joints. The member ductility depends directly on the material ductility and on the cross-sectional ductility.

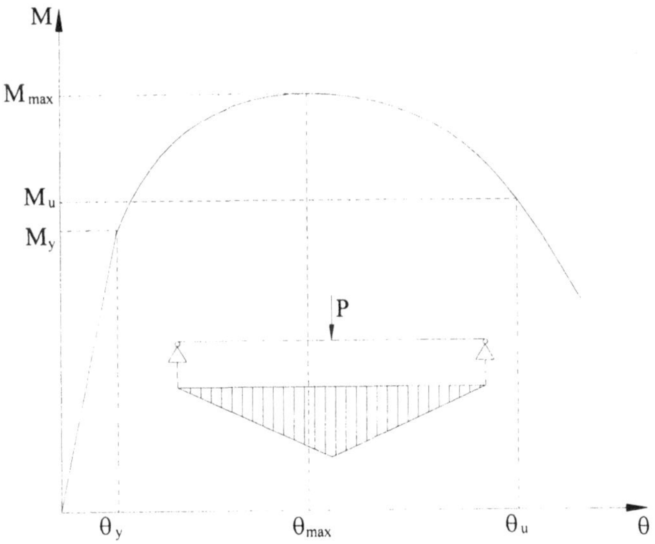

Figure 6.1.3: Idealized moment-rotation curve for an element

The member ductility is expressed in terms of rotational ductility, according to the following formula (see Figure 6.1.3):

$$\mu_\varepsilon = \frac{\theta_u}{\theta_y} \tag{6.1.6}$$

where: θ_u is the ultimate plastic rotation
θ_y is the yielding rotation

In literature there are several suggestions for the definition of the ultimate and yielding rotation, and various methods for the computation of the rotation capacity: theoretical (Kato, Mazzolani-Piluso, Kemp, etc); empirical, based on the statistical interpretation of the obtained tests (Kato-Akiyama, Mitani-Makino, Nakamura); methods based on the interpretation of the plastic failure mechanisms (Gioncu, Ivany, Piluso).

- *Joint Ductility*

The joint moment-rotation characteristic curve and the rotational ductility are similar to the ones defined for members. Anyway, the behaviour of the joints is more complex, depending on all the components that are part of the joint: end plates, bolts, welds, stiffeners, backing plates, etc. In recent years there were made appreciable efforts for the understanding of the behaviour connections and joints. The Annex J of EUROCODE 3 introduces the component method for joint strength and stiffness evaluation.

ECCS Manual on Design of Steel Structures in Seismic Zones (1994) classifies the plastic behaviour of the joints as follows (see Figure 6.1.4):

- type A: full strength joints
- type B: full strength joints with limited rotation capacity
- type C: partial strength joints with limited rotation capacity
- type D: partial strength joints with sufficient rotation capacity
- type E: partial strength joints with premature failure of joint components

In the case of Moment Resisting Frames, the properties of beam-to-column joints significantly influence the structural global characteristics: resistance, stiffness and ductility.

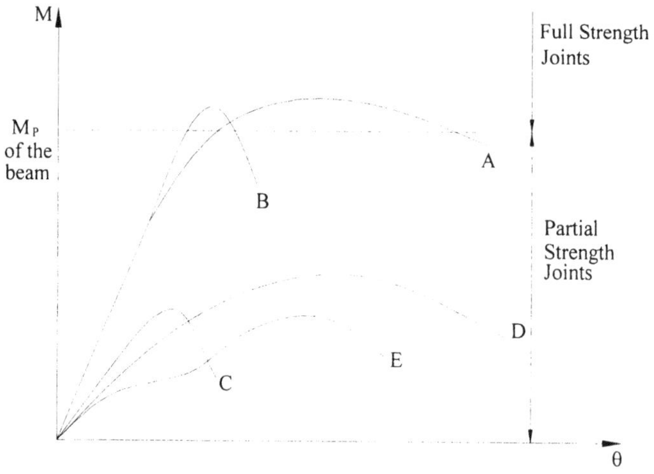

Figure 6.1.4 Moment-Rotation curves of beam-to-column connections

- *Collapse Mechanism and Structural Rigidity*

The collapse mechanism plays a very important role in the seismic analysis of structures, because it influences the value of the available global ductility and the energy dissipation capacity of the structure. The case in which a frame fails by a global type mechanism (see Figure 6.1.13) can be adopted as reference, because, in this case, the structure is considered able to exhibit enough ductility to withstand severe earthquakes. In modern seismic codes the problem of reduction of the available ductility, due to undesirable collapse mechanisms, is faced by providing design criteria which have the goal to guarantee the complete exploitation of the plastic reserves of the structures. It is, therefore, requested to avoid the local failure modes.
In order to obtain global collapse mechanisms, at each beam-to-column joint, the following condition has to be satisfied:

$$\sum M_{R,C} > \sum M_{R,B} \tag{6.1.7}$$

where: $\sum M_{R,C}$ - sum of the resisting moments of the columns connected to the joint.
$\sum M_{R,B}$ - sum of the resisting moments of the beams connected to the joint.

For the situations of this type, the design is based on the Strong Column Weak Beam (SCWB) approach, by which the plastic hinges should form in columns rather than in beams.

- *Soil conditions*

The seismic response of a building depends on structural characteristics, such as local and global rigidity, its strength, the horizontal and vertical regularity, etc. Besides these structural characteristics, a very important factor is the interaction soil-structure, fact that can affect the overall behaviour of the structure. There can be distinguished three different eigenperiods (structure-soil - earthquake) that may influence the structural behaviour:
- Fundamental period of vibration of the structure. This directly depends on the structural members, joints, regularity, etc.
- Fundamental period of the soil profile. This period plays an important role in ground motion and building vibration, especially at their resonance, being a very important index for a site-specific seismic evaluation.
- The control period of the earthquake. It depends mainly on the earthquake characteristics. Here it should be mentioned that the control period is directly derived from the response spectra, and it is strongly influenced by the soil profile from where it is recorded.

These three periods are different for each soil profile, ground motion and building. In this way, each site is a particular case, the structural damage differing even for very close structures. The soil dynamic properties may influence the damages of a structure during a ground motion. The research in the domain has revealed that there exist three main situations that characterise the interaction between soil and structure (Men and Cui, 1996):
- Stiff structures on soft ground. In this case the values of the displacements induced in structures are smaller than that of the soil, but the internal forces became bigger.
- Softer structures on stiffer ground. This case is characterised by bigger displacements as compared to the displacements of the ground.
- Soft structures on soft soil and stiff structures on stiff ground. In this case the degradations are mainly due to the resonance effect.

ANALYSED FRAMES

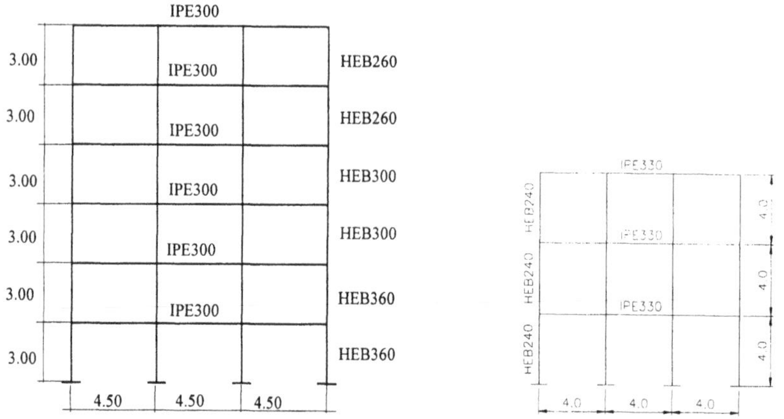

Figure 6.1.5. Geometry of the analysed frames

The parametrical study is developed on two series of frames: a 6 storey – 3 bay series frame (C36) with members from Fe430 steel and a 3 storey – 3 bay frame series (C33) with members from Fe360 steel, shown in Figure 6.1.5. The first one is taken from the ECCS manual (1994) and was designed according to Mazzolani-Piluso method in order to assure a global failure mechanism.

Characteristic values of the vertical loads acting on the beams are equal to 31.4 kN/m and 12 kN/m for the permanent G_k and live Q_k actions, respectively. The combination of actions corresponding to the frame subjected only to vertical loads is $1.35\Sigma G_k + 1.5\Sigma Q_k$. For the seismic design situation, the load combination is $\Sigma G_k + \Sigma\psi_2 Q_k + \gamma_1 A_{Ed}$, where ψ_2 is the coefficient for the quasi-permanent value of the variable action, γ_1 is the importance factor and A_{Ed} is the design value of the earthquake action.

The member cross-sectional dimensions and joint properties were obtained by means of an equivalent static design procedure, using an elasto-plastic analysis. Member characteristics for the C36 and C33 frames are given in Table 6.1.1.

Table 6.1.1
DESIGN VALUES OF THE SECTION CHARACTERISTICS.

Frame	Column	Beam	$M_{pl,b,Rd}$ x10^6 Nxmm	$M_{pl,c,Rd}$ x10^6 Nxmm	$M_{pl,c,Rd}/M_{pl,b,Rd}$	material
C36	HEB360 HEB300 HEB260	IPE300	157.1	670.8 467.3 320.7	4.27 2.97 2.04	Fe430
C33	HEB240	IPE330	171.8	225.0	1.31	Fe360

Eight frame typologies have been considered in the study: rigid frame with full-strength and infinitely rigid connections (RIG) and seven different semi-rigid frames (SR1, SR2, DU1, DU2, DU3, DU4 and DU5), as shown in Figure 6.1.6 and Figure 6.1.7 for C33 and C36 frame series respectively. The rigid connections are supposed to have the moment capacity much bigger than that of the beams, and the rotational stiffness equal to infinity.

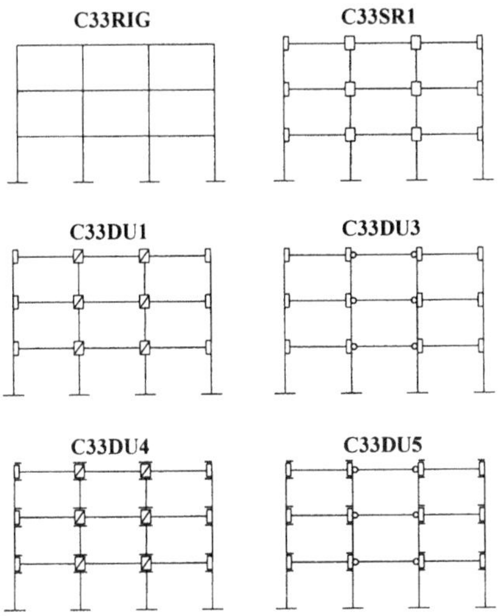

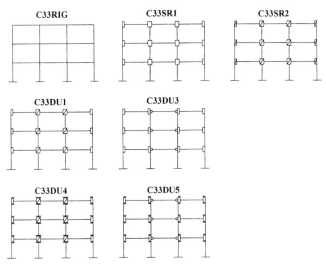

Figure 6.1.6 Frame typologies for C33 frame series

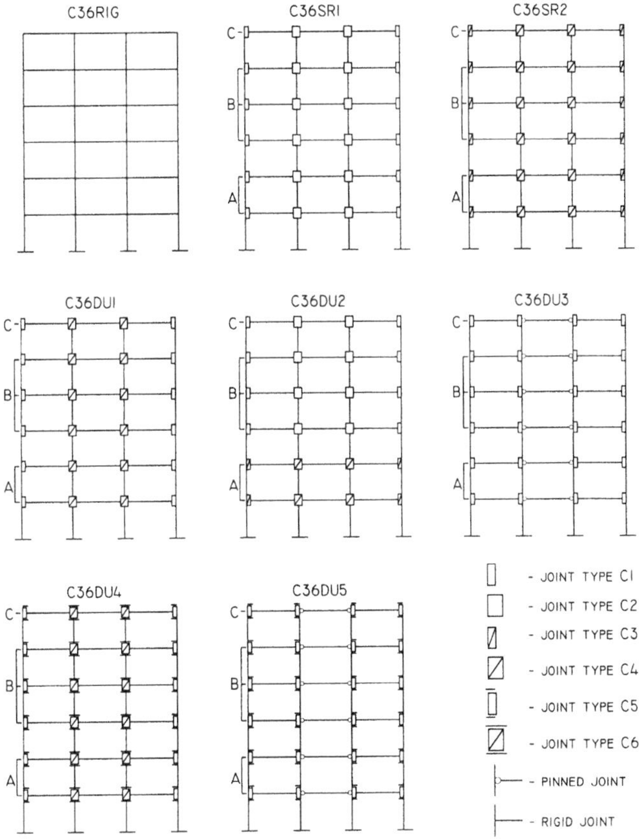

Figure 6.1.7. Frame typologies for C36 frame series

TABLE 6.1.2
CHARACTERISTICS OF THE SEMI-RIGID JOINTS.

Joint Type	$Mj_{,Rd}$ $x10^6$ Nxmm	m^* $(M_{j,Rd}/M_{pl,b,Rd})$	$S_{j,ini}$ $x10^{11}$ Nxmm/rad	S_j $x10^{11}$ Nxmm/rad	S^*_j
C36.1A	153.4	0.98	0.628	0.314	8.05
C36.1B	153.4	0.98	0.561	0.281	7.21
C36.1C	153.4	0.98	0.493	0.246	6.31
C36.2A	133.8	0.85	0.421	0.211	5.40
C36.2B	109.3	0.70	0.340	0.127	3.26
C36.2C	91.4	0.58	0.305	0.152	3.90
C36.3A	153.4	0.98	1.235	0.618	15.85
C36.3B	153.4	0.98	1.274	0.640	16.41
C36.3C	153.4	0.98	1.292	0.646	16.57
C36.4A	153.4	0.98	1.235	0.618	15.85
C36.4B	153.4	0.98	1.274	0.640	16.41
C36.4C	153.4	0.98	1.292	0.646	16.57
C36.5A	219.3	1.40	1.186	0.977	25.05
C36.5B	280.8	1.79	1.509	1.509	38.70
C36.5C	172.3	1.10	0.936	0.468	12.00
C36.6A	280.8	1.79	∞	∞	∞
C36.6B	280.8	1.79	∞	∞	∞
C36.6C	243.4	1.55	∞	∞	∞
C33.1	130.9	0.76	0.501	0.250	4.05
C33.2	74.63	0.43	0.307	0.153	2.48
C33.3	155.8	0.91	1.362	0.681	11.02
C33.4	152.5	0.89	1.362	0.681	11.02
C33.5	140.9	0.82	0.907	0.454	7.34
C33.6	210.4	1.22	∞	∞	∞

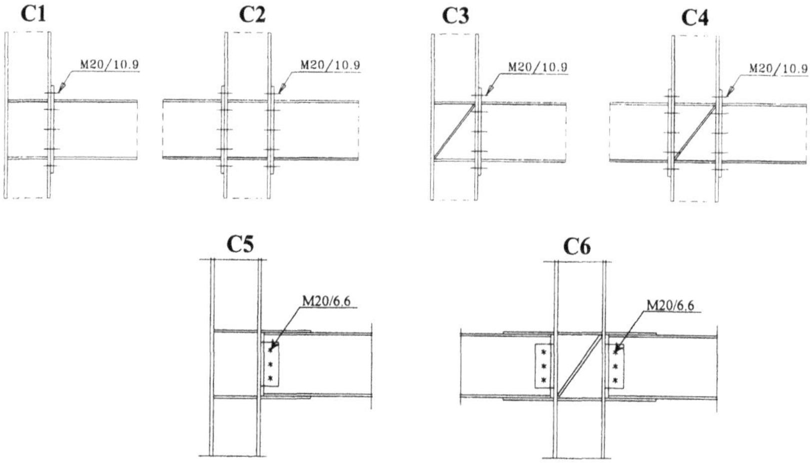

Figure 6.1.8. Detailing of semi-rigid joints.

The semi-rigid joints have been designed according to Annex J of Eurocode 3. Three different beam-to-column joints have been considered: extended end-plate bolted joints with transversal stiffeners, extended end-plate bolted joints with transversal and diagonal stiffeners and welded with reinforcing cover plates on beam flanges and web cleat. The joints are shown in Figure 6.1.8, while the design values of moment capacity and stiffness are given in Table 6.1.2, where:
A – connections at levels 1-2
B - connections at levels 3-5
C - connections at the roof level

Distinction is made between double-sided joints subjected to balanced and unbalanced moments, as is the case under seismic horizontal forces. Taking into account the web panel of the C2 joints is unstiffened to shear, compared to C4, the drop of stiffness is very important in the case of unbalanced bending moments.

Perfect elasto-plastic material behaviour was considered in the dynamic time-history analyses that have used the DRAIN2DX computer program for a set of selected accelerograms.

SELECTION AND SCALING OF RECORDS

The accelerograms from different historical earthquakes were considered. The ground motions records come from California (US), Greece (GR), Japan (J), Former Yugoslavia (FY), Romania (RO) and Italy (I). The accelerograms used in numerical simulations were selected from the available records on basis of the following criteria:

- the strongest record (max PGA)
- record corresponding to soft soil conditions ($T_{C,\,max}$)
- record corresponding to stiff soil conditions ($T_{C,\,min}$)

The deterministic descriptors of the frequency content of the ground motion time-history are the two control or corner periods defined from the maximum values of the response spectra as follows:

$$T_C = 2\pi \frac{SV_{max}}{SA_{max}} \tag{6.1.8}$$

$$T_D = 2\pi \frac{SD_{max}}{SV_{max}} \tag{6.1. 9}$$

The most significant and stable period of response spectra is T_C control period. This criterion was used in order to classify ground motion records as belonging to different soil conditions.

The effective peak acceleration (EPA) was used in order to scale the selected ground motions. To better understand EPA, it should be considered as normalising factor used to generate smoothed elastic response spectra for ground motions of normal duration. The following, somehow arbitrary, definition of EPA holds for the broad and intermediate frequency band, ground motions having a relatively small or medium control period (T_c<0.6s):
EPA represents the average of the maximum ordinates of elastic acceleration response spectra within the period range of 0.1 to 0.5 seconds, divided by a standard (mean) value of 2.5 for 5% damping (Naiem and Anderson, 1993).

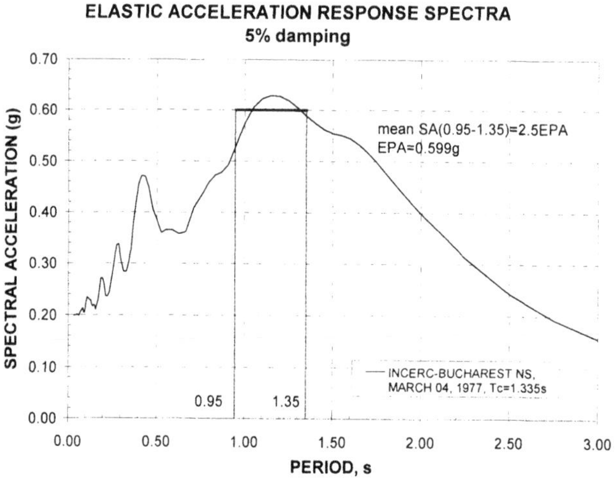

Figure 6.1.9 Definition of EPA

For the narrow frequency band (motions of long predominant period), the above definition must be definitely changed. In that case, the 0.4s averaging interval must be centred on the period range corresponding to the dominant peak of the spectra SA.

This last definition was used here in order to determine the EPA for the considered ground motions, taking into account its ability to adjust itself to motions of any frequency content (see Figure 6.1.9):

$$EPA = \frac{mean \cdot of \cdot SA\Big|_{T(SA\max)-0.2s}^{T(SA\max)+0.2s}(\xi = 0.05)}{2.5} \tag{6.1. 10}$$

The EPA characterises the intensity of a ground motion by averaging the effects of shaking on the most exposed – to that spectral content – structures.

The selected ground motions are scaled with respect to a reference EPA value of an EC8 elastic design response spectrum corresponding to a value of PGA of 0.25g (see Figure 6.1.10):

$$EPA_{ref} = \frac{mean \cdot of \cdot SA\Big|_{T_C}^{T_B}(\xi = 0.05)}{2.5} \tag{6.1. 11}$$

$$SA_{ref} = a_g \cdot S \cdot \eta \cdot \beta_0 = 0.25 \cdot 1.0 \cdot 1.0 \cdot 2.5 = 0.625g \tag{6.1. 12}$$

$$EPA_{ref}=0.625/2.5=0.25g\ (245.17\ cm/s^2) \tag{6.1. 13}$$

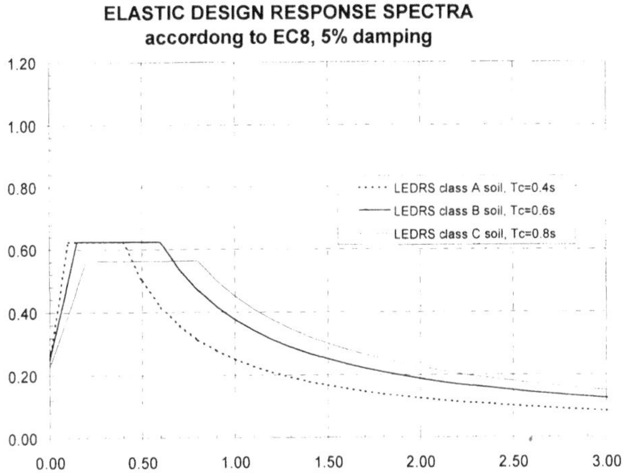

Figure 6.1.10 EC8 Elastic design response spectrum

The main factors (PGA, EPA, T_C etc), that characterise the ground motions considered are listed in Table 6.1.3, where the accelerograms are classified according to their control period values: $T_C <$ 0.4sec.; 0.4sec. $< T_C <$ 0.8sec.; and $T_C >$ 0.8sec.

Due to the huge amount of results in terms of ductility derived from these analyses, in the present study are comprised only the results given by three selected accelerograms. Considering the fact that all accelerograms have been scaled to the same EPA value, the selection criterion was the control period T_C. Secondly the variation of the initial EPA was considered. The accelerograms selected are:

- Incerc-Bucharest NS, March 04, 1977 –Vrancea - (PGA=0.199g, T_C=1.335s, EPA=0.240g)
- Kobe NS (PGA=0.836g, T_C=0.622s, EPA=0.704g)
- Brienza NS, November 23, 1980 (PGA=0.21g, T_C=0.19s, EPA=0.22g).

Their elastic spectral acceleration (for 5% damping) is given in Figure 6.1.11.

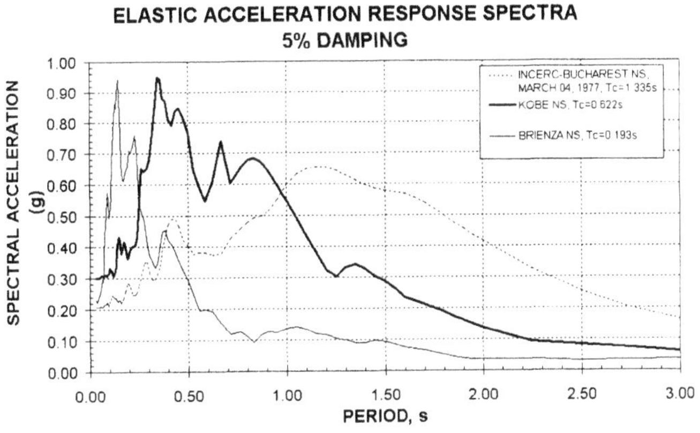

Figure 6.1.11: Acceleration response spectra of considered records.

As it can be seen from Figure 6.1.11, the selected accelerograms cover a very large range of the control period values T_C 0.19sec.-1.335sec.. These accelerograms have been considered representative for all the accelerograms under study. The fundamental periods of vibration of the considered frames are given in Table 6.1.4. It can be seen that Vrancea control period is comparable to the vibration periods of the frames. From here we can draw a presumptive conclusion that the Vrancea type earthquake, among the selected ground motions, seems to be the most destructive one.

TABLE 6.1.3
GROUND MOTIONS CHARACTERISTICS

Record	PGA (g)	T_C (sec)	EPA (g)	scaling factor	scaled PGA (g)	EPV (cm/sec)
I Ground motions with T_C<0.4sec.						
Brienza NS, November 23, 1980 (I)	0.21	0.190	0.22	1.156	0.248	9.06
La County Fire St, Newhall 90 Deg. Jan 17, 1994 (USA)	0.583	0.346	0.629	0.397	0.231	52.42
Hercegnovi – Sola D. Pavi'i' WE, April 15, 1979 (FY)	0.230	0.390	0.194	1.286	0.296	11.30
II Ground motions with 0.4sec. < T_C < 0.8sec.						
Thessaloniki Trans, 1978 (GR)	0.143	0.450	0.140	1.786	0.255	9.71
Focsani NS, August 30, 1986 (RO)	0.303	0.492	0.198	1.259	0.381	24.21
Egion Peloponnese, Tr., 1995 (GR)	0.543	0.497	0.447	0.559	0.304	31.98
Petrovac – Hotel Oliva NS, April 15, 1979 (FY)	0.438	0.531	0.539	0.464	0.203	51.06
Kobe NS (J)	0.836	0.622	0.704	0.355	0.297	97.17
Eren-Bucharest N10W, Aug 30, 1986 (RO)	0.159	0.664	0.133	1.879	0.299	16.39
La County Fire St, Newhall 360 Deg, Jan 17, 1994 (USA)	0.589	0.736	0.700	0.357	0.210	92.27
Sturno EW, November 23, 1980 (I)	0.29	0.750	0.30	0.834	0.245	60.98
Kalamata Peloponnese Long, 1986 (GR)	0.220	0.768	0.197	1.266	0.278	26.64
III Ground motions with T_C > 0.8sec						
Ulcinj-1 – Hotel Olimpik WE, April 15, 1979 (FY)	0.240	1.180	0.185	1.353	0.324	48.96
Ntt NS (J)	0.156	1.204	0.138	1.814	0.284	30.27
Incerc-Bucharest NS, Aug 30, 1986 (RO)	0.090	1.256	0.081	3.072	0.278	19.20
Calitri NS, November 23, 1980 (I)	0.15	1.270	0.18	1.401	0.214	35.86
Incerc-Bucharest NS, March 04, 1977 (RO)	0.199	1.335	0.240	1.042	0.207	52.18
Pico Canyon Rd., Newhall N46E Jan 17, 1994 (USA)	0.419	1.393	0.358	0.699	0.293	83.74
Muntele Rosu NS, August 30, 1986 (RO)	0.081	1.499	0.067	3.709	0.299	18.63

Table 6.1.4

FUNDAMENTAL PERIODS OF VIBRATION OF THE ANALYSED FRAMES

Frame	Fundamental period (sec)	Frame	Fundamental period (sec)
C36RIG	1.33	C33RIG	0.95
C36SR1	1.69	C33SR1	1.15
C36SR2	1.47	C33SR2	1.04
C36DU1	1.52	C33DU1	1.09
C36DU2	1.61	--	--
C36DU3	1.86	C33DU3	1.31
C36DU4	1.35	C33DU4	1.01
C36DU5	1.63	C33DU5	1.20

In order to analyse the influence of soil conditions on the structural seismic response of considered frames, two accelerograms coming from the same earthquake event were selected: INCERC 1986 (EPA=0.081, T_C=1.256 and PGA=0.090) and EREN 1986 (EPA=0.133, T_C=0.664 and PGA=0.159). These accelerograms were both recorded in Bucharest, but the soil conditions of the seismic stations were different: the INCERC site conditions corresponds to a soft soil, while EREN site to a medium one. Their elastic response spectra is shown in Figure 6.1.12, and is compared to the Eurocode 8 linear elastic design response spectra (LEDRS) for stiff (A), medium (B) and soft (C) soil conditions.

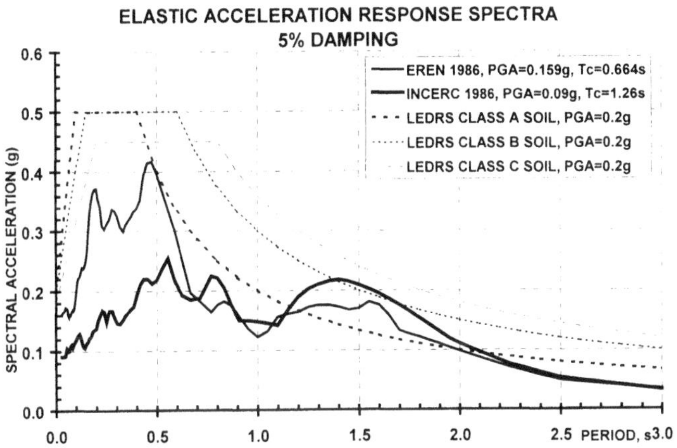

Figure 6.1.12: Elastic Response Spectra for INCERC and EREN 1986, Vrancea earthquake

In analyses, the accelerograms have been multiplied by a scaling factor in order to reach the desired imposed acceleration.

MONITORED PARAMETERS

Ductility Demand

The main aim of the investigation was to study the ductility demand for frames with semi-rigid joints. For this purpose, all the accelerograms have been scaled so as to have the same value for EPA. The EPA value that has been chosen corresponds to a high intensity earthquake according to EC8. Practically, all structures have been subjected to earthquakes with accelerograms of the types described in Table 6.1.3, scaled in order to have the same value of EPA: 0.35g.

The accounting of the ductility demand for the structures under study was performed at the level of member and beam-to-column joint – local ductility – and by means of inter-storey drift and q factors at the level of frames - global ductility.

Local Ductility

For accounting the ductility demand at the element level (beams, columns and joints) the maximum plastic rotations of elements have been monitored (rotational ductility). They have been compared to the maximum plastic rotations that the element can sustain. The monotonic maximum plastic rotations (θ_p) were computed by DUCTROT computer program. Plastic rotation capacity of members in conditions of cyclic behaviour (θ_p^{cor}) has been computed by adjusting the monotonic values by correction coefficients that depend on the slenderness of member cross section and the axial force. A material partial safety factor γ=1.5 was used (Gioncu, 1997). Computed values of plastic rotation supply of frame members are given in Table 6.1.5.

TABLE 6.1.5
PLASTIC ROTATION SUPPLY OF MEMBERS

Frame type	member	θ_p	θ_p^{cor}
C36	IPE300	0.0946	0.054
	HEB360	0.1085	0.049
	HEB300	0.0926	0.042
	HEB260	0.1267	0.065
C33	IPE330	0.0934	0.053
	HEB240	0.0963	0.044

For the semi-rigid joints, two values of limiting plastic rotations have been considered:

- 0.02 rad. for welded joints
- 0.03 rad. for bolted joints

The ultimate joints rotations considered correspond to Intermediate respectively Special Moment Resisting Frames, according to American code AISC '97. For the limitations chosen it was considered that the welded joints are more rigid than the bolted ones. Actually, the exact value of the plastic rotation of a certain connection depends directly on all the components that are part of that connection. However, the last research in the domain has revealed the fact that the joints without diagonal stiffeners in the column panel under cyclic symmetrical loading can withstand larger strengths and smaller rotation capacities, while in the case of joints under anti-symmetrical loading, the resistance is diminished but with a substantial increase in rotational capacity (see chapter 3.2). This fact is due to the work of the column panel zone in case of anti-symmetrical loaded joints.

The authors' opinion regarding the above values of the rotational limitations is that they are too severe, taking into account the fact that in the analysis are considered reduced moment capacity and rotational stiffness (computed for anti-symmetrical loaded joints – by EUROCODE 3, Annex J) and simultaneously a reduced rotation capacity. However, these limitations have been used in reference with the above-mentioned code.

Global Ductility

The global ductility has been monitored by two parameters, as described in subchapter 2.2, in terms of inter-storey and the q-factor values.

A limiting value of 3% for inter-storey drift has been considered. This value has been taken based on research studies, so as to correspond to the total consumption of the rotational capacity in members or of the bearing capacity of a structure:

$$\Delta_i = \frac{\delta_i - \delta_{i-1}}{h_i} \leq 0.03 \tag{6.1.14}$$

where δ_i is the displacement at floor i, and h_i is the height of the storey under consideration. This limit has been widely used in other studies, and is supported by field evidence correlating inter-storey drift with damage (Elnashai et al., 1996, 1998).

The ultimate amplitude of the acceleration a_u used for the computation of q factors (computed by equation 6.1.3.) was considered the minimum value corresponding to three theoretical states of collapse:

$$a_u = \min(a_\Delta, a_\theta, a_m) \tag{6.1.15}$$

where:

- a_θ - ultimate acceleration amplitude corresponding to the total consumption of the local rotational ductility (attainment of the ultimate plastic rotation)
- a_Δ – ultimate acceleration amplitude corresponding to the total consumption of the global ductility limit (in terms of inter-storey drift, Δ_{max}=3%)
- a_m – ultimate acceleration amplitude corresponding to attainment of the failure mechanism.

In case of Moment Resisting Frames (Mazzolani and Piluso, 1994) a global failure mechanism and three partial mechanisms can be considered as shown in Figure 6.1.13

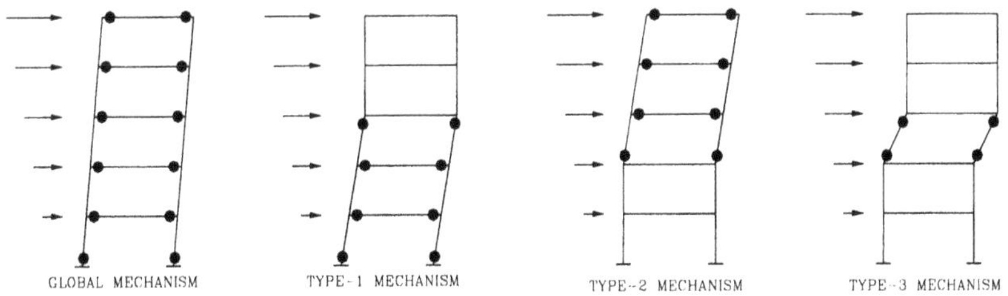

Figure 6.1.13 Collapse mechanism typologies of MR frames under seismic horizontal forces

According to the present study, the collapse mechanism occurs when the plastic hinge configuration forms a one-degree cinematic mechanism. For this purpose, all plastic hinges that appeared during the time-history analysis have been considered, independently of the moment when they appeared. This is a conventional mechanism only; in fact the collapse mechanism occurs when the number and configuration of the plastic hinges at a certain moment are so that they generate a partial or global mechanism.

Factors Expressing the Structural Seismic Performance

Beside the parameters that account for the structural ductility, other two parameters have also been considered in order to evaluate the global performances of a structure under seismic circumstances: the ultimate EPA that can be sustained by a structure for a given ground motion, and the Global Damage Index (GDI) that is a measure of structural degradation. These parameters complete the description of the global behaviour of structures.

Seismic Performance Factor

It has to be noted that the q factor does not express the structural capacity to resist to an earthquake, - for instance two structures characterised by the same q factor can experience different values of the acceleration corresponding to the first plastic hinge and collapse, respectively. For this purpose, an absolute parameter such as the ultimate EPA that a structure can withstand has been considered. Similarly to the definition of the ultimate acceleration used to compute the q factor, the ultimate EPA value is considered the minimum of:

- EPA_θ – ultimate EPA value corresponding to attainment of the rotational local ductility limit (consumption of the ultimate plastic rotation in members or joints)
- EPA_Δ – ultimate EPA value corresponding to the consumption of the global ductility limit (in terms of inter-storey drift, Δ_{max}=3%)
- EPA_m – ultimate EPA value corresponding to attainment of the failure mechanism.

$$EPA_\theta = \min.(EPA_\theta, EPA_\Delta, EPA_m) \qquad (6.1.16)$$

A very important parameter that can be used to characterise the seismic performance of a given structure in conditions of a given ground motion is the performance factor – η, defined by:

$$\eta = \frac{EPA_U}{EPA_S} \qquad (6.1.17)$$

where:

EPA_S is the Effective Peak Acceleration induced by the seismic excitation in the structure

For a structure well designed in a certain seismic region, the value of the seismic performance factor should be always greater than 1.

Damage Indices

The damage indices were introduced in order to account for local and global structural degradations. The values of the Local Damage Indices have been computed by:

$$I_{D_L} = \frac{\theta_p}{\theta_{p.\lim}} \qquad (6.1.18)$$

while the Global Damage Indices are computed by the formula:

$$I_{Dg} = \frac{\sum I_{D_{Li}}^{2}}{\sum I_{D_{Li}}} \tag{6.1.19}$$

where:

- I_{DL} is the local damage index,
- θ_p is the plastic rotation demand for beams, columns and joints;
- $\theta_{p.lim}$ is the corresponding minimum guaranteed plastic rotation supply;
- I_{Dg} is the global damage index.

A damage index equal to 1 represents the total consumption of the rotational capacity, while a damage index equal to 0, corresponds to an undamaged member or joint. Practically the damage index can be neither higher than 1 nor lower than 0.

NUMERICAL RESULTS

All the factors characterising the local and global ductility, as well as the factors expressing the structural seismic performance have been derived from dynamic inelastic analyses of C36 and C33 series of frames. The seismic action was introduced by means of the selected accelerograms, already presented.

Results in Terms of Local Ductility

The results in terms of local ductility are given in Figure 6.1.14 for C33 frame series and Figure 6.1.15. for C36 frame series. In charts are presented the maximum plastic rotations for each storey (3 storeys for C33 frames and 6 storeys for C36 frames), differentiated for columns, beams and joints. Distinction has been made between the internal joints (connecting the external columns to beams) and external joints (connecting the internal columns to beams), as the frames typology differs. The plastic rotation for the pinned joints used at frames DU3 and DU5 have not been taken into account, considering that these joints exhibit only elastic rotations. On the charts are also drawn the limiting values of plastic rotations, as they have been computed for beams and columns (see Table 6.1.5), and the theoretical values considered for joints (0.02 radians for welded joints and 0.03 radians for bolted joints).

A first conclusion drawn after analysing these charts is the alternation of the plastic hinges between beams and joints. As expected, in the case of full strength joints, the rotational ductility demand is present only in beams, while in the case of partial-strength joints the situation is, of course, reversed. For the case in which the moment capacities of beams and joints are closer, (C33DU4 and C36DU4), a sharing of the plastic rotation demand between the beams and joints can be observed.

Analysing the influence of ground motions, it can be easily seen that for the Vrancea type earthquake, the plastic rotation demand for beams, columns and joints exceeds considerably the corresponding values of the other two ground motions. The limiting value of rotation considered for joints is exceeded for almost all the frame typologies for the Vrancea earthquake.

A very usual frame typology that has been largely used in USA was the one with pinned joints on the middle spans and rigid joints on the outer spans, represented in this study by frame typology DU5. The frame typology DU3 is similar to the typology DU5, with the mention that the non-pinned joints are semi-rigid. In both cases it can be noticed that the rotational ductility demand is considerably higher than in the rest of the frames. The explanation of this fact is that for a certain storey there are less potential plastic regions, which can dissipate the input energy. In this way, the earthquake input energy is dissipated only within the available dissipative regions.

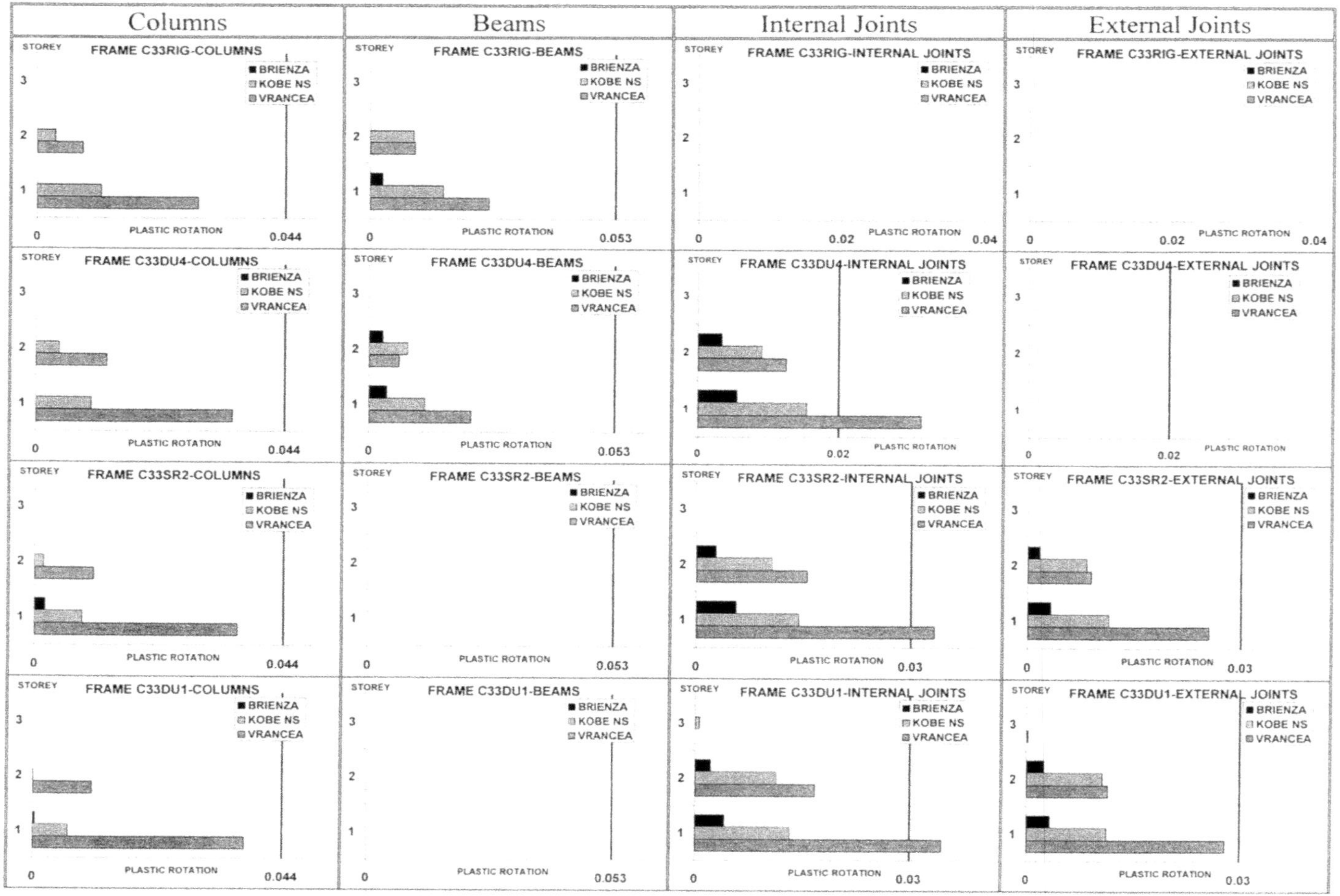

Columns
Beams
Internal Joints
External Joints
FRAME C33RIG-COLUMNS
FRAME C33RIG-BEAMS
FRAME C33RIG-INTERNAL JOINTS
FRAME C33RIG-EXTERNAL JOINTS
FRAME C33DU4-COLUMNS
FRAME C33DU4-BEAMS
FRAME C33DU4-INTERNAL JOINTS
FRAME C33DU4-EXTERNAL JOINTS
FRAME C33SR2-COLUMNS
FRAME C33SR2-BEAMS
FRAME C33SR2-INTERNAL JOINTS
FRAME C33SR2-EXTERNAL JOINTS
FRAME C33DU1-COLUMNS
FRAME C33DU1-BEAMS
FRAME C33DU1-INTERNAL JOINTS
FRAME C33DU1-EXTERNAL JOINTS
STOREY
PLASTIC ROTATION
BRIENZA
KOBE NS
VRANCEA
0
0.02
0.03
0.04
0.044
0.053
1
2
3

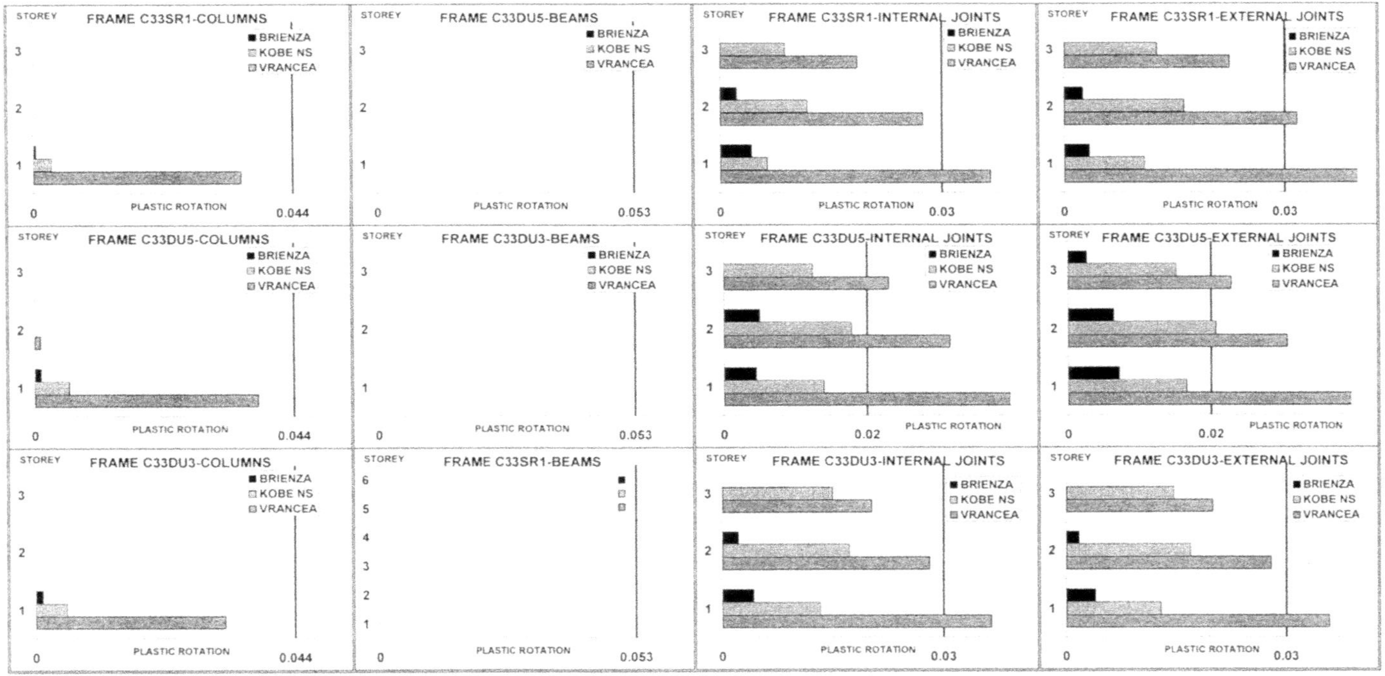

Figure 6.1.14: Results obtained in terms of rotational ductility for C33 frame series

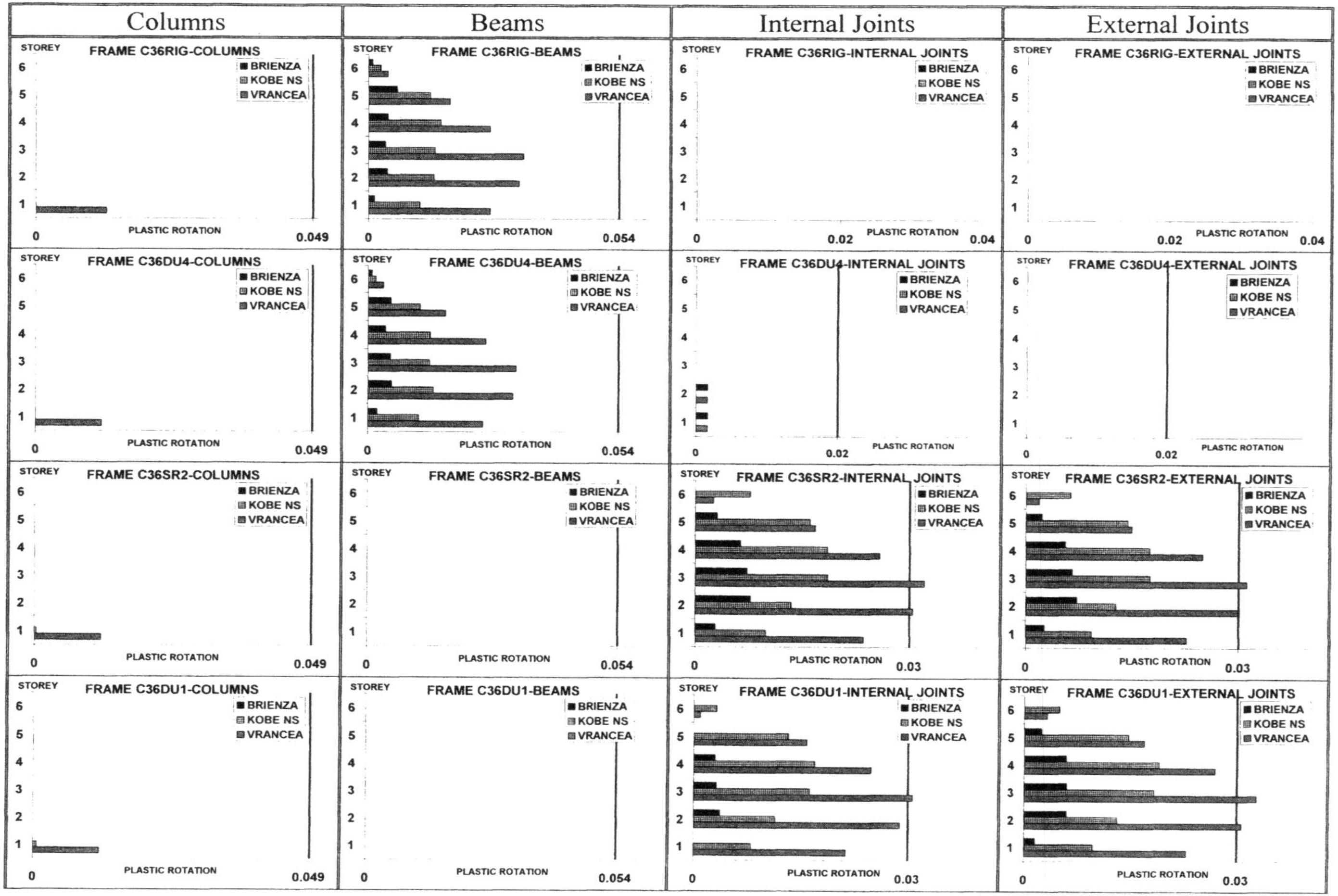
Columns
Beams
Internal Joints
External Joints
FRAME C36RIG-COLUMNS
FRAME C36RIG-BEAMS
FRAME C36RIG-INTERNAL JOINTS
FRAME C36RIG-EXTERNAL JOINTS
FRAME C36DU4-COLUMNS
FRAME C36DU4-BEAMS
FRAME C36DU4-INTERNAL JOINTS
FRAME C36DU4-EXTERNAL JOINTS
FRAME C36SR2-COLUMNS
FRAME C36SR2-BEAMS
FRAME C36SR2-INTERNAL JOINTS
FRAME C36SR2-EXTERNAL JOINTS
FRAME C36DU1-COLUMNS
FRAME C36DU1-BEAMS
FRAME C36DU1-INTERNAL JOINTS
FRAME C36DU1-EXTERNAL JOINTS
STOREY
PLASTIC ROTATION
BRIENZA
KOBE NS
VRANCEA
0.049
0.054
0.02
0.03
0.04

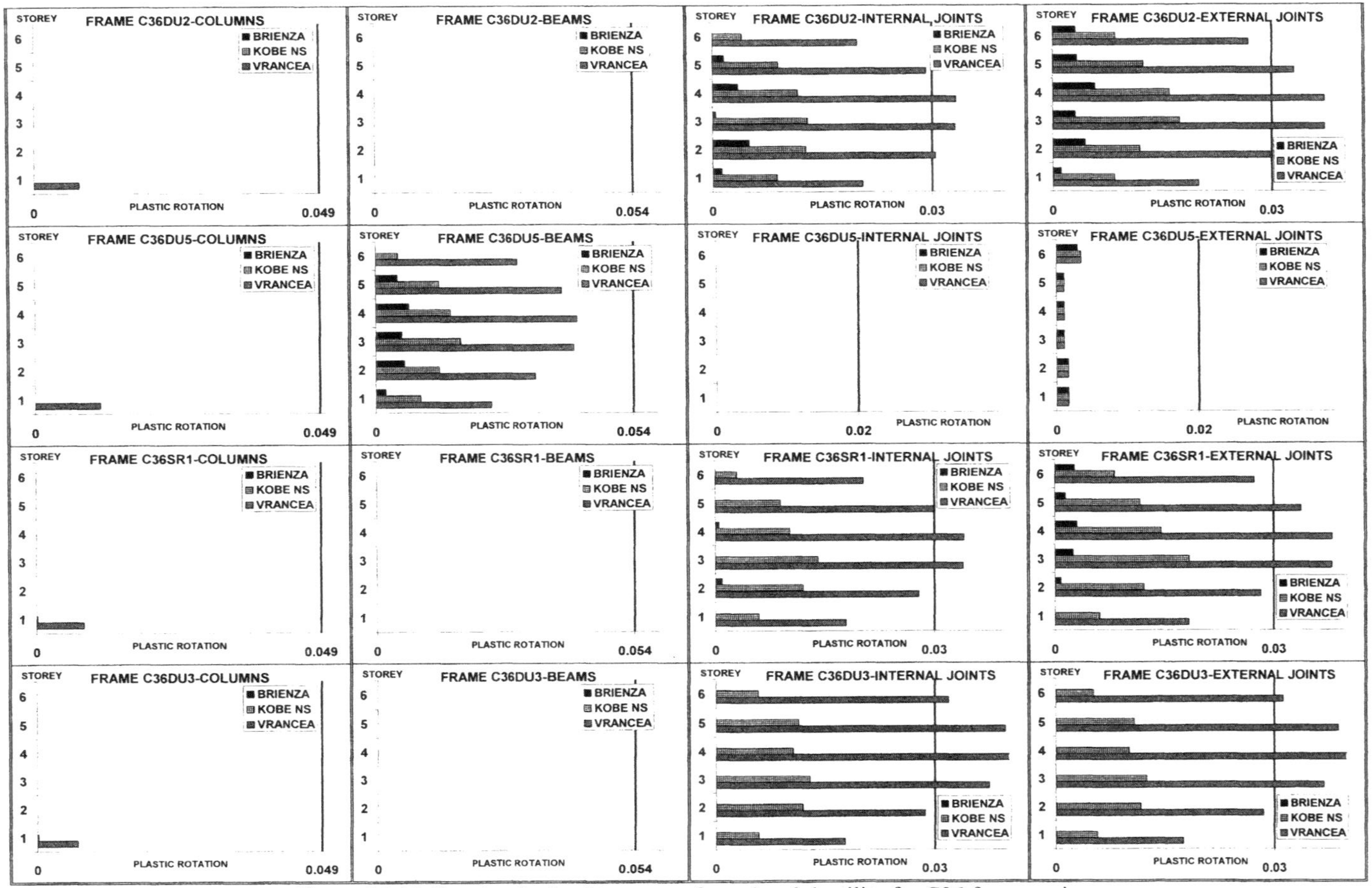

Figure 6.1.15: Results obtained in terms of rotational ductility for C36 frame series

Although the strength and rigidity of the considered connections are very different for the frames under study, in columns there are no big variations of the plastic rotations as the frame typology changes. It results that the semi-rigidity of connections does not influence much the required rotational local ductility in columns. Also can be seen that in the case of C36 frame series (frame designed on the Strong Column Weak Beam concept) the plastic rotations in columns are very small. On the other hand, in the case of beams, the ductility demand of beams is directly affected by the semi-rigidity of joints.

The ductility demand for the frames seems to increase with their flexibility, larger rotation capacities of joints and/or beams being required for the flexible frames. In the case of frames DU3 and DU5, this situation is explicable, due to diminution in the dissipative zones. In the case of frame C36, the local ductility requirement in terms of rotation for the middle storeys is higher than that for lower storeys, a fact that indicates that the middle storeys are more solicited. Generally, there are not big differences between the required local ductility for internal and external joints. This indicates that the input energy on a storey is dissipated proportionally by the internal and external joints. However, by performing a time-history analysis on a structure, it can be concluded that the affected regions are changing within the structure with the evolution of plastic hinges.

Global Ductility Parameters

Displacement Ductility

Similarly to the required rotational ductility, the values of the required global ductility, expressed in terms of inter-storey drift for each storey are given in Figure 6.1.16.

It can be seen that the maximum inter-storey drifts are concentrated in the middle storeys for all frames, with the only exception for Vrancea earthquake for frame C33. Connecting this conclusion with the similar one coming from the ductility in terms of rotation, it means that these storeys dissipate the higher amount of energy, but also the second-order effects are greater in that part of structure.

As expected, the Vrancea earthquake was the most destructive, the limiting inter-storey drift value (3%) being always exceeded, while the Brienza earthquake, with the control period T_C much smaller compared to that of the frames, was the less destructive one.

The required displacement ductility is strongly influenced by the semi-rigidity of joints, the charts clearly showing that the more flexible the frame, the bigger the required displacement. This is valid for both frame series. For the structures with the pinned joints in the middle spans, this effect is more visible.

q Factor

The q factor expresses the ability of a structure to work and dissipate energy in the post-elastic domain. In Figure 6.1.17 are presented the q factors obtained for the frame typologies under consideration.

A parameter that is fundamental for the values of the q factor is the number of dissipative zones. The bigger the number of dissipative zones the bigger the q factor. This can be graphically seen in Figure 6.1.17, the q factors for C36 series (a) being generally higher than that for C33 series (b). Also, in the case of frames with pinned joints on the middle spans where the number of dissipative zones is reduced, the q factor is also reduced.

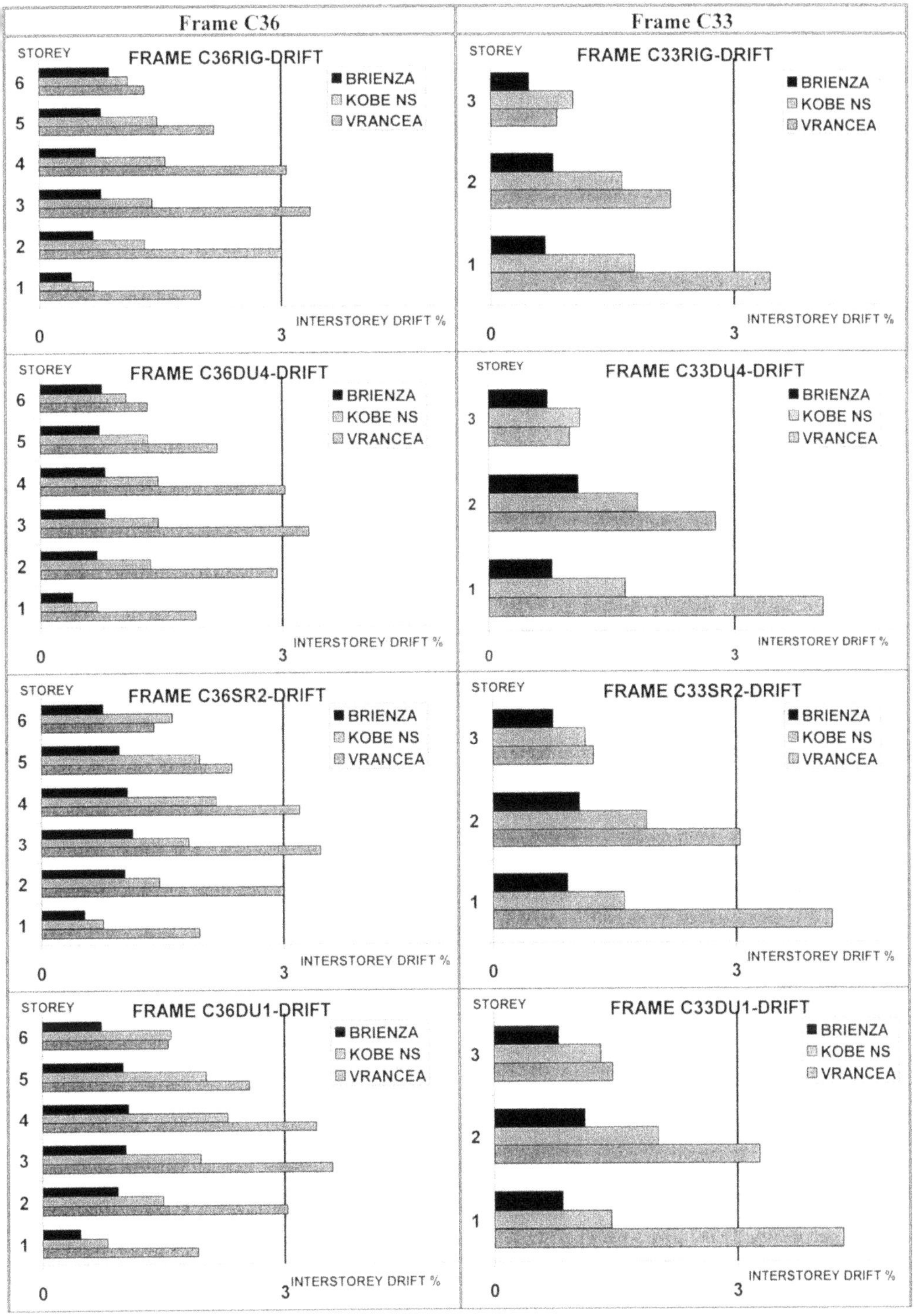
Frame C36
Frame C33
FRAME C36RIG-DRIFT
FRAME C33RIG-DRIFT
FRAME C36DU4-DRIFT
FRAME C33DU4-DRIFT
FRAME C36SR2-DRIFT
FRAME C33SR2-DRIFT
FRAME C36DU1-DRIFT
FRAME C33DU1-DRIFT
STOREY
BRIENZA
KOBE NS
VRANCEA
INTERSTOREY DRIFT %
0
3
1
2
3
4
5
6

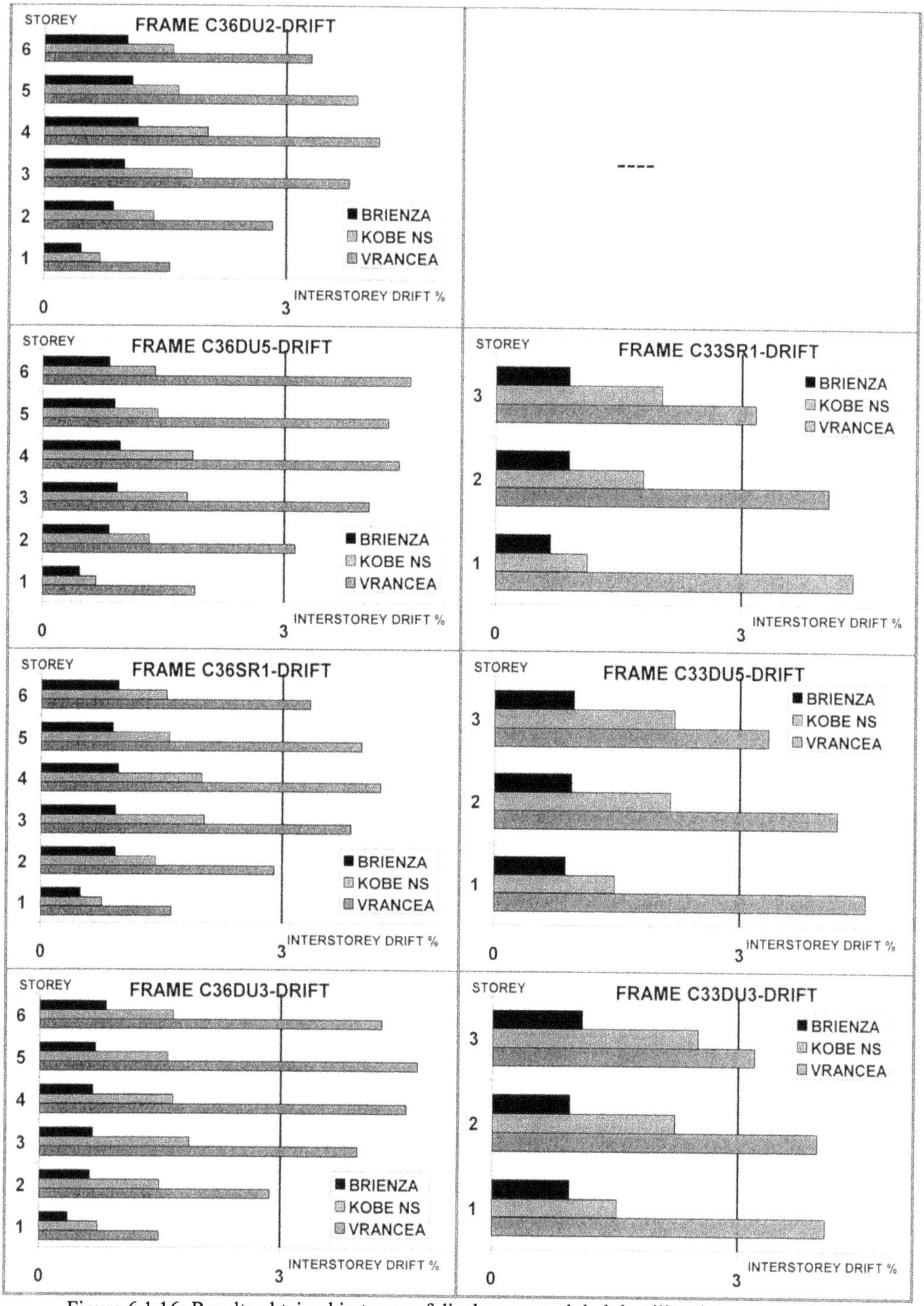

Figure 6.1.16: Results obtained in terms of displacement global ductility (inter-storey drift)

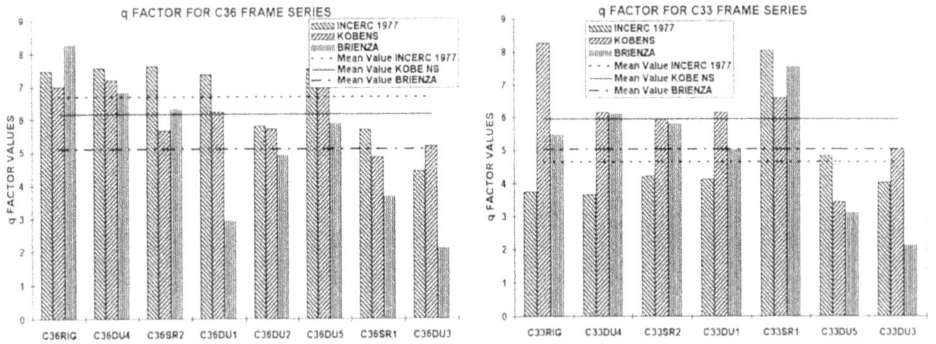

(a) (b)

Figure 6.1.17 q factors for frames under consideration: C36 series – (a) and C33 series –(b)

On the other hand, the q factors depend on the selected ground motion and the frame typology. Generally, the perfect rigid frames show a good behavioural factor. It cannot be said that for the three earthquakes considered in these analyses there are significant differences between the q factor values obtained for C33 and C36 frame series.

It should be reminded that for the design of structures a q-factor equal to 6 was used, value that is for many cases higher than the value resulted from analyses, fact that means that the frames cannot exhibit the designed energy.

Factors Expressing the Structural Seismic Performance

Seismic Performance Factor

The EPA criterion (equation 6.1.17) is very important in characterisation of the seismic performance of a particular structure for a given seismic excitation. The η values shows the structural capacity to resist a seismic action in respect to the failure criteria accepted in design: global ductility (Δ), local ductility (θ) or the collapse mechanism (m). Figure Figure 6.1.18 shows the η values, considering all the failure criteria used to define EPA_u up to a value of η=8.

A main conclusion drawn from these charts is the fact that the destructivity of a seismic motion cannot be characterised only by the PGA or EPA of the structure, but also of the interaction: control period T_C of the motion - fundamental period of the structure. This can be seen in the case of earthquake types Vrancea and Brienza: although both of them have almost the same PGA, the ultimate η factors are about 1.5 (mean) for Vrancea and about 7 (mean) for Brienza. However, the ultimate EPA values are smaller than 0.5g for Vrancea earthquake type (T_C=1.335sec.), about 0.75g for Kobe earthquake (T_C=0.622sec.) and about 1.4g for Brienza (T_C=0.190 sec.). The ground motions with the values of T_C closer to that of frames show the smallest value of ultimate EPA.

In case of Kobe earthquake type the mechanism condition was the limiting criterion. In some cases, the values of η factors was smaller than 1, this indicating the high intensity of the earthquake.

It can also be seen that the three criteria used to establish the ultimate EPA values are very close, which means that the criteria chosen are very close to the failure limit.

The frames with theoretical rigid joints (RIG) proved the best values of the ultimate EPA values, but such frames cannot be realised in practice. The real rigid frames (frames with real characteristics of rigid joints – DU4) show a slightly smaller value of the ultimate EPA. The frames with semi-rigid joints do not necessarily lead to smaller values for EPA; sometimes the semi-rigid or dual frames showing higher values for ultimate EPA than rigid frames as is the case of C33DU1 and C36 DU5 frames.

In the case of C33 frame series, the dual frames with pinned joints on middle spans showed the smallest EPA values.

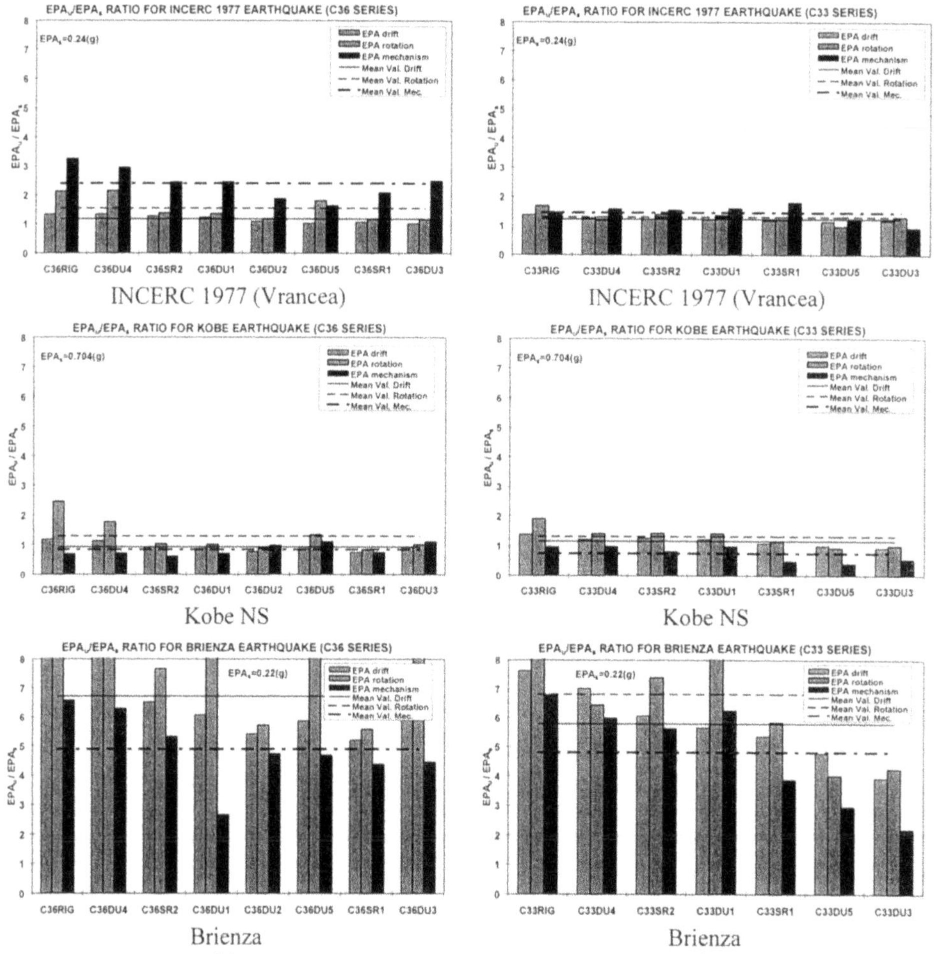

(a) (b)

Figure 6.1.18: Ultimate EPA values for frame series: C36 series - (a) and C33 series -(b)

Global Damage Indices (GDI)

The local damage indices (LDI) indicate the degree of destruction within the structural elements i.e. beams, columns and joints. Also, the global damage index (GDI) indicates the overall state of damages of a structure. In figure Figure 6.1.19 are presented GDI values computed for the minimum failure criterion, as used to define the q factors.

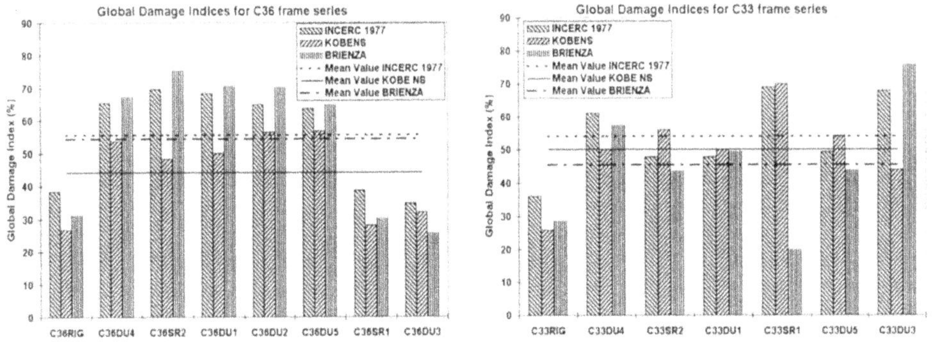

(a) (b)

Figure 6.1.19: Global damage indices for frame series: C36 series - (a) and C33 series -(b)

For a better identification of the global and partial damage indices, their values are given in Table 6.1.6. It is to be noted that the global damage index is not a linear sum of the partial indices.

TABLE 6.1.6

THE GLOBAL AND PARTIAL (COLUMNS, BEAMS AND JOINTS) DAMAGE INDICES.

Frame	Global Damage Indices					Frame	Global Damage Indices			
	$I_{D,col}$	$I_{D,beam}$	$I_{D,joint}$	$I_{D,G}$			$I_{D,col}$	$I_{D,beam}$	$I_{D,joint}$	$I_{D,G}$
INCERC-BUCHAREST NS, MARCH 04, 1977 ROMANIA										
C36RIG	0.168	0.397	0	0.384		C33RIG	0.442	0.256	0	0.360
C36DU4	0.162	0.398	0	0.386		C33DU4	0.352	0.192	0.762	0.494
C36SR2	0.125	0	0.710	0.696		C33SR2	0.459	0	0.488	0.478
C36DU1	0.122	0	0.698	0.651		C33DU1	0.428	0	0.505	0.479
C36DU2	0.047	0	0.656	0.651		C33SR1	0.451	0	0.649	0.613
C36DU5	0.018	0.383	0.206	0.348		C33DU5	0.266	0	0.758	0.680
C36SR1	0.049	0	0.661	0.656		C33DU3	0.327	0	0.763	0.690
C36DU3	0.024	0	0.642	0.639						
KOBE NS JAPAN										
C36RIG	0.027	0.269	0	0.267		C33RIG	0.228	0.281	0	0.258
C36DU4	0.037	0.284	0	0.281		C33DU4	0.318	0.272	0.860	0.543
C36SR2	0.043	0	0.488	0.483		C33SR2	0.263	0	0.646	0.561
C36DU1	0.064	0	0.506	0.499		C33DU1	0.270	0	0.583	0.503
C36DU2	0.011	0	0.564	0.564		C33SR1	0.055	0	0.513	0.501
C36DU5	0.014	0.351	0.207	0.320		C33DU5	0.086	0	0.470	0.440
C36SR1	0.021	0	0.540	0.538		C33DU3	0.166	0	0.743	0.700
C36DU3	0.055	0	0.577	0.568						
BRIENZA, 24 NOV. 1980, N-S										
C36RIG	0.018	0.314	0	0.311		C33RIG	0.325	0.243	0	0.286
C36DU4	0.018	0.304	0	0.301		C33DU4	0.319	0.206	0.681	0.439
C36SR2	0.021	0	0.758	0.755		C33SR2	0.361	0	0.473	0.437
C36DU1	0.020	0	0.709	0.706		C33DU1	0.378	0	0.547	0.495
C36DU2	0.019	0	0.706	0.703		C33SR1	0.118	0	0.601	0.576
C36DU5	0.020	0.275	0.211	0.257		C33DU5	0.089	0	0.796	0.759
C36SR1	0.022	0.314	0	0.311		C33DU3	0.076	0	0.215	0.199
C36DU3	0.027	0	0.655	0.650						

The best behaviour appears in the case of frames with perfectly rigid joints (RIG); however, the frames with real rigid joints (DU4) show significantly higher GDI values. Taking into account that the structural elements (beams and columns) are the same for both typologies, the only parameter that changes the values of GDI are the real characteristics of the joints.

GDI values of these frames are tributary to the imposed ultimate rotation of the joints. If the rotation capacity of the real joints can be larger than the limits accepted in the present analysis (see chapter 3), the values of GDI will be, for sure, smaller.

From this table there can be seen the same alternation of the damage indices between beams and joints as it was shown in terms of rotational ductility. Also, it can be noticed that in the case of frames with partial-strength joints, the main factor that affects the GDI is the damage index for joints.

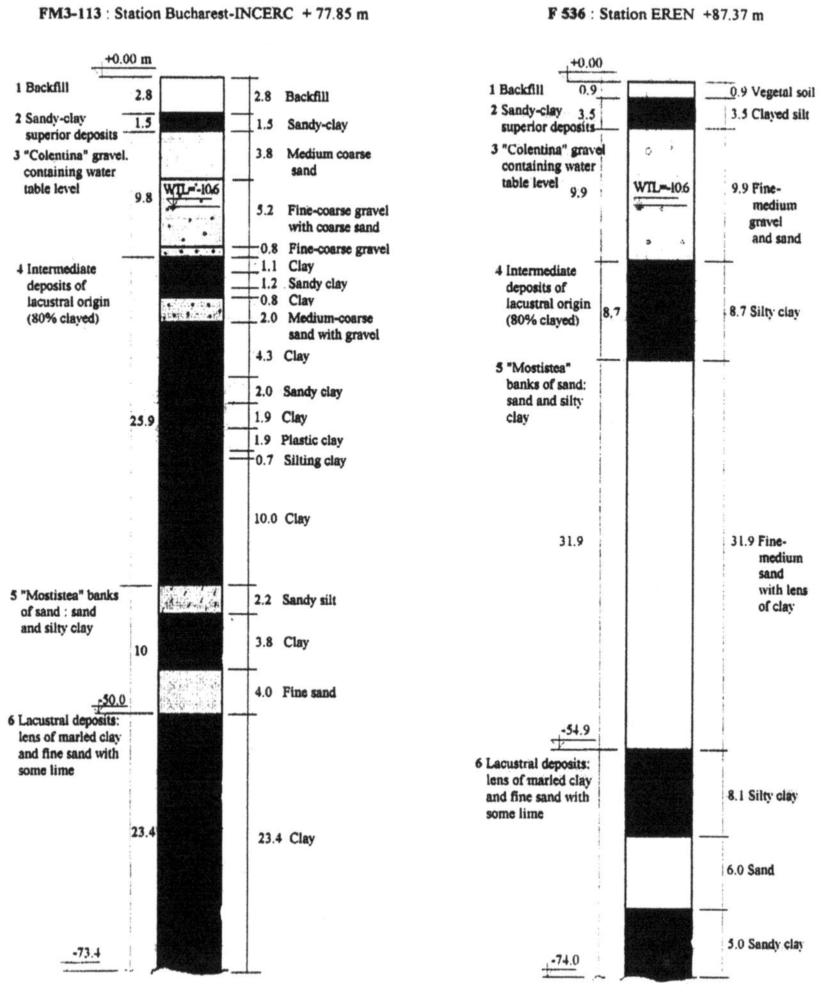

(a) (b)

Figure 6.1.20 Soil conditions in INCERC seismic station (a) and EREN (b)

The Influence of Soil Conditions

In order to see the influence of soil conditions, two accelerograms were selected, generated by the same source (Vrancea 1986), both of them recorded in Bucharest, at two seismic stations located in different soil conditions: EREN – medium soil conditions, and INCERC – soft soil conditions. In order to compare the structural response, four frame typologies have been selected (RIG, DU1, DU2 and SR1) with the joint typology from the Figure 6.1.7. All the frames are selected from C36 frame series.

The soil stratification corresponding to the two seismic stations is given in Figure 6.1.20 for INCERC (a) and EREN (b), the difference between the two being obvious.

The structural responses are given in terms of q factors, global and partial damage indices. The ductility demand is expressed in terms of both inter-storey drif, top sway displacement and plastic rotations are shown in charts of required ductility, function of accelerogram multiplier. The influence of soil conditions on the collapse mechanism is shown in Figure 6.1.25.

TABLE 6.1.7

Q FACTORS OBTAINED FOR EREN AND INCERC1986 GROUND MOTIONS.

Frame	Earthquake	λ_e	λ_θ	λ_Δ	λ_m	λ_u	q fct.	1st pl. hinge
C36RIG	EREN	1.74	12.6	4.80	7.03	4.80	2.76	Beam
	INCERC	1.32	6.72	5.60	6.55	5.60	4.23	Beam
C36SR1	EREN	0.46	6.36	6.80	5.66	5.66	12.3	Connection
	INCERC	0.54	6.26	5.55	7.39	5.55	10.3	Connection
C36DU1	EREN	0.87	6.14	5.90	8.47	5.90	6.78	Connection
	INCERC	1.01	5.99	5.41	7.58	5.41	5.38	Connection
C36DU2	EREN	0.52	4.77	5.15	4.22	4.22	8.12	Connection
	INCERC	0.63	5.16	5.02	5.73	5.02	7.98	Connection

The values of q-factors are presented in Table 6.1.7 together with the accelerogram multipliers from which they have been computed. The rigid frames show the smallest q factor values. The premature appearance of the first plastic hinge influences significantly the values of the q factor in case of SR and DU frames, although the ultimate accelerogram multipliers are on the same range with RIG frames. It can be stressed that a higher q factor does not necessarily mean that the frame ability to work in the plastic range is better.

TABLE 6.1.8

LOCAL AND GLOBAL DAMAGE INDICES CORRESPONDING TO ULTIMATE ACCELERATION.

		Maximum values			Global values			GDI
		$I_{D,c}$	$I_{D,b}$	$I_{D,j}$	$I_{D,c}$	$I_{D,b}$	$I_{D,j}$	$I_{D,g}$
RIG	EREN	0.41	0.31	--	0.39	0.22	--	0.26
	INCERC	0.37	0.30	--	0.32	0.21	--	0.24
SR1	EREN	0.29	--	0.86	0.28	--	0.59	0.57
	INCERC	0.21	--	0.95	0.20	--	0.68	0.66
DU1	EREN	0.53	--	0.89	0.45	--	0.64	0.62
	INCERC	0.36	--	0.86	0.33	--	0.58	0.56
DU2	EREN	0.20	--	0.80	0.18	--	0.51	0.49
	INCERC	0.18	--	0.98	0.16	--	0.62	0.60

Table 6.1.8 gives the values of maximum local and global damage indices computed for the ultimate acceleration. A first conclusion is that in case of RIG frames the local damages are going mainly in columns, while in case of semi-rigid frames they are concentrated in joints. As said earlier, the joint damage indices are tributary to their rotation capacity, fact that explains their greater values for semi-rigid joints.

The required local ductility in terms of rotation is given in charts - plastic rotation versus acceleration multiplier. The initial ground motions were scaled at the same value of EPA (0.25g), practically, the accelerogram multipliers λ adjust this value. Separate graphs are drawn for columns - see Figure 6.1.21 - and beams/joints – see Figure 6.1.22 - (as they are working in plastic range: RIG: beams, SR1, DU1 and DU2: joints). On the graphs are also pointed the attainment of the inter-storey drift and ultimate plastic rotation limits, as well as the formation of the plastic mechanism.

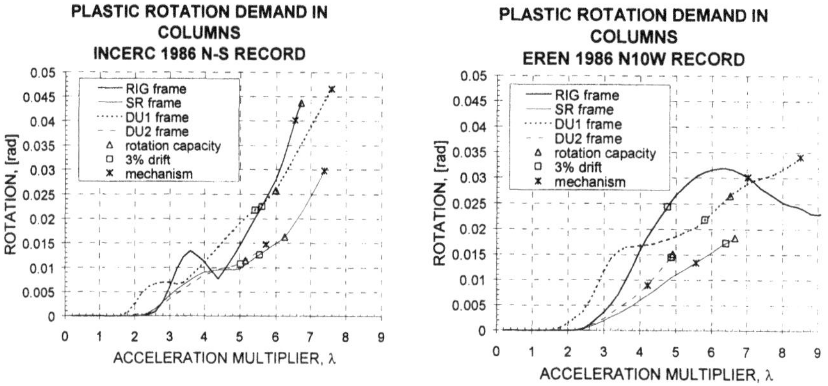

Figure 6.1.21: Plastic rotation demand in columns versus acceleration multiplier

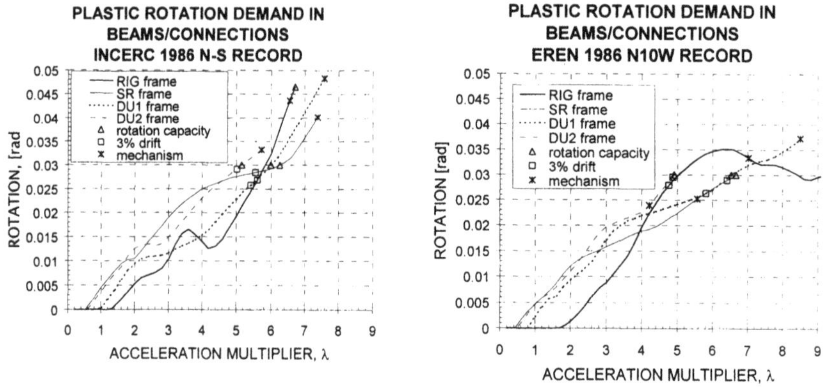

Figure 6.1.22: Plastic rotation demand in beams/connections versus acceleration multiplier

The general shapes of curves are different for the two earthquakes, for both columns and beams/joints, the curves increasing sharply for soft soils (INCERC) and smoothly for medium soils (EREN). The rotational ductility demand is greater in the case of soft soils both for columns and beam/joints. It may be concluded that the required local ductility is smaller in the case of earthquake with control period different from that of frames (EREN), and it is higher in the case of earthquake with control period closer to fundamental period of vibration of structures (due to resonance effect).

The required global ductility in terms of inter-storey drift displacement is given in Figure 6.1.23, and the global ductility in terms of top-sway displacement in Figure 6.1.24.

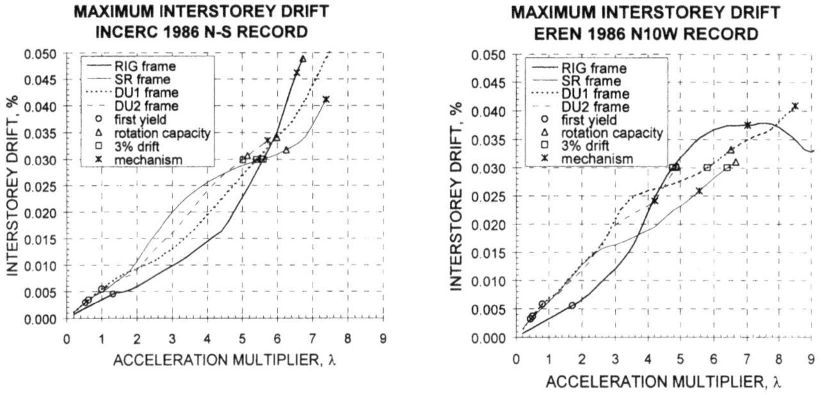

Figure 6.1.23: Maximum inter-storey drift versus acceleration multiplier

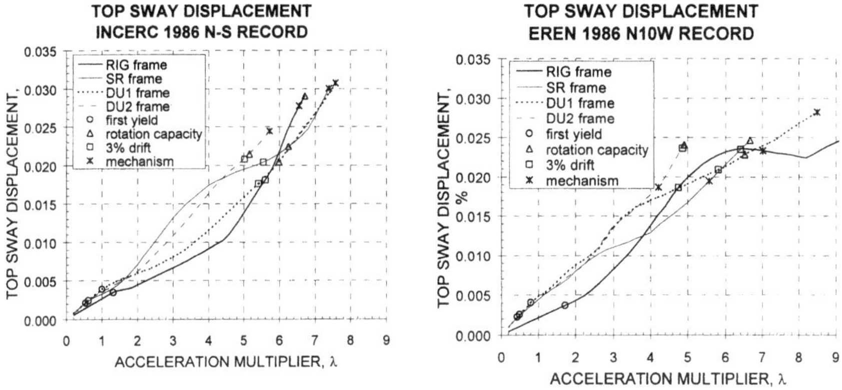

Figure 6.1.24: Top sway displacement versus acceleration multiplier

Even in terms of displacement, the structural response is different. Actually, the curves are going on similar shapes until a certain point (approx. 4 λ) from which in the case of soft soils (INCERC record) they are increasing sharply with a small increase of λ, while for the case of medium soils (EREN record) the λ values increase sharply with the increase of deformations. With other words, one can say that the soft soil conditions increase the frames' sensibility to deformations. Also, the soil conditions affect also the behavioural parameters considered in study: the points where the maximum values of inter-storey drift, the maximum plastic rotation or formation of mechanism being different. Concerning the first plastic hinge appearance, it may be seen that in some cases it appears for very small values of λ. This may influence the values of q factors computed.

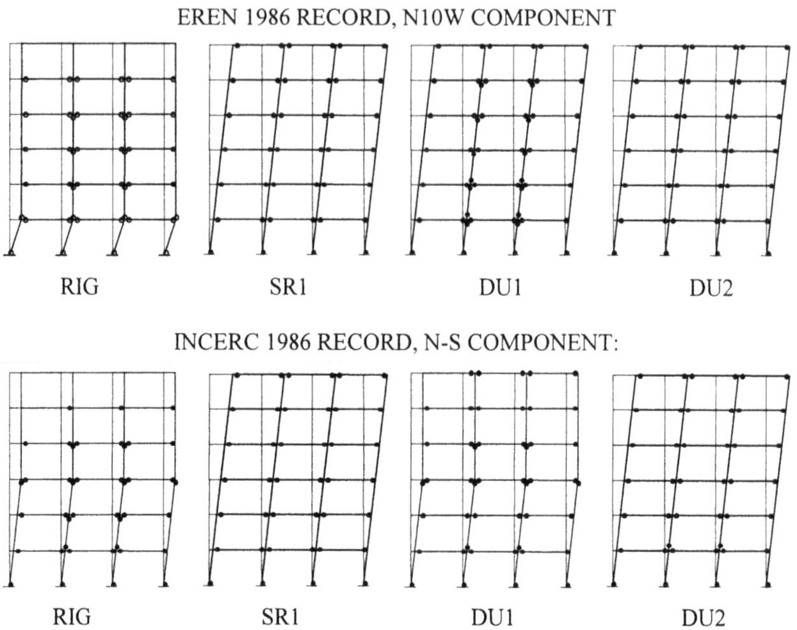

Figure 6.1.25 Influence of soil conditions on collapse mechanisms

The global mechanism is present for DU2 and SR1 frames only, frames that have the largest fundamental periods. In case of DU1 frame, the global mechanism is present for EREN record, while a storey mechanism was formed for INCERC record. A storey mechanism is formed in the case of RIG frame, both for the case of INCERC and EREN earthquakes although it was designed in order to assure a global mechanism. The use of semi-rigid joints can improve the structural collapse mechanism, so that to dissipate as much as possible from the earthquake input energy.

CONCLUSIONS

The previous chapter presented results of numerical analyses performed on C36 and C33 series of frames, for all monitored parameters: local rotational ductility, global ductility (in terms of inter-storey drift and q factor) and parameters expressing the structural performances (ultimate EPA values and the Global Damage Indices). The influence of the soil conditions on the global structural performance response given of frame was studied by means of two accelerograms generated by the same source, on two different sites. In order to have an overview of the behaviour of the frames under study, two tables with some qualitative descriptions are hereafter presented: the first characterising the influence of joint properties on global behaviour, while the second shows the influence of soil conditions on structural behaviour.

In Table 6.1.9 the following notations have been used:
RLD – Required Local Ductility (rotational)
RGD – Required Global Ductility (inter-storey drift)
q-fct. - Behavioural factor q
I_D – Damage Index
η – the seismic performance factor

TABLE 6.1.9

INFLUENCE OF JOINT PROPERTIES ON THE GLOBAL STRUCTURAL BEHAVIOUR.

Types of joints	Output Results - Frame series **C36**	Output Results - Frame series **C33**	Comments
Rigid Frame (theoretical) **RIG**	RLD – low RGD – low q-fct - high η - high I_D - low	RLD – low RGD – low q-fct - high η - high I_D - low	*Good response for all parameters. Ideal behaviour, but it remains theoretical. All the plastic hinges are in beams*
Real Rigid Frame (real rigid joints) **DU4**	RLD – low RGD – low q-fct - high η - high I_D - high	RLD – medium RGD – low η - med-high q-fct - high I_D - high	*Plastic hinges are formed in beams and internal joints Higher values of damages. This is the consequence of real rigid joints.*
Frames with semi-rigid joints: (semi-rigid joints) **SR1-SR2**	RLD – SR1: med.-high SR2: low RGD – SR1: high SR2: medium q-fct - SR1 - med -low - SR2 - high η - SR1-medium - SR2 - med-high I_D - SR1 - low - SR2 - high	RLD – SR1: med.-high SR2:medium RGD –SR1: high SR2:medium q-fct - SR1 - med -high - SR2 - medium η - SR1-medium - SR2 - med-high I_D - SR1 - high - SR2 - medium	*There are important behavioural differences between the two typologies. Frame behaviour is influenced by the joint properties.*
Dual frames without pinned joints (rigid and SR joints) **DU1 - DU4**	RLD – DU1: med.-high DU4: medium RGD – DU1: low DU4: low q-fct - DU1 - medium - DU4 - high η - DU1-med.-low - DU4 - med-high I_D - DU1 - high - DU4 - high	RLD – DU1: medium DU4: medium RGD – DU1: medium DU4: low q-fct - DU1 - medium - DU4 - medium η - DU1 - high - DU4 - high I_D - DU1 - med-high - DU4 - med-high	*Good general behaviour. The great values of the damage indices are due to the joint degradation. Frame behaviour is influenced by the joint properties and by their disposal within the structure.*
Dual frames with pinned joints (rigid and SR joints) **DU3 - DU5**	RLD – DU3: med.-high DU5: high RGD – DU3: high DU5: high q-fct - DU3 - low - DU5 - high η - DU3-medium - DU5 - medium I_D - DU3 - medium - DU5 - high	RLD – DU3: high DU5: medium RGD – DU3: high DU5: high q-fct - DU3 - low - DU5 - low η - DU3 - low - DU5 - low I_D - DU3 - high - DU5 - high	*General bad behaviour of the frames. Great values of Required Local and Global Ductilities, and also of the Damage indices.* *Pinned joints have a bad influence on frame behaviour*
Horizontal Dual Frames (semi-rigid joints) **DU2**	RLD – DU3: med.-high RGD – DU3: medium q-fct - low η - med-low I_D - high	**--**	*Intermediary results, between the two types of SR frames. Can be a good alternative.*

TABLE 6.1.10

INFLUENCE OF SOIL CONDITIONS ON GLOBAL BEHAVIOUR.

Frame	INCERC record (soft soil) T_C=1.256s.	EREN record (medium soil) T_C=0.664s.	Comments
C36RIG T=1.33sec.	UD – high URR – high CM – mech. type 1	UD – low URR – low CM – mech. type 3	*The frame behaviour is better on medium soils.*
C36SR1 T=1.69sec.	UD – high URR – high CM – global mech.	UD – low URR – low CM – global mech.	*The frame behaviour seems to be better for medium soils, but for this case the limitations are stricter.*
C36DU1 T=1.52sec.	UD – high URR – high CM – mech. type 1.	UD – medium URR – medium CM – global mech.	*The frame behaviour is better on medium soils. The soil conditions influence the collapse mechanism.*
C36DU2 T=1.61sec.	UD – limited URR – limited CM – global mech.	UD – limited URR – limited CM – global mech.	*Limited values of both deflections and rotations are observed. No big differences up to collapse.*

In Table 6.1.10 the following notations have been used:
UD – Ultimate Deflections (inter-storey drift and top-sway displacement)
URR – Ultimate Required Rotations (columns and beams/columns)
CM - Collapse Mechanism

Finally, the scope of this study was to analyse the influence of beam-to-column joint properties on the global performance and ductility demand of multi-storey steel building frames. In the following, the main conclusions of the study are listed:

1. Considering the seismic response parameters:

- The effects of a particular seismic motion on a given structure depend on all the parameters that are part of the system: earthquake (T_C, EPA, PGA), soil (T_S) and structure (T). From this reason, all results and comparisons on the scaled accelerograms should be considered with certain reserves.
- For structures designed in seismic areas, the seismic performance factor (η) can be a good tool in evaluation their seismic vulnerability in regard with the seismic motion characteristics. It can give also information about the structural performances under different earthquakes (e.g. earthquakes with different return periods).
- The definition of the collapse mechanism in these analyses was considered just like an equivalent limit state. There are still uncertainties in the definition of the collapse mechanism in the case of frames under dynamic time-history analyses.

2. Considering the influence of frame typology:

- Perfect rigid frames (RIG) proved a very good behaviour, but they still remain unrealistic (joints with infinitely rotational stiffness cannot be designed). The real rigid frames display big damages, damages due mainly to joint degradation. This may be related to the relatively low value of ductility (plastic rotation capacity of 0.02 rad. for welded and 0.03 rad. for bolted joints, respectively) that have been imposed for computing the damage index. It is expected an improvement in the frame performances if the joint rotation capacity increases.
- Frames having semi-rigid joints seem to have a good general behaviour. The problem is to find the right joint properties that "fit" best for the given frame. Anyway, the "semi-rigid" frames analysed in the present study could be an alternative to the "rigid" ones. The only problem is to establish the limits of their use.

- Dual frames with rigid / semi-rigid joints represent an open problem: it seems possible to find configurations which can improve the structural performances, but further studies are necessary.
- Dual frames with pinned joints in middle span have generally a bad behaviour and they should be avoided. They have a bad behaviour of all the three parameters monitored.
- A Moment resisting Frame should have as many dissipative zones as possible, in order to be able to dissipate a larger amount of seismic input energy.

3. Concerning the joint properties and their use on Moment Resisting Frames:

- Joints with larger rotation capacities should be used in zones of high seismicity, especially in frames with partial-strength joints. The larger the rotation capacity, the larger the available ductility of frames.
- The joint detailing and resistance can control the formation of the plastic hinges in a structure.
- The changes in joint semi-rigidities do not necessarily lead to a significant change in columns' required ductility, but they can lead to a change in the required ductility of beams.
- The use of semi-rigid joints can also influence the collapse mechanism of structures, in many cases in a better manner, in order to have a global mechanism, which makes the structure more ductile (practically it uses its entire capacity to dissipate energy).

4. Concerning the soil influence:

- Structures with fundamental periods near to the soil period are prone to resonance. This is sustained also by this study, in which the earthquakes with control periods T_C closer to the structural fundamental period resulted more destructive for these situations.
- Taking into account the soil dynamic characteristics, it can be stressed that the use of semi-rigid joints can change the fundamental period of the frame from that of the soil, to a safer period range. In this way, the ductility supply could also change.

Both the required and the available ductility of a structure depend on many parameters. By the use of semi-rigid joints, we could improve both the available and the required ductility, but what is most important, we may control the structural behaviour.

However, the conclusions of the present study have to be regarded with prudence and, anyway, this must be limited to the particular analysed frames.

REFERENCES

AISC 97, (1997). *Seismic Provisions for Structural Steel Buildings*. American Institute of Steel Construction, Inc. Chicago, Illinois, USA.

Astaneh A., Nader N. (1992). Seismic Behaviour and Design of Semi-rigid Steel Frames. *Report No. UCB/EERC 92/06, EERC,* University of California, Berkeley.

Bertero V. (1997). General Report on Codification, Design and Application. *STESSA '97 Behaviour of Steel Structures in Seismic Areas,* Proceedings of the Second International Conference 3-8 Aug. 1997, Kyoto, Japan.

Dubina D. et. al. (1999). Global Performances of Steel Moment Resisting Frames with Semi-rigid Joints. *Proceedings of the 6th international colloquium on Stability and Ductility of Steel Structures,* Timisoara, Romania, 367-377.

Dubina D. et al. (1998). Suitability of Semi-rigid Joint Steel Building Frames in Seismic Areas. *Eleventh European Conference on Earthquake Engineering,* Paris.

Elnashai A. S. et al. (1996). Experimental and Analytical Investigation into the Seismic Behaviour of Semi-rigid Steel Frames. *ESEE Research Report No. 96 - 7 December 1996,* Imperial College, London.

Elnashai A. S. et al. (1998). Response of Semi-Rigid Steel Frames to Cyclic and Earthquake Loads. *Journal of Structural Engineering,* ASCE, Vol. 124 No8, 857-868.

ENV 1993-1-1 (1993). *EUROCODE 3: Design of Steel Structures. Part 1.1. General Rules and Rules for Buildings.* Brussels: CEN, European Committee for Standardisation.

ENV 1998-1-1 (1993). *EUROCODE 8: Earthquake Resistant Design of Structures. Part 1.: General Rules and Rules for Buildings - Seismic Actions and General Requirements for Structures.* Brussels: CEN, European Committee for Standardisation.

Gioncu et. al. (1997). Simplified Approach for Evaluating the Rotation Capacity of Double T Steel Sections. *STESSA '97 Behaviour of Steel Structures in Seismic Areas,* Proceedings of the Second International Conference 3-8 Aug. 1997, Kyoto, Japan.

Kato B. et. al. (1997). Seismic Damage of Steel Beam-to-Column Rigid Connections in the Hyogoken-Nanbu Earthquake. *STESSA '97 Behaviour of Steel Structures in Seismic Areas,* Proceedings of the Second International Conference 3-8 Aug. 1997, Kyoto, Japan.

Lungu D. et. al. (1997). Effective Peak Ground Acceleration (EPA) Versus Peak Ground Acceleration (PGA) and Effective Peak Ground Velocity (EPV) Versus Peak Ground Velocity (PGV) for Romanian Seismic Records. *Eurocode 8 - Worked examples*

Mahin A. S. (1998). Lessons from Damage of Steel Buildings During the Northridge Earthquake. *Engineering Structures*, vol No. 20, 261-270.

Mazzolani F. M., Piluso V. (1996). *Theory and Design of Seismic Resistant Steel Frames.* London, E&FN Spon.

Mazzolani F.M., Piluso V. (1994). *ECCS Manual on Design of Steel Structures in Seismic Zones.* Brussels ECCS, European Convention for Constructional Steelwork.

Men F., Cui J. (1996). A Simplified Reasoning of Rigidity Effect of Foundation Soil on Seismic Damage to Building. *11th WCEE,* Acapulco, Mexico. Paper No.246. Elsevier Science.

Nakashima M. et. al. (1998). Classification of Damage to Steel Buildings in the 1995 Hyogoken-Nambu Earthquake. *Engineering Structures*, vol No. 20, 271-281.

SAC Joint Venture, (1996). *Connection Test Summaries.* SAC 96-02. Sacramento, California.

SEAOC Vision 2000 Committee Report, (1995). *Performance Based Seismic Engineering.* Sacramento, California: Structural Engineers Association of California.

Watanabe et. al. (1998). Performance and Damages to Steel Structures During the 1995 Hyogoken-Nambu Earthquake. *Engineering Structures,* vol No. 20, 282-290.

6.2

INTERACTION BETWEEN LOCAL AND GLOBAL PROPERTIES

Ioannis Vayas

INTRODUCTION

Building structures are designed against earthquakes so that they undergo inelastic deformations during a strong seismic event. This presumes a stable hysteretic behaviour of structural elements during cyclic loading. Accordingly, structures have to be designed for ductility, in addition to stiffness and strength. Ductility under seismic conditions expresses the ability of the structural elements to exhibit inelastic cyclic deformations, without considerable stiffness or strength degradation. The provision of ductility allows for inelastic dissipation of energy, which leads to the development of lower forces that have to be resisted by the structure compared to those that would develop under an elastic response. Under certain conditions it is therefore possible to result in a more economic design.

In order to achieve a ductile response, measures have to be taken both at the level of individual members and at the level of the structure. Relative member strengths, structural geometric configuration, position of resisting elements etc. are important design considerations. It is therefore important during design to associate local and global properties of the structure. For moment resisting frames considered here, local properties are expressed through the ability of the members to form plastic hinges at localised positions and therefore to the rotation capacity of the plastic hinges under cyclic conditions. On the other hand, global properties are a function of a number of parameters as the structural layout, the influence of 2nd order effects, the relative strength of members etc. In the present work, the main parameters affecting local and global properties as well as their mutual relation are studied.

Chapter 2 provides information for the local properties in respect to the material, the cross sections, the members and the joints. The local properties are considered in as much as they are needed for further evaluations. A deeper insight is given elsewhere in this publication. For the purpose of computations, three levels of local ductility, high, medium and low, associated to different rotation capacities are introduced. Additionally, a proposal is made that allows for a low-cycle-fatigue evaluation of the local properties.

The global properties are studied in chapter 3. Simple bilinear systems and complete frames are investigated. The response to monotonic, dynamic and seismic loading is examined by means of extensive parametric studies. The main parameters under consideration are the level of local ductility of members, the properties of connections, the presence or not of irregularities, the level of vertical loading affecting the structural stability and the fatigue behaviour of members and connections.

NOTATION

D	damage index
E	modulus of elasticity
E_d	energy dissipated during a cycle
E_p	energy dissipated by a bilinear material
EI_b	stiffness of a connected beam
H_y	lateral yield force
H_{max}	lateral maximal force
I	second moment of area
K	elastic joint stiffness
L_b	length of the connected beam
M_{bRd}	design strength of a connected beam
M_{jRd}	design strength of a joint
M_p	plastic moment
N	number of stress range cycles
P	vertical load
P_{crit}	critical vertical load
S^1	stiffness from 1st order analysis
S^2	stiffness from 2nd order analysis
S_j	elastic joint stiffness
W_e	elastic energy
W_p	plastic energy
ΔH	additional horizontal force from 2nd order effects
$\Delta\sigma$	stress range
$\Delta\sigma_R$	fatigue strength
$\Delta\varphi_p$	fatigue deformability
Ω	overstrength
a	constant of the fatigue curve
c	outstand width of the flange of a cross section
d	height of the web of a cross section
f_y	yield stress
f_{yst}	static yield strength
f_u	tensile strength
k_y	yield curvature
k_p	plastic curvature
k_u	ultimate curvature
l	span of a beam
l_p	plastic hinge length
m	slope of the fatigue strength curve
$\overline{m}$	relative joint moment capacity
n	number of cycles
α	plastic hinge to beam span ratio
α_y	yield acceleration
α_u	ultimate acceleration
δ_p	lateral plastic deformation
δ_y	lateral yield deformation
δ_u	lateral ultimate deformation

$\varepsilon = \sqrt{235 / f_y}$, f_y in N/mm^2

ε_y yield strain

ε_u ultimate strain

$\bar{\lambda}_{LT}$ relative slenderness to lateral torsional buckling

μ_δ ductility in terms of displacement

μ_ε ductility in terms of strain

μ_E ductility in terms of energy

μ_κ ductility in terms of curvature

μ_φ ductility in terms of rotation

φ_{mon} rotation capacity for monotonic loading

φ_y yield rotation

φ_u ultimate rotation

LOCAL PROPERTIES

General

Moment frames resist imposed actions primarily through bending of their members. Bending also develops at beam-to-column joints, so that the relevant connections shall be formed as moment connections. Local ductility is accordingly understood as the ability of members connections and joints to rotate due to bending beyond the plastic moment. Though steel is ductile as a material, local ductility is limited due to the appearance of instabilities such as local buckling, lateral torsional buckling etc. Additionally, ductility is adversely influenced under cyclic loading conditions due to cracking at the parent material, or near welds as a consequence of low-cycle-fatigue. Evidently, the global non-linear performance relies heavily on the corresponding performance of the zones where the inelastic action develops. In the following the local properties for monotonic and cyclic loading are presented. Local ductility is treated in detail elsewhere in the present publication, so that this chapter will focus on a more or less qualitative description. For further calculations, local ductility and response to low-cycle-fatigue will be will be considered as parameters.

Properties of steel

The mechanical properties of steel may be determined by means of a tensile coupon test. Such a test provides the stress-strain characteristic of the material (Figure 6.2.1). For normal structural steels, following values may be determined:

- The yield strength f_y which corresponds to first significant yielding
- The static yield strength f_{yst} which corresponds to yielding at a very low loading rate
- The ultimate tensile strength f_u which is the maximal engineering stress
- The modulus of elasticity E
- The yield strain ε_y and
- The ultimate strain ε_u

The yield and ultimate strength at normal temperature is a function of the element thickness and is given by the relevant Euronorm EN 10025 for structural steels in Table 6.2.1. The static yield strength is 5% to 10% lower than the yield strength listed in the table.

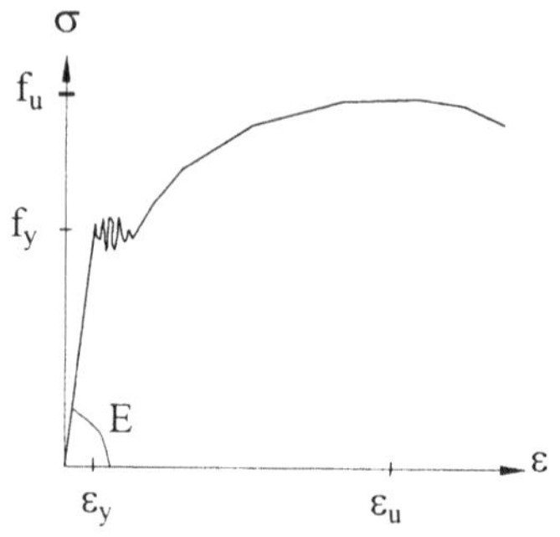

Figure 6.2.1: Stress – strain diagram for structural steel

The overstrength of steel may be determined as following:

$$\Omega = \frac{f_u}{f_y} \tag{6.2.1}$$

The yield ratio, YR, is the inverse of the overstrength. According to Table 6.2.1 and eq. (6.2.1), the overstrength of structural steel is approximately equal to 1.5, the yield ratio is then approximately equal to 0.65.
The yield strain may be determined as following:

$$\varepsilon_y = \frac{f_y}{E} \tag{6.2.2}$$

while the ultimate strain is usually ranging between 20% and 30%.

The ductility of steel may be determined according to:

$$\mu_\varepsilon = \frac{\varepsilon_u}{\varepsilon_y} \tag{6.2.3}$$

so that according to (6.2.2) it takes values between 120 and 250.

Ductility of steel may be also expressed in terms of material toughness as determined from Charpy impact tests. Material toughness strongly depends on **temperature**, the transition temperature being the temperature at which steel becomes brittle (Figure 6.2.2). Accordingly, steel is ductile above the transition temperature and brittle below it. The relevant material specifications include values of the material toughness in dependence on the application. Accordingly, the Japanese Code requires a value of Charpy V-notch energy of 27 J at 0° C, while the German Code requires 80 J at 20° C for steels with moderate weldability and 70 J at 0° C for steels with high weldability determined as the average between three specimen.

Besides temperature, the applied **strain rate** is another important parameter influencing the ductility of steel. Figure 6.2.3 shows the relationship between yield strength and tensile strength of steel and the strain rate.

TABLE 6.2.1
NOMINAL VALUES OF YIELD STRENGTH AND ULTIMATE STRENGTH FOR STRUCTURAL STEEL ACCORDING TO EN 10025

Nominal steel grade	Thickness t [mm] *)			
	$t \leq 40$ mm		40 mm $< t \leq 100$ mm **)	
	f_y (N/mm^2)	f_u (N/mm^2)	f_y (N/mm^2)	f_u (N/mm^2)
EN 10025:				
Fe 360	235	360	215	340
Fe 430	275	430	255	410
Fe 510	355	510	335	490
pr EN 10113:				
Fe E 275	275	390	255	370
Fe E 355	355	490	335	470

*) t is the nominal thickness of the element.
**) 63 mm for plates and other flat products in steels delivery condition TM to prEN 10113-3.

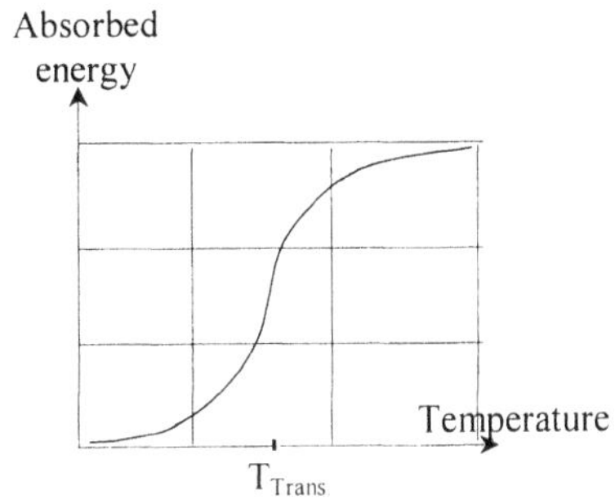

Figure 6.2.2: Material toughness as a function of test temperature

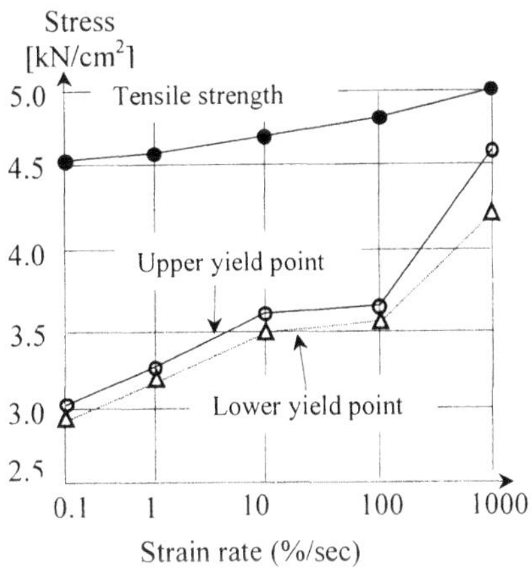

Figure 6.2.3: Relationship between yield stress or tensile strength and strain rate [13]

For the yield stress and the tensile strength this relationship may be expressed as following:

$$f_y = f_{y0} + k_y \log(\dot{\varepsilon}/\dot{\varepsilon}_o) \quad (6.2.4)$$

$$f_u = f_{u0} + k_u \log(\dot{\varepsilon}/\dot{\varepsilon}_o) \quad (6.2.5)$$

where:
f_y, f_u = yield stress and tensile strength at the strain rate $\dot{\varepsilon}$
f_{y0}, f_{u0} = yield stress and tensile strength at the strain rate $\dot{\varepsilon}_0$.
Under quasi-static loading, the strain rate may be taken equal to
$\varepsilon_0 = 10^{-4}$/sec The values of the parameters k were found to be equal to (Kato, 1997):
k_y = 2.1 kN/cm^2 and
k_u = 0.74 kN/cm^2

It may be seen that the yield stress is much more affected by the strain rate than the tensile strength. Consequently the yield ratio of normal steel increases from 0.6 under static conditions to 0.8 at high strain rates, while the overstrength decreases from 1.5 to 1.25. The ultimate elongation at failure does not appear to be affected by the strain rate. Accordingly, with increasing strain rate the ductility of the material, expressed by eq. (6.2.3), decrease.

The strain rate affects also largely the material toughness. Results of Charpy impact tests reported in Kato (1997) are shown in Figure 6.2.4. Consequently, fracture is brittle at high strain rate and at low temperatures in contrast to room temperature where it is ductile.

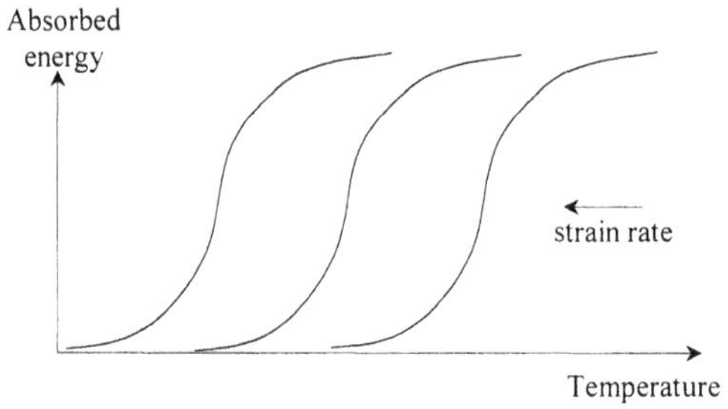

Figure 6.2.4: Energy Absorption and transition temperature vs. temperature

Considering that the strain rate during an earthquake is approximately equal to 0.1/sec, the yield stress, tensile strength and yield ratio for a S 235 steel are equal to:

$$f_y = 23.5 + 2.10 \log(0.1/0.0001) = 29.8 \text{ kN/cm}^2 \quad (6.2.6)$$

$$f_u = 35.5 + 0.74 \log(0.1/0.0001) = 37.7 \text{ kN/cm}^2 \quad (6.2.7)$$

$\Omega = f_u / f_y = 37.7 / 29.8 = 1.27$

$YR = 1 / \Omega = 0.79$

Another important parameter that affects the material properties is **triaxiality**. The relevant expressions will be derived subsequently, avoiding for the sake of simplicity a tensor formulation. The stress-strain relation in the principal axes may be written as:

$$\sigma_i = \lambda \cdot e + 2 \cdot G \cdot \varepsilon_i \quad \text{i=1,2,3} \tag{6.2.8}$$

where:

$e = \varepsilon_1 + \varepsilon_2 + \varepsilon_3$ $\qquad$ $\lambda = \dfrac{\nu \cdot E}{(1+\nu)\cdot(1-2\nu)}$

$G = \dfrac{E}{2\cdot(1+\nu)}$ = shear modulus $\qquad$ ν = Poisson's ratio = 0,3 for steel

The linear and quadratic invariants of the stresses may be written as:

$$S_1 = \sigma_1 + \sigma_2 + \sigma_3 \tag{6.2.9a}$$

$$S_2 = \frac{1}{2}\cdot(\sigma_1^2 + \sigma_2^2 + \sigma_3^2) \tag{6.2.9b}$$

The stresses may be split into a hydrostatic component and a deviator as following:

$$\sigma_i^H = \sigma_M = (\sigma_1 + \sigma_2 + \sigma_3) / 3 \tag{6.2.10a}$$

$$\sigma_i^D = \sigma_i - \sigma_M \tag{6.2.10b}$$

After numerical treatment, the invariants of the deviator may be written as:

$$S_1^D = 0 \quad \text{and} \quad S_2^D = S_2 - \frac{1}{6}\cdot(S_1)^2 \tag{6.2.11}$$

The von Mises yield criterion that applies to steel is written as:

$$S_2^D = \frac{1}{3}\cdot f_y^2 \tag{6.2.12}$$

Eq. (6.2.12) implies that only deviatoric stress states lead to material yielding. On the contrary, hydrostatic stress states lead only to volume changes. This is illustrated in Figure 6.2.5, where hydrostatic stress states are represented by the space diagonal and the yield criterion by a circle.

The stress states and the yield criterion may be formulated at various positions of the structural elements. For example four cases may be distinguished (Figure 6.2.6):

- Unrestrained conditions (Uniaxial stess states, point 1 in Figure 6.2.6)
  It is $\sigma_1 = \sigma$ and $\sigma_2 = \sigma_3 = 0$

  After numerical manipulation it may be found that $S_2^D = \frac{1}{3}\cdot\sigma^2$

  The yield criterion leads to:

  $$\sigma = f_y \tag{6.2.13}$$

- Partly restrained conditions (Plane stress states, point 2 in Figure 6.2.6)
  It is $\varepsilon_1 = \varepsilon$, $\varepsilon_2 = 0$ and $\sigma_3 = 0$

  Using eq. (6.2.8), the stress conditions are determined from:

$\sigma_1 = \dfrac{\sigma}{(1+\nu)\cdot(1-\nu)}$ $\qquad$ $\sigma_2 = \dfrac{\nu\cdot\sigma}{(1+\nu)\cdot(1-\nu)}$ $\qquad$ $\sigma_3 = 0$

After numerical manipulation it may be found that $S_2^D = \dfrac{\sigma^2 \cdot (1-\nu+\nu^2)}{3 \cdot (1-\nu)^2 \cdot (1+\nu)^2}$

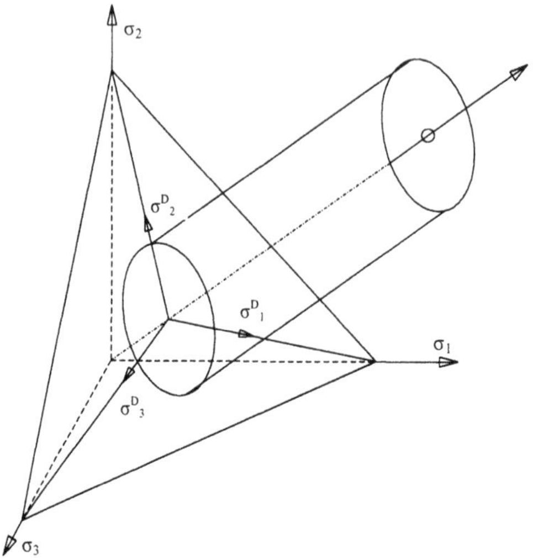

Figure 6.2.5: Three-dimensional principal stress states

The yield criterion leads to:

$$\sigma = \frac{(1-\nu)\cdot(1+\nu)}{\sqrt{1-\nu+\nu^2}} \cdot f_y \quad or \quad \sigma = 1{,}02 \cdot f_y \tag{6.2.14}$$

- Pure shear (Point 3 in Fig. 6.2.6)

It is $\sigma_1 = \sigma$ $\sigma_2 = -\sigma$ and $\sigma_3 = 0$

After numerical manipulation it may be found that $S_2^D = \sigma^2$

The yield criterion leads to:

$$\sigma = f_y / \sqrt{3} \tag{6.2.15}$$

- Restrained conditions (Tri-axial stress state , point 4 in Fig. 6.2.6)

It is $\varepsilon_1 = \varepsilon$ and $\varepsilon_2 = \varepsilon_3 = 0$

Using eq. (6.2.8), the stress conditions are determined from:

$$\sigma_1 = \frac{(1-\nu)\cdot\sigma}{(1+\nu)\cdot(1-2\cdot\nu)} \qquad \sigma_2 = \sigma_2 = \frac{\nu\cdot\sigma}{(1+\nu)\cdot(1-2\cdot\nu)}$$

After numerical manipulation it may be found that $S_2^D = \dfrac{\sigma^2}{3\cdot(1+\nu)^2}$

The yield criterion leads to:

$$\sigma = (1+\nu)\cdot f_y \quad or \quad \sigma = 1{,}3 \cdot f_y \tag{6.2.16}$$

Under restrained conditions, it is accordingly required that the stress shall increase to 1,30 f_y for yielding to start. On the other side, as previously found, the yield stress increases at high strain rates and the overstrength ratio becomes 1,27. It may be therefore concluded that the combination of triaxiality and high strain rates "consumes" all the ductility of steel, so that brittle failure occurs.
It may be concluded that:

- Steel is ductile at normal temperature and when quasi-statically loaded,
- Steel becomes non-ductile at low temperature or if dynamically loaded with high speed
- Brittle failure occurs at high strain rates and under restrained conditions.

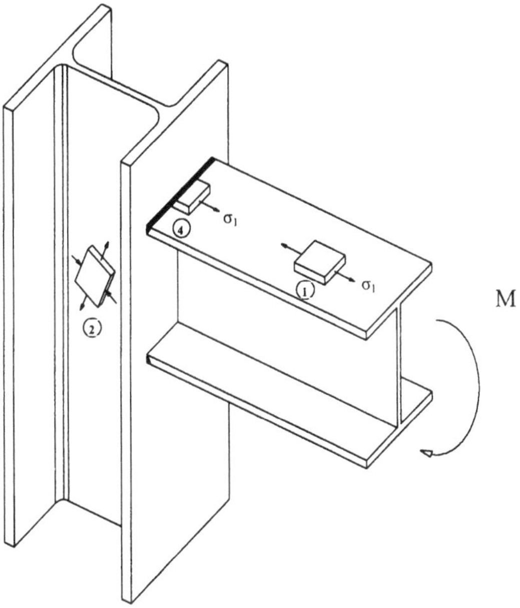

Figure 6.2.6: Stress and strain conditions at various positions

Behaviour of cross sections for monotonic loading

The cross sectional response is traditionally described by means of moment-curvature curves. Such curves may be derived by application of layer models, where the cross section is divided in fibers. Under the assumption that plain sections remain plain, the strain distribution is linear. From the σ-ε diagram of steel, the stresses and, after appropriate integration, the bending moments and normal forces in the cross section may be determined.

For steel sections prone to local buckling, this model needs an extension in so far that the stress in the fibers cannot be derived only from the σ-ε diagram of steel. The stress distribution over the width of a plate element becomes nonlinear after buckling, the stress of the buckled parts being reduced. This effect may be numerically treated by an introduction of effective widths for the compressed plated parts of the cross section, where instead of gross areas and reduced stresses, equivalent effective areas and full stresses are used. This method, called effective width method, is proposed by most steel design Codes for the determination of resistance of thin-walled elements. However, the application of this method is restricted to the loading part of the response curve only and does not allow the evaluation of ductility.

By introduction of a strain-oriented formulation and inclusion of strain hardening as described in Vayas (1994), (1997), the effective width method is extended beyond the limit load, allowing thus the derivation of the cross sectional response in both the loading and the unloading branch (Figure 6.2.7). As a result, local ductility in terms of curvature as given by eq. (6.2.17) may be calculated.

$$\mu_k = \frac{k_u}{k_y} \tag{6.2.17}$$

where:
k_u = ultimate curvature,
k_y = yield curvature at the development of the plastic moment

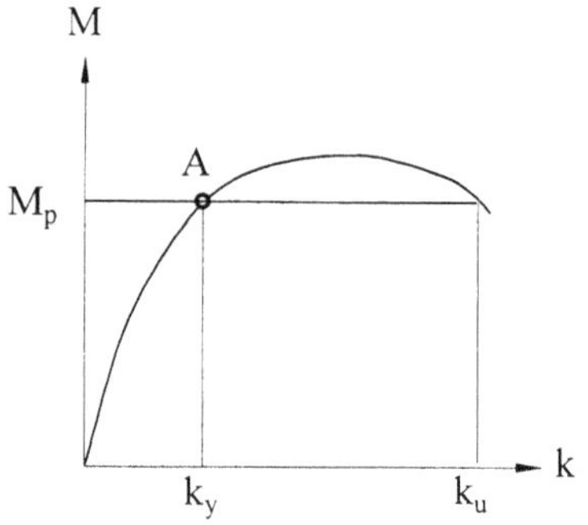

Figure 6.2.7: Definition of ductility in terms of curvature

The calculations show that ductility in terms of curvature is a characteristic of the cross section, influenced by:

- the compactness of the compressed plated parts of the cross section (flange, web) and
- the level of the applied axial force.

Behaviour of cross sections for cyclic loading

For the derivation of moment-curvature curves for cross sections subjected to cyclic loading, the same procedure as described before may be applied. For a certain strain distribution, the stress distribution at the cross section can be found provided that the constitutive material law for cyclic loading is known. Such laws depend on the type of steel, the sequence of cycling (high-to low, constant, low-to-high amplitudes), the existence or not of load reversals etc. The problem becomes more complicated in the presence of local buckling, where the stress is non-uniform. Adopting a simplified constitutive law, which includes isotropic hardening but neglects Bauschinger effects, moment-curvature curves for I-sections have been derived in Vayas (1990) (Figure 6.2.8).

The results show a stiffness and strength deterioration under increasing cyclic reversals, especially after the initiation of local buckling. This is due to the fact that the combination of local buckling with plasticity leads to out-of-plane displacements of the plated parts, so that the overall response becomes both weaker and softer. A measure of deterioration is given by the ratio:

$$\mu_E = \frac{E_d}{E_p} \tag{6.2.18}$$

where:
E_d = energy dissipated during a cycle (which is expressed by the area enclosed in a full cycle)
E_p = corresponding energy of a bilinear linear elastic-perfectly plastic response.
By defining certain values of the above determined energy ratio, compactness requirements for cross sections may be established in cases of cyclic loading.

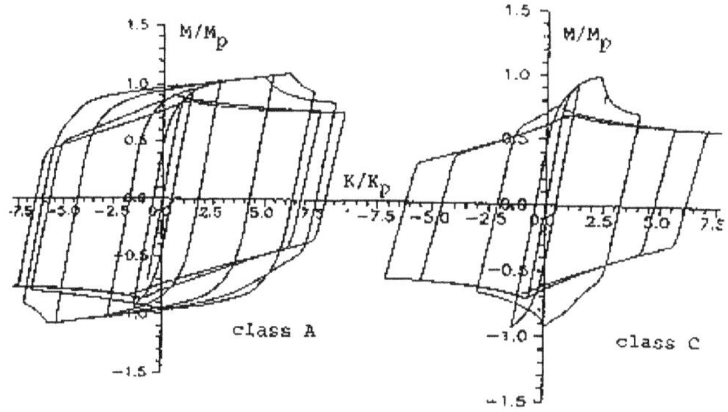

Figure 6.2.8: Moment-curvature curves for cyclic loading [21]

Behaviour of members for monotonic loading

The response of steel members is adversely influenced by several possible instability modes of failure such as flexural, lateral torsional, distorsional or shear buckling, contrary to cross sections which are prone only to local buckling. Ductility is therefore a **member** rather than a cross section property (Vayas et al. 1996, 1999). For moment resisting frames subjected to seismic loading, inelastic behaviour is preferably to be restricted in the beams. The ductility considerations refer therefore to beam elements, in which the influence of axial forces is small. The possible modes of failure for beam elements are local and lateral torsional buckling.

The rotation capacity of steel beams has been extensively studied both experimentally and theoretically (Gioncu 1997, Spangemacher 1992, Vayas et. al 1999). A detailed reference list is given in Gioncu (1997). Most experimental investigations have been performed on single span beams subjected to one or two concentrated loads, referred to usually as three or four point bending tests. The response of the beam, by means of the mid-span moment - end rotation curve, is shown in Figure 6.2.9.

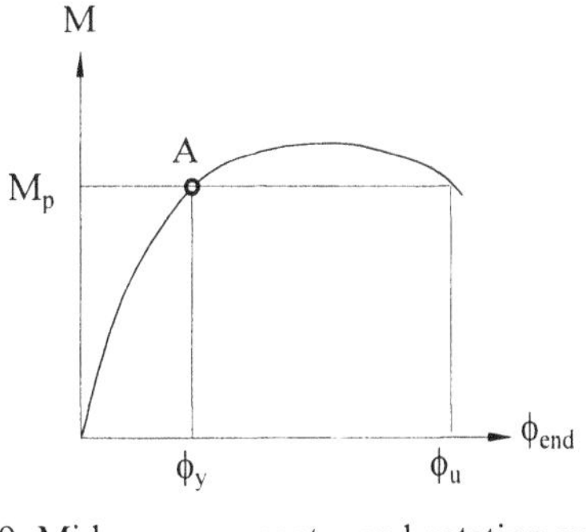

Figure 6.2.9: Mid-span moment – end-rotation curve of a beam

This response may be numerically treated by application of the plastic zone method or the plastic hinge method. The former allows for a gradual extend of the plastic deformations, starting from the most stressed fiber of the most stressed cross section and spreading over the depth of the cross section and along the member longitudinal axis. Accordingly, the stiffness along the member changes with increasing loading, as at cross sections where yielding has occurred the initial elastic stiffness (EI) gradually reduces. The deflection line is therefore continuous during loading and unloading and rotation takes place only at true, and not plastic, hinges.

In the plastic hinge method plasticity is supposed to concentrate at the positions of plastic hinges, while the remaining parts behave elastic. After the plastic hinge formation, point A in Figure 6.2.9, the deflection line becomes discontinuous as the beam section is starting to rotate in the span. The rotation capacity is defined as the amount of plastic rotation that exhibits the beam without unloading beyond the plastic moment. As shown in Figure 6.2.9, the rotation capacity is equal to:

$$\phi_p = \phi_u - \phi_y \tag{6.2.19}$$

Local ductility in terms of rotation is expressed by:

$$\mu_\varphi = \frac{\varphi_u}{\varphi_y} \tag{6.2.20}$$

where:
φ_u = ultimate rotation at mid-span
φ_y = yield rotation at mid-span when the plastic moment M_p develops

The rotation capacity may be expressed in terms of rotation ductility according to:

$$\phi_p = (\mu_\phi - 1) \cdot \phi_y = (\mu_\varphi - 1)\frac{1}{2}\frac{M_p}{EI}l \tag{6.2.21}$$

where:
EI = stiffness of the cross section and
l = span of the beam

The relationship between local ductility in terms of rotation and curvature may be written as:

$$\mu_\varphi = 1 + 2\,(\mu_\kappa - 1)\,\alpha \tag{6.2.22}$$

where:
$k_p = k_u - k_y$ = plastic curvature
$\alpha = l_p / l$ = plastic hinge length to beam span ratio

μ_κ in eq. (6.2.22) is a property of the cross section, μ_φ is a property of the member. As previously stated, ductility is reduced due to the appearance of instabilities, depends therefore on the relevant slenderness to local and lateral torsional buckling ($\bar{\lambda}_{LT}$) as illustrated in Figure 6.2.10.

A proposal for classification based on the member rather than the cross section, properties is made by Vayas et. al. (1999). Figure 6.2.11 gives a qualitative illustration of the proposal, Figure 6.2.12 present the curves for $\bar{\lambda}_{LT}$=0.

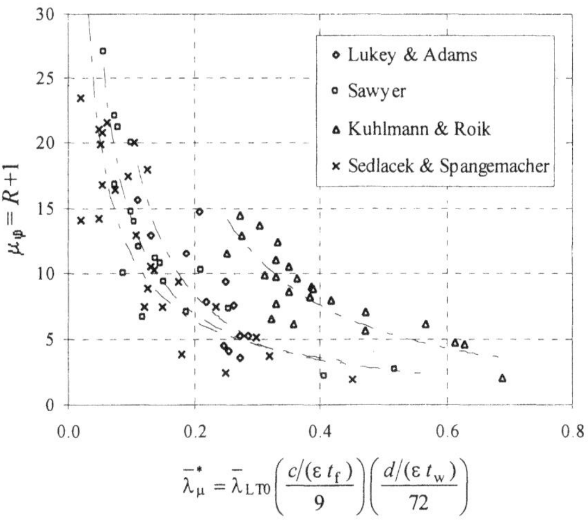

Figure 6.2.10: Rotation ductility vs. generalized slenderness from tests

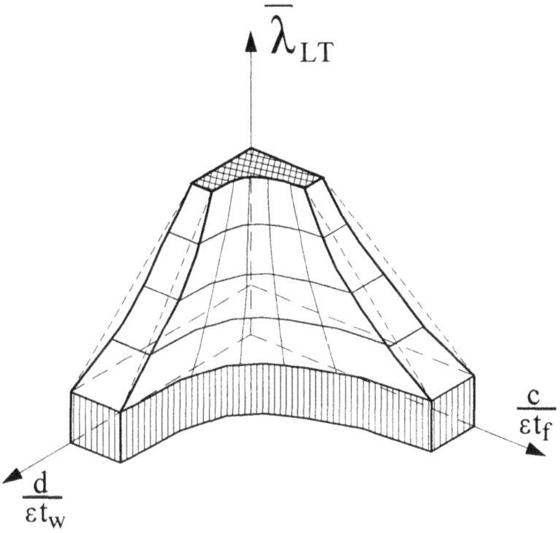

Figure 6.2.11: Limit curves for member classification

In the present context local ductility is regarded as a parameter. Accordingly, three levels of local ductility designated as H(igh), M(edium) and L(ow) are introduced. The corresponding values of local ductility μ_φ are taken equal to 10, 5 and 2.

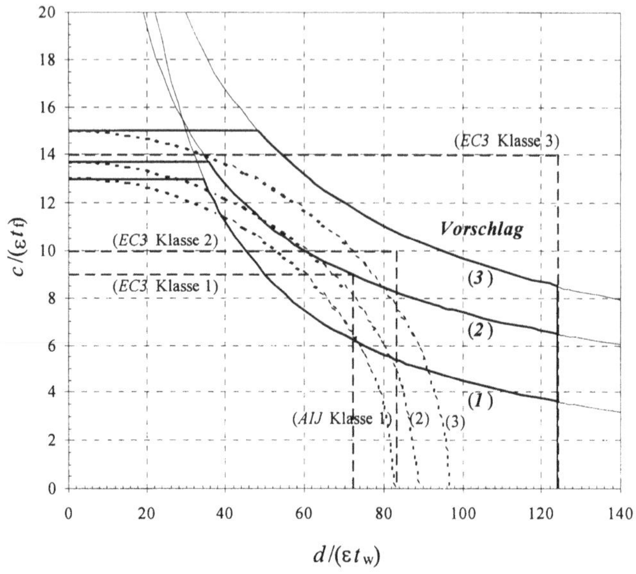

Figure 6.2.12: Limit curves for $\bar{\lambda}_{LT} = 0$

Behaviour of members for cyclic loading – Low-cycle fatigue evaluation

The behaviour of steel members under cyclic loading has been extensively studied both experimentally and analytically. The investigations refer to isolated members and subassemblies between beams, columns and their joints. The response is a function of:

- the dimensions of the member cross section
- the member slenderness and
- the level of axial force.

Depending on the deformation amplitude, and therefore on the ductility level, a reduction in strength, stiffness and energy dissipation may be observed. This is due to buckling and plasticity as explained before, but sometimes also to crack formation and development as a consequence of low-cycle fatigue. The quantification of the member response requires the definition of appropriate failure criteria. Such criteria may be the attainment of a specific reduction ratio for the stiffness or strength, the formation of the first crack, the ratio of the actual energy dissipation in a cycle compared to the corresponding energy dissipation under the assumption of bilinear elastoplastic behaviour etc. The last definition may be more appropriate, since it includes both strength and stiffness degradation. The results may be presented in the form of low-cycle fatigue curves (Ballio et. al. 1994, Bernuzzi et. al. 1997, Ferreira et. al. 1998, Vayas, Ciutina 1999).

In order to derive a low-cycle fatigue expression, the relevant rules for high-cycle fatigue will be recalled. In high-strength fatigue the material behaviour is elastic, so that the fatigue criterion is expressed in terms of stress vs. number of cycles. The fatigue strength curves for nominal stresses are defined by:

$$\log N = \log a - m \log \Delta\sigma_R$$

where:
$\Delta\sigma_R$ = fatigue strength,
N = number of stress range cycles,
m = slope of the fatigue strength curves and
log a = a constant.

The value of the slope constant m is equal to 3 or 5 in dependence on the level of the applied stress range. For a certain stress range there exists a cut-off limit, under which no failure occurs independent on the applied number of cycles. Between the number of cycles N_1 and N_2 for different stress ranges $\Delta\sigma_2$ and $\Delta\sigma_2$, following relation may be derived from eq. (6.2.23):

$$\frac{N_1}{N_2} = \left(\frac{\Delta\sigma_2}{\Delta\sigma_1}\right)^m \qquad (6.2.24)$$

The fatigue verification is performed by means of a damage index D calculated from eq. (6.2.25):

$$D = \frac{n}{N} \qquad (6.2.25)$$

where:
n is the number of cycles of stress range $\Delta\sigma$ and
N is the number of cycles of stress range $\Delta\sigma$ that cause failure.

The value of the damage index is limited to:

$$0 \leq D \leq 1 \qquad (6.2.26)$$

0 corresponds to no damage, i.e. the stress range is below the cut-off limit
1 corresponds to complete damage and
intermediate values of D correspond to partial damage.

For variable applied stress ranges, the damage assessment is performed in accordance to a cumulative law. As such, the linear Palmgren-Miner rule in accordance to eq. (6.2.27) is usually applied.

$$D = \Sigma \frac{n_i}{N_i} \qquad (6.2.27)$$

where:
n_i = number of cycles of stress range $\Delta\sigma_i$ and
N_i = number of cycles of the same stress range that cause failure

For the determination of the design spectrum in the fatigue assessment, a method for counting the cycles for a certain stress history has to be employed. As such, the *rainflow* or *reservoir* method was proven to be appropriate. According to that method, the stress history diagram is supposed to be filled with water as a reservoir. Subsequently the water is drained, starting from the lowest point. The stress ranges are counted as the differences in height of the various reservoirs.

The above-described procedure is appropriate for application in high-cycle fatigue problems. For low-cycle fatigue problems the material enters into the inelastic range, so that a stress-based fatigue evaluation is not possible. In order to take advantage of the fatigue assessment evaluation as described before, a strain-oriented formulation of the fatigue problem is necessary. This requires the substitution of stress as fatigue parameter with strain. Using the plastic rotation as a generalized strain parameter (Vayas, Ciutina 1999), the low-cycle fatigue verification may be performed analogously to eq. (6.2.23) according to (Figure 6.2.13):

$$\log N = \log a - m \log \Delta\varphi_p \tag{6.2.28}$$

where

$\Delta\varphi_p$ = fatigue deformability (corresponding to the fatigue strength),
N = number of plastic rotation range cycles,
m = slope of the fatigue plastic rotation range curves and
log a = a constant.

The slope m in eq. (6.2.28) as determined from tests is usually equal to 2. In high cycle fatigue, m in eq. (6.2.23) ranges between 3 and 5 (Eurocode 3).

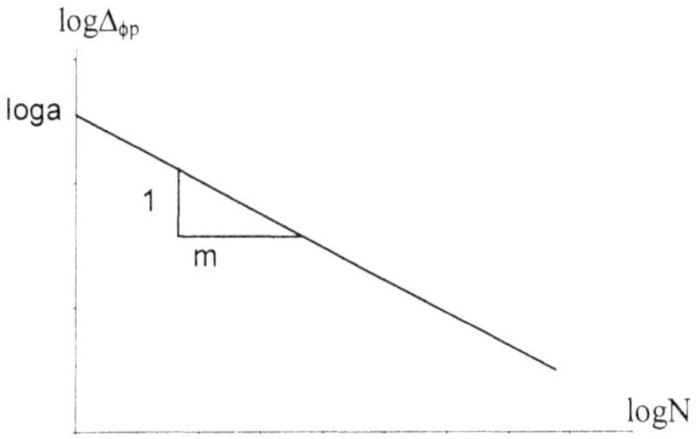

Figure 6.2.13: Low-cycle fatigue curve

Analogously to eq. (6.2.24), following expression holds between the number of cycles to failure and the corresponding plastic rotation ranges:

$$\frac{N_1}{N_2} = \left(\frac{\Delta\varphi_{p2}}{\Delta\varphi_{p1}}\right)^m \tag{6.2.29}$$

Reference values for the damage evaluation may be provided from the results of monotonic tests. In counting the value of N for monotonic loading, it shall be remembered that one cycle corresponds to loading in one direction, unloading, loading in the other direction at the same amplitude and reloading to zero. Accordingly, N=1/4 and $\Delta\varphi = 2\varphi_{mon}$ for monotonic loading. However, in counting the cycles in the reservoir method, one cycle corresponds to loading in one direction to the value $\Delta\varphi$ and unloading to zero. That means that monotonic loading may be taken into account by the pair N_{mon}=1/2 and $\Delta\varphi = \varphi_{mon}$. Accordingly, if the rotation capacity under monotonic loading φ_{mon} is known, the number of cycles for a certain range of plastic rotation may be determined from eq. (6.2.30):

$$N = \frac{1}{2}\left(\frac{\varphi_{mon}}{\Delta\varphi_p}\right)^m \tag{6.2.30}$$

The damage index for application of n cycles with a plastic rotation range of $\Delta\varphi_p$ is then determined from eq. (6.2.26), while the damage index for a certain plastic rotation spectrum may be calculated from eq. (6.2.27). For the evaluation of the plastic rotation spectrum, the reservoir method as described for high-cycle fatigue may be applied.

Behaviour of joints for monotonic loading

The response of moment frames is largely influenced by the behaviour of the beam-to-column connections and joints. The term "joint" has a broader meaning than the term "connection" in that it includes both the connection and the panel zone of the column. The connection morphology depends on the type of the column and the beam sections as well as on the construction practice. For all types of joints, bolted or welded, three zones, the tension zone, the compression zone and the shear zone, may be distinguished (Figure 6.2.14). By considering possible failure modes and the deformability in the three zones, the stiffness, strength and ductility of the joint may be evaluated. Apart from the geometric characteristics, the joint response is also influenced by the loading conditions. This is due to the fact that the panel zone is activated in shear by the **difference** of moments on opposite sides of the column, so that the joint may become softer if the applied moments on the two sides are of different sign.

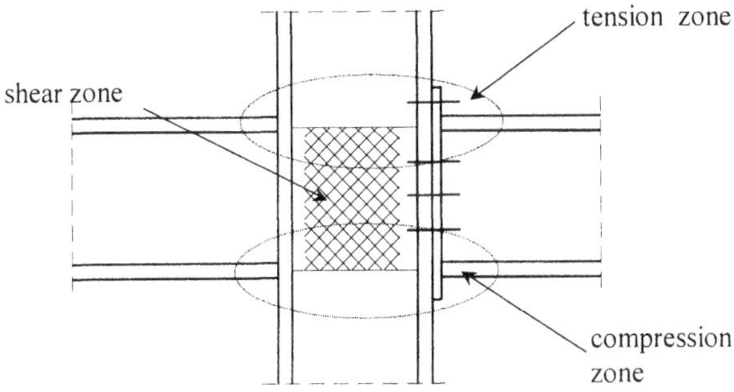

Figure 6.2.14: Zones of a beam to column joint

The behaviour of joints may be described by means of moment rotation curves (Figure 6.2.15). It is mainly characterized by three quantities, the moment capacity M_{jRd}, the stiffness S_j and the rotation capacity φ_{cd}.

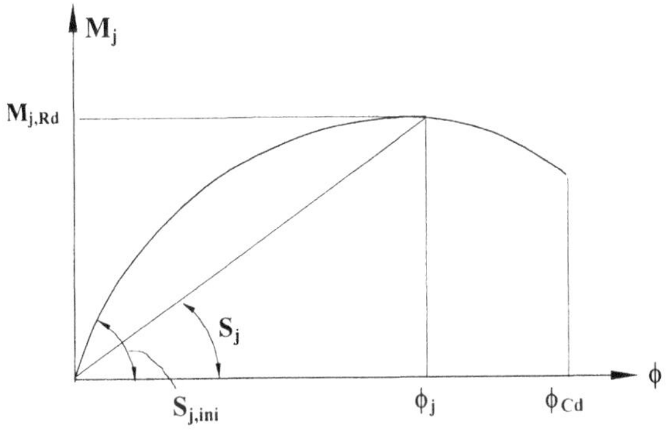

Figure 6.2.15: Moment rotation curve of a joint

Joints are classified into categories in accordance to their moment capacity and their stiffness [4]. In respect to the moment capacity, joints are classified as:

- full strength joints if $\overline{m} \geq 1$ and
- partial strength joints if $\overline{m} \leq 1$ where

$$\overline{m} = \frac{M_{jRd}}{M_{bRd}} \tag{6.2.31}$$

M_{jRd} = design strength of the joint,
M_{bRd} = design strength of the connected beam.

In respect to stiffness, joints in moment resisting frames examined here are classified as:

- rigid if $$S_j \geq 25\frac{EI_b}{L_b} \tag{6.2.31a}$$
- semi-rigid if $$S_j \leq 25\frac{EI_b}{L_b} \tag{6.2.31b}$$
- pinned if $$S_j \leq 0.5\frac{EI_b}{L_b} \tag{6.2.31c}$$

where:
S_j = elastic joint stiffness
EI_b = stiffness of the connected beam
L_b = length of the connected beam

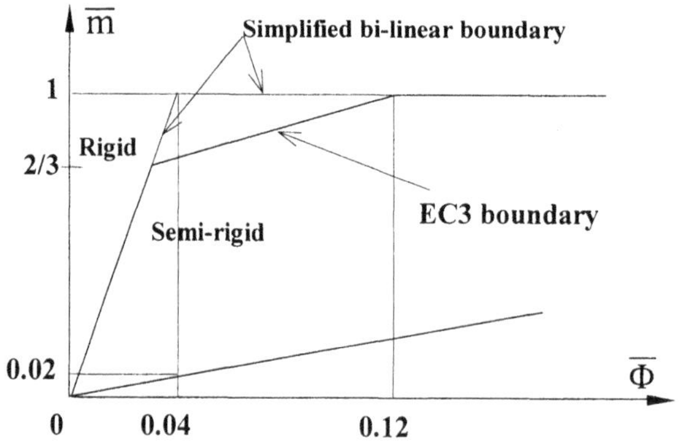

Figure 6.2.16: Classification of joints

Joints were formed as either rigid or pinned in the past. The application of semi-rigid joints seems to be advantageous under certain conditions. For joints subjected to monotonic loading, following failure modes were mostly recorded:

- Tensile rupture of bolts
- Yielding of end plates or column flanges along yield lines
- Yielding of column web due to tension or compression

- Buckling of column web due to compression
- Yielding or buckling of column web due to shear

The studies reveal that the application of capacity design rules favouring ductile failure modes, e.g. column flange or end plate yielding prior to bold failure in tension zone, or avoidance of web buckling in the compression zone, result in an enhancement of overall joint ductility. However, considering the large number of parameters influencing the joint behaviour, it may be stated that at the present there doesn't exist a generally recognised design method allowing for an accurate prediction of the strength, stiffness and rotation capacity of joints. Of special importance is the rotation capacity of joints if inelastic behaviour is not confined in the beam but takes place in the connection, e.g. if partial strength connections are used. However, if inelastic deformations in the connection are to be avoided, the moment resistance of the connection shall well exceed the corresponding resistance of the connected beam taking into account strain hardening. In accordance with Eurocode 3, this condition is considered as fulfilled if it is:

$$\overline{m} \geq 1.2 \tag{6.2.32}$$

If the plastic hinge is going to be formed in the joints, their rotation capacity becomes important. The relevant requirements are similar to those at beams as discussed before, since the failure mechanism does not change. These requirements may be easily fulfilled by application of semi-rigid joints, while rigid joints do not generally possess adequate rotation capacity. Semi-rigidity has also the positive effect that joints attract lower moments due to vertical loading. However there is no reduction in sway moments. In frames with semi-rigid joints an increase in sway displacements due to the stiffness reduction is expected. This results a reduction in global ductility due to P-Δ effects as will be seen later.

Behaviour of joints for cyclic loading

The study of the behaviour of beam-to-column joints under cyclic loading has been the subject of a large number of experimental and theoretical researches. The main target of this research was the investigation of the **ductility supply**, i.e. the ability of connections and joints, regarded as a subassemblage with the connected beams and columns to dissipate inelastic energy. Here again instabilities lead to less ductile behaviour. For example shear yielding, but not shear buckling, in the panel zone of the column provides stable hysteresis loops (Vayas et al 1995) as shown in Figure 6.2.17.

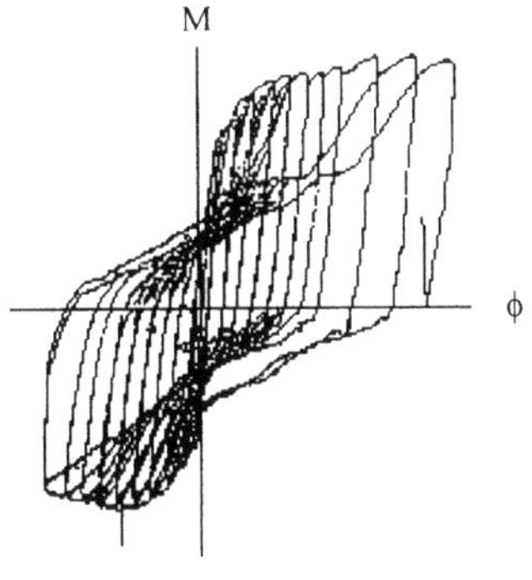

Figure 6.2.17: Moment-rotation-curves of knee joints with shear on the panel zone

Observations made at steel buildings after the recent earthquakes of Northridge and Kobe revealed extensive cracks in the beam to column connections (Kato et. al. 1997, Yamada 1996). This shows that low-cycle fatigue procedures as outlined before may be employed for the analysis of the joint behaviour. In addition to the presentation on the properties of steel, reasons for this unfavourable behaviour are:

a) high strains at the lower beam flange due to the shift of the neutral axis of the beam towards the upper flange that results in the composite action
b) the presence of cope holes that weaken the beam at its most critical section near the joint over a very short length
c) stress concentrations at steel backing bars and run-off tabs
d) inadequate quality of weld execution, especially for field welds.

GLOBAL PROPERTIES

General

The global properties describe the ability of a frame to deform in the inelastic range. For moment resisting frames considered here, inelastic deformations correspond to the formation of plastic hinges in members and/or joints. It is therefore obvious that the number and the rotation capacity of the developing plastic hinges directly affect global ductility. Global ductility is high, if inelastic behaviour may spread out over a large part of the frame. As the plastic hinges form, the frame response becomes softer so that, due to the presence of vertical loads, stability effects may become important. Another reason for a reduction in lateral stiffness and increase importance of 2nd order effects is if beams are not rigidly connected to the columns. As observed with local ductility, instability effects lead to a reduction of global ductility too. For frames with semi-rigid joints, this may be compensated by a potentially better redistribution of moments in the frame. Summarizing, global ductility is affected by:

- The redundancy of the structural system
- The level of local ductility (rotation capacity)
- The rigidity of beam to column connections and
- The level of vertical loading.

In the following, global ductility and the influence of the parameters affecting it will be studied for moment resisting frames subjected to monotonic and cyclic loading.

Static behaviour under monotonic loading

Bilinear systems

♦ Cantilever column

The response of a single cantilever column to monotonic horizontal loading will be studied in this section. EI is the stiffness of the column and l its height. The column is subjected to concentrated vertical and horizontal forces P and correspondingly H at its top. The moment-curvature curve of the cross section is considered to be bilinear. As soon as the plastic moment at the column base, possibly reduced due to the presence of the axial force, is reached a plastic hinge form. Ignoring the influence of deformations on equilibrium, the lateral deformations subsequently grow due to rotation of the plastic hinge, without increase of the

$$\varphi_p = (\mu_k - 1)\kappa_y l_p = (\mu_k - 1)\frac{M_p}{EI} l_p \tag{6.2.33}$$

The total deformation may be written as:

$$\delta_u = \delta_y + \delta_p = \frac{M_p \cdot l^2}{3 \cdot EI} + \varphi_p \cdot l = \frac{M_p \cdot l^2}{3 \cdot EI} + (\mu_k - 1) \cdot \frac{M_p}{EI} \cdot l_p \cdot l \tag{6.2.34}$$

Global ductility may be defined as the ratio between the ultimate and the yield displacements, so that following relationship between global ductility and local ductility in terms of curvature may be established:

$$\mu_\delta = \frac{\delta_u}{\delta_y} = 1 + 3\alpha(\mu_\kappa - 1) \tag{6.2.35}$$

The relationship between local ductility in terms of curvature and rotation may be established taking into account the yield rotation at the top, $\varphi_y = \frac{M_p l}{2EI}$, and the plastic rotation at the top given by eq. (6.2.33). It may be written as:

$$\mu_\varphi = \frac{\varphi_u}{\varphi_y} = 1 + \frac{\varphi_p}{\varphi_y} = 1 + 2\alpha(\mu_\kappa - 1) \tag{6.2.36}$$

The relationship between global and local ductility in terms of rotation may be written as:

$$\mu_\delta = 1.5(\mu_\varphi - \frac{1}{3}) \tag{6.2.37}$$

- Sway column clamped at both ends

For such a column following expressions may be written:

Yield rotation at mid-height:

$$\varphi_y = \frac{M_p \cdot l}{6 \cdot EI} \tag{6.2.38}$$

Relationship between local ductility in terms of curvature and rotation:

$$\mu_\varphi = \frac{\varphi_u}{\varphi_y} = 1 + \frac{\varphi_p}{\varphi_y} = 1 + 6\alpha(\mu_\kappa - 1) \tag{6.2.39}$$

Relationship between global ductility and local ductility in terms of rotation:

$$\mu_\delta = 1 + \frac{\delta_p}{\delta_y} = 1 + \frac{\varphi_p l}{M_p l^2 / 6EI} = \mu_\varphi = 1 + 6\alpha(\mu_\kappa - 1) \tag{6.2.40}$$

The relationship between global and local ductility for the two systems is illustrated in Figure 6.2.18. It may be observed that the cantilever column corresponds roughly to frames with weak beams, while the clamped column to a frame with strong beams. For intermediate cases, the relationship between global and local ductility is between the two extremes.

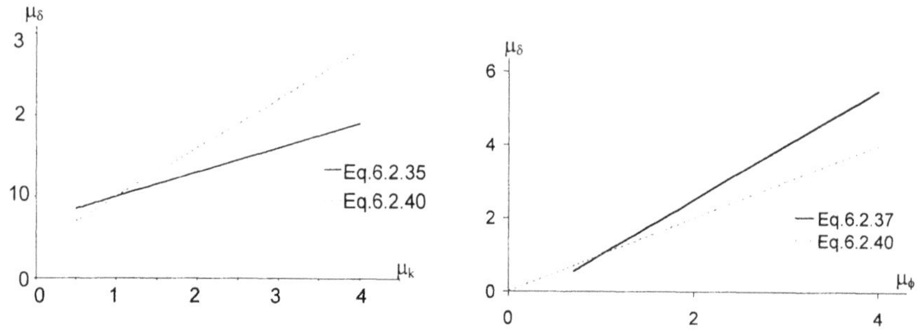

Figure 6.2.18: Relationship between global ductility and local ductility in terms of curvature and displacement for a cantilever (solid line) and a sway clamped column (dotted line)

The definition of global ductility in terms of displacements may become ambiguous for more complicated systems, like for irregular multi-storey frames. This is due to the fact that the top displacement may not express, as a global parameter, possible discontinuities in displacements, e.g. due to the appearance of soft storeys. In order to overcome this deficiency, a definition for global ductility based on energy is proposed. Considering the plastic energy dissipated and the elastic energy stored in a cantilever column, following expressions may be derived:

Elastic energy:

$$W_e = \int \frac{1}{2}\frac{M^2}{EI}dx = \frac{1}{2}\frac{1}{3}\frac{M_p^2}{EI}l \qquad (6.2.41)$$

Plastic energy:

$$W_p = \sum M_p \varphi_p = M_p(\mu_\kappa - 1)\kappa_y l_p = \frac{M_p^2}{EI}(\mu_\kappa - 1)\alpha l \qquad (6.2.42)$$

Considering equal energy between the elastic and the elastic-plastic system, global ductility in terms of energy may be defined as:

$$\mu_E = \sqrt{1 + \frac{W_p}{W_e}} \qquad (6.2.43)$$

Accordingly, following relationship between global ductility and local ductility may be derived:

$$\mu_E = \sqrt{1 + 6\alpha(\mu_\kappa - 1)} = \sqrt{1 + 3(\mu_\varphi - 1)} \qquad (6.2.44)$$

For a sway column with both ends clamped, the elastic energy is still expressed by eq. (6.2.41), while the plastic energy by twice the value of eq. (6.2.42), as two plastic hinges form at the column ends. Accordingly, eq. (6.2.44) is substituted as following:

$$\mu_E = \sqrt{1 + 12a(\mu_\kappa - 1)} = \sqrt{1 + 2(\mu_\varphi - 1)} \qquad (6.2.45)$$

The above relationships are graphically illustrated in Figure 6.2.19.

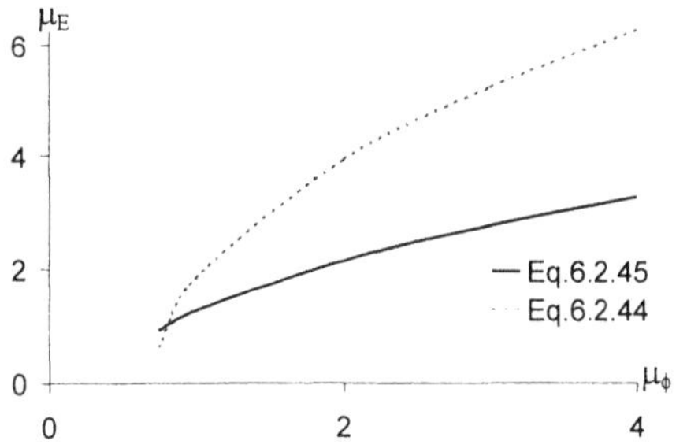

Figure 6.2.19: Relationship between global ductility in terms of energy and local ductility in terms of rotation

The influence of deformations on equilibrium was not considered so far. By application of 2nd order analysis, additional moments arise. For a cantilever column with a lateral deformation δ determined from 1st order analysis, the moment at the column base is increased by $V \cdot \delta$. 2nd order effects may also be taken into account by considering a fictitious additional horizontal force at the top and formulating the equilibrium conditions for the undeformed system. The additional horizontal force may be taken equal to:

$$\Delta H = \frac{P\delta}{l} \tag{6.2.46}$$

Such as approach in which the axial deformations and the resulting bending moments are neglected, corresponds to consideration of P-Δ (global instability) and neglect of P-δ (local instability) effects. The relevant mistake amounts to maximally 20% (Rubin 1973). The relation between the stiffness with and without consideration of P-Δ effects may be written as:

$$S^2 = S^1 - \frac{P}{l} \tag{6.2.47}$$

Recalling that the critical load corresponds to zero stiffness, eq. (6.2.47) may be substituted by:

$$S^2 = S^1\left(1 - \frac{P}{P_{cr}}\right) \tag{6.2.48}$$

As expected, the reduction in stiffness due to P-Δ effects is influenced by the level of vertical forces. P-Δ effects influence the behaviour also in the plastic range. After the attainment of the plastic moment at the base, equilibrium may be reached only if the external horizontal force is reducing. The stiffness of the unloading branch is equal to:

$$S_{un} = -\frac{\Delta H}{\delta} = -\frac{P}{l} \tag{6.2.49}$$

The load-deformation curves for 1st order and 2nd order analysis are illustrated in Figure 6.2.20b.

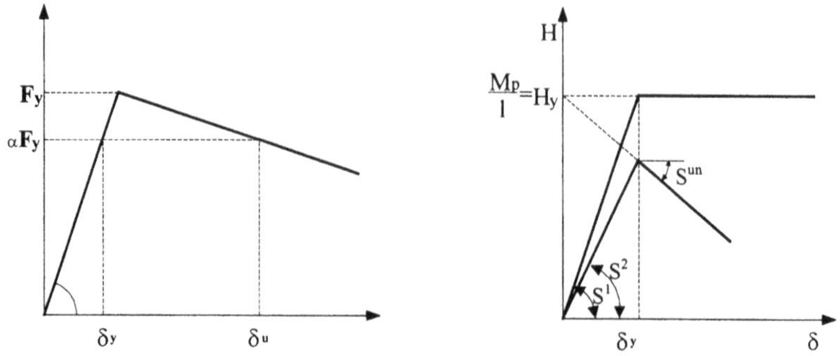

Figure 6.2.20: Column response for 1st and 2nd order analysis

As shown in Figure 6.2.20, global ductility in terms of displacement, as expressed by eq. (6.2.35), is impossible to be defined for simple bilinear systems subjected to P – Δ effects. This is because the horizontal line through the yield state Y does not cut, due to unloading, the response curve. In such a case, global ductility may be determined only if another definition for the yield-state is used. This may be done by defining a fictitious yield force lower than the actual yield force (Figure 6.2.20a). In that way, the horizontal line through the new yield state does cut the response curve at an ultimate limit state U. However, if global ductility is expressed in terms of energy according to eq. (6.2.43), the relations derived before are approximately valid even if unloading occurs. The approximation refers to the fact that the 2nd order moments do not vary linearly along the column as the displacements grow.

Frames

The response of frames under monotonic loading may be studied by appropriate pushover analysis. In order to simulate the conditions for a seismic excitation, the vertical loading is kept constant, while the horizontal loads are considered to increase according to a certain pattern, this pattern corresponding to the frame's fundamental mode. As the horizontal loads increase in the sense of a load multiplier, plastic hinges develop and the frame response becomes softer. The loads may increase up to the point where the stiffness becomes zero. Further increase in deformations result in unloading due to 2nd order effects, as discussed in the previous section. The maximum load does therefore not necessarily coincide with the mechanism formation. The frame response, as expressed by the base shear - top displacement curve, is illustrated in Figure 6.2.21.

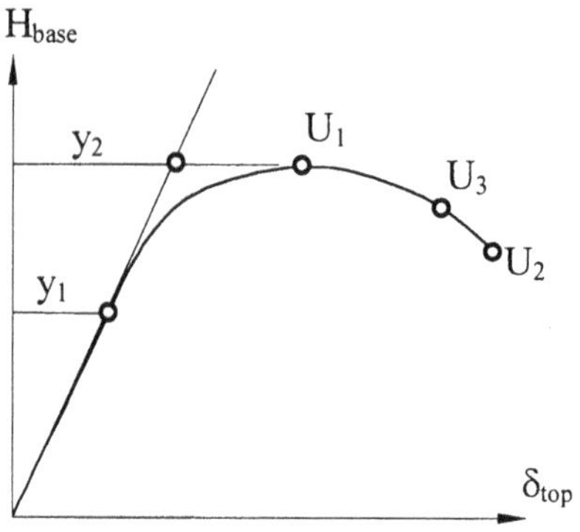

Figure 6.2.21: Frame response for 2nd order analysis

The determination of global ductility requires a definition for the yield and the ultimate states. Regarding the yield state two definitions are possible:

- Yield state Y1: corresponds to the appearance of the first plastic hinge
- Yield state Y2: corresponds to the intersection between the elastic line with the horizontal line at maximum loading.

For the ultimate state, following definitions are possible:

- Ultimate state U1: corresponds to the state at maximum loading
- Ultimate state U2: corresponds to the mechanism formation
- Ultimate state U3: corresponds to the exhaustion of the rotation capacity at a plastic hinge.

Once the limit states Y and U are defined, global ductility may be evaluated in terms of displacements by eq. (6.2.35) or in terms of energy by eq. (6.2.43). **Overstrength** is defined as the ratio between the base shears (horizontal multipliers) at the ultimate and the yield limit state:

$$\Omega = \frac{H_u}{H_y} \quad \text{or} \quad \Omega_{max} = \frac{H_{max}}{H_y} \qquad (6.2.50)$$

For single storey frames following expression was approximately derived for the global ductility in terms of displacements [15]:

$$\mu_\delta = 1 - 2\frac{1 - P/P_{cr}}{1 - 4P/P_{cr}}(\Omega_{max} - 1) + \mu_\varphi \qquad (6.2.51)$$

The above expression indicates how local ductility, frame overstrength and P-Δ effects affect global ductility. For multi-storey frames, it was also shown that the type of collapse mechanism heavily influences the value of global ductility. Under various assumptions for structural regularity and consideration of three possible types of failure mechanism (global, partial, single storey), it was possible to derive analytical expressions for the global ductility of multi-storey frames (Mazzolani, Piluso 1996).

In the following, a method is presented that allows for the determination of global ductility, taking into account all parameters of importance. The method is not restricted to specific geometric, loading or other conditions. It is based on a pushover 2nd order analysis of the frame. In this method, limit states Y1 and U3 were chosen to define the yield and the ultimate limit state. This selection was made for the following reasons:

- The yield limit state is associated to the design criterion and accordingly to the type of global analysis. If elastic global analysis is used, the limit state corresponds to the 1st plastic hinge formation and therefore Y1 should be used. If plastic global analysis is applied, limit state Y2 is more appropriate. In the present work, state Y1 is defined as the yield limit state, as this corresponds to the design criterion for seismic loading where a dynamic response spectrum analysis is applied.
- Limit state U3 has a clear physical interpretation, in that it is associated to the exhaustion of local ductility. If this ductility is exhausted, any further deformation is not possible due to failure of the corresponding member or joint. On the other hand, a selection of the instant of the mechanism formation as the ultimate state, would lead to zero ductility for bilinear systems. However, such a system is able to deform further, as long as local ductility allows it. It must be noted, that the above definition is restricted in that the load at the ultimate state must be higher than the yield load. Otherwise the deformation capacity is, due to the appearance of instability, of no help.

Non-linearity in respect to the material behaviour may be considered during a pushover analysis by two methods, the method of plastic zones and the method of plastic hinges. The plastic zones method allows for the evaluation of the gradual plastification across the members. On the contrary, in the plastic hinge method cross sections behave either fully elastic or rigid plastic. Structural deformations in the plastic zones method are continuous, while in the plastic hinge method discontinuities appear at plastic hinges where plastic rotations occur.

The plastic zones method, although appearing as more accurate than the plastic hinge method, has certain drawbacks such as:

- No rotations are recorded, so that the rotation capacity cannot be checked.
- A fine subdivision of members along their length, especially at positions where plastic behaviour is expected, is required so that the numerical effort is considerable.
- Its results are highly dependent on the element subdivision, without ensuring that a finer subdivision provides more accurate results.

In the present work, the pushover analysis is based on the application of the plastic hinge, 2nd order analysis, using the SOFiSTiK general purpose programme. The connections in beam-to-column joints are modelled by means of bilinear rotation springs. By varying the spring stiffness, all possible connection configurations, from a simple to a rigid connection, can be simulated. Springs of high rigidity are also introduced at potential plastic hinge positions within the span of members. The introduction of springs at the position of plastic hinges, allows for a direct determination of the relevant plastic rotation, which is equal to the calculated rotation of the spring. The yield moment of the spring is equal to the ultimate moment of the connection or the beam plastic moment for partial strength or full strength connections respectively.

According to the above description, global ductility for a certain frame under monotonic loading was evaluated as following (Figure 6.2.22):

1. The frame is modelled as a 2D-structure consisting of linear members and non-linear springs and vertical loading is applied.
2. The dynamic characteristics of the structure (frequencies, modes of vibration) are determined.
3. Horizontal loading, distributed along the height in accordance to the fundamental mode, is gradually applied.
4. The frame response is determined by plastic hinge, 2nd order analysis.
5. The appearance of the first plastic hinge, which defines the yield limit state Y, is recorded.
6. The lateral loading is further increased up to the point, where the rotation capacity of one plastic hinge is exhausted. This defines the ultimate limit state U.
7. If the rotation capacity is not exhausted, until the maximum load at which the overall stiffness becomes zero, mechanism deformations including P-Δ effects is applied. The limit state U corresponds again to the exhaustion of the rotation capacity at one plastic hinge. However, if due to P-Δ effects the reduction in load is so high that the overstrength becomes less than 1 (i.e. if $H_u < H_y$ in Figure 6.2.22) , the limit state U corresponds to the situation at which Ω is equal to 1 (i.e. $H_u = H_y$).

The rotation capacity is determined by consideration of equivalent beams subjected to three-point bending, as described in a previous section. Starting from the moment distribution at the ultimate limit state, each member is considered as isolated from the frame and transformed to an equivalent single span beam (Figure 6.2.23). If a member has two plastic hinges at its ends, it will be transformed to two equivalent single span beams. The rotation capacity of each plastic hinge is subsequently found following the procedure outlined in the relevant section for beams.

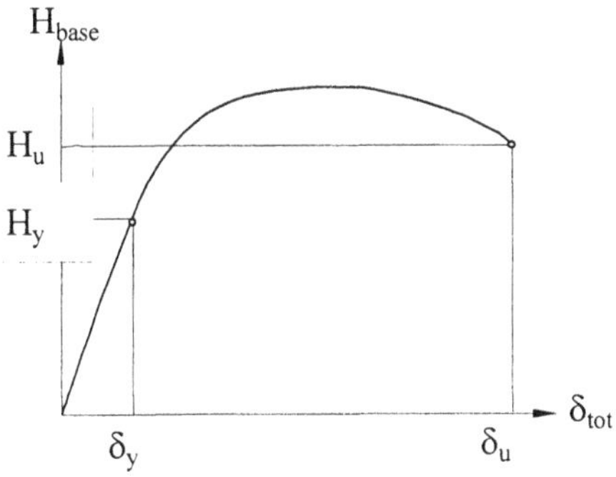

Figure 6.2.22: Evaluation of global ductility for a frame

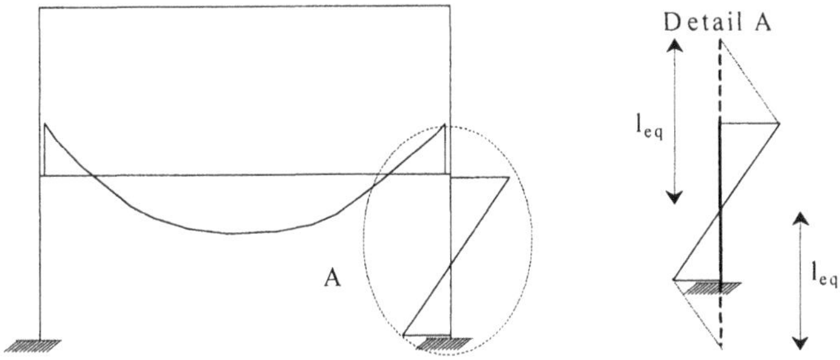

Figure 6.2.23: Determination of equivalent single span beams

Several frames were analysed for the following parameter variation:

- Frame geometry
- Level of local ductility
- Level of vertical loading
- Semi-rigidity of joints

Figure 6.2.24 shows the range of values for each parameter under consideration. The specific frames were selected for this investigation, due to the fact that these frames were used before in another study (Guerra et. a;. 1990), aiming at the determination of the behaviour factor q according to various proposals. The geometric properties, the member dimensions and other frame data were therefore adopted from that study. For each frame the parameters μ_E, Ω, as well as the product $\mu_E\Omega$ were recorded. The results are summarised in Figures 6.2.25 to 6.2.28.

Results for the seven frames are illustrated in Figure 6.2.25. They refer to frames with rigid joints and 40% level of vertical loading. It may be seen that all frames exhibit similar overstrength ratios, ranging between 1.0 and 1.5. The ductility ratio and its product with the overstrength ratio decrease with the number of storeys. This shows that low-rise buildings have an inherent redistribution capacity in contrast to higher buildings, where storey shears and overturning moments due to lateral loading accumulate over the storeys, leading to increased demand for strength at the lower storeys. For the specific frames, where the member cross sections were kept the same over the height of the building, the higher storeys were accordingly over-proportioned and did not yield when the rotation capacity at the lower levels was exhausted. As a result, the members at higher levels did not participate in the energy dissipation, leading to a drop in overall ductility.

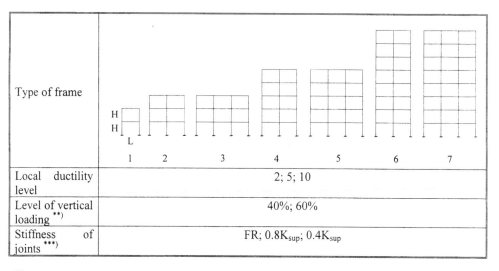

Type of frame	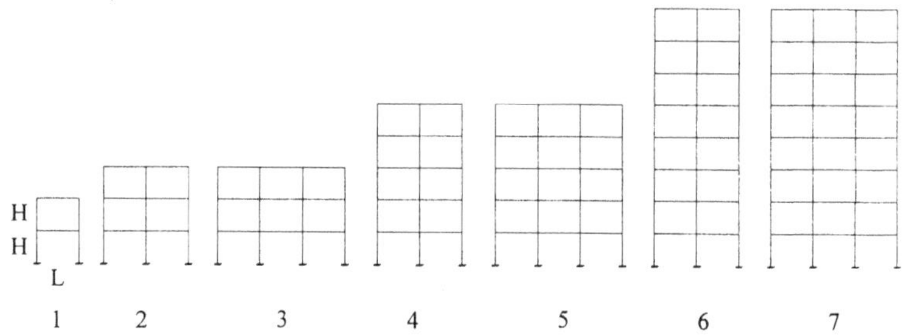
Local ductility level	2; 5; 10
Level of vertical loading **)	40%; 60%
Stiffness of joints ***)	FR; $0.8K_{sup}$; $0.4K_{sup}$

**) Percentage of the beam strength used for vertical loading

***) FR = full rigid joint ($K \geq 25\, EI_b / L_b$),

$K_{sup} = 25\, EI_b / L_b$

K = elastic joint stiffness,

EI_b = stiffness of the connected beam,

L_b = length of the connected beam

Frame	L(m)	H(m)	T (sec)	Beam	Column
1	5	3	0.62 (0.76)	IPE300	HEB180
2	4	4	0,99 (1.21)	IPE330	HEB240
3	4	4	1.12 (1.37)	IPE330	HEB240
4	4	3	1.14 (1.39)	IPE360	HEB280
5	4	3	1.15 (1.42)	IPE360	HEB280
6	4	3	1.26 (1.89)	IPE450	HEB320
7	4	3	1.41 (1.73)	IPE450	HEB360

T = fundamental period for 40% (60%) vertical loading and rigid joints

Yield stress for beams and columns f_y = 235 Mpa

Figure 6.2.24: Investigated frames and range of parameter variation

Figure 6.2.26 presents the overstrength ratios and their products with the global ductility ratios for frame 2 and three levels of joint rigidity. It may be seen that global ductility is generally improved by local ductility. However, this improvement is not linearly varying with local ductility. This is due to the fact that for larger local ductility levels the lateral deformations increase over-proportionally, leading to higher P-Δ effects. Consequently the frame response may be in the descending branch of the force-deformation curve (Fig. 6.2.21) causing a drop in the overstrength ratio.

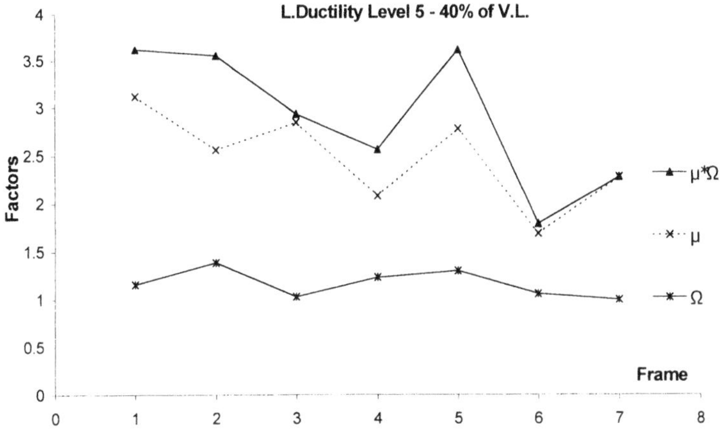

Figure 6.2.25: Global ductility and overstrength for frames with rigid joints (40% vertical loading)

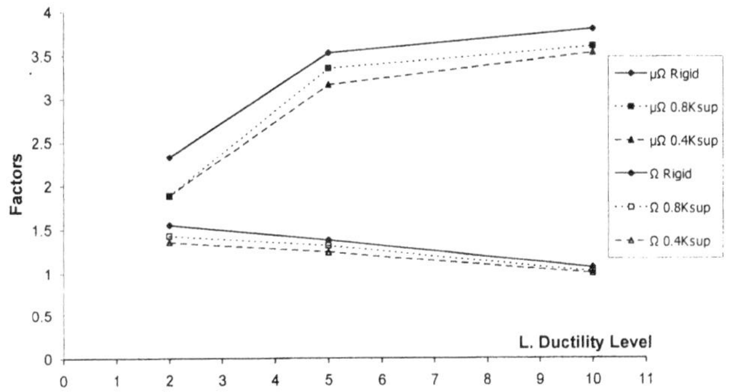

Figure 6.2.26: Global ductility and overstrength vs level of local ductility of frame 2 with rigid and semi-rigid joints (40% vertical loading)

Figure 6.2.27 shows the product between overstrength ratio and global ductility for all frames for 60% vertical loading, local ductility level 2 and three levels of joint rigidity. It may be seen that this product, and therefore overall ductility, decreases with increasing joint flexibility. This is due to the fact that as the joints become more flexible, the lateral deformations, and therefore the importance of 2nd order effects, are increase leading to lower values of overall ductility. However, it may be noted that, as will be illustrated later, dynamic analyses showed that the results of pushover analysis are in respect to the flexibility of joints over-conservative.

In respect to the level of vertical loading it may be observed that global ductility is insensitive to it for low levels of local ductility. However, this influence becomes more important at higher ductility levels due to 2nd order effects. For higher vertical loading, yielding starts at lower levels of the horizontal multiplier. This results in an increase in overstrength, because the denominator decreases, and therefore an increase in its product with the global ductility. However, this increase is generally not compensated from the aforementioned decrease in the yield load. Consequently, the final value of the horizontal multiplier is generally smaller for higher levels of vertical loading.

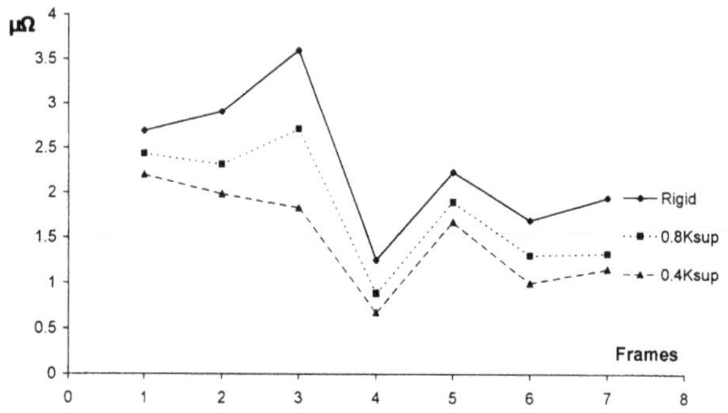

Figure 6.2.27: Global ductility of frames for rigid and semi-rigid joints (local ductility level 2, 60% vertical loading)

Figure 6.2.28 shows for three levels of local ductility the product between overstrength ratio and global ductility as a function of the aspect (height-to-width) ratio of the frame. It may be seen that global ductility decreases with increasing aspect ratios. This is due to the aforementioned increase of overturning moments and the increased strength demand at lower storeys. This situation, combined with invariable member sections over the height of the building for the specific frames, leads to a reduction in global ductility. It may be seen that the beneficial effect of increasing local ductility is different for the various values of aspect ratio. For low aspect ratios, the increase is more pronounced from an increase of local ductility from level 2 to level 5, while for higher aspect ratios this occurs for an increase from level 2 to level 10. This shows that for monotonic loading considered here, higher values of local ductility are needed for slender frames to improve considerable their global ductility.

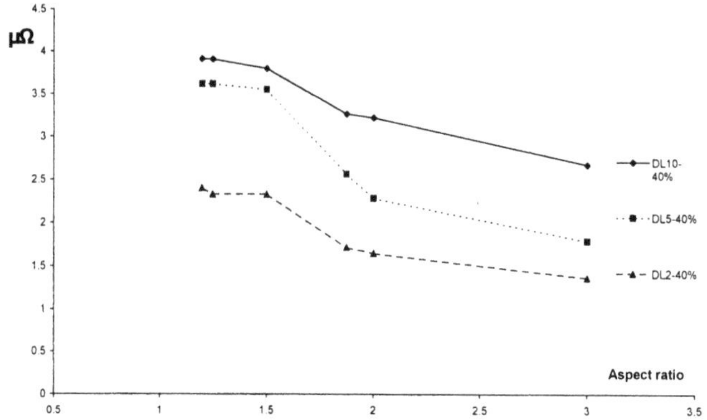

Figure 6.2.28: Global ductility vs. aspect (height-to-width) ratio of the frame for various levels of local ductility

Summarising the investigation on the behaviour of frames under monotonic loading, it may be said that:

- Based on a pushover 2nd order analysis, an energy-based method for the determination of global ductility for moment resisting frames was introduced.
- The method is not restricted to a specific geometrical or other properties of the frame (irregularity, member properties, configuration of joints etc.).
- The application of this method to several frames leads, at least within the range of parameters considered, to following conclusions:
  - Overstrength decreases with increasing number of storeys.
  - Global ductility is increased, although not linearly, with an increase in local ductility (rotation capacity).
  - Global ductility is reduced with increasing joint flexibility due to 2nd order effects.
  - For the same reason global ductility is reduced with increasing height-to-width ratio of the frame.

Dynamic behaviour of frames under monotonic loading

In the previous section, the static frame response under monotonic loading was investigated by pushover analysis, where the lateral load distribution was kept constant. In order to study the dynamic frame response under monotonic loading, an extension of the above described pushover analysis is required. This refers to the need to update the dynamic characteristics of the frame during loading. Due to the appearance of plastic hinges the frame becomes softer, so that its natural frequencies and its mode shapes change. Accordingly, the application of a pushover analysis must be accompanied by a change in shape of the lateral load distribution. This study was performed by use of a specially developed software programme (Vayas, Sophokleous 1996).

Figure 6.2.29 illustrates as an example the progressive plastic hinge formation in a five-storey frame up to the exhaustion of the local ductility. The relevant position is indicated in this figure by a circle not in black. The variation of the 5 first natural periods of the frame at progressive loading is shown in Figure 6.2.30. It may be seen that most affected is the fundamental period, which increases progressively. The change is generally smooth, except from the instant at which all plastic hinges form at its base where a sudden increase in the fundamental period is observed. The final value of the fundamental period, near collapse, is 8-times the initial one on the intact frame. The increasing distance between the fundamental period from the other natural periods indicates that its importance increases at higher load levels.

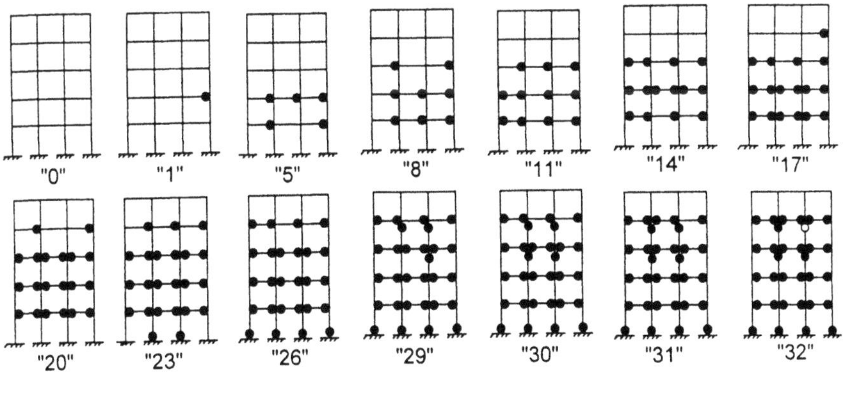

Figure 6.2.29: Progressive plastic hinge formation in a frame

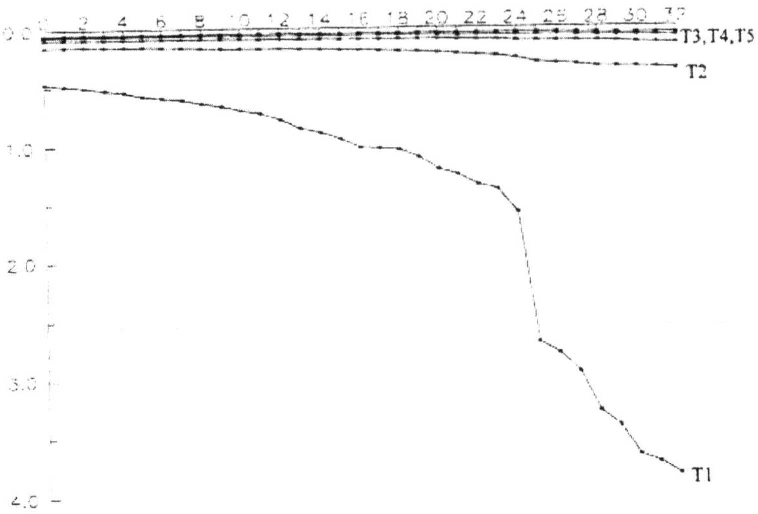

Figure 6.2.30: Variation of the natural periods of the frame with progressive plastification

The change in the fundamental mode shape, and accordingly in the lateral load distribution, at progressive loading is illustrated in Figure 6.2.31. This load distribution is also compared to a uniform and a triangular distribution of lateral loads. It may be observed that at higher loading levels the mode shape shifts to the uniform distribution.

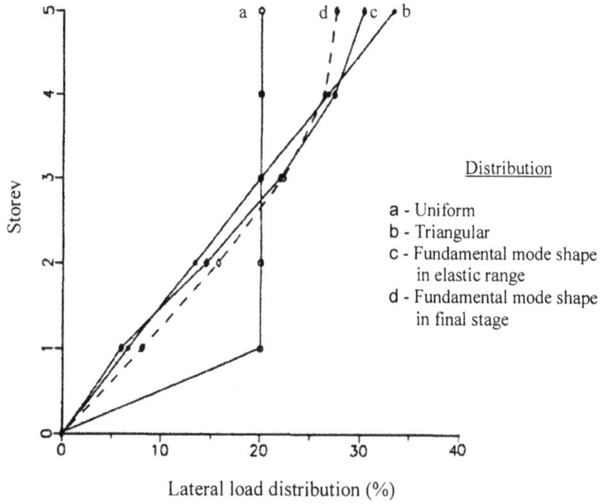

Figure 6.2.31: Lateral load distribution according to various laws

The above procedure was applied to investigate the behaviour of irregular frames to dynamic loading, since it is believed that such frames are more susceptible to such type of loading than regular frames. Starting from the five storey, three-bay frame (frame 5 in Figure 6.2.24) as reference, which will be subsequently designated as S, M or R, several types of irregularity over the height of the frame were introduced. They refer to:

- Irregularity in stiffness
- Irregularity in mass
- Irregularity in strength
- Irregularity in geometry

Irregularity in stiffness

In order to study the influence of the irregularity in stiffness over the height of the building, the stiffness of the columns at one specific storey was reduced by 50%. Figure 6.2.32 presents the results in respect to overstrength and global ductility, where S describes the reference frame and S1 to S5 the frames with reduced stiffness in the 1st to the 5th storey. It may be seen that the changes are of minor importance. This is primarily due to the fact that the stiffness distribution is taken into account by the dynamic analysis in the determination of the horizontal loads, so that no additional local ductility requirements are imposed on the members of the softer storey.

Irregularity in mass

In order to study the influence of the irregularity in mass over the height of the building, the vertical loading, and accordingly the mass, on the beams of one specific storey of the reference frame was increased by 50%. Figure 6.2.33 presents the results in respect to overstrength and global ductility, where M describes the reference frame and M1 to M5 the frames with increased mass

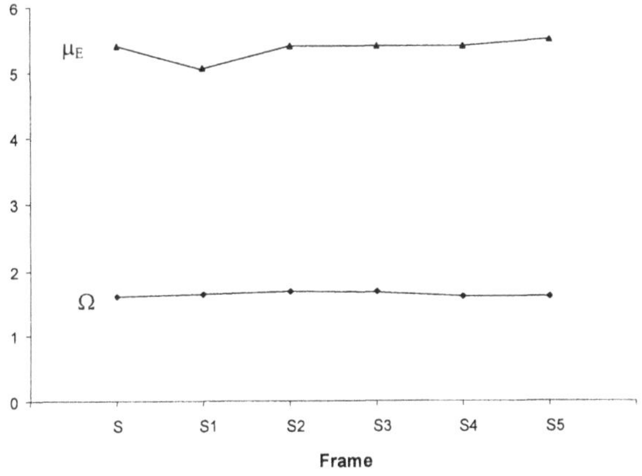

Figure 6.2.32: Global ductility and overstrength of frames with stiffness irregularity

in the 1st to the 5th storey. It may be seen that the overstrength is increasing for frames M1 to M3. This is due to premature yielding at those frames due to the high existing moment from vertical loading, which results in a reduction in the yield load H_y and therefore an increase in the overstrength factor Ω. However, for frames M4 and M5 the overstrength does not increase because, as explained before, the members at higher storeys do not exhibit nonlinear behaviour so that both the yield and the ultimate loads remain unchanged.

In respect to global ductility, it may be observed that it is reduced for frames M1 and M2. This is due to early yielding at lower storeys which is causing increased ductility requirements on the relevant members.

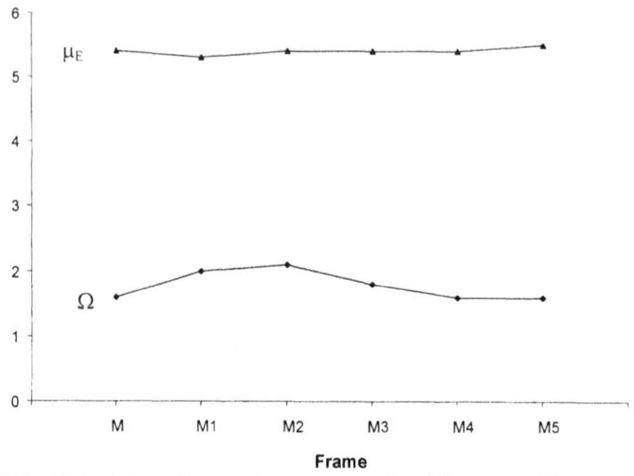

Figure 6.2.33: Global ductility and overstrength of frames with mass irregularity

Irregularity in strength

In order to study the influence of the irregularity in strength over the height of the building, the resistance all columns at one specific storey of the reference frame was reduced by 50%. Figure 6.2.34 presents the results in respect to overstrength and global ductility, where R describes the reference frame and R1 to R5 the frames with reduced resistance in the 1st to the 5th storey. A considerable reduction in global ductility for the irregular frames compared to the regular one may be observed. This is caused by the concentration of ductility demand at the members of the weak storey.

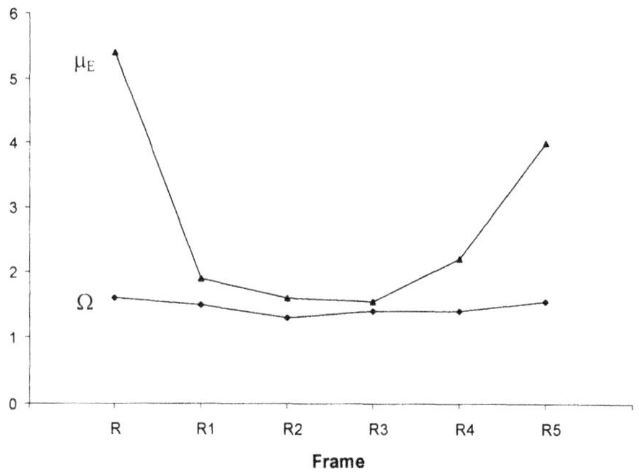

Figure 6.2.34: Global ductility and overstrength of frames with strength irregularity

The initial and final lateral load distribution for the investigated frames is shown in Figure 6.2.35. It has to be observed that, opposite to an irregularity in stiffness or mass, a reduction in strength does not change the initial mode shape, and accordingly the lateral load distribution, of the structure. The initial distribution is therefore identical for all frames. However, the deviation between initial and final distribution is much larger compared with irregular frames in respect to stiffness or mass. This shows that the stiffness properties due to inelastic deformations change much more when the strength is irregularly distributed over the height. The deviation between initial and final distribution is increasing from frame R5 to frame R1. For all frames, the final distribution resembles to a uniform distribution above the weak storey. This is because the final lateral load distribution follows more or less the deformation pattern of the frame. Accordingly there is a jump in the lateral forces at the weak storey where the drift becomes larger. It may be observed, that such a change in the distribution of lateral loads due to inelastic behaviour is not taken into account when linear analysis methods are applied.

Irregularity in geometry

In order to study the influence of geometrical irregularity in the global ductility and overstrength, several frames with variable geometric properties shown in Figure 6.2.36 were studied. The reference frame is a five storey four bay regular frame indicated in the Figure as frame G3. Geometric irregularity was introduced by omission of complete bays from the reference frame.

For frames with such geometric irregularities, a definition of global ductility by means of top displacements is obviously not appropriate. On the contrary, the expression of global ductility by means of energy appears to represent better the frame behaviour, since it account for the energy dissipation within the frame without consideration of the form of the lateral deformations over the height. The results for the investigated frames are shown in Figure 6.2.37. The figure shows also the plastic hinge pattern as well as the position where the rotation capacity was exhausted. Frames G2 to G5 where gradual irregularities were introduced exhibit similar behaviour. In those frames plastic hinges appear in the beams and at column bases. In the tower-like frames G1, G7, G8 and G12, the lower part remains elastic due to its higher strength. These frames behave more or less like regular frames above the level of the complete frame, as for example frame G1 which could be regarded as a 4-storey one bay regular frame. When the tower, where the plastic deformations appear, constitutes a small portion of the overall frame, as in frame G12, the global ductility is low. Unexpectedly, frame G11 in which plasticity was spread-out over the entire structure showed a low value of global ductility is low.

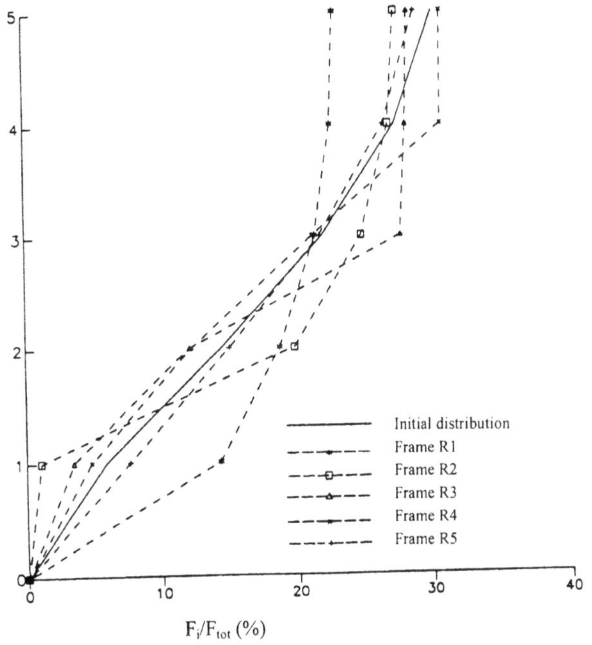

Figure 6.2.35: Initial and final lateral load distribution of the frames

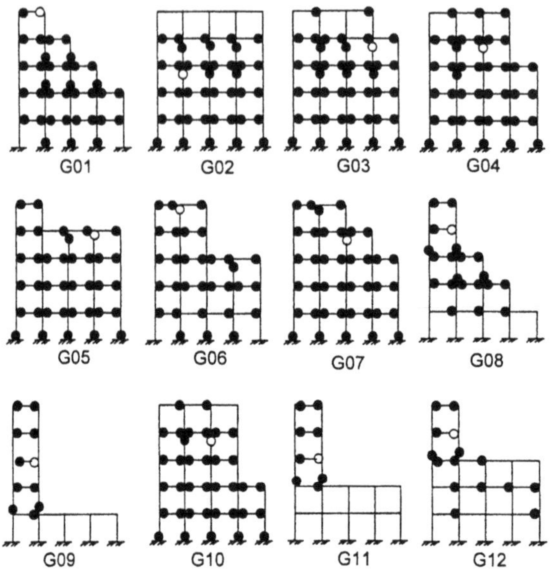

Figure 6.2.36: Frames with irregularity in geometry

Unexpectedly, frame G11 in which plasticity was spread-out over the entire structure showed a low value of global ductility. As a conclusion it may be said that, contrary to the provisions of Eurocode 8, no general rule may be given for the influence of geometric irregularity in respect to global ductility. This has to be evaluated for each specific case.

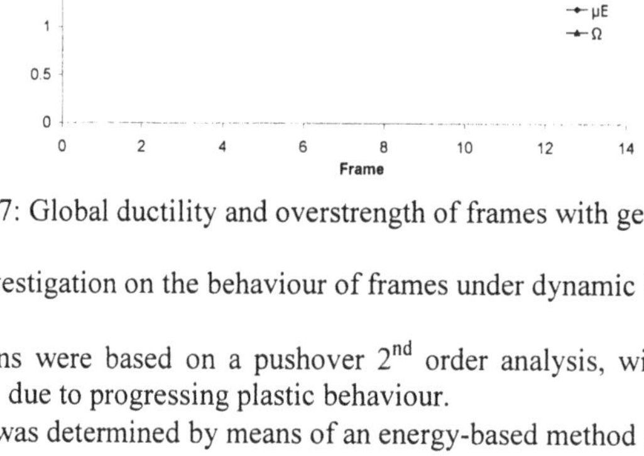

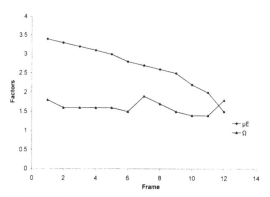

Figure 6.2.37: Global ductility and overstrength of frames with geometric irregularity

Summarising the investigation on the behaviour of frames under dynamic loading, it may be said that:

- The investigations were based on a pushover 2nd order analysis, with up-date of the dynamic frame properties due to progressing plastic behaviour.
- Global ductility was determined by means of an energy-based method as for monotonic loading.

- The application of this method to several irregular frames lead to following conclusions:
  - Irregularity in stiffness and mass over the height of the frame do not appear to considerably affect global ductility.
  - On the contrary, irregularity in strength, and especially weak storeys, appear to be of detrimental influence on global ductility.
  - No general rule may be given on how geometric irregularities affect global ductility. This must be examined from case to case.

Behaviour of frames under seismic loading

The study of the response of frames to seismic loading cannot be entirely relied upon investigations by means of pushover analysis, even if the latter takes into account the dynamic characteristics of the structure as described before. One of the main reasons is that the frame response under earthquakes is both influenced by its own dynamic characteristics and the characteristics of the seismic record. Additionally, and in contrast to monotonic loading, the frame is subjected to cyclic inelastic deformations of variable amplitude that may cause low-cycle fatigue problems in specific elements.

In order to investigate the response under seismic loading, non-linear dynamic analyses by means of the general purpose DRAIN-2DX software package were performed. Several seismic records were initially considered in this study, including records from Greece (Aigion 1995, Kalamata 1985, Korinthos 1981, Xylocastro 1981, Thessaloniki 1978), Romania (Vrancea 1977, 1986 and 1990), the United States (El Centro 1945, Northridge 1994) and Japan (Kobe 1995). From the above, some records covering a wide range of response were selected for further elaboration.

Two sets of analysis were performed. One refers to the evaluation of the structural response in respect to serviceability and ductility criteria, the other in respect to low-cycle fatigue criteria. The relevant studies will be presented in the following.

Evaluation in respect to serviceability and ductility criteria

The study refers to the investigation of the non-linear response of frames 1 to 7 shown in Figure 6.2.24. The joints are considered as rigid, the level of vertical loading is varied between 40% and 60% as outlined before. The fundamental periods of the frames are between 0.6 and 1.9 sec as shown in the Figure. The records considered in this study are those of Aigion, Thessaloniki and Kobe. Their acceleration spectra scaled to the peak ground acceleration, together with the design of the Greek seismic Code for a value of the behaviour factor q = 4, are shown in Figure 6.2.38. For all frames, the yield and the ultimate acceleration α_y, respectively α_u, were determined by appropriate scaling of the relevant records. The yield acceleration corresponds to the appearance of the first plastic hinge, the ultimate acceleration to certain limit states criteria as described below.

- *Serviceability criterion (A)*

The limit state is defined as the situation where the inter-storey drift exceeds 2% of the relevant storey height.

Compliance with this limit state roughly corresponds to the serviceability requirement included in Eurocode 8. Indeed, Eurocode 8 requires a drift limitation under serviceability conditions (moderate earthquakes) between 0.4 and 0.6%, depending on the susceptibility of non-structural elements to damage. Under ultimate conditions (strong earthquakes), the calculated drift values must be enlarged. Eurocode 8 does not introduce two types of earthquake, a moderate and a strong one. It rather proposes a ratio between the drift values for the strong to the moderate seismic record equal 2.5. The enhancement of the above drift limits by the factor 2.5 leads to drift limits of 1% and 1.5%. The limit values of the Code are therefore more conservative than the 2% value adopted here.

- *Residual drift criterion (B)*

The limit state is defined as a situation where the residual non-recoverable part of the inter-storey drift exceeds 1% of the storey height.

The residual drift angle is widely used as a criterion for the assessment of a building's condition after strong earthquakes. Indeed, if the residual drift exceeds roughly 3% the damage is supposed to be so heavy that the building must be demolished. For larger values than the adopted value 1% of the residual drift, the damage is considered to be moderate to heavy.

- *Ductility criterion (C)*

The limit state is defined as a situation where the rotation at any developing plastic hinge becomes larger than 0.03 radians.

The above limit value of rotation is supposed to correspond to an exhaustion of the rotation capacity of structural members or connections. In that sense this criterion expresses the limits imposed by local ductility.

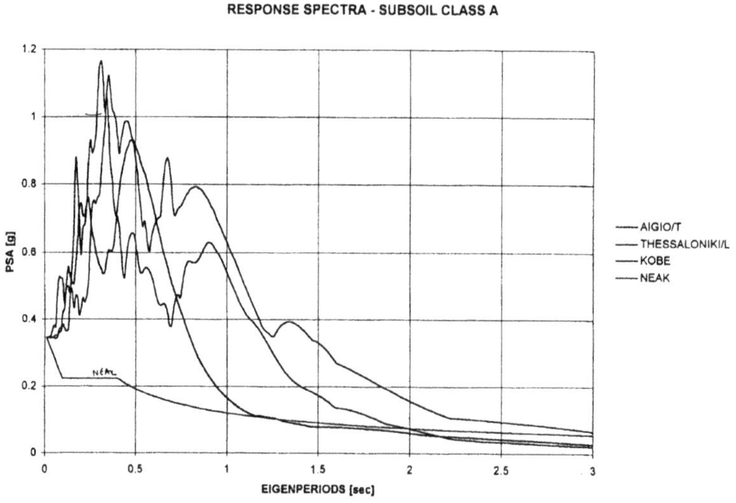

Figure 6.2.38: Acceleration response spectra of records considered and design Code spectrum

The yield accelerations α_y for two levels of vertical loading are presented in Figure 6.2.39. It may be observed that this acceleration is heavily influenced by the type of record and the frame characteristics. The Kobe record with the highest response values over a wide range of periods provides generally the lowest values of accelerations, contrary to the Aigion record whose peak response values concentrate over a narrow frequency range. The yield accelerations are as expected lower for higher level of vertical loading, as the structural reserves are lower in this case.

The values of the ultimate accelerations in accordance to the previously defined limit states criteria are illustrated in Figures 6.2.40 to 6.2.42.

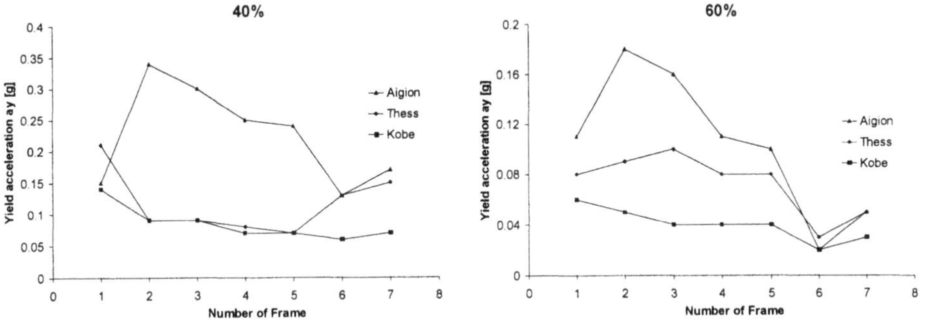

Figure 6.2.39: Yield acceleration a) for 40% and b) for 60% of vertical loading

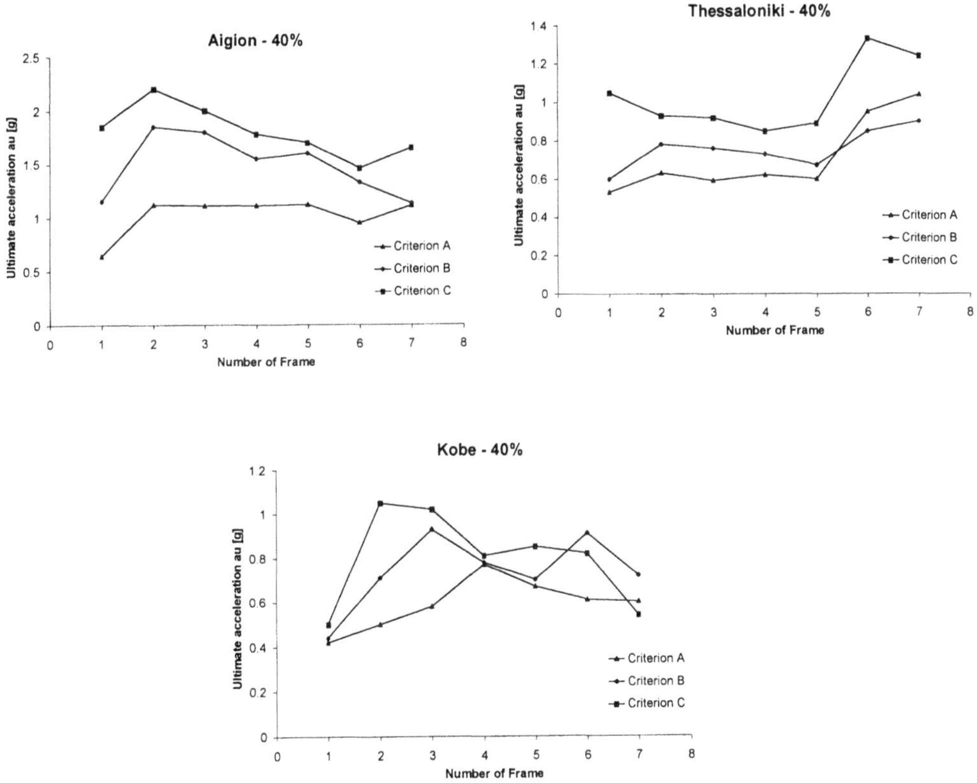

Figure 6.2.40: Ultimate accelerations for 40% level of vertical loading for a) Aigion, b) Thessaloniki and c) Kobe records

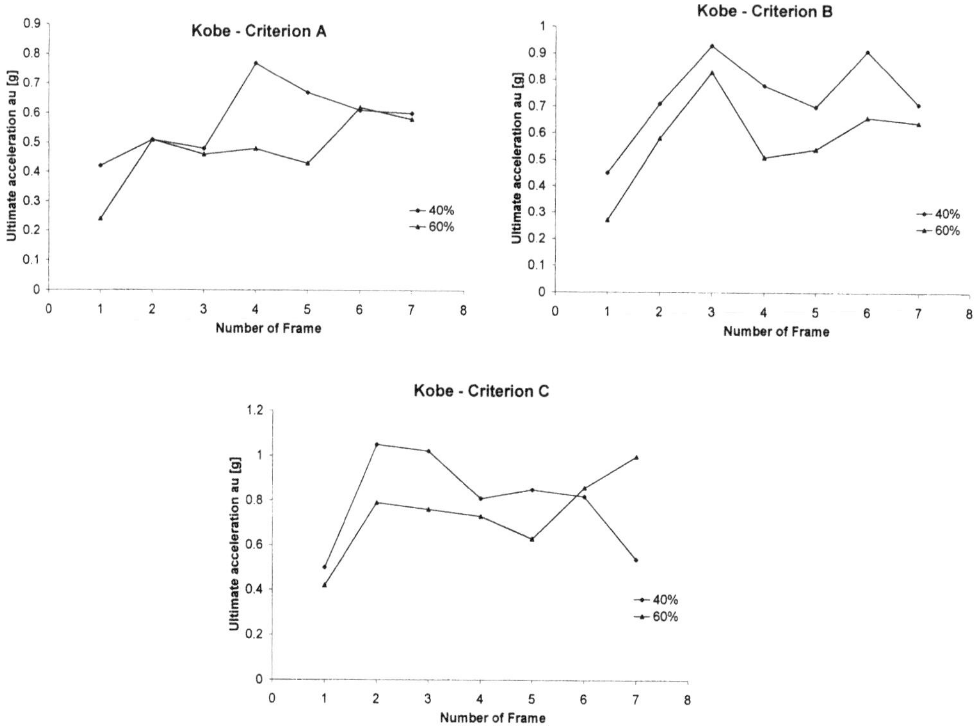

Figure 6.2.41: Ultimate accelerations for 40 and 60% of vertical loading for a Kobe record

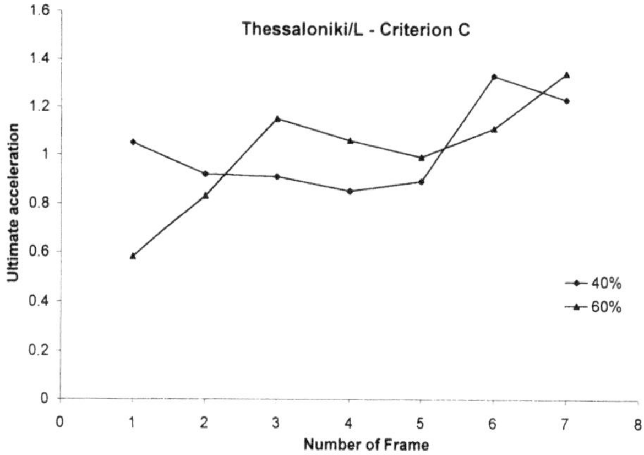

Figure 6.2.42: Ultimate accelerations for 40 and 60% of vertical loading for a Thessaloniki record

Following observations may be done:

a) Like the yield accelerations, the ultimate accelerations are strongly dependent on the characteristics of the structure and the type of record. For the frames investigated here, the Aigion record provides generally the highest larger acceleration values, almost twice as large as the Kobe record.
b) The three failure criteria provide different values of the ultimate accelerations. For the frames and the records under consideration, the serviceability criterion A leads generally to the lowest values. This shows that the drift limitation, as currently proposed in Eurocode 8, will be the decisive design criterion for moment resisting frames. This is confirmed by practical design applications.
c) Higher levels of vertical loading lead generally, although not always, to lower ultimate accelerations.
d) Contrary to the results of pushover analysis, frames with higher aspect ratios did not generally exhibit lower limit accelerations. This shows that pushover analysis may be over-conservative for more slender frames.

The ratio between the ultimate and the yield acceleration, expressing according to Eurocode 8 the behaviour factor q, is shown in Table 6.2.2. As outlined before, the value of this ratio is lowest for the serviceability limit state criterion A. However, the results of table 6.2.2 may be misleading due to the fact that what finally counts is the absolute value of the limit acceleration a structure can sustain during an earthquake and not its ratio to a yield acceleration. For the frames considered here, the absolute value ranges between approximately 1.0g for the Aigion record and 0.6g for the Kobe record. Taking into account that the actual PGA in Aigion was 0.54g and 0.86g in Kobe, the destructiveness of the Kobe earthquake may be concluded.

TABLE 6.2.2
MEAN VALUE OF THE RATIO BETWEEN ULTIMATE AND YIELD ACCELERATIONS FOR THREE RECORDS AND TWO LEVELS OF VERTICAL LOADING

Frame	1	2	3	4	5	6	7
A	4.5	6.4	6.3	8.6	8.1	16.8	13.0
B	7.2	8.5	9.7	9.5	9.4	18.2	14.1
C	6.9	10.5	10.2	11.8	11.2	22.7	18.0

Evaluation in respect to low-cycle-fatigue criteria

As well known, progressive damage is caused in a structure during a seismic event. The damage accumulation can be accounted for by application of low-cycle-fatigue lows. Since only inelastic deformations are causing damage, it is appropriate to adopt the plastic rotation as the controlling parameter. The fatigue curve is described then by eq. (6.2.28), the damage index by eq. (6.2.25). The value of the plastic rotation for monotonic loading, as well as the value m of the slope of the fatigue curve may be taken as parameters. It may be reminded that the adoption of m=1 corresponds to the algebraic summation of the occurring plastic deformations, an assumption that constitutes the basis of many design methods. For each part of the structure exhibiting inelastic deformations a damage index may be calculated.

Figure 6.2.43 presents maximum values of the damage index for frames 3 and 6 of Figure 6.2.24. The slope of the fatigue curve is taken equal to m=2 and limit plastic rotation for monotonic loading equal to 0.03 radians. The level of vertical loading is equal to 40%, the stiffness of the beam-to-column joints varies between rigid and 40% of the full rigid. The records of Aigion, Vrancea and Kobe are considered, the corresponding PGAs are equal to 0,54g, 0.21g and 0.86g. The Figures indicate that the frames collapse due to the Kobe earthquake. The damage due to the Aigion earthquake is lower than the damage due to the Vrancea earthquake, although that its PGA is more than twice as large.

This indicates that PGA cannot be the only decisive parameter in a seismic design. The effect of semi-rigid joints is not conclusive in this study. In some cases semi-rigidity improves, in other cases it worsens the behaviour. For more slender frames, it seems that rigid joints are more beneficial than semi-rigid ones.

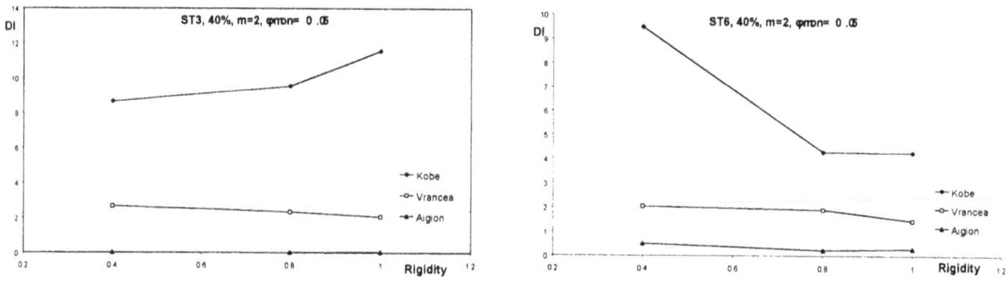

Figure 6.2.43: Damage index for frame with rigid and semi-rigid joints

Comparison between pushover and dynamic analyses

The study refers to non-linear dynamic analyses of the previously investigated frames in a manner that allows to a comparison of the results with the corresponding results of pushover analysis. For such a comparison, several seismic records from Greece (Thessaloniki 1978, Aigion 1985), Romania (Vrancea 1977), the US (El Centro 1945) and Japan (Kobe 1995 were considered. The normalised acceleration and relative input energy spectra of the five records are presented in Figures 6.2.44 and 6.2.45. For the non-linear dynamic analysis, the records were scaled in respect to acceleration to allow for comparisons with the results of the pushover analysis. The scaling factors were determined according to following considerations:

- Level of strength at first yielding as the ratio between the corresponding base shear to the sum of vertical loads from pushover analysis: $\varepsilon_y = H_y / P_{tot}$
- Contribution of ductility and overstrength from pushover analysis: $\mu_E\,\Omega$.

If the peak ground acceleration of a specific record in respect to the gravity acceleration is equal to α_{PGA}, the scale factor is determined from eq. (6.2.52):

$$sf = \frac{\varepsilon_y \mu_E \Omega}{\alpha_{PGA}} \tag{6.2.52}$$

The final peak ground acceleration of all records for a specific frame is then equal to:

$$(sf)\,\alpha_{PGA} = \varepsilon_y\,\mu_E\,\Omega \tag{6.2.53}$$

As a result of the non-linear analysis, the damage index is determined at all sections that exhibit inelastic deformations. The reference rotation capacity $\bullet_{mon}$ applied in the fatigue evaluation corresponds to the rotation capacity for monotonic loading. This was determined from the pushover analysis for the various levels of local ductility in accordance to eq. (6.2.21). The equivalent member length as shown in Figure 6.2.24 was determined for the moment diagram that also results in from pushover analysis.

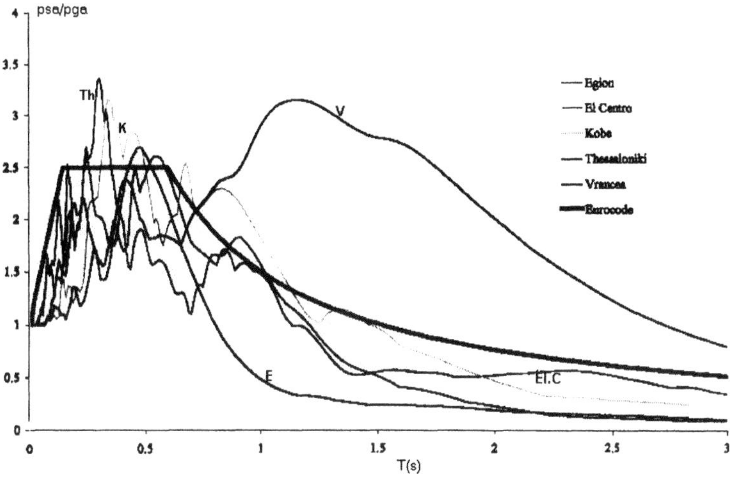

Figure 6.2.44: Normalised acceleration response spectra

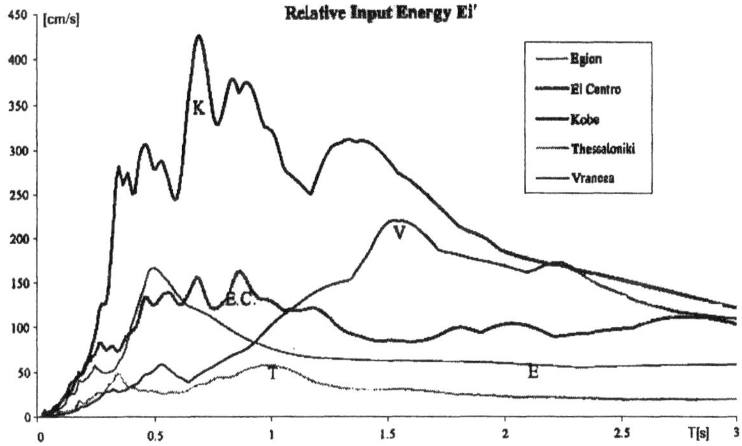

Figure 6.2.45: Relative input energy spectra

In case that pushover and non-linear dynamic analysis lead to the same results, the above scaling procedure leads to damage indices equal to 1. This means that the maximum damage index of a structure subjected to an earthquake with PGA scaled to the acceleration as determined by monotonic analysis is equal to 1. Monotonic analysis is conservative if the resulting damage index is higher than 1, not conservative if it is lower than 1. In the following, some characteristic results will be presented.

Figure 6.2.46 presents the influence of the value of the exponent m of the low-cycle fatigue curve. The figure refers to the three-storey three-bay frame with rigid joints, 40% level of vertical loading and local ductility level 5. As expected, a higher value of the exponent results in a reduction of the damage index. The adoption of m=1, which corresponds to the algebraic summation of the plastic deformations leads accordingly to conservative results. As observed before, different records although scaled to the same PGA produce considerably different damage.

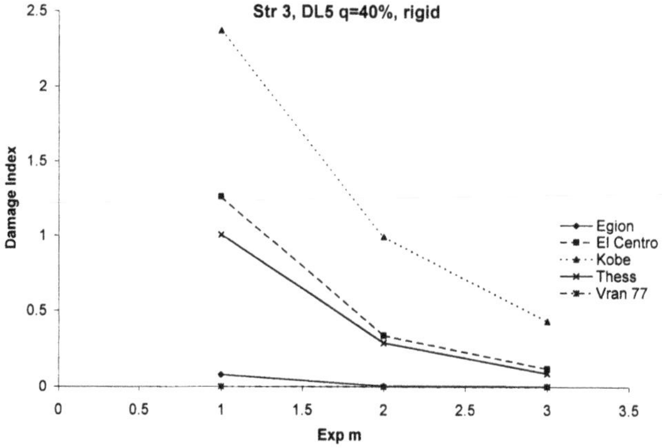

Figure 6.2.46: Damage index vs. value of the exponent m of the low-cycle fatigue curve

Figure 6.2.47 presents the influence of the local ductility level. The figure refers to the three-storey three-bay frame subjected to the Kobe record. The joints are rigid and semi-rigid, the level of vertical loading is 60% and the value of the exponent of the fatigue curve is m = 2. It may be observed that all values of the damage index are well below 1, indicating that pushover analysis is overconservative especially with increasing flexibility of the joints. Higher values of local ductility result in a reduction of damage. The beneficial influence of an increase in local ductility is even larger, if it is considered that for higher local ductility the peak ground acceleration was scaled to a larger value.

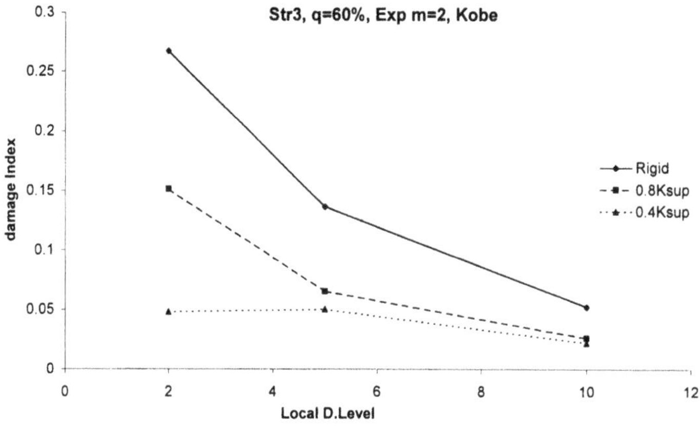

Figure 6.2.47: Damage vs. level of local ductility for rigid and semi-rigid joints

Figures 6.2.48 and 6.2.49 present results for the same frame subjected to four different records as a function of the local ductility level. The joints are rigid, the level of vertical loading is 60% and 40%. Higher values of local ductility result in again lower damage, which combined with the higher scaling factor indicates the highly beneficial influence of local ductility on the frame behaviour. The low values of the damage index (Figure 6.2.48) shows the conservatism of pushover analysis with increasing level of vertical loading.

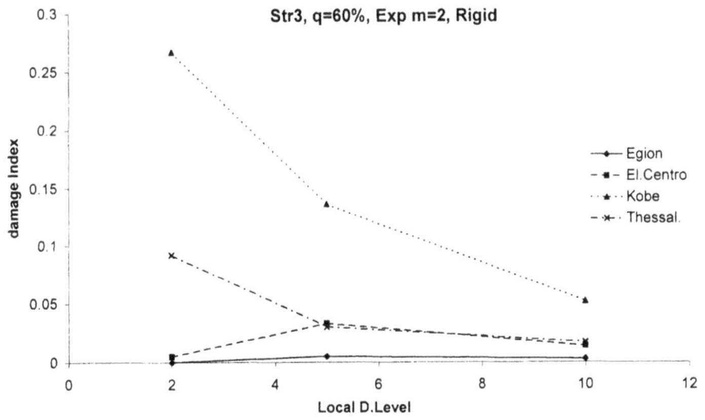

Figure 6.2.48: Damage vs. level of local ductility for 60% vertical loading

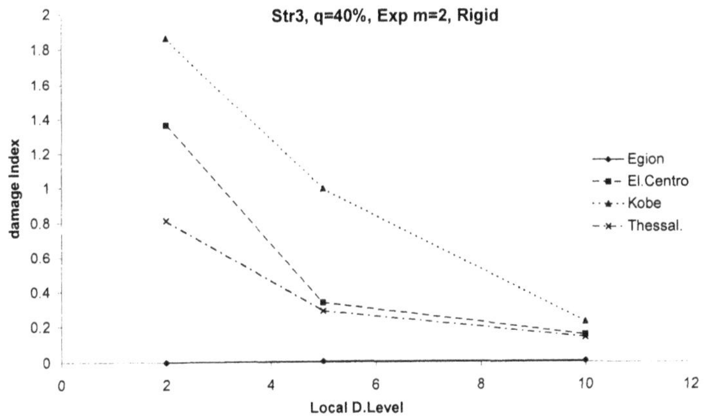

Figure 6.2.49: Damage vs. level of local ductility for 40% vertical loading

The effect of vertical loading is illustrated for another frame in Figure 6.2.50. The figure refers to the two-storey three-bay frame with rigid joints, ductility level 2 and value of the exponent of the fatigue curve m = 2. It may be again observed that higher vertical loading leads to lower damage index. The Figure is characteristic for most frames that have been investigated. It may be accordingly concluded that pushover analysis becomes conservative as the level of vertical loading increases.

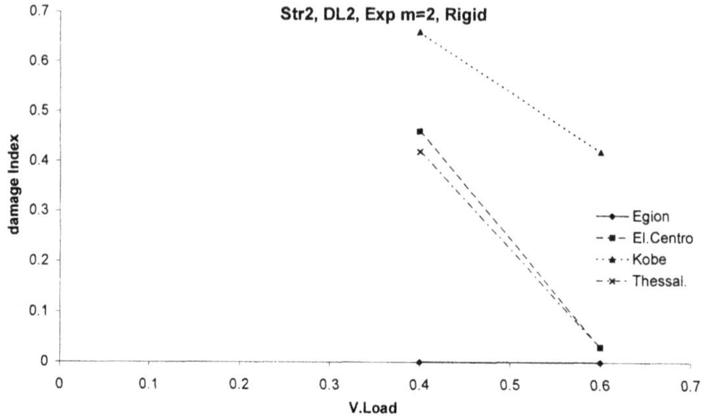

Figure 6.2.50: Damage vs. level of vertical loading

The effect of the joint rigidity is presented in Figures 6.2.51 to 6.2.53. Figures 6.2.51 and 6.2.52 refer to the three-storey three-bay frame, Figure 6.2.53 to the one-storey two-bay frame. A reduction in damage with decreasing joint rigidity may generally be observed. However, it must be reminded that this reduction may be balanced by the lower scaling factors for semi-rigid joints. The Figures only indicate that pushover analysis is conservative for frames with flexible joints.

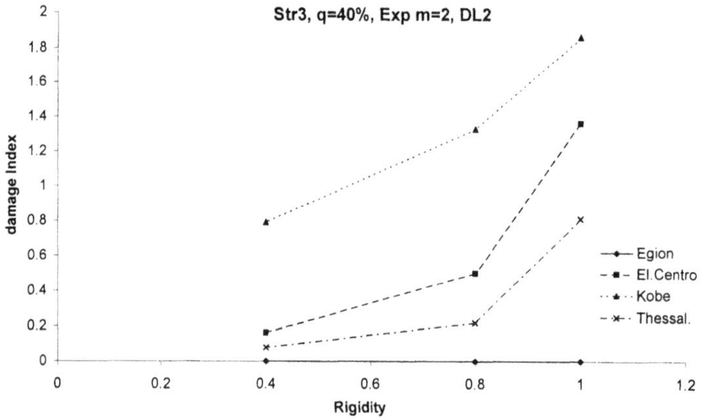

Figure 6.2.51: Damage index vs. joint rigidity

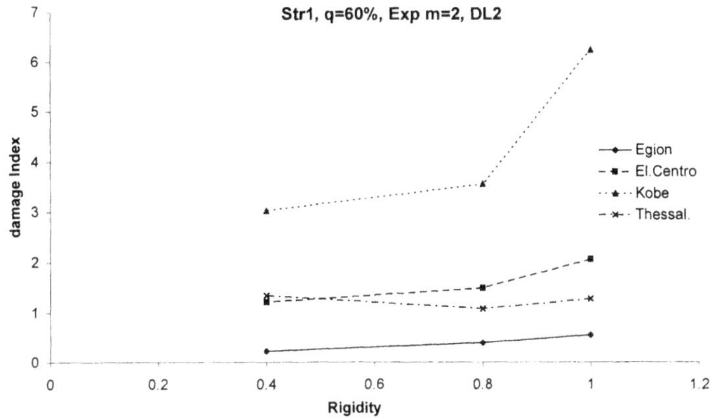

Figure 6.2.52: Damage vs. joint rigidity

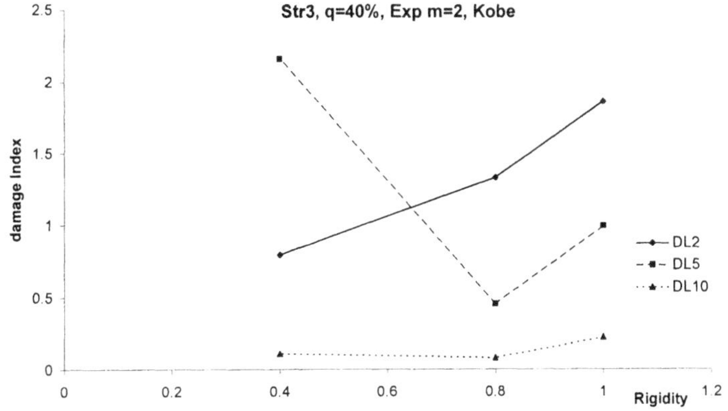

Figure 6.2.53: Damage index vs. joint rigidity

The examination of the above results leads to following conclusions:

a) The structural response to earthquakes is highly affected by the characteristics of both the structure and the seismic record. The results of pushover analysis cannot be directly extrapolated for seismic loading.
b) Pushover analysis is generally conservative for frames with high overall slenderness (height-to-width ratio), high level of vertical loading and flexible joints.
c) Local ductility and better fatigue behaviour is beneficially influencing the structural response.
d) Higher joint flexibility is sometimes positively, sometimes negatively influencing the structural response.
e) Observations from results of non-linear analyses constitute general trends of the structural behaviour. They do not necessarily apply for a specific frame subjected to a specific seismic record.

SUMMARY AND CONCLUSIONS

The present study refers to the investigation of the structural response of moment resisting frames under seismic loading. The local behaviour of structural members to monotonic loading is taken into account by introduction of various levels of local ductility in terms of inelastic rotation, characterising the rotation capacity of members and connections. The local response to cyclic loading is characterised by low-cycle fatigue curves, expressed in terms of inelastic rotation.

The response of the frames was investigated by means of various types of analysis and structural evaluation, as following:

- Pushover monotonic analysis. Structural evaluation by means of energy criteria
- Pushover dynamic analysis that corresponds to monotonic analysis with up-dated dynamic characteristics of the structure. Structural evaluation by means of energy criteria
- Non-linear dynamic analysis. Structural evaluation by means of serviceability, ductility and low-cycle fatigue criteria

Pushover analysis constitutes a useful tool to study the structural behaviour. It overestimates however the effects of some parameters, leading generally to very conservative results. Non-linear analysis predicts better the structural response as it is highly affected by the characteristics of both the structure and the seismic record. Its results are therefore restricted to the frame configurations and seismic records which have been considered. As a result of both types of analyses, some main trends may be observed that lead to following conclusions for the response of moment resisting frames:

a) Higher local ductility (rotation capacity at potential plastic hinge positions) and enhanced fatigue resistance as a result of appropriate detailing rules is beneficially influencing the structural response.
b) Higher joint flexibility may be in certain cases beneficial to the structural response
c) Irregularity in the distribution of the stiffness and the mass over the height does not considerably affect the structural behaviour.
d) Irregularity in strength, and especially weak storeys, considerably reduce the ability of the structure to resist seismic loading.
e) Serviceability criteria often prevail over resistance and ductility criteria.

REFERENCES

Ballio, G., Castiglioni, C. (1994). Seismic behavior of steel sections, *J. of Constructional Steel Research,* **29**, 21-54.

Bernuzzi, C., Calado, L., Castiglioni, C. (1997). Ductility and load carrying capacity predictions of steel beam-to-column connections under cyclic reversal loading. J. *of Earthquake Engineering*, 401-432.

Ciutina, A., Vayas, I. (1998). Degradation of framed structures due to low cycle fatigue induced by earthquakes, *Report Lab. of Steel Structures*, NTU Athens.

Eurocode 3, (1992). *Design of Steel Structures. Part 1-1 General Rules and Rules for Buildings*, CEN, European Committee for Standardization, ENV 1993-1-1

Eurocode 3, (1995). *Annex J. Joints in building frames.* CEN, European Committee for Standardization, ENV 1993-1-1

Eurocode 8, (1994). *Design provisions for earthquake resistance of structures*. CEN, European Committee for Standardisation, ENV 1998-1-1.

European Convention for Constructional Steelwork (ECCS). *Recommended testing procedure for assessing the behaviour of structural steel elements under cyclic loads*. ECCS Publ.No 45, Rotterdam, The Netherlands

European Norm EN 10025. *Hot rolled products of non-alloy structural steels* – Technical delivery conditions

Ferreira, J. et. al. (1998) Low cycle fatigue strength assessment of cruciform welded joints, *J. of Constructional Steel Research*, **47**, 223-244.

Gioncu, V., Petcu, D. (1997). Available Rotation Capacity of Wide-Flange Beams and Beam-Columns. Parts 1 & 2, *J. of Constructional Steel Research,* **43**, No 1-3, 161-244.

Guerra, C., Mazzolani, F., Piluso, V. (1990). Evaluation of the q-factor in steel framed structures:state-of-art. *Ingegneria sismica*, Anno VII n. 2, 42-63.

Kannan, A., Powel, G. (1975) DRAIN-2D. *A general purpose computer program for dynamic analysis of inelastic plane structures*, EERC 73-6 and EERC 73-22 reports, Berkeley, USA.
Kato, B. et. al. (1997). Kobe earthquake damage steel moment connections and suggested improvement, *Japanese Society of Steel Construction*, Techn. Rep. 39.

Kurobane, Y., Azuma, K., Ogawa, K. (1997) *Brittle fracture in steel building frames- Comparative study of Northridge and Kobe Earthquake damage*, Int. Inst. of Welding, Annual Assembly, San Francisco, Calif..

Mazzolani,M., Piluso, V. (1996). *Theory and Design of seismic Resistant Steel Frames*, E&FN SPON

Rubin, H. (1973). Das Q-Delta-Verfahren zur vereinfachten Berechnung verschieblicher Rahmensysteme nach dem Traglastverfahren der Theorie II. Ordnung, *Bauingenieur* 48, **8**, 275.

SOFiSTiK. (1999) 3-D *Stabwerk nach Theorie III. Ordnung*, SOFiSTiK Software GmbH

Spangemacher, R., Sedlacek G. (1992). Zum Nachweis ausreichender Rotationsfähigkeit von Fliessgelenken bei der Anwendung des Fliessgelenkverfahrens, *Stahlbau*, **61**, 329-339.

Vayas, I. (1997). Stability and Ductility of Steel Elements, *J. of Constructional Steel Research*, **44**, No 1-2, 23-50.

Vayas, I., Psycharis, I. (1994). Local cyclic behaviour of steel members, *Intern. Workshop on behaviour of steel structures in seismic areas*, STESSA '94, Mazzolani, Gioncu (eds), E & FN SPON, 231-241.

Vayas, I., Psycharis, I.(1990). Behaviour of thin-walled steel elements under monotonic and cyclic loading, European Conference on Structural Dynamics, Eurodyn' 90, Bochum, preprints 213-217

Vayas, I. (1996). Stregnth and Ductility of Axially Loaded Members with Outstand Plated Elements, *Proc. Coupled Instabilities in Metal Structures*, CIMS '96, Imperial College press, 189-198

Vayas, I., Psycharis, I. (1993). Ein dehnungsorientiertes Verfahren zur Ermittlung der Duktilität bzw. Rotationskapazität von Trägern aus I-Profilen, *Stahlbau*, **62**, 333-341.

Vayas, I. (1997). Investigation of the cyclic behaviour of steel beams by application of low-cycle fatigue criteria, *Intern. Conference on behaviour of steel structures in seismic areas*, STESSA '97, Kyoto, Mazzolani, Akiyama (eds), Edizioni 10/17, 350-357.

Vayas, I., Ciutina, A. (1999). Low-cycle-fatigue gestützter Erdbebennachweis von Rahmen aus Stahl, *Bauingenieur*, (in print).

Vayas, I., Sophokleous, A. (1996). Einfluß von Unregelmäßigkeiten auf die Duktilität von Rahmentragwerken, *Bauingenieur*, **71**, 329-339.

Vayas, I., Rangelov, N., Georgiev, Tz. (1999). Schlankheitsanforderungen zur Klassifizierung von Trägern aus I-Profilen, *Stahlbau*, (in print).

Vayas, I., Pasternak, H., Schween, T. (1995). Cyclic Behavior of beam-to-column steel joints with slender web panels, ASCE, *J. of Structural Engineering.*, **121**, No 2, 240-248.

Yamada M., Kawabata, T. Yamanaka, K.(1989). Biege-Ermüdungsbruch von Stahlstützen mit I- und Kastenquerschnitt, I Versuche, *Stahlbau*, 58, 361-364.

Yamada, M. (1996). Das Hanshin-Awaji-Erdbeben, Japan, *Bauingenieur*, **71**, 15-19, 73-80

Chapter 7

Failure Mode and Ductility Demand

7.1

DESIGN OF SEMI-RIGID STEEL FRAMES FOR FAILURE MODE CONTROL

Rosario Montuori, Vincenzo Piluso

INTRODUCTION

The failure mode control is universally recognized as a fundamental requirement for designing seismic resistant structures, because local failure modes, such as partial and soft-storey mechanisms, lead to damage concentration and, as a consequence, to the premature collapse of structures.
For this reason, the development of a collapse mechanism of global type is promoted by modern seismic codes. In fact, this mechanism typology involving the beam ends and the base sections of first storey columns maximizes the number of dissipative zones. Unfortunately, as demonstrated by numerous numerical studies dealing with the seismic inelastic performances of steel frames, this ambitious pattern of yielding cannot be realized by means of simplified design rules. In other words, simplified criteria such as the member hierarchy criterion and the three-quarter rule (Lee, 1996) are able to prevent the formation of storey mechanisms, but are not able to assure the development of a collapse mechanism of global type.
In order to assure a collapse mechanism of global type, a sophisticated theoretical procedure, based on the extension of the kinematic theorem of plastic collapse to the concept of mechanism equilibrium curve, has been recently proposed (Mazzolani and Piluso, 1996a; 1997) with reference to rigid full-strength beam-to-column joints. The problems to be faced for the extension of the design procedure to semi-rigid frames have been also outlined (Mazzolani and Piluso, 1996b).
With reference to the ideal case of full-strength rigid beam-to-column joints, i.e., the case where beam-to-column joints do not constitute a frame imperfection, the sections of the columns required to develop a collapse mechanism of global type, are strictly dependent on the plastic moment of the beam sections.
In such a case, the design of frames failing in global mode can require economically severe design solutions when a large beam section is necessary due to long spans or heavy vertical loads. This observation constituted the motivation of the work presented in this paper. In fact, the aim of the work is to look for structural solutions forcing the yielding to occur at the beam ends rather than in the columns even in the case of large moment capacity beams required to cover long spans, without resorting to columns having very large sections. One of the possible structural solutions is constituted by the use of semi-rigid partial-strength connections. In fact, in this case the flexural strength of the connections can be properly balanced to develop a pattern of yielding involving the beam-to-column connections only, with the exception of the base sections of the first storey columns where the formation of plastic hinges is expected. The application of this alternative approach, which will be examined in this paper, requires additional design issues to be accurately examined. This is the case of the check of the resistance of beam-to-column joints under the load combinations provided by the codes and the case of the fulfilment of the serviceability requirements concerning the limitation of the interstorey drifts under seismic forces corresponding to frequent earthquakes. In particular, this design

requirement is very important in the case of partial-strength connections due to the corresponding rotational deformability which can lead to a significant increase of the lateral displacements with respect to the case of rigid frames. Regarding the relationship between the joint flexural resistance and the corresponding rotational stiffness reference will be made to bolted extended end plate connections. All these aspects are taken into account within the proposed design procedure which will be presented in the following Sections. Finally, the results of a wide parametric analysis will be used to show the practical situations where the use of partial-strength semi-rigid connections can lead to a significant saving in structural weight, without renouncing to a collapse mechanism of global type, i.e. excluding column hinging.

FAILURE MODE CONTROL

Generality

As already stated, a design procedure assuring the development of a collapse mechanism of global type has been recently proposed with reference to geometrically regular steel frames with full-strength beam-to-column connections (Mazzolani and Piluso, 1996a; 1997) and, successively, extended to the case of set-back steel frames (Mazzolani and Piluso, 1996c).
The design procedure is based on the extension of the kinematic theorem of plastic collapse to the concept of mechanism equilibrium curve. Three collapse mechanism typologies are considered (Figure 7.1.1), so that $3n_s$ collapse mechanisms are analysed, being n_s the number of storeys.

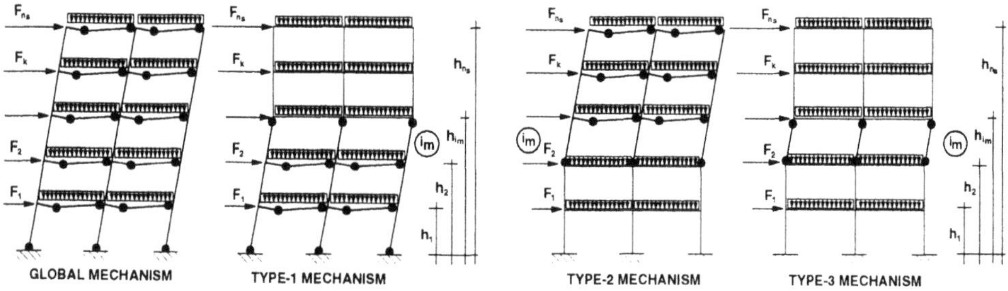

Figure 7.1.1: Collapse mechanism typologies.

It is assumed that the beam sections are known, because they are designed to resist vertical loads. Therefore, the unknowns of the design problem are the column sections, i.e. the plastic moments of columns. The column plastic moments are derived by imposing that, for a properly selected range of the frame top sway displacement, the mechanism equilibrium curve corresponding to the global failure mode has to lie below those corresponding to all the other kinematically admissible mechanisms.
The mathematical details of the design procedure for failure mode control of rigid frames are given elsewhere (Mazzolani and Piluso, 1997). Some additional issues have to be considered in the case of partial-strength beam-to-column connections and in the case of concentrated forces transmitted to the girders (main beams) by the secondary beams constituting the floor deck. These issues regard the location of plastic hinges within the beam span, the internal work due to the beams and the external work due to the vertical loads acting on the beams, as it will discussed in the following sections. The basic concepts of the design procedure for failure mode will also be briefly summarized.

Location of plastic hinges in beams

The aim of the design procedure is to maximise the number of dissipative zones by promoting the yielding of the beam ends and/or the yielding of the beam-to-column connections. In addition, depending on the magnitude of vertical loads, the hinging of an intermediate beam section can also arise. All these possibilities are accepted, because the primary aim of the failure mode control is to limit the column hinging to the base sections of the first storey columns.

The first issue to be faced is the location of the dissipative zones within the beam span. Regarding the location of the first plastic hinge, it can be observed that the seismic actions can be modelled by means horizontal forces whose distribution along the height of the structure can be selected according to an appropriate combination of the of vibration modes.

The magnitude of these forces can be represented by a common multiplier α, so that the bending moment diagram is obtained as the superposition of the one due to vertical loads, which are assumed to be constant, and the one due to horizontal forces (Figure 7.1.2).

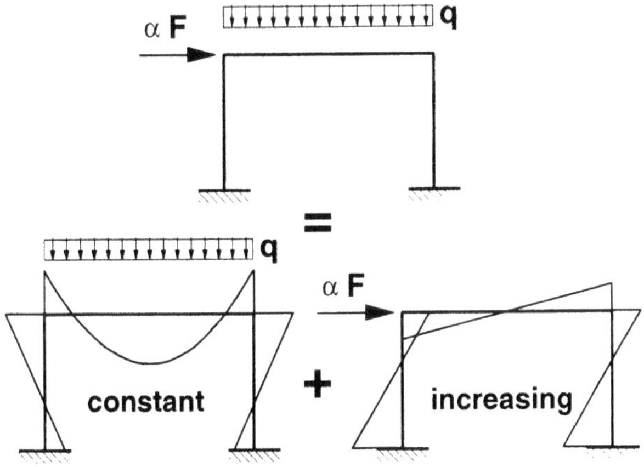

Figure 7.1.2: Superposition of bending moment

It can be recognised that increasing horizontal forces, i.e. the multiplier α, the first plastic hinge develops at the beam end or in the connection where the bending moments due to seismic forces and to vertical loads have the same sign (Figure 7.1.2).

The locations of the second plastic hinge depends on the magnitude of vertical loads and on the flexural resistance of connections. The flexural resistance of connections can be expressed through the following nondimensional parameters:

$$\overline{m}_L = \frac{M_{j,Rd}^{(left)}}{M_p} \qquad \overline{m}_R = \frac{M_{j,Rd}^{(right)}}{M_p} \tag{7.1.1}$$

where $M_{j,Rd}^{(left)}$ and $M_{j,Rd}^{(right)}$ are, respectively, the design flexural resistance of left and right beam-to-column joints and M_p is the design plastic moment of the beam section. With reference to the case in which the vertical loads acting on the beam are constituted by N concentrated forces, the beam span is divided in N+1 equal fields. It can be observed that the expressions for computing the bending moments in any field of beam are (Figure 7.1.3):

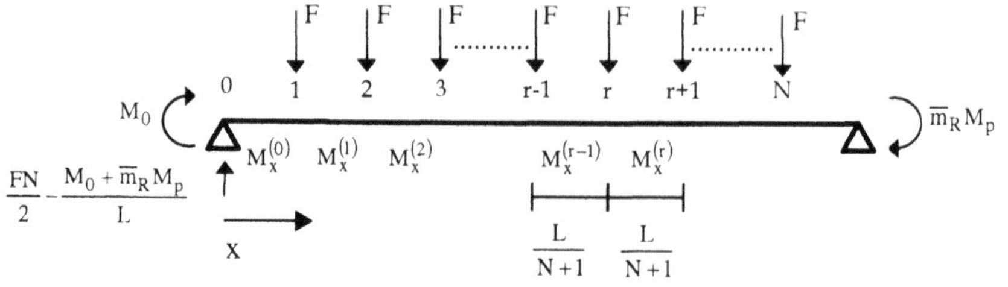

Figure 7.1.3: Adopted notation

$$M_x^{(0)} = M_0 + \left(\frac{FN}{2} - \frac{M_0 + \overline{m}_R M_p}{L} \right) x \tag{7.1.2}$$

$$M_x^{(1)} = M_x^{(0)} - F\left(x - \frac{L}{N+1} \right) = M_0 + \left(\frac{FN}{2} - F - \frac{M_0 + \overline{m}_R M_p}{L} \right) x + \frac{FL}{N+1} \tag{7.1.3}$$

$$M_x^{(2)} = M_x^{(1)} - F\left(x - \frac{2L}{N+1} \right) = M_0 + \left(\frac{FN}{2} - 2F - \frac{M_0 + \overline{m}_R M_p}{L} \right) x + \frac{FL}{N+1}(1+2) \tag{7.1.4}$$

and so on.
Therefore the following recurrent relationship can be adopted:

$$\begin{aligned} M_x^{(r)} &= M_x^{(r-1)} - F\left(x - \frac{rL}{N+1} \right) = M_0 + \left(\frac{FN}{2} - rF - \frac{M_0 + \overline{m}_R M_p}{L} \right) x + \frac{FL}{N+1} \sum_{t=1}^{r} t = \\ &= M_0 + \left(\frac{FN}{2} - rF - \frac{M_0 + \overline{m}_R M_p}{L} \right) x + \frac{FL}{N+1} \frac{(r+1)r}{2} \end{aligned} \tag{7.1.5}$$

where the index r denotes the beam section corresponding to the r-th vertical force.
In addition, the r-th derivative is given by:

$$\frac{dM_x^{(r)}}{dx} = F\frac{(N-2r)}{2} - \frac{M_0 + \overline{m}_R M_p}{L} \tag{7.1.6}$$

By a simple equilibrium consideration, it can be pointed out that the bending moment is decreasing for $x>L/2$, so that the second plastic hinge can develop only in a section satisfying the following relation:

$$\frac{rL}{N+1} \leq \frac{L}{2} \qquad \text{i.e.} \qquad r \leq \frac{N+1}{2} \tag{7.1.7}$$

According to the previous relationship, the more distant section from left end of the beam where the second plastic hinge can develop is identified by the index:

$$r_{max} = \frac{N}{2} \qquad \text{if N is even}$$
$$(7.1.8)$$
$$r_{max} = \frac{N+1}{2} \qquad \text{if N is odd}$$

The development of the second plastic hinge in section 0 (i.e. at the first beam end or in the corresponding beam-to-column connection) requires the fulfilment of the following conditions in the sections regarding the sections where concentrated forces are applied:

$$M_x^{(1)}\left(x = \frac{L}{N+1}\right) < M_p \quad \ldots\ldots \quad M_x^{(r)}\left(x = \frac{rL}{N+1}\right) < M_p \quad \ldots\ldots \quad M_x^{(r_{max})}\left(x = \frac{r_{max}L}{N+1}\right) < M_p \tag{7.1.9}$$

In addition, the yielding condition of section 0 is given by:

$$M_0 = M_p\overline{m}_L \tag{7.1.10}$$

Combining the r-th relationship (7.1.9), with the yielding condition (7.1.10), the following limitation is obtained:

$$F < \frac{2M_p\left[r\overline{m}_R + N + 1 - \overline{m}_L(N+1-r)\right]}{rL(N+1-r)} = F_0^{(r)} \tag{7.1.11}$$

which has to be verified for $r = 1,\ldots,r_{max}$. As a consequence the second plastic hinge is located in section 0 provided that the following condition is satisfied:

$$F < \min\left\{F_0^{(1)},\ldots\ldots,F_0^{(r_{max})}\right\} \tag{7.1.12}$$

The second plastic hinge can occur in an intermediate beam section, below the r-th vertical force, provided that the bending moment $M_x^{(r-1)}$ in the previous field is increasing, i.e. $\frac{dM_x^{(r-1)}}{dx} > 0$, and the one $M_x^{(r)}$ in the successive field is decreasing, i.e. $\frac{dM_x^{(r)}}{dx} < 0$. In addition, the following conditions have to be verified:

$$M_0 < \overline{m}_L M_p \tag{7.1.13}$$

which excludes the yielding of the beam-to-column connection or the beam end, and

$$M_x^{(r)}\left(x = \frac{rL}{N+1}\right) = M_p \tag{7.1.14}$$

corresponding to the yielding condition of the t-th beam section.
By imposing the increasing of $M_x^{(r-1)}$ and the decreasing of $M_x^{(r)}$, the following limitation can be, respectively, obtained:

$$F > \frac{2\left(M_0 + \overline{m}_R M_p\right)}{L(N - 2r + 2)} \tag{7.1.15}$$

$$F < \frac{2\left(M_0 + \overline{m}_R M_p\right)}{L(N - 2r)} \tag{7.1.16}$$

Taking into account the yielding condition (7.1.14), the value of $M_x^{(r)}$ given by the equation (7.1.5), provides the bending moment occurring in section 0 when the second plastic hinge is located in the r-th section:

$$M_0 = \frac{M_p(N+1+r\overline{m}_R)}{(N-r+1)} - \frac{rFL}{2} \tag{7.1.17}$$

By substituting the above value of M_0 into the relationships (7.1.13), (7.1.15) and (7.1.16) the following three limitations are obtained:

$$F > \frac{2M_p\left[N+1+r\overline{m}_R - (N+1-r)\overline{m}_L\right]}{rL(N+1-r)} = F_{M_0}^{(r)} \tag{7.1.18}$$

$$F > \frac{2M_p(N+1)(1+\overline{m}_R)}{L(N-r+1)(N-r+2)} = F_i^{(r)} \tag{7.1.19}$$

$$F < \frac{2M_p(N+1)(1+\overline{m}_R)}{L(N-r)(N-r+1)} = F_s^{(r)} \tag{7.1.20}$$

which can be summarised as follows:

$$\max\left\{F_{M_0}^{(r)}, F_i^{(r)}\right\} < F < F_s^{(r)} \tag{7.1.21}$$

The comparison between the relations (7.1.11) and (7.1.18) and between relations (7.1.19) and (7.1.20) shows that:

$$F_0^{(r)} = F_{M_0}^{(r)} \qquad \text{for} \quad r = 1,2,3,\ldots\ldots, r_{max} \tag{7.1.22}$$

$$F_i^{(r)} = F_s^{(r-1)} \qquad \text{for} \quad r = 2,3,4,\ldots\ldots., r_{max} \tag{7.1.23}$$

Therefore, accounting for the above relationships, the second plastic hinge develops in the first end of the beam or in the connection corresponding to section 0 provided that the following condition is fulfilled:

$$F < \min\left\{F_{M_0}^{(1)}, F_{M_0}^{(2)}, \ldots, F_{M_0}^{(r_{max})}\right\} \tag{7.1.24}$$

which is obtained from (7.1.12) and (7.1.22).
On the contrary, the second plastic hinge develops in an intermediate beam section *($r<r_{max}$)*, provided that the following condition holds:

$$\max\left\{F_s^{(r-1)}, F_{M0}^{(r)}\right\} < F < F_s^{(r)} \tag{7.1.25}$$

The particular case in which the second plastic hinge develops below the last vertical force *($r=r_{max}$)* preceding (when N is even) or coincident with (when N is odd) the midspan, is identified by the following condition:

$$F > \max\left\{F_i^{(r_{max})}, F_{M0}^{(r_{max})}\right\} \tag{7.1.26}$$

In order to simplify the requirement (7.1.24) corresponding to the occurrence of the second plastic hinge in section 0, it is useful to compare the expressions of $F_{M0}^{(r)}$ and $F_{M0}^{(r+1)}$ aiming at the identification of the minimum value of of $F_{M0}^{(r)}$ for $(r=1.....r_{max})$. It is easy to recognise that:

$$F_{M0}^{(r+1)} > F_{M0}^{(r)} \quad (r=1.....r_{max}-1) \quad \text{for} \quad \frac{r(r+1)}{(N+1)(N-2r)}\overline{m}_R + \frac{(N-r)(N+1-r)}{(N+1)(N-2r)}\overline{m}_L > 1 \tag{7.1.27}$$

Similarly, aiming at the simplification of the condition (7.1.25), corresponding to the occurrence of the second plastic hinge in an intermediate beam section, it is useful to compare the expression of $F_s^{(r)}$ and $F_{M0}^{(r+1)}$ to recognise the minimum value of terms contained in the first member. It is easy to show that:

$$F_s^{(r)} > F_{M0}^{(r+1)} \quad (r=1.....r_{max}-1) \quad \text{for} \quad \frac{r(r+1)}{(N+1)(N-2r)}\overline{m}_R + \frac{(N-r)(N+1-r)}{(N+1)(N-2r)}\overline{m}_L > 1 \tag{7.1.28}$$

Therefore, relationships (7.1.27) and (7.1.28) suggest the introduction of a function *f* defined as:

$$f(r, \overline{m}_R, \overline{m}_L) = \frac{r(r+1)}{(N+1)(N-2r)}\overline{m}_R + \frac{(N-r)(N+1-r)}{(N+1)(N-2r)}\overline{m}_L \tag{7.1.29}$$

Making use of the function *f* just introduced, relationships (7.1.27) and (7.1.28) can be expressed as:

$$F_s^{(r)} > F_{M0}^{(r+1)} > F_{M0}^{(r)} \quad (r=1.....r_{max}-1) \quad \text{for} \quad f(r, \overline{m}_R, \overline{m}_L) \geq 1 \tag{7.1.30}$$

By considering that function *f* is increasing with r, it can be stated that if condition (7.1.30) is satisfied for a given value $r = r^*$ then it is fulfilled also for $r = r^*+1, r^*+2,\ldots, r_{max}-1$.
Accounting for the above considerations, the location of the second plastic hinge can be identified in a simple way. In fact, it is possible to compute the minimum value of r, namely r_{min}, which satisfies relationship (7.1.30). Therefore, due to increasing of function *f*, if results $f(r, \overline{m}_R, \overline{m}_L) > 1$ for $r > r_{min}$ while $f(r, \overline{m}_R, \overline{m}_L) < 1$ for $r < r_{min}$. As a consequence, relationship (7.1.30) provides:

$$F_s^{(r)} > F_{M_0}^{(r+1)} > F_{M_0}^{(r)} \quad \text{for } r = t, \ldots\ldots, r_{max} - 1 \quad \text{and} \quad F_s^{(r)} < F_{M_0}^{(r+1)} < F_{M_0}^{(r)} \quad \text{for } r = 1, \ldots\ldots, t-1 \qquad (7.1.31)$$

It is useful to observe that, according to the second one of relationships (7.1.31), for $r < r_{min}$, the condition (7.1.25) has no meaning because, in this case, $F_s^{(r)} < F_{M_0}^{(r)}$. It means that intermediate beam sections preceding r_{min} cannot yield. As a consequence, the condition (7.1.24) corresponding to the yielding of section 0 can be written as:

$$F < \min\left\{F_{M_0}^{(r_{min})}, F_{M_0}^{(r_{min}+1)}, \ldots, F_{M_0}^{(r_{max})}\right\} \qquad (7.1.32)$$

which, taking into account the first one of relationships (7.1.31), becames:

$$F < F_{M_0}^{(r_{min})} \qquad (7.1.33)$$

In addition, the condition (7.1.25) to be verified to develop the second plastic hinge in an intermediate beam section, accounting for the first one of relationships (7.1.31), can be written as:

$$F_{M_0}^{(r)} < F < F_s^{(r)} \quad \text{with} \quad r = r_{min} \qquad \text{and} \qquad F_s^{(r-1)} < F < F_s^{(r)} \quad \text{with} \quad r = r_{min}, r_{min-1}, \ldots, r_{max-1} \qquad (7.1.34)$$

because, as already demonstrated, the second plastic hinge cannot develop in sections identified by $r < r_{min}$.

As a conclusion, if the magnitude of vertical loads satisfies condition (7.1.33) the second plastic hinge develops in section 0 (connection or beam end). On the contrary, the second plastic hinghe occurs in an intermediate beam section whose location is identified by the index $r = r_h$ which can be defined as the characteristic value of r assuring the fulfilment of condition (7.1.34). Obviously, this index depends on the magnitude of vertical loads.

In addition, it can be observed that relationship (7.1.33) shows that $F_{M_0}^{(r_{min})}$ can be considered as a limit force. In fact if F has a value smaller than $F_{M_0}^{(r_{min})}$ then the second plastic hinge is located at the left end of the beam or in the corresponding connection, on the contrary if F has a value greater than $F_{M_0}^{(r_{min})}$ the second plastic hinge can develop in an intermediate section of the beam. In the case of full-strength beam-to-column joints ($\overline{m}_L = 1$ and $\overline{m}_R = 1$), considering the equivalent uniform distributed load $q = F(N+1)/L$ the result obtained by Mazzolani and Piluso (1997) is derived as a particular case of the most general one herein presented. In fact, for $N \to \infty$ the following limit holds:

$$\lim_{N \to \infty} \frac{F_{M_0}^{(r_{min})}(N+1)}{L} = \frac{4M_P}{L^2} = q_{lim} \qquad (7.1.35)$$

Mechanism equilibrium curves

As it has been pointed out by Mazzolani and Piluso (1997), the mechanism equilibrium curve can be expressed as:

$$\alpha_c = \alpha - \gamma\delta \qquad (7.1.36)$$

where α is the kinematically admissible multiplier of horizontal forces, γ is the slope of the mechanism's equilibrium curve and δ is the top sway displacement of the moment resisting frame. In particular, it results:

$$\alpha = \frac{tr\left(\mathbf{C}^T\mathbf{R}_c\right) + 2tr\left(\mathbf{B}^T\mathbf{R}_b\right) - tr\left(\mathbf{q}^T\mathbf{D}_v\right)}{\mathbf{F}^T\mathbf{s}} \tag{7.1.37}$$

and

$$\gamma = \frac{\mathbf{V}^T\mathbf{s}\dfrac{\delta}{H_0}}{\mathbf{F}^T\mathbf{s}} \tag{7.1.38}$$

where the following notation has been used:

- **C** is a matrix of order $n_c \times n_s$ (n_c is the number of columns and n_s is the number of storeys) whose elements are the plastic moment columns which are reduced due to the contemporary action of the axial load, i. e. C_{ik} is the plastic moment of the i-th column of the k-th storey;
- $\mathbf{R_c}$ is a matrix of order $n_c \times n_s$ whose coefficient $R_{c,ik}$ account for the participation of the i-th column of the k-th storey to the collapse mechanism, in fact it is given by:
  $R_{c,ik} = 2$ when both the column ends are yielded;
  $R_{c,ik} = 1$ when only one column end is yielded;
  $R_{c,ik} = 0$ when the column does not participate to the collapse mechanism;
- **B** is a matrix of order $n_b \times n_s$ (n_b is the number of bays) which contains the plastic moment of the relevant dissipative zones of the j-th bay of the k-th storey, its elements are defined as:

$$B_{jk} = \frac{\left(\overline{m}_{L,jk} + \overline{m}_{R,jk}\right)}{2} M_{b,jk} \qquad \text{for } x_{jk} = 0 \tag{7.1.39}$$

$$B_{jk} = \frac{\left(1 + \overline{m}_{R,jk}\right)}{2} M_{b,jk} \qquad \text{for } x_{jk} > 0 \tag{7.1.40}$$

  where x_{jk} is the abscissa of section where the secon plastic hinge is located.
- $\mathbf{R_b}$ is a matrix of order $n_b \times n_s$ whose coefficient $R_{b,jk}$ accounts for the participation of j-th beam of k-th storey to the collapse mechanism, it is given by $R_{b,jk} = L_j/(L_j - x_{jk})$ when the beam is involved in the mechanism, while is equal to zero in the opposite case;
- **q** is a matrix of order $n_b \times n_s$ whose general term q_{jk} is the vertical uniform load acting on the j-th beam of the k-th storey when vertical loads are uniformly distributed, while it is the equivalent uniform load, i.e. $q_{jk} = F_{jk}\left(N_{jk} + 1\right)/L_j$ (F_{jk} is the value of the concentrated forces acting on the j-th beam of the k-th storey, N_{jk} is the corresponding number of concentrated forces and L_j is the span of the j-th beam) when vertical loads are constituted by concentrated forces;
- $\mathbf{D_v}$ is a matrix of order $n_b \times n_s$ whose coefficient $D_{v,jk}$ accounts for the external work of vertical loads acting on the j-th beam of the k-th storey, it is given by $D_{v,jk} = L_j x_{jk}/2$ when the beam is involved in the collapse mechanism, while it is equal to zero in the opposite case;
- **F** is the vector of the design horizontal forces (order n_s);
- **s** is the shape vector of horizontal displacements (order n_s);

- $\mathbf{V}$ is the vector of storey vertical loads, i.e. V_k is the total vertical load acting on the k-th storey given by $V_k = \sum_{j=1}^{n_b} q_{jk} L_j$;
- H_0 is the number of the interstorey heights of the storeys involved in the collapse mechanism.

It is useful to remember that relationships (7.1.36) is obtained by equalizing the internal work with the external work which includes the second order work of vertical loads, i.e. the work of vertical loads due to second order vertical displacements.
Regarding Eq.(7.1.37), it is useful to point out that the term related to the external work due to the vertical loads, i. e. $tr(\mathbf{q}^t\mathbf{D}_v)$, remains formally coincident with that occurring in the case of uniformly distributed vertical loads.

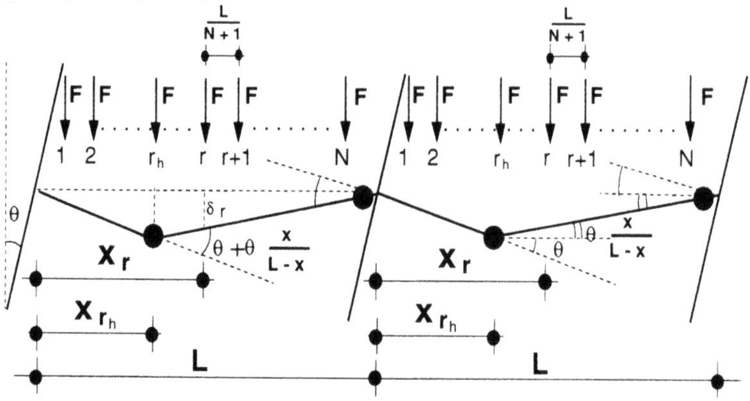

Figure 7.1.4: Action of vertical loads

In fact, it can be observed that, the abscissa where the generic r-th force is applied is given by:

$$x_r = \frac{L}{N+1} r \qquad r = 1,2,\ldots, N \qquad (7.1.41)$$

and the corresponding vertical displacement provided that the beam is involved in the collapse mechanism, is given by (Figure 7.1.4):

$$\begin{cases} \delta_r = \theta x_r & \text{for} \quad x_r \le x_{r_h} \\ \delta_r = \dfrac{\theta x_r}{L - x_r}(L - x_r) & \text{for} \quad x_r > x_{r_h} \end{cases} \qquad (7.1.42)$$

where r_h and x_{r_h} represent, respectively, the section and the corresponding abscissa where the second plastic hinge occurs (Figure 7.1.4).
The combination of relationships (7.1.41) and (7.1.42) gives:

$$\delta_r = \frac{\theta L}{N+1} r \qquad \text{for} \quad r \le r_h$$

$$\delta_r = \frac{\frac{\theta L}{N+1} r_h}{L - \frac{L}{N+1} r_h}\left(L - \frac{Lr}{N+1}\right) = \frac{\theta L r_h (N+1-r)}{(N+1-r_h)(N+1)} \qquad \text{for} \quad r > r_h \tag{7.1.43}$$

As a consequence the external work due to the vertical forces acting on the generic beam can be expressed as:

$$L_F = \sum_{r=1}^{N} \delta_r F = \theta F\left[\sum_{r=1}^{r_h} \frac{Lr}{N+1} + \sum_{r=r_h+1}^{N} \frac{L r_h (N+1-r)}{(N+1-r_h)(N+1)}\right] = \frac{FL\theta}{N+1}\left[\sum_{r=1}^{r_h} r + \sum_{r=r_h+1}^{N} \frac{r_h (N+1-r)}{(N+1-r_h)}\right] =$$

$$= \frac{FL\theta}{N+1}\left[\frac{r_h(r_h+1)}{2} + \frac{r_h(N+1)}{(N+1-r_h)} \sum_{r=r_h+1}^{N} 1 - \frac{r_h}{(N+1-r_h)} \sum_{r=r_h+1}^{N} r\right] \tag{7.1.44}$$

The above expression can be simplified pointing out that:

$$\sum_{r=r_h+1}^{N} 1 = (N - r_h) \qquad \text{and} \qquad \sum_{r=r_h+1}^{N} r = \sum_{r=1}^{N} r - \sum_{r=1}^{r_h} r = \frac{N(N+1)}{2} - \frac{r_h(r_h+1)}{2} \tag{7.1.45}$$

by substituting Eq. (7.1.45) into Eq. (7.1.44), the following simple expression can be obtained:

$$L_F = \sum_{r=1}^{N} \delta_r F = \frac{FLr_h}{2}\theta \tag{7.1.46}$$

By multiplying and dividing for L(N+1) the above expression and introducing the equivalent uniform load $q = F(N+1)/L$ it can be rearranged as:

$$L_F = \frac{L(N+1)}{L(N+1)} \frac{FLr_h}{2}\theta = \frac{F(N+1)}{L} \cdot \frac{Lr_h}{N+1} \cdot \frac{L}{2} \cdot \theta = q \cdot x \cdot \frac{L}{2} \cdot \theta \tag{7.1.47}$$

Summary of design procedure for failure mode control

In the previous Section, the issues to be faced in evaluating the mechanism equilibrium curves of steel frames with partial strength beam-to-column joints subjected to seismic horizontal forces and to concentrated vertical loads acting on the beams have been discussed. As soon as these issues have been solved, the design procedure for failure mode control, can be easily applied (Mazzolani and Piluso, 1997).

From the conceptual point of view, as already stated, the goal of the design procedure is the dimensioning of column sections in order to assure that the equilibrium curve corresponding to the global mechanism lies below those corresponding to all other kinematically admissible mechanisms within a selected range of the top sway plastic displacement (Figure 7.1.5). Therefore, $3n_s$ design conditions have to be satisfied:

$$\alpha^{(g)} - \gamma^{(g)}\delta_u \leq \alpha_{im}^{(t)} - \gamma_{im}^{(t)}\delta_u \qquad \text{with } t = 1,2,3 \qquad \text{and} \qquad im = 1,2,3.........,n_s \qquad (7.1.48)$$

where

- $\alpha^{(g)}$ is the kinematically admissible mulptiplier of horizontal foeces corresponding to the global mechanism;
- $\gamma^{(g)}$ is the slope of the global mechanism;
- $\alpha_{im}^{(t)}$ is the kinematically admissible multiplier of horizontal forces corresponding to the im-th mechanism of t-th type;
- $\gamma_{im}^{(t)}$ is the slope of the mechanism equilibrium curve corresponding to the im-th mechanism of t-th type;
- δ_u is the ultimate displacement to be selected according to the plastic rotation supplay ($\theta_{p.lim}$) of the dissipative zones (i.e. $\delta_u = \theta_{p.lim} H$, being H the frame height).

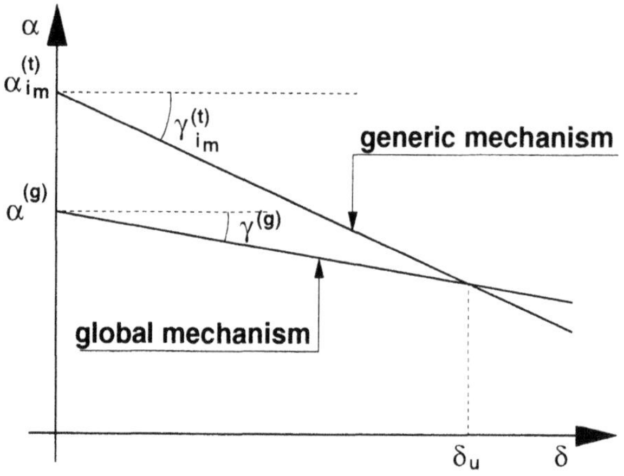

Figure 7.1.5: Design conditions

The beam section are assumed to be known, because they are designed to resist vertical loads. On the contrary, the column sections are the unknown of the design problem. The algorithm to compute the smallest column sections fulfilling the design requirements (7.1.48) has been detailed described by Mazzolani and Piluso (1997) and has been codified into a computer program namely SOPDOSF (Second Order Plastic Design of Steel Frames).

DESIGN OF EXTENDED END PLATE CONNECTIONS

The structural weight of a semi-rigid steel frame failing according to the collapse mechanism of global type is strictly dependent on the plastic moment of the beam sections, which are designed to resist vertical loads, and on the degree of flexural resistance of beam-to-column joints which is expressed through the nondimensional parameter $\overline{m}$.

Two cases can be identified corresponding to full-strength beam-to-column joints ($\overline{m} \geq 1$) and to partial-strength beam-to-column joints ($\overline{m} < 1$). It is evident that, for any given beam section, the weight of the columns required to obtain a collapse mechanism of global type reduces as the joint resistance $\overline{m}$ reduces. This weight reaches its minimum value for $\overline{m} = 0$, which corresponds to the

ideal case of beams pin-jointed to the columns. In fact, in such a case, columns behave as cantilevers and just the yielding of their base section is required for developing a "global" mechanism.
However, this is just a theoretical consideration, because the value of $\overline{m}$ has to be chosen considering not only the minimization of the structural weight (within frames failing in global mode), but also the need to comply with the strength and serviceability requirements under seismic actions corresponding to frequent earthquakes. All these checks require the knowledge of the rotational stiffness K_φ of beam-to-column joints.
The evaluation of the rotational behaviour of beam-to-column joints, i.e. the computation of their rotational stiffness and flexural resistance, can be carried out by means of the component method which has been recently codified in Annex J of Eurocode 3 (CEN, 1997). With reference to the case of bolted extended end plate connections, considered in this work, an alternative approach is constituted by the use of relationships relating the joint flexural resistance and the rotational stiffness to the geometrical and mechanical parameters describing the structural detail of the connection (Faella et al., 1997; 1999). By means of an exhaustive parametric analysis, referred to the structural detail depicted in Figure 7.1.6 it has been pointed out that the nondimensional flexural resistance $\overline{m}$ of the joint is related to the corresponding rotational deformability according to the following relationship:

$$\overline{m} = C_1 \eta^{-C_2} \tag{7.1.49}$$

where

$$\eta = \frac{L}{K h_b} \tag{7.1.50}$$

and

$$K = \frac{K_\varphi L}{E I_b} \tag{7.1.51}$$

being h_b, I_b and L the beam depth, the beam moment of inertia and the beam span, respectively. The parameter K, given by Eq. (7.1.51), is usually referred as nondimensional rotational stiffness of the joint, so that η has to be considered a rotational deformability parameter.

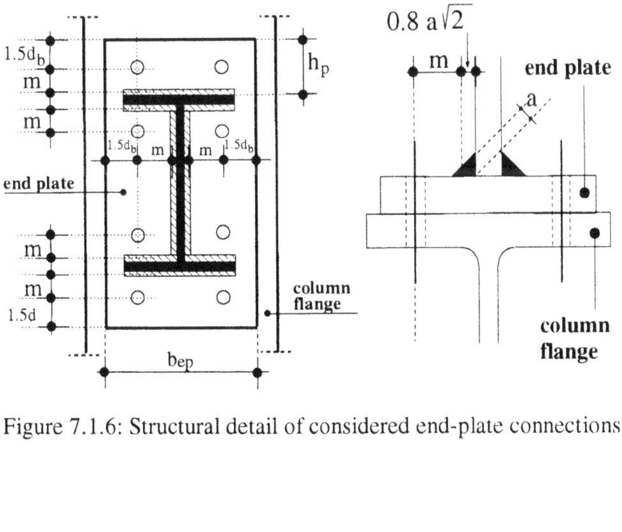

Figure 7.1.6: Structural detail of considered end-plate connections.

The coefficients C_1 and C_2 have been derived from the regression analysis of the data obtained through numerical simulations developed by means of the JMRC (Joint Moment Rotation Curve) program which codifies the component approach according to a modelling of the joint components recently proposed to improve the accuracy of the predicted moment-rotation curves (Faella et al., 1999).The values of C_1 and C_2 depend on the coupling of the beam shape (IPE series) and the column shape (HEA, HEB or HEM series), on the bolt location through the m/d_b ratio (Figure 7.1.6) (being d_b the bolt diameter) and on the joint location (internal or external joints) (Faella et al., 1997;1999).
For any given value of $\overline{m}$ and a given beam of span L and for a selected value of the m/d_b ratio, Eqs. (7.1.49), (7.1.50) and (7.1.51) allow the computation of the joint rotational stiffness to be used for the mathematical representation of the moment-rotation curve of beam-to-column joints. This allows the global structural analysis to be performed for the load combinations prescribed by the codes, being the frame members already defined as required for failure mode control. Therefore, all resistance and serviceability requirements can be properly checked.
In addition, the structural detail of the joint can be completely defined considering that the end-plate thickness is related to the joint rotational deformability parameter η according to the following relationship:

$$\eta^{0.25} = \frac{C_3}{\tau - C_4} + C_5 \geq C_6 \tag{7.1.52}$$

where:

$$\tau = \left(\frac{t_{eq}^3 h_b}{I_b} \right)^{1/4} \tag{7.1.53}$$

and:

$$\frac{1}{t_{eq}^3} = \frac{1}{t_{fc}^3} + \frac{1}{t_{ep}^3} \tag{7.1.54}$$

being t_{fc} the thickness of the column flanges and t_{ep} the end plate thickness. In addition, the coefficients C_3, C_4, C_5 and C_6, as well as C_1 and C_2, are regression coefficients depending on the coupling of the beam shape and the column shape, the m/d_b ratio and the joint location (Faella et al., 1997;1999). The above relationships allow the computation of the end-plate thickness required for obtaining the flexural resistance $\overline{m}$ and the corresponding rotational deformability η, while the bolt diameter is obtained considering that, in order to avoid brittle failure modes, they are designed to withstand a bending moment equal to 1.20 times the plastic moment of the connected beam.

CHECKING EUROCODE PROVISIONS

As already stated, starting from the knowledge of the beam sections, which are designed to resist vertical loads, the column sections can be derived for any given value of $\overline{m}$ by imposing the occurrence of a collapse mechanism of global type. In addition, by means of the relationships described in the previous Section, the moment-rotation curve of the beam to column joints can be obtained and the frame input data required for elastic global structural analysis are completely established.
The global structural analysis is necessary to check the fulfilment of the provisions given both in Eurocode 3 (CEN,1993) and in Eurocode 8 (CEN, 1998). The following load combinations are considered:

$$1.35G + 1.50Q \qquad \text{and} \qquad G+0.3Q \pm E \tag{7.1.55}$$

and considering the influence of geometrical imperfections:

$$1.35G + 1.50Q \pm I \qquad \text{and} \qquad G+0.3Q \pm (E+I) \tag{7.1.56}$$

where G and Q are the dead load and the live load, respectively, while E is the earthquake load (i.e. the seismic forces scaled down according to the design value of the q-factor). In addition, I represents the horizontal forces equivalent to the frame imperfections, depending on the magnitude of vertical loads in the considered load combination.
For each load combination, the following checks are required:

- check of beam resistance

$$C_{b.k} = \frac{M_{b.Sd}}{M_{b.Rd}} \leq 1 \tag{7.1.57}$$

where $M_{b.Sd}$ is the design value of the beam bending moment for the considered load combination and $C_{b.k}$ is the checking code for the *k-th* beam.

- check of column resistance

$$C_{c.1.k} = \frac{M_{c.Sd}}{M_{c.N.Rd}} \leq 1 \tag{7.1.58}$$

where $M_{c.Sd}$ is the design value of the column bending moment and $M_{c.N.Rd}$ is the design flexural resistance of the column reduced to account for axial force $N_{c.Sd}$, which for double T sections can be computed as:

$$M_{c.N.Rd} = M_{c.Rd} \qquad \text{for} \qquad N_{c.Sd} \leq 0.10 N_{c.Rd} \tag{7.1.59}$$

$$M_{c.N.Rd} = 1.10 M_{c.Rd} \left(1 - \frac{N_{c.Sd}}{N_{c.Rd}}\right) \qquad \text{for} \qquad N_{c.Sd} > 0.10 N_{c.Rd} \tag{7.1.60}$$

where $M_{c.Rd}$ is the design flexural resistance of the column and $N_{c.Rd}$ is the design axial resistance.

- check of column stability

$$C_{c.2.k} = \frac{N_{c.Sd}}{\chi_{min} N_{c.N.Rd}} + k_x \frac{M_{c.Sd}}{M_{c.Rd}} \leq 1 \tag{7.1.61}$$

where k_x is the coefficient accounting for bending moment distribution along the member axis and χ_{min} is the stability coefficient depending on member slenderness. The maximum value between $C_{c.1.k}$ and $C_{c.2.k}$ is $C_{c.k}$ and represents the checking code for the *k-th* column.

- check of beam-to-column joint resistance

$$C_{j.k} = \frac{M_{j.Sd}}{M_{j.Rd}} \leq 1 \tag{7.1.62}$$

where $M_{j.Sd}$ is the design value of the joint bending moment computed for the considered load combination and $C_{j.k}$ is the checking code for the *k-th* beam-to-column joint.

Regarding the influence of second order effects, they have been accounted for by means of the moment amplification method.
In addition, the global structural analysis provides also the values of the interstorey drifts to be used for checking serviceability requirements. Regarding this issue, it is useful to remember that Eurocode 3 suggests the limitation of the maximum interstorey drift to *h/300*, being *h* the interstorey height.
The analysis of all the load combinations provided by Eurocodes and the required resistance and stability checks of members and/or connections have been performed by means of a computer program, namely ESFA97 (Elastic Steel Frame Analysis), for the analysis and design of rigid and semi-rigid steel frames according to Eurocodes.

DESIGN PROCEDURE

The proposed design procedure for semi-rigid frames failing in global mode is summarized by the flow-chart depicted in Figure 7.1.7.

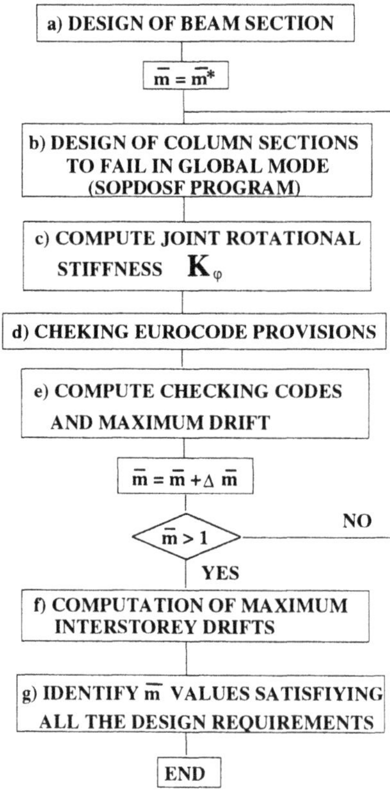

Figure 7.1.7: Flow chart of the proposed design procedure.

It consists in the following steps:

a) preliminary design of the beam section considering the vertical loads corresponding to the load combination *1.35G+1.50Q*. The plastic moment required to withstand vertical loads can range between $qL^2/16$ and $qL^2/8$, while, the minimum value of the nondimensional flexural resistance of the joints, for a selected beam section, is given by:

$$\overline{m}^* = \frac{qL^2}{8M_{b.Rd}} - 1 \tag{7.1.63}$$

which can be used as initial value of the joint resistance for the loop described in steps from b) to e);

b) design the column sections for the selected beam and the current value of $\overline{m}$, using the SOPDOSF program, required for developing a collapse mechanism of global type ;
c) compute the joint rotational stiffness K_φ corresponding to $\overline{m}$ for the selected m/d_b ratio according to section regarding the design of extended end plate;
d) perform, for the designed frame, all resistance and stability checks, according to Eurocode 3 and considering all the load combinations according to previous section (program ESFA97);
e) compute checking codes C_b and C_c as the maximum value attained by $C_{b.k}$ (Eq. (7.1.57)) and $C_{c.k.}$, (Eqs.(7.1.58) and (7.1.61)) respectively, considering all the members (beams and columns, respectively) and all the load combinations; similarly, the checking code of joints is evaluated as the maximum value attained by $C_{j.k}$ (Eq. (7.1.62)) considering all the beam-to-column joints and all the load combinations. It is evident that $C_j \leq 1$ means that the check of the flexural resistance of joints is satisfied for all the beam-to-column joints and all the load combinations. Similarly, $C_b \leq 1$ and $C_c \leq 1$ means that all checks regarding beams and columns, respectively, are satisfied. The maximum interstorey drift is also computed. In addition, if $\overline{m} < 1$ then increase the value of $\overline{m}$ and return to step b), otherwise perform the following steps;
f) identify the structural solutions satisfying all the design requirements. These solutions correspond to the frames designed (through the SOPDOSF program) for $\overline{m}$ values leading also to the fulfilment of Eurocode provisions. It means that, for such frames, all the checking codes C_b, C_c and C_j are less than 1.0, in addition, the corresponding maximum interstorey drift is acceptable;
g) among all the structural solutions satisfying the design requirements, which have been identified in the previous step, the selection of that leading to the minimum weight completes the design procedure.

In order to show the great amount of informations obtained by means of the above design procedure, reference can be made, as an example, to the four storey-three bay frame depicted in Figure 7.1.8.

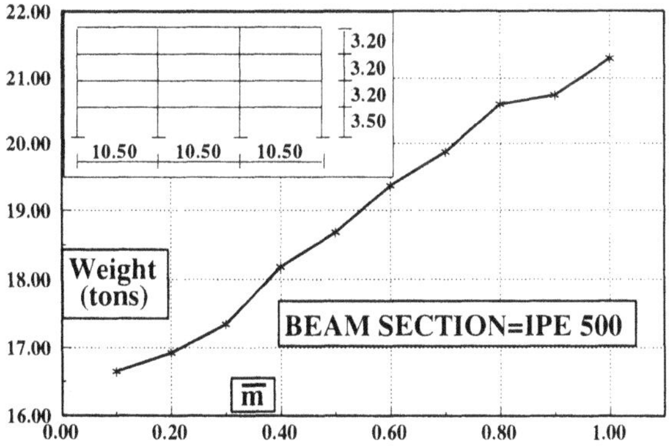

Figure 7.1.8 Structural weight vs. joint flexural resistance

The bay span is 10.50 m, the interstorey height is 3.20 m, with the exception of the first storey whose height is equal to 3.50 m. The characteristic values of the dead and live loads acting on the beams are 21 kN/m and 10.5 kN/m, respectively. The beam section required to withstand the load combination *1.35G+1.50Q* is an IPE500 shape, made of Fe430 steel. The curve depicted in Figure 7.1.8 is obtained as result of the step b) of the proposed design procedure. It shows the influence of the flexural resistance of beam-to-column joints on the structural weight of the frame designed to fail according to the global mode. Obviously, the weight decreases as the value of the nondimensional joint resistance $\overline{m}$ decreases.
However, the structural solutions corresponding to the points of Figure 7.1.8 can be adopted only if the resistance and stability checks of members and/or joints, described, are satisfied under the load combinations prescribed by the codes.
As already stated, the global structural analysis under the load combinations corresponding to serviceability conditions requires the complete modelling of the rotational behaviour of beam-to-column joints. To this scope the rotational stiffness of joints has been computed with reference to an m/d_b ratio equal to 2. This choice is based on the observation that for any given value of $\overline{m}$ the joint rotational stiffness increases as the m/d_b ratio decreases, so that it provides the minimum frame lateral displacements for the given value of $\overline{m}$. In addition, in order to simplify the structural detail of beam to column joints, continuity plates have been omitted. With reference to such connection detail (unstiffened joint, m/d_b = 2 with bolt class 10.9) the following values can be used for the constants appearing in Eqs. (7.1.49) and (7.1.52):

$$C_1 = 3.6069 \qquad C_2 = 1.7982 \qquad C_3 = 0.006 \qquad C_4 = 0.047 \qquad C_5 = 1.034 \qquad C_6 = 1.182 \qquad (7.1.64)$$

The knowledge of the joint rotational behaviour allows the structural analysis under the load combinations previously described to be performed. As a consequence, resistance and stability cheks of memebers and/or joints can be carried out.
These checks are the result of the step d) of the design procedure and are summarized in the lower part of Figure 7.1.9b. The fulfilment of all the resistance and stability checks is guaranteed for the $\overline{m}$ values leading to checking codes C_b, C_c and C_j less than 1.0.
Therefore, for example with reference to the case of 4 storey – 3 bay frames with IPE 500 as beam section, it can be concluded that the beam-to-column joints require a non-dimensional resistance $\overline{m} \geq 0.50$. Obviously, the solution corresponding to $\overline{m} = 0.50$ is the one leading to the minimum weight without renouncing to the global mechanism.
However, the final selection of the structural solution to be adopted has to account for serviceability requirements and, therefore, for the magnitude of the interstorey drifts which has to be properly limited. For this reason, the upper part of pictures represented in Figure 7.1.9b provides, on the right hand scale, also the maximum value of the ratio between the interstorey drift δ and the corresponding interstorey height h.
In the considered case of 4 storey – 3 bay frames with IPE 500 as beam section, for $\overline{m} \geq 0.50$ the maximum drift is $\delta_{max} \cong h / 170$. This value can be reduced by increasing $\overline{m}$. The limit value of the interstorey drift depends on code provisions. However, if the above maximum value can be accepted, the structural solution corresponding to $\overline{m} = 0.50$ can be adopted. The column sections corresponding to this solution are given in Table 7.1.1.
It is important to underline that Figure 7.1.9b provides also the results corresponding to the case of full-strength ($\overline{m} = 1$) rigid beam-to-column joints. In fact, the solid circle and the solid square in the lower part of pictures depicted in Figure 7.1.9b correspond to the resistance and stability checks for this case, while the solid triangle in the upper part gives the corresponding interstorey drift.
The column sections corresponding to the particular case of 4 storey – 3 bay frames with IPE 500 as beam section, are given in Table 7.1.2. The structural weight corresponding to full-strength rigid joints is equal to 21.3 tons while that corresponding to the most convenient solution with partial strength

semi-rigid joints is 18.7 tons. Therefore, without renouncing to a collapse mechanism of global type, the use of semi-rigid joints leads to a saving in structural weight of about 12%.

Table 7.1.1. Column sections corresponding to $\overline{m} = 0.50$ (Weight 18.7 tons), beam section = IPE 500

Storey	Columns			
4	HEB 280	HEB 320	HEB 320	HEB 280
3	HEB 320	HEB 400	HEB 400	HEB 320
2	HEB 320	HEB 400	HEB 400	HEB 320
1	HEB 340	HEB 450	HEB 450	HEB 340

Table 7.1.2. Column sections corresponding to the case of full strength $\overline{m} = 1$ rigid beam-to-column joints (Weight 21.3 tons), beam section = IPE 500

Storey	Columns			
4	HEB 400	HEB 450	HEB 450	HEB 400
3	HEB 500	HEB 500	HEB 500	HEB 500
2	HEB 500	HEB 550	HEB 550	HEB 500
1	HEB 550	HEB 600	HEB 600	HEB 550

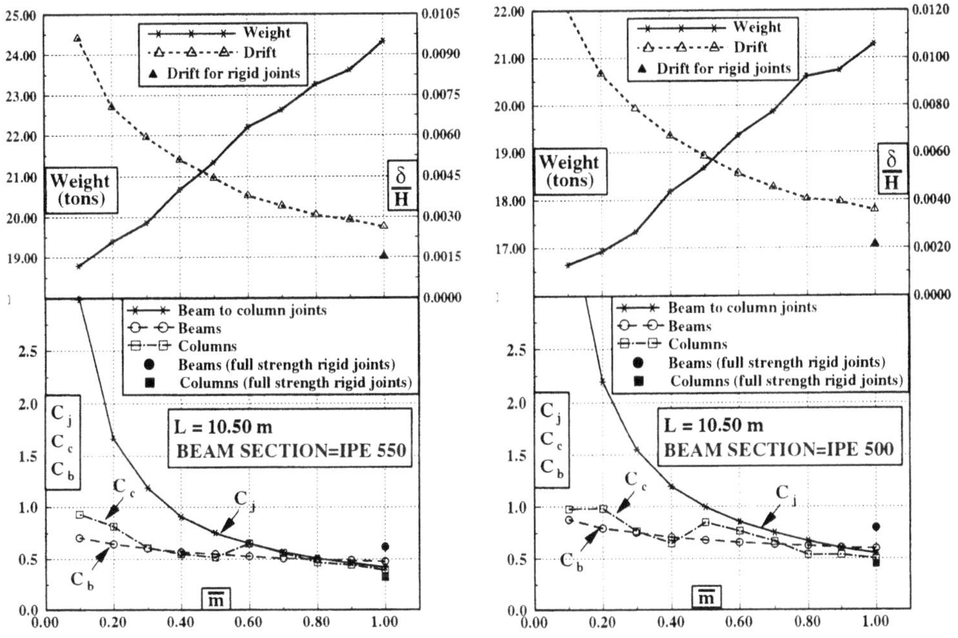

(a) (b)

Figure 7.1.9: Result of the design procedure

In Figure 7.1.9a, Figure 7.1.10a and Figure 7.1.10b the results corresponding to the same frame, but, realized with different beam sections are depicted. In particular it is important to observe that with an IPE 400 (Figure 7.1.10b) or an IPE 450 (Figure 7.1.10a) as beam section, only frames with full strength semi-rigid joints can be realized. In fact, in these cases full-strength rigid joints cannot be adopted, because the beam checking section code, i.e. C_b is grater than 1.0.

Therefore, if an IPE 400 is used as beam section, then it is possible to realize the 4 storey 3-bay frame with a weight of 15.6 tons, reaching a saving in structural weight of about 26%, but , this solution leads to a drift $\delta_{max} \cong h/133$, which could be not compatible with code provisions.

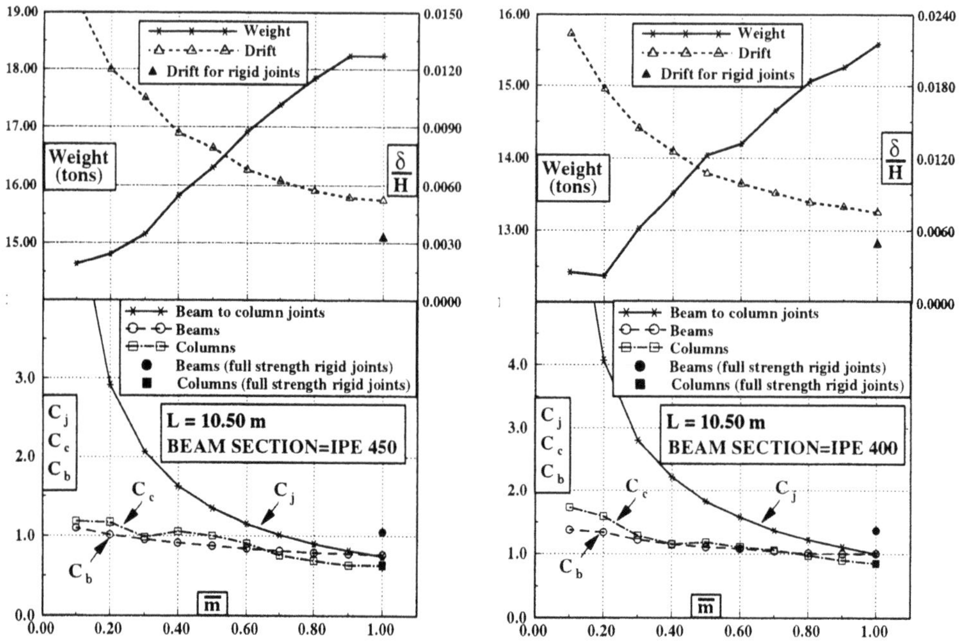

(a) (b)

Figure 7.1.10 Result of the design procedure

It is interesting to observe that the width of the range of $\overline{m}$ values that can be adopted, provided that interstorey drifts are compatible with code provisions, increases as the beam section increases. In fact this range reachs its maximum width when an IPE 500 section is used. In this case $\overline{m}$ can assume values ranging from $\overline{m}$=0.37 to $\overline{m}$=1.0 (Figure 7.1.9a). The corresponding frames lead to a reduction of the frame lateral displacements, but the weight of these structural solutions is increased with respect to the alternative design solutions depicted in Figure 7.1.9b, Figure 7.1.10a, Figure 7.1.10b.

PARAMETRIC ANALYSES

In order to investigate the structural situations where the use of partial-strength semi-rigid joints can lead to significant saving in structural weight, a wide parametric analysis has been developed with reference to low-rise frames.

The number of storeys varied from 2 to 6, the number of bays from 2 to 4. The bay span varied from 4.50 m to 13.50 m.

For each structural scheme, the proposed design procedure has been applied varying also the size of the beams. The result of the analyses carried for each structural scheme is the ratio between the weight P_k of the structural solution adopting partial-strength semi-rigid joints and that P_∞ of the structural solution adopting full-strength rigid joints. This comparison has been developed considering different limit values for the maximum interstorey drift. The results of the parametric analysis have been summarized in Figure 7.1.11 where δ and h are the maximum interstorey displacement and the interstorey height, respectively, so that each curve refers to a given limit value of the interstorey drift for serviceability conditions.

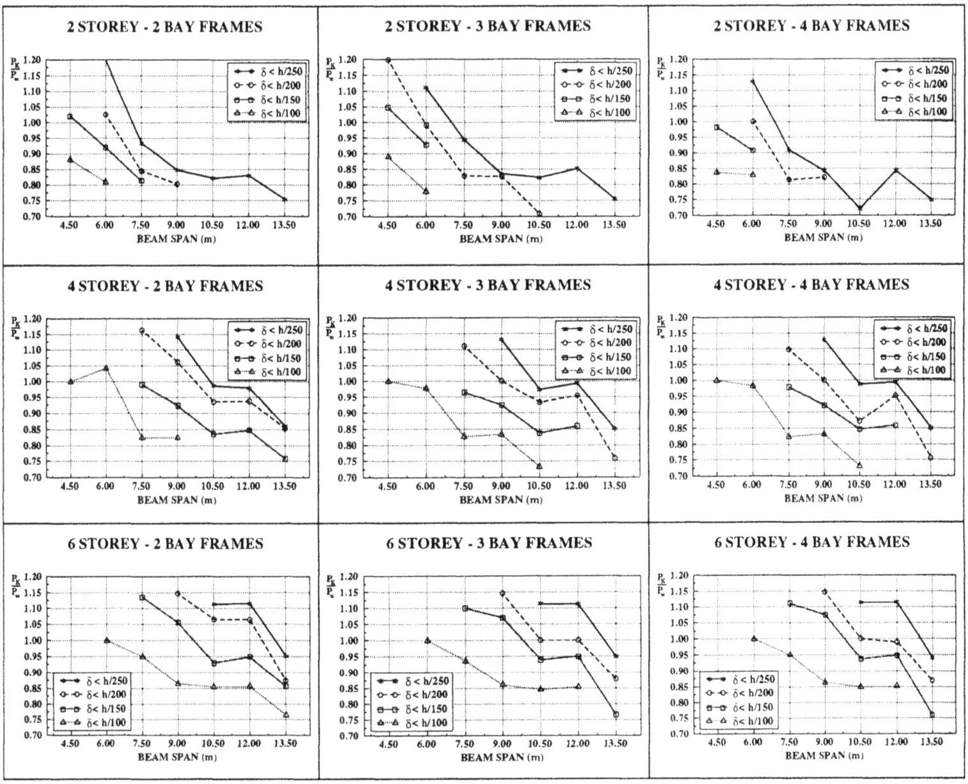

Figure 7.1.11 Saving in structural weight

It can be observed that the amount of saving in structural weight, deriving from the use of partial strength semi-rigid joints without renouncing to the global failure mode, increases as the bay span increases. In addition, the influence of the limit value for the maximum interstorey drift is particularly relevant. In fact, depending on the severity of this limit value, cases can arise in which semi-rigid joints cannot be applied due to excessive lateral displacements, so that the use of rigid joints is mandatory. In addition, in some cases the use of semi-rigid joints requires the increase of the beam section with respect to that adopted in the rigid frame solution. These cases are characterized by P_k/P_∞ values greater that 1.0, so that, in these cases, the use of rigid joints is more convenient.
However, it has to be underlined that, as expected, in the case of long span frames the use of partial-strength semi-rigid frames for developing a global failure mode can be significantly convenient leading to saving in structural weight which can reach 30%.

DYNAMIC INELASTIC ANALYSES

As evidenced in the previous Sections, the proposed design procedure leads to frames failing in global mode exploiting partial-strength beam-to-column joints. This is particularly convenient in the case of long span frames where the use of full-strength connections, without renouncing to the global failure mode, would require very large column sections leading to heavy structural solutions.
However, the structural solutions identified in the previous Section can be practically used provided that, under severe earthquakes, the plastic rotation demands in beam-to-column joints are compatible

with the corresponding supply. For this reason, it is of primary interest to compare these plastic rotation demands with those occurring when the structural solution with full-strength rigid joints is applied. To this scope, dynamic inelastic analyses have been carried out with reference both to the structural solution with partial-strength semi-rigid joints and the corresponding solution with full-strength rigid joints.

As an example, with reference to 4storey-3bay frames having different bay span and for two different limit values of the interstorey drift under service conditions, Figure 7.1.12 shows the plastic rotation demands occurring under an earthquake record generated to match the elastic design response spectrum of Eurocode 8 for stiff soil conditions.

The black circles refer to frames with full-strength rigid connections while the white circles refer to frames with partial-strength semi-rigid connections. All frames have been designed to fail according to the global mechanism and such pattern of yielding has been confirmed by the dynamic inelastic analyses. For any considered case, in Figure 7.1.12 the beam section adopted is also given. Regarding the serviceability requirements, two cases have been considered which are characterized by limit value of the maximum interstorey drift under the load combinations $G+0.3Q \pm (E+I)$ equal to $h/150$ and to $h/200$, respectively.

In addition, Figure 7.1.12 provides the maximum plastic rotation demands for three values of the peak ground acceleration (PGA=0.35g, PGA=1.5*0.35g and PGA=1.5*0.35g). It can be observed that, depending on the bay span and the magnitude of the ground motion, in some cases the plastic rotation demand in semi-rigid frames is greater than that occurring in rigid frames, in other cases the opposite occurs.

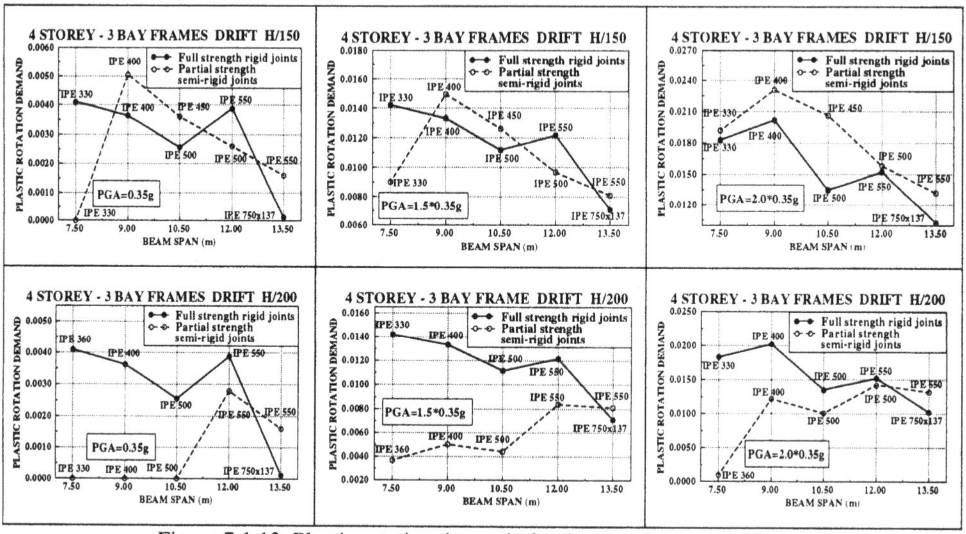

Figure 7.1.12: Plastic rotation demands for the case 4 storey – 3 bays.

However, the most important outcome of these preliminary analyses is that the magnitude of the plastic rotation demands, occurring for the two alternative structural solutions of the same design problem, are comparable. This means that partial-strength semi-rigid joints can be effectively used provided that their plastic rotation supply is similar to that developed by full-strength rigid joints.

In addition, it is useful to underline that the above results can be practically obtained provided that the rotational behaviour of the joints is accounted for in all the steps of the design process. In other words, they refer to frames whose members have been dimensioned considering the influence of the joint semi-rigidity.

CONCLUSIONS

A new design procedure for seismic resistant semi-rigid frames failing in global mode has been presented. The originality of the design strategy is its ability to force yielding to occur in the beams rather than in the columns exploiting partial-strength connections, so that a collapse mechanism of global type is obtained.
A parametric analysis has been developed to show the structural situations where the use of partial-strength semi-rigid joints can lead to significant saving in structural weight, without renouncing to a collapse mechanism of global type. These structural situations are characterized by long spans and heavily loaded beams.
Finally, some preliminary dynamic inelastic analyses of the designed frames have shown that, under seismic action, the use of partial-strength semi-rigid joints does not lead to an increase of the plastic rotation demands, provided that semi-rigidity, i.e., the rotational behaviour of beam-to-column joints, is taken into account in all the steps of the design process.

REFERENCES

CEN (1993). *Eurocode 3: Design of Steel Structures.* CEN/TC250/SC3.

CEN (1997) *Eurocode 3 Part 1.1. Revised Annex J. Joints in Building Frames.* Final approved draft, January, CEN/TC250/SC3-PT9.

CEN (1998). *Eurocode 8: Structures in Seismic Regions.* CEN/TC250/SC8.

Faella, C., Piluso, V. and Rizzano, G. (1996). A New Design Approach for Braced Frames with Extended End Plate Connections. IABSE Colloquium on Semi-Rigid Structural Connections, Istanbul, September.

Faella, C., Piluso, V. and Rizzano, G. (1996). Some Proposal to Improve EC3-Annex J Approach for Predicting the Moment-Rotation Curve of Extended End Plate Connections, *Costruzioni Metalliche*, N.4, July-August.

Faella, C., Piluso, V. and Rizzano, G. (1997). A New Method to Design Extended End Plate Connections and Semi-Rigid Braced Frames. *Journal of Constructional Steel Research*, Vol. 41, No.1, pp.61-91

Faella, C., Piluso, V. and Rizzano, G. (1999). *Structural Steel Semi-rigid Connections*, CRC Press, Boca Raton., Florida.

Lee, H.-S. (1996). Revised Rule for Concept of Strong-Column Weak-Girder Design. *Journal of Structural Engineering*, ASCE, Vol. 122, pp. 359-364.

Mazzolani, F.M. and Piluso, V. (1996a). *Theory and Design of Seismic Resistant Steel Frames.* London, New York, E and FN Spon, an Imprint of Chapman and Hall.

Mazzolani, F.M. and Piluso, V. (1996b). Plastic Design of Semirigid Frames for Failure Mode Control. IABSE Colloquium on Semi-Rigid Structural Connections, Istanbul, September 1996.

Mazzolani, F.M. and Piluso, V. (1996c). Behaviour and Design of Set-Back Steel Frames. First European Workshop on Asymmetric and Set-Back Structures, Anacapri, October.

Mazzolani, F.M. and Piluso, V. (1997). Plastic Design of Seismic Resistant Steel Frames. *Earthquake Engineering and Structural Dynamics*, Vol. 26, pp.167-191.

Montuori, R (1997). Seismic-Resistant Semirigid Steel Frames: Design for Failure Mode Control (in italian). Thesis presented for obtaining the Civil Engineer Degree. Tutors: Faella, C., Piluso, V. and Rizzano, G. University of Salerno, Italy.

7.2

INFLUENCE OF CONNECTION MODELLING ON SEISMIC RESPONSE OF MOMENT RESISTING STEEL FRAMES

Gaetano Della Corte, Gianfranco De Matteis, Raffaele Landolfo

INTRODUCTION

During the recent earthquakes of Northridge and Kobe steel structures exhibited an unexpected non-ductile behaviour, essentially due to bad joint performance. Therefore, an unquestionable new development in research activities is dealing with the actual response of steel connections under severe dynamic loading conditions. Through this direction, the first step is the correct identification of mechanical properties of joints, which influence the global performance of steel building frames. In particular, besides to accurately evaluate strength and stiffness, particular attention must also be paid in determining the available rotation capacity of connections. A plastic rotation supply equal to 0.02-0.03 rad is commonly used as a benchmark to judge the seismic effectiveness of member sections and connections. This value is used in numerical simulations, in order to assess rotation demand to rotation capacity ratio. But, recent earthquakes point out the occurrence of even higher ductility demand, it being particularly influenced by the structural scheme and the intensity of the ground motion.

Experimental response of typical beam-to-column joint typologies evidences several behavioural aspects. First of all, in case of monotonic loading, the effect of both non-linearity and kinematic hardening is particularly significant. On the other hand, in case of alternate loading, low-cycle fatigue phenomena are an important concern, which may induce premature failure and, in some cases, deterioration of the main mechanical properties. Besides, cyclic hardening is not negligible. Finally, it is to be considered that the qualitative shape of hysteretic loops is strongly dependent on the connection typology, which may exhibit either fully or poorly dissipative behaviour, with a corresponding growth of pinching effects.

It is worthy to remember that mechanical behaviour is rate-dependent and this is another difficult task of seismic engineering. However, in the following, there is no account for rate dependent phenomena, assuming that it is accurate to introduce in analytical simulations hysteretic force-deformation relationships obtained starting from quasi-static cyclic tests.

With reference to joint modelling, it is nowadays accepted the idea of considering simple elastic-perfectly plastic models in case of monotonic loading, as testified by recent code prescription (see for example ENV 1993, 1992). On the contrary, a similar approach may be not sufficiently accurate in aseismic design of dissipative structures. In this latter case, the capacity of the structure to survive a strong earthquake motion is directly related not only to its strength but also to its plastic deformation capacity (ENV 1998, 1994). Therefore it is necessary an adequate evaluation of inelastic structural behaviour, aiming at a correct and reliable prediction of collapse. Hence, it seems proper to wonder whether the adoption of simplified restoring force characteristics is allowed or a more sophisticated modelling of joint mechanical behaviour is necessary for correct prediction of seismic demands to structural components.

The first step for answering the above question has been the developing of a versatile mathematical model interpreting joint mechanical behaviour. It is a semi-empirical model, based upon a number of parameters to be defined according to experimental results, and it provides a flexible tool for the dynamic inelastic analysis of steel moment resisting frames. By implementing the hysteretic model as new subroutine into a general computer program typically used to perform global structural analyses, i.e. DRAIN 2D, the actual dissipation capacity of the connection is accounted for and its influence on the structural global behaviour is evaluated. Then, a parametrical study devoted to assess the influence of the behavioural parameters of the above hysteretic model has been performed. With reference to a SDoF system, the sensitivity of global response to actual connection behaviour is firstly estimated, emphasising the differences with respect to results obtained by using typical simplified models, i.e. elastic-perfectly plastic and slip type restoring force characteristics. Finally, the influence of some behavioural aspects characterising the cyclic joint response is investigated with reference to some moment resisting frames designed according to Eurocode 3 and 8. Analyses, which have been carried out varying some significant parameters of the proposed model, have permitted to emphasise the sensitivity of predicted values of ductility and hysteretic energy demands to different modelling assumptions.

THE PROPOSED HYSTERETIC MODEL

Basic concepts

For describing joint behaviour, the relationship between bending moment M transmitted by beam to column and the dual kinematic parameter ϕ (relative rotation between beam and column) is used. For the sake of simplicity, only systems with a symmetrical M-ϕ monotonic behaviour are considered in the following, but it is conceptually simple to extend the proposed approach to non-symmetrical systems. It is opportune to state preliminarily some fundamental concepts, extensively used throughout this Section. The basic unit for the description of a generic loading history is the deformation excursion, which is constituted by a loading branch and by the subsequent unloading branch of the M-ϕ path. The deformation range of excursion ($\bar{\phi}$) is the deformation range between the beginning and the peak deformation of the excursion (ATC, 1992). As it will be explained herein after, the beginning of a loading and the end of an unloading branch always stand on a straight line passing through the origin with a slope equal to the one concerned with the hardening of the system (Figure 7.2.1).

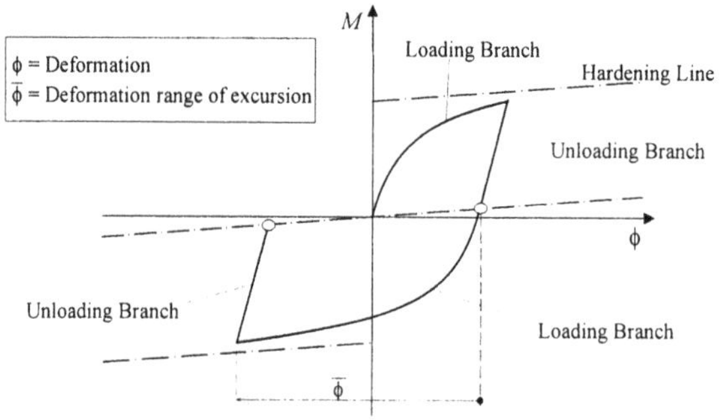

Figure 7.2.1: Basic definitions

The hysteretic model for 'perfect' dissipative systems

The loading branch without pinching

For the description of the loading branch in absence of pinching, a mathematical formulation proposed by Richard and Abbott (1975) has been adopted. This formulation, expressed in terms of Moment (M) versus Rotation (ϕ) relationship, can be written as follows:

$$M = \frac{(k_0 - k_h)\phi}{\left[1 + \left|\frac{(k_0 - k_h)\phi}{M_0}\right|^n\right]^{1/n}} + k_h\phi \qquad (7.2.1)$$

In such a formula, k_0 represents the initial stiffness of the system; the parameters M_0 and k_h are the intersection with the M axis and the slope of a straight-line (hardening line) asymptote of the M-ϕ curve, respectively; finally, n is a shape parameter, which regulate the sharpness of transition from elastic to fully plastic behaviour (increasing values of n correspond to an increasing sharpness). Figure 7.2.2*a* illustrates the graphical representation of Richard-Abbott mathematical formulation, also reporting the meaning of adopted symbolism.

The loading branch with pinching

For describing pinching, two limit curves are introduced, representing a lower bound and an upper bound to possible M-ϕ values, respectively. Both curves have a Richard-Abbott type law, with parameters k_{0p}, M_{0p}, k_{hp}, n_p (lower bound curve) and k_0, M_0, k_h, n (upper bound curve). A point (M, ϕ) of the real path is considered to belong also to a Richard-Abbott type curve, whose parameters are defined as follows:

$$k_{0t} = k_{0p} + (k_0 - k_{0p})t \qquad (7.2.2.a)$$

$$M_{0t} = M_{0p} + (M_0 - M_{0p})t \qquad (7.2.2.b)$$

$$k_{ht} = k_{hp} + (k_h - k_{hp})t \qquad (7.2.2.c)$$

$$n_{pt} = n_p + (n - n_p)t \qquad (7.2.2.d)$$

where the parameter t, ranging in [0,1], defines the transition law from the lower bound to the upper bound curve. Its mathematical formulation is to be defined in such a way to reproduce the shape of the curve as experimentally observed. In the proposed model, in order to describe pinching-type behaviour, parameter t is defined by the following relationship:

$$t = \left[\frac{(\phi/\phi_{\lim})^{t_1}}{(\phi/\phi_{\lim})^{t_1} + 1}\right]^{t_2} \qquad (7.2.3)$$

where t_1, t_2 and ϕ_{lim} are three parameters to be defined on the basis of experimental data. Figure 7.2.2*b* illustrates, qualitatively, the description of pinching with reference to one single excursion from the origin.

In case of a generic deformation history, the parameter ϕ_{lim} is to be related to the maximum experienced deformation in the direction of the loading branch to be described. Therefore, it could be evaluated through the parameter λ, defined by the following relationship:

$$\phi_{lim} = \lambda(|\phi_0| + \phi_{max}) \tag{7.2.4}$$

where: $|\phi_0|$ is the absolute value of deformation corresponding to the starting point of the current excursion; ϕ_{max} is the maximum absolute value of deformation experienced, in the whole previous loading history, in the direction of loading branch to be described (Figure 7.2.3*a*).

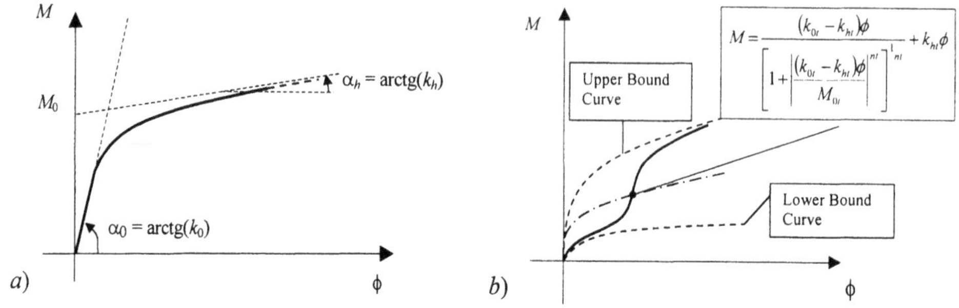

Figure 7.2.2: The loading branch without pinching (*a*) and with pinching (*b*)

The unloading branch

The unloading branch is assumed to be linear with a slope equal to the initial stiffness k_0 up to the intersection with the straight line obtained drawing the parallel to the hardening line going through the origin. This allows the Bauschinger effect to be considered.

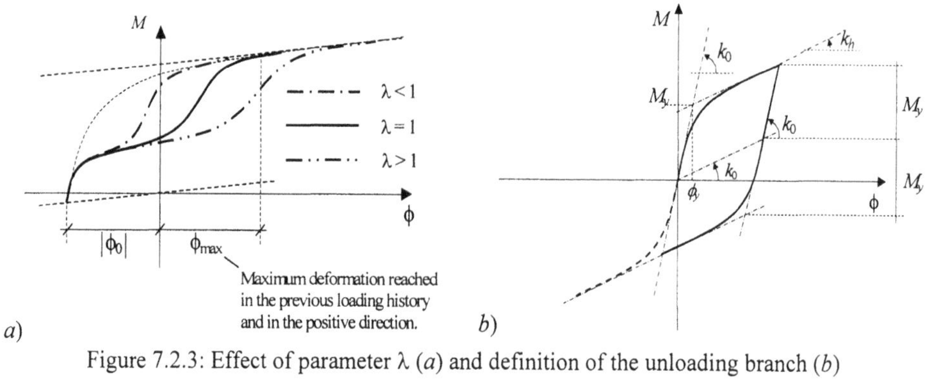

Figure 7.2.3: Effect of parameter λ (*a*) and definition of the unloading branch (*b*)

Description of damage and cyclic hardening

Damage due to plastic fatigue

Cyclic action in the inelastic range produces accumulation of plastic deformation, until ductility of the system is locally exhausted and failure occurs due to fracture. In some cases, the repetition of loading is accompanied by degradation of the system, that is deterioration of its mechanical properties.

Using the original idea proposed by Park and Ang (1985), a collapse index is here introduced as follows:

$$IC = \frac{\bar{\phi}}{\bar{\phi}_{u,0}} + \beta \frac{E_h}{M_y \bar{\phi}_{u,0}} \tag{7.2.5}$$

In relation (7.2.5), $\bar{\phi}$ is the current value of the deformation range of excursion and $\bar{\phi}_{u,0} = \phi_{u,0}$ is the corresponding ultimate value in the case of one single excursion from the origin (monotonic loading). E_h is the hysteretic energy accumulated in all previously experienced excursions, except the current one. M_y represents the conventional yielding resistance of the system, as defined in Figure 7.2.3*b*. Coefficient β is an empirical parameter related to damage rate. By considering the system to be in a collapse state ($IC = 1$), it is possible to evaluate the ultimate deformation value in the current range of excursion:

$$\bar{\phi}_u = \bar{\phi}_{u,0}\left(1 - \beta \frac{E_h}{M_y \bar{\phi}_{u,0}}\right) = \bar{\phi}_{u,0}(1 - D_d) \tag{7.2.6}$$

where $D_d = \beta \frac{E_h}{M_y \bar{\phi}_{u,0}}$ is a damage factor accounting for the ductility reduction, due to accumulated plastic deformation.

Starting from this formulation, it is possible to consider damage of other mechanical properties, such as strength ($M_{0,\text{red}}$) and stiffness ($k_{0,\text{red}}$), by adopting similar relationships. For example, in case of strength:

$$M_{0,red} = M_0(1 - D_M) = M_0\left(1 - \beta_M \frac{E_h}{M_y \bar{\phi}_{u,0}}\right) \tag{7.2.7}$$

It is worthy to remember that parameter β_M may be different from β. In fact, some systems do not exhibit deterioration of mechanical properties, but they fail due to low-cycle fatigue. In this case β_M is to be set equal to zero, while a different value should be used for β.

Damage due to unstable branch

Some systems may exhibit unstable branches beyond plastic deformation at some extent. Here, this branch is assumed to be linear and therefore it is completely individuated by the value of the deformation range of excursion corresponding to its activation ($\bar{\phi}_{inst}$) and by the value of its slope (k_{inst}). Figure 7.2.4*a* illustrates this concept in case of a single excursion starting from the origin.
In case of a generic deformation history, it is necessary to fix a rule for determining $\bar{\phi}_{inst}$. In Figure 7.2.4*b*, hardening and softening branches individuate a quadrilateral region in the M-ϕ plane. All points external to this area are impossible states for the system. Contrary to the hardening lines, softening branches are not fixed at the beginning of the deformation history. In fact, their activation is due to a value for the deformation range of excursion being reached. Such a limit is assumed to be equal to the value measured in the case of one single excursion from the origin (Figure 7.2.5*a*). Once it has been activated, the decreasing branch remains fixed in that direction and defines the boundary of incompatible states (Kumar *et al.*, 1996). Consequently, it implies a decrease of the resistance of the system. This strength degradation is supposed to be ruled by the following relationship:

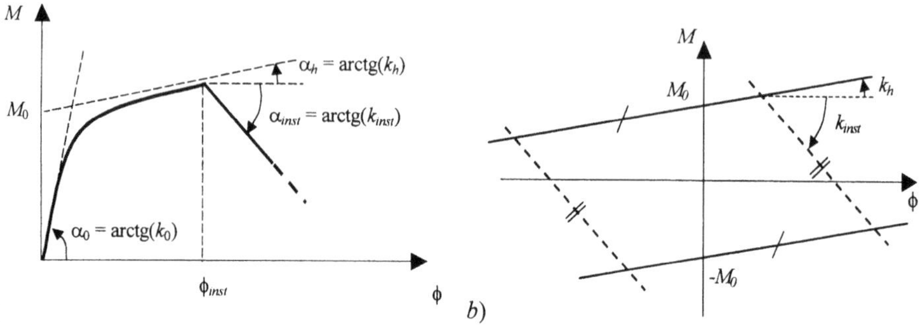

a) b)

Figure 7.2.4: Unstable branch for monotonic loading (*a*) and region of possible *M*-ϕ states for cyclic loading (*b*)

$$M_{0,red} = M_0\left(1 - \frac{k_{inst}\left[\left(\phi^{+}_{max} - \phi^{+}_{inst}\right) + \left(\phi^{-}_{max} - \phi^{-}_{inst}\right)\right]}{M_0}\right) \tag{7.2.8}$$

where: ϕ^{+}_{max} is the maximum positive deformation; ϕ^{+}_{inst} is the value of deformation corresponding to the activation of the decreasing branch in the positive direction, being related to the instability limit value of the deformation range of excursion as well as to the deformation history; ϕ^{-}_{max} and ϕ^{-}_{inst} are the analogous terms for the negative direction; $M_{0,red}$ is the reduced value of parameter M_0. Figure 7.2.5.b illustrates these concepts, representing the unstable branch and the corresponding system damage.

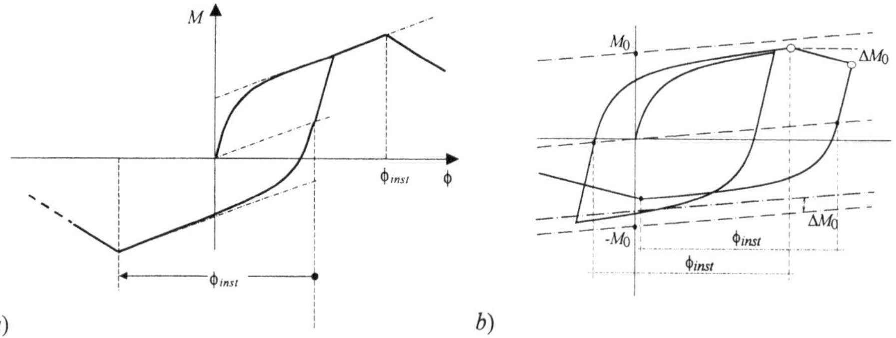

a) b)

Figure 7.2.5: Deformation range of excursion activating the unstable branch (*a*) and effect of strength degradation due to unstable branches (*b*)

In the current model, damage due to unstable branch is simply added up to damage due to accumulation of plastic deformation. However, it is important to observe that these two damages have different mechanical cause and meaning. The decreasing branch is, in fact, associated with a change in the macroscopic geometry of the system (i.e. local buckling of some parts of the joint) producing a lowering of resistance, while damage due to accumulation of plastic deformation corresponds to micro-structural modifications of the material.

Cyclic hardening

Hardening due to cyclic plastic deformation is considered to be isotropic. Besides, experimental results of constant deformation amplitude tests regarding systems not exhibiting strength deterioration show that cyclic hardening grows up in few cycles and then becomes stable. Therefore, the following assumption has been made:

$$M_{0,inc} = M_0 \qquad \text{if } \phi_{max} \le \phi_y$$

$$M_{0,inc} = M_0\left(1 + H_h \frac{\phi_{max} - \phi_y}{\phi_y}\right) \qquad \text{if } \phi_{max} \ge \phi_y \tag{7.2.9}$$

where: M_0 and $M_{0,inc}$ are the initial and the increased value of strength, respectively; ϕ_{max} is the maximum value of deformation reached in the loading history (in either positive or negative direction); ϕ_y is the conventional yielding value of deformation (see Figure 7.2.3.b); H_h is an empirical coefficient defining the level of isotropic hardening (Filippou *et al.* 1983). The above formulation practically corresponds to translate the asymptotic line of equation 7.2.1 as a function of the plastic deformation extent.

The application of the model

General

In order to investigate the capability of the proposed model to interpret joint mechanical behaviour, the simulation of some existing experimental tests of beam-to-column steel joints has been performed. Analysed typologies have been chosen among the most commonly used in steel frame structures. Therefore, welded, end plate and angle connections have been considered. It is interesting to observe that each one of these typologies is identified by a different characteristic shape of hysteretic cycles. Consequently, selected values of model parameters could be assumed as reference values for predicting the behaviour of joints belonging to the same categories. Tables 7.2.1-7.2.5 synthesise the values assumed for the model parameters for each joint typology.

Welded connections

Figure 7.2.6 refers to a joint in which the principal source of inelastic behaviour is the web column panel in shear (Matsui *et al.*, 1992). It shows a very stable behaviour, representing the 'perfect' dissipative system.

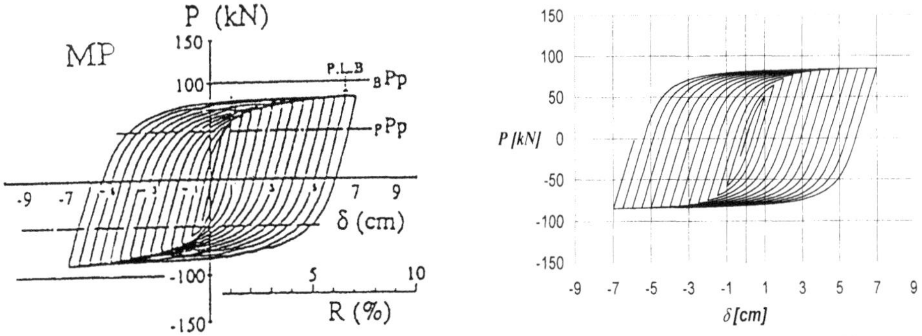

Figure 7.2.6: Typical US welded-flange bolted-web connection (Matsui & Sakai, 1992)

TABLE 7.2.1
MODEL PARAMETER VALUES FOR THE SIMULATION OF FIGURE 7.2.6 TEST

	Upper Bound Curve				**Lower Bound Curve**			
Parameter	k_0 (kN/cm)	M_0 (kN)	k_h (kN/cm)	n	k_{0p} (kN/cm)	M_{0p} (kN)	k_h (kN/cm)	n_p
Value	75	86	0	1.5	-	-	-	-
	Transition Curve			**Hardening**	**Damage**		**Unstable branch**	
Parameter	t_1	t_2	λ	H_h	β	$\phi_{u,0}$ (cm)	ϕ_{inst} (cm)	k_{inst} (kN/cm)
Value	-		-	-	-	-	-	-

Figure 7.2.7 refers to the cyclic behaviour of a beam section in bending. Even if it is not representative of a joint response, it has been chosen for illustrating the case when the system exhibits an unstable branch. In particular, it is interesting to observe that the first activation of the unstable branch corresponds to a deformation excursion range practically equal to the value of deformation activating the decreasing branch in the case of monotonic loading. Besides, in the following deformation history, the unstable branch does not change, constituting an asymptote for the actual M-ϕ path. Even if the model is not able to take into account the smooth tendency to approach the softening branch exhibited by the actual system, the comparison seems to be satisfactory. The parameter $\phi_{u,0}$, which is necessary for defining the damage function, was assumed equal to 3 in., evaluated as the intersection of the monotonic loading curve with the horizontal line M=M_0.

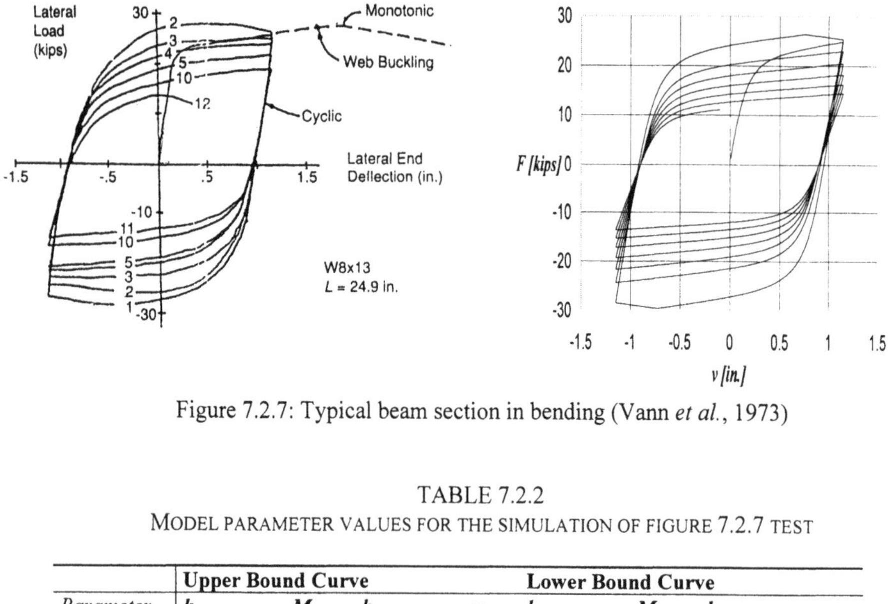

Figure 7.2.7: Typical beam section in bending (Vann *et al.*, 1973)

TABLE 7.2.2
MODEL PARAMETER VALUES FOR THE SIMULATION OF FIGURE 7.2.7 TEST

	Upper Bound Curve				**Lower Bound Curve**			
Parameter	k_0 (kN/cm)	M_0 (kN)	k_h (kN/cm)	n	k_{0p} (kN/cm)	M_{0p} (kN)	k_h (kN/cm)	n_p
Value	126	22	2.7	2	-	-	-	-
	Transition Curve			**Hardening**	**Damage**		**Unstable branch**	
Parameter	t_1	t_2	λ	H_h	β	$\phi_{u,0}$ (cm)	ϕ_{inst} (cm)	k_{inst} (kN/cm)
Value	-		-	0.055	0.08	3	1.7	2.7

End-plate connections

Figure 7.2.8 refers to the case of an extended end plate joint. The small pinching effect in the hysteretic cycle is due to flexural plastic deformation of the end plate and plastic elongation of bolts in tension, producing the opening of gaps between the beam and the column flange. The value of parameter $\phi_{u,0}$ was arbitrarily assumed equal to 100 mm.

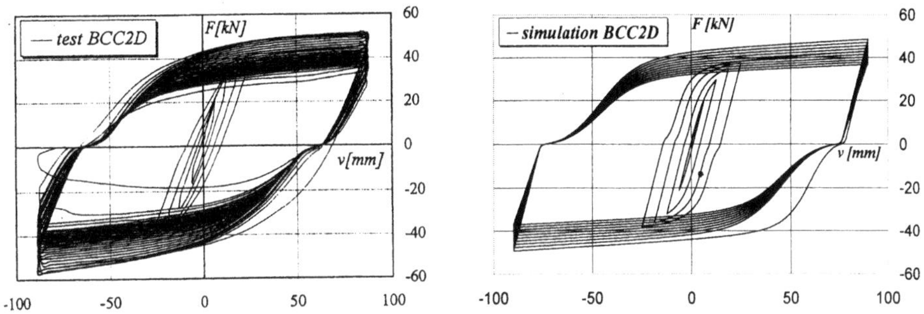

Figure 7.2.8: Typical extended end-plate connection (Calado & Castiglioni, 1995-1996)

TABLE 7.2.3

MODEL PARAMETER VALUES FOR THE SIMULATION OF FIGURE 7.2.8 TEST

	Upper Bound Curve				**Lower Bound Curve**			
Parameter	k_0 (kN/mm)	M_0 (kN)	k_h (kN/mm)	n	k_{0p} (kN/mm)	M_{0p} (kN)	k_h (kN/mm)	n_p
Value	3.6	40	0.055	2	3.6	3.5	0.055	8
	Transition Curve			**Hardening**	**Damage**		**Unstable branch**	
Parameter	t_1	t_2	λ	H_h	β	$\phi_{u,0}$ (mm)	ϕ_{inst} (mm)	k_{inst} (kN/mm)
Value	5	0.5	0.25	0.03	0.01	100	-	-

Figure 7.2.9 refers to the case of a flush end-plate steel joint. It is interesting to observe that the transition from the lower bound to the upper bound curve is quite linear. In this case the effect of pinching is more pronounced and is essentially due to plastic elongation of bolts (this specimen failed with bolt fracture). Also in this case parameter $\phi_{u,0}$ was arbitrarily fixed as 100 mm.

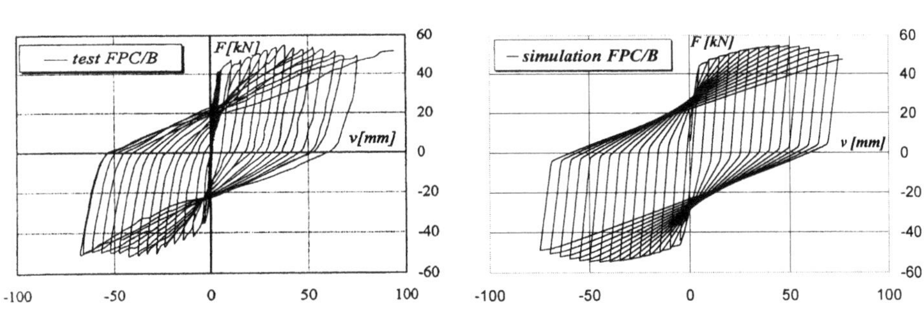

Figure 7.2.9 Typical flush end-plate connection (Bernuzzi *et al.*, 1996)

TABLE 7.2.4
MODEL PARAMETER VALUES FOR THE SIMULATION OF FIGURE 7.2.9 TEST

	Upper Bound Curve				**Lower Bound Curve**			
Parameter	k_0 (kN/mm)	M_0 (kN)	k_h (kN/mm)	n	k_{0p} (kN/mm)	M_{0p} (kN)	k_h (kN/mm)	n_p
Value	12	45	0.2	4	12	5	0.2	4
	Transition Curve			**Cyclic Hardening**	**Damage**		**Unstable branch**	
Parameter	t_1	t_2	λ	H_h	β	$\phi_{u,0}$ (mm)	ϕ_{inst} (mm)	k_{inst} (kN/mm)
Value	30	0.03	1	0.02	0.03	100	-	-

Angle connections

Figure 7.2.10 refers to a typical top and seat angle connection, which exhibits strong pinching effects. In this case, the cause of pinching is due to both plastic flexural deformations of angles, producing the opening of gaps, and the ovalisation of bolt holes, producing slip between beam flange and angles. The availability of experimental data has allowed also a quantitative comparison between experimental and simulation results to be carried out in terms of both dissipated energy and peak moment per cycle (Figure 7.2.11). Parameter $\phi_{u,0}$ was assumed equal to 0.1 rad.

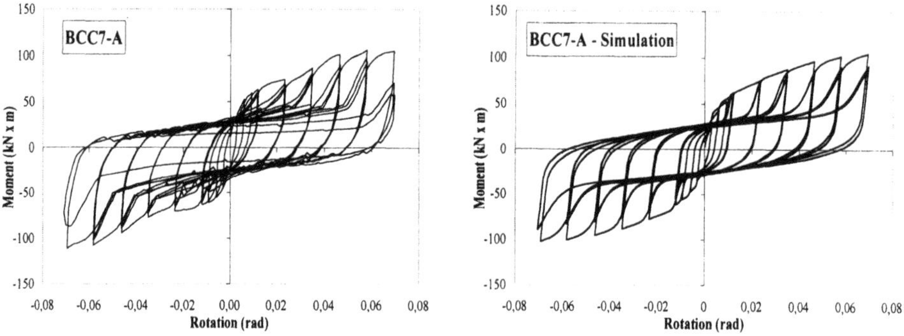

Figure 7.2.10: Typical angle connection (Calado *et al.*, 1999)

TABLE 7.2.5
MODEL PARAMETER VALUES FOR THE SIMULATION OF FIGURE 7.2.10 TEST

	Upper Bound Curve				**Lower Bound Curve**			
Parameter	k_0 (kNm)	M_0 (kNm)	k_h (kNm)	n	k_{0p} (kNm)	M_{0p} (kNm)	k_h (kNm)	n_p
Value	25000	85	750	0.6	25000	45	250	0.6
	Transition Curve			**Cyclic Hardening**	**Damage**		**Unstable branch**	
Parameter	t_1	t_2	λ	H_h	β	$\phi_{u,0}$ (rad)	ϕ_{inst} (rad)	k_{inst} (kNm)
Value	50	0.5	1	-	0.03	0.1	-	-

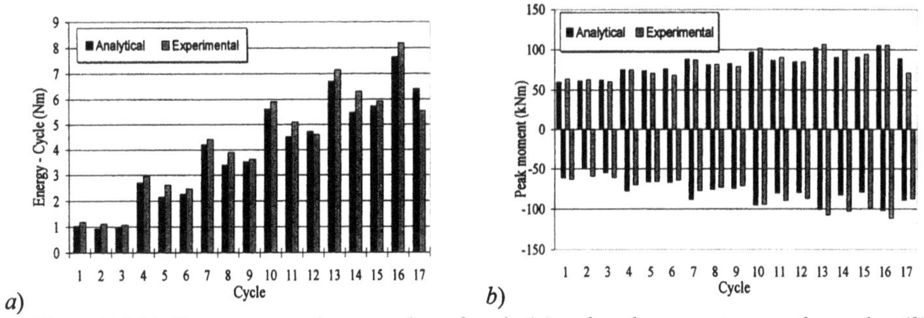

Figure 7.2.11: Energy per cycle vs number of cycle (*a*) and peak moment per cycle number (*b*)

THE PARAMETRIC ANALYSIS OF SDOF SYSTEMS

General

In the previous Section the description of a general model interpreting hysteretic behaviour of structural components has been presented. It was developed with specific reference to the mechanical behaviour of steel beam-to-column joints. However, it takes into account phenomenological aspects that are also common to other types of components, even made with different base material.

The mechanical behaviour of steel frames is fundamentally related to the mechanical behaviour of its joints, which are the locations where plastic deformations are concentrated. Therefore, the type of hysteretic behaviour of connections determines the type of hysteretic relation between global frame parameters, such as relation between story shear force and displacement.

On the other side, the inelastic performance of a multi-degree of freedom system is a difficult subject that involves the evaluation of the effect of many factors over the hysteretic modelling of connections. In particular, the type of accelerogram in relation with the numerous possible modes of vibration of a steel frame is to be taken into account when analysing such systems. Besides, analysis results may be strongly influenced by frame geometrical configuration, which represents another important parameter of the study.

Therefore, in order to investigate the influence of different connection hysteretic rules on the prediction of inelastic performance of frames, the authors decided to start analysing a single-mass subject to uni-directional ground motions, in such a way to minimise the factors to be considered.

Previous studies

First important studies concerning with seismic response of SDoF systems date back to the last fifties (Housner, 1959). Ten years later Newmark and Hall (1969) published a fundamental work on linear elastic response spectra. Since that date a large amount of research effort has been spent for the evaluation of seismic response of linear SDoF systems, with particular attention to the influence of input motion, in some cases reflecting the influence of site conditions. It is interesting to observe that the study of inelastic SDoF systems developed contemporary to the elastic case. In fact, in 1969 Veletsos presented first studies on IRS (inelastic response spectra). In 1975 Murakami and Penzien computed probabilistic non-linear constant strength response spectra for SDoF systems with four types of hysteretic behaviour and subjected to 100 artificially generated earthquakes. In 1979 Riddel and

Newmark computed constant ductility IRS of 10 recorded earthquake ground motions considering the effects of both damping and hysteretic behaviour.

More recently, Minami and Osawa (1988) conducted parametric studies on elastic-plastic response spectra for different hysteretic models classified according to their strain energy-absorbing capacities. By analysing obtained results it seems that fundamental inelastic response parameters, i.e. kinematic and cumulated ductility, are slightly influenced by hysteretic assumptions for models belonging to the same group. While significant differences can be observed when radically changing the type of dissipative behaviour (for example going from a bilinear full-dissipative model to a partial-dissipative pinching-type one).

Krawinkler and Nassar (1992) studied average IRS of bilinear and stiffness degrading SDoF systems. They concluded that IRS are only slightly modified by the type of hysteretic model considered. Extensive parametric studies were conducted by Fajfar *et al.* (1989, 1992, 1994) in almost ten years of research activity. The influence of both damping and hysteretic modelling on four types of interrelated constant ductility non-linear response spectra (strength, displacement, input and hysteretic energy) was considered. Their relevant results show again that hysteretic model influence significantly inelastic response only when changing radically the type of dissipative behaviour (with or without pinching of hysteretic cycles), while the influence is slight in case of substantial similar shape of hysteretic cycles. Besides an important influence of damping modelling was observed. These conclusions are drawn in comparison with results of influence of input earthquake motions, which is more significant.

Cosenza and Manfredi (1994) carried out a study on the influence of stiffness and strength degradation on constant ductility strength reduction factor spectra. It was concluded that damage phenomena are somewhat influencing the design strength, but this is strictly related with the type of input motion considered. However, this influence seems to be slight when compared with the one concerned with other factors, e.g. the adopted collapse criterion.

Equation of motion

The well known equation of motion of SDoF systems writes:

$$m\ddot{x} + b\dot{x} + F_s = -ma_g \qquad (7.2.10)$$

where m is the mass, b the viscous damping coefficient, F_s the restoring strain-related force, x the displacement of the mass relative to the ground, and a_g the time-dependent ground acceleration. With the following positions:

$$\mu = \frac{x}{x_y}; \quad \varphi = \frac{F_s(x)}{k_0 x_y}; \quad \omega_0 = \sqrt{\frac{k_0}{m}}; \quad \nu = \frac{b}{2\sqrt{k_0 m}}; \quad PGA = max[a_{g,max}; abs(a_{g,min})] \qquad (7.2.11)$$

x_y being a conventional yield displacement, k_0 the initial stiffness of the system and PGA the peak ground acceleration, equation (7.2.10) can be rewritten as:

$$\ddot{\mu} + 2\nu\omega_0\dot{\mu} + \omega_0^2\varphi = -\frac{k_0}{m}\frac{mPGA}{k_0 x_y}\frac{a_g}{PGA}. \qquad (7..2.12)$$

By introducing the 'resistance level' R^* defined as follows:

$$R^* = \frac{mPGA}{k_0 x_y} = \frac{mPGA}{F_y} \tag{7.2.13}$$

equation of motion finally writes:

$$\ddot{\mu} + 2\nu\omega_0\dot{\mu} + \omega_0^2\varphi = -R^*\omega_0^2 \frac{a_g}{PGA} \tag{7.2.14}$$

For a given hysteretic model (i.e. relationship $\varphi = \varphi(\mu)$), a fixed value of ω_0, ν, R^* and a chosen non-dimensional ground acceleration history (a_g/PGA), the step-by-step numerical integration of equation (7.2.14) has been performed by means of the linear acceleration Newmark's method (1959). Thus we obtained ductility and non-dimensional hysteretic energy demand ($e_h = E_h/R_y x_y$, E_h being the actual hysteretic energy), defined as the maximum values reached throughout the whole deformation history.

The analysed hysteretic models

The hysteretic models investigated in this study can be grouped in two different types according to their dissipative capacities: (1) Full-dissipative-type models (without pinching); (2) Partial-dissipative-type models (with pinching).

Group (1) comprises classical elasto-plastic bilinear model (EPB) and the new specifically developed hysteretic fully non-linear model (EPNL) described in the previous Section. Figures 7.2.12*a* and 7.2.12*b* show a qualitative application of these two models.

Group (2) comprises classical elasto-plastic bilinear with slackness model (EPBS) and the new specifically developed fully non-linear with pinching model (EPNLP) described in the previous Section. Figures 7.2.13*a* and 7.2.13*b* show a qualitative application of these two models.

The parametric study is addressed to evaluate the effect of non-linearity, hardening ratio, cyclic damage and pinching on plastic engagement, measured through kinematic ductility μ and non-dimensional hysteretic energy e_h.

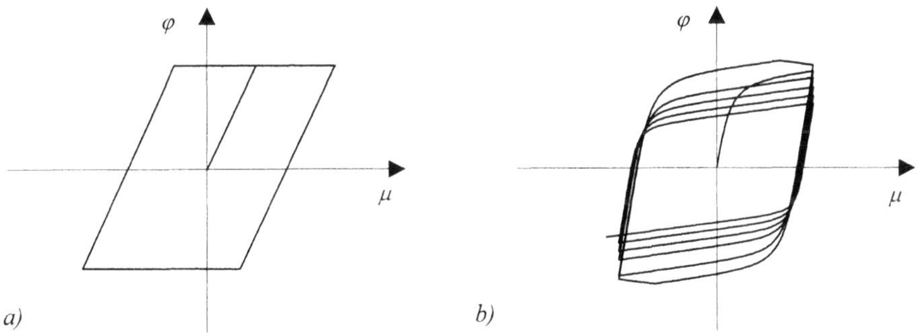

Figure 7.2.12: *a)* Elasto-plastic bilinear (EPB) and *b)* Elasto-plastic fully non-linear model (EPNL)

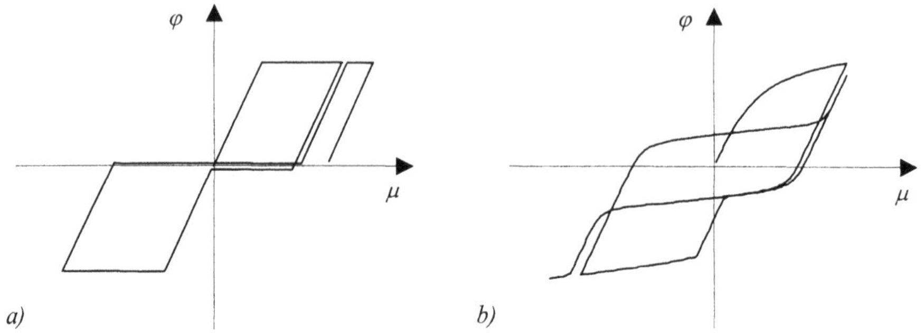

Figure 7.2.13: *a)* Elasto-plastic bilinear with slackness (EPBS) and *b)* Elasto-plastic fully non-linear with pinching model (EPNLP)

Result format

For presentation of results, the following parameter R has been introduced:

$$R = R^* \frac{S_{a,el}(T, \nu = 0.05)}{PGA} = \frac{mPGA}{F_y} \frac{S_{a,el}(T, \nu = 0.05)}{PGA} = \frac{mS_{a,el}(T, \nu = 0.05)}{F_y} \qquad (7.2.15)$$

in which $S_{a,el}$ is the pseudo-acceleration of the equivalent elastic system of period $T = T_0$ ($= 2\pi/\omega_0$) and damping ratio $\nu = 5\%$. In other words, R is the ratio of maximum elastic force that would stress the system ($mS_{a,el}$) and its actual strength (F_y). Five values of the reduction factor R were assumed in the parametric studies: 1, 2, 4, 6 and 8. However, for saving the space, only some results related to $R = 1$ (elastic system) and $R = 6$ (high inelastic deformation demand) are presented in the following.

Analyses have been carried out considering 9 acceleration time histories, recorded on rigid and medium soil conditions, corresponding to earthquakes registered in different World Regions. All accelerograms have been scaled to the same *PGA* and then amplified by means of parameter R in order to consider increasing plastic engagement.

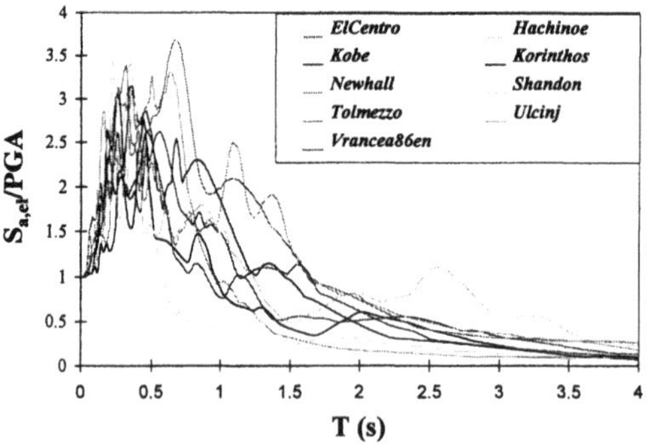

Figure 7.2.14: Normalised linear elastic pseudo-acceleration spectra for considered earthquake records ($\nu = 5\%$).

For every value of R, non-linear ductility and non-dimensional hysteretic energy demand spectra have been computed as mean values among all the earthquakes. The range of variation for the initial period of vibration ($T = T_0$) was assumed equal to 0.4 – 3 s, with time step equal to 0.2 s. Results are presented by varying the values of several parameters of the above hysteretic models. The meaning of these parameters and their influence on plastic engagement are discussed in the following.

Result Presentation

Figures 7.2.15 and 7.2.16 refer to full-dissipative systems (EPB and EPNL) without damage of mechanical properties and in absence of strain hardening. They show the influence of the shape parameter *n*. As it was expected, figure 7.2.15*a* shows that there is no plastic engagement for $R = 1$ and that displacements of non-linear systems are slightly smaller than displacements concerned with EPB model. This is due to the hysteretic energy dissipated by non-linear system as also confirmed by energy demand graphs (figure 7.2.16*a*). For high plastic engagement ($R = 6$), the influence of shape parameter is even less important, allowing the EPB model to be used for the evaluation of plastic engagement.

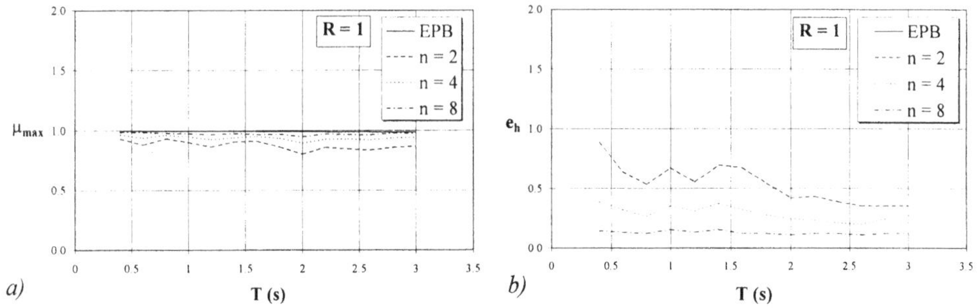

Figure 7.2.15: Ductility (*a*) and non-dimensional hysteretic energy (*b*) demand for full-dissipative type models characterised by various degrees of non-linearity (*n*), for $R = 1$.

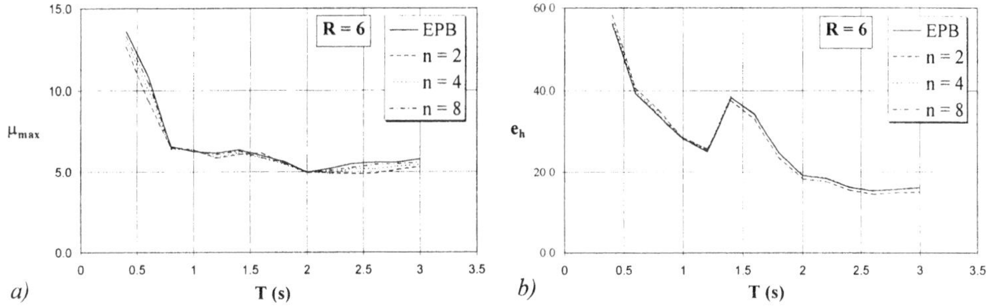

Figure 7.2.16: Ductility (*a*) and non-dimensional hysteretic energy (*b*) demand for full-dissipative type models characterised by various degrees of non-linearity (*n*), for $R = 6$.

Figure 7.2.17 illustrates effects of kinematic strain hardening on the response of EPNL model ($n = 4$). Hardening ratio *h* equal to 0, 5, 10 and 20% have been considered. Generally, a decrease of ductility demand with increasing values of *h* can be observed. Hysteretic energy is however increasing with *h* probably owing to the compensation effect due to wider hysteretic cycles for higher hardening. Besides, it can be noticed that the influence of strain hardening is reduced as far as the period of vibration increases, this being partially due to the shape of the single spectrum characterised by higher gradient for lower periods of vibration. It is also interesting to observe that there are only slight differences in ductility demand when changing the value of *h* from 0.05 to 0.20 but differences are much more pronounced when going from $h = 0$ to $h > 0$. This means that the value chosen for the kinematic hardening ratio is not so important as the assumption of h different from zero itself.

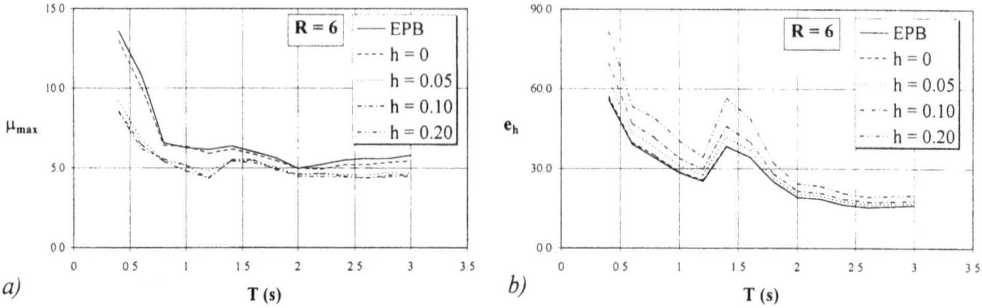

Figure 7.2.17: Ductility (*a*) and non-dimensional hysteretic energy (*b*) demand for full-dissipative type models characterised by various degrees of kinematic hardening (h), for $R = 6$.

Influence of strength degradation on ductility and non-dimensional hysteretic energy demand is illustrated in figure 7.2.18, where $n = 4$ and $h = 0$ have been considered for EPNL model. Values for parameter β equal to 0.025, 0.05 and 0.15 have been assumed, in order to consider a low, medium and high rate of strength degradation, respectively. Besides, for all cases, a fixed ultimate ductility $\mu_{ult} = x_{ult}/ x_y$ equal to 10 has been adopted. From figure 7.2.18 it can be observed that the influence of strength degradation on both ductility and non-dimensional hysteretic energy demand is quite limited. Only for high rate of strength degradation ($\beta = 0.15$) differences in prediction may be occasionally significant.

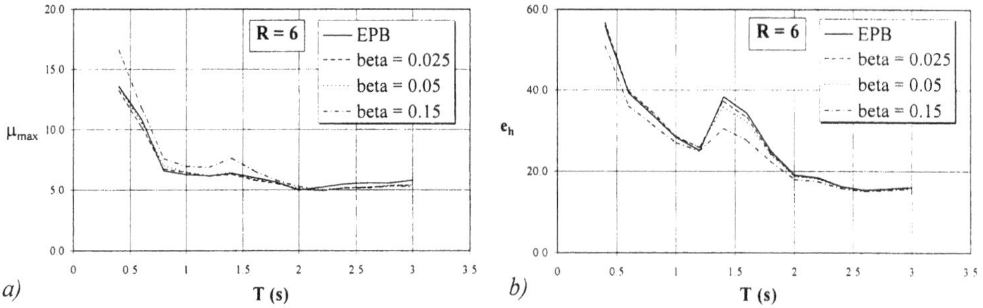

Figure 7.2.18: Ductility (*a*) and non-dimensional hysteretic energy (*b*) demand for full-dissipative type models characterised by various degrees of strength degradation (β), for $R = 6$.

Figure 7.2.19 shows the effect of softening branches in the hysteretic restoring force-deformation relationship (EPNL model with $n = 4$, $h = 0$, $\beta = 0$). Two values of ductility activating softening branches have been considered, namely $\mu_{inst} = 3$ and $\mu_{inst} = 5$. An increment in ductility demand with decreasing values of μ_{inst} can be observed in figure 7.2.19*a*, while figure 7.2.19*b* shows the decrease of non-dimensional hysteretic energy with decreasing values of μ_{inst}, which may be due to the compensation effect of increasing ductility and smaller size of hysteretic cycles. It can be noticed that any influence for long initial periods of vibration vanishes and all curves in figure 7.2.19*a* approach the same value of ductility demand, practically coincident with the value of R.

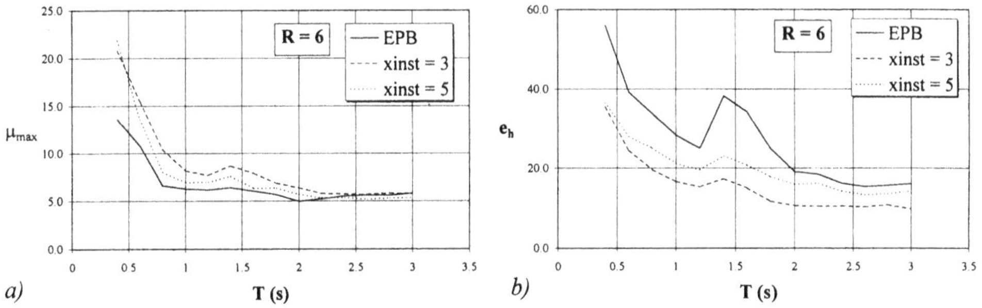

Figure 7.2.19: Ductility (*a*) and non-dimensional hysteretic energy (*b*) demand for full-dissipative type models characterised by various levels of softening ductility (x_{inst}), for $R = 6$.

Finally, figure 7.2.20 shows the variation of ductility and non-dimensional hysteretic energy demand going from full-dissipative (without pinching) to partial-dissipative (with pinching) type of hysteretic behaviour. The two examined models were characterised by the same envelope curve (same initial stiffness k_0, yielding resistance F_y, shape factor n, hardening ratio h) but they differ for the shape of the hysteretic cycles. The analysed cases are referred to a constant value of parameter F_{0p}, chosen equal to 0.2 times the yielding resistance F_y, which corresponds to a strong pinching effect. It can be observed that ductility demand increases when going from EPNL to EPNLP model, while hysteretic energy demand decreases, due to compensation of increased ductility and reduced size of hysteretic cycles. Once again, the effect of pinching on ductility demand decreases for increasing values of the period of vibration.

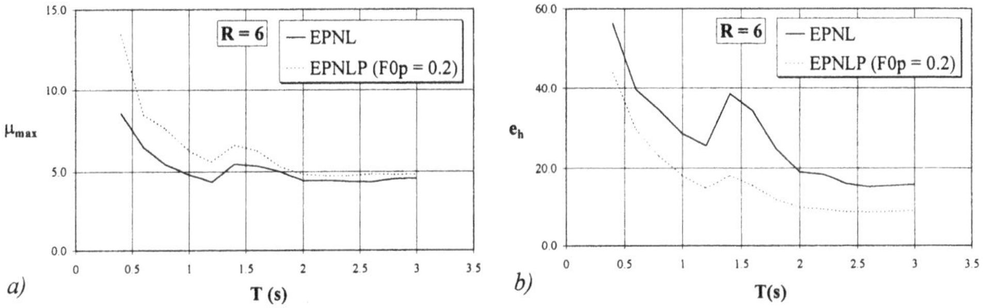

Figure 7.2.20: Ductility (*a*) and energy (*b*) demand for full-dissipative (without pinching) and partial dissipative (with pinching) type models, for $R = 6$.

It is interesting to observe from figures 7.2.21 and 7.2.22 the influence of the value chosen for viscous damping ratio in evaluating inelastic response spectra. Obviously, both ductility and non-dimensional hysteretic energy demand increase when decreasing values of damping ratios are considered. Differences are of the same order of magnitude than those due to the most influencing hysteretic modelling parameters.

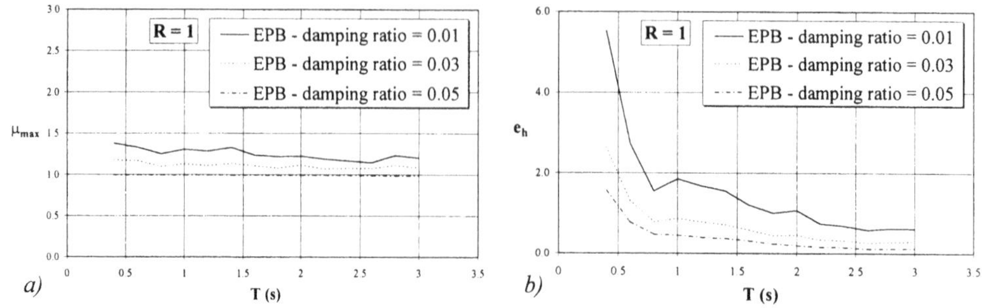

Figure 7.2.21: Ductility (*a*) and non-dimensional hysteretic energy (*b*) demand for EPB model characterised by different values of damping ratio (ν), for $R = 1$.

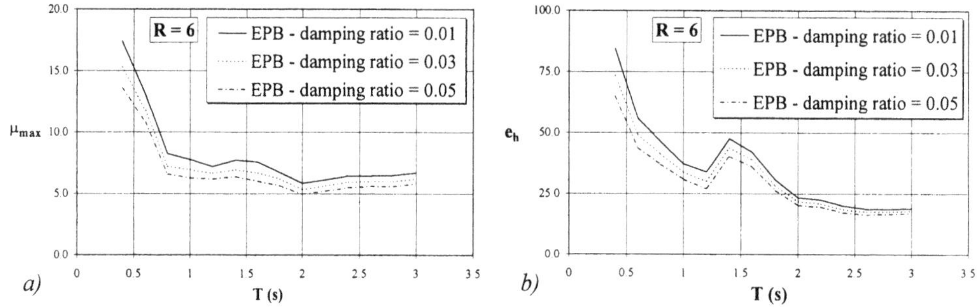

Figure 7.2.22: Ductility (*a*) and non-dimensional hysteretic energy (*b*) demand for EPB model characterised by different values of damping ratio (ν), for $R = 6$.

Main outcomes

Obtained results on SDoF system show that kinematic hardening, pinching of hysteretic cycle and strength degradation due to softening branches in the hysteretic response are the three phenomenological aspects that have a major impact on seismic demand to the system. On the contrary, influence of fully non-linear behaviour is negligible, thus confirming the validity of piece-wise linear models. As far as strength degradation due to plastic fatigue (without decreasing branches) is concerned, a slight influence on plastic engagement has been observed. Moreover, the effect due to the above factor becomes more significant for short periods of vibration. The latter is however a general trend, all investigated parameters presenting a very small effect on plastic engagement for high periods of vibration.

As a general conclusion, since the most of parameters affecting the hysteretic response of the structural component have a limited influence on the global performance of the system, elastic-perfectly plastic models, with or without slackness, seem to be suitable for simplified seismic analyses of SDoF systems. Nevertheless, the adoption of sophisticated hysteretic rules could be justified for non-linear global analyses of MDOF structures, by a detailed schematisation of all structural components, only one or few of these being the possible responsible for the collapse of the whole.

THE PARAMETRIC ANALYSIS OF MOMENT RESISTING FRAMES

General

Hysteretic model previously described has been applied to study the dynamic inelastic performance of a typical steel moment resisting frame designed according to Eurocode 3 and Eurocode 8. A medium seismicity zone ($PGA = 0.20g$) and sub-soil conditions type A have been accounted for (De Matteis *et al.* 1998). As design force reduction factor, a q-factor value equal to 6 was assumed. Vertical loads are those codified for civil buildings. It must be emphasised that integral code prescription on serviceability limit state was applied, in particular considering a limit value of $0.006h$ for interstorey drift. Consequently, frame members were strongly over-resistant as respect to the analogous that could be obtained by considering ultimate limit state checking and capacity design rules only. Figure 7.2.25 illustrates the analysed frame geometry, members sizes and furnishes some generalities of modelling assumptions.

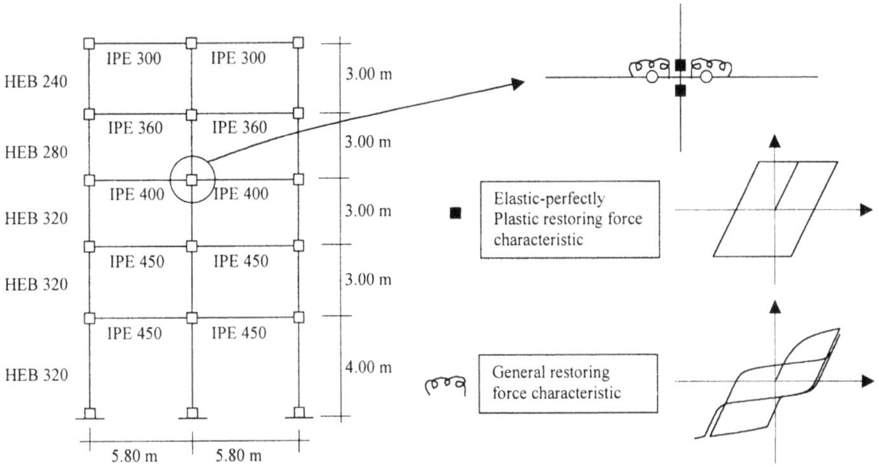

Figure 7.2.23: Analysed frame geometry and fundamental modelling assumptions

Dynamic time-history inelastic analyses have been carried out by means of a general purpose computer program, developed in co-operation with University of Catania (Italy). Only two accelerograms have been considered in the analyses, one recorded during Kobe earthquake and the other one during El Centro earthquake. Records were scaled to the same value of maximum acceleration. In particular, values of 0.5g, 1.0g, and 1.5g for PGA have been considered in the analyses. Besides, since the main objective of the study was the influence of some model parameters on ductility and energy demand, no collapse criteria have been fixed. However, the analyses are focused on the response of joints and therefore, implicitly, local type collapse criteria have been referred to.

The parametric study has been organised as in the following:

(1) Study of the influence of hysteretic cycle shape

(2) Study of the influence of strength degradation

(3) Study of the influence of initial values of stiffness and strength

In each group of analyses, the investigated parameter is varied in a realistic range, but when it is combined with other parameter values this sometimes could lead to unrealistic joint characteristics.

However, the above methodology allows the effect of the analysed parameter to be clearly quantified. In the following, study cases and obtained results for each group of analyses are described in details.

Study planning

Study of the influence of hysteretic cycle shape

In this group of analyses the influence of pinching effects is investigated. Therefore, the ratio of transition parameter t_1 to transition parameter t_2 has been varied within a realistic range. Initial strength and stiffness values have been considered to be constant in all cases, by assuming values corresponding to possible rigid, full-strength joints.

Table 7.2.6 synthesises the values associated to model parameters, where $M_{pl,b}$ is the cross-section plastic moment, EI_b the flexural stiffness and L_b the length of connected beam.

TABLE 7.2.6
STUDY ON THE HYSTERETIC CYCLE SHAPE

	Upper bound curve				**Lower bound curve**				**Damage**
Parameter	k_0	M_0	k_h	N	k_{0p}	M_{0p}	k_h	n_p	β
Value	50 EI_b/L_b	$M_{pl,b}$	0	0.5	k_0	$0.3M_0$	0	0.5	0
	Transition parameter								
Parameter	t_1	t_2	λ						
Value	50	0.5 0.1 0.02 0.005	1						

Considered values of t_1 and t_2 correspond to ratios t_1/t_2 equal to 100, 500, 2500, 10000, which refer to hysteretic cycles characterised by correspondingly decreasing pinching effects.

Study of the influence of strength degradation

The influence of strength degradation for full-dissipative type systems has been analysed with reference to two typical hysteretic rules, one approaching the elastic-perfectly plastic model (n=10) and the other one being strongly non linear (n=0.5). For each model type, 4 values of damage parameter β have been assumed (ranging from 0.025 to 0.15), modelling an increasing degradation rate (Table 7.2.7).

TABLE 7.2.7
STUDY ON THE STRENGTH DEGRADATION (FULL-DISSIPATIVE SYSTEMS)

Parameter	k_0	M_0	k_h	n	β
Value	50 EI_b/L_b	$M_{pl,b}$	0	10 0.5	0.025 0.05 0.10 0.15

Study of the influence of initial stiffness level

Three levels of initial stiffness have been considered: $k_0/k_r = 2$, $k_0/k_r = 1$, $k_0/k_r = 0.5$, k_r being the limit value of initial stiffness which, according to Eurocode 3, marks the transition from rigid and semi-rigid steel joints. This study is referred to a pinching type hysteretic model, because semirigidity is usually a characteristic of bolted joints where pinching is also associated. Table 7.2.8 shows parameter values considered in this group of analyses.

TABLE 7.2.8
STUDY ON THE INITIAL STIFFNESS

	Upper bound curve				**Lower bound curve**				**Damage**
Parameter	k_0	M_0	k_h	n	k_{0p}	M_{0p}	k_h	n_p	β
Value	$50\ EI_b/L_b$ $25\ EI_b/L_b$ $12.5\ EI_b/L_b$	$M_{pl,b}$	0	0.5	k_0	$0.3M_0$	0	0.5	0

	Transition parameter		
Parameter	t_1	t_2	λ
Value	50	0.5	1

Study of the influence of initial strength level

The effect of joint moment resistance has been investigated considering three strength levels: $M_0/M_{pl,b} = 1, 0.75, 0.5$. In this case, a pinching type hysteretic model has been considered. Table 7.2.9 shows parameter values considered in this group of analyses.

TABLE 7.2.9
STUDY ON THE INITIAL STRENGTH

	Upper bound curve				**Lower bound curve**				**Damage**
Parameter	k_0	M_0	k_h	n	k_{0p}	M_{0p}	k_h	n_p	β
Value	$50\ EI_b/L_b$	$1.00\ M_{pl,b}$ $0.75\ M_{pl,b}$ $0.50\ M_{pl,b}$	0	0.5	k_0	$0.3M_0$	0	0.5	0

	Transition parameter		
Parameter	t_1	t_2	λ
Value	50	0.5	1

The obtained results

As previously emphasised the main objective of the current study was to establish a hierarchy among phenomenological aspects that are experimentally apparent for beam-to-column steel joints in terms of their effects on the seismic response of moment resisting frames. The aim is pursued by individuating hysteretic modelling parameters having a major impact on fatigue life of joints. At present there is no agreement in the international scientific community on what is the most effective method for evaluating fatigue life of members and joints in a simple and practical way. Nevertheless, it is customarily recognised the importance of two factors in determining collapse due to low-cycle fatigue: ductility and hysteretic energy demand.

The empirical method proposed by Park and Ang (1985) combines these two factors in a linear way, it suggesting the following collapse index to be assumed:

$$IC = \frac{\phi}{\phi_{u,mon}} + \beta_{PA} \frac{E_h}{M_y \phi_{u,mon}} \tag{7.2.16}$$

The key parameter is β_{PA}, which is related to material and typology of structural component considered, and must be set on the basis of experimental evidence. Corresponding values for steel structure components are very uncertain due to lack of experimental data. On the other hand, current research seems to be oriented to other methods for evaluating the fatigue life of structures (Krawinkler & Zohrei 1983, Cosenza *et al.* 1993).

Notwithstanding, ductility and hysteretic energy demand surely influence collapse due to low-cycle fatigue and, because of the clearness of their meaning and simplicity of calculation, they are used here as parameters measuring susceptibility of frame seismic response to hysteretic model.

Therefore, for each considered model parameter, obtained results have been synthesised in two types of diagrams: a first one, in which maximum plastic and total rotation demand to joints are shown, and a second one, in which maximum and total hysteretic energy demand are illustrated. These diagrams are drawn with reference to both considered natural acceleration records and for three different peak ground acceleration values.

The influence of hysteretic cycle shape

Figures 7.2.24 and 7.2.25 show the influence of pinching effect on both maximum ductility and energy demands to joints. Firstly, it can be observed that in case of low values of PGA, i.e. in case of trivial plastic engagement, differences among models are insignificant. On the contrary, in case of strong plastic deformations, a remarkable influence of pinching modelling is noticed. In particular, for PGA = 1.5g, maximum required plastic rotation may be over 50% greater going from a pinching type model (t_1/t_2=100) to a full-dissipative type one (t_1/t_2=10000). Also, variations in hysteretic energy demands are smaller, but this could be due to a sort of compensation effect of increasing ductility with decreasing sizes of cycles.

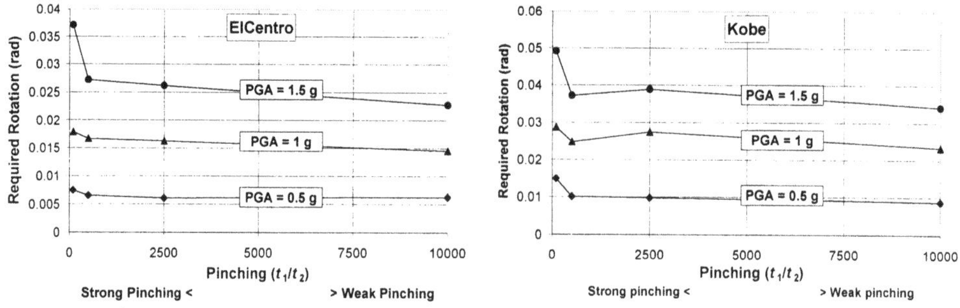

Figure 7.2.24: Maximum plastic rotation demand (influence of hysteretic shape)

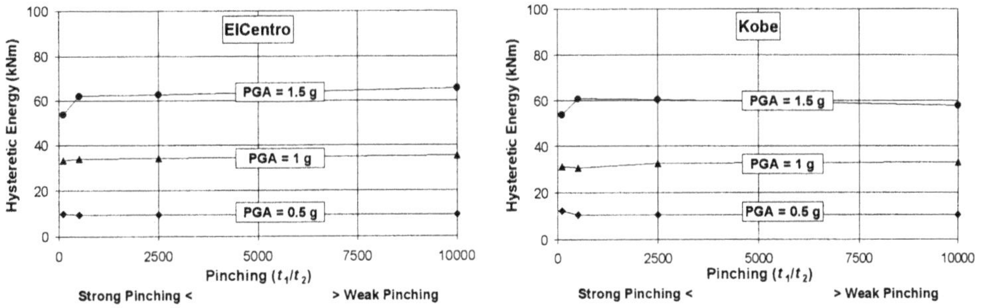

Figure 7.2.25: Maximum hysteretic energy dissipation required at joints (influence of hysteretic shape)

Influence of strength degradation

Figures 7.2.26 and 7.2.27 illustrate the influence of strength degradation for a joint behaviour practically equivalent to the elastic-perfectly plastic one (n = 10). Results indicate that maximum required plastic rotation is practically independent from β (at least in the range of considered values), while slight variations are observed for energy demands. Nevertheless, it is important to be reminded that results related to energy are surely less important than the ones related to ductility, the former having to be multiplied by a very low factor ($10^{-2} \div 10^{-1}$) for applying collapse criteria, as for instance Park and Ang model.

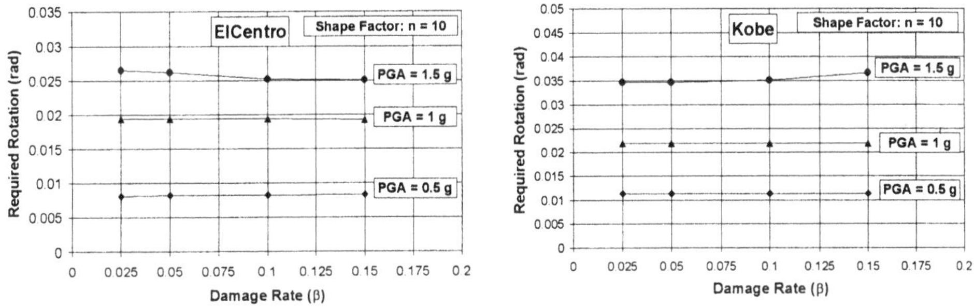

Figure 7.2.26: Maximum plastic rotation demand (influence of strength degradation)

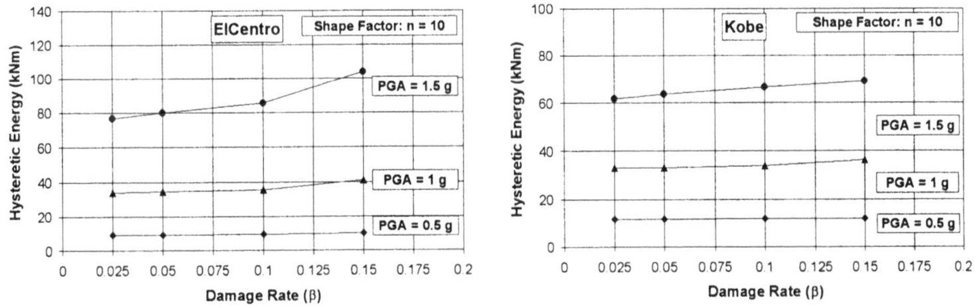

Figure 7.2.27: Maximum hysteretic energy dissipation demand (influence of strength degradation)

Analogously, figures 7.2.28 and 7.2.29 show the influence of strength degradation with reference to a fully non-linear model (n = 0.5). As respect to the previous case, higher plastic rotation and energy dissipation demand may be observed. Anyway, the effect of parameter β seems to be slight on both plastic rotation and energy dissipation demand.

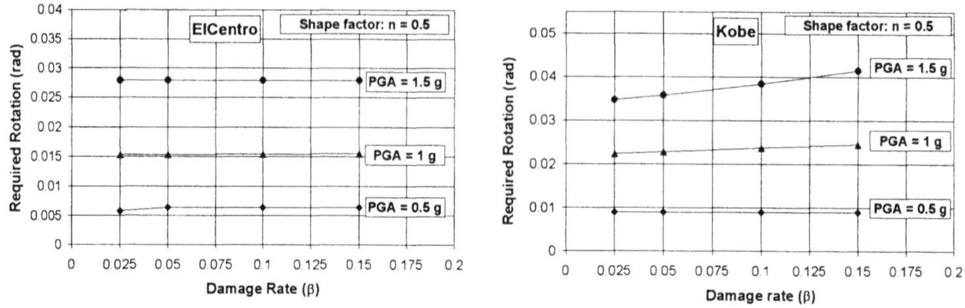

Figure 7.2.28 Maximum plastic rotation demand (influence of strength degradation)

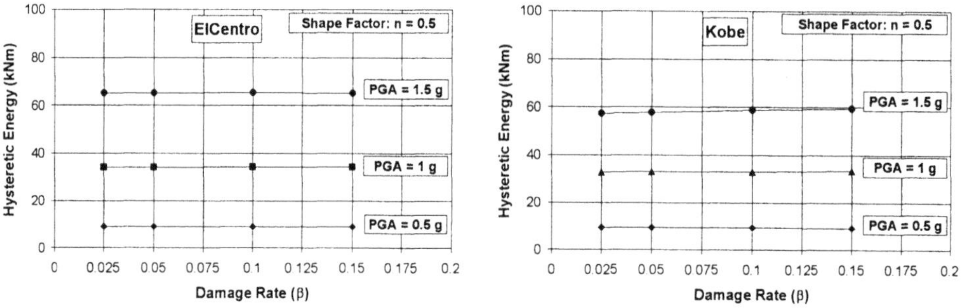

Figure 7.2.29: Maximum hysteretic energy dissipation demand (influence of strength degradation)

Influence of initial stiffness level

In all previous cases, initial stiffness level was so high to confuse plastic and total rotation. On the contrary, in this group of analyses, initial stiffness level is varied, not permitting the above confusion, the difference being the elastic part of rotation, evaluated with reference to conventional plastic resistance M_0. Therefore, in figures 7.2.30 e 7.2.31 both total (full line) and plastic (dashed line) rotations are depicted. It can be noticed that in case of low values of PGA the required plastic rotation is almost constant, but for strong input motions, rotations increase as far as values of initial stiffness decreases.

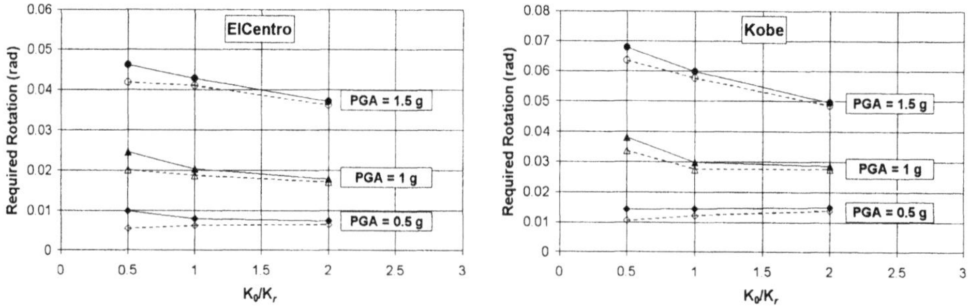

Figure 7.2.30: Maximum plastic rotation demand (influence of initial stiffness)

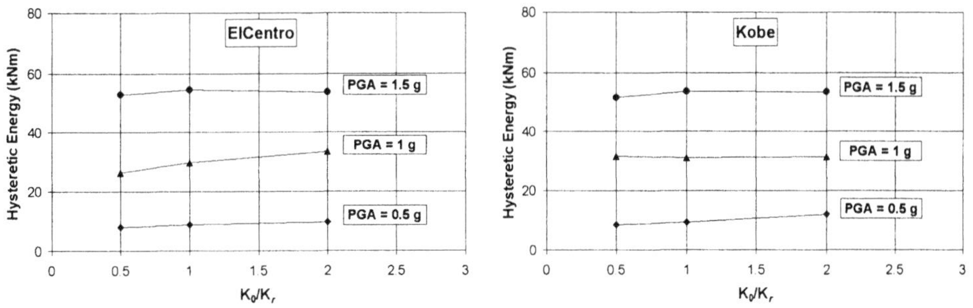

Figure 7.2.31: Maximum hysteretic energy dissipation demand (influence of initial stiffness)

Influence of initial strength

Figures 7.2.32 and 7.2.33 show the variation of plastic demand with reference to several levels of initial joint flexural strength. As it could be expected, plastic rotation increases with decreasing value of joint plastic strength level ($M_0/M_{pl,b}$), that is going from full strength ($M_0/M_{pl,b}$ = 1) to partial strength ($M_0/M_{pl,b}$<1) joints. The same trend may be drawn for the energy dissipation demand, even if the compensation effect due to the reduction size of hysteretic cycles sizes with decreasing plastic resistance is to be noticed. As a matter of the fact, this parameter seems to be the most influencing, but it is to be observed that it is also the easiest to be determined by analytical computations as respect to the other ones.

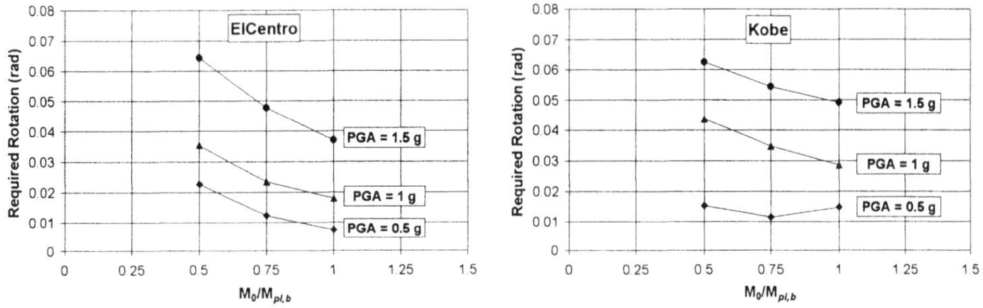

Figure 7.2.32: Maximum plastic rotation demand (influence of initial strength)

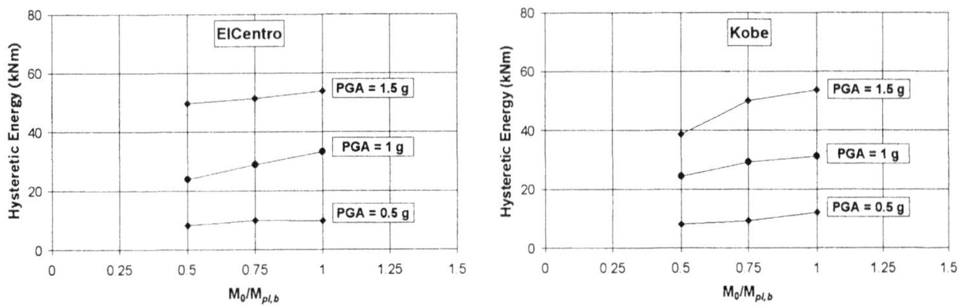

Figure 7.2.33: Maximum hysteretic energy dissipation demand (influence of initial strength)

Main outcomes

The above parametric analysis, even if not exhaustive with reference to a very complex problem, allow some important considerations about influence of joint modelling on seismic performance of steel moment resisting frames to be drawn.

(1) Pinching modelling may significantly affect joint ductility demand. Therefore, the possibility to use connection models accounting for such a phenomenological aspect could be very profitable for a correct prediction by numerical analyses. (2) Strength degradation due to plastic fatigue is not very important in determining plastic demands to joints. (3) Initial stiffness level is quite influential for the whole response of steel frames in terms of both ductility and energy dissipation demand. (4) Initial strength level strongly affects the structural response of moment-resisting frames.

Finally, it is useful to be reminded that the above analyses are referred to specific frame geometry and acceleration records. Nevertheless, the structural regularity of the considered frame allows the drawn conclusions to be extended to similar regular configurations. On the other side, the influence of the acceleration record is determinant, not allowing the obtained results to be easily generalised.

GENERAL CONCLUSIONS

The current study has dealt with the influence of connection modelling on the seismic response of steel moment resisting frames. A mathematical model interpreting cyclic behaviour of beam-to-column steel joints has been firstly presented. The comparison with experimental results has shown that such a model is satisfactory to represent correctly the main phenomenological aspects as they are apparent on the basis of experimental evidence.

Then, the effect of joint modelling has been investigated on SDoF systems. Several types of hysteretic rules have been considered. Obtained results have shown that kinematic hardening, pinching of hysteretic cycle and strength degradation due to softening branches are the phenomenological aspects having a major impact on seismic demand to the system. On the contrary, the influence of non-linearity seems to be negligible, while strength degradation due to plastic fatigue (without decreasing branches) is not determinant. Also, the effect due to the above factors becomes generally more significant for short periods of vibration. The latter is however a general trend, all the investigated parameters presenting a very small effect on plastic engagement for high periods of vibration.

Finally, the effect of joint hysteretic model has been examined with reference to a moment resisting frame configuration. In general, the results seem to be in good agreement with the ones concerned with SDoF system. In particular, pinching of hysteretic behaviour significantly affect ductility demand to joint, while strength degradation due to plastic fatigue provides a slight influence on the whole response of steel frames. Also, as it was expected, initial strength and stiffness are determinant for the correct prediction of seismic structural response.

However, it is useful to observe that the current study does not consider the effect due to geometry of the frame. Besides, the adopted methodology of analysis does not allow the interaction effect among different parameters to be assessed. Eventually, the performed analyses show that the influence of the acceleration record is considerable, in some cases it being more important than the one concerning with joint modelling itself.

REFERENCES

ATC, (1992). *Guidelines for cyclic seismic testing of components of steel structures.* Applied Technology Council, Publication N. 24.

Attiogbe, E. & Morris, G. (1991). Moment-Rotation functions for steel connections. *Journal of Structural Engineering,* ASCE, **117**(6), 1703-18.

Bernuzzi, C., Zandonini, R. and Zanon, P. (1996). Experimental analysis and modelling of semi-rigid steel joints under cyclic reversal loading. *J. Constructional Steel Research*, **38**(2), 95-123.

Calado, L., Castiglioni, C.A. (1995). Low Cycle Fatigue Testing of Semi-Rigid Beam-to-Column Connections, *Proc. of 3rd International Workshop on Connections in Steel Structures,* Trento, 371-380.

Calado, L., Castiglioni, C.A. (1996). Steel Beam-to-Column Connections under Low-Cycle Fatigue Experimental and Numerical Research, *Proc. of XI World Conference on Earthquake Engineering,* Acapulco, Mexico.

Calado, L. De Matteis G., Landolfo R. (1999). Experimental analysis on angle beam-to-column joints under reversal cyclic loading, *Proc. of XVII Congress CTA,* Napoli 3-5 October.

Cosenza, E., Manfredi, G. & Ramasco, R. (1993). The use of damage functionals in earthquake resistant design: a comparison among different procedures. *Structural Dynamics and Earthquake Engineering,* **22**, 855-868.

Cosenza, E., Manfredi G. 1994. Classificazione e comportamento sismico di modelli ciclici degradanti. *Proc. of Danneggiamento ciclico e prove pseudo-dinamiche,* Napoli, 2-3 giugno.

Della Corte G., De Matteis G., Landolfo R. (1999*a*). A mathematical model interpreting the cyclic behaviour of steel beam-to-column joints, *Proc. of XVII CTA Congress,* Napoli 3-5 ottobre.

Della Corte, G., De Matteis G. & Landolfo R. (1999*b*). Modellazione di nodi trave-colonna e risposta sismica di telai di acciaio, *Proc. of IX National Conference "L'Ingegneria Sismica in Italia",* Torino 20-23 September.

De Matteis G. (1998). *The effect of cladding panels in steel buildings under seismic actions.* PhD Thesis, University of Naples 'Federico II'.

De Matteis, G., Landolfo, R. & Mazzolani, F.M. (1998). Dynamic response of infilled multistorey steel frames, *Proc. of XI European Conference on Earthquake Engineering,* Paris la Defense, France, 6-11 September.

ECCS, (1986). *Recommended Testing Procedure for Assessing the Behaviour of Structural Steel Elements under Cyclic Loads.* European Convention for Constructional Steelwork, Publication No. 45, Brussels.

ENV 1993. 1992. *Eurocode 3: Design of Steel Structures.* Commission of the European Communities.

ENV 1998. 1994. *Eurocode 8: Structures in Seismic Regions.* Commission of the European Communities.

Fajfar P., Vidic T., Fischinger M. (1989). Seismic demand in medium- and long-period structures. *Earthquake Engineering and Structural Dynamics*, **18**, 1133-1144.

Fajfar P., Vidic T. (1994). Consistent inelastic design spectra: hysteretic and input energy. *Earthquake Engineering and Structural Dynamics*, **23**, 523-537.

Filippou, F.C., Popov, E.P., Bertero, V.V. (1983). *Effects of Bond deterioration on Hysteretic behaviour of reinforced concrete joints.* Report No. UCB/EERC-83/19, Earthquake Engineering Research Center, University of California, Berkeley.

Housner, G. W. 1959. Behaviour of structures during earthquakes. *J. Engrg. Mech. Div.*, ASCE, **85**(4), 109-129.

Krawinkler, H. & Zohrei, M. (1983). Cumulative Damage in Steel Structures subjected to Earthquake Ground Motions. *Journal on Computers and Structures,* **16**(1-4).

Krawinkler, H. Nassar, A. (1992). *Strength and ductility demands for SDoF and MDOF systems.* J. A. Blume EEC, Report **95**, Stanford University.

Kumar, S., Usami, T. (1996). An Evolutionary-Degrading Hysteretic Model for Thin-walled Steel Structures. *Engineering Structures,* **18** (7), 504-514.

Matsui, C. Sakai, J. (1992). Effect of collapse mode on ductility of steel frames, *Proc. of the Tenth World Conference on Earthquake Engineering,* Madrid.

Mazzolani, F.M., Piluso, V. (1996). *Theory and design of seismic resistant steel frames.* E & FN SPON, London.

Mazzolani, F.M. (1999) Reliability of moment resistant connections of steel building frames in seismic areas: the first year of activity of the RECOS project. *Proc. of 2nd European Conference on Steel Structures, Eurosteel 1999,* Praha, Czech Republic. May 26-29.

Minami, T., Osawa, Y. (1988). Elastic-plastic response spectra for different hysteretic rules. *Earthquake Engineering and Structural Dynamics*, **16**, 555-568.

Murakami, M. Penzien, J. (1975). *Non-linear response spectra for probabilistic seismic design and damage assessment of reinforced concrete structures.* Report No. UCB/EERC-75-38, Earthquake Engrg. Res. Ctr., Univ. of California, Berkeley, Calif.

Nader M.N. & Astaneh A. (1991). Dynamic behaviour of flexible semirigid and rigid steel frames. *Journal of Constructional Steel Research* **18**, 179-192.

Newmark, N.M. Hall, W. J. (1969). Seismic design criteria for nuclear reactor facilities. *Proc. of the Fourth World Conf. on Earthquake Engrg.,* Chilean Association on Seismology and Earthquake Engrg., Santiago, Chile, **2**(B-4), 37-50.

Park, Y.-J. & Ang, A.H.-S. (1985) Mechanistic seismic damage model for reinforced concrete. *Journal of Structural Engineering,* ASCE, **111**(4), 722-739.

Richard, R.M. Abbott, B.J. (1975). Versatile Elastic-Plastic Stress-Strain Formula. *J. Engrg. Mech. Div., ASCE,* **101** (4), 511-515.

Riddell, R., Newmark, N. M. (1979). *Statistical analysis of the response of nonlinear systems subjected to earthquakes.* Struct. Res. Series No. 468, Dept. of Civ. Engrg., Univ. of Illinois, Urbana, Ill.

Veletsos, A. S. (1969). *Response of ground-excited elastoplastic systems.* Report No. 6, Rice Univ., Houston, Tex.

Vann, W.P, Thompson, L.E., Whalley, L.E., Ozier, L.D. (1973). Cyclic behaviour of rolled steel members. *Proc. 5th World Conf. Earthquake Engineering,* **1**, 1187-1193, International Association for Earthquake Engineering, Rome, Italy.

Vidic, T. Fajfar, P. Fishinger M. (1994). Consistent inelastic design spectra: strength and displacement. *Earthquake Engineering and Structural Dynamics*, **23**, 507-521.

7.3

INFLUENCE OF THE STRUCTURAL TYPOLOGY ON THE SEISMIC PERFORMANCE OF STEEL FRAMED BUILDINGS

Gianfranco De Matteis, Raffaele Landolfo, Dan Dubina, Aurel Stratan

INTRODUCTION

Moment-resisting frames are more and more used in the field of steel structure buildings as a suitable and convenient solution resisting to both vertical and horizontal actions (Mazzolani & Piluso, 1996). They may be preferred to alternative structural schemes where pin-jointed steel frames are stabilised against lateral forces by means of purposely designed bracing systems (usually, steel diagonal bracing, concrete core and concrete walls). Framed structures allow, in fact, the optimisation of architectural space, also providing higher dissipative features in case of earthquake loading. Middlemost these two solutions, dual structures, where horizontal loading are resisted in part by moment resisting frames and in part by bracing systems acting in the same plane, are being practised.

In order to optimise lateral stiffness and system energy dissipation capability, traditional space moment-resisting frames would present all the beam-to-column connections moment-resisting and full-strength type. In fact, this solution should provide the greatest number of dissipative zones which, according to the strong column - weak beam design philosophy, have to be located at the beam ends (ENV 1998, 1994). On the other side, it is common in nowadays practice to design steel moment resisting frames as perimeter frames and/or dual rigid-pinned frames. Aiming at optimising the structural configuration, building layouts where some frames (interior and/or perimeter) have a reduced number of bays with moment-resisting connections are adopted.

Reason for using these alternative solutions is to reduce the number of three and four-way rigid connections in space moment-resisting frames, which are quite expensive and present a questionable mechanical performance, unless particular precautions in details are taken. Therefore, it results to be preferable to eliminate weak-axis moment connections. Consequently, the number of moment-resisting connections is reduced as much as possible, resulting in structures with low redundancy. This structural solution is frequently used as seismic-resistant system especially in Southern California. However, performance of steel MRFs during Northridge earthquake was much lower than the implicitly assumed for this category of systems. All the causes of this unexpected poor behaviour are difficult to be established, but connections are strongly incriminated. In fact, the reduced number of moment connections in perimeter moment resisting frames, due to reduced redundancy, provides higher ductility demands to rigid beam-to-column connections, resulting in their failure.

A possible way to mitigate this problem is to use semirigid and partial resistant connections instead of complete pinned ones. The expected benefits of these dual structural schemes are the increased redundancy with respect to rigid/pinned solutions, as well the possibility to control damage distribution throughout the structure. This corresponds to an alternative category of dual system, where, for each frame, duality must be intended in terms of joint mechanical properties. This structural configuration has became almost popular in United States and Japan (Shen, 1996, Kishi *et al.*, 1996). Another way would be the employment of semirigid joints everywhere in the structure. But, in this case, one of the most important concerns is the difficulty to yield the serviceability conditions imposed by technical codes. In fact, two main effects can be identified: the reduction of lateral stiffness, which reduces the performance of the structure under horizontal loading, and the increase of the period due to the different soil-to-structure interaction (Nader and Astaneh, 1992; Astaneh, 1994). The latter effect is beneficial due to decreasing earthquake equivalent loading on the structure, but, generally, the former one is dominant.

Eventually, the possibility of combining economy and seismic performance through the rational distribution of joints within the frame with different mechanical properties constitutes an alternative design philosophy, which is attractive but not yet completely established. In fact, the variation of joint mechanical properties throughout building frames could be intended as an additional source of structural irregularity. Therefore, particular attention must be paid in the evaluation of the actual seismic performance of the whole structure. Hence, preventively, dynamic non-linear time-history analyses should be performed aiming at checking the above concerns and at establishing proper design methodologies.

STUDY PLANNING

The influence of the structural typology on the seismic response of moment-resisting frames has been examined through a number of dynamic inelastic analyses carried out by means of the computer code DRAIN 2DX (Prakash *et al.*, 1993). Both space moment resistant frames and perimeter moment resistant frames are considered. In both cases, parametrical analyses are concerning with moment resistant frames where members are assembled with different combinations of rigid, semi-rigid and pinned joints. Besides, in every case, several joint rigidity and resistance levels are accounted for. In order to account for the influence of accelerogram typology, several historical earthquakes have been selected. Results, summarised in terms of ductility demand as a function of the peak ground acceleration, allow the effect of the joint semi-rigidity on the global behaviour of the structure to be emphasised. Finally, the variation of the q-factor as well as the distribution of the structural damage within the structure for all examined schemes are outlined.

The current study is subdivided in two Sections. (1) The first one is dealing with a parametrical study, where a typical symmetric spatial frame building is considered. Frame members are designed by considering rigid full-strength beam-to column joints, by following a conventional procedure, which is only partially in accordance with EC8 provisions. Then, joints with different mechanical characteristics are introduced in the same frame, and the effect of dual frame structures, which consists of rigid external bays and pinned or semi-rigid internal ones, is analysed. As a comparison solution with all semi-rigid joint is investigated. In both cases several collapse criteria are considered. (2) The second Section refers to a study case, it being focused on a more realistic frame building configuration that is designed completely according to EC8. The effect due to joint stiffness is accounted for in designing members. Contrary to frame analysed in the first Section, dual structures are constituted of rigid internal bays and pinned or semi-rigid connections at external columns. In this case, pushover analyses are performed as well. They provide useful information for a better understanding of the differences in performance among analysed structural typologies.

THE PARAMETRIC STUDY

Examined Cases

In the first parametrical study, the symmetric 3-bay 5-storey building designed by Mazzolani & Piluso (1997) has been considered. Thus, the response of a single frame (Figure 7.3.1) has been evaluated by means of inelastic time-history analyses. Member sizes, which are have been carried out according to EC8 (ENV 1998, 1994), by considering stiff soil conditions, a peak ground acceleration equal to a_d=0.35g and a design behavioural factor q_d = 6. Member hierarchy criterion has been therefore applied. As far as the serviceability conditions are concerned, the design value of the inter-storey drift has been limited to h/250 (where h is the inter-storey height). Despite this less restrictive limitation, the design has been governed by serviceability conditions rather than by member strength requirements.

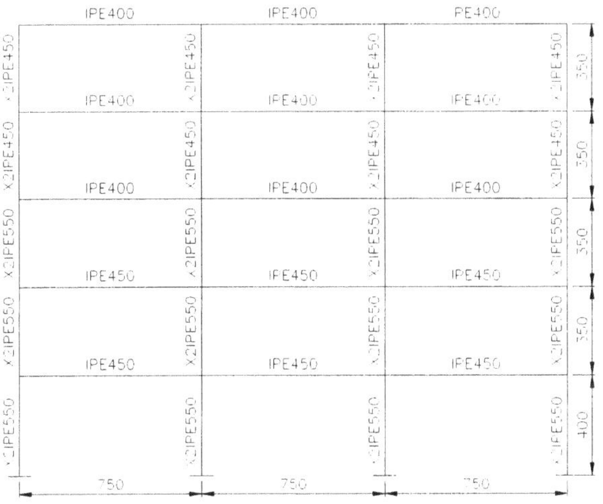

Figure 7.3.1: The analysed frame

With reference to the above frame structure, in order to investigate the effect of joint behaviour on the global structural response, several behavioural parameters for beam-to-column connections have been considered, by using both semirigid and pinned joints as alternative to rigid joints. As a first step, semirigid joints have been located in the middle bay only, obtaining a dual joint structural configuration (Figure 7.3.2*a*). In the following, these configurations are labelled with the letter *'S'*. For the sake of comparison, configuration where central bay is characterised by pinned joints *'P'* has been considered too. In the second step semirigid joints are present everywhere in the structure, this corresponding to a homogeneous joint structural configuration (Figure 7.3.2*b*). In the following, these configurations are labelled as *'T'*. For the sake of comparison, the configuration where all joint are fully rigid and total strength *'R'* has been considered too. Therefore, the seismic performance of four structural schemes have been determined and compared each other.

In all cases, several joint rigidity and resistance levels have been considered. With reference to the stiffness, joint rigidity values (k_j) equal to k_{lim}, 0.8 k_{lim} and 0.6 k_{lim} have been assumed, where k_{lim} is the joint rigidity value corresponding to the boundary between rigid and semirigid joint behaviour as suggested by the above code for unbraced frames (ENV1993, 1994). In the following, the semirigidity level is indicated by the first number of contracted names. Therefore, S1x and T1x refer to k_j = 1 k_{lim}, S2x and T2x refer to k_j = 0.8 k_{lim} and S3x and T3x refer to k_j = 0.6 k_{lim}.

At the same way, several moment capacity levels (M_j) have been considered, by analysing both partial strength and full strength joints. Values have been referred to moment capacity of the connected beam (M_b). M_j / M_b ratios equal to 1.2, 1, 0.8 and 0.6 have been considered. In contracted names, moment capacity level is associated to the second number, by using S/T*x*1, S/T*x*2, S/T*x*3, S/T*x*4, in a decreasing order for resistance level. All examined cases in terms of both joints stiffness and moment capacity level are summarised in TABLE 7.3.1.

TABLE 7.3.1
EXAMINED CASES

STRUCTURAL CONFIGURATION	**Joint Stiffness (k_j/ k_{lim})**		**Joint Moment Capacity (M_j/ M_b)**	
	Semirigid (S xx)	*Pinned (P)*	*Semirigid (Sxx)*	*Pinned (P)*
Dual joint configuration	1 (S1*x*), 0.8 (S2*x*) 0.6 (S3*x*)	0	1.2 (S*x*1) 1 (S*x*2), 0.8 (S*x*3) 0.6 (S*x*4)	
	Semirigid (Txx)	*Rigid (R)*	*Semirigid (T)*	*Rigid (R)*
Homogeneous joint configuration	1 (T1*x*), 0.8 (T2*x*) 0.6 (T3*x*)	∞	1.2 (T*x*1) 1 (T*x*2) 0.8 (T*x*3) 0.6 (T*x*4)	∞

(Symbols in brackets indicate nomenclature used for contracted names)

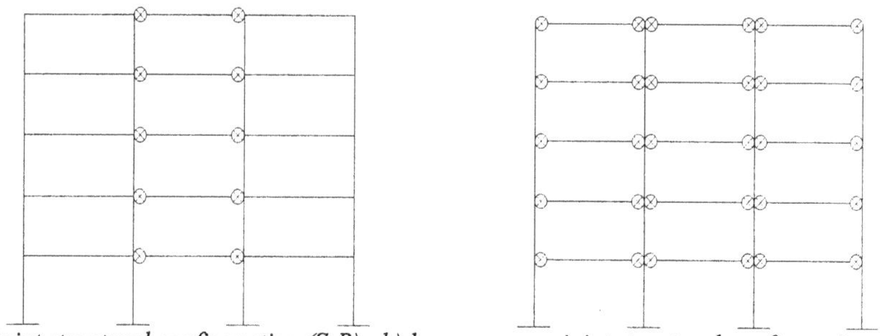

a) dual joint structural configuration (S-P) *b) homogeneous joint structural configurations (T-R)*
Figure 7.3.2: Analysed frame configurations

Methodology of analysis

General

The performed parametrical study has been carried out by means of 6240 time-history analyses. The structural performance of 26 structural typologies under 6 earthquake records has been examined, by considering progressively increasing peak-ground acceleration values, starting from 0.05*g* to 2.0*g*. The value 0.05*g* has been chosen because of, according to the design procedure, the first yield of the structure theoretically should appear at $0.35/q_d$=0.058*g*, so probably at this step the structure should behave elastically. The value of 2.0*g* has been chosen after a preliminary analysis, by checking that at this earthquake level the structure has reached a conventional collapsing state, as stated below. Integration step equal to 0.01s has been adopted, after checking that unbalanced forces produce numerical errors less than 10% in term of the plastic rotation demand to member sections.

Selection of records

In order to take allowance for the seismic input, six records corresponding to different historical earthquakes have been considered. Accelerograms have been selected in such a way the corresponding elastic spectra was similar to the one assumed by Eurocode 8 (ENV 1998, 1994) for subsoil class A. In order to eliminate the differences in terms of peak ground acceleration, all the records have been scaled at the same acceleration value. In Figure 7.3.3, in case of viscous damping coefficient equal to 5%, normalised elastic response spectra of the examined earthquakes are depicted together with the design one (EC8-class A). The figure allows the shape and the frequency content of analysed spectra to be compared each other. It is clearly shown as the average shape of analysed earthquakes is very similar to the one assumed by the code, especially for vibration periods higher than 0.5s.

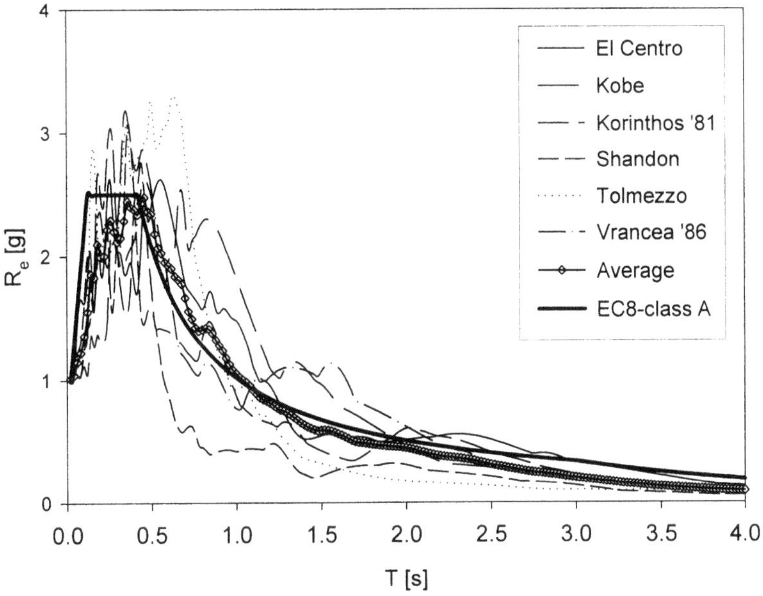

Figure 7.3.3: Elastic spectra of examined earthquakes

Dispersion factor of spectral amplifications of each record respect to the design one, has been evaluated in the period range 1.4s-4.0s, which should be representative of the structural behaviour all over the seismic event, being the fundamental period of vibration of all examined frame configurations ranging between 1.4s and 1.5s. Percentage values of standard deviation for examined records are reported in TABLE 7.3.2 together with the one corresponding to the average spectrum. Dispersion among earthquakes seems to be acceptable. In fact, standard deviation value is about 10% for the average. Nevertheless, one should mention that for some earthquakes, namely Shandon and Tolmezzo, it is higher than 20%.

TABLE 7.3.2

STANDARD DEVIATION (%) OF ELASTIC RESPONSE SPECTRA

El Centro	*Kobe*	*Korinthos*	*Shandon*	*Tolmezzo*	*Vrancea*	*Average*
6.98	13.60	12.91	23.11	25.98	17.13	10.19

Selection of collapse criteria

In order to compare the ultimate structural performance of analysed structures, collapse criteria should be preventively defined allowing ultimate peak ground acceleration values of considered records to be determined. A local criterion based on the comparison between plastic rotation supply and plastic rotation demand in either the most engaged member section or beam-to-column connection has been applied. In order to account for the structural damage due to repeated actions, two different simplified approaches have been adopted: the first one is simply based upon the maximum plastic deformation demand, the second one is an energy-type approach concerned with the cumulated plastic deformation computed along the time-history analysis. Conventionally and conservatively, in both cases, in order to asses the collapsing state, plastic rotation demand may be then compared with plastic rotation supply deduced on the basis of either experimental results or analytical methods.

For each structure and for each record, acceleration multiplier versus maximum plastic rotation curves have been therefore plotted. They have been referred to the most engaged plastic hinge among those formed in beam sections, column sections and connections, by considering both maximum plastic excursion and cumulated plastic excursion. These curves can be simply used for determining the maximum peak ground acceleration that the structure is able to withstand when the available plastic rotation capacity for beam sections, column sections is fixed. Consequently, actual q-factor of the structure may be promptly defined as the ratio between the acceleration level corresponding to the attainment of conventional collapse (a_{ult}) and the yielding acceleration corresponding to the appearance of the first plastic hinge in the structure (a_y).

In the current study, in order to determine the ductility supply, the stable part of available plastic rotation capacity (R_{st}) for member sections have been assessed as the average of values suggested by the theoretical method proposed by Mazzolani & Piluso (1996) and the empirical one proposed by Mitani & Makino (1980). Besides, in order to account for the unstable part in the evaluation of total of rotation capacity (R), the simplified method proposed by Gioncu & Pectu (1997) has been considered too. Corresponding ultimate plastic rotation values θ_{pl} and $\theta_{pl,st}$ are summarised in TABLE 7.3.3, where rotation supply has been computed for each member section. Then, as final values of available plastic rotations, minimum values, which correspond to section IPE 450 for beam and to section 2xIPE 550 for columns, have been considered.

TABLE 7.3.3
MEMBER PLASTIC ROTATION SUPPLY (RAD)

Member Section	**Stable part ($\theta_{pl,st}$)**			**Total (θ_{pl})**
	Mazzolani-Piluso method	*Mitani-Makino method*	*Average*	*Gioncu's method*
IPE 450	0.0610	0.0513	0.0562	0.0837
IPE 400	0.0663	0.0529	0.0596	0.0850
2XIPE 550	0.0379	0.0353	0.0366	0.0415
2XIPE 450	0.0618	0.0372	0.0495	0.0747

The problem of determining available plastic rotation capacity for beam-to-column joints is more complicated due to the lack of both satisfactory theoretical approaches and comprehensive experimental results. Usually, plastic rotation supply ranging from 0.02 to 0.03 rad are considered to be realistic values (Tsai & Popov, 1988) and they are generally assumed for performing seismic analyses by checking that they are not exceeded by maximum plastic excursions experienced all over the deformation history. On the other side, further studies suggest to relate the connection rotation capacity to the beam depth, also providing in some cases higher values than the ones above mentioned (Roeder

& Fountch, 1996). Anyway, once this value has been established, the collapse state of plastic hinge involving the connection may be assessed through plotted θ_{pl} versus PGA curves.

Finally, in order to account for a global parameter also in estimating the structure capability to withstand increasing intensity earthquakes, interstorey drift criterion has been applied as well, by assuming a ultimate value equal to $0.03h$, where h is the interstorey height.

The obtained results

The influence of joint resistance

The performed numerical analysis has produced thousands of results, allowing all selected parameters to be investigated in terms of influence on the seismic structural performance of the building. Herein, only the main results are shown and discussed, while a more comprehensive examination is reported by Fulop (1998).

In Figure 7.4.4, the influence of joint strength is shown in case of dual joint structural configuration and rigid beam-to-column connections (k_j/k_{lim}=1). Plotted curves represent the average of rotation demand for all examined earthquakes. It can be seen the increasing influence of the connection as the moment capacity decreases, while no significant modification of the plastic rotation requirement in plastic hinges on the beams and columns can be observed. Obtained results show that generally beam sections present the highest plastic rotations, unless partial strength connection with M_j=0.6M_b are considered. This is due to the effect of dual joint configuration, which induce a concentration of plastic rotation within members of outer bays rather than in the central one. Really speaking, this does not mean that

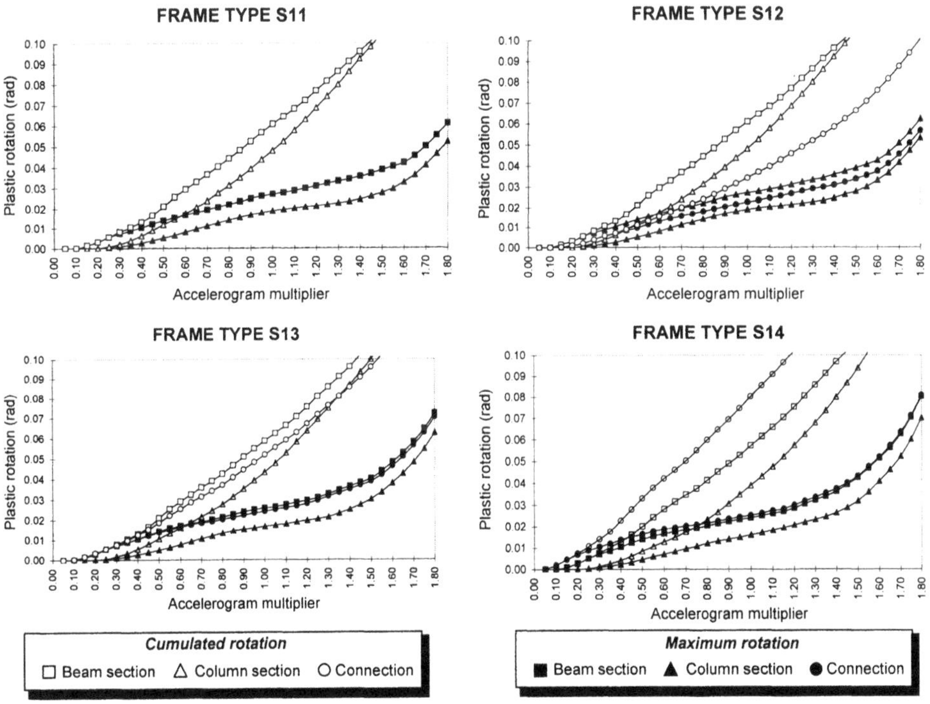

Figure 7.3.4: The influence of joint resistance for dual joint structural configurations

the collapse of the frame is always due to the beam ('beam controlled collapse'), because of the different ductility supply of beam sections, column sections and connections. For instance, collapse of frame types S11 and S12 seems to be ascribed to the column sections rather than beams, being rotation requirements slightly higher for beams than for columns and, concurrently, rotation capacity supply substantially lower for the latter. Similarly, in case of frame type S13, the collapse is ruled by connections if they have a limited rotation capacity respect to beam and column sections. Structural performances are clearly worst in case of homogeneous joint configuration, especially for partial strength connections. In Figure 7.3.5, plots for $M_j=1M_b$ and $M_j=0.8M_b$ are represented. It is clear that, in both cases, connections overrule the collapse of the structure.

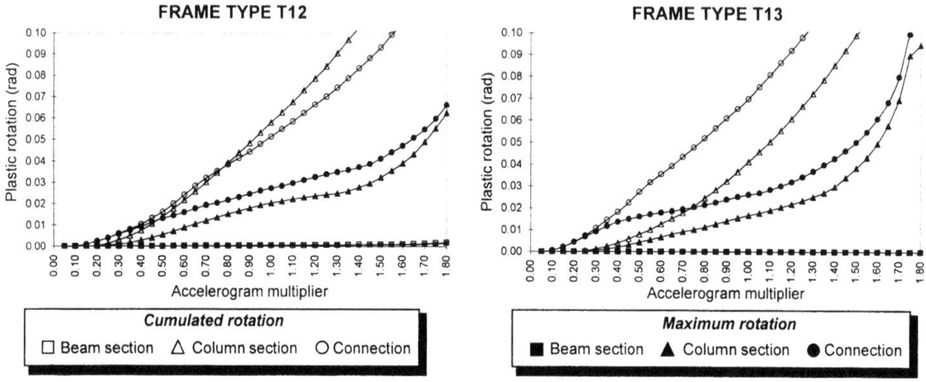

Figure 7.3.5: The influence of joint resistance for homogeneous joint structural configurations

In Figure 7.3.6, in case of dual joint configurations, the required rotation capacity for connections and column sections providing a collapse ruled by beam sections is plotted. It allows the minimum connection and column rotation capacity supply in order to have collapse due to the rotation capacity limitation of the beams. The above figure is referred to the case of structures with joint stiffness $k_j = k_{lim}$ and joint moment capacity $M_j=1M_b$, $0.8M_b$, and $0.6M_b$, respectively, while beam rotation capacity has been limited to the stable part. It is possible to verify that only in case of partial strength connections, namely $M_j=0.8M_b$ and $0.6M_b$, required plastic rotation for connections are higher than 0.03 rad. Besides, generally, connection rotation capacity demand decreases as far as the joint stiffness decreases, emphasising that semirigid connections could be particularly effective in dual joint configurations. Finally, it is to be observed the influence of the seismic input on the collapse mechanism, being the required plastic rotation in columns very high in case of Korinthos, Shandon and Vrancea records.

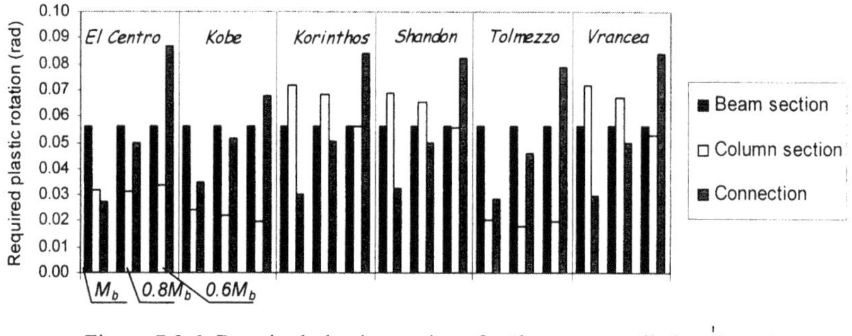

Figure 7.3.6: Required plastic rotations for 'beam controlled collapse'

The comparison between dual and homogeneous joint structural configurations

In Figure 7.3.7, the structural performance is synthesised for all examined structures and earthquakes. In fact, for each adopted collapse criterion (exhausting of available stable rotation capacity, exhausting of available total rotation capacity, interstorey drift higher than 3%), the corresponding values of ultimate accelerogram multiplier are reported. Conventionally, for all the connections, available plastic rotation capacity values have been fixed equal to 0.045 rad. Depicted graphs allows the influence of all main behavioural parameters to be easily recognised with respect to both the accelerogram type and the collapse criterion.

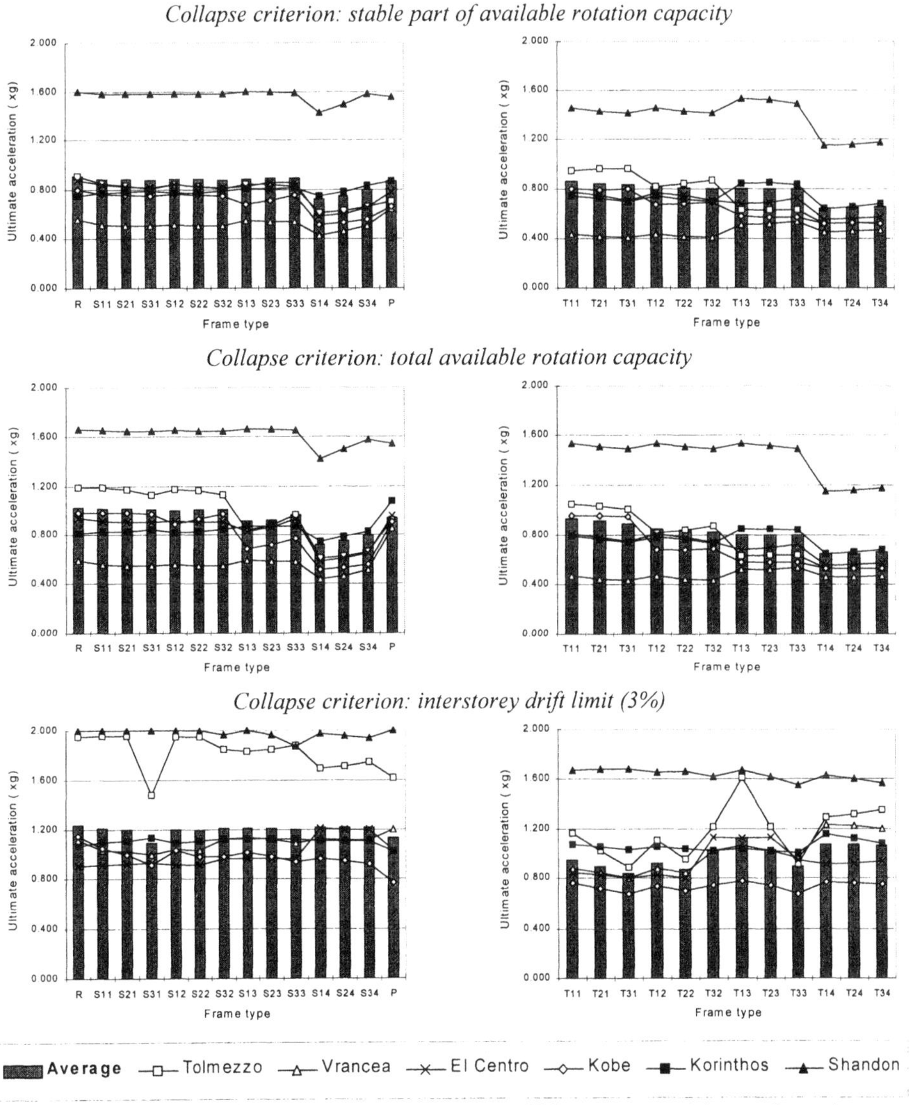

a) dual joint configurations *b) homogeneous joint configurations*

Figure 7.3.7: Intensity earthquake structural resistance of examined frames

Obtained results seem to be quite homogeneous for all examined collapse criteria, leading to slightly increasing resistance values going from stable rotation capacity to interstorey drift limitation. On the contrary, record typology has a strong influence on the evaluation of structural performance, resulting a ratio up to three between ultimate acceleration of best behaving earthquake (Shandon) and the worst (Vrancea). Anyway, in all cases, analysed structures are able to withstand all examined earthquakes up to peak ground accelerations much higher than the one assumed for building design (a_d=0.35 g).

The comparison between dual and homogeneous configuration shows that generally structural performance is better in the former case. This is particularly evident if reference to interstorey drift limitation criterion is made. Corresponding graphs emphasise lateral stiffness reduction effect due to semirigid joints. Besides, obtained results confirm the important rule of connection moment capacity, especially in case of homogeneous joint configuration, generally providing undermining of seismic response for weaker connections. As a consequence, in case of partial strength connections, in homogeneous joint configurations, connection ductility becomes the key parameter controlling the collapse of the structure, while in dual joint configurations beam sections of outer bays firstly exhaust their available rotation capacity. Truly, for a connection plastic rotation capacity equal to 0.045 rad, there is no significant drop in the performance for partial strength joints, if reference stable part of rotation capacity is made. This is because the plastic rotation capacity of the joint is comparable with the one of the beam, 0.045 rad against 0.056 rad. If total rotation capacity is considered, the differences are much more significant due to the bigger difference between the two values, 0.045 rad and 0.084 rad. Nevertheless, results are sometimes contradictory, showing an increasing of structural performance in case of moment capacity M_j=0.8M_b, but this outcome is strongly affected by the individual earthquake record.

On the contrary, within a structural typology, for both dual and homogeneous joint configurations, the reduction of connection stiffness does not negatively affect frame structural performance. Conversely, dual joint configurations show a slightly increase of ultimate accelerogram multiplier as connection rigidity decreases, but result dispersion due to record type in not negligible. This is due to two main beneficial effects: firstly, the shifting of the structure toward higher vibration periods; secondly, the moving of damage localisation from the central bay to the outer ones, where connection are rigid and full strength and the collapse is ruled by beam element sections. The first of this effect is also significant in case of homogeneous joint configuration when interstorey drift criterion is adopted.

In Figure 7.3.8, in order to examine the different localisation of damage within frame members and connections, for both dual and homogeneous joint configurations, global damage index distribution is depicted. It is referred to Vrancea earthquake at peak ground acceleration equal to 0.8g. For each element (member sections and connections), local damage index $I_{D,L}$ has been evaluated as the ratio between the attained plastic rotation θ_p and the stable part of plastic rotation supply $\theta_{pl,st}$. Then, according to Park *et al.* (1985), a global damage index $I_{D,G}$ within the whole structure and in each bay $I_{D,G,i}$ has been computed as the weighted average of the local damage indexes, by using the following relationship type:

$$I_{D,G} = \frac{\sum I_{D,L_i}^2}{\sum I_{D,L_i}} \tag{7.3.1}$$

It can be clearly noticed as in homogeneous joint configurations global damage indexes strongly increase as far as connection resistance decreases, while they slightly decrease as far as the connection stiffness decreases. With reference to dual joint configurations, the connection behaviour influences damage distribution within the structure only in case of very poor connections, i.e M_j=0.6M_b. In fact, in all the other cases, damage is concentrated in outer bays only, where connections are rigid and full strength. Consequently, global damage index is substantially less than the corresponding one in homogeneous joint configurations.

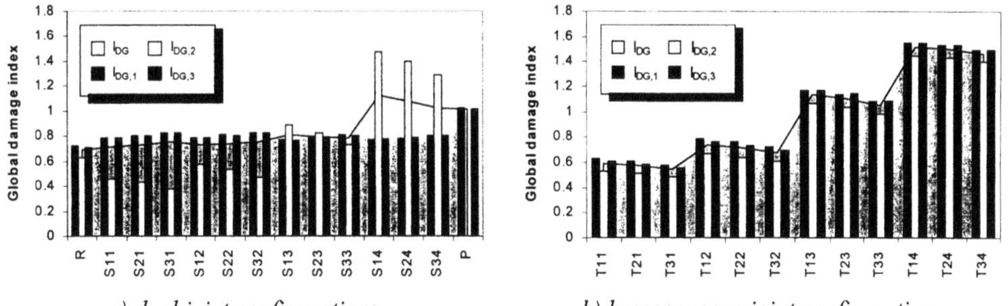

a) dual joint configurations *b) homogeneous joint configurations*

Figure 7.3.8: Damage distribution among bays

The effect of structural overstrength

Finally, a meaningful aspect to be mentioned is the effect due to overstrength. In Figure 7.3.9, q-factor values determined through the ratio between yielding acceleration and ultimate accelerogram multiplier evaluated with reference to stable part of plastic rotation supply are depicted. Over and above the influence of the record, it can be observed that higher q-factor values correspond to frames with partial strength connections, even if they provide a worst structural response in terms of maximum intensity earthquake that the structure is able to withstand (see Figure 7.3.7).

This contradictory result evidences that q factor, as it is usually defined, is not enough to describe by itself the seismic performance of the structure, another fundamental parameter to be considering being the structural overstrength due to design Ω_d. It can be defined as the ratio between the actual yielding acceleration a_y and the theoretical one $a_{d,y} = a_d / q_d = 0.058g$. For the sake of example, obtained values for dual joint structures are presented in Figure 7.3.10*a*, where an important overstrength can be observed in each frame typology. In particular, results are remarkably influenced by the examined record, varying from 6.7 to 1.8 in case of rigid full-strength joints.

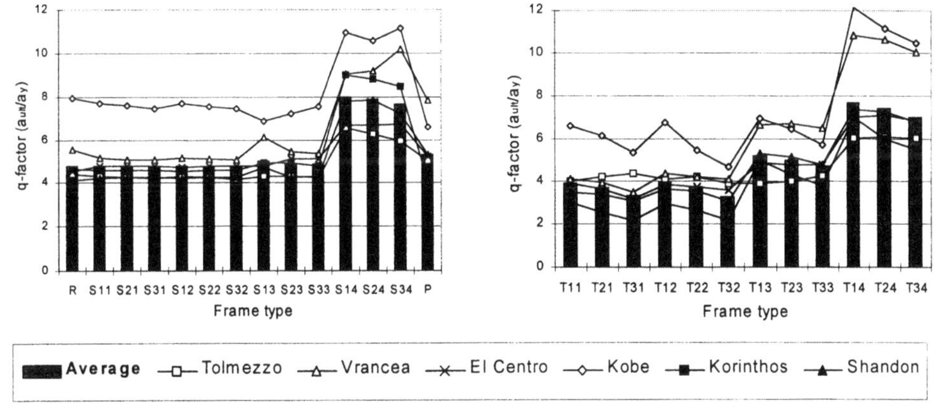

a) dual joint configurations *b) homogeneous joint configurations*

Figure 7.3.9: q-factor values for examined frames

Structural overstrength is basically caused by two effects: the first is due to differences between the design spectra and the actual one corresponding to the examined record; the second one is an intrinsic factor of the structure, it being imputable to the design procedure. In particular, this intrinsic factor is remarkable for the analysed structure, due to the following aspects: (1) the effect of seismic forces on the structure have been evaluated by means of a simplified analysis, while numerical time-history analysis account for actual mass modal participation; (2) in design evaluation of base shear force, reference to the approximate fundamental vibration period suggested by EC8-Part 1.2-Appendix C has been made. Since the effect due to earthquake response spectrum may be easily calculated as ratio of spectral ordinate at the fundamental period of the structure for the considered record and the design value, in Figure 7.3.10*b*, overstrength rate due structural design is presented. Even if results are much less scattered, one can see that intrinsic overstrength of the structure due to the design remains remarkable and tends to 1 only in case of very partial strength connections. Similar conclusions may be drawn up for homogeneous joint structural configurations.

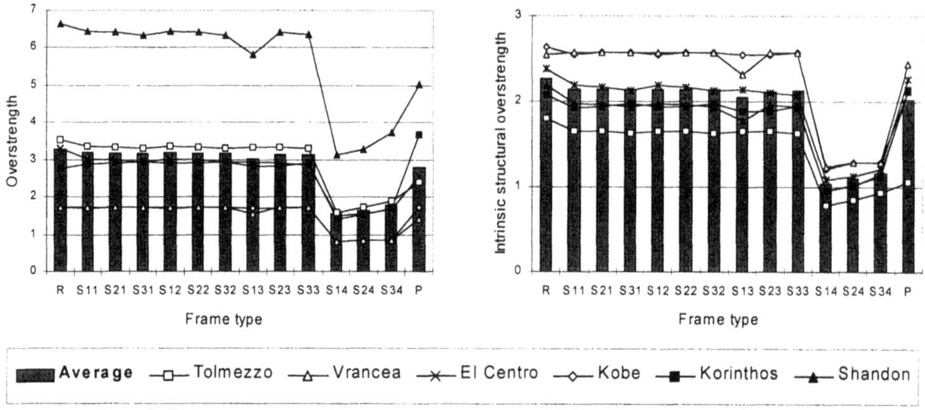

a) total design overstrength *b) intrinsic structural overstrength*

Figure 7.3.10: Overstrength for dual joint structural configurations

Actually, once the overstrength has been computed, the actual structural seismic performance may be quantified through the coefficient $q' = \Omega_d \cdot a_{ult} / a_y$, which may be intended as a modified *q*-factor value. Corresponding values are reported in Figure 7.3.11. It can be observed that now partial strength connection structures provide the worst performance, confirming that this factor is able to correctly quantify the ultimate resistance of the structure. It is also important to observe that results show that analysed structures are able to withstand earthquake whose peak ground acceleration is up to 25 times higher than the one that is expected to produce the first yielding. Only in case of the most damaging earthquake (Vrancea), such modified *q*-factor is similar to the one which has been hypothesised in design (q_d=6), being in this case total design overstrength value almost equal to one. This result emphasises the high structural capability of moment-resisting frames independent of the joint behavioural parameters, provided that member and connections guarantee significant local ductility levels.

Main outcomes

The wide parametrical analysis presented in this Section allows important remarks on the influence of connection behaviour on the seismic response of moment-resisting frames to be drawn. (1) Generally, when semirigid or partial strength connections are employed, dual joint structural configurations

behave better than analogous homogeneous joint configurations. (2) The effect of joint semi-rigidity on the global structural behaviour is not strongly influencing. (3) Connection moment capacity is able to remarkably undermine the seismic performance of the structure in homogeneous joint configurations.
In such a case, joint ductility is the key parameter. It must be consistent with plastic rotation demand, which is a function of the structure itself as well as of the seismic records. On the contrary, dual joint configurations, allowing that the major damage is concentrated in frame bays where rigid, full strength connections are employed, may lead to a limited penalisation of the frame behaviour due to the employment of poor connections. (4) The structural overstrength for usual frame structures designed according to recent code provisions is considerable and should not be disregarded in analysing results coming from numerical analyses. Also, it covers the essence of many other effects, which should be investigated on properly designed structures.

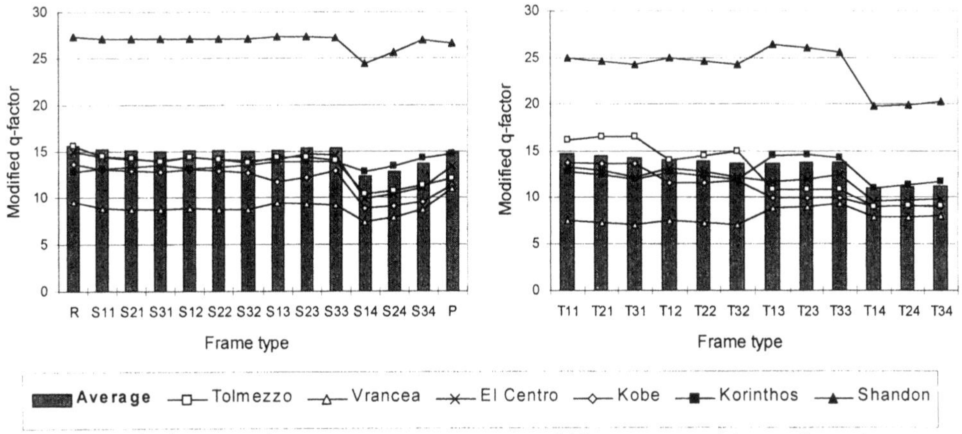

a) dual joint configurations *b) homogeneous joint configurations*

Figure 7.3.11: Modified q-factor values for examined frames

CASE STUDY

The structural scheme

Aiming at checking the above conclusion with reference to a more realistic frame structure, a different 3-bay x 3-bay 5-storey steel building structure is analysed in the current Section (Figure 7.3.12). The structure is supposed to be located in a high seismicity zone, with stiff subsoil conditions. The loads considered in the design are: dead floor load: G_{FLOOR}=4.75 kN/m^2, dead load for exterior surface: G_{CL}=1.70 kN/m^2, live load: Q=3.0 kN/m^2. Seismic specific parameters for design are:

- PGA=0.35g
- Behaviour factor q=6
- Subsoil class - A
- Interstorey drift limit d_{lim}=0.006 h

There are several possibilities that may be followed when designing this structure. The two "classical" schemes would be with all frames fully rigid in both directions, and the dual rigid-pinned scheme that would eliminate weak-axis moment connections. The second structural scheme is the one that will probably be chosen by most of designers due to its simplicity (3 and 4-way connections are difficult to design, also, their performance is questionable). At the same time it would reduce the number of expensive rigid connections. Due to fewer moment connections and reduced redundancy, higher demands will be imposed on remaining rigid beam-to-column connections in this case. A possible way to override this problem would be to use partially restraint weak-axis connections instead of pinned ones.

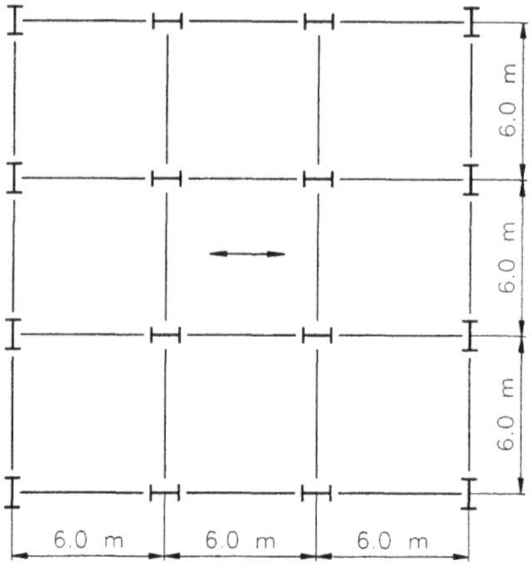

Figure 7.3.12: Horizontal layout of the building

Several structural variants have been accounted for. The two limiting cases herein considered are the fully rigid frame (Figure 7.3.13*a*), and the rigid/pinned dual structure, with the pinned joints in the external bay, corresponding to weak-axis connections (Figure 7.3.13*b*). In order to investigate the influence of semirigid (SR) – partial resistant (PR) joints, two additional structural schemes are considered. The first one corresponds to the dual rigid/(SR-PR) configuration, with rigid joints for internal columns, and SR-PR joints for the external weak-axis connections (Figure 7.3.13*c*). The second structural scheme corresponds to a fully SR/PR joint configuration (Figure 7.3.13*d*). For all the above structural configurations, stiffness and moment capacity of joints used in the parametric study are synthesised shown in TABLE 7.3.4.

TABLE 7.3.4
ANALYSED SCHEME SOLUTIONS

Reference name	**Stiffness (k_j/k_{lim})**		**Moment capacity (M_j/M_b)**	
	Interior conn.	*Exterior conn.*	*Interior conn.*	*Exterior conn.*
RIG	Infinity	Infinity	1.2	1.2
D11	Infinity	0.6	1.2	1.0
D12	Infinity	0.6	1.2	0.8
D13	Infinity	0.6	1.2	0.6
D24	Infinity	0.25	1.2	0.4
F11	0.6	0.6	1.0	1.0
F12	0.6	0.6	0.8	0.8
F13	0.6	0.6	0.6	0.6
DUP	Infinity	0	1.2	-

Two types of analysis methods were employed: a static pushover analysis and an inelastic time-history one. The first one is simpler and is believed to provide reliable results for structures oscillating predominantly in the first mode of vibration. The time history analysis is greatly influenced by the modelling parameters and also by the earthquake records used.

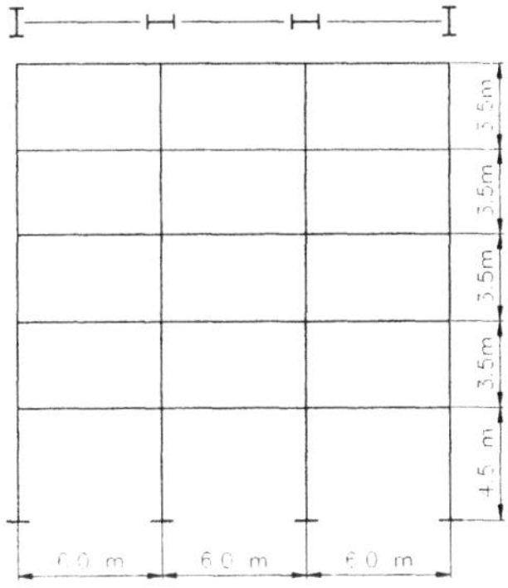

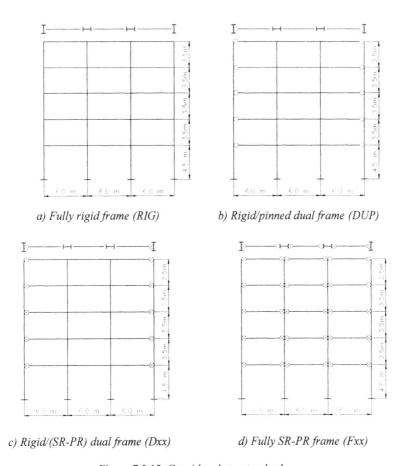

a) Fully rigid frame (RIG)

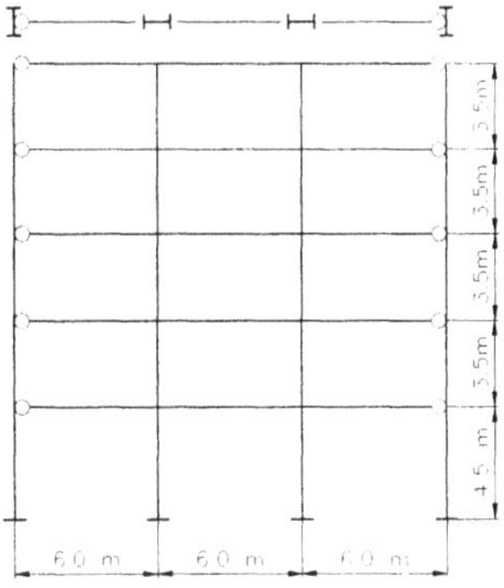

b) Rigid/pinned dual frame (DUP)

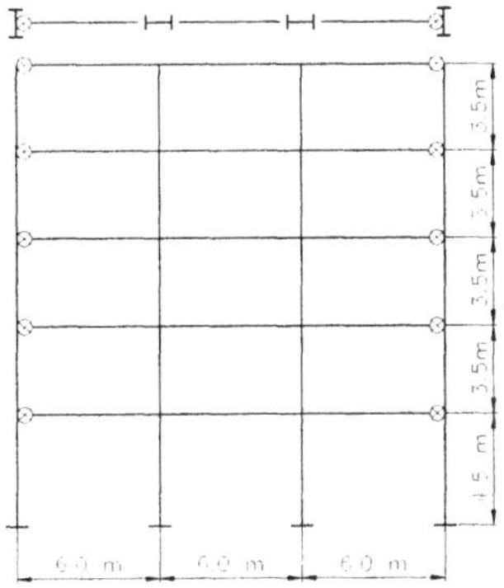

c) Rigid/(SR-PR) dual frame (Dxx)

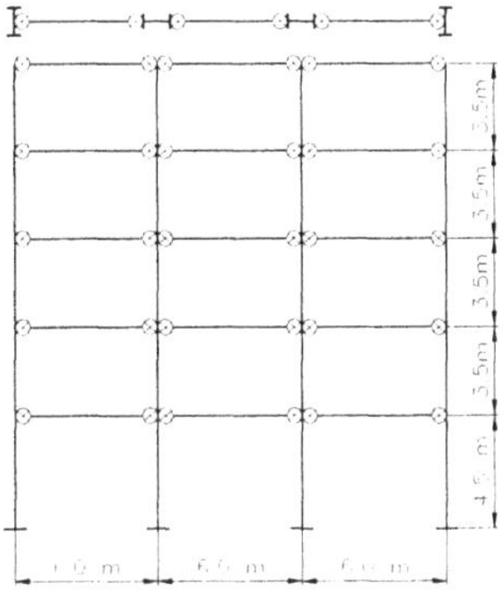

d) Fully SR-PR frame (Fxx)

Figure 7.3.13: Considered structural schemes

The adopted design procedure and obtained member sizes

It is recognised that the actual behaviour of the semirigid/partial resistant joints shall be accounted for when designing the structure. Therefore, an appropriate design procedure was followed for each structural scheme considered. The consequences on the resulted structures in relation with the adopted joint properties are given in the following. The member design were carried out according to Eurocode 8 (1994) and Eurocode 3 (1990). As it was expected, the design has been ruled by serviceability conditions (interstorey drift limitation) rather than by member strength requirements.

The fundamental period of vibration computed by modal analysis instead of the one given by empirical formulation of EC8 was used. The fundamental period of vibration, the member dimensions resulted from analyses, and the structural weight of steel members are presented in TABLE 7.3.5.

Consequences on the design of the use of rigid, semirigid, partially resistant joints, and pinned joints for the different structural configurations analysed can be summarised as follows:

- Increase of member dimensions for fully SR and dual rigid/pinned structural schemes

- Selected PR joint characteristics do not influence member dimensions
- Dual rigid/(SR-PR) frame configuration can be accommodated by member sizes of the fully rigid frame
- As shown in Figure 7.3.14 a somewhat increase in structural weight will result for fully SR and dual rigid/pinned frames (not higher than 15%).
- Increase of lateral stiffness of moment-resisting frames cannot be accomplished by increasing column section only, while keeping beam section constant.
- The decrease of design shear force due to the increase of the fundamental period of vibration in the case of SR joints is not sufficient for the drift control. As a result, lateral stiffness of the frame must be increased, resulting in greater beam sections. Due to this situation, because of the same global stiffness, all schemes yielded practically the same fundamental period of vibration.

TABLE 7.3.5

CHARACTERISTICS OF THE CONSIDERED SCHEMES

	T_1, s	Columns 1-3	Columns 4-5	Beams 1-3	Beams 4-5	weight/frame (to)
RIG	0.94	HEB550	HEB360	IPE500	IPE400	20.4
D1x	0.96	HEB550	HEB360	IPE500	IPE400	20.4
D24	0.98	HEB550	HEB400	IPE500	IPE400	20.8
F1x	0.94	HEB550	HEB360	IPE550	IPE450	21.6
DUP	0.94	HEB550	HEB400	IPE600	IPE500	23.3

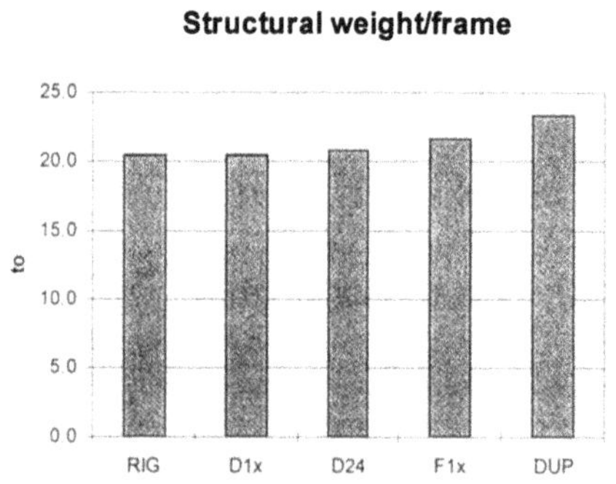

Figure 7.3.14: Structural weight of designed frames

The analysed earthquake records

Likewise to the above parametric study, earthquake records have been selected with reference to the their linear response spectrum in order to be similar with the design spectrum of Eurocode 8 for stiff subsoil conditions. Only four of the six selected records have been chosen. Records corresponding to earthquakes of Tolmezzo and Shandon have been disregarded, they being the ones providing the highest scatter as respect to others. Also in this case, peak ground acceleration of the records used has been scaled at the same value, in order to be consistent with the same seismic hazard as defined by the code.

Collapse criterion

As far as the collapse criterion is concerned, only the one concerned with 3% interstorey drift has been chosen in this case. This criterion is relatively simple and allows the uncertainties in predicting reliably the ductility and cyclic behaviour of connections under dynamic loading to be overcome. Also, it is to be considered that the main aim of the study is to compare the seismic performance of different structures. Hence, the primary characteristic of the collapse criterion to be adopted should be its stability in relation to the analysed structures rather than its effectiveness in matching the structural performance itself. Nevertheless, in order to assess the structural damage, absolute values of maximum and accumulated plastic rotations in members and connections are investigated as well.

The obtained results

Pushover analysis

The pushover analysis was conducted under a triangular distribution of lateral loads, keeping vertical ones constant. The lateral load is defined as the sum of horizontal seismic forces applied at each storey, unit lateral load being equal to unity. The pushover curve is obtained by performing a displacement-controlled analysis in terms of the roof displacement δ. It must be underlined that the pushover analysis provides only an approximation to the displacements and internal forces in the corresponding dynamic inelastic system. Moreover, due to second-order effects, the lateral load will not be equal to the base shear force determined as the sum of shear forces in the first storey columns.

Plots of the base shear and lateral load versus roof displacement are shown in Figure 7.3.15. Also, first yield and attainment of the interstorey drift limit are indicated. Slope of the descending branch of the lateral load versus top displacement curve is given by the parameter γ_S. It should be mentioned that this slope is strongly influenced by the modelling options, and especially by the strain hardening. In case of significant strain hardening the slope may result positive, i.e. effect of strain hardening may compensate second-order effects. Anyway, this is not the case here, since no strain hardening was considered in the analysis.

It has been shown (Mazzolani and Piluso, 1996) that a measure of the frame sensitivity to second-order effects may be given by the stability coefficient $\gamma = \gamma_s \cdot \delta_r$, where δ_r represents the roof displacement under the unit lateral force according to a linear elastic analysis (*P*-Δ effects not included). Other parameters that may be used to characterise the inelastic performance of a structure are:

- Ductility: $\mu = \dfrac{\delta_u}{\delta_y}$
- Ductility related reduction factor: $R_\mu = \dfrac{V_e}{V_y}$

The quantitative value of the ductility reduction factor strongly depends on the energy dissipation capacity of the structural system which is closely related to the ductility of the system. It should be mentioned that in a special case, when the displacements of the inelastic and the corresponding elastic systems are equal (δ_u=δ_e – this assumption is approximately valid in the medium- and long-period range of the spectrum), the ductility related reduction factor is equal to the ductility factor, therefore being $R_\mu = \mu$.

- Redundancy: $\rho = \dfrac{V_y}{V_1}$
- Total reduction factor: $R = R_\mu \cdot \Omega$
- Design overstrength: $\Omega_d = \dfrac{V_1}{V_d}$
- Behaviour factor: $q = R_\mu \cdot \rho$
- Total overstrength: $\Omega = \rho \cdot \Omega_d$

where:
δ_u – top sway displacement in the inelastic system at collapse
δ_y –yield displacement in the idealised bilinear base shear – top sway displacement relationship
V_d – design base shear force
V_1 – base shear force at first yield
V_y – base shear force at the yield state (at collapse)
V_e – base shear force in the indefinitely elastic system

A typical plot of base shear (V) versus roof displacement (δ) of a MDOF inelastic system, the bilinear idealisation, and the corresponding indefinitely elastic system are given in Figure 7.3.16. In the same figure, the basic terms previously defined are reported as well.

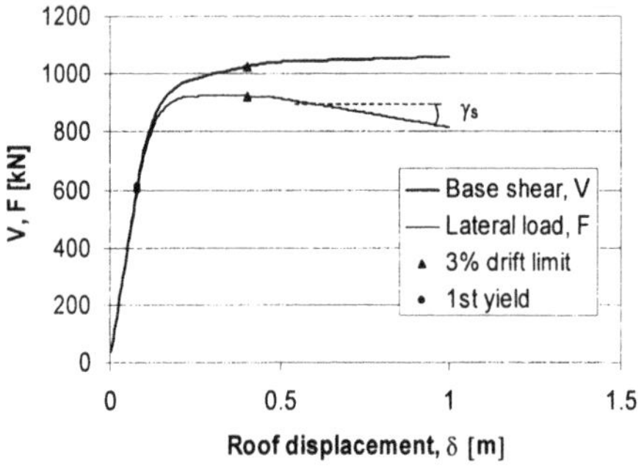

Figure 7.3.15: Base shear and lateral load versus roof displacement (RIG)

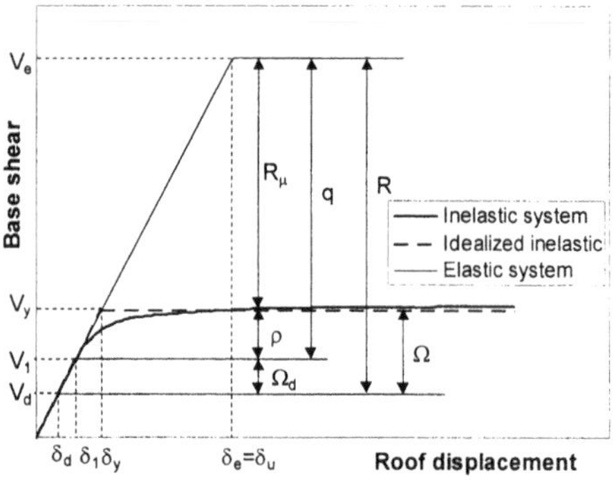

Figure 7.3.16: Typical base shear – roof displacement relationship

Figure 7.3.17 shows the base shear and lateral load versus top displacement relationships for some analysed frames. The base shear V is normalised to the seismically effective weight W, and the roof displacement δ is normalised to the structure height H. Earlier yielding, lower strength capacity and greater global ductility can be recognised for frames with PR connections. At the same time, the intermediate position of the dual (D13) frame and the steep softening branch of the F-δ curve in the case of dual-pinned (DUP) frame can be observed. Anyway, it can be noticed that the formation of the collapse mechanism occurs generally after the attainment of the 3% interstorey drift limitation.

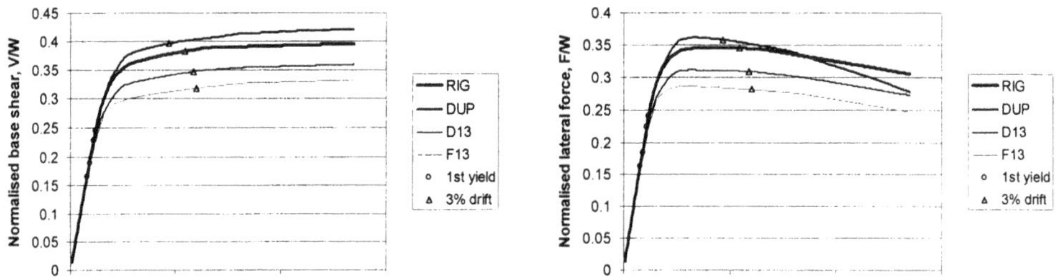

Figure 7.3.17: Normalised base shear and lateral force versus normalised top displacement relationship

The component reduction factors are given in Figure 7.3.18. The lowest ductility reduction factor R_μ is observed for the dual-pinned frame, as well as for D11 and F11 frames. Ductility reduction R_μ and redundancy ρ show a tendency to increase for frames including partial resistant connections (both for dual and semirigid schemes). On the other hand, the design overstrength Ω_d has an opposite tendency. This situation is explained by earlier yielding of frames with partial resistant joints and by the promotion of global collapse mechanism for these frames.

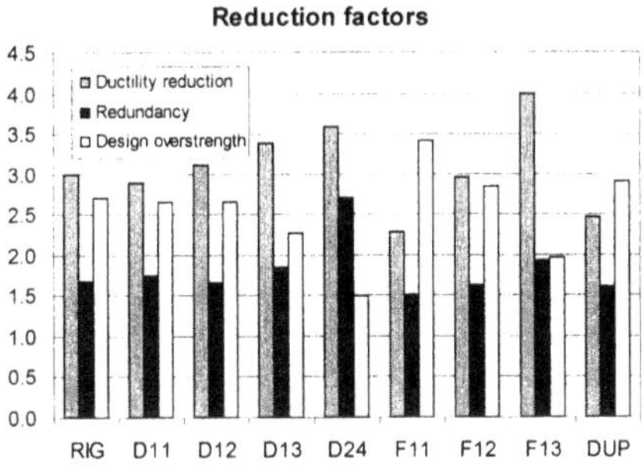

Figure 7.3.18: Component reduction factors

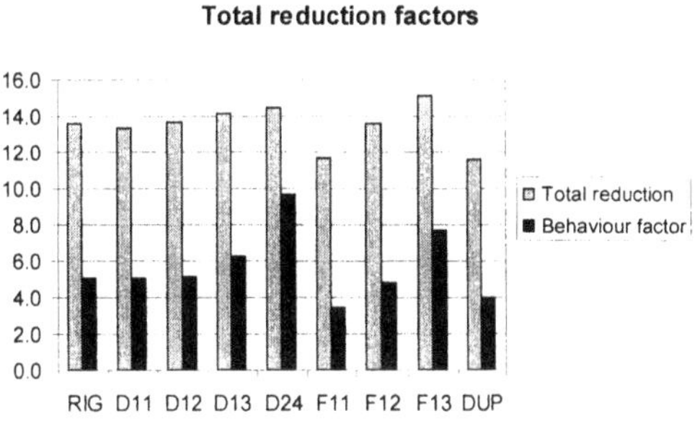

Figure 7.3.19: Total reduction factors (R)

Total reduction factors R and q (Figure 7.3.19) show the same tendency described for the ductility reduction factor, even though it is somewhat attenuated in the case of total reduction factor R, due to opposite tendency of the design overstrength factor Ω_d, whose variation is depicted in Figure 7.3.20.

TABLE 7.3.6

NUMBER OF STORIES INVOLVED IN THE COLLAPSE MECHANISM

Frame typology	RIG	D11	D12	D13	D24	F11	F12	F13	DUP
	4	4	4	4	5	2	4	4	2

Finally, in Figure 7.3.21, the variation of the stability coefficient, which is a measure of the structure sensitivity to second-order effects, is reported. The poor behaviour of DUP and F11 frames is observed also in terms of this coefficient. It has to be mentioned that these two frames exhibited the worst behaviour due to premature formation of partial collapse mechanism (only two stories for both cases, see TABLE 7.3.6).

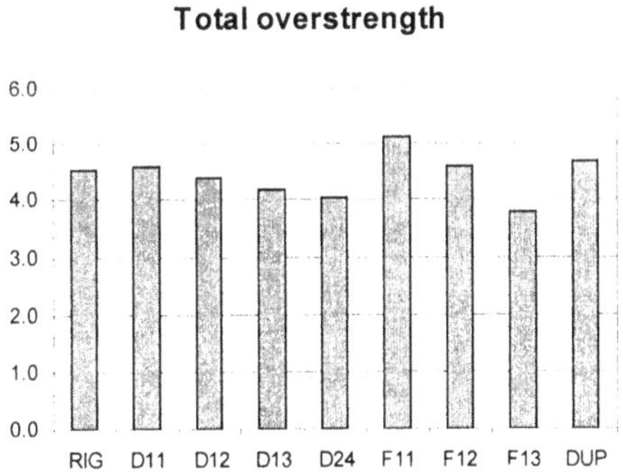

Figure 7.3.20: Total overstrength factor (Ω)

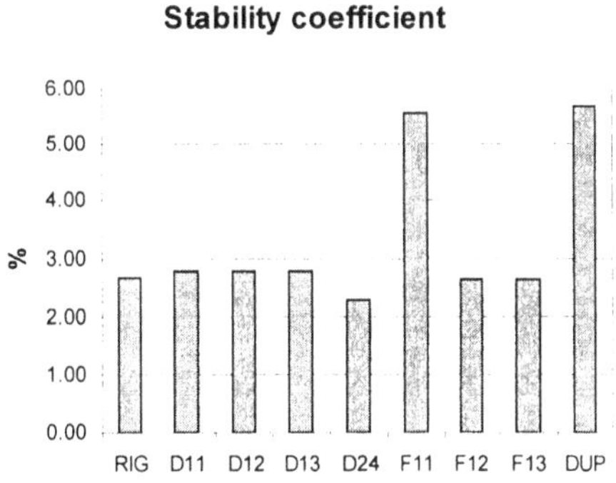

Figure 7.3.21: Stability coefficient (γ)

Time-history analysis

The parameters that may be used to characterise the structural behaviour are basically the same as those defined for the pushover analysis. However, in some cases slight modifications should be introduced, while some other parameters, i.e. stability coefficient, have no meaning at all. Eventually, the following parameters will be considered:

- Ductility related reduction factor: $R_\mu = \dfrac{\lambda_u}{\lambda_1}\dfrac{V_1}{V_y}$
- Redundancy: $\rho = \dfrac{V_y}{V_1}$
- Design overstrength: $\Omega_d = \dfrac{V_1}{V_d}$
- Total overstrength: $\Omega = \rho \cdot \Omega_d$
- Total reduction factor: $R = R_\mu \cdot \Omega$
- Behaviour factor: $q = R_\mu \cdot \rho = \dfrac{\lambda_u}{\lambda_1}$

in which λ_u is the ultimate accelerogram multiplier and λ_1 is the accelerogram multiplier at first yielding.

Some other useful information may be provided by values of yield and ultimate peak ground acceleration. Besides, plastic rotations evaluated in structural components (beams, columns and connections), both the maximum and accumulated values, have been inspected.

In Figure 7.3.22, the comparison among the analysed structures is depicted in terms of peak ground acceleration at first yielding. It may be observed a tendency to increase for fully SR structures, especially in case of full strength joints (F11) and for dual rigid/pinned frames (DUP). This is due to member sections, they being bigger in these cases as respect to the others. Obviously, frames with partial strength connections show a decrease of PGA producing the first yielding.

Similarly, in Figure 7.3.23, the comparison is shown in terms of ultimate PGA. An increase of ultimate PGA is accomplished for PR joints, for every structural configuration. A possible explanation of this is the promotion of global collapse plastic mechanism with higher energy dissipation. Anyway, it should be considered that ultimate PGA is strongly dependent on the chosen collapse criterion. Therefore care shall be taken when interpreting such results.

Influence of earthquake shows a considerable scatter of results for both yield and peak ground accelerations. All the structures yield earlier under Kobe earthquake and much later under Vrancea earthquake. No trend can be find in the case ultimate PGA. At the same time, ultimate multiplier of the ground motion is much more uniform for different earthquakes. It means that the type of ground motion provide less influence on collapse of the structure.

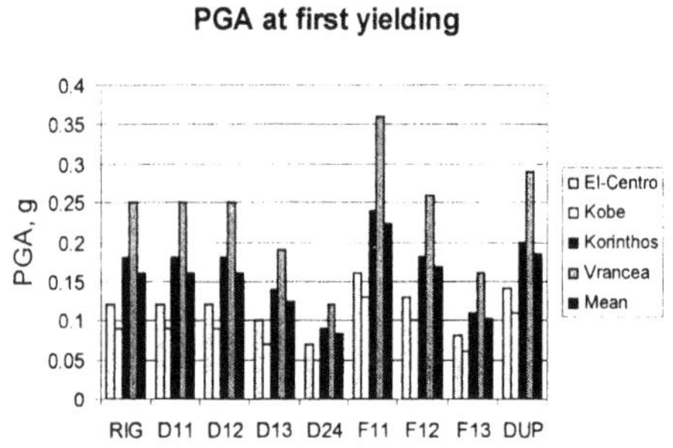

Figure 7.3.22: Yielding PGA

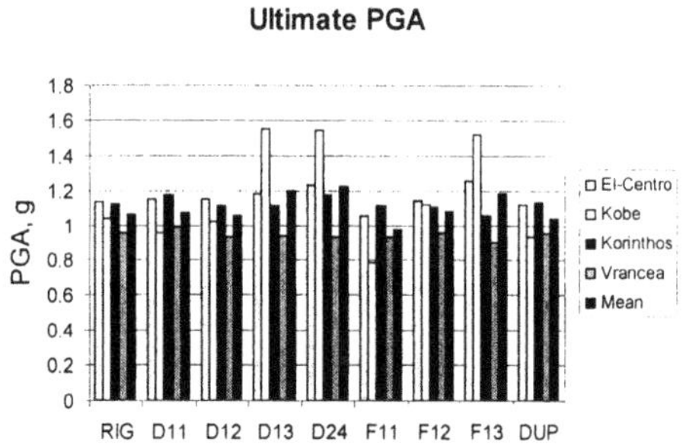

Figure 7.3.23: Ultimate PGA

The resulted values of the q-factor are presented in Figure 7.3.24. Higher values in the case of PR joints may be generally observed (high ultimate PGA versus low yield PGA). The trend of q-factor values for dual configurations is to increase with the decrease of joint resistance.

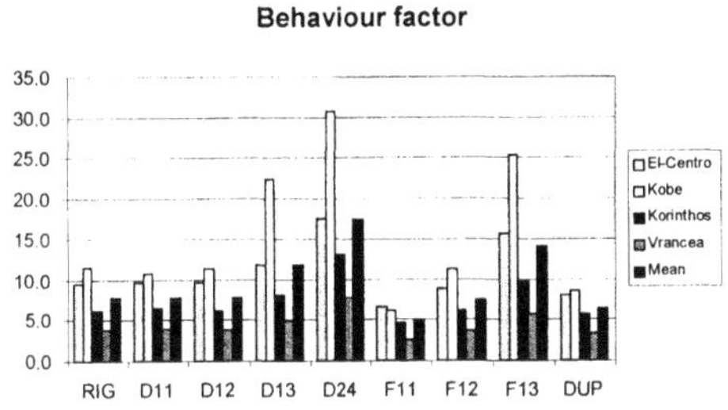

Figure 7.3.24: The behaviour factor q

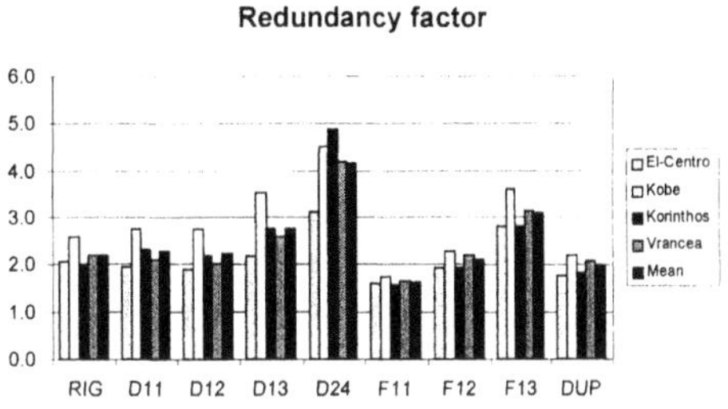

Figure 7.3.25: Redundancy factor ρ

The following tendency is observed for the redundancy factor (Figure 7.3.25):

- Higher redundancy for PR joint frames
- Lower one for dual rigid/pinned and low stiffness/high strength joints

The design overstrength (Figure 7.3.26) is given mainly by design governed by serviceability conditions, difference between the design spectra and spectra of the particular earthquake, design governed by other limit states, and commercially available structural shapes:

- Values close to 1.0 are observed for dual rigid/(SR-PR) frames
- High values are observed for SR-FR frames, as well as for dual rigid/pinned ones

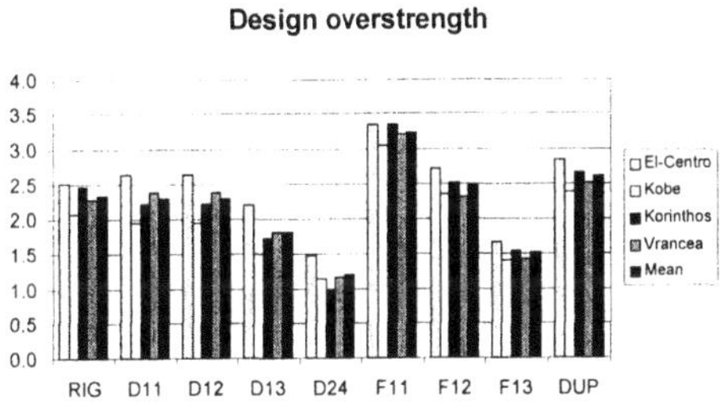

Figure 7.3.26: Design overstrength factor Ω_d

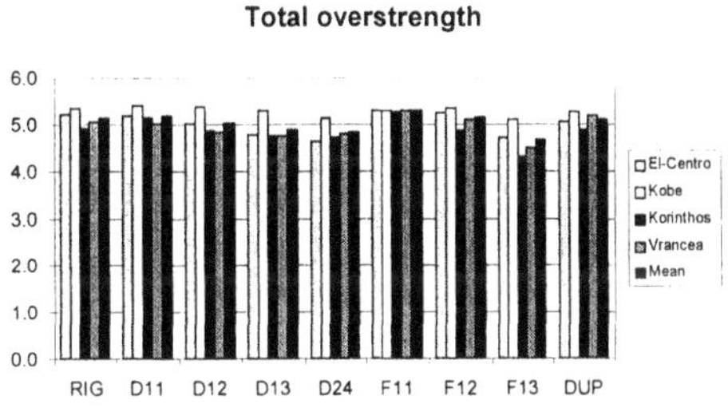

Figure 7.3.27: Total overstrength factor Ω

In the case of total overstrength shown in Figure 7.3.27, an important impact of the selected acceleration records is observed (different trends of the reliability/redundancy and design overstrength factors must be reminded). Anyway, an increase of the total overstrength is observed for PR joints. Redundancy (ρ) and design overstrength (Ω_d) are somewhat influenced by connection strength. Nevertheless, the total overstrength (Ω) is almost constant independent of the connection typology. This is mainly a peculiarity of dual structural configurations.

The results (see Figure 7.3.28 and Figure 7.3.29) in terms of total reduction factor and behaviour factor are greatly influenced by the earthquake. This influence is mainly connected with the trend of yield PGA. Lower values for these factors result in the case of Vrancea earthquake, due to quite late first yielding. It should be reminded that these two global parameters should not be seen as performance indicators. However, an increase of these factors for frames with partial resistant connections (both dual and fully semirigid) is observed.

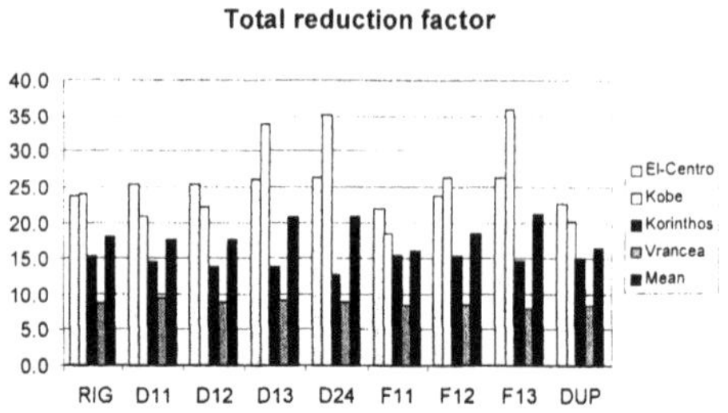

Figure 7.3.28: Total reduction factor R

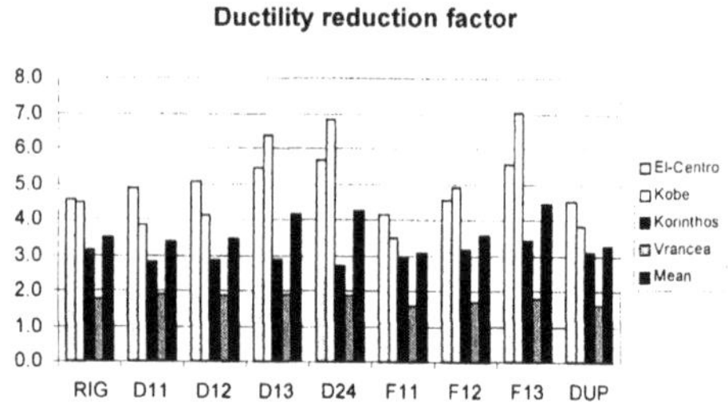

Figure 7.3.29: Ductility factor R_μ

Maximum and accumulated rotations in beams, connections and columns for a ground acceleration of 1g are shown in Figure 7.3.30 and Figure 7.3.31, as average value for all the earthquakes. The acceleration level is close to the average of ultimate PGAs, corresponding to 3% interstorey drift limit. Maximum rotations provide information upon the ductility demand in one cycle, while accumulated ones are related to the overall ductility demand during history of cycles generated by the ground motion.

Generally, seismic design codes specify the strong column – weak beam design philosophy. It aims at promoting the formation of plastic hinges in beams rather than in columns, for assuring a greater dissipation capacity through collapse mechanism of global type. Anyway, it is recognised that code specifications can not ensure the above criterion through simple relations. From this point of view, it can be observed that maximum plastic rotation demands in columns is particularly high in the case of dual rigid/pinned (DUP) and SR/full strength (F11) frames. When partial resistant connections are employed (for both D and F series), the demand in inelastic deformations is shifted toward the latter. The same phenomenon is observed for the accumulated plastic rotations, and it is even more evident.

Use of semirigid connections in the case of homogeneous scheme (F) is shown to have a negative influence on the structural performance, unless connections are partial strength also. This is a result of the design procedure, governed by the serviceability condition (lateral stiffness). Due to the reduced stiffness of SR joints, bigger beam sections result, and unless PR connections are involved, the beam/column plastic moment ratio will be higher.

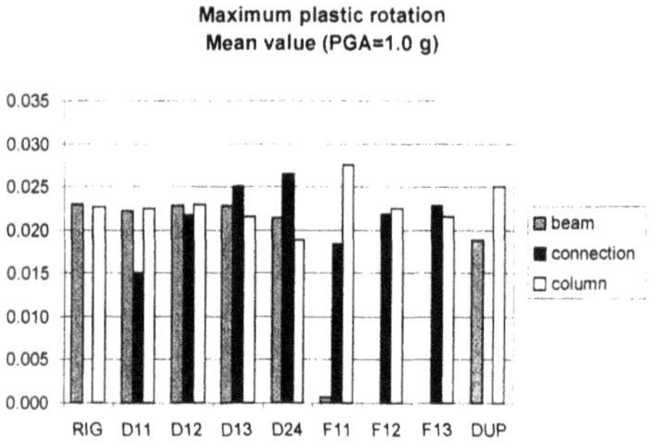

Figure 7.3.30: Maximum plastic rotations at 1g (rad)

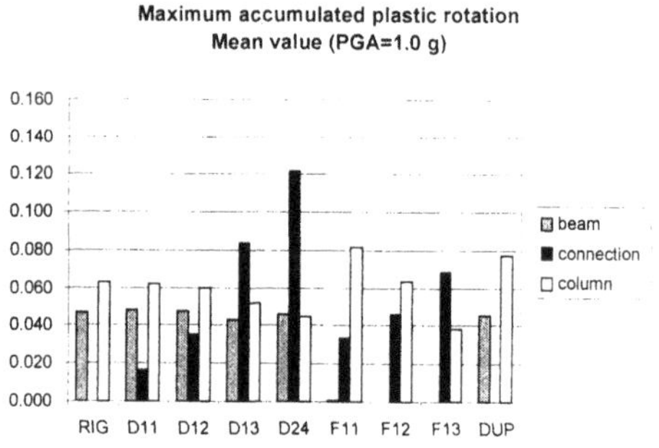

Figure 7.3.31: Accumulated plastic rotations at 1g (rad)

In the case of dual schemes (D), element cross-sections are only slightly changed (for D24), and only stiffness/moment capacity of external joints is changed. The effect of reduced stiffness of exterior joints is to shift plastic demand on interior rigid connections, but when combined with reduced moment capacity, the effect is inverse. Providing that these joints have sufficient and stable rotation capacity, global performance of dual schemes is better.

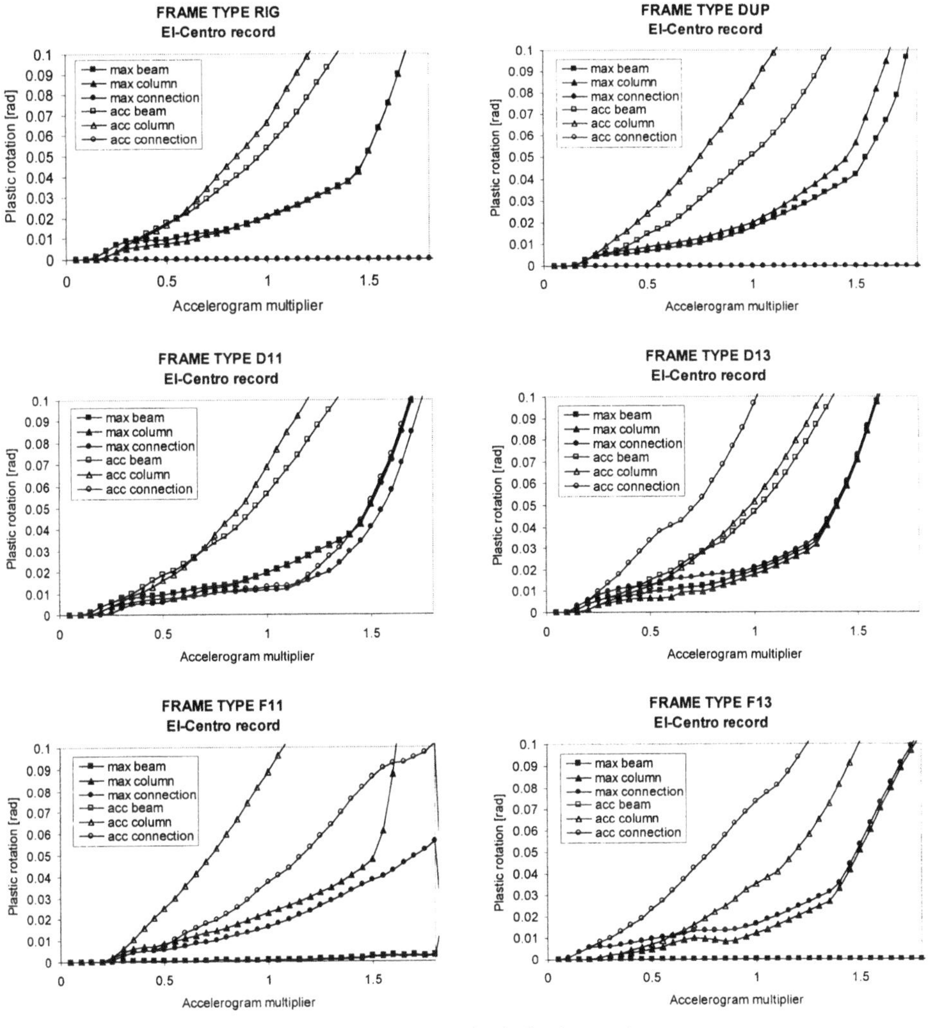

Figure 7.3.32: Required plastic rotations

Plastic rotation demands in beam, columns and connections for representative frames and El-Centro record are presented in Figure 7.3.32. Distribution of plastic rotation demands can be followed on a larger range of acceleration multipliers. The results clearly show that low stiffness – high strength connections are not a good solution for MRFs. Dual structural schemes provide a good alternative not only to dual rigid/pinned (DUP), but also to homogeneous (F) schemes.

Eventually, some results concerning the comparison of the pushover and the time-history analyses are shown in Figure 7.3.33 and Figure 7.3.34. It can be observed that the trend of the investigated parameters is the same in both analysis methods. Time-history analysis provides higher values for base shear at 3% interstorey drift, ductility reduction, and redundancy. An opposite situation is observed in the case of yield base shear and design overstrength. At the same time, ductility reduction and redundancy parameters yield smaller values in the case of pushover analysis, thus providing values on the safe side.

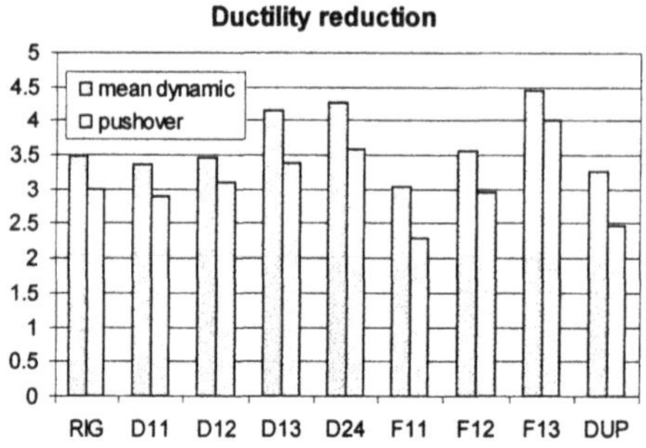

Figure 7.3.33: Ductility reduction – comparison of analysis methods

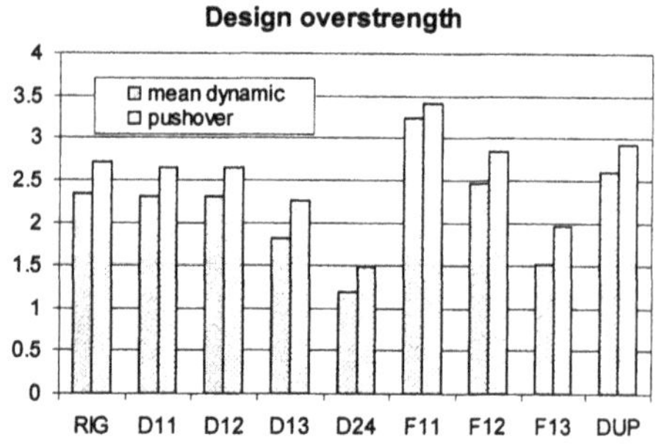

Figure 7.3.34: Design overstrength – comparison of analysis methods

Main outcomes

The above study allows the following main conclusions to be drawn. (1) Design of steel moment resisting frames for seismic loading is governed by serviceability conditions, consequently resulting in a considerable design overstrength. If semi-rigid connections are considered in the design procedure, increased element sections will result. At the same time, partial-resistant connections can be involved without any change in structural sizes. For dual schemes, even SR-PR connections yielded no modifications of elements. (2) Higher structural damage is expected for frames with low-stiffness/high-strength joints. Therefore, semirigid/full-strength joints (if they would be practically possible) are not suitable for earthquake applications. But, taking into account the realistic structural detailing, semirigid beam-to-column joints are generally partial resistant. (3) In the case of dual rigid/(SR-PR) frames, damage is shifted from rigid connections and columns to the exterior SR-PR connections. This type of frames behaves better under earthquake loads than dual rigid/pinned frames. (4) Provided that SR/PR joints with adequate ductility are available, these can be used as a tool for damage control in dual schemes.

Nevertheless, despite the above conclusions are interesting and attractive, it is to be reminded that due to the particular structural layout herein considered (weak axis external columns, three bays and five storeys), further investigation is needed. The type of ground motion, number of storeys, structural layouts different from those studied here may yield different results. Also, performance (stiffness, capacity and available ductility) of practical rigid, as well as SR/PR connections are parameters that are needed in order to get additional insight into the performance of dual structural schemes.

The pushover analysis, being much simpler to be implemented in design offices with respect to the time-history one, may be recommended as an alternative to the equivalent static analysis methods. It is able to provide additional information on the structure inelastic response and deformation capacity, usually providing results on the safe side. Anyway, care must be taken when the influence of higher modes is important. For instance, it seems that in the case of "far field" earthquakes, the horizontal loads introduced in the pushover analysis have to correspond to the first eigenmode distribution, while for "near field" ones, the second eigenmode distribution would be more suitable. Obviously, in order to prove these assumptions, further research efforts are necessary.

CONCLUSIONS

Results obtained from the parametrical study, as well as those from the case study (design according to Eurocode 8) show that moment resisting frames in dual schemes with combination of rigid/full-strength and semirigid/partial-strength joints could provide an adequate seismic performance.

Structural schemes and general rules could not be defined yet. From this point of view, the dual configuration of the frame has to be studied and optimised taking into account the type of ground motion, soil conditions and the moment-rotation characteristic of joints. Consequently, the accurate prediction of structural seismic behaviour is a complex matter and could not be performed according to simplified lateral force procedure stipulated in most of the present design codes. The pushover analysis can provide a convenient compromise, supplying an analysis tool for optimisation of structural scheme performance.

It has to be mentioned that the structural behaviour factor (q) should not be considered as a measure of the structure performance, especially for semirigid or dual schemes. The definition of yield is important for the computation of the relevant parameters. However, the simple definition based on the first plastic hinge seems to be inappropriate for the dual structural schemes, where different connection typologies are present in the same lateral force resisting system. Influence of the earthquake ground motion shows to be considerably more important than other effects.

It is clear that steel moment resisting frames, where bays with rigid/full strength joints and alternating bays with pinned joints are used (pre-Northridge perimeter frames), present a worse seismic performance than similar dual schemes, where pinned joints are replaced by semirigid/partial resistant ones.

REFERENCES

Astaneh A. (1994). Seismic behaviour and design of steel semi-rigid structures. In Proc. *Behaviour of Steel Structures in Seismic Areas (STESSA '94)*, Eds. F.M. Mazzolani, and V. Gioncu, E & FN Spon, pp. 547-556.

ENV 1993, Eurocode 3, (1994). *Joints in Building Frames,* CEN/TC250/SC3-PT9.

ENV 1998, Eurocode 8, (1994). *Design Provisions for Earthquake Resistance Structures*, Commission of the European Communities.

Fischinger M. and Fajfar P. (1994). *Seismic Force Reduction Factors,* Earthquake Engineering. Balkema ed., Rotterdam.

Fulop L. A. (1998). *Seismic Performance of Moment-Resisting Frames with Semirigid Joints*, Diploma project, University of Naples Federico II, Univesitatea Politehnica Timisoara, Supervisors: Prof. F. M. Mazzolani and Prof. D., Dubina, Co-ordinator: Dr. G. De Matteis.

Gioncu V. (1999). Framed Structures. Ductility and Seismic Response. General Report In Proc. *6th International Colloquium on Stability and Ductility of Steel Structures.* September 1999, Timisoara, Romania (in press).

Gioncu V. and Pectu D. (1997). Available rotation capacity of wide flange beam-columns. *J. Constructural Steel Research,* Vol. **43** (1-3), 161-217.

Kishi N., Chen W.F., Goto Y. and Hasan R. (1996). Behaviour of tall buildings with mixed use of rigid and semi-rigid connections. *Computer & Structures*, No. **6**, Elsvier Science Ltd, 1193-1206.

Krawinkler H. (1995). *Systems Behaviour of Structural Steel Frames Subjected to Earthquake Ground Motion,* Background Reports: Metallurgy, Fracture Mechanics, Welding, Moment Connections and Frame Systems Behaviour. Report No. SAC-95-09. SAC Joint Venture, California, USA.

Mazzolani F.M. and Piluso V (1996). *Theory and design of seismic resistant steel frames,* Chapman & Hall, London, UK.

Mazzolani F.M. and Piluso V. (1997). The influence of the design configuration on the seismic response of moment-resisting frames. In Proc. of *Behaviour of steel Structures in Seismic Areas - STESSA '97*, Kyoto, Japan, 444-453.

Mitani I. and Makino M. (1980). Post local Buckling Behaviour and Plastic Rotation Capacity of Steel Beam-Columns. In Proc. 7th *World Conference on Earthquake Engineering,* Istanbul.

Nader M.N., Astaneh A. (1991). Dynamic behaviour of flexible semirigid and rigid steel frames. *Journal of Constructional Steel Research*, **18**, 179-192.

Park Y.J., Ang A.H.S. and Wen, Y.W., (1985). Seismic damage Analysis of Reinforced Concrete Buildings, *Journal of Structural Engineering*, ASCE, **111**.

Prakash V., Powell G.H. and Campbell S. (1993). *Drain 2DX Base program description and user guide*, Version 1.10.

Roeder C. W. and Fountch D.A. (1996). Experimental results for seismic resistant steel frame connections. *Journal of Structural Engineering*, ASCE, Vol. **122:6**.

Shen J. (1996). A new dual system for seismic design of steel buildings. In Proc. of *Advances in Steel Structures*, ICASS'96, Hong-Kong, Vol 2, Edited by S. L. Chan and J. G. Teng, Pergamon, Elsevier Science ltd, 1027-1033.

Tsai K. and Popov E.P. (1988). *Steel beam-column joint is seismic moment resisting frames*, Report no. UCB/EERC-88/19, University of California, Berkeley.

7.4

INFLUENCE OF BUILDING ASYMMETRY

Peter Fajfar, Damjan Marusic, Iztok Perus

GENERAL

Asymmetric buildings are considered to be more vulnerable in strong earthquakes than symmetric ones. Substantial research efforts have been devoted to understanding the seismic response behaviour of asymmetric building systems during the last decades. Nevertheless, the progress has been rather slow. This is manifested in the relative scarcity of conclusions of general validity. The main reason seems to be a very large number of parameters, which influence linear and especially nonlinear response.

In this chapter, a part of the research performed at the University of Ljubljana is summarised. The main aims of the research are to contribute to the understanding of the linear and nonlinear seismic response of asymmetric structures and to develop a relatively simple method for the simulation of this response. Research was performed in parallel on simple single-storey and multi-storey buildings. The influence of the most important structural parameters has been studied on idealised single-storey structures. Elastic response spectrum analysis, as well as linear, and nonlinear time-history analyses were performed on a number of test structures. The multi-storey buildings studied were 5-storey steel frame structures with mass eccentricity. Included were frames with semi-rigid and partial-strength connections. The basic symmetric structure was designed according to Eurocodes 3 and 8. The influence of the structural system, of the magnitude of eccentricity, and of the intensity of ground motion has been studied by performing a series of linear and nonlinear time-history analyses. In all cases, the main response parameters studied are maximum displacements at the edges. Ratios of these displacements and displacements in corresponding symmetric structures indicate the influence of torsion. Comparisons are made between elastic and inelastic responses. Furthermore, the results obtained by applying bi-directional and uni-directional horizontal ground motion are compared.

From the results of analyses several conclusions have been drawn. They are based on a very limited number of test examples. Nevertheless, at least some of them may be generally valid and they may contribute to a better understanding of seismic response of asymmetric buildings. They also provide information needed for the achievement of the second objective of the study, i.e. development of a simple nonlinear analysis procedure for asymmetric buildings.

SINGLE-STOREY BUILDINGS

Parametric studies have been performed on idealised single-storey structures. Analytical expressions have been derived for the determination of elastic response of structures with eccentricity in both directions subjected to ground motions defined by response spectra. Elastic response spectrum analysis, as well as linear, and nonlinear time-history analyses were performed on a number of test structures.

The response parameters studied are displacements, mainly their maximum values at the edges. Comparisons are made between elastic and inelastic responses. The difference between the behaviour of torsionally stiff and torsionally flexible buildings is discussed. The influence of the type of eccentricity is studied. Furthermore, the response of structures subjected to bi-directional and uni-directional horizontal excitation is compared.

Mathematical models

Models with elements oriented in two directions are used for realistic simulation of the actual behaviour, even if some studies have demonstrated, for a restricted range of input parameters, that appropriate results can be obtained also with elements only in the direction of loading. The distribution of strength in the majority of the models investigated so far has been determined according to a specific code. A strength distribution strictly following a code cannot be attained in a real structure. Because of this reason, and because of our ambition to explore the inelastic response in a more general way, independently of specific codes, we decided to investigate some limit cases regarding strength distribution. Like the majority of researchers, we used a model with three elements in each of two orthogonal directions. The basic symmetric model is shown in Figure 7.4.1. Six lateral load-resisting elements are connected with a slab, which is rigid in horizontal plan and has no stiffness in the vertical direction. Mass is uniformly distributed at the slab level. The periods of vibration amount to $T_X = 0.3$ s, $T_Y = 0.4$ s, and $T_Z = 0.254$ s, where index Z denotes vertical axis and T_Z corresponds to torsional vibration. The hysteretic behaviour of all elements is ideal elasto-plastic with 5% post-yield stiffness.

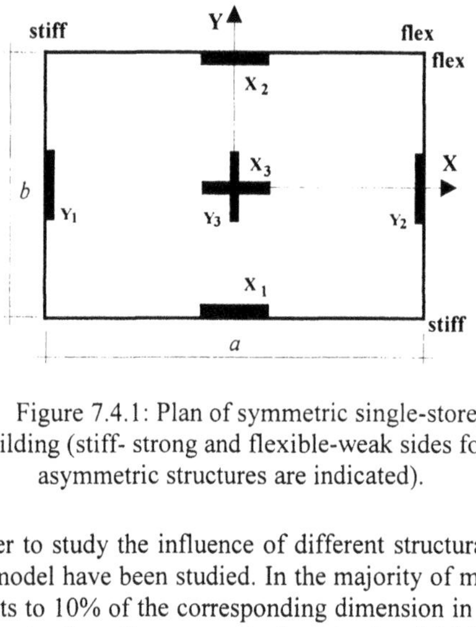

Figure 7.4.1: Plan of symmetric single-storey building (stiff- strong and flexible-weak sides for asymmetric structures are indicated).

Data for the basic symmetric model
$a = 6$m, $b = 4$m
stiffness K, strength $F_y = 0.256Mg$
M is total mass

elements Y_1, Y_2, Y_3
stiffness $k = K/3$, strength $f_y = F_y/3$
yield displacement $u_y = 1.02$ cm

elements X_1, X_2, X_3
stiffness $1.778\,k = 0.593\,K$,
strength $1.5\,f_y = 0.5\,F_y$
yield displacement $u_y = 0.86$ cm

periods: T_X=0.3 s, T_Y=0.4 s, T_Z=0.254 s

radius of inertia: $r = 0.347$a

In order to study the influence of different structural parameters, several (asymmetric) variants of the basic model have been studied. In the majority of models, the eccentricity in both horizontal directions amounts to 10% of the corresponding dimension in plan. Mass eccentricity is produced by shifting the mass centre from the geometrical centre (model M). Stiffness and/or strength eccentricity is obtained by changing the characteristics of elements. The static force-displacement relations for different elements, representing envelopes of the hysteretic loops, are shown in Figure 7.4.2. S and R models are eccentric regarding stiffness and strength, respectively. In the SR model, a linear relation between stiffness and strength is assumed. Consequently, stiffness and strength eccentricity are equal. In the SRa model, it is more realistically assumed that the changes of strength are smaller than the changes of stiffness. In model M30 the mass eccentricity is increased to 30%. All models mentioned above are torsionally stiff. A torsionally flexible building (M1) has been studied as well.

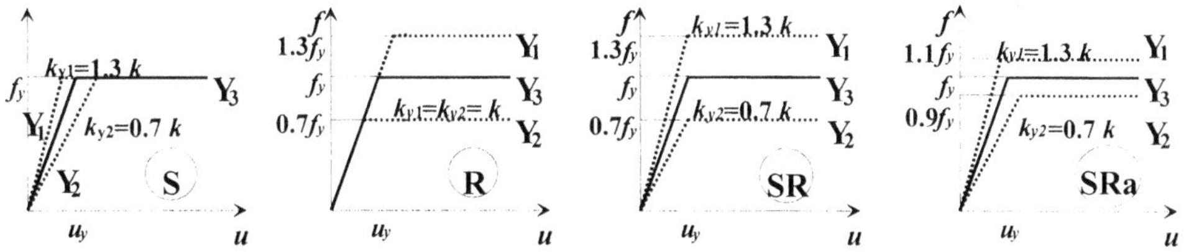

Figure 7.4.2: Force - displacement relations for Y-elements in asymmetric structures. The same ratios apply for X-elements.

An overview of the investigated models is given below. Only those characteristics, which are different from the basic symmetric structure, are given. Eccentricities are defined as the ratios of mass, stiffness and/or strength eccentricity and the corresponding dimension in the plan (a or b). Index u denotes period of the uncoupled system.

- Model M — mass ecc. — $e_{MX}= e_{MY}= 0.1$
- Model S — stiffness ecc. — $e_{SX}= e_{SY}= -0.1$
- Model R — strength ecc. — $e_{RX}= e_{RY}= -0.1$
- Model SR — strength and stiff. ecc. — $e_{SX}= e_{RX}= e_{SY}= e_{RY}= -0.1$
- Model SRa — strength and stiff. ecc. — $e_{SX}= e_{SY}= -0.1$, $e_{RX}= e_{RY}= -0.084$
- Model M30 — mass ecc. — $e_{MX}= e_{MY}= 0.3$
- Model M1 — mass ecc. — $e_{MX}= e_{MY}= 0.1$, T_{Zu}=0.5 s

Response spectrum analysis

For the case of the elastic response spectrum analysis of a single-storey building, modelled as a system with three degrees of freedom, analytical expressions have been derived for displacements at an arbitrary point of the floor (Perus and Fajfar, 1999).

$$u_{xj,x}(l_y) = m^*_{j,x} S_{dj,x}\left(1 - l_y \frac{\varphi_{zj} r}{\varphi_{xj}}\right), \quad u_{yj,x}(l_x) = m^*_{j,x} S_{dj,x}\left(\frac{\varphi_{yj}}{\varphi_{xj}} + l_x \frac{\varphi_{zj} r}{\varphi_{xj}}\right) \tag{7.4.1}$$

$$u_{xj,y}(l_y) = m^*_{j,y} S_{dj,y}\left(\frac{\varphi_{xj}}{\varphi_{yj}} - l_y \frac{\varphi_{zj} r}{\varphi_{yj}}\right), \quad u_{yj,y}(l_x) = m^*_{j,y} S_{dj,y}\left(1 + l_x \frac{\varphi_{zj} r}{\varphi_{yj}}\right)$$

where the first index (x or y) represents the direction of displacement, the second index (j) the mode, and the third one (x or y after comma) the direction of loading. The influence of three vibration modes can be combined by the CQC rule, and the influence of two ground motion directions can be combined by the SRSS rule. Normalised coordinates, defining the location of the point under consideration, are defined as $l_x = L_x/r$ and $l_y = L_y/r$, where r represents radius of inertia of the floor mass. L_x and L_y are coordinates measured from the centre of masses. φ_{xj}, φ_{yj} and φ_{zj} are components of the jth mode shape. Index z stands for rotation. $S_{dj,x}$ and $S_{dj,y}$ are values in displacement spectra for X and Y direction, respectively, corresponding to the period T_j. Effective masses are determined as follows

$$m^*_{j,x} = \frac{\varphi^2_{xj}}{\varphi^2_{xj} + \varphi^2_{yj} + \varphi^2_{zj} r^2}, \quad m^*_{j,y} = \frac{\varphi^2_{yj}}{\varphi^2_{xj} + \varphi^2_{yj} + \varphi^2_{zj} r^2} \tag{7.4.2}$$

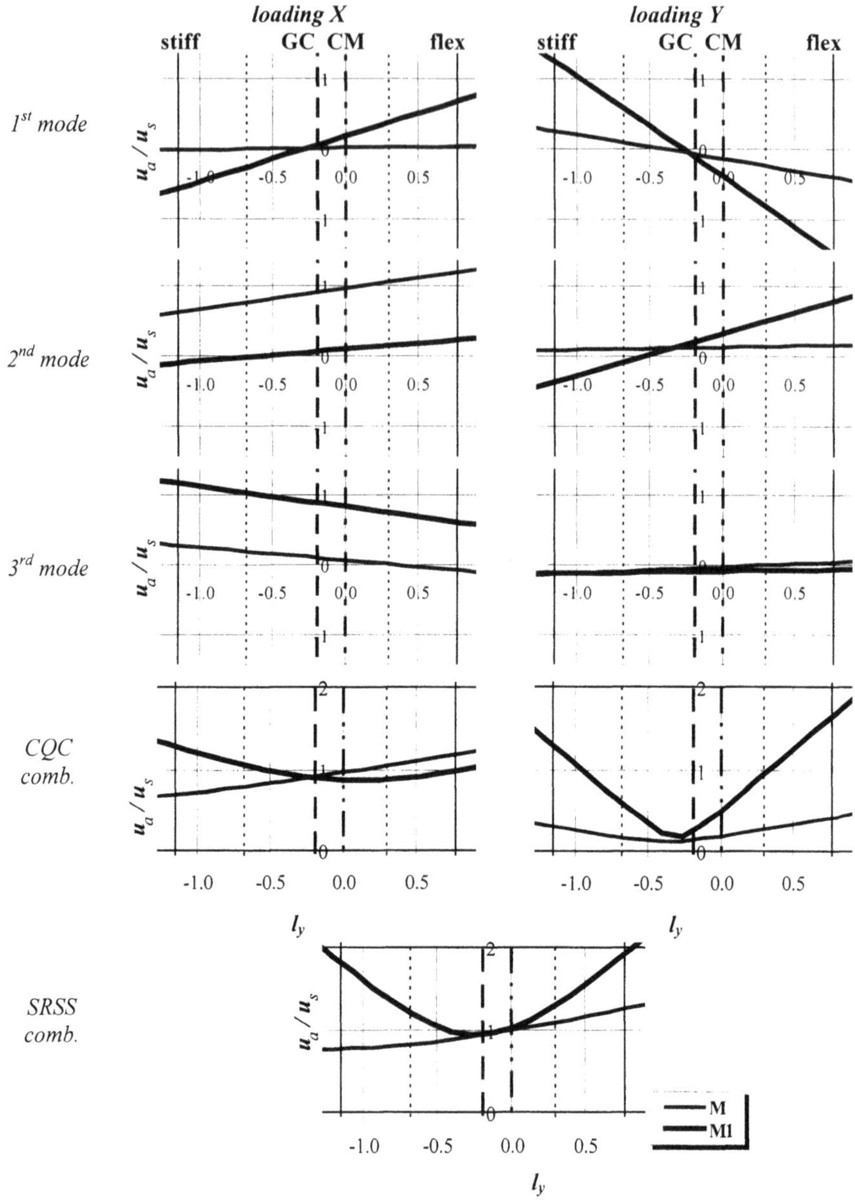

Figure 7.4.3: Ratio of displacements in X-direction in asymmetric structures (models M and M1) and symmetric structure as a function of the normalised distance from the mass centre. Solid vertical lines denote stiff and flexible edge. Thin dashed vertical lines denote locations of elements for model M1.

The response spectrum approach will be used to demonstrate the difference between the response of torsionally stiff and torsionally flexible structures. It will be used also for the demonstration of the influence of the location at the floor plan on displacement. Furthermore, the difference between mass and stiffness eccentricities will be discussed. In all cases, seismic loading is defined by a constant acceleration spectrum and is applied independently in both horizontal directions. For each loading direction, the displacements for three modes are combined using the CQC combination rule. The influence of the simultaneous excitation in two horizontal directions is estimated by combining the results obtained for loadings in X- and Y-direction by the SRSS (square root of sum of squares) rule.

TABLE 7.4.1

NATURAL PERIODS, MODE SHAPES AND EFFECTIVE MASSES FOR DIFFERENT MODELS. ORIGIN OF THE COORDINATE SYSTEM IS ALWAYS IN MASS CENTRE.

model	**M** 			**M1** 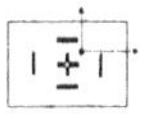			**S** 		
mode	1	2	3	1	2	3	1	2	3
T_j	0.411	0.310	0.240	0.554	0.373	0.291	0.412	0.311	0.245
φ_{xj}	-0.077	1.000	0.703	-0.451	0.286	1.000	-0.084	1.000	0.805
φ_{yj}	1.000	0.136	-0.335	1.000	1.000	-0.109	1.000	0.158	-0.361
φ_{zj}	0.090	-0.152	1.000	0.797	-0.252	0.162	0.099	-0.173	1.000
$r\ \varphi_{zj}$	0.187	-0.316	2.082	1.659	-0.525	0.337	0.206	0.360	2.082
$m^*_{j,x}$	0.006	0.894	0.100	0.051	0.060	0.888	0.007	0.866	0.127
$m^*_{j,y}$	0.961	0.017	0.022	0.253	0.737	0.010	0.953	0.022	0.025
	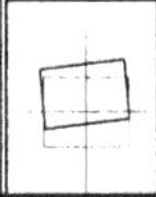	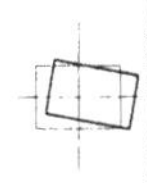	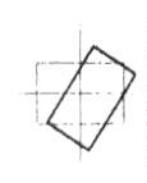	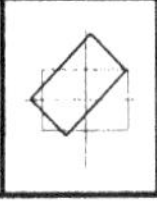	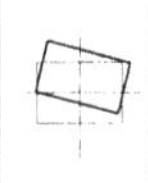	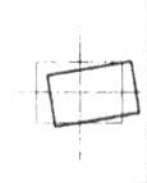	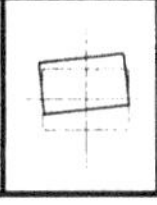	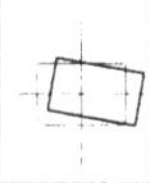	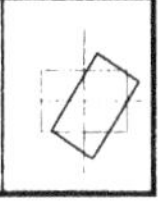

First, the mass eccentric model M will be analysed. The eccentricity in both horizontal directions amounts to 10% of the corresponding dimension in plan. It is assumed that the total mass and the mass moment (and, as a consequence also the radius of inertia) do not change. (Physically, such a situation can be obtained by a redistribution of mass.) Results of free vibration analysis are summarised in Table 7.4.1. Note that the free vibration response is predominantly translational for the first two modes and predominantly torsional for the third mode. Consequently, the structure is torsionally stiff. The forced vibration response in terms of displacements can be determined by Eqns. 7.4.1 and 7.4.2. In Figure 7.4.3, the ratio between displacements in the X-direction of the mass eccentric and of the symmetric structure are plotted for each vibration mode and for each of two horizontal directions of excitation separately. The ratio is a function of l_y, which defines the distance from the mass centre (normalised by the radius of inertia). Similar results can be obtained for displacements in Y-direction (see Figure 7.4.4). Results indicate that by far the largest contribution to the displacement response comes from the second mode in the case loading in X-direction. Compared to the symmetric building, displacements are larger at the flexible side (i.e. at the side to which mass centre was moved), and smaller at the stiff side. It is well known that such a behaviour is typical for torsionally stiff structures.

The next example is a torsionally flexible structure, characterised by the first mode, which is predominantly torsional. Such a structure can be created from the torsionally stiff structure M in two different ways: (a) by decreasing the torsional stiffness by changing the location of load bearing elements, or (b) by increasing the mass moment. The first possibility is very usual in practice, whereas the second option is less realistic. It can be produced by concentrating masses at the perimeter of the building or even outside of it. In the option (a) the load bearing elements are moved toward the centre (model M1). Their new location is at the distances 0.254a and 0.254b from the geometrical centre (GC). All structural characteristics remain the same as those in the model M, with the exception of the torsional stiffness, which has decreased. Periods, mode shapes, and effective masses are summarised in Table 7.4.1. The results plotted in Figure 7.4.3, clearly indicate that several modes and both directions of loading importantly contribute to the final displacements. Due to torsion, displacements at the flexible side increase, like in the case of torsionally stiff structures, but the increase is much larger. The most notable (well known) difference can be seen at the stiff side. Displacements are larger than in the case of symmetric structure at this side as well. Note, however, that displacements at the edges are, in the case of structures with the investigated distribution of load resisting elements in plan, relevant only for non-structural damage. The structural elements are located at smaller coordinates where the torsional influence is much lower than at the edges (Figures 7.4.3 and 7.4.4).

In both structures discussed so far, eccentricity was introduced by shifting the mass centre, whereas the structural elements have been distributed symmetrically. Now, we will investigate a structure, in which eccentricity is a consequence of the asymmetric distribution of stiffness. It is well known that, under certain conditions, periods and mode shapes of mass- and stiffness-eccentric structures are equal if eccentricities are equal. We will investigate a torsionally stiff model (model S) which is only approximately equivalent to the mass eccentric model M. Mass and mass moment (and, as a consequence, radius of inertia) are the same in both models. Note, however, that the mass moment corresponds to the centre of masses, which is located in the geometrical centre for model S and is shifted in the case of model M. As a result of the modified stiffnesses of load bearing elements of model S the stiffness centre is displaced in both directions from the geometrical and mass centre toward stiff edges. Stiffness eccentricities in both directions amount to 10%, like in the case of the mass-eccentric system M. The modification of element stiffnesses has not changed the total translational stiffness in both directions. The torsional stiffness, however, has changed a little bit, and is responsible for small differences in free vibration results (see Table 7.4.1). The displacement curves for M and S model are practically the same (compare Figures 7.4.4 and 7.4.5). However, the normalised coordinates at the edges are not the same. They are measured from the mass centre which has a different location for each model.

Linear and non-linear time-history analyses

In addition to response spectrum analysis, discussed in the previous chapter, linear and non-linear time-history analyses have been performed as well. Two horizontal components of eight strong motion records (Sylmar and Newhall from Northridge 1994, Kobe J.M.A. from Kobe 1995, El Centro 1940, Petrovac, Ulcinj 1, Ulcinj 2 and Bar from Montenegro 1979) were used for time-history analyses. Each pair of accelerograms was applied twice (in the second run X- and Y-directions were interchanged). So, sixteen time-histories were computed for each system. The fundamental periods of the investigated structures are in the short-period range of spectra. Consequently, we normalised the records regarding the peak ground acceleration. After scaling, for each record the maximum value of ground acceleration in horizontal plane (considering vectorial sum of both components) was equal to 0.4 g. Normalised acceleration spectra are shown in Figure 7.4.6. The stronger components (i.e. for each accelerogram the component with larger peak ground acceleration) were grouped together. These are N-S components of all records except Ulcinj 2, where the E-W component is stronger.

The comparison of results obtained by response spectrum analysis and linear time-history analysis (where unlimited elastic behaviour was assumed) can provide some information on the influence of the two approximations used in response spectrum approach, i.e. CQC rule for combination of modes and SRSS rule for combination of the two horizontal directions of ground motion. In response spectrum analyses a constant acceleration spectrum was used, which, in the relevant period range, roughly corresponds to the mean spectrum of accelerograms used in time-history analyses. This is another approximation, which also contributes to the difference between the two approaches.

The maximum values of edge displacements, obtained by linear and nonlinear time-history analyses for different models, are marked in Figures 7.4.4 and 7.4.5. The marked values were calculated for each model as follows. For each of 16 bi-directional ground excitations, maximum response in terms of

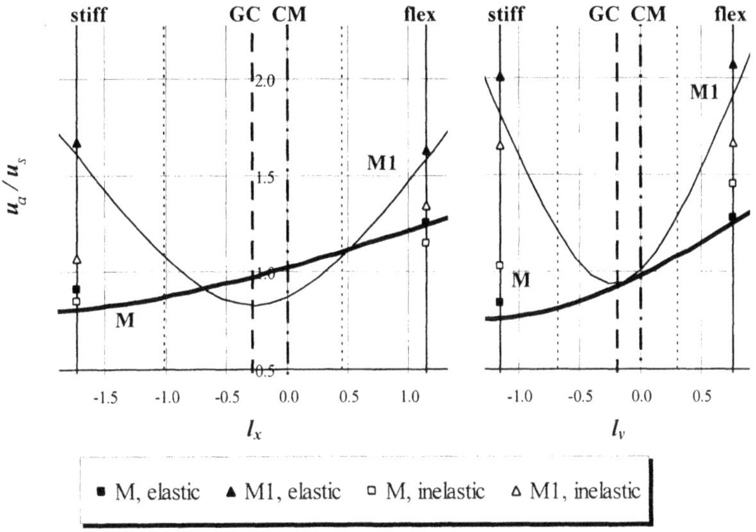

Figure 7.4.4: Ratio of displacements in asymmetric structures and symmetric structure, obtained by elastic response spectrum analysis. Solid vertical lines denote edges of M and M1. Thin dashed vertical lines denote locations of elements for M1. Maximum edge displacements obtained by linear and nonlinear time-history analyses are marked.

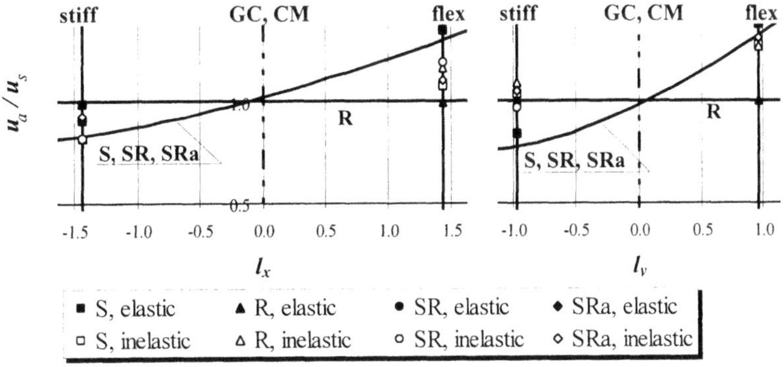

Figure 7.4.5: Ratio of displacements in asymmetric and symmetric structures for models with stiffness and/or strength eccentricity. Maximum edge displacements obtained by linear and non-linear time-history analyses are marked. Note that elastic results for S, SR and SRa are identical.

displacements at all four edges of the building were determined, separately for elastic and inelastic behaviour. Then, the mean values were calculated. In figures the ratios between the mean values of asymmetric model and of the basic symmetric model are given. In the case of the symmetric model (Figure 7.4.1), the absolute values of inelastic displacements are 2.06 cm and 4.62 cm in X- and Y-direction, respectively. The corresponding ductilities amount to 2.4 and 4.5, respectively. In the case of unlimited elastic behaviour, the displacements amount to 1.92 cm and 3.29 cm in X- and Y-direction, respectively. A fair correlation of elastic time-history results with the results of the response spectrum analysis can be observed. Inelastic results also do not differ dramatically.

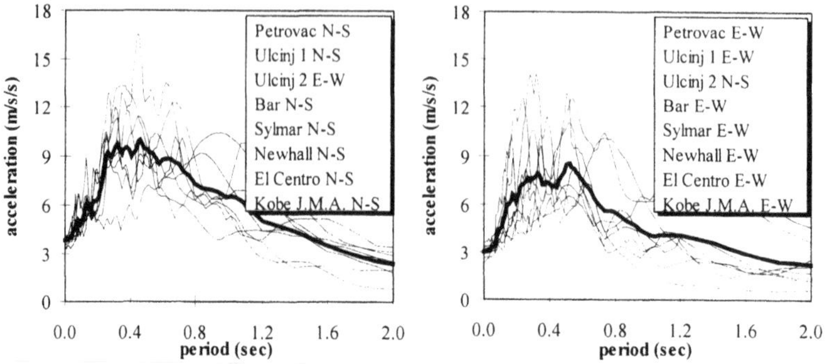

Figure 7.4.6: Normalised elastic acceleration spectra for 5% damping. Thick lines correspond to mean spectra.

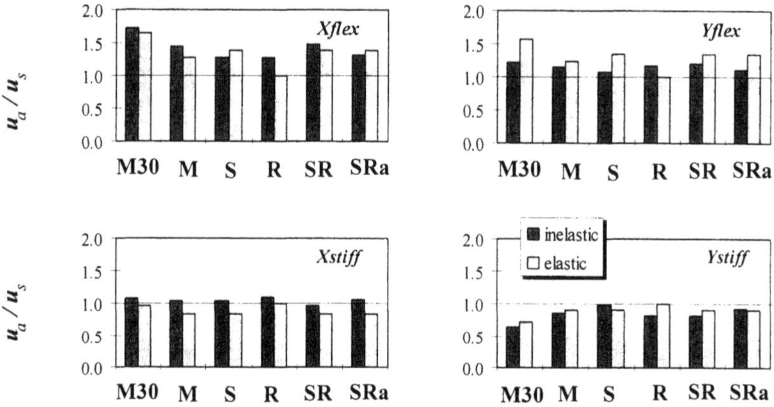

Figure 7.4.7: Mean values of the ratio of maximum displacements at edges of asymmetric and symmetric structures for different models obtained by elastic and inelastic time-history analyses.

Influence of torsion on different torsionally stiff models, expressed as the ratio of edge displacements of asymmetric and symmetric structures, can be seen also in Figure 7.4.7, separately for elastic and inelastic behaviour. The ratio applies to mean values. The values of the coefficient of variation, indicating the scatter of results obtained for 16 time-histories, vary from about 0.1 to about 0.4 with an average of about 0.3. From the figure it can be seen that torsion increases displacements at the flexible edge and decreases them at the stiff edge (with a few exceptions at the stiff side in X-direction). Influence of torsion is larger in the case of a large eccentricity (model M30). There are relatively small differences between models with different types (but the same value) of eccentricity. Moderate influence of inelastic behaviour can be noticed. The influence is the largest in the case of R model, which is symmetric if the response is elastic. In almost all inelastic cases the torsional influence is larger in X- than in Y-direction. Consequently, in the majority of cases, inelasticity increases displacements in X-direction and decreases them in Y-direction. The observed phenomenon might be a consequence of the fact that the elements in Y-direction (i.e. transverse regarding to X) experience larger displacements and larger inelastic deformations.

Uni-directional versus bi-directional loading

In addition to simultaneous bi-directional excitation, analyses were performed also by applying uni-directional ground motion separately in X- and Y-direction. The computational procedure was similar as in the case of bi-directional excitation. The results are summarised in Figure 7.4.8, separately for elastic and inelastic behaviour. Two values are used for uni-directional results. The first one corresponds to the larger of the two values obtained by performing two separate analyses with uni-directional ground motion in X- and Y-direction, respectively. The second one is obtained by combining both values, determined by separate analyses with uni-directional input. For the combination, the SRSS rule is used, i.e. $u = \sqrt{(u_X^2 + u_Y^2)}$, where u denotes displacement. u_X and u_Y are displacements obtained by applying ground excitation only in X- and only in Y-direction, respectively.

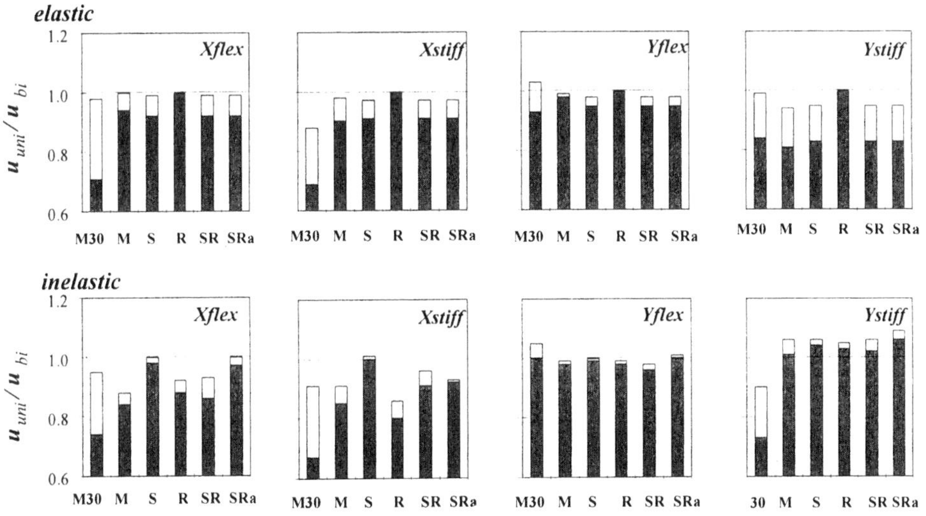

Figure 7.4.8: Ratio of mean values of maximum displacements for different models obtained by uni- and bi-directional excitation. The total height and the dark part of columns correspond to the SRSS combination and to the maximum uni-directional value, respectively.

Figure 7.4.8 demonstrates that, in all elastic cases and in the great majority of inelastic cases, uni-directional ground motion underestimates displacements. The underestimation is large in the case of large eccentricity (model M30). In general, the estimates based on uni-directional ground motions can be improved if the SRSS rule is applied, most notably in the case of large eccentricity. The SRSS rule has been widely used in linear analyses. In our study it provides very good results for elastic structures. The results for inelastic structures are not as favourable, but still adequate for most practical purposes. It is interesting that the best correlation is achieved for the most critical part of the structure, i.e. the flexible side in the Y-direction.

Under uni-directional loading, elements in transversal direction remain in elastic region, unless the eccentricity is very large, and provide torsional stiffness also in the case if the elements in the direction of loading yield. Under bi-directional loading, however, elements in both directions may yield simultaneously and a torsional mechanism may occur. The plastic mechanisms, including the torsional one, occur a limited number of times and that they last only a short time.

Conclusions

From the results of analyses several conclusions can been drawn. They are based on a very limited number of test examples. Nevertheless, at least some of them seem to be generally valid and they may contribute to a better understanding of seismic response of asymmetric buildings.

Torsionally stiff versus torsionally flexible structures: It is well known that there is an important difference between the seismic response of torsionally stiff (the first mode is predominantly translational) and torsionally flexible (the first mode is predominantly torsional) buildings. In the case of torsionally flexible buildings, not only the maximum displacements at the flexible side of the building increase (compared to symmetric structures), but also those at the stiff side of the building. Such behaviour is qualitatively different from that observed in torsionally stiff buildings and from that obtained in the case of static loading in mass centre. In torsionally flexible buildings torsional influence may be very large, thus there is a general agreement that it is preferable not to use such systems. Buildings, in which torsional flexibility occurs due to small torsional stiffness and strength, are more dangerous than those in which torsional flexibility occurs due to large mass moment.

Elastic versus inelastic response: For a nonlinear analysis, data on strength of elements are needed. Asymmetry can occur not only due to asymmetric distribution of mass and/or stiffness, but also due to asymmetric distribution of strength. Strength and stiffness of structural elements can be related in different ways. Consequently, for a general study of inelastic seismic response of asymmetric structures, a number of possible combinations of different types of eccentricities have to be studied. Moreover, different possibilities can be chosen for modelling force – deformation envelopes and hysteretic behaviour of structural elements. As a consequence, the number of parameters, relevant for the inelastic response, is much larger than in the case of a linear analysis. Even for a simple single-storey structure, a study has to be restricted to a highly limited number of parameters and to limit cases. Qualitatively, the response of elastic and inelastic models considered in our study is similar, with the exception of the model with only strength eccentricity, which behaviour in elastic range is symmetric. Quantitatively, moderate differences can be observed. In the majority of cases, inelastic behaviour increases displacements in the direction, which experiences smaller displacements and smaller inelastic deformations (i.e. X-direction), and decreases them in the other (Y) direction, both at the flexible and stiff side.

Type and magnitude of eccentricity: There are relatively small differences between investigated models with different types (but the same, moderate magnitude) of eccentricity. The torsional influence increases with an increase of eccentricity.

Uni-directional versus bi-directional excitation: In all investigated elastic cases and in the great majority of inelastic cases, uni-directional ground motion underestimates displacements. The underestimation is large in the case of large eccentricity. In general, the estimates can be improved if results of two independent uni-directional analyses with ground motion applied separately in two horizontal directions are combined using the SRSS rule. Most notable improvement can be achieved in the case of large eccentricity. The best correlation is achieved for the most critical part of the structure, which is the flexible edge in the direction, which experiences larger displacements and larger inelastic deformations (i.e. Y-direction). Based on this observation, it will be possible to extend the applicability of simplified nonlinear analysis procedures (based on pushover analysis) to asymmetric structures. The analysis will consist of two independent pushover analyses in two horizontal directions. In all cases, the transversal elements must be included.

MULTI-STOREY BUILDINGS

Test structures and ground motion

Five-storey steel frame buildings are used as test structures. The basic symmetric structures were designed by Mazzolani and Piluso (1997) according to Eurocodes 3 and 8. The storey heights are 4 and 3.5 m for the first storey and other storeys, respectively. Masses amount to 331 tons at the top storey and to 315 tons at all other storeys. The design spectrum for stiff soil, normalised to peak ground acceleration 0.35 g, was used. The behaviour (reduction) factor q = 6 was applied. A conservative estimate of the natural period was made, which resulted in design base shear of about 10% of the total weight of the building. The first structure is a space moment- resisting frame (S in Figure 7.4.9) in which all the beam-to-column connections are moment-resistant. In the second structure, there are only few moment-resisting bays at the perimeter (F1 in Figure 7.4.9). In this structure, the cost of connections is reduced. On the other hand, the cross-sections of the columns carrying the lateral loads have to be increased. As a result, the displacement shape of F1 building resembles to a cantilever beam and not to a frame (Figure 7.4.16). The third structure (F2) consists of the same moment-resisting frames as the second structure. However, the layout in the plan is different. In the new structure, the moment-resisting frames are located in the interior of the plan (Figure 7.4.9). Thus, the torsional stiffness and torsional resistance are greatly reduced. In such a way, a torsionally flexible structure is created. The fundamental period of torsional vibration amounts to 2.13 s, compared to 0.73 s in the original structure. The initial translational stiffness of all structures is practically the same. Accordingly, the fundamental translational periods are almost equal: 1.25 s for S structure, and 1.26 s for F structures. Semi-rigid variants of S and F1 structures were also studied. For structures denoted as S_R and $F1_R$ the strength of beam-to-column connections is reduced to 60% of the beam strengths (partial-strength rigid connections), whereas in structures S_S and $F1_S$, only the connection stiffness is reduced to 60% of the joint stiffness, which, according to Eurocode 8, represents the lower limit for rigid joints (full-strength semi-rigid connections). The fundamental periods of S_S and $F1_S$ structures are about 12% and 10% higher than the corresponding periods of rigid structures.

Beams are hot-rolled I-profiles. The sections of all columns in S structure, and of some columns in F structures are composed of two hot-rolled I or H-profiles welded along the longitudinal axis to form X-shaped section. Other columns are hot-rolled H profiles. Asymmetry is introduced in plan by taking into account 5% accidental eccentricity, as proposed by Eurocode 8. The influence of larger mass eccentricity was also investigated.

Six different ground motion records (with 2 horizontal components each) obtained during the 1979 Montenegro earthquake (3 records), the 1994 Northridge earthquake (2 records) and the 1995 Kobe earthquake (1 record) were used. For each record, the component with the higher peak ground velocity (v_g) was scaled to the same value and applied in Y-direction. Acceleration spectra are shown in Figure 7.4.10.

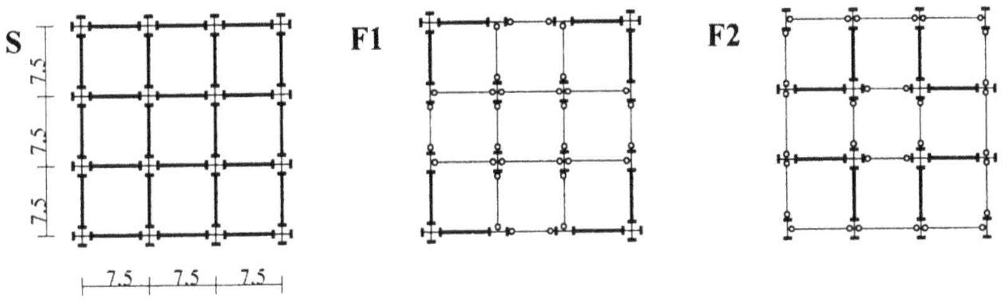

Figure 7.4.9: Plans of investigated buildings

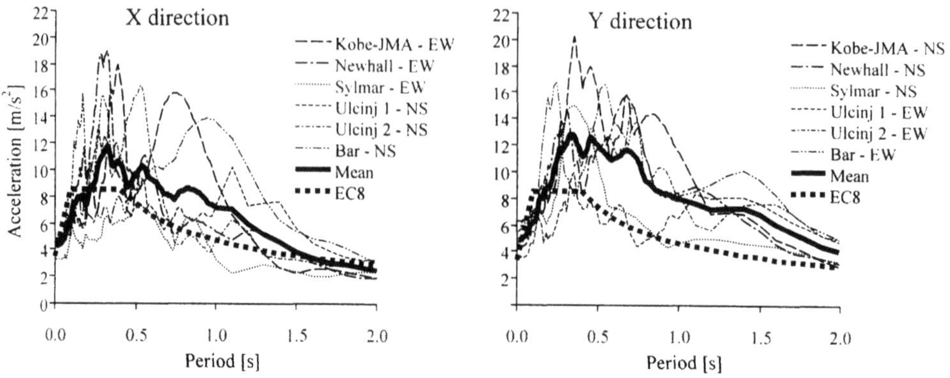

Figure 7.4.10: Acceleration spectra for 5% damp. (spectra for different ground motions, stronger comp. in Y-dir. scaled to $v_g = 70$ cm/s; mean spectrum; Eurocode 8 design spectrum with $q=1$)

Mathematical modelling

A pseudo three-dimensional global mathematical model, usually applied for linear analyses of multi-storey structures, was used. The model is composed of planar frames connected with rigid diaphragms. In this modelling approach, each column belonging to two frames (in two directions) is modelled independently in both directions, and is subjected to independent uni-axial bending in two directions, rather than to bi-axial bending. The compatibility of axial deformations in columns, belonging to two frames is not considered. For structural elements, a relatively simple mathematical model, typical for steel frame structures, is applied. All beams and columns are modelled as planar elastic beam elements with two nonlinear rotational springs at the two ends. The moment – rotation relationship for each spring is assumed to be elasto-plastic with zero post-yield stiffness. Torsional stifnesses and strengths of all elements are neglected. An initial study (Marusic and Fajfar, 1999) has shown that influences of axial force - bending moment interaction in columns and of the second order theory are small. Thus, they have not been taken into account. Initial moments due to vertical loading are considered. Secondary beams are oriented in Y-direction. Consequently, the initial moments influence the frames in X-direction. The damping matrix is proportional to mass matrix and initial stiffness matrix. The target damping is 5% in the first two modes. The flexibility of semi-rigid connections is taken into account in a simplified approximate way by reducing the second moment of area (moment of inertia) of beams.

Parametric study

A parametric study of dynamic behaviour of structural systems was performed with the CANNY-E program (Li, 1996). The dynamic response to simultaneous ground excitations in two orthogonal horizontal directions was studied. The parameters, varied in the study, include eccentricity of the mass centre and the intensity of ground motion. In addition to the symmetric structures, eccentric structures with 5, 10, and 15% eccentricity (the values are defined as the eccentricity divided by the dimension of the building in the plan and multiplied by 100) were analysed. In the case of semi-rigid structures, only 10% eccentricity was considered. The same eccentricity applies to both directions. Different intensities, defined by peak ground velocity v_g, were used. All accelerograms applied in Y-direction were scaled to the same v_g. Each accelerogram in X-direction was scaled with the same factor as the corresponding accelerogram in Y-direction. v_g amounts to 10, 40, 70 (only this velocity was used in the case of semi-rigid structures), and 100 cm/s. In the first case ($v_g = 10$ cm/s), the structures remain in the elastic range. The extent of plasticity increases with increasing velocity. For all investigated cases dynamic analyses were performed for 6 input ground motions. Maximum values for each ground motion were determined. Finally, mean values and standard deviations of selected response quantities were computed. The same procedure was repeated for uni-directional ground excitation in X- and in Y-direction. In addition, pushover analyses were performed. Horizontal loads with inverted triangular distribution were applied in mass centres, separately in X- and in Y-direction. Target displacement of the mass centre at the top of each building was equal to the mean value of the dynamic top displacement of the symmetric structure. The final results of pushover analyses were computed as SRSS combinations of results obtained from two independent analyses for loading in each of two horizontal directions.

TABLE 7.4.2

MEAN VALUES OF TOP DISPLACEMENTS OF SYMMETRIC STRUCTURES [cm]

	v_g = 10 cm/s		v_g = 40 cm/s		v_g = 70 cm/s		v_g = 100 cm/s	
Building	X - dir.	Y - dir.	X - dir.	Y - dir.	X - dir.	Y - dir.	X - dir.	Y - dir.
S	4.0	5.3	15.2	19.6	22.6	30.3	29.8	37.9
F1, F2	4.2	5.6	15.5	22.0	23.0	34.6	31.8	43.8

TABLE 7.4.3

MEAN VALUES OF TOP DISPLACEMENTS OF SYMMETRIC SEMI-RIGID STRUCTURES (v_g = 70 cm/s) [cm]

Building	S_R		S_S		$F1_R$		$F1_S$	
	X - dir.	Y - dir.	X - dir.	Y - dir.	X - dir.	Y - dir.	X - dir.	Y - dir.
	19.9	28.5	26.1	33.4	21.8	31.6	26.0	38.7

Results

Selected results are shown in Figures 7.4.11-7.4.19 and in Tables 7.4.2 and 7.4.3. The relationships between base shear and top displacement at geometrical centre, determined by pushover analyses (Figures 7.4.11 and 7.4.12), show that the overstrength factor, i.e. the ratio between the actual and design strength, is large (about 2.5 to 3). The influence of eccentricity on stiffness and strength is negligible in all cases. There is a small influence of initial moments due to vertical loading (compare X- and Y-direction, initial moments apply only to frames in X-direction), especially in the case of S-structure. Semi-rigid structures exhibit reduced strength or stiffness.

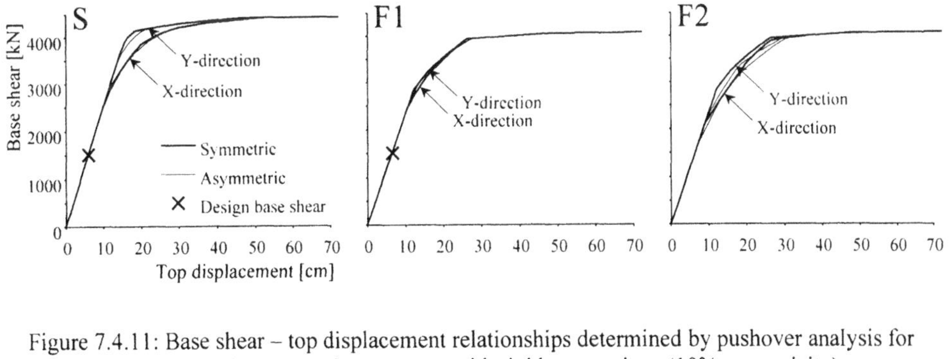

Figure 7.4.11: Base shear – top displacement relationships determined by pushover analysis for symmetric and asymmetric structures with rigid connections (10% eccentricity).

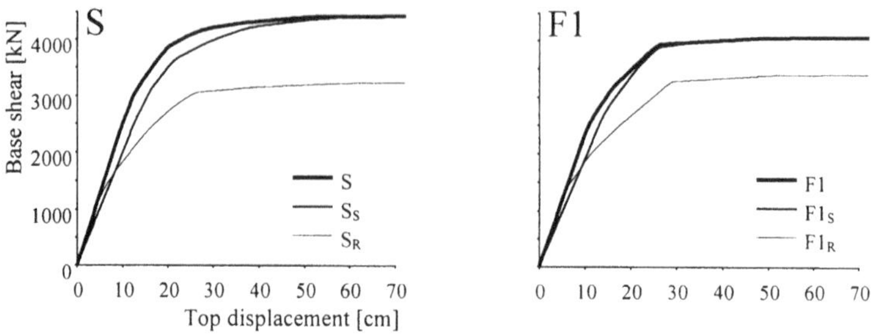

Figure 7.4.12: Base shear – top displacement relationships determined by pushover analysis for symmetric structures with rigid, semi-rigid, and partial-strength connections (X-direction).

Maximum dynamic displacements at the top of symmetric structures are given in Tables 7.4.2 and 7.4.3. (Note that all results of dynamic analyses, presented here, are mean values obtained from six ground motions, if not indicated otherwise. The values of the coefficient of variation, indicating the scatter of results, vary from about 0.1 to about 0.5 with an average of about 0.3.) Top displacements of asymmetric and corresponding symmetric structures, obtained by dynamic analyses and pushover analyses are compared in Figure 7.4.13 and 7.4.14. Ratios are given for four edges. In Figure 7.4.15, ratios of displacements, obtained by dynamic and pushover analysis are shown. Note that the displacements at the mass centre are the same for both analyses. Envelopes of displacements and storey drifts for selected cases (rigid and semi-rigid variants of structures S and F1, 10 % eccentricity, ground motion scaled to $v_g = 70$ cm/s) are shown in Figures 7.4.16 and 7.4.17. For comparison, the results obtained by pushover analyses are also plotted in Figure 7.4.16. For the same structures, rotations at element ends in the most displaced frames (at the flexible edge in Y-direction) are presented (Figure 7.4.18). Because of symmetry, only one half of the frame is plotted. The results correspond to Kobe ground motion, stronger component scaled to $v_g = 70$ cm/s. Only values larger than yield rotation are shown. Comparisons of results obtained by uni- and bi-directional input are summarised in Figure 7.4.19. The uni-directional values correspond to the larger of the values u_X and u_Y obtained by performing two separate analyses with uni-directional ground motion in X- and Y-direction, respectively, and to the values, combined according to the SRSS rule. In Figure 7.4.19 mean values of the ratios of uni-versus bi-directional displacements are shown for different subgroups of data obtained for structures with rigid connections. Standard variations of results are also shown.

Observations and conclusions

Based on the results of the study, several conclusions can be drawn. The numerical values given in the conclusions apply only to the investigated structures. Most of the qualitative observations, however, are considered to be generally valid. They mostly confirm the conclusions obtained within our research on single-storey and other multi-storey buildings. Similar conclusions have been obtained also by some other researchers.

The global behaviour of the structures, designed according to Eurocode 8 (without taking into account asymmetry, but with substantial overstrength), suggests that the appropriately detailed buildings would in most cases survive ground motions, which are much more severe than design ground motion. There are some important differences between the seismic response of S and F1 structures. They will be not discussed in this section that deals with torsional effects. Some more detailed observations and conclusions are as follows.

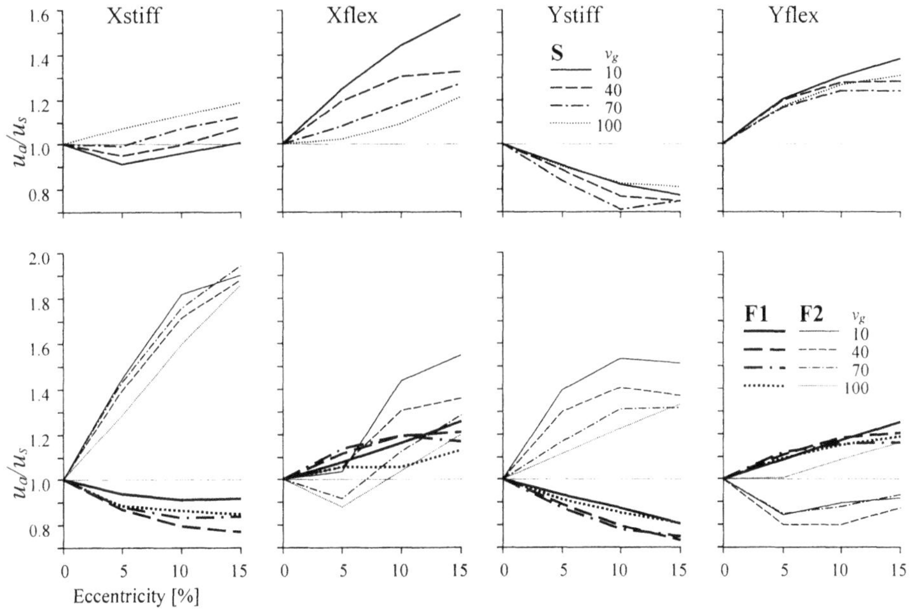

Figure 7.4.13: Ratio of top displacements in asymmetric and symmetric structures with rigid connections (mean values).

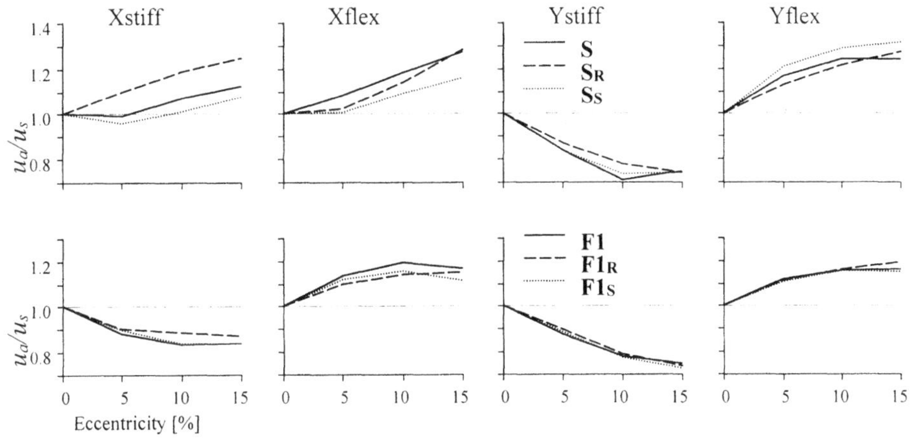

Figure 7.4.14: Ratio of top displacements in asymmetric and symmetric rigid and semi-rigid structures (mean values).

Influence of semi-rigidity: Deformation demand in structures with full-strength semi-rigid connections is larger than in corresponding structures with rigid connections. In the case of partial-strength rigid connections deformation demand is slightly smaller than in original structures (this observation cannot be generalized), whereas the ductility demand is increased. Moderate changes in damage distribution occur. As far the influence of torsion is concerned, no important difference between the structures with rigid and semi-rigid connections can be observed. All conclusions, which follow, apply to both types of structures.

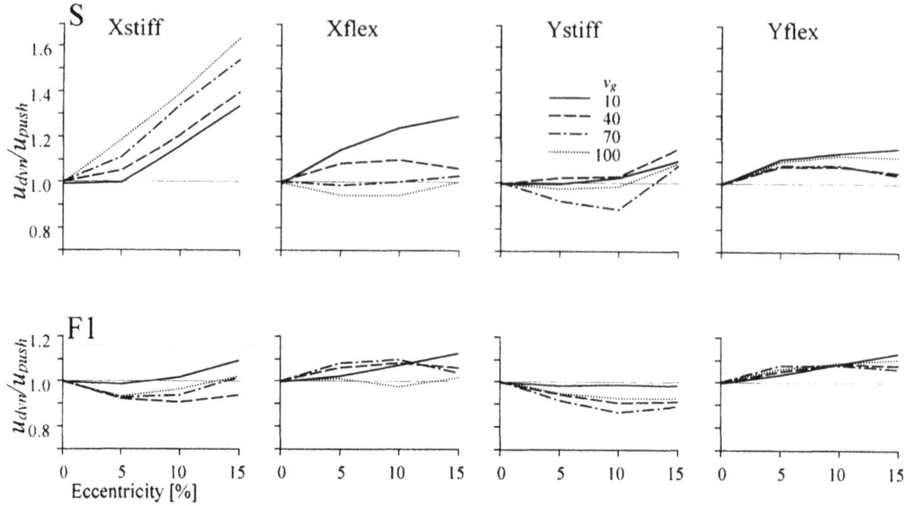

Figure 7.4.15: Ratio of top displacements obtained by dynamic analyses (mean values) and pushover analyses.

Influence of asymmetry: The behaviour of asymmetric torsionally stiff and torsionally flexible buildings is qualitatively different (see discussion in the chapter on single-storey buildings). As regards the structural systems, the largest influence of torsion can be observed in the torsionally flexible system F2. Such a system, in which the load resisting elements are located in the interior of the plan, is quite usual in the practice. For this system, in the most critical case the increase of top displacements due to torsion amounts to almost 100%. Note that displacements are measured at the perimeter. At the location of structural elements the increase of displacements due to torsion is much lower (see discussion in the chapter on single-storey buildings). Within the group of torsionally stiff buildings, larger influence of torsion can be observed in the space moment-resisting structure S. In this structure, the maximum increase of top displacements due to torsion amounts to about 23, 40, and 50% for 5, 10, and 15% eccentricity, respectively. In general, the influence of torsion increases with an increase in eccentricity. However, the relation is not linear. For S building, the torsional amplification at the flexible side generally decreases with an increase of the intensity of ground motion. The same trend can be observed at all edges of the F2 building. In the case of F1 building, the influence of ground motion intensity is small.

Uni- versus bi-directional excitation: Figure 7.4.19 demonstrates that uni-directional excitation slightly underestimates displacements if the maximum values obtained from two separate uni-directional ground motions are considered. Improved, mostly conservative estimates can be obtained with the SRSS combination of uni-directional results.

Pushover versus dynamic analysis: In the case of torsionally stiff structures, a pushover analysis yields qualitatively correct results (amplification) at flexible edges. However, the torsional effects at these edges are usually underestimated (see Figures 7.4.15 and 7.4.16). At stiff edges, a pushover analysis may provide severely underestimated results (X-direction of S building). On the other hand, overestimates can be also observed (F1 building). The error at the flexible side, which is more important, can be eliminated or at least reduced by increasing the eccentricity of the static lateral loading, i.e. moving the point of application from the mass centre toward the flexible edge (as is usual in the case of linear analyses according to codes). In principle, with a different adjustment of eccentricity the results at the stiff side could be matched as well. Some guidelines for the values of

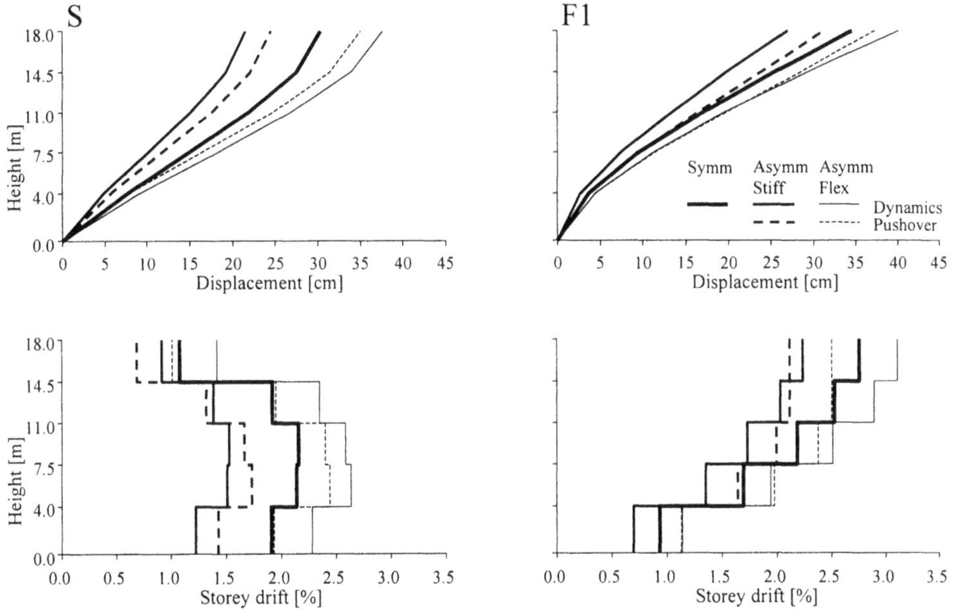

Figure 7.4.16: Mean values of envelopes of displacements and storey drifts at stiff and flexible edge in Y-direction for symmetric and asymmetric (10% mass eccentricity) structures (v_g=70 cm/s). Pushover results (SRSS combination) for asymmetric structures.

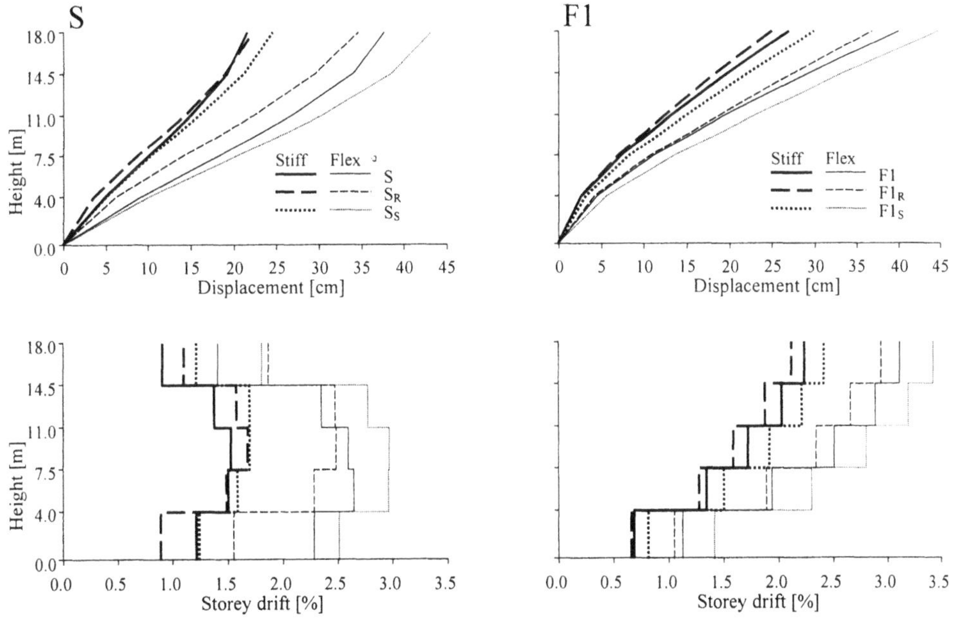

Figure 7.4.17: Mean values of envelopes of displacements and storey drifts at stiff and flexible edge in Y-direction for rigid and semi-rigid asymmetric (10% mass eccentricity) structures (v_g=70 cm/s).

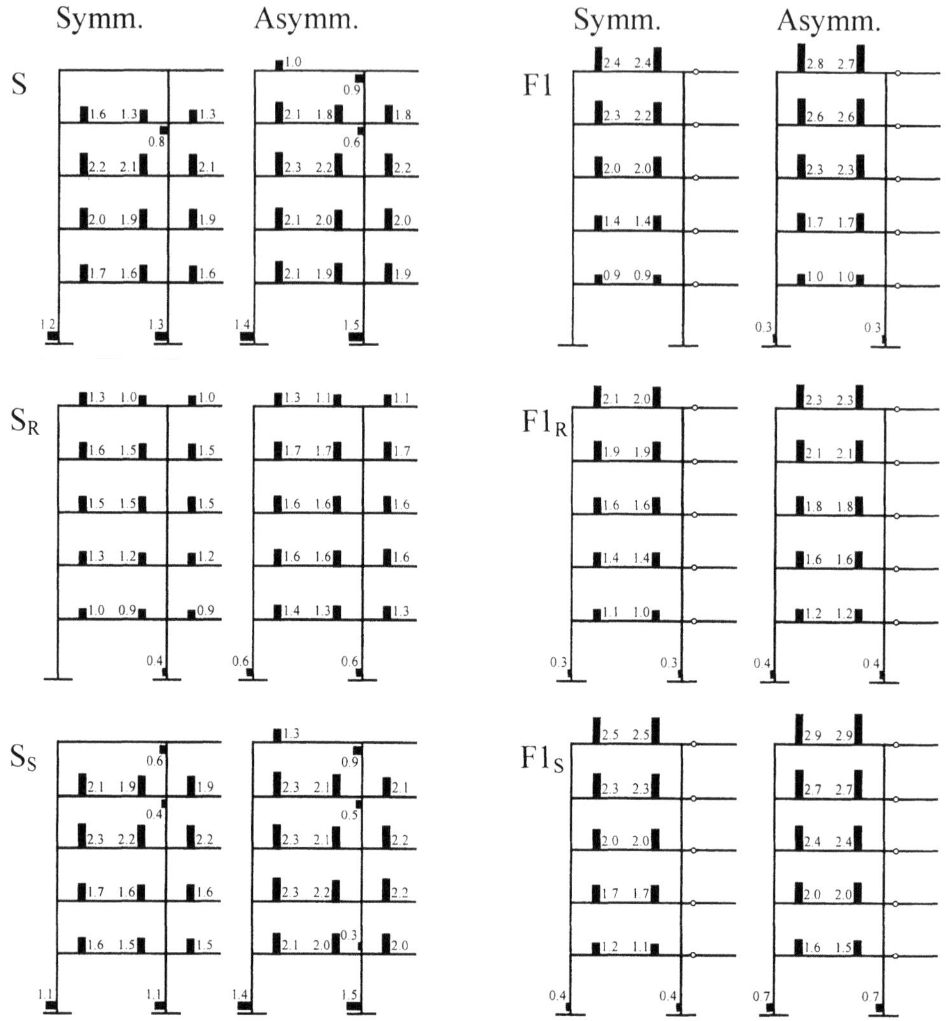

Figure 7.4.18: Maximum rotations at element ends (in 10^{-2} rad) of frames in Y-direction for symmetric and asymmetric structures (10% eccentricity, edge frames at the flexible side) for Kobe-JMA ground motion (bi-directional excitation with NS component in Y-direction, scaled to v_g = 70 cm/s).

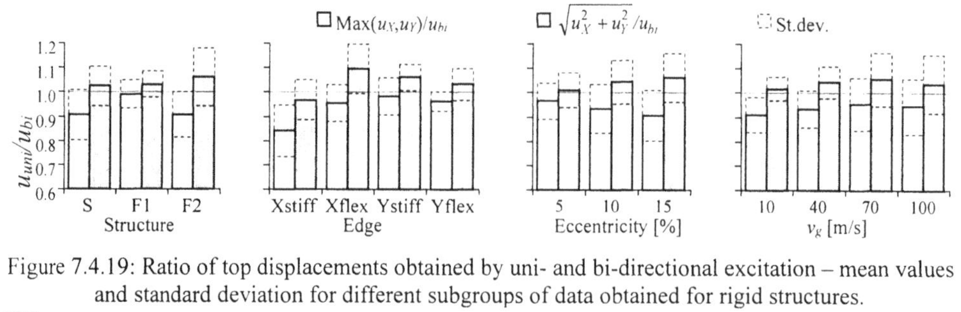

Figure 7.4.19: Ratio of top displacements obtained by uni- and bi-directional excitation – mean values and standard deviation for different subgroups of data obtained for rigid structures.

effective eccentricities have still to be determined. Pushover analysis does not take into account the higher mode effect and thus underestimates the seismic response in the upper parts of structures. In the case of torsionally flexible buildings, the dynamic and static responses are qualitatively different and pushover analysis, in principle, cannot provide adequate results.

Appendix: Simplified modelling of semi-rigid and partial-strength connections

The straightforward approach for mathematical modelling of semi-rigid and partial-strength beam-to-column connection is the use of a special nonlinear connection element. For the time being the number of appropriate computer programs for three-dimensional nonlinear dynamic analysis is very limited. The CANNY-E program, which has been used in our study, unfortunately does not provide an appropriate connection element. For this reason we modelled the semi-rigid and partial-strength connections by using reduced stiffness and strength of beams, respectively. Such an approach greatly simplifies the modelling and yields results of adequate accuracy, considering that our study is concerned mainly with the global structural behaviour.

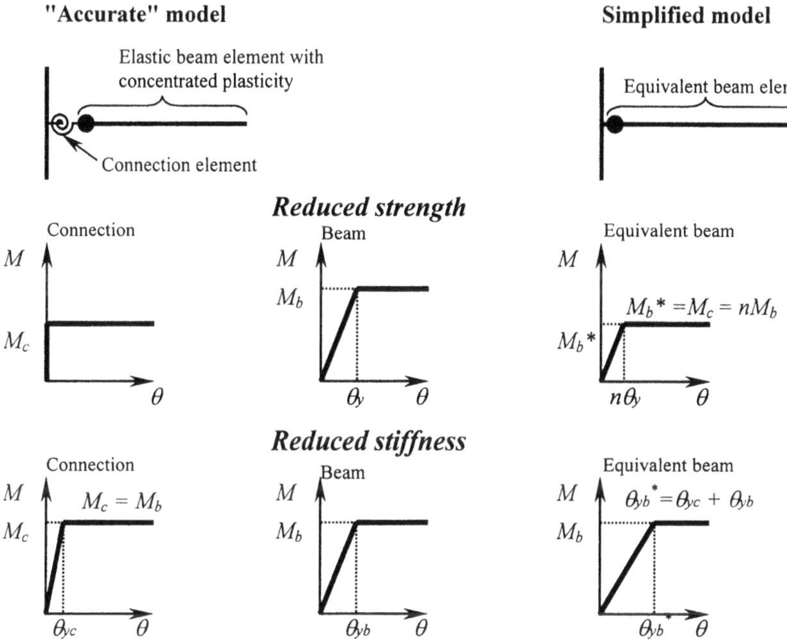

Figure 7.4.20: "Accurate" and simplified modelling of partial-strength and semi-rigid connections

The simplified modelling approach is illustrated in Figure 7.4.20. If the connection strength is smaller than the beam strength, the strength of the equivalent beam (i.e. the yield moment of the plastic hinge at the end of the elastic beam) is simply reduced to the value equal to the connection strength. If the connection is flexible, the elastic stiffness of the equivalent beam is reduced. The reduced beam second moment of area (moment of inertia) I_b^* can be determined as

$$I_b^* = I_b \frac{m}{m+6} \tag{7.4.3}$$

where I_b is the second moment of area of the original beam and m is defined as

$$m = \frac{S_{j,\,ini}}{EI_b/L_b} \tag{7.4.4}$$

where $S_{j,ini}$ is the initial rotational stiffness of the beam-to-column joint and L_b is the length of the beam. Eqn. 7.4.3 is based on assumption that the connection has zero length. A similar formula was proposed by Nader and Astaneh (1992). Their formula exactly fits one coefficient of the "exact" combined stiffness matrix of the connection element and beam, whereas formula 7.4.3 approximately fits all four coefficients connected with rotations. Details can be found in Marusic (2000).

The simplified modelling approach with the equivalent beam stiffness has been already used in elastic analysis (e.g. Nader and Astaneh, 1992, Chan, 1994). It can be applied also in the case of inelastic response. The approach was validated by means of an example. A planar frame, taken from the S structure, was modelled using equivalent beams (simplified model) and analysed with CANNY-E program. For comparison, the same frame was modelled using connection elements ("accurate" model) and analysed with the DRAIN-2DX program (Prakash et al, 1993). For the simplified model, there is practically no difference between the results obtained with CANNY-E or DRAIN-2DX. Pushover curves are compared in Figure 7.4.21a. Inelastic response to the Kobe-JMA NS accelerogram is shown in Figures 7.4.21b and 7.4.21c. Note that in the case of the "accurate" model, the rotations of the connection element and of the beam-end are superimposed. Hysteretic behaviour (moment versus rotation) of the beam-end at the first storey is shown in Figure 7.4.22. The results, presented in Figures 7.4.21 and 7.4.22 demonstrate that reasonably accurate response can be obtained by using the simplified modelling approach.

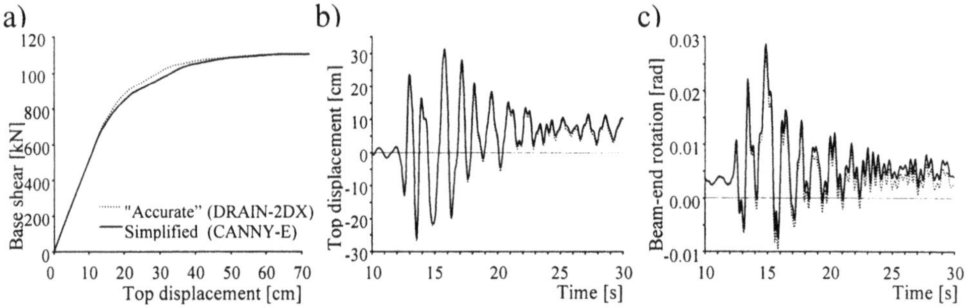

Figure 7.4.21: Comparison of results obtained by "accurate" and simplified model (a – pushover, b and c – dynamic response due to Kobe-JMA record NS component).

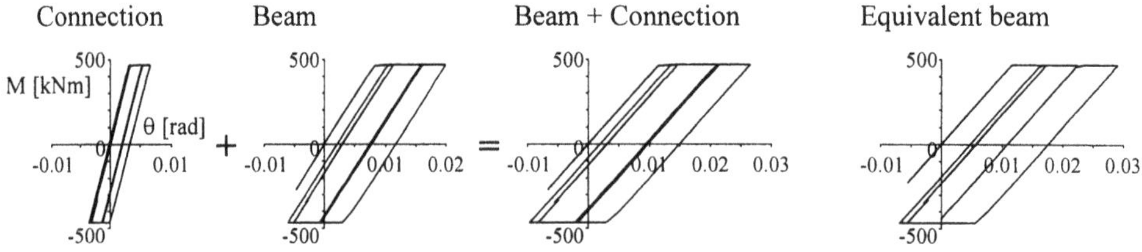

Figure 7.4.22: Comparison of moment-rotation relationship at the end of the beam for "accurate" and simplified models.

CONCLUSIONS

Based on the study of elastic and inelastic torsional response of a limited number of regular asymmetric single- and multi-storey buildings, the following conclusions can be drawn.

Type of structures: In torsionally flexible buildings for which, by definition, the first mode is predominantly torsional, very large amplifications of displacements may occur at all edges of the building due to torsion. There is a general agreement that it is preferable not to use such systems. In the case of torsionally stiff buildings, displacements increase mainly at the flexible (weak) sides of the building. The increase is larger in the case of a space moment-resisting frame structure than in the case of a structure with only few moment-resisting bays at the perimeter. It is questionable, however, is this conclusion can be generalized.

Influence of semi-rigidity: As far the influence of torsion is concerned, no important difference between the structures with rigid and semi-rigid connections can be observed.

Elastic versus inelastic response: For a nonlinear analysis, data on strength of elements are needed. Asymmetry can occur not only due to asymmetric distribution of mass and/or stiffness, but also due to asymmetric distribution of strength. Strength and stiffness of structural elements can be related in different ways. Consequently, for a general study of inelastic seismic response of asymmetric structures, a number of possible combinations of different types of eccentricities have to be studied. Moreover, different possibilities can be chosen for modelling force – deformation envelopes and hysteretic behaviour of structural elements. As a consequence, the number of parameters, relevant for the inelastic response, is much larger than in the case of a linear analysis. Even for a simple single-storey structure, a study has to be restricted to a highly limited number of parameters and to limit cases. Qualitatively, the response of elastic and inelastic models considered in our study is equal, with the exception of the model with only strength eccentricity, which behaviour in elastic range is symmetric. Quantitatively, moderate differences can be observed. Based on the studied test structures no clear general trends can be determined. In many cases, the influence of torsion decreases with an increase of the intensity of ground motion and with corresponding increased inelastic deformations.

Type and magnitude of eccentricity: A structure may be eccentric due to distribution of mass, stiffness and/or strength, or any combination of them. A relatively small difference between a very limited number of investigated simple models with different types, but the same magnitude of eccentricity was observed. The influence of torsion generally increases with an increase in eccentricity.

Uni-directional versus bi-directional excitation: In all investigated elastic cases and in the great majority of inelastic cases, displacements obtained from uni-directional ground motion are underestimated. The underestimation may be large in the case of large eccentricity. In general, improved, mostly conservative estimates can be obtained if results of two independent uni-directional analyses with ground motion applied separately in two horizontal directions are combined using the SRSS rule. Most notable improvement can be achieved in the case of large eccentricity. The best correlation is achieved for the most critical part of the structure, which is the flexible side in the direction, which experiences larger displacements and larger inelastic deformations.

Pushover versus dynamic analysis: In the case of torsionally flexible buildings, dynamic and static responses are qualitatively different, and pushover analysis, in principle, cannot provide adequate results. For torsionally stiff structures, pushover analysis yields qualitatively correct (amplified) results. They are usually underestimated results at flexible edges. This error can be eliminated or at least reduced by increasing the eccentricity of the static lateral loading, i.e. by moving the point of application from the mass centre toward the flexible edge (as is usual in the case of linear analyses according to codes). At stiff edges, the results of a pushover analysis may be considerably smaller or larger than the dynamic results. Pushover analysis does not take into account the higher mode effect and thus underestimates the seismic response in the upper parts of structures.

Implications for development of a simplified nonlinear analysis: Recently, especially after the 1994 Northridge and 1995 Kobe earthquakes, a consensus has been reached that present codes need significant improvements and expansion. The structural engineering community is now developing a new generation of design and rehabilitation procedures that incorporate performance based engineering concepts. For lower performance levels, e.g. "life safety" or "collapse prevention", the use of a nonlinear analysis procedure is inevitable. The long-term solution is nonlinear dynamic (time-history) analysis. However, for the time being and perhaps at least for the next decade, the appropriate method for a large number of buildings seems to be a nonlinear static (pushover) analysis combined with response spectrum approach. Such methods are used for example in new guidelines developed in the USA. Such a method is also the N2 method, developed at the University of Ljubljana (Gaspersic and Fajfar, 1996, Fajfar 1999). The simplified nonlinear methods have been developed for planar structures and their applicability has to be extended to asymmetric structures that experience torsional rotations. Here, inter alia, the problems arise at which point to apply static loading and how to combine the influence of the two horizontal excitations.

Based on the observations, reported above, it is possible to extend the applicability of simplified non-linear analysis procedures (based on pushover analysis) to asymmetric torsionally stiff structures. Two independent pushover analyses in two horizontal directions should be performed and results should be combined by the SRSS rule. The transverse elements must be included in the mathematical model. Further investigations are needed in order to provide some guidelines on the effective eccentricity of the point of application of lateral loading. In any case, it seems reasonable to use two limit cases of eccentricity in each direction. The larger one is intended to control the flexible edge and the smaller one the stiff edge. Furthermore, it is wise not to consider any computed favourable effect of torsion.

Eurocode 8: It has been confirmed that the appropriately detailed buildings designed according to the current version of Eurocode (without taking into account actual asymmetry, but with substantial overstrength), would in most cases survive ground motions, which are much more severe than design ground motion.

Further research needs: Extensive research is still needed on all topics discussed above. An additional problem, which has been not yet understood and has been hardly touched in our investigation, is the considerable difference between the torsional effects in X- and Y-direction.

REFERENCES

Chan, S.L. (1994). Vibration and Modal Analysis of Steel Frames with Semi-Rigid Connections. *Engineering Structures*, **16(1)**, 25-31.

Fajfar, P., Gaspersic, P. (1996). The N2 Method for the Seismic Damage Analysis of RC Buildings. *Earthquake Engineering and Structural Dynamics*, **25(1)**, 31-46.

Li, K.N. (1996). *Three-dimensional Nonlinear Dynamic Structural Analysis Computer Program Package CANNY-E.* CANNY Consultants Pte Ltd., Singapore.

Marusic, D., Fajfar, P. (1999). Comparative Study of Seismic Response of Steel Frame Structures, *Structural Dynamics - EURODYN'99*, Vol. **2**, 1159-1164 (Frýba L. and Náprstek J., eds.). Balkema, Roterdam, Netherlands.

Marusic, D. (2000). *Influence of Torsion on Seismic Response of Steel Buildings* (Ph.D. thesis in preparation). University of Ljubljana, Ljubljana, Slovenia.

Mazzolani, F.M., Piluso, V. (1996). *Theory and Design of Seismic Resistant Steel Frames*. E & FN Spon, London, UK.

Mazzolani, F.M., Piluso, V. (1997). The Influence of the Design Configuration on the Seismic Response of Moment-Resisting Frames, *Behaviour of Steel Structures in Seismic Areas: STESSA '97* (Mazzolani F.M. and Akiyama H., eds.), Edizioni, Salerno, Italy.

Nader, M.N., Astaneh-Asl, A. (1992). *Seismic Behavior and Design of Semi-Rigid Steel Frames*. Report No. UBC/EERC-92/06, University of California, Berkeley, USA.

Perus, I., Fajfar, P. (1999). On the Seismic Response of Idealised Asymmetric Single-Storey Structures. *Proceedings of the 2nd European Workshop on the Seismic Behaviour of Asymmetric and Irregular Structures*, Istanbul, Turkey.

Prakash, V., Powell, G.H., Campbell, S. (1993). *DRAIN-2DX, Base Program Description and User Guide, Ver. 1.10*. Report No. UBC/SEMM-93/17, University of California, Berkeley, USA.

Chapter 8

Design Methodology

8.1

GENERAL DEFINITIONS AND BASIC RELATIONS

Peter Fajfar

INTRODUCTION

Seismic design is still as much an art as a science. It strongly relies on empirical findings. However, some basic relations exist which define a conceptual framework of current design procedures and which represent the background of code requirements. The understanding of these relations is essential for the development of codes, for research in support of code development and for correct applications of codes.

The basic quantities, which control the seismic response, are accelerations and displacements. Seismic demand can be expressed in terms of both quantities. It can be compared with the capacity of structure expressed by the same quantities (note that acceleration can be related to strength). The structural response may be linear (in the case of a minor earthquake) or nonlinear (in the case of a strong ground motion). Relations exist between acceleration and displacement, not only for elastic but also for inelastic structural behaviour. Moreover, relations exist among elastic and inelastic response quantities. Here, a very important parameter is the so-called reduction factor (called in Europe behaviour factor q).

For an ideal elasto-plastic single-degree-of-freedom (SDOF) system, the basic relations can be presented graphically in the AD (acceleration - displacement) format. In the AD chart the basic relations can be visualised. The chart can be used for both traditional force-based design as well as for the new deformation-controlled (or displacement-based) design. The relations apply to SDOF systems. However, they can be used also for a large class of multi-degree-of-freedom (MDOF) systems, which can be adequately represented by equivalent SDOF systems.

In this chapter the basic relations, used in seismic design, will be defined, discussed and presented in a graphical form. Different definitions of behaviour (reduction) factors, used by different authors throughout this book, will be discussed.

HISTORICAL DEVELOPMENT

Early provisions for aseismic design, which were developed in Italy, Japan, and the U.S.A. in first decades of this century, required buildings to be designed to withstand a lateral force of about 10 per cent of their own weight. This value was determined purely empirically by observing the consequences of earthquakes. It is interesting to note that the order of magnitude of design seismic forces has remained more or less unchanged over decades in spite of tremendous progress in earthquake engineering and is still used in many areas of high seismicity. What has changed is the explanation of the fact that the great majority of well designed and constructed buildings survived strong ground motions, even if they were designed only for a fraction of the forces that would develop if the structure behaved entirely linearly elastic.

Investigations into structural behaviour during earthquakes at the beginning of 20th century led researchers to the conclusion that structures with a lateral strength of about 10% of their weight or more survive the strongest earthquakes. It was concluded, probably as a consequence of this finding (and of reasoning in terms of statics), that peak ground acceleration during strong earthquakes can reach a value of about 0.1 g. The first strong motion records obtained in California from 1933 onwards, demonstrated that peak ground acceleration can reach values larger than 0.3 g. After this, it was possible to explain the earthquake resistance of many buildings only by taking into account the energy dissipation capacity (ductility) of structures. However, in the case of an earthquake with accelerations in the range of 1 g, like those recorded in some recent earthquakes, a building with a base shear capacity of 10% of its weight would probably collapse unless an extremely large ductility was provided. The majority of structures do not possess such high ductility. Nevertheless, again the majority of the buildings meeting code requirements survived such severe ground motions. There are two possible explanations for this observation. First, that the actual demand was lower that it appeared to be, and/or second, that the actual capacity was greater than that which had been predicted in design. In fact, both factors are usually influencing the structural behaviour, because maximum acceleration is a poor measure of earthquake potential to damage medium- and long-period structures, and because the actual strength usually exceeds that required by codes.

In short, most structures can be designed taking into account reduced forces which are typically much lower than the linear elastic response forces. This fact is realised by the seismic design codes which use the force reduction factors (for example "behaviour factor" q in Eurocode 8, or "response modification factor" R in US codes) to define seismic design forces. The reduction factors are predominantly based on empirical observations of the behaviour of common structural systems during earthquakes. Consequently, on average they yield acceptable results. However, to describe a complex phenomenon of the response reduction for a particular structure by a single average number may be confusing and misleading. Therefore, the main factors that influence the response reduction should be identified, analysed, and quantified.

PHYSICAL BACKGROUND OF BEHAVIOUR FACTORS

The structural system is simulated by an equivalent SDOF model with a bilinear force - displacement envelope (Figure 8.1.1). The following relations apply

$$F_y = m\,\omega^2\,D_y \qquad (8.1.1)$$

$$\mu = \frac{D}{D_y} \qquad (8.1.2)$$

where F_y is the strength of the system (yield strength), m its mass, ω the natural frequency, D_y the yield displacement, D the maximum displacement, and μ the corresponding ductility factor.

A structure with a given strength F_y, which relies on energy dissipation through inelastic deformation, should have a limit deformation capacity which exceeds seismic demand in the case of severe earthquakes. It is well known that seismic demand in terms of ductility (which is related to deformation according to Eqn. 8.1.2) and seismic demand in terms of the strength F_y are interrelated. The problem can therefore be stated in a different way. Assuming that a certain deformation capacity is provided (or, in the case of a serviceabality limit state, a certain deformation is tolerated), the strength F_y of the

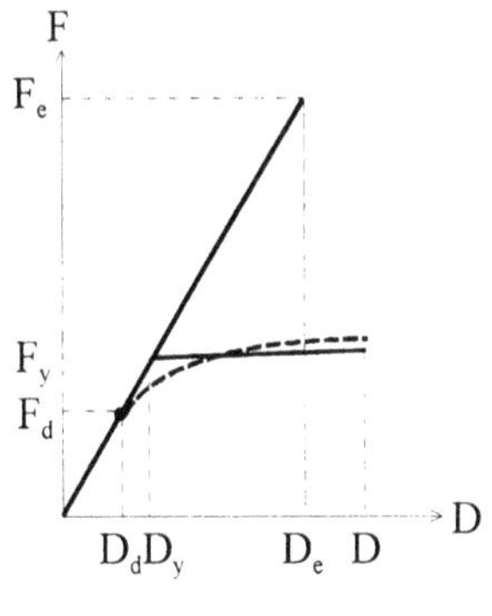

Figure 8.1.1: Idealized force - displacement relationships

system should at least be equal to the required strength. This approach is actually used in design procedures and can be written in the form

$$F_y = \frac{m\,A_e}{q_\mu} \tag{8.1.3}$$

where A_e is the value in the elastic (pseudo)acceleration spectrum and q_μ is the strength reduction factor, which is equal to the elastic strength demand $F_e = m\,A_e$ divided by the inelastic strength demand F_y

$$q_\mu = \frac{F_e}{F_y} \tag{8.1.4}$$

q_μ depends mainly, but not exclusively, on the prescribed target ductility and on the period of the system. Expressions similar to Eqn. 8.1.3 can be found in various seismic codes. However, an important difference should be noted between Eqn. 8.1.3 and the expressions in the codes. In Eqn. 8.1.3, F_y represents the actual strength, whereas the seismic forces in the codes correspond to design strength F_d which is, as a rule, lower than the actual strength. This difference is mainly due to overstrength, which is an inherent property of properly designed, detailed, constructed and maintained highly redundant structures.

Taking into account the overstrength

$$q_s = \frac{F_y}{F_d} \tag{8.1.5}$$

the following relation applies

$$q = \frac{F_e}{F_d} = \frac{F_e}{F_y}\frac{F_y}{F_d} = q_\mu q_s \tag{8.1.6}$$

Thus the total force reduction q, which is equal to the elastic strength demand F_e divided by the code prescribed seismic force F_d, can be defined as the product of the ductility dependent factor q_μ and the overstrength factor q_s. The factors q_μ and q_s are discussed in more detail in next sections.

The maximum relative displacement of the system, D, can be determined from Eqns. 8.1.1 – 8.1.3

$$D = \frac{\mu A_e}{q_\mu \omega^2} = \frac{\mu}{q_\mu} D_e \tag{8.1.7}$$

where D_e is the maximum relative displacement of the system with unlimited elastic behaviour subjected to the ground motion defined by elastic spectrum A_e. An alternative form of Eqn. 8.1.7 is

$$D = \mu \, q_s \, D_d \tag{8.1.7a}$$

where D_d is the maximum relative displacement of the system under design loading F_d.

DUCTILITY DEPENDENT FACTOR q_μ

The ductility dependent reduction factor q_μ has been the subject of extensive research. An excellent overview was made by Miranda and Bertero (1994). The reduction factor q_μ is, in the medium-period (velocity controlled) and long-period (displacement controlled) region, only slightly dependent on the period T, and is roughly equal to the prescribed target ductility μ. In the short-period (acceleration controlled) region, however, the q_μ factor depends strongly on both T and μ. In the limit case of an infinitely rigid structure ($T = 0$) there is no reduction due to ductility (q_μ=1). A moderate influence of hysteretic behaviour and damping can be observed in the whole period region. The transition period from the period dependent part to the more or less period independent part of q_μ spectrum is roughly equal to the transition period between the acceleration controlled short-period region and the velocity controlled medium-period region T_C. This period is an important characteristic of the ground motion and is often referred to as the the characteristic period or the "predominant" period. It roughly corresponds to the period at which the largest energy is imparted to the structure.

In our derivations and discussions, simple bilinear q_μ spectra will be used

$$q_\mu = (\mu - 1)\frac{T}{T_C} \qquad T \le T_C \tag{8.1.8}$$

$$q_\mu = \mu \qquad T \ge T_C \tag{8.1.9}$$

Eqns. 8.1.7 and 8.1.9 suggest that, in the medium- and long-period ranges, the equal displacement rule applies, i.e. the displacement of the inelastic system is equal to the displacement of the corresponding elastic system with the same period. Several researchers have empirically demonstrated that this rule is approximately valid for a large number of structural systems.

OVERSTRENGTH FACTOR q_S

Strength exceeding that required by codes (overstrength) is a major factor contributing to the seismic resistance of structures. The name "overstrength factor" is used by different authors for different quantities. This fact may cause confusion and misunderstanding.

According to Paulay and Priestley (1992) and several other authors the overstrength factor is basically defined on the section (of a member) level as the ratio between the maximum strength and the nominal or ideal strength. The nominal or ideal strength of a section is based on the established theory predicting a prescribed limit state with respect to failure of this section. It is derived from the dimensions and code-specified nominal material strength properties. Factors that may contribute to strength exceeding the nominal or ideal value include steel strength greater than the specified yield strength, strain hardening and strain rate effects. The overstrength factor according to this definition is typically relatively small and its quantification is difficult or impossible.

Other overstrength factors are also used by the same authors, e.g. flexural overstrength factor and system overstrength factor. The flexural overstrength factor measures the flexural overstrength in terms of the required strength for earthquake forces alone at one node point of the structural model. The system overstrength factor represents the sum of the overstrengths of a number of interrelated members. (All definitions are taken from Paulay and Priestley, 1992).

According to several other authors the overstrength factor is defined on the level of the whole structure as a ratio between the actual structural yield level (note that, in actual structures, modelled as a MDOF system, this is not the first yield level but the yield level in an idealised bilinear force – deformation diagram) and the code prescribed strength demand arising from the application of prescribed loads and forces. It results from the following groups of sources: (a) redistribution of internal forces in the inelastic range in the case of ductile, statically indeterminate (redundant) structures; difference between the design level and required member strength (e.g. allowable versus yield stresses, load factors); member oversize (due to discrete member sizes and due to desired uniformity of members for constructibility); minimum requirements according to code provisions regarding proportioning and detailing; multiple loading combinations (e.g. effects of gravity loads); deflection constraints on system performance; architectural considerations. (b) conservatism in mathematical models; effects of structural elements that are not considered as a part of lateral load resisting system; effects of nonstructural elements. (c) higher material strength than those specified in the design, strain hardening and strain rate effects. The influence of the majority of (a) group factors can be easily at least approximately quantified by a nonlinear pushover analysis, while the (b) group sources are less reliable or require sophisticated mathematical modelling and may be neglected in practical design.The (c) group factors are the same as those determining the basic overstrength factor as defined by Paulay and others, and they are difficult to be quantified.

It is obvious that overstrength may come from a variety of sources and that, in real structures, it varies widely, depending on the material and the type of the structural system, the structural configuration, the number of stories, detailing, and the kind and the date of the code to which the structure was designed.

IMPLEMENTATION IN CODES

The code suggested values of q-factors are essentially of an empirical origin. So, in addition to ductility, they generally automatically imply overstrength, although this is usually not explicitly realised. According to Eurocode 8, Part 1.1 (EC8, 1994), »the behavior factor q is an approximation of the ratio of the seismic forces, that the structure would experience if its response was completely elastic with 5% viscous damping, to the minimum seismic forces that may be used in design – with a conventional linear model – still ensuring a satisfactory response of the structure«. Typically, the majority of codes use constant behaviour factors (up to $q = 8$) throughout the entire spectrum. Having in mind the fact that the ductility dependent factor q_{μ} decreases in the short-period region, while the overstrength factor typically increases in this region (Figure 8.1.2), it appears that a constant, period independent overall behaviour factor q, as used in the majority of codes, might be a very rough but reasonable approximation.

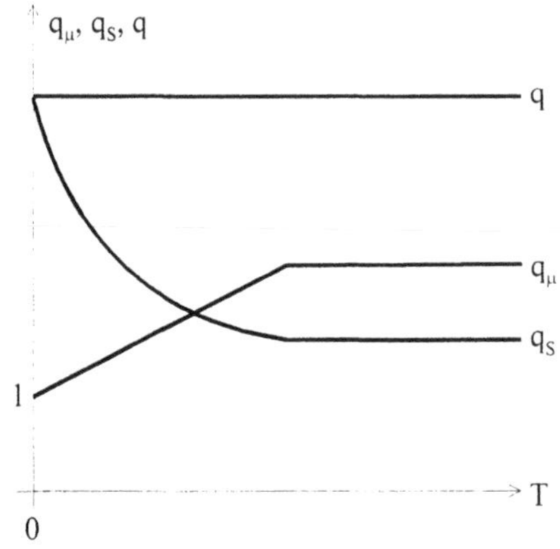

Figure 8.1.2: Idealized qualitative relationship between reduction factors and period

However, using a single number (q-factor) to account for the implicit interaction of at least two parameters (ductility and overstrength) my be misleading. An appropriate solution appears to be to define the q-factors in aseismic codes according to Eqn. 8.1.6 as the product of the equivalent global ductility factor q_{μ} and the overstrength factor q_S. An appropriate and not excessively demanding way for the determination of the overstrength is a nonlinear static (pushover) analysis.

Recently, it has been widely recognised that, in the case of severe earthquakes, deformations are usually more important than forces. Consequently, different deformation controlled (displacement based) design approaches have been proposed. In the majority of codes, however, not enough attention has been paid to displacements which are generally the main source of damage. The displacement demand in codes is usually computed as

$$D = q_D D_d \tag{8.1.10}$$

where the value of the factor q_D is equal to or similar to the value of the behaviour factor q. By comparing Eqns. 8.1.10 and 8.1.7a and considering Eqns. 8.1.6 and 8.1.8 it can be seen that Eqn. 8.1.10 underestimates the displacement demand in the short-period region, especially in the case of larger inelastic deformation.

VISUALISATION OF BASIC QUANTITIES AND RELATIONS IN ACCELERATION – DISPLACEMENT FORMAT

A convenient possibility for the presentation of demand spectra is the acceleration - displacement (AD) format, in which spectral accelerations are plotted against spectral displacements, with the periods T represented by radial lines. In the AD format, accelerations and displacements, which control the seismic response, are plotted in the same diagram. Furthermore, seismic demand can be directly compared with the capacity of the structure expressed by the same quantities (note that acceleration can be related to strength). This idea has been employed in the capacity spectrum method (Freeman, 1998) which recently became very popular in the USA and also in Japan. In this method, the capacity of a structure is compared with the demands of earthquake ground motion on the structure by means of a graphical procedure. The graphical presentation makes possible a visual evaluation of how the structure will perform when subjected to earthquake ground motion. The method is easy to understand. The capacity of the structure is represented by a force - displacement curve. By converting forces to accelerations the capacity diagram is obtained. In the capacity spectrum method, the demands of the earthquake ground motion are defined by highly damped elastic spectra. This is a contraversial approach. Recently, it was modified by the author (Fajfar, 1999) in order to allow the presentation of seismic demand in terms of inelastic spectra rather than equivalent elastic spectra. The intersection of the capacity diagram and the demand spectrum provides an estimate of the inelastic acceleration (strength) and displacement demand (Figure 8.1.5). In a chart, plotted in AD format, the basic relations can be visualised. The chart can be used for both traditional force-based design as well as for the increasingly popular deformation-controlled (or displacement-based) design. In this chapter, the basic relations, given in previous chapters, will be presented in AD format.

Starting from the elastic acceleration spectrum, we will determine the inelastic spectra in AD format. For an elastic SDOF system the following relation applies

$$D_e = \frac{T^2}{4\pi^2} A_e \tag{8.1.11}$$

where A_e and D_e are the values in the elastic acceleration and displacement spectrum, respectively, corresponding to the period T and a fixed viscous damping ratio. A typical smooth elastic acceleration spectrum for 5% damping (used as elastic spectrum for medium soil conditions in Eurocode 8), normalised to a peak ground acceleration of 1.0 g, and the corresponding elastic displacement spectrum, are shown in Figure 8.1.3a. Both spectra can be plotted in the AD format (Figure 8.1.3b).

Eqn. 8.1.3 can be written in terms of accelerations rather that forces

$$A = \frac{A_e}{q_\mu} \tag{8.1.12}$$

Eqns. 8.1.7 and 8.1.12 define inelastic acceleration and displacement spectra. They depend on the reduction factor q_μ which is defined by Eqns. 8.1.8 and 8.1.9.

Starting from the elastic spectrum shown in Figure 8.1.3b, and using Eqns. 8.1.7 – 8.1.9 and 8.1.12, the demand spectra (for the constant ductility factors μ) in AD format can be obtained (Figure 8.1.4).

In order to directly compare the seismic demand with the capacity of the structure, the force – deformation relationship of SDOF system has to be transformed into the so-called capacity diagram which is defined in the AD format. The capacity diagram can be easily obtained by dividing forces in the force – displacement relationship by the mass

$$A = \frac{F}{m} \tag{8.1.13}$$

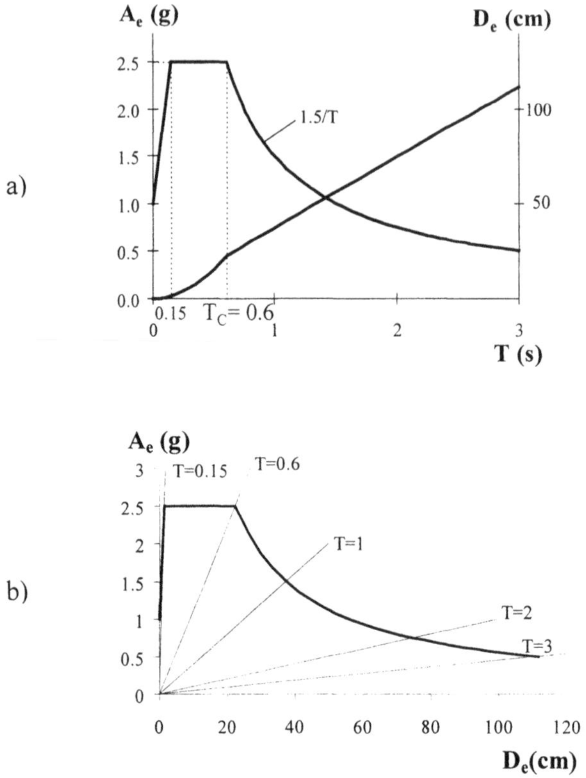

Figure 8.1.3. Typical elastic acceleration (A_e) and displacement spectrum (D_e) for 5 % damping, normalized to 1.0 g peak ground acceleration. a) Traditional format. b) AD format.

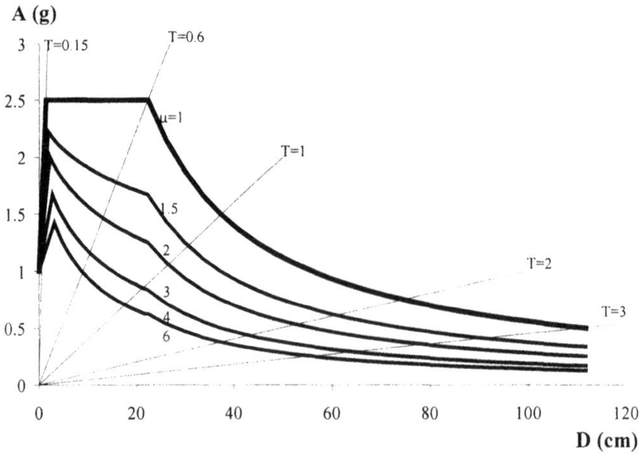

Figure 8.1.4: Demand spectra for constant ductilities in AD format normalized to 1.0 g peak ground acceleration

The basic quantities and relations, used in seismic design, can be visualised in Figure 8.1.5 which applies for medium- and long-period structures (for short-period structures see Figure 8.1.6). The visualization of the procedure may help in better understanding the relations between the basic quantities. Both the demand spectra and the idealised elasto-plastic capacity diagram have been plotted in the same graph in the AD format. The intersection of the radial line corresponding to the elastic period of the idealised bilinear system T with the elastic demand spectrum A_e defines the acceleration demand (strength) required for elastic behaviour and the corresponding elastic displacement demand A_e. The yield acceleration A_y represents both the acceleration demand and the capacity of the inelastic system. The reduction factor q_μ can be determined as the ratio between the accelerations corresponding to the elastic and inelastic systems

$$q_\mu = \frac{A_e(T)}{A_y} \tag{8.1.14}$$

If the elastic period T is larger than or equal to T_C, the equal displacement rule applies and the inelastic displacement demand D is equal to the elastic displacement demand D_e (see equations 8.1.7 and 8.1.9, and Figure 8.1.5a). From triangles in Figure 8.1.5a it follows that the ductility demand μ is equal to q_μ. If the elastic period of the system is smaller than T_C, the reduction factor is smaller than μ and the displacement demand is larger than in the case of the corresponding elastic structure with the same period (Figure 8.1.5b).

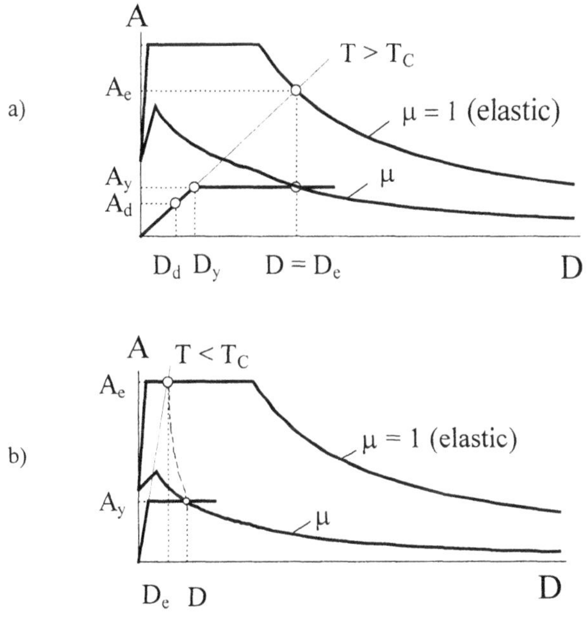

Figure 8.1.5: Elastic and inelastic demand spectra versus capacity diagram. Period of the system is (a) in medium- or long-period range or (b) in short-period range

In both cases ($T < T_C$ and $T \geq T_C$) the inelastic demand in terms of accelerations and displacements corresponds to the intersection point of the capacity diagram with the demand spectrum corresponding to the ductility demand μ. At this point, the ductility factor determined from the capacity diagram and the ductility factor associated with the intersecting demand spectrum are equal.

The expected global performance can be assessed by comparing displacement capacity and demand. The determination of seismic capacity is not discussed here. It relies largely on empirical or semi-empirical values. The effect of cumulative damage can easily be taken into account by using the so-called "equivalent ductility factors" (Fajfar 1992). The idea behind the equivalent ductility factor is to reduce the monotonic deformation capacity of an element and/or a structure as a consequence of cumulative damage due to the dissipation of hysteretic energy. Cumulative damage effects caused by several inelastic cycles may be important especially for existing structures with poor detailing subjected to long duration ground motions.

If necessary, the displacement demand can be modified as well, e.g. in order to take into account larger displacements expected for systems with narrow hysteresis loops or negative post-yield stiffness. The effect is equivalent to an appropriate modification of the reduction factor q_{μ}.

Both force-based and deformation-based approaches can be easily visualised with the help of Figure 8.1.5. In these two approaches, different quantities are chosen at the beginning. Let us assume that the approximate mass is known. The usual force-based design typically starts by assuming the stiffness (which defines the period) and the approximate global ductility capacity. The seismic forces (defining the strength) are then determined, and finally displacement demand is calculated. In direct displacement-based design, the starting points are typically displacement and/or ductility demands. The quantities to be determined are stiffness and strength. The third possibility is a performance evaluation procedure, in which the strength and the stiffness (period) of the structure being analyzed are known, whereas the displacement and ductility demands are calculated. It should be noted that, in all cases, the strength corresponds to the actual strength and not to the design base shear according to seismic codes, which is in all practical cases less than the actual strength.

DISPLACEMENT SPECTRUM IN LONG- AND VERY LONG-PERIOD RANGE

The spectra in Figures 8.1.3 – 8.1.5 have been intentionally cut off at the period $T = 3$ s. At longer periods the displacement spectrum is typically constant. Consequently, the acceleration spectrum in the long-period range typically decreases with the square of the period T. Depending on the earthquake and site characteristics, the constant displacement range of the spectrum may begin at even at shorter periods, e.g. at about 2 s (Tolis and Faccioli, 1999). In the very long-period range, spectral displacements of all systems (elastic and inelastic) decrease to the value of the peak ground displacement d_g which is, for the time being, very uncertain. Elastic displacement spectrum, extended to the long- and very long-period range, may be idealized according to Figure 8.1.6 (Fajfar, 1999, Bommer et al, 2000). Reliable values for corner periods periods T_D, T_E and T_F , and for the peak ground displacement d_g have still to be determined. It should be noted that these values strongly depend on the characteristics of ground motion, especially on the characteristic period T_C.

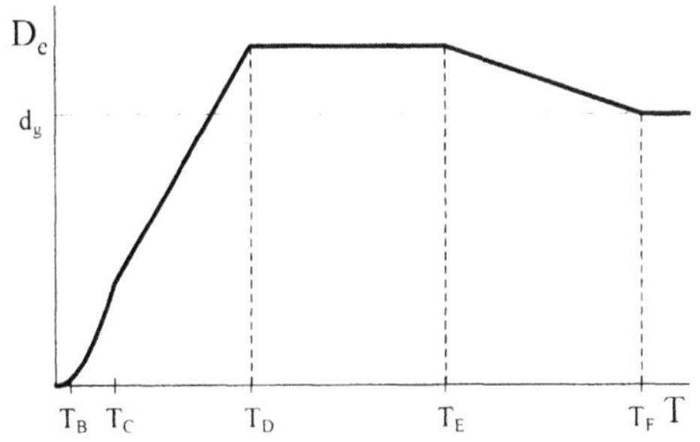

Figure 8.1.6: Idealised elastic response spectrum for displacements

According to the equal displacement rule, inelastic displacements are approximately equal to elastic ones for all periods larger than T_C. Based on this assumption, elastic displacement spectrum applies in the long and very long-period range also for inelastic structures.

MDOF STRUCTURES

So far, all discussions have been restricted to SDOF systems. For real structures, mostly MDOF models are used. The response spectrum approach, used in previous chapters, is by definition not applicable for inelastic MDOF systems. However, the seismic behaviour of a large class of MDOF structural systems can be closely approximated by equivalent SDOF models. In such cases, all considerations of previous subchapters, with small modifications, apply also to MDOF systems.

The starting point is a force - displacement relationship of the MDOF system obtained by a pushover analysis (static analysis under monotonically increasing lateral loads). In the case of building structures the representatives for force and displacement are usually base shear and lateral displacement at the top, respectively. The force - displacement relationship of the equivalent SDOF system is obtained by a simple transformation of forces and displacements. The transformation factors between the quantities in SDOF and MDOF systems depend on the procedure used for the determination of the equivalent SDOF system. One, perhaps the most convenient possibility for the transformation is the relation (Fajfar, 1999)

$$Q = \Gamma Q^* \tag{8.1.15}$$

where Q^* represents the quantities in the equivalent SDOF system (force and displacement), and Q represents the corresponding quantities in the MDOF system (base shear and top displacement). The constant Γ, which is usually called modal participation factor, is defined as

$$\Gamma = \frac{\sum m_i \Phi_i}{\sum m_i \Phi_i^2} \tag{8.1.16}$$

where m_i and Φ_i are mass and the component of the assumed displacement shape Φ in the i-th storey, respectively. Note that the assumed displacement shape Φ is normalised – the value at the top is equal to 1. Note also that any reasonable shape can be used for Φ. Only in a special case Φ represents the first mode shape. The value in the numerator represents the mass of the equivalent SDOF system

$$m^* = \sum m_i \Phi_i \tag{8.1.17}$$

Since the same transformation is used both for forces and displacements (Eqn. 8.1.15), the initial stiffness of the equivalent SDOF system remains the same as of the MDOF system.

The actual force – displacement relationship has to be idealised into an elastic – perfectly plastic form. Here, engineering judgement has to be used. For an idealised equivalent SDOF system all relations, defined in previous subchapters, apply.

The transformation from a MDOF to an equivalent SDOF system is based on the assumption that the displacement shape Φ is time-independent. In addition, the lateral load distribution, used for the determination of the base shear - top displacement relationship, has to be assumed. Eqn. 8.1.15 has been obtained by assuming lateral loads P_i which are in each storey proportional to the product of storey mass m_i and the component of assumed lateral displacement Φ_i

$$P_i = m_i \ \Phi_i \tag{18}$$

The most essential is the first assumption on the displacement shape that remains constant throughout the time history. This assumption is reasonable only for structures that vibrate predominantly in a single mode. It may not be acceptable for structures with important higher mode and torsional effects, and for irregular structures with specific strength and/or stiffness distributions over the height and/or over the plan. For such structures, modification factors have to be introduced in analyses (see for example Seneviratna and Krawinkler, 1997). However, in a qualitative sense, the basic relations still apply for the global behaviour.

BEHAVIOUR FACTORS USED IN THIS BOOK

A great variety of interpretations of behaviour factors is present in earthquake engineering community. An excellent overview of different approaches for the determination of behaviour factors was prepared by Mazzolani and Piluso (1996). The variety, which is causing some confusion when results obtained by different authors are compared, is witnessed also in this book. In this subchapter we will try to compare and to comment behaviour factors used throughout this book.

First, it should be noted that all definitions and relations given in previous subchapters (of Chapter 8.1) are based on an idealised elasto-plastic SDOF system subjected to equivalent static load (determined from response spectrum). The situation becomes more complicated if one attempts to determine behaviour factors for an actual structure modelled as a MDOF system (and not as an equivalent SDOF system) and/or if a dynamic time-history analysis is used.

If a multi-linear force-displacement diagram, corresponding to a MDOF system, is used instead of a bilinear diagram, typically two characteristic forces (or accelerations) exist in addition to the design force. One corresponds to the first yield and the other one to the ultimate limit state. In such a case the overstrength may be divided in two components as done in Chapter 7.4. In that chapter, the ratio between the base shear at first yield V_I and the design base shear V_d is called »design overstrength«, whereas the ratio between the base shear at ultimate limit state V_y and V_I is called »redundancy«. The product of both is called »total overstrength». In Chapter 7.4, a distinction is made between »total reduction factor R«, which is defined as the ratio between the base shear in the indefinitely elastic system V_e and V_d, and the »behaviour factor q«, which is defined as a ratio between V_e and V_I. Considering the definition of behaviour factor q in EC8 (see subchapter Implementation in codes), it follows that the »total reduction factor R« in Chapter 7.4 corresponds to the behaviour factor q in EC8, whereas the »behaviour factor q« in Chapter 7.4 is typically smaller than the behaviour factor q in EC8.

It should be noted that there is no consensus about the definition of the »ultimate limit state«. Different criteria like exceeding of a predefined storey drift or element rotation are used. The formation of the plastic mechanism has been also used as an ultimate limit state in the research reported in this book. Fortunately, the choice of the definition of the ultimate limit state does not influence much the corresponding force since the force - deformation curve is flat in the relevant region. Consequently, the definition of the ultimate limit state does not influence much the q-factor as well. However, it strongly influences the deformations at the ultimate limit state. Interestingly, no much attention has been paid to the definition of the »first yield«, although different possibilities exist for this definition as well (Sanchez and Plumier, 1999).

The first yield state is taken as a reference for determination of q-factor in the majority of chapters in this book. In Chapter 8.4, however, the q-factor is defined as the ratio between V_e and V_y. This definition corresponds to the ductility dependent behaviour factor q_μ (Eqn.8.1.4). The same ratio is used as q-factor for SDOF systems in Chapter 7.3.

Additional complexity is introduced if q-factors are determined based on dynamic (time-history) analysis. An approach, in principle consistent with the approach based on static analysis, has been proposed by Aribert and Grecea (»the base shear approach«, see Chapter 8.4). However, as already mentioned, the definition of the q-factor in this approach (ratio of V_e and V_y) is different from the definition in EC8 (ratio of V_e and V_d) and also from definitions used by other authors in this book.

Among the researchers studying steel structures, the most popular definition of q-factor based on dynamic analysis seems to be the ratio between the maximum ground acceleration leading to ultimate limit state and that corresponding to the first yield. This definition, which has been widely used in research presented in this book (e.g. Chapters 7.4, 8.4, 8.5), seems to yield results which are usually similar to those obtained from the definition based on the ratio of shear forces. However, many cases exist, where the approach does not work (see discussion in Chapter 8.4). Again, the definition is not consistent with the EC8 definition since the reference level is the first yield and not the design level.

The q-factors obtained by dynamic analysis, especially those obtained with the ratio of ground accelerations, are highly sensitive to the characteristics of ground motions. The values for the same structure obtained with different ground motions, reported in this book, show a large scatter. Differences are due to different time-histories and different spectral shapes. Note that for structures with the fundamental period in the medium- or long-period range, maximum ground acceleration is a parameter which generally exhibits a very poor correlation with structural damage.

A basically different approach for the determination of q-factor is based on the energy concept (»Energy approach«, Chapter 8.3). In this approach, the essential assumption is related to the energy dissipation capacity of the structure. The values of the computed q-factors strongly depend on this assumption.

Based on the above discussion, it can be concluded that q-factors, used in different parts of this book, are defined and determined in different ways. Thus, comparisons can be made only with great care. Furthermore, it should be noted that none of the q-factors, as defined throughout this book, corresponds to the EC8 behaviour factor.

REFERENCES

Bommer, J.J., Elnashai, A.S, Weir, A.G. (2000). Compatible Acceleration and Displacement Spectra for Seismic Design Codes. *Proc. 12th World Conference on Earthquake Engineering*, Auckland.

EC8 (1994). *Eurocode 8* – Design Provisions for Earthquake Resistance of Structures – Part 1-1: General Rules – Seismic Actions and General Requirements for Structures, ENV 1998-1-1, CEN.

Fajfar, P. (1992). Equivalent Ductility Factors Taking into Account Low-Cycle Fatigue. *Earthquake Engineering and Structural Dynamics* **21**, 837-848.

Fajfar, P. (1999). Capacity Spectrum Method Based on Inelastic Demand Spectra, *Earthquake Engineering and Structural Dynamics* **28**, 979-993.

Fajfar, P. (1999). Slovenian Comments to Eurocode 8 (submitted to CEN). Slovenian Working Group for Seismic Codes, March 1999, Ljubljana.

Fajfar, P., Krawinkler, H., editors (1997). *Seismic Design Methodologies for the Next Generation of Codes,* Balkema, Rotterdam.

Freeman, S.A. (1998). Development and use of capacity spectrum method, *Proc. of the 6th US National Conference on Earthquake Engineering*, Seattle, CD-ROM, EERI, Oakland.

Mazzolani, F.M., Piluso, V. (1996). *Theory and Design of Seismic Resistant Steel Frames*, E&FN Spon, London.

Miranda, E., Bertero, V. V. (1994). Evaluation of Strength Reduction Factors for Earthquake Resistant Design, *Earthquake Spectra* **10**, 357-379.

Paulay, T., Priestley, M.J.N. (1992). *Seismic Design of Reinforced Concrete and Masonry Buildings,* J. Wiley & Sons.

Sanchez, L., Plumier, A. (1999). Particularities Raised by the Evaluation of Load Reduction Factors for the Seismic Design of Composite Steel-Concrete Structures. *Stability and Ductility of Steel Structures* (D.Dubina and M.Ivanyi, eds.), 41-48, Elsevier.

Seneviratna, G.D.P.K., Krawinkler, H. (1998). *Evaluation of Inelastic MDOF Effects for Seismic Design.* The John A. Blume Earthquake Engineering Center Report No. 120, Stanford University.

Tolis, S.V., Faccioli, E. (1999). Displacement design spectra. *Journal of Earthquake Engineering* **3,** 107-125.

Vidic, T., Fajfar, P., Fischinger, M. (1994). Consistent Inelastic Design Spectra: Strength and Displacement. *Earthquake Engineering and Structural Dynamics* **23**, 502-521.

8.2

ENERGY APPROACH

Jordan Milev, Peter Sotirov, Zdravko Bonev, Nikolay Rangelov, Tzvetan Georgiev

INTRODUCTION

It is not efficient to design buildings that will not suffer any damages during any earthquake in the areas with high degree of seismic risk. The design philosophy of the current earthquake codes, including Eurocode 8 is as follows:

1) Buildings should be prevented from collapsing during the most severe earthquakes. However some non-destructive damages may be permitted and some of the energy input during the severe earthquake is dissipated through inelastic deformations.
2) Building should be designed to remain almost within the elastic range for the earthquakes, which are expected to occur several times during the lifetime of the building.

The basic problem in earthquake resistant design is to grasp the nature of seismic input on buildings. Despite high irregularity and spatial distribution of ground motions, the energy input into a building is a stable quantity. Akiyama (1985) defined the total energy input exerted by an earthquake, E, which was absorbed by the structure as

$$E = W_p + W_e + W_h \tag{8.2.1}$$

where: W_e :kinetic energy and elastic strain energy;
W_h :damping energy caused by miscellaneous damping effects other than inelastic deformations;
W_p :the cumulative inelastic strain energy defined by Akiyama (1985) as damage of the structure;
E : total seismic energy input;

Akiyama (1985) defined that the "collapse" occurred when one story of a multi-story building loses its restoring force and collapses. And the term "almost elastic range" refers to a case when the displacements of all stories do not exceed the yield story displacement. The condition under which the structure can remain almost elastic is

$$W_Y \geq W_e = E_s - W_h \tag{8.2.2}$$

where: W_Y :elastic vibration energy adsorbed by the entire structure while it is in the almost elastic range;

E_s :energy input of the earthquakes, which are expected to occur several times during the lifetime of the building;

On the other hand, the condition under which the structure can survive without collapse is

$$W_u \geq W_p = E_u - W_h - W_e \tag{8.2.3}$$

where: W_u :the cumulative inelastic strain energy adsorbed by the entire structure while the weakest story is brought to a collapse state;

E_u :energy input of the most severe earthquakes;

In the multi-story buildings it is very important to know how the energy input induced by an earthquake is distributed over the entire structure. Past earthquakes damage reveals that damage concentration is the main cause of the collapse of buildings. Damage distribution in each story depends primarily on the strength distribution along the height of the structure. The alternative expression of Eqn. (8.2.3) is

$$W_{ui} \geq W_{pi} \tag{8.2.4}$$

where: W_{ui} :energy that can be adsorbed by i-th story until it reaches a collapse state;

W_{pi} :energy adsorbed by i-th story;

ENERGY APPROACH FOR BEHAVIOUR FACTOR EVALUATION

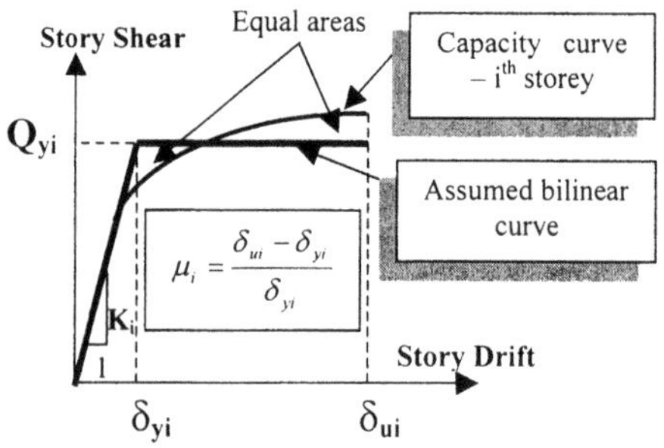

a) storey monotonic behaviour

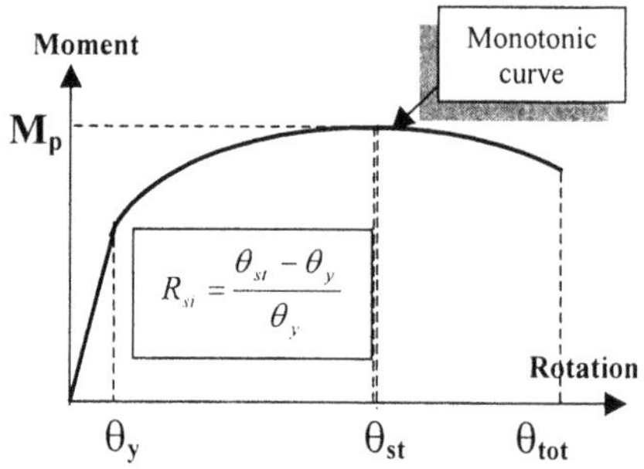

b)Plastic rotation capacity of beam-columns

Figure 8.2.1: Monotonic behaviour

The evaluation of the total energy input from an earthquake to the damage is made by Housner (1956) and Akiyama (1985). They assume that energy input attributable to the damage of an elastic-plastic system is the same as that producing damage in the relevant elastic system. The energy input attributable to the structural damage, E_D is expressed by:

$$E_D = E - W_h = W_p + W_e \approx \frac{1}{2} M S_v^2 \tag{8.2.5}$$

where: M : total mass of the structure;

S_v : spectral pseudo-velocity response of a damped SDOF for the considered earthquake.

Following Akiyama (1985) and Mazzolani & Piluso (1986), if the storey restoring forces of a multi-storey structure are assumed with the elastic-perfectly plastic characteristics (Fig. 8.2.1[a]), then the inelastic strain energy, W_{pi}, which is adsorbed by the i-th story is expressed by

$$W_{pi} = Q_{yi} \delta_{yi} \eta_i , \tag{8.2.6}$$

where:

η_i : cumulated ductility ratio of the story (see Fig. 8.2.2);

$Q_{yi} = \alpha_i \sum_{j=i}^{N} m_j g$: yield shear force of the i-th story;

α_i : yield shear force coefficient;

m_j : mass of the j-th story;

N : number of stories;

g :gravity acceleration;

$\delta_{yi} = \frac{Q_{yi}}{K_i}$: elastic story deformation under Q_{yi};

K_i :the stiffness of the i-th story.

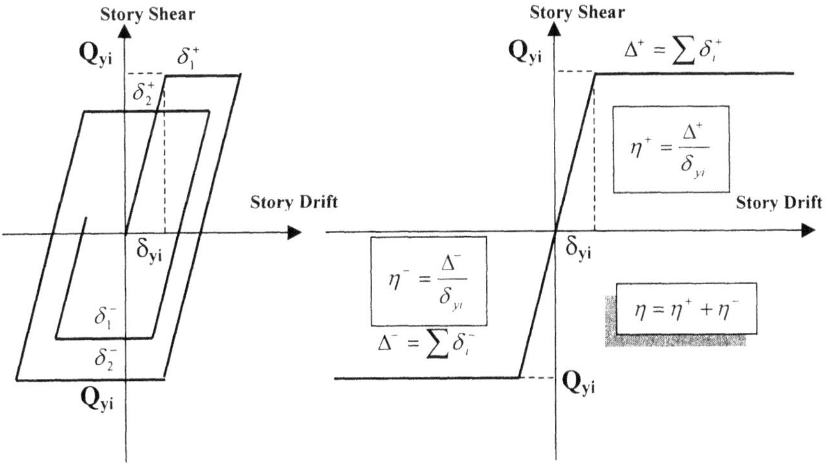

Figure 8.2.2: Cumulated ductility ratio definition

The equivalent stiffness K_{eq}, meaning that the spring constant of the SDOF system with the same natural period, T_1, as the period of the fundamental mode of the relevant multi-story building, is introduced by the following equation

$$K_{eq} = \frac{4\pi^2 M}{T_1^2} \tag{8.2.7}$$

Then by substituting the expressions for Q_{yi} and δ_{yi} in Eqn. (8.2.6) and by denoting $c_i = \frac{4\pi^2}{K_i T_1^2} \frac{\left(\sum_{j=i}^{N} m_j\right)^2}{M}$, the equation for the adsorbed inelastic strain energy in the i-th story becomes

$$W_{pi} = \frac{Mg^2 T_1^2}{4\pi^2} c_i \alpha_i^2 \eta_i \tag{8.2.8}$$

The inelastic strain energy adsorbed by the entire structure, W_p, is a sum of the inelastic strain energy dissipated in the stories

$$W_p = \sum_{J=1}^{N} W_{pi} \tag{8.2.9}$$

The damage distribution coefficient is introduced $\gamma_i = \frac{W_p}{W_{pi}}$. Then the inelastic strain energy dissipated by the entire structure W_{ps} could be expressed by the inelastic strain energy W_{pi} absorbed by the i-th story

$$W_p = \gamma_i W_{pi} \tag{8.2.10}$$

Substituting Eqn. (8.2.8) into Eqn. (8.2.10) the expression for W_p becomes

$$W_p = \frac{Mg^2 T^2}{4\pi^2} \gamma_i c_i \alpha_i^2 \eta_i \tag{8.2.11}$$

The elastic vibration strain energy in the first story is within the range

$$0 \le W_{e1} \le \frac{Q_{y1}\delta_{y1}}{2} \tag{8.2.12}$$

The upper bound given by Eqn. (8.2.12) is assumed, approximating the elastic vibration energy (see Fig. 8.2.2),

The expressions for Q_{y1} and δ_{y1} in Eqn. (8.2.12) are applied and the following approximate equation for the elastic vibration energy of the first story is derived

$$W_{e1} = \frac{(\alpha_1 Mg)^2}{2K_1} \tag{8.2.13}$$

It is obvious that the relation between the elastic vibration energy of the first story and the elastic vibration energy of the entire structure is as follows

$$\frac{W_{e1}}{W_e} = \frac{\delta_{y1}}{\delta_{eq}} = \frac{K_{eq}}{K_1} \tag{8.2.14}$$

Employing Eqn. (8.2.13), (8.2.14) and (8.2.7) the following expression for W_e is derived

$$W_e = \frac{1}{2} M \left(\frac{T_1}{2\pi} \alpha_1 g \right)^2 \tag{8.2.15}$$

The energy absorbed by the entire structure until the collapse occurs in the observed story (i-th story) can be expressed using the Eqs. (8.2.11) and (8.2.15)

$$W_i = W_e + W_p = \frac{1}{2} Mg^2 \left(\frac{T}{2\pi} \right)^2 \left[1 + 2c_i \gamma_i \eta_i \left(\frac{\alpha_i}{\alpha_1} \right)^2 \right] \tag{8.2.16}$$

The condition under which the observed story (i-th story) is prevented for collapse is

$$W_i \geq E_D \tag{8.2.17}$$

Then by substituting Eqs. (8.2.16) and (8.2.5) in (8.2.17), taking into consideration the relationship between spectral pseudo-velocity and spectral pseudo-acceleration response, $S_v = \frac{T}{2\pi} S_a$, the Eqn. (8.2.17) becomes

$$\alpha_1 g \geq \frac{S_a}{\sqrt{1 + 2c_i \gamma_i \eta_i \left(\frac{\alpha_i}{\alpha_1} \right)^2}} \tag{8.2.18}$$

Considering that $Q_{y1} = \alpha_1 gM$ is the required yield base shear strength and $Q_{e1} = S_a M$ is the elastic maximum base shear force corresponding to the spectral acceleration response Sa the Eqn. 8.2.18 becomes

$$Q_{y1} = \frac{Q_{e1}}{q_i} \tag{8.2.19}$$

where:

$$q_i = \sqrt{1 + 2c_i\gamma_i\eta_i\left(\frac{\alpha_i}{\alpha_1}\right)^2} \qquad (8.2.20)$$

is the lower bound of the q-factor derived by the assumption that the collapse in the i-th story of the building is prevented.

Therefore the value of the q-factor that assures that all stories will be prevented from the collapse is

$$q = \min(q_1, q_2, ..., q_N) \qquad (8.2.21)$$

Akiyama (1985) made three main assumptions, which are based on the large amount of experimental data and parametric studies evaluation, for calculation of the necessary parameters in Eqn. (8.2.20):

(1) The optimum yield shear force coefficient distribution is established;

$$\frac{\alpha_{opt,i}}{\alpha_1} = 1 + 1.593.X - 11.85X^2 + 42.6X^3 - 59.5X^4 + 30X^5 \quad \text{for } 0.2 < X \leq 1.0$$
$$\frac{\alpha_{opt,i}}{\alpha_1} = 1 + 0.5X \quad \text{for } X \leq 0.2 \qquad (8.2.22)$$

where: $X = x_i/H$, x_i is height of the floor of the relevant storey and H is the total height of the building.

Note: The Eqn. (8.2.22) is derived in the case of shear type structures. It is necessary to add ***1.3X⁴*** *to the right side of Eqn. (8.2.22) in the case of flexural-shear type structure.*

(2) The fundamental damage distribution law is introduced;

$$\gamma_i = \frac{\sum_{J=1}^{N} s_j p_j^{-n}}{s_i p_i^{-n}}$$
$$p_i = \frac{\alpha_i}{\alpha_{opt,i}} \qquad (8.2.23)$$
$$s_i = \left(\sum_{j=i}^{N}\frac{m_j}{M}\right)^2\left(\frac{\alpha_{opt,i}}{\alpha_1}\right)^2\left(\frac{K_1}{K_i}\right)$$

where: n=12 for "weak column-strong beam" structures and n=6 for "weak beam- strong column" structures.

(3) The energy input attributable to the structural damage is expressed by

$$E_D = \frac{E}{\left(1 + 3h + 1.2\sqrt{h}\right)^2} \tag{8.2.24}$$

where: h : damping constant.

The original Kato-Akiyama method for q-factor evaluation for the given earthquake requires the following steps:

(1) The story stiffness K_i are calculated. The Muto's method is usually used in Japanese practice;
(2) The story yield shear force coefficients α_i are calculated. The simplified method of AIJ is usually applied in Japanese practice;
(3) The limit value of the cumulated story ductility ratio η_{lim} is assumed.
(4) The energy input attributable to the damage E_D is calculated by applying Eqn. (8.2.5) and by using the scaled pseudo-velocity response spectra of studied earthquake;
(5) The cumulative inelastic strain energy W_p of the entire structure is calculated by applying Eqs. (8.2.5) and (8.2.15);
(6) The damage distribution coefficients γ_i are calculated by applying Eqs. (8.2.23) and (8.2.24). Then the story inelastic strain energy W_{pi} for each story is calculated by applying Eqn. (8.2.10).
(7) The cumulated story ductility ratios η_i are calculated by applying Eqn. (8.2.6). If the maximum η_i is equal (with some tolerance) to the assumed η_{lim} then go to step 8. Otherwise the pseudo-velocity response spectrum of studied earthquake is scaled again and steps (4)-(7) are repeated until the ultimate accelerogram multiplier is found;
(8) The q-factor of the structure is calculated by applying Eqs. (8.2.20 and (8.2.21).

MODIFICATION OF AKIYAMA-KATO METHOD FOR THE q-FACTOR EVALUATION

The original Akiyama-Kato method does not require any non-linear analysis. However the evaluation of each story capacity needs several assumption concerning the simplified expression of the inelastic strain energy absorbed by the entire structure and the optimum distribution of this work between different storeys. A modification of the original Akiyama-Kato method has been already proposed by Milev (1998). The most important revisions of the original Akiyama-Kato method being as follows:

(1) The i-th story stiffness K_i and yield shear forces Q_i are calculated by performing the non-linear push-over analysis on the considered structure. The distribution of horizontal forces along the height of the structure is considered on the basis of story shears obtained by elastic multi-modal analysis of the structure.
(2) The limit value of the cumulated story ductility ratio η_{lim} is assumed.
(3) The cumulative inelastic strain energy W_p for the entire structure and the story inelastic strain energy W_{pi} of each story are calculated by performing the non-linear time history analysis with the scaled acceleration time history (recorded or artificial);
(4) The damage distribution coefficients γ_i are calculated by applying Eqn. (8.2.10).
(5) The cumulated story ductility ratios ηi are calculated by applying Eqn.(8.2.6). If the maximum η_i is equal (with some tolerance) to the assumed η_{lim} then go to step 6. Otherwise the

acceleration time history is scaled again and steps (3)-(5) are repeated until the ultimate accelerogram multiplier is found;

(6) The q-factor of the structure is calculated by applying Eqs. (8.2.20) and (8.2.21).

The modification of Akiyama-Kato method does not require the simplified expression of the inelastic strain energy absorbed by the entire structure and the optimum distribution of this work between different storeys. However it needs both non-linear static pushover and dynamic time history analysis.

LIMIT VALUE OF THE CUMULATED STORY DUCTILITY RATIO

Both the original and modified Akiyama-Kato methods require to judge the limit value of the story cumulated ductility ratio η_{lim}. The assumption made by Akiyama (1988) is that the collapse occurs when the energy dissipated at the observed story during the entire time history in each direction (positive or negative) is equal to the critical energy dissipated under monotonic loading. Under this hypothesis

$$\eta_{\lim} = 2\mu_{\lim} \tag{8.2.25}$$

where: μ_{lim}: limit value of storey monotonic ductility in one direction,

For the shear type frame, the columns only are deformed. Therefore the storey monotonic ductility μ_{lim} is related to the rotation capacity of the columns, (R_{st}=R_{col}) as follows:

$$\mu_{\lim} = \frac{2}{3}R_{st} + 2 \tag{8.2.26}$$

The factor “2/3” in Eqn. 8.2.26 considers that the beams deform, too. The additional “2” takes into account energy dissipation capacity during the post critical behaviour. (see Mazozlani & Piluso (1996) for more details).

For the “weak beams-strong column” frames mainly the beams are supposed to dissipate energy during the earthquake. According to Akiyama & Yamada (1997) the story ductility is related to the beam rotation capacity (R_{st}=R_b) as follows:

$$\mu_{\lim} = R_{st} \tag{8.2.27}$$

Considering Eqs. (8.2.25), (8.2.26) and (8.2.27) the following values for η_{lim} are assumed in this study:

$$\begin{aligned} &\eta_{\lim} = 2R_{st} && \text{in the case of "weak beams - strong columns" frames} \\ &\eta_{\lim} = \frac{4}{3}R_{st} + 4 && \text{in the case of shear type frames} \end{aligned} \tag{8.2.28}$$

STUDIED FRAMES

Frame design

Some parametric studies are performed for behaviour factors evaluation of some typical for the Bulgarian construction practice steel-moment resisting frames. Six real frames with two types of haunching (A-type and B-type, see Fig. 3.3.6) are designed and their seismic behaviour is studied. A total number of 12 studied frames has been carefully designed, having 1 to 3 bays and 2 to 6 stories. The following notation is used:

RF $N_b\, N_s\, K_h$,

where:

RF = 'Recos Frame';

N_b = Number of bays;

N_s = Number of stories;

K_h = Key denoting the shape of the beam flanges at the beam-to-column connections (i.e. the type of haunching):

"*A*" = linear haunching (Fig. 3.3.6), and

"*B*" = haunching aimed at provoking the plastic hinge location at a specific point (Fig. 3.3.6).

In fact, the "*B*-type" haunching is an equivalent to the "dog bone" idea.

The frames are as follows:

- one span two stories frame – RF12A and RF12B
- two spans three stories frame – RF23A and RF23B
- two spans four stories frame – RF24A and RF24B
- three spans four stories frame – RF34A and RF34B
- three spans five stories frame – RF35A and RF35B
- three span six stories frame – RF36A and RF36B

Results for the frames type B only are presented herein.

The dimensions adopted are as follows: all bays = 6.0 m, all story heights = 3.6 m, all erection splices located at 1.0 m from column centre line, frames @ 6.0 m longitudinally.

Both columns and beams have been designed with welded built-up I-sections. The column cross-sections are assumed constant over the entire height, different for internal and for external columns. The beam-to-column connections are fully shop-welded with continuity plates (stiffeners) in the columns, and therefore should be classified as rigid full-strength connections. The column bases are considered fixed. The erection splices are made with splice plates (single in the flanges and double in the web), site-welded by fillet welds (see Fig. 3.3.6). Such a detail may also be regarded as rigid connection, therefore no semi-rigidity is actually introduced.

The design has been carried out according to the current BG specifications: BG Steel design code (1987), BG Code for actions on structures (1991) and BG Seismic code (1987). Additionally, the provisions of the capacity design methodology have been taken into account. Thus all the frames designed fail into global plastic mechanism. Moreover all are applicable for real construction needs.

Generally, the strength and stability criteria according to BG steel design code are similar to those of EC3 for Class 3 cross-sections. Additionally, the capacity design rules have also been taken into consideration.

No irregularities have been considered and all the frames are symmetrical. The most important feature is the column–tree configuration (erection splices located within the beam spans), as well as the horizontal haunching of beams (normally, there is no space for the more efficient vertical haunching at the joints). Mild steel is considered of grade equivalent to grade S235, having $f_y = 235$ N/mm^2. The material safety factor adopted is $\gamma_m = 1.1$.

TABLE 8.2.1
SECTIONAL DIMENSIONS OF STUDIED FRAMES

Frame	**Exterior**	**Interior**	**Floor beam**			**Roof beam**		
« R F »	**column**	**column**	**haunch**	**span**	**l_h**	**haunch**	**span**	**l_h**
2 1 B	fl 240.12 w 320.8		fl 220.10 w 380.6	fl 140.10 w 380.6	1000	fl 220.10 w 380.6	fl 140.10 w 380.6	1000
3 2 B	fl 240.12 w 320.8	fl 240.12 w 380.8	fl 220.10 w 380.6	fl 140.10 w 380.6	1000	fl 220.10 w 380.6	fl 140.10 w 380.6	1000
4 2 B	fl 300.16 w 300.10	fl 300.16 w 350.12	fl 260.12 w 350.8	fl 120.12 w 350.6	1000	fl 220.10 w 260.6	fl 120.10 w 260.6	1000
4 3 B	fl 280.15 w 320.10	fl 280.15 w 370.10	fl 220.10 w 380.6	fl 140.10 w 380.6	1000	fl 220.10 w 380.6	fl 140.10 w 380.6	1000
5 3 B	fl 300.18 w 400.12	fl 300.18 w 450.14	fl 220.10 w 380.8	fl 140.10 w 380.6	1000	fl 220.10 w 380.8	fl 140.10 w 380.6	1000
6 3 B	fl 300.18 w 400.12	fl 300.18 w 450.14	fl 240.14 w 400.8	fl 120.14 w 400.6	1000	fl 240.12 w 300.6	fl 120.12 w 300.6	1000

The "B-type" frames have been designed as follows:

(i) The beam cross-section is designed for the maximum span bending moment (it has been adopted that LT buckling is prevented by the secondary floor structure);

(ii) The length of the haunch, l_h, is determined on the basis of the bending moment diagram;

(iii) The haunch cross-section is designed following the capacity design rules with reference to the beam capacity at point "l_h", i.e. at the planned location where the dissipative plastic hinge is provoked to develop;

(iv) Finally the columns are designed based both on stability criteria and the capacity design rules with regard to the resistance of the adjacent haunch cross-sections.

The results are summarised in the table 8.2.1 ('fl'='flange', 'w'='web').

Modelling of the frames

Both static inelastic pushover analysis and dynamic inelastic time history analysis were performed with the frames described above. The general purpose computer program DRAIN-2DX (see Parkash at all (1994)) is used.

Beams and columns are modelled with the plastic hinge beam-column element 02 (see Powel and Campbell (1994)). The in-plane rigid floor is modelled as the cross sectional area of the beam is set to infinity. The damping in the structure is considered as 5% of the critical with damping matrix, which is linear combination of stiffness and mass matrix.

EARTHQUAKE RECORDS

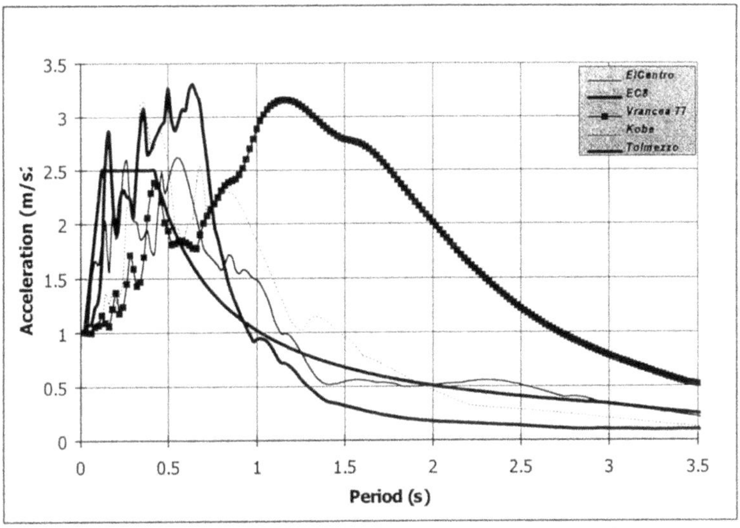

Figure 8.2.3: Response spectra of the selected earthquakes

Four earthquake records were selected in this study. They were normalised to the gravity acceleration (g) in order to eliminate the influence of the peak ground acceleration (PGA). Some of their parameters are presented in Table 8.2.2 and they are refereed as El Centro, Kobe JMA, Tolmezzo and Vrancea77.

The elastic pseudo acceleration response spectra of the chosen earthquakes are presented in Fig. 8.2.3.

TABLE 8.2.2
PARAMETERS OF SELECTED EARTHQUAKE RECORDS

Name of the event	**Record Step**	**PGA/g**	**Time with max. acceleration**	**Number of Steps**	**Record Length [sec]**
El Centro	0.02	0.348	2.12	2691	53.8
Kobe JMA	0.02	0.834	4.92	2820	56.38
Tolmezzo	Var.	0.313	3.92	1025	19.988
Vrancea77	0.02	0.199	6.12	2008	40.16

ANALYSIS PROCEDURE

TABLE 8.2.3
STORY STIFFNESS, STORY YIELD FORCES AND HORIZONTAL FORCE DISTRIBUTION (HFD) OF THE STUDIED FRAMES

Frame	Story	HFD	Story stiffness, K_i [kN/cm]	Story yield force, Qyi [kN]
RF21B ***(T_1=0.662 sec)***	1	1.00	83.64	215.77
	2	0.44	46.77	150.00
RF32B ***(T_1=0.990 sec)***	1	1.00	137.30	324.74
	2	0.50	80.25	266.25
	3	0.33	73.16	177.50
RF42B ***(T_1=1.254 sec)***	1	1.00	161.44	357.86
	2	0.44	85.99	309.72
	3	0.34	77.00	250.56
	4	0.28	74.21	174.00
RF43B ***(T_1=1.260 sec)***	1	1.00	235.61	500.14
	2	0.45	129.18	434.97
	3	0.34	117.25	352.30
	4	0.27	112.36	243.00
RF53B ***(T_1=1.430 sec)***	1	1.00	326.1	584.60
	2	0.41	159.6	522.00
	3	0.28	132.9	441.12
	4	0.31	128.2	358.13
	5	0.24	105.7	261.00
RF63B ***(T_1=1.550 sec)***	1	1.00	377.98	685.35
	2	0.49	195.42	629.10
	3	0.29	168.27	550.80
	4	0.26	163.55	480.60
	5	0.26	162.29	402.30
	6	0.21	154.47	270.00

The applied procedure is as follows:

1) The SRSS of story shear are determined based on the modal analysis of the frames;
2) The distributions of fictional story forces (DFSF), which correspond to the obtained story shears are calculated. Their normalized to unity distributions are listed in tables 8.2.3.
3) The procedure of modified Kato-Akiyama method, described above is applied. Story stiffness and story yield forces, obtained by nonlinear static pushover analysis with assumed story forces distribution, are listed in table 8.2.3. The ultimate limit value of the story cumulated ductility ratio η_{lim} is calculated for the case of "weak beams-strong columns" frames by applying Eqn. 8.2.28.

TABLE 8.2.4
BEAMS AND COLUMNS PLASTIC ROTATION CAPACITY

		R_{st}			θ_{st} [rad]		
Frame		*Mazzolani -Piluso*	*Mitani-Makino*	*Average*	*Mazzolani -Piluso*	*Mitani-Makino*	*Average*
RF21B	Beam	7.59	5.18	6.40	0.0530	0.0360	0.0445
	Column	3.66	3.84	3.75	0.0176	0.0184	0.0180
RF32B	Beam	7.59	5.18	6.40	0.0530	0.0360	0.0445
	Int. col.	3.43	3.84	3.63	0.0147	0.0161	0.01556
	Ext. col.	3.02	3.03	3.03	0.007	0.007	0.007
RF42B	Beam	12.48	7.37	9.90	0.093	0.055	0.074
	Int. col.	6.73	5.81	6.3	0.0269	0.0311	0.029
	Ext. col.	8.57	7.13	7.85	0.0247	0.0206	0.0226
RF43B	Beam	7.59	5.18	6.40	0.0530	0.0360	0.0445
	Int. col.	5.84	6.27	6.05	0.0246	0.0264	0.0255
	Ext. col.	4.43	5.72	5.10	0.0108	0.0139	0.0124
RF53B	Beam	7.59	5.18	6.40	0.0530	0.0360	0.0445
	Int. col.	7.57	8.10	6.40	0.0263	0.0282	0.0273
	Ext. col.	11.35	8.49	9.92	0.0255	0.0191	0.0223
RF63B	Beam	12.00	8.41	10.20	0.0823	0.0577	0.0700
	Int. col.	13.42	8.50	10.96	0.0251	0.0159	0.0205
	Ext. col.	7.90	8.10	8.00	0.0257	0.0263	0.0260

RESULTS OF ANALYSIS

The summary of the analysis results is presented in Fig. 8.2.4. The behaviour factor values (see Fig. 8.2.4^{a}) obtained by the energy method for different earthquakes show very small variation. The difference is in the reasonable limits. The behavior factors increase with increase of story number. That fact is expected for the case of "weak beams-strong columns" frames because with the increase of story number the energy dissipation capacity of the frame increases. The average values of ultimate accelerogram multipliers as a fraction of gravity acceleration are presented in Fig. 8.2.4^{b}.

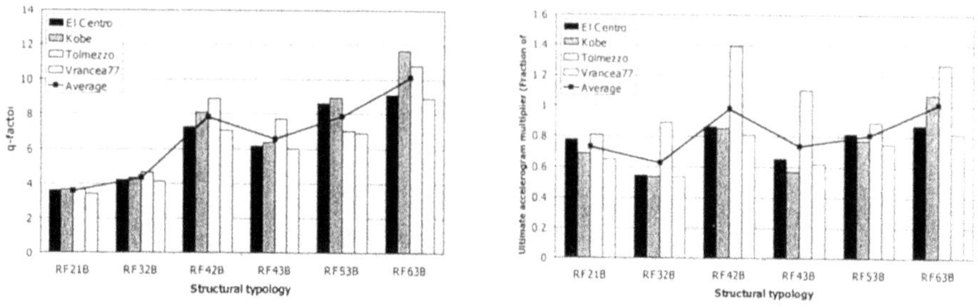

a) Behaviour factor values b)Ultimate accelerogram multipliers

Figure 8.2.4: Energy approach analysis results for frames with the B-type haunching

COMPARISON WITH EXPERIMENTAL DATA

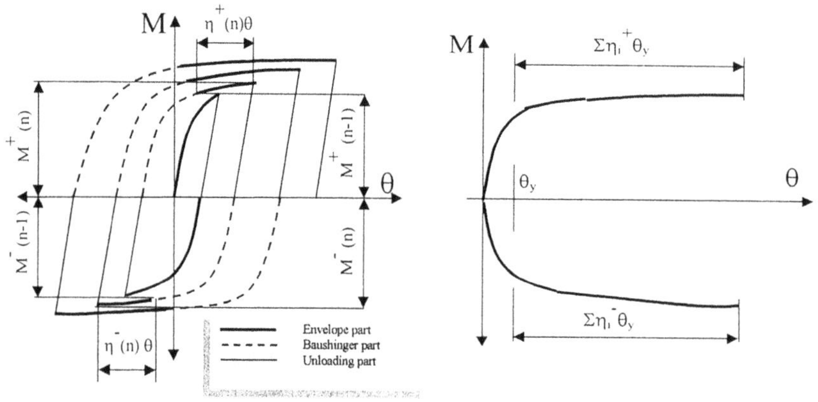

Figure 8.2.5: Decomposition of moment-rotation diagram

The hysteretic moment-rotation curves of columns, beams and panel zones can be decomposed into envelope parts, Baushinger parts and unloading parts (see Fig. 8.2.5). The bold solid lines in the Fig. 8.2.5 represent the inelastic part in each loading cycle under assumption that the load at the beginning of unloading of the foregoing loading cycle corresponds to the load at the onset of plastification after one cycle has elapsed. Those inelastic parts in each cycle are connected in each direction of loading (positive and negative). According to Akiyama (1985), the two curves obtained, which correspond to the positive and negative directions, coincide with the inelastic part of the moment-rotation curve under monotonic loading. The slope of the unloading parts agree with the initial elastic slope. After developing of inelastic deformation in one direction, the initial path in the subsequent reverse loading is not linear. It takes a non-linear curve with considerable deterioration in stiffness (see the dashed parts in Fig. 8.2.5). This phenomenon is termed as Baushinger effect.

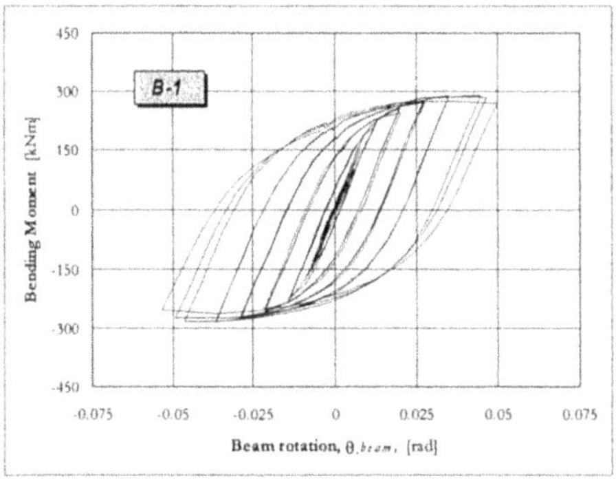

a) Specimen B-1

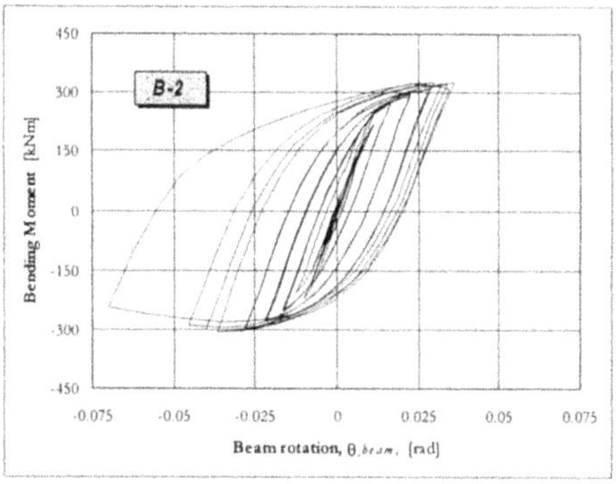

b) Specimen B-2

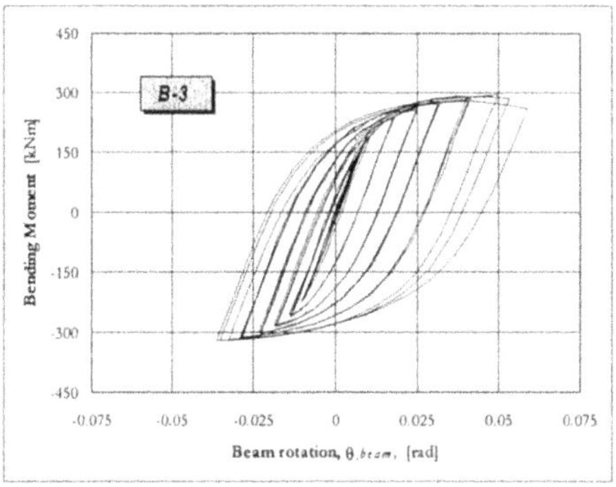

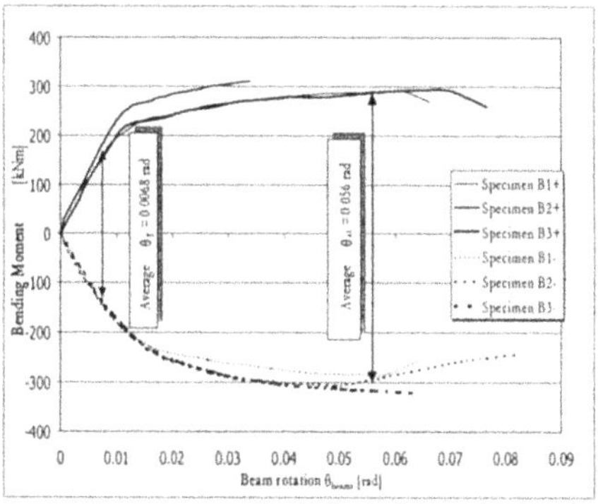

c) Specimen B-3 d) Decomposition of M-θ diagrams

Figure 8.2.6: Experimental data elaboration

TABLE 8.2.5

THEORETICAL AND EXPERIMENTAL PLASTIC ROTATION CAPACITY OF TEST SPECIMENS

	R_{st}			θ_{st} [rad]		
Frame	*Mazzolani -Piluso*	*Mitani-Makino*	*Experimental*	*Mazzolani -Piluso*	*Mitani-Makino*	*Experimental*
RF42B	12.48	7.37	7.24	0.093	0.055	0.056

Six full scale beam-column specimens were tested in the test laboratory of UACEG - Sofia. Those specimens were extracted from frames RF42B and RF42A (see paragraph 3.3 for details). Only the specimens with B-type haunching are discussed herein. Experimental beam moment-rotation relationships of specimens B-1, B-2 and B-3 are presented in Fig. 8.2.6[a]-8.2.6[c] Moment-rotation diagram decomposition for all specimens in both direction of loading (positive and negative) are presented in Fig. 8.2.6[d]. The monotonic rotation capacity of tested beams Rst and corresponding plastic rotation capacity θst obtained with the theoretical method of Mazzolani-Piluzo (1996) and the empirical method of Mitani-Makino (see Mazzolani-Piluzo (1996)for more details) as well as average experimental values of Rst and θst are listed in Table 8.2.5.

The analysis procedure, which is described above, is applied again with the frame RF42B employing all four earthquake records (El Centro, Kobe JMA, Tolmezo and Vrancea77). The ultimate limit value of the story cumulated ductility ratio η_{lim} is calculated for the case of "weak beams-strong columns" frames by using Eqn. 8.2.28. However the average experimental value for Rst=7.24 is assumed. Comparison, between behavior factor values obtained by assuming average experimental value for Rst=7.24 and previously considered value Rst=9.90 as average of the values calculated by both theoretical methods (Mazzolani-Piluso and Mitani-Makino), is presented in Table 8.2.6.

TABLE 8.2.6
BEHAVIOUR FACTORS OF FRAME RF42B BASED ON ANLYTICAL PREDICTION AND EXPERIMENTAL RESULTS FOR Rst

Earthquake	**Theoretical**		**Experimental**	
	Rst	***q-factor***	***Rst***	***q-factor***
El Centro	9.90	7.25	7.24	6.85
Kobe JMA	9.90	8.08	7.24	7.02
Tolmezo	9.90	8.59	7.24	7.24
Vrancea 77	9.90	7.12	7.24	6.71
Average	-	**7.76**	-	**6.96**

CONCLUSIONS

This study is focused on the energy approach for behaviour factor evaluation of steel moment resisting frames. A modification of Akiyama-Kato method is employed. Some parametric study, with frames designed according to current Bulgarian codes and considering the capacity design approach, are performed. Besides, the theoretically calculated and experimentally obtained beam plastic rotation capacity is set to one of analysed parameters. The following conclusions could be drawn on the bases of the current study:

- The scatter of behaviour factor values, which are calculated by applying the energy approach, is in reasonable limit (within 15%) for different earthquake records;

- The behaviour factor values increases with increasing of story number of the frames. In the case of properly designed "strong columns-weak beams" frames, this trend could be explained by increasing of frame energy dissipation capacity with increasing of story number;

- One of the most important assumption in applying of energy approach is the limit value of member rotation capacity. The comparison between theoretical recommendation and experimental data shows that the member rotation capacity could be theoretically predicted with reasonable accuracy. The difference between analytical and experimental values of member rotation capacity affects the q-factor values in the limits of 10-15%. This scatter is in the same range as one influenced by the different earthquake records.

However it is necessary to underline that more experimental and theoretical studies are necessary for collapse state definition based on energy approach.

REFERENCES

Akiyama, H., (1985). "Earthquake Resistant Limit-State Design for Buildings", University of Tokyo Press, 1985

Akiyama, H., (1988). "Earthquake Resistant Design Based on the Energy Concept", Proceedings of the 9th World Conference on Earthquake Engineering, Tokyo, Kyoto.

Akiyama, H., and Yamada S., (1997) "Seismic Input and Damage of Steel Moment Frames", General Report of STESSA'97, Behaviour of Steel Structures in Seismic Areas, Kyoto, Japan.

Housner, G., (1959). "Behaviour of Structures during the Earthquakes", Journal of ASCE, EM4.

Milev, J., (1998). "Modelling of Shear Walls for Seismic Analysis of Frame-Wall Structures", PhD Thesis, (in Bulgarian).

Mazzolani, F., and Piluso, V., (1996) "Theory and Design of Seismic Resistant Steel Frames", E&FN SPON, 1996

Nakata, S., (1995). "An Example of the Structural Design of Reinforced Concrete Building", Lecture Note for the Earthquake Engineering Training Course, IISEE, BRI, Tsukuba, Japan.

Parkash, V., Powell, G. H., and Campbell, S., (1994). "DRAIN-2DX. Base Program Description and User Guide", Report No UCB/SEMM/94/07,1994

Powell, G. H., and Campbell, S., (1994). "DRAIN-2DX. Element Description and User Guide for Element Type01, Type02, Type04, Type06 and Type15", Report No UCB/SEMM/94/08.

8.3

THE BASE SHEAR FORCE APPROACH

Jean-Marie Aribert, Daniel Grecea

INTRODUCTION

Generally codes require only an elastic static analysis at geometrical first order to design steel frames subject to seismic actions. Advantage of the very significant dissipative phenomena in steel structures is taken by means of a q-factor reducing the seismic forces which would be obtained assuming a perfectly elastic behaviour.
Simulating frame responses by non-linear dynamic analyses, it was realised that most of the definitions of this q-factor in the relevant literature are conventional without direct relation with the internal forces and moments in the structures. Even some definitions are very optimistic, leading to an unsafe determination of the internal forces and moments and also of the forces applied to the foundations.
Determination of the behaviour q-factor is a complex problem due to several parameters affecting its value, in particular the following ones :

- the partial character of the energy dissipation in the structure when local storey mechanisms occur ;
- the second order geometrical effects, so-called P-Δ effects, necessary developed when large energy dissipation is required ;
- the structural irregularity of the vertical configuration as well as the plan one ;
- the occurrence of local buckling in the dissipative beams reducing their rotation capacity ;
- the buckling risk of columns subject to axial force and bending moment, whose drastic consequence requires to limit the energy dissipation (accepting only a few plastic hinges at the column ends), etc.

In Mazzolani & Piluso (1996), a synthesis chapter can be found about the different methods existing in the literature to evaluate the q-factor. Comparisons show clearly a large scattering of the results which may be partially explained by absence of a general philosophy on the q-factor definition, consistent with its determination procedure. For steel structures, it is understandable easily that the first order elastic static analysis may not be well adapted to the dynamic behaviour. In fact, this type of analysis appears elementary to express suitably the real dissipative behaviour due to the occurrence of a very large number of plastic hinges associated with important P-Δ effects.
This chapter intends to present a new method to define and to evaluate the q-factor for steel frames, based on elastic-plastic dynamic analyses. This definition is related essentially to the maximum inelastic shear force at the structure base deduced from the time-history response. Advantages of this new definition are pointed out as well as some possible adjustment of the q-value as a function of the ground acceleration. Then, the practical use of this q-factor is explained when design is considered in association with elastic response spectra. To check the structural resistance, equivalent static global analyses of rigid-plastic type are introduced, preventing against the risk of local storey mechanisms when the structure is able to be fully dissipative. The above approach may be generalised to the case of

portal frames with semi-rigid and partial-strength joints provided that a global analysis of elastic-plastic type is adopted to control the required rotation of the joints. Several numerical examples are calculated to illustrate the new concepts proposed in this chapter.

DETERMINATION OF q-FACTOR

Comments on existing methods

Partially according with the classification given by Mazzolani & Piluso (1996), existing methods well-known in the literature to evaluate the q factor can be grouped in three main categories, as mentioned hereafter.

Methods based on the inelastic response of simple degree of freedom systems (SDOF)

The basic method is a static one, using the ductility factor theory (Cosenza & all. (1986)). However, the method can be also deduced from the dynamic analysis whose results are interpreted by means of an inelastic response spectrum in pseudo-acceleration (Giuffre & Giannini (1982) ; Palazzo & Fraternali (1987)).
But the q values obtained by these simplified methods cannot be easily transposed to real multidegree of freedom systems (MDOF). Generally it is required that the structure satisfies conditions of “structural regularity“ and ”energy global dissipation”, which means the development of plastic hinges generally in all the beams during the dynamic response under a strong ground accelerogram.
It is clear that these methods cannot take into account some local limitations, for instance the attainment of rotation capacity at the ends of beams or columns, the risk of buckling in columns subject to high compression and flexural bending, etc.

Methods based on an energy approach

These methods appear more attractive because they do not require to satisfy conditions of structural regularity and energy global dissipation.
One of these methods (Como & Lanni (1983)) is based on the evaluation on the one hand of the elastic strain energy W_e stored in the state of first yielding, on the other hand of the total energy W_u stored and dissipated by elastic-plastic deformations up to failure.
But in practice, this evaluation needs to use the approximate concept of equivalent horizontal forces, statically applied and distributed according to a combination of a selected number of vibration modes.
Another method (Akiyama (1988))consists in verifying that under the design major seismic action, the capacity of the structure to dissipate energy by cumulated plastic deformations remains greater than the earthquake input energy into the structure; the last one can be obtained by $1/2MS_V^2$ where M is the total mass of the structure and S_V is the spectral elastic response in pseudo-velocity. But the evaluation of the dissipative capacity needs to accept several assumptions, in particular concerning the simplified expression of the hysteretic plastic work at each storey and an optimum distribution of this work between the different storeys (established empirically from parametric studies).

Methods based on inelastic dynamic analyses of multidegree of freedom systems (MDOF)

Using the non-linear dynamic analysis to provide the time-history response of a MDOF structure submitted to natural or artificial ground motions, these methods are incontestably the most precise ones, even though it is laborious to perform them. As explained hereafter, the main difficulty is related to the interpretation mode of the dynamic results.

The well-known approach of Ballio & Setti (1985), which is the oldest in literature, consists in performing a non-linear dynamic analysis and obtaining the maximum response of the structure during its time-history for different levels of ground motion; practically the considered ground acceleration a(t) is multiplied by a factor λ that is step-by-step increased. For each analysis with a fixed value λ, the response of a multi-storey structure is characterised by a significant displacement, generally the upper storey deflection δ (Figure 8.3.1). As long as $\lambda \leq \lambda_e$, the response remains elastic (segment OE in Figure 8.3.2); for higher values of λ, the real elastic-plastic displacements δ become generally smaller than the calculated ones assuming an ideal elastic behaviour, so that the curve (δ, λ) determined step-by-step has the position EIU shown in Figure 8.3.2. An interesting reference value λ_u^* of the seismic action multiplier can be defined by the intersection point U^* between the curve (δ, λ) and the linear elastic line extending the segment OE. At this stage, Ballio & Setti consider that the structure is almost reaching a state of global dynamic instability beyond which the plastic structural dissipation is not enough to offer resistance to important deflections.

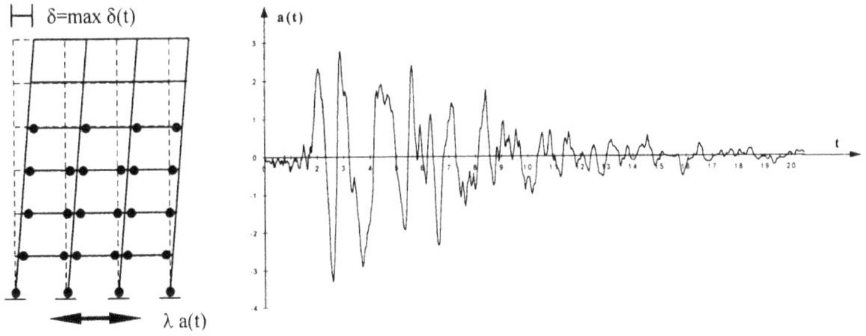

Figure 8.3.1 : Dynamic response of a structure subject to a given accelerogram

Usually the behaviour factor q is evaluated as the ratio:

$$q_B = \lambda_u^* / \lambda_e \tag{8.3.1.a}$$

which means the ratio between the ground acceleration close to the structural collapse and that corresponding to the first yielding.

At first sight the definition given in (8.3.1.a) is attractive because:

- It guaranties the satisfaction of criterion so-called "displacement equality" (which means equality between the second order elasto-plastic displacement $\delta^{(inel)*}$ and the first order elastic one $\delta^{(e,th)}$ corresponding theoretically to the same level of acceleration λ^*); it is worth noting that this criterion is often presented as a postulate in design codes.
- It is consistent with the ductility factor theory (noting that : $\lambda_u^* / \lambda_e = \delta^{(inel)*} / \delta^{(e)}$).

Contrary to what is sometimes affirmed in the literature, relation (8.3.1.a) cannot be a suitable definition because :

- On the one hand, it keeps up some confusion of the external ground acceleration with the inelastic spectral response in acceleration of the structure, whose consequence may be the more significant as the structure is governed by several degrees of freedom. Similarly, the relevant definition is not connected efficiently to the internal forces and moments in the structure, neither to the forces applied to the foundation.
- On the other hand, according to the accelerogram type and the fundamental period T of the studied structure, it may be met cases where the definition of point U^* is not clear (due to quasi-parallelism

of the (δ, λ) and OE curves), also cases where there is no real intersection (due to reduced rotation capacity of some elements, beams and columns ; see farther relationship (8.3.1.c)). Even, there are cases where the (δ, λ) curve is above the OE line as soon as $\lambda>\lambda_e$.

To compensate the inaccurate determination of intersection U^* and to introduce a more general definition of q (which is not associated with the criterion of displacement equality), Sedlacek & Kuck (1993) have proposed to change $\overline{OEU^*}$ for another straight line defined by equation :

$$\delta = k\frac{\delta^{(e)}}{\lambda_e}\lambda$$

where the k factor may have different conventional values, for instance k=1.5 (this value seems to cover a more realistic domain of application). Then, the new point of intersection leads to a different value λ_{uS} of the accelerogram multiplier, but the behaviour q-factor is still given by a relation similar to (8.3.1.a), namely :

$$q_S = \lambda_{uS}/\lambda_e \qquad (8.3.1.b)$$

When using this modification, it should not be forgotten that now the second order elasto-plastic displacement $\delta_S^{(inel)}$ is equal k times the first order theoretical elastic displacement $\delta^{(e,th)}$ calculated for the same level of acceleration λ_{uS}.
It is obvious that definitions (8.3.1.a) and (8.3.1.b) do not take into account limitations imposed by the rotation capacity of elements, beams and columns. This capacity may be evaluated on the safety side using the concept of class of cross-section; but on the one hand Mazzolani & Piluso (1996), on the other hand Gioncu & Petcu (1995), have proposed more precise methods based respectively on semi-empirical approach and on plastic yield line mechanisms. Consequently, Mazzolani and all. have considered the following definition :

$$q_M = \lambda_{uM}/\lambda_e \qquad (8.3.1.c)$$

when the rotation capacity is attained for an acceleration level λ_{uM} lower or greater than λ_u^* shown in Figure 8.3.2. Before presenting a new definition, it should be noted that definitions (8.3.1.b) and (8.3.1.c) are still open to criticism because there are not related to the real inelastic responses in pseudo-acceleration at level λ_{uS} and λ_{uM}, respectively.

New definition of q-factor

Complementary to Figure 8.3.2, Figure 8.3.3 represents the variation of the horizontal base shear force V of the structure versus the accelerogram multiplier λ. This force reaches the values $V^{(e)}$ and $V^{(inel)}$ corresponding to the first yielding state and to the ultimate limit state, respectively. Numerical studies have shown that generally $V^{(e)}$ is clearly less than $V^{(inel)}$ so that the definitions (8.3.1.a) to (8.3.1.c) of q-factor used through an elastic global analysis, lead inevitably to the underestimation of internal forces and moments and consequently to the underestimation of the forces acting on the foundation.

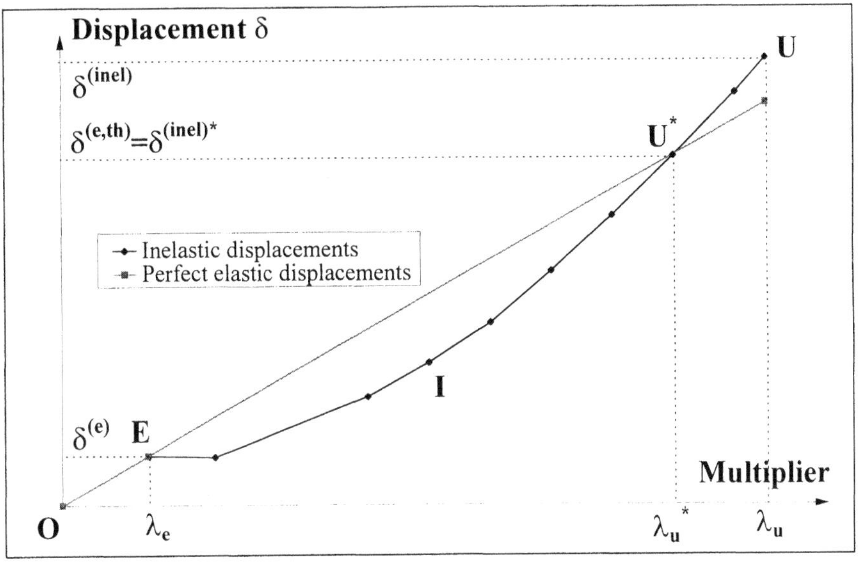

Figure 8.3.2 : Maximum displacement versus accelerogram multiplier

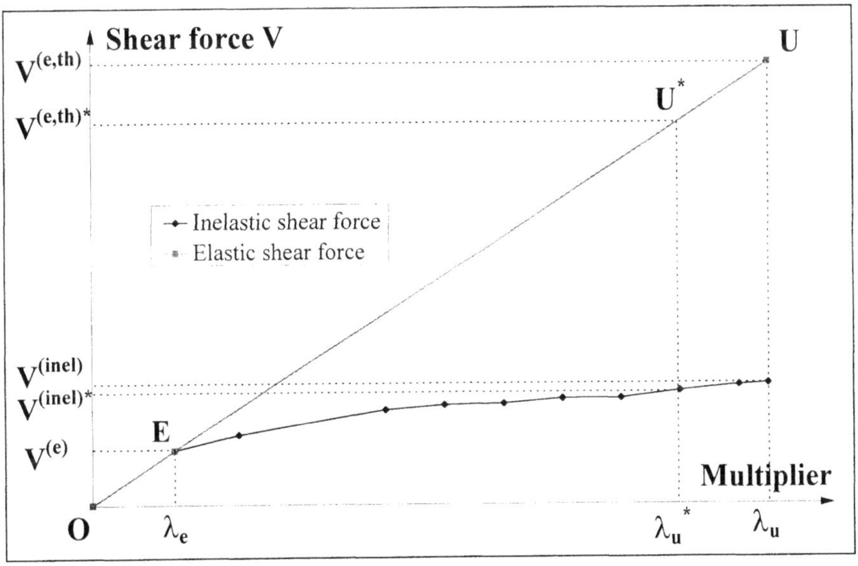

Figure 8.3.3 : Maximum base shear force versus accelerogram multiplier

The new definition of the q-factor proposed by the present authors is based on the ratio between the elastic theoretical base shear force $V^{(e,th)}$ corresponding to an elastic theoretical behaviour at the ultimate value λ_u of the multiplier and the real inelastic base shear force $V^{(inel)}$:

$$q = V^{(e,th)} / V^{(inel)} \quad (8.3.2)$$

or more precisely :

$$q = \left(V^{(e)}/\lambda_e\right) / \left(V^{(inel)}/\lambda_u\right) \quad (8.3.3)$$

The main advantage of this definition is the suitable evaluation of the reduced horizontal force applied to the foundation, which allows to expect a suitable evaluation of the internal forces and moments in the structure, provided that an appropriate global analysis is used, taking into account the dissipative effects.

Another advantage of definition (8.3.2), more explicit in (8.3.3), is the fact that there is no need to know the value λ_e corresponding to the first yielding stage (it is only necessary to know the slope $V^{(e)}/\lambda_e$) ; determination of λ_e is sometimes difficult (because of a bifurcation not always clear of the curve EIU), and even significance of λ_e in dynamic behaviour remains questionable.

As important comment, it should be underlined a complementary aspect of definition (8.3.2) (which seems having been neglected by the users of definitions (8.3.1.a) to (8.3.1.c)) : the q-factor defined by (8.3.2) has to be associated with a precise value of the maximum soil acceleration, $a_N^{(u)} = \lambda_u \cdot \max|a(t)|$, which appears to be practically the definition of a nominal acceleration.

For practical design, it is more convenient to adopt the reference values (q^*, $a_N^{(u)*}$) corresponding to point U^* in Figure 8.3.3 assuming, if necessary, an ideal behaviour of the dissipative elements (rotation capacity, stability). This state of dissipation is very interesting because of the equality between the first order elastic displacements and the second order inelastic ones ($\delta^{(e,th)} = \delta^{(inel)*}$). For levels of nominal accelerations different from $a_N^{(u)*}$, due to another seismic intensity of the site or another rotation capacity of the beams, columns or joints, the q-factor will be adjusted moving apart this reference factor q^*, as explained in a further paragraph.

A few results from a parametric investigation

A few results extracted from a parametric study carried out by Grecea (1999), dealing with the behaviour q-factor of frames with different configurations and with rigid and full strength joints are examined in this paragraph.

The main objective of the study was to dispose of significant values of q-factor, and to establish a sort of definition of q as a function of nominal acceleration a_N, parameter of plastic redistribution capacity α_u/α_y, fundamental period T_1, and interstorey drift sensitivity coefficient θ_j :

$$q^* = q\left(a_N^{(u)*}, \alpha_u/\alpha_y, T_1, \theta_j\right) \quad (8.3.4)$$

α_u and α_y are, respectively, the maximum and first yielding values of the horizontal seismic forces multiplier determined by a static elastic-plastic analysis ; θ_j is defined farther on (see relationship (8.3.13)).

The frames adopted in the parametric study are shown in Figure 8.3.4 with the associated characteristics given in Table 8.3.1. In Table 8.3.1, w means the storey load.

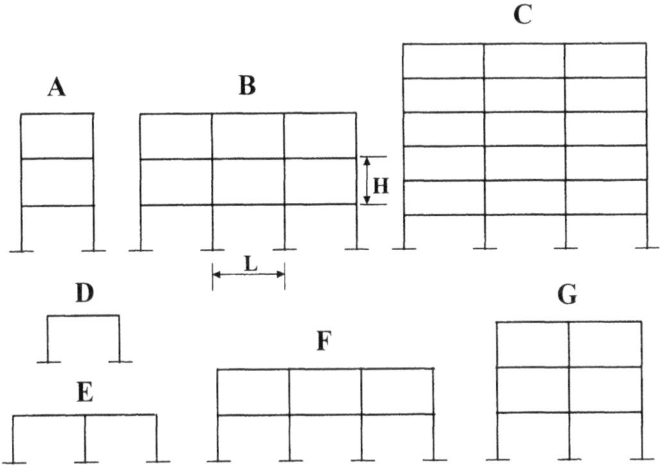

Figure 8.3.4 : Investigated frames

TABLE 8.3.1
CHARACTERISTICS OF ANALYSED FRAMES

Frame	L [m]	H [m]	w [kN/m]	m_j [kg]	Beams	Columns
A	5.0	4.0	22	11000	IPE 300	HEB 240
B	4.0	4.0	32	38400	IPE 330	HEB 240
C	4.5	3.0	35	47250	IPE 330	HEB 360 (1,2)
						HEB 300 (3-5)
						HEB 260(6)
D	5.0	6.0	40	20000	IPE 300	HEB 240
E	5.0	6.0	40	40000	IPE 300	HEB 240
F	5.0	4.0	40	60000	IPE 300	HEB 240
G	5.0	4.0	41.6	41600	IPE 300	HEB 300

The concerned frames were subject to three different accelerograms namely these of Vrancea - 1977 (Figure 8.3.5), Kobe - 1995 (Figure 8.3.6) and El Centro - 1940 (Figure 8.3.7). Inelastic dynamic analyses of the frames were performed using the DRAIN-2DX computer code, developed at Berkeley University by Prakash & all. (1993).
The obtained results from the parametrical study are presented in Table 8.3.2, for all the frames and for the three accelerograms.

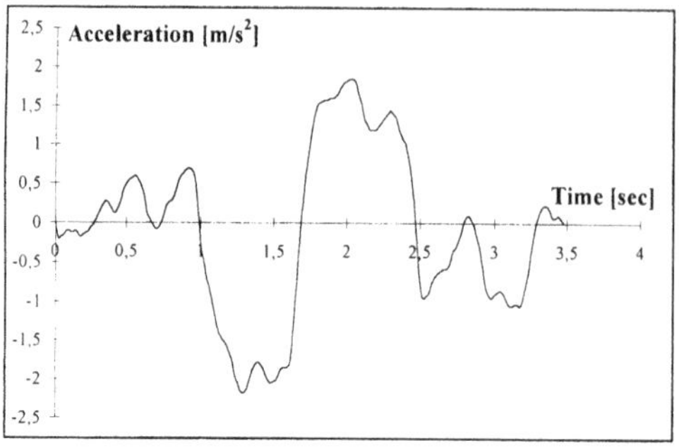

Figure 8.3.5.a : Accelerogram of Vrancea (1977)

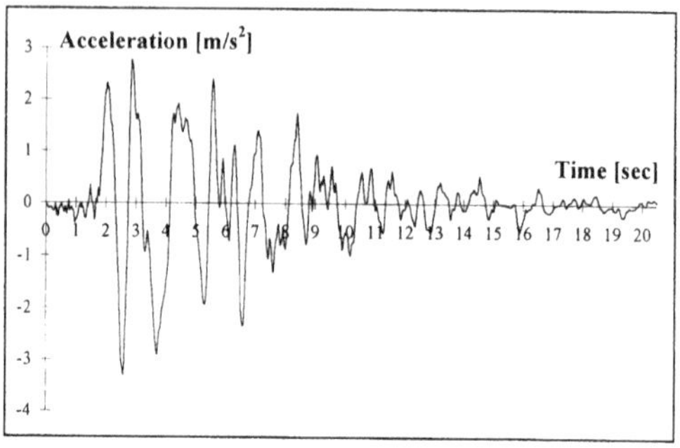

Figure 8.3.5.b : Accelerogram of Kobe (1995)

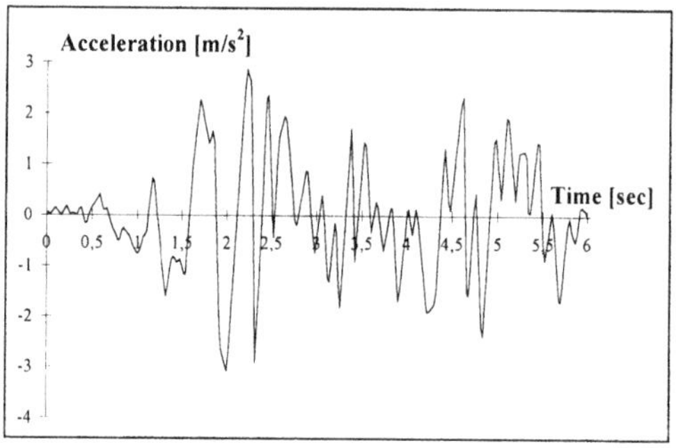

Figure 8.3.5.c : Accelerogram of El-Centro (1940)

TABLE 8.3.2

RESULTS OF q-FACTOR FROM PARAMETRIC INVESTIGATION

Accelerogram of :	Frame	A	B	C	D	E	F	G
	T_1 (sec)	0.883	0.917	1.330	0.675	0.761	0.865	1.081
	α_u/α_y	1.38	1.27	1.41	1.28	1.31	1.32	1.40
VRANCEA	q^*	1.6	1.5	3.4	1.2	1.1	1.6	3.0
	$a_N^{(u)*}$	3.1	2.6	4.8	4.0	3.3	2.2	2.9
	θ_j	0.020	0.030	0.130	0.021	0.035	0.056	0.099
KOBE	q^*	3.1	3.8	6.7	4.0	2.4	3.5	3.9
	$a_N^{(u)*}$	4.6	5.6	7.9	4.5	2.9	3.3	3.1
	θ_j	0.070	0.160	0.170	0.038	0.034	0.115	0.114
EL-CENTRO	q^*	5.8	5.8	4.0	1.9	1.9	5.8	4.0
	$a_N^{(u)*}$	15.1	11.1	12.0	5.1	5.6	7.7	9.3
	θ_j	0.127	0.131	0.129	0.034	0.045	0.139	0.115

Analysing these results, the following remarks may be underlined :

- Values of q-factor determined according to the new method are smaller than those given usually by codes and by other methods of the literature.
- Values of q-factor are clearly influenced by the type of frame. For Vrancea accelerogram, q-factor is varying from 1.1 or 1.2 for a simple frame like D or E until 3.4 for a dissipative frame like C.
- But values of q-factor are also strongly influenced by the shape of the accelerogram and of its associated spectrum, with regard to the fundamental period of the structure. So, for Kobe accelerogram, q-factor is varying from 2.4 for frame E until 6.7 for frame C. In case of El Centro accelerogram, q-factor is varying from 1.9 for frame D and E until 5.8 for frames A, B and F.
- The reference value q^* should be associated with the nominal ground acceleration $a_N^{(u)*}$ which can be regarded as the consequence of only q^* (the dissipation level depending eventually on the spectrum shape and the fundamental period of the frame).
- The plastic redistribution parameter α_u/α_y and the interstorey drift sensitivity coefficient θ_j have a slight influence on the values of the q-factor, but not always evident due to the prevailing influence of the spectrum shape and the fundamental period of structure.
- According to Table 8.3.2, for q^* values higher than 2.5 or 3.0, P-Δ effects cannot be neglected because of $\theta_j > 0.10$ (for example, frames C and G subject to accelerogram of Vrancea, frames B, F, G subject to accelerogram of Kobe, and frames A, B, C, F, G subject to El Centro accelerogram).

Adjustment of q-factor

When the nominal acceleration $a_N^{(u)}$ is different from the reference one $a_N^{(u)*}$, the q-factor has to be adjusted. The parametric investigation of Grecea (1999) demonstrated that the hereafter linear interpolation gives always results very often on the safe side :

$$q = q^* \frac{a_N^{(u)}}{a_N^{(u)*}} \tag{8.3.5}$$

provided that $a_N^{(u)}$ lies within the following interval :

$$0.5a_N^{(u)*} \le a_N^{(u)} \le 1.5a_N^{(u)*} \tag{8.3.6}$$

Of course, the upper bound of (8.3.6) is reached seldom because it implies high dynamic stability of the structure and large rotation capacity of its elements.
An illustration of relationship (8.3.5) in the case of frame C is shown in Figure 8.3.6.
It is worth nothing that linking relationships (8.3.4) and (8.3.5) means in fact :

$$q\left(a_N^{(u)}, \alpha_u/\alpha_y, T_1, \theta_j\right) = \frac{a_N^{(u)}}{a_N^{(u)*}} q^*\left(\alpha_u/\alpha_y, T_1, \theta_j\right). \tag{8.3.7}$$

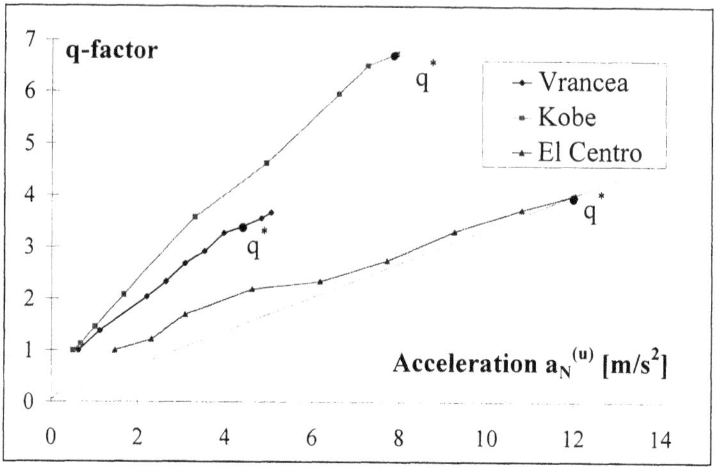

Figure 8.3.6 : Variation of q-factor (frame C)

USE OF THE NEW q-FACTOR WITH ELASTIC RESPONSE SPECTRUM

Evaluation of equivalent static forces

From the elastic response spectrum, it is possible to evaluate the elastic base shear force $V^{(e,th)}$, corresponding to the nominal acceleration $a_N^{(u)}$, as given by relationship (8.3.8) where $R_e(T_1)$ is the normalized elastic response, T_1 the fundamental period and M the total mass of the structure. When the accelerogram is very representative of the normalized spectrum, the so-obtained value $V^{(e,th)}$ is generally close to the elastic one deduced from the dynamic analysis. Let be :

$$V^{(e,th)} = M \cdot R_e(T_1) \cdot a_N^{(u)} \tag{8.3.8}$$

Using the adjusted q-factor according to relationship (8.3.5) and the elastic response spectrum, the reduced inelastic base shear force $V^{(inel)}$ corresponding to the nominal acceleration $a_N^{(u)}$ is given by :

$$V^{(inel)} = M \cdot R_e(T_1) \cdot a_N^{(u)} / q \tag{8.3.9}$$

It should be noted that relationship (8.3.9) may be slightly approximative because the inelastic response spectrum is never perfectly deduced by anamorphosis from the elastic one, specially when the latter is very irregular.

The distribution of the seismic horizontal forces is given by relationship (8.3.10) where x_{j1} represents the displacement of storey j in the fundamental mode of vibration :

$$F_j^{(inel)} = \frac{m_j x_{j1}}{\sum_j m_j x_{j1}} V^{(inel)} \tag{8.3.10}$$

This relation is valid only for regular structures, which means that the participant mass M_1 of the fundamental mode is close to the total mass M of the structure. If it is not the case, it should be taken account of several vibration modes in such a manner that the sum of participant modal masses exceeds 90% of the total mass of the structure ($\sum_i M_i \geq 0.90M$) ; but such a generalization will not be developed in the present chapter.

Evaluation of inelastic displacements

The inelastic displacement of storey j (possibly with 2nd order effects) is simply given by the familiar elastic relationship (8.3.11) when $a_N^{(u)}$ is strictly equal to $a_N^{(u)*}$:

$$\delta_j^{(inel)} = \delta_j^{(e,th)} = \frac{T_1^2}{4\pi^2} \frac{x_{j1} M}{\sum_{k=1}^{n} m_k x_{k1}} a_N^{(u)} R_e(T_1) \tag{8.3.11}$$

In the general case where $a_N^{(u)} \neq a_N^{(u)*}$, a linear variation may be proposed again :

$$\delta_j^{(inel)}(a_N^{(u)}) = \delta_j^{(e,th)}(a_N^{(u)}) \cdot \frac{a_N^{(u)}}{a_N^{(u)*}}, \tag{8.3.12}$$

as demonstrated using the parametric investigation already mentioned for relationship (8.3.5). An illustration of relationship (8.3.12) for frames B and C is shown in Figure 8.3.7.
The evaluation of $\delta_j^{(inel)}$ is useful to control the interstorey drift at the « serviceability limit state (SLS) » under seismic action, often transformed in equivalent ultimate limit state (ULS) by most of the seismic codes. Also, this evaluation is considered farther at the ultimate limit state to know if the second order effects on the global rigid-plastic analysis or the elastic-plastic one are negligible or not.

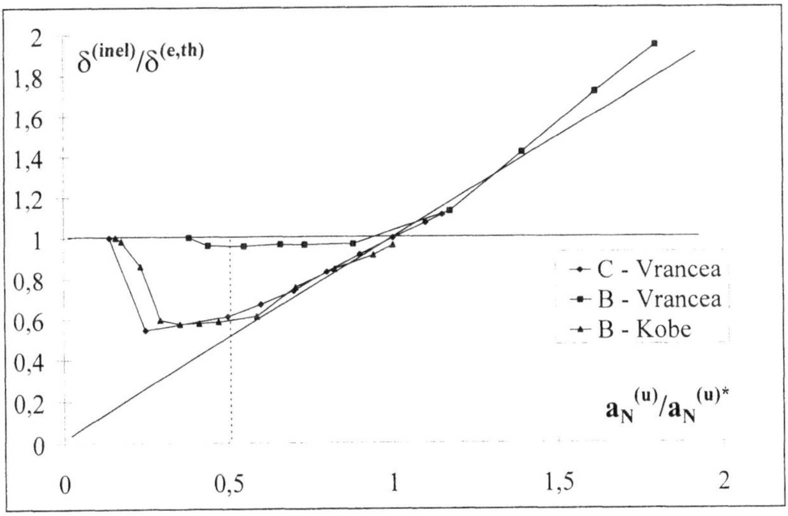

Figure 8.3.7 : Variation and evaluation of inelastic displacements

Evaluation of the P-Δ effects

P-Δ effects can be evaluated by means of the interstorey drift sensitivity coefficient θ_j defined as follows :

$$\theta_j = \frac{\Delta\delta_j}{h_j} \cdot \frac{W_j}{H_j} \tag{8.3.13}$$

where $\Delta\delta_j$ is the horizontal displacement at the top of the storey j, relative to the bottom of the storey, h_j is the storey height, H_j is the total horizontal shear force at the bottom of the storey and W_j is the total vertical axial force acting at the bottom of the storey. For values of $\theta_j \leq 0.1$, second-order effects can be neglected.

Checking of the structural resistance

At this stage of the presentation, it can be explained what procedure may be used as *equivalent static global analysis* of the structure in order to check its resistance capacity to the seismic action. For the moment, it is assumed only the case of a structure able to be « *fully dissipative* », meaning by this term that the rotation capacity of dissipative elements is high (for example, cross-sections of Class 1) and the dissipative phenomena due to plastic hinges are well distributed over the whole structure.
Rigid-plastic analysis (in the usual static sense) may be used, taking account of second order effects if θ_j given by relationship (8.3.13) is greater than 0.10. For each type of mechanism considered, a plastic multiplier α_p of the seismic horizontal forces, supposed to be proportional to the fundamental response as follows :

$$F_j^{(p)} = \alpha_p \, m_j \, x_{j1} \, , \tag{8.3.14}$$

can be determined.
For a good distribution of the dissipative phenomena, the *global mechanism* consisting of plastic hinges at the ends of all the beams and at the base of the first storey columns should be favoured. In the contrary, it has to make sure of no development of premature *local storey mechanisms*. Explicitly, a local storey mechanism means a plastic hinge mechanism which is concentrated on one storey or a few ones including possible plastic hinges at column ends, leading necessarily to a reduced dissipative capacity. Finally, the two following conditions have to be satisfied :

- <u>*First condition*</u> : The plastic multiplier of the global mechanism is less than all the multipliers of the local mechanisms :

$$\alpha_p^{(glob)} < \alpha_p^{(loc)} \, , \quad \forall(loc) \tag{8.3.15}$$

- <u>*Second condition*</u> : The base shear force evaluated by relationship (8.3.9) is less than all the multipliers of the local mechanisms :

$$V^{(inel)} < V_p^{(loc)} = \sum_{j=1}^{n} \alpha_p^{(loc)} m_j x_{j1} \qquad , \quad \forall(loc) \, . \tag{8.3.16}$$

In addition, for a base shear force equal to $V^{(inel)}$, under cyclic hysteretic behaviour, the occurrence of plastic hinges at the ends of several columns cannot be excluded, especially at the base of the structure.

It is obviously essential to avoid any *column local mechanism* due to the occurrence of another plastic hinge between those at the column ends ; for that reason it is necessary to limit the axial force of the column N_{Sd} and/or the column non-dimensional slenderness $\bar{\lambda}$ (using a buckling length equal to the system length), satisfying for instance the following condition given by French code PS 92 (1995), clause 13.7.1.1.1, for columns with double curvature bending :

$$N_{Sd}/N_{pl.Rd} + 0.8\bar{\lambda} \leq 1 \quad \text{if} \quad N_{Sd}/N_{pl.Rd} \geq 0.15 \tag{8.3.17.a}$$

$$\bar{\lambda} \leq 1.6 \quad \text{if} \quad N_{Sd}/N_{pl.Rd} < 0.15 \tag{8.3.17.b}$$

In addition to (8.3.15), (8.3.16) and (8.3.17), other classical checks remain to be done (stability of members with combined axial force and moment, lateral-torsional buckling of beams, overstrength of beam-to-column joints in dissipative zones, etc.).
When dissipative elements have a limited rotation capacity (cross-sections in class 2 or 3), the structure cannot be classified as fully dissipative and the behaviour q-factor has to be determined as a function of this rotation capacity. An elastic-plastic analysis may substitute for the rigid-plastic analysis in order to control the required rotations in the dissipative elements corresponding to the new value of the shear force $V^{(inel)}$ (provided that the structure was designed so that : $V^{(inel)} < V_p^{(glob)}$).

Example : 6 storey - 3 bay frame - Accelerogram of Vrancea

The 6 storey - 3 bay frame presented in Figure 8.3.8 is analysed when being subject to the seismic action characterised by the accelerogram of Vrancea (Romania), recorded to the Building Research Institute INCERC Bucharest (the 4th of March 1977), already shown in Figure 8.3.5.a and expressed by its associated elastic response spectrum in Figure 8.3.9. All the cross-sections of members are Class 1.

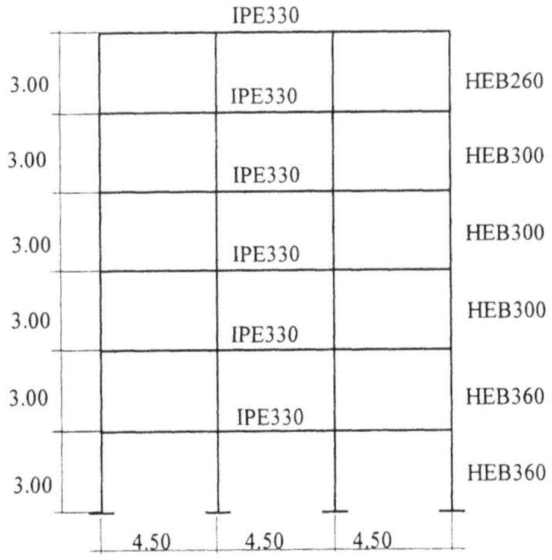

$m_j = 47250$kg
T_1=1,33 sec.

$$\{X_1\} = \begin{Bmatrix} 1,000 \\ 0,896 \\ 0,738 \\ 0,532 \\ 0,300 \\ 0,101 \end{Bmatrix}$$

Steel grade : S 275
$M_{pl,b}$ = 221,1 kNm (IPE 330)
$M_{pl,c}$ = 737,0 kNm (HEB 360)
513,7 kNm(HEB 300)
352,6 kNm(HEB 260)

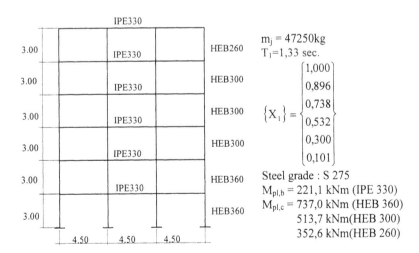

Figure 8.3.8 : Frame geometry and characteristics

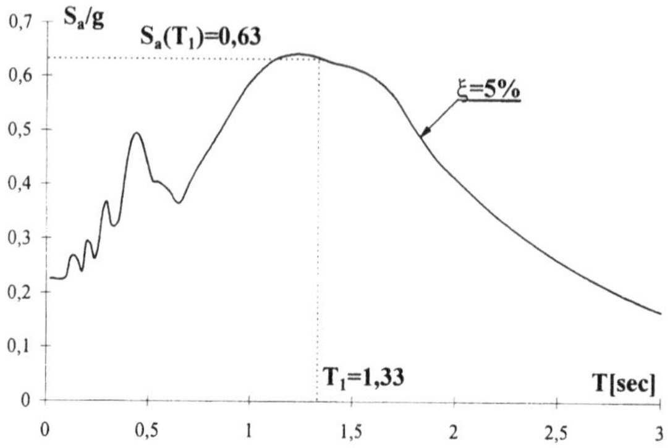

Figure 8.3.9 : Elastic response spectrum of Vrancea earthquake

From the non-linear dynamic analysis the following reference values have been obtained (with a small correction due to the fact that the inelastic response spectrum is not strictly similar with the elastic spectrum) :

$$q^{*} = 3.4 \qquad \text{for} \qquad a_N^{(u)*} = 4.8\,\text{m/s}^2\,.$$

In order to illustrate the design procedure, first of all is considered the case where $a_N = 5.0\,\text{m/s}^2$. The behaviour factor according to (8.3.5) is equal now to q = 3.5 and relationship (8.3.9) gives :

$$V^{(\text{inel})} = 6 \times 47250 \times \left(\frac{0.63 \times 9.81}{2.2}\right) \times \frac{5.0}{3.5} = 1138\ \text{kN}\,.$$

The inelastic displacement of the 6th floor deduced from (8.3.11) and (8.3.12) is :

$$\delta_6^{(\text{inel})} = 1.10\ \text{m},$$

leading to the highest interstorey drift sensitivity coefficient $\theta_3 = 0.16$ and consequently needing to take into account the geometrical 2nd order effects.

As comparison, dynamic analysis gives, for $a_N = 5.0\,\text{m/s}^2$, the following values for the base shear force and the inelastic displacement which appear not far from the previous ones :

$$\overline{V}^{(\text{inel})} = 1145\ \text{kN} \qquad \text{and} \qquad \overline{\delta}_6^{(\text{inel})} = 0.95\ \text{m}.$$

Figure 8.3.10.a shows other results obtained from the dynamic analysis as the distribution of the real horizontal inelastic forces, that of the storey displacements and the position of the plastic hinges. Figure 8.3.10.b gives the horizontal force distribution in accordance with (8.3.14) corresponding to the global mechanism with a second order rigid-plastic analysis (based on the distribution of the displacements associated with $\delta_6^{(\text{inel})} = 1.10\ \text{m}$). Also is taken into account a reduction of the plastic resistance moment due to the axial force in the base columns. Finally, the base shear force corresponding to the global mechanism is equal to :

$$V_p^{(\text{glob})} = 675\ \text{kN}\,.$$

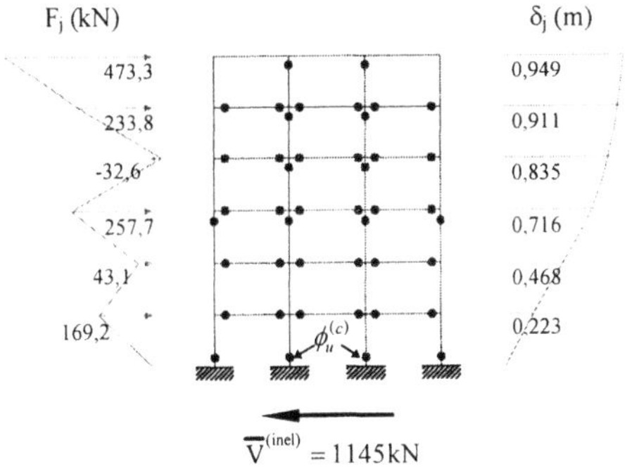

Figure 8.3.10.a : Results of non-linear dynamic analysis for a_N =5.0 m/s^2

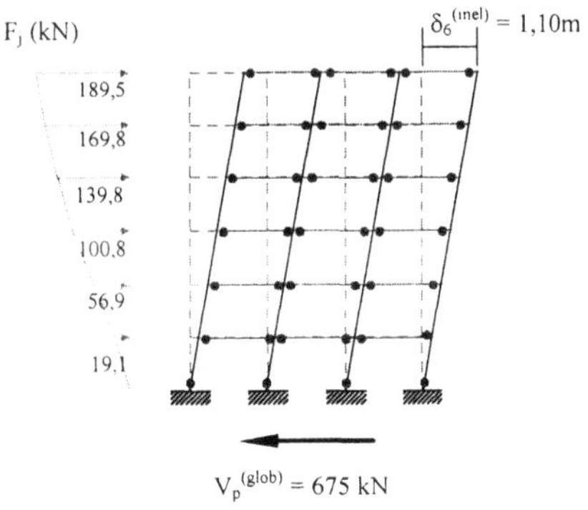

Figure 8.3.10.b : Second order rigid-plastic analysis of the global mechanism

Examination of different local mechanisms using also second order rigid-plastic analyses is illustrated partially in Figure 8.3.11.

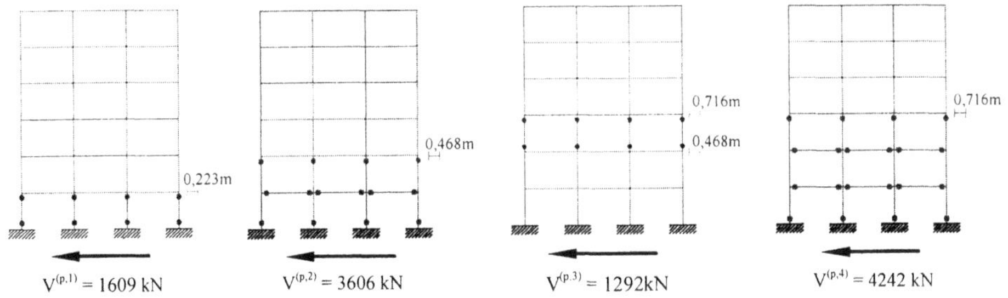

Figure 8.3.11 : Second order rigid-plastic analyses of local mechanisms

It is easy to check that conditions expressed by inequalities (8.3.15) and (8.3.16) are well satisfied. As for the risk of local mechanism in the base columns, it is out of question because :

- for external columns : $N_{Sd}/N_{pl.Rd} = 0.11 < 0.15$,

  and according to 17.b : $\overline{\lambda} = 0.46 < 1.6$

- for internal columns : $N_{Sd}/N_{pl.Rd} = 0.22 > 0.15$,

  but according to 17.a : $N_{Sd}/N_{pl.Rd} + 0.8\overline{\lambda} = 0.22 + 0.8 \times 0.46 = 0.59 < 1$

The above method of design can be easily adapted for a lower level of nominal acceleration, for instance a_N = 3.5 m/s^2. According to relationship (8.3.5) acting on the safe side, the behaviour factor takes the value q = 2.5, but the value of $V^{(inel)}$ remains unchanged. In return, the inelastic displacements are reduced, so :

$$\delta_6^{(inel)} = 1.10 \times \overline{3.5}^2 / \overline{5.0}^2 = 0.54 \text{ m}.$$

Obviously inequalities (8.3.15) and (8.3.16) are still well satisfied. These results are confirmed by the non-linear dynamic analysis which gives in particular :

$$\overline{V}^{(inel)} = 999 \text{ kN} \qquad \text{and} \qquad \overline{\delta}_6^{(inel)} = 0.49 \text{ m}.$$

GENERALIZATION TO FRAMES WITH SEMI-RIGID AND PARTIAL STRENGTH JOINTS

As already mentioned, the structure should be considered partially dissipative when the rotation capacity of dissipative members is limited (due to cross-sections in Class 2 or 3) or partial strength joints are used in dissipative zones.

The reference values (q^*, $a_N^{(u)*}$) are not reachable generally and the maximum values (q, a_N) should be determined as function of the allowable rotation capacity directly from dynamic analysis.

A few results from a parametric investigation

Grecea (1999) carried out a parametric investigation on some of the structures in Figure 8.3.4 (structures A, B, C), including partial strength joints with different characteristics and considering the three accelerograms of Figures 8.3.5.a,b,c, in order to evaluate their influence on the q-factor. Moment resistance $M_{j,R}$ and initial rotational stiffness $S_{j,ini}$ of the studied joints are given in Table 8.3.3 where $M_{b,pl,R}$ is the plastic resistance moment of the connected beam to the joint and K_{sup} is the limit rotational stiffness corresponding to the distinction between rigid joint behaviour and semi-rigid one according to Annex J of Eurocode 3. Figure 8.3.12 shows skeleton moment-rotation curves of joints (for a quarter space) assuming here perfect parallelograms without stiffness degradation when repeated cyclic bending moments are applied. This type of joint behaviour was introduced in the inelastic dynamic analyses performed using again DRAIN-2DX computer code of Prakash (1993). As often as not, partial strength joints have got a limited rotation capacity ϕ_u. Here three values of rotation capacity were adopted, namely 0.015 ; 0.030 and 0.045 radians which seem to be realistic to cover most of applications.

Tables 8.3.4 and 8.3.5 collect the results of the q-factor for structure C subject to the accelerogram of Vrancea (1977) and Kobe (1995), respectively.

TABLE 8.3.3
JOINT CHARACTERISTICS FOR THE PARAMETRIC STUDY

Moment resistance $M_{j,R}$	Rotational stiffness $S_{j,ini}$
$1.0\ M_{b,pl,R}$	K_{sup}
	$0.8K_{sup}$
	$0.6K_{sup}$
$0.8\ M_{b,pl,R}$	K_{sup}
	$0.8K_{sup}$
	$0.6K_{sup}$
$0.6\ M_{b,pl,R}$	K_{sup}
	$0.8K_{sup}$
	$0.6K_{sup}$

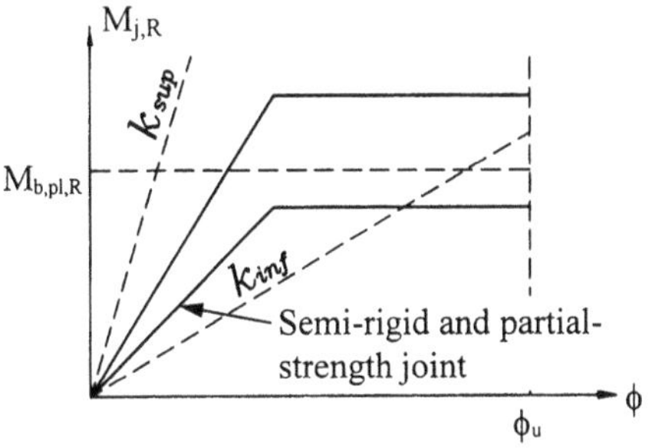

$K_{sup} = 25EI_b / L_b$

Figure 8.3.12 : Skeleton moment-rotation curves

TABLE 8.3.4
VALUES OF q-FACTOR FOR ACCELEROGRAM OF VRANCEA (1977)

$M_{j,R}$ / $M_{b,pl,R}$	Criterion	$S_{j,ini}$ = Ksup			$S_{j,ini} = 0.8K_{sup}$			$S_{j,ini} = 0.6K_{sup}$		
		q	a_N	θ_j	q	a_N	θ_j	q	a_N	θ_j
1.0	$\phi_{0.015}$	1.5	1.30	0.042	1.6	1.30	0.039	1.6	1.30	0.043
	$\phi_{0.030}$	2.0	2.20	0.061	2.0	2.20	0.061	2.1	2.20	0.065
	$\phi_{0.045}$	2.8	3.00	0.090	2.9	3.00	0.094	3.0	3.00	0.096
0.8	$\phi_{0.015}$	1.8	1.20	0.044	1.8	1.20	0.049	1.8	1.20	0.049
	$\phi_{0.030}$	2.3	2.10	0.074	2.3	2.00	0.074	2.4	2.00	0.075
	$\phi_{0.045}$	2.9	2.90	0.097	2.9	2.80	0.101	3.0	2.80	0.102
0.6	$\phi_{0.015}$	2.0	1.10	0.060	2.0	1.10	0.060	2.0	1.10	0.058
	$\phi_{0.030}$	2.4	2.00	0.086	2.4	2.00	0.087	2.6	2.00	0.090
	$\phi_{0.045}$	2.9	2.60	0.111	3.0	2.60	0.112	3.0	2.60	0.115

TABLE 8.3.5
VALUES OF q-FACTOR FOR ACCELEROGRAM OF KOBE (1995)

$M_{j,R}$ / $M_{b,pl,R}$	Criterion	$S_{j,ini} = K_{sup}$			$S_{j,ini} = 0.8K_{sup}$			$S_{j,ini} = 0.6K_{sup}$		
		q	a_N	θ_j	q	a_N	θ_j	q	a_N	θ_j
1.0	$\phi_{0.015}$	1.5	0.50	0.044	1.6	0.60	0.046	1.8	0.80	0.054
	$\phi_{0.030}$	2.3	1.00	0.054	2.8	1.20	0.066	2.7	1.20	0.071
	$\phi_{0.045}$	3.4	1.60	0.069	3.5	1.60	0.073	3.2	1.60	0.079
0.8	$\phi_{0.015}$	1.5	0.50	0.043	1.7	0.50	0.047	1.9	0.70	0.052
	$\phi_{0.030}$	2.8	1.00	0.068	2.9	1.00	0.070	2.8	1.10	0.074
	$\phi_{0.045}$	3.5	1.40	0.069	3.8	1.50	0.072	3.5	1.60	0.078
0.6	$\phi_{0.015}$	1.9	0.50	0.042	2.3	0.50	0.049	2.3	0.60	0.056
	$\phi_{0.030}$	3.2	1.00	0.065	3.5	1.00	0.070	3.2	1.10	0.075
	$\phi_{0.045}$	4.2	1.40	0.075	4.6	1.50	0.079	4.0	1.70	0.082

From examination of Tables 8.3.4 and 8.3.5 it is clear that the initial rotational stiffness $S_{j,ini}$, even for the lowest value $0.6K_{sup}$, has no influence practically on the q-factor. Moreover, decrease in the joint moment resistance tends to increase slightly the q-factor, maybe favouring the occurrence of global dissipative mechanism. So, these results incite to present the average values of the q-factor for the two accelerograms, as a function only of the joint resistance moment and of the rotation capacity. These average results are presented in Tables 8.3.6 and 8.3.7.

TABLE 8.3.6

AVERAGE VALUES OF q-FACTOR FOR ACCELEROGRAM OF VRANCEA (1977)

$M_{j,R}$	$1.0M_{b,pl,R}$			$0.8M_{b,pl,R}$			$0.6M_{b,pl,R}$		
ϕ	$\phi_{0.015}$	$\phi_{0.030}$	$\phi_{0.045}$	$\phi_{0.015}$	$\phi_{0.030}$	$\phi_{0.045}$	$\phi_{0.015}$	$\phi_{0.030}$	$\phi_{0.045}$
q	1.6	2.0	2.9	1.8	2.3	2.9	2.0	2.4	3.0
a_N	1.30	2.20	3.00	1.20	2.00	2.80	1.10	2.00	2.60
θ_j	0.04	0.06	0.09	0.05	0.07	0.10	0.06	0.09	0.11

TABLE 8.3.7

AVERAGE VALUES OF q-FACTOR FOR ACCELEROGRAM OF KOBE (1995)

$M_{j,R}$	$1.0M_{b,pl,R}$			$0.8M_{b,pl,R}$			$0.6M_{b,pl,R}$		
ϕ	$\phi_{0.015}$	$\phi_{0.030}$	$\phi_{0.045}$	$\phi_{0.015}$	$\phi_{0.030}$	$\phi_{0.045}$	$\phi_{0.015}$	$\phi_{0.030}$	$\phi_{0.045}$
q	1.6	2.6	3.4	1.4	2.8	3.6	2.2	3.3	4.3
a_N	0.60	1.20	1.60	0.60	1.00	1.50	0.50	1.00	1.50
θ_j	0.05	0.06	0.07	0.05	0.07	0.07	0.05	0.07	0.08

Example : 6 storey - 3 bay frame - Accelerogram of Vrancea

An application to the 6 storey - 3 bay frame of Figure 8.3.8 but including partial strength joints with resistance moment $M_{j,R} = 0.8M_{b,pl,R}$ and with rotation capacity $\phi_u = 0.030$ radians is presented bellow.
The q-factor deduced from non-linear dynamic analyses is q = 2.1 with the associated acceleration a_N = 2.1 m/s^2. An equivalent static second order elastic-plastic analysis is performed using PEP-micro computer code developed in CTICM by Galea & Bureau (1995) up to reach step-by-step the base shear force equal to $V^{(inel)}$ = 759 kN. It leads to 6th storey displacement equal to $\delta_6^{(inel)}$ = 0.029m and a maximum required rotation 0.029 radians.
From the inelastic dynamic analysis, the real solution is in fact : $V^{(inel)}$ = 773.3 kN, with a 6th storey displacement equal to $\delta_6^{(inel)}$ = 0.025m and a maximum required rotation 0.030 radians. These results are illustrated in Figure 8.3.13 for the accelerogram multiplier λ = 0.94 which corresponds to the required rotation 0.030 radians.
The comparison between the internal forces evaluated by the static elastic-plastic analysis and those calculated by the inelastic dynamic analysis, into the base columns is presented in Table 8.3.8 (bending moment M, horizontal shear V and axial force N).

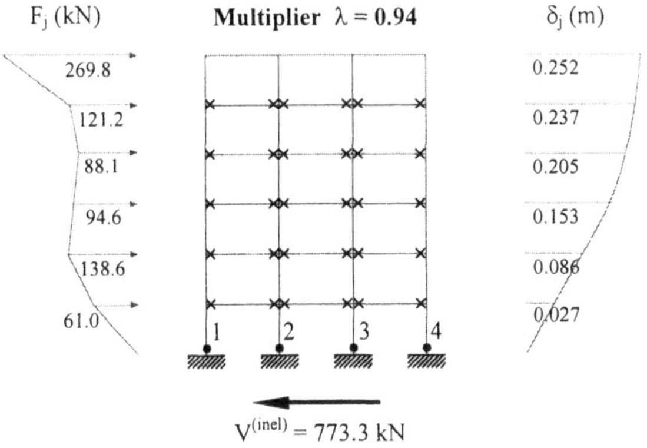

Figure 8.3.13 : Distribution of seismic forces and storey displacements from inelastic dynamic analysis for required rotation 0.030 radians

TABLE 8.3.8
INTERNAL FORCES AT THE BASE COLUMNS

Base floor columns	PEP-micro - Static elastic-plastic analysis			DRAIN-2DX - Inelastic dynamic analysis (a_N=2.1m/s^2 for req. rotation 0.030rd)		
	M [kNm]	V [kN]	N [kN]	M [kNm]	V [kN]	N [kN]
1	737.0	171.2	68.5	737.0	177.0	72.0
2	711.9	207.5	931.6	712.2	211.0	927.2
3	701.4	208.8	947.9	700.2	210.8	960.6
4	714.2	171.0	887.0	713.8	174.5	895.2

CONCLUSION

Using a new definition of the q-factor, appropriate equivalent static methods are proposed to design steel structures in seismic zones.
It appears a good agreement between these methods and the non-linear dynamic analyses (considering the inelastic base shear force and the inelastic displacements).
Fully dissipative frames can be designed by means of 1st and 2nd order rigid-plastic analyses whereas partially dissipative frames need to use an elastic-plastic analysis.
Another advantages should be pointed out :

- easy adjustment of the q-factor in accordance with nominal acceleration a_N ;
- possible taking account of frames where semi-rigid and partial strength joints are active in dissipative zones.

As perspective, there is need to continue the calibration of reference values (q^* , $a_N^{(u)*}$) for other types of structure and shapes of response spectrum so that simplified analytical relationships to evaluate these values a priori could be established.

A tentative generalisation is in progress for the case of irregular structures (filtering the contribution of the different vibration modes to the inelastic base shear force).

References

Mazzolani F.M. and Piluso V. (1996). *Theory and design of seismic resistant steel frames*, E & FN Spon, London, UK

Cosenza E., De Luca A. and Faella C. (1986). Criteria for evaluation of overall ductility of steel frames. *Costruzioni Metalliche n°5.*

Giuffre A. and Giannini R. (1982). La duttilita delle strutture in cemento armato. ANCE, AIDIS, Roma.

Palazzo B. and Fraternali F. (1987). L'uso degli spettri di collasso nell'analisi sismica: proposta per una diversa formulazione del coefficiente di struttura. 3 Convegno Nazionale l'Ingegneria Sismica in Italia, Roma.

Mazzolani F.M. and Piluso V. (June/July 1994). A new method to design steel frames failing in global mode including P-Δ effects. International Workshop on Behaviour of Steel Structures in Seismic Areas, Timisoara, Romania.

Cosenza E., De Luca A., Faella C. and Piluso V. (August 1988). A rational formulation for the q factor in steel structures. 9th World Conference on Earthquake Engineering, Tokyo, Japan.

Como M. and Lanni G. (1983). Aseismic toughness of structures. *Meccanica 18*, pp. 107-114

Akiyama H. (August 1988). Earthquake resistant design based on energy concept. 9th World Conference on Earthquake Engineering, Tokyo, Japan, Vol V, 8.1.2.

Ballio G. (Sept. 1985). ECCS approach for the design of steel structures against earthquakes. Symposium on Steel in Buildings, Luxembourg, IABSE-AIPC-IVBH Report, Vol. 48, pp. 373-380.

Setti P. (1985). Un metodo per la determinazione del coefficiente di strutura per le costruzioni metalliche in zona sismica. *Costruzioni Metalliche n°3.*

Aribert J.M. and Grecea D. (1997). A new method to evaluate the q-factor from elastic-plastic dynamic analysis and its application to steel frames. *Behaviour of Steel Structures in Seismic Areas*, Edited by F.M. Mazzolani & H. Akiyama, Kyoto, Japan.

Aribert J.M. and Grecea D. (1998). Experimental behaviour of partial-resistant beam-to-column joints and their influence on the q-factor of steel frames. XIème Conférence Européenne de Génie Parasismique, 6-11 Septembre 1998, Paris, France.

Sedlacek G. and Kuck J (31.8.1993) Determination of q-factors for Eurocode 8, Aachen, Germany.

Gioncu V. and Petcu D. (Sept.21/23, 1995). Numerical investigations on the rotation capacity of beam and beam-column. International Colloquium "Stability of Steel Structures", Budapest, Hungary.

Grecea D. (1999). Caractérisation du comportement sismique des ossatures métalliques - Utilisation d'assemblages à résistance partielle. Thèse de Doctorat. INSA de Rennes, France.

Prakash V., Powell G.H. & Campbell S. (1993). DRAIN-2DX, base program description and user guide. Version 1.10.

AFNOR (1995). Règles PS 92 applicables aux bâtiments, NFP 06.013, France.

Galea Y. and Bureau A. (1995) PEP-micro. Analyse plastique au second ordre de structures planes à barres. Manuel d'utilisation, Version 2b, CTICM, France.

8.4

COMPARISON AMONG METHODS

Jordan Milev, Peter Sotirov, Zdravko Bonev, Nikolay Rangelov, Tzvetan Georgiev

INTRODUCTION

The design philosophy adopted by current seismic codes, including Eurocode 8, is based on the assumption that some of the energy input during a strong earthquake is dissipated through inelastic deformations. The limit values of these plastic deformations should be restricted in order to prevent the building from collapse. This restriction is usually made on the basis of local and global ductility, and energy dissipation capacity of the structure. Nowadays there are two main methods for performing seismic resistant design of structures, considering their inelastic behaviour:

(1) Inelastic time history analysis, which is able to provide the response history of the structure subjected to generated or recorded earthquakes;
(2) The behaviour factor concept, which is based on the use of elastic modal analysis by assuming an inelastic design spectrum. This is obtained in almost all seismic codes by modifying the elastic design pseudo acceleration response spectrum by means of a factor (q-factor in Eurocode 8), which takes into account the energy dissipation capacity of the whole structure up to collapse;

However, the inelastic time history analysis is limited into the research purposes and into the study of seismic response of very important facilities such nuclear power plants because it requires an accurate and sophisticated modeling of the cyclic behaviour of the structural members, which is provided nowadays by specialized software only. Moreover such analysis is practically cumbersome from the computational point of view. In comparison with those the behavior factor (q-factor) approach is a simple tool for designers. It allows performance of an equivalent static analysis instead of inelastic dynamic one.

The maximum allowable values of the behaviour factor, q_{code}, are specified in Eurocode 8 for various structural types. The value of q-factor is the ratio of the ordinates of the linear elastic design response spectrum (LEDRS) used to define the seismic hazard at a site to those of the inelastic design response spectrum (IDRS), or

$$q_{code} = \frac{S_a^{LEDRS}}{S_a^{IDRS}} \tag{8.4.1}$$

where: S_a^{LEDRS} : the elastic design spectral pseudo acceleration (see Fig. 8.4.1);

S_a^{IDRS} : the inelastic design spectral pseudo acceleration (see Fig. 8.4.1);

Behaviour factor, q, for the specified acceleration time history can be presented as follows:

$$q = \frac{S_a^c}{S_a^y} \tag{8.4.2}$$

where: S_a^c : the spectral pseudo acceleration for the ground motion intensity corresponding to the collapse state of the structure (see Fig. 8.4.1);

S_a^y : the spectral pseudo acceleration for the ground motion intensity corresponding to the yield state of the structure (see Fig. 8.4.1);

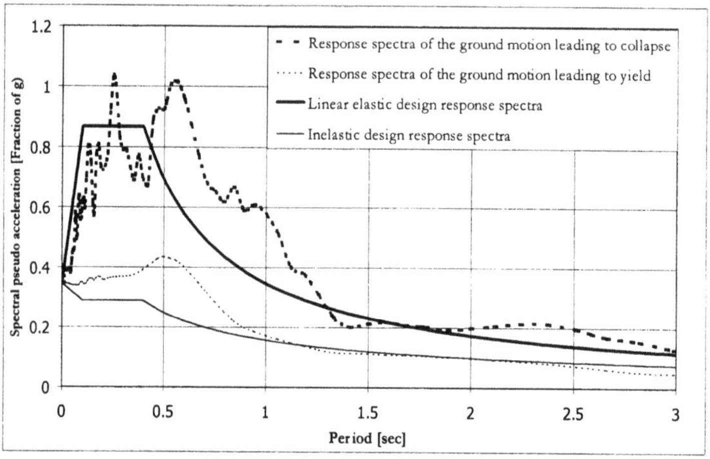

Figure 8.4.1. Design, yield and collapse pseudo acceleration response spectra

Therefore the evaluation of the q-factor for the specified acceleration history requires inelastic dynamic analysis for evaluation of two spectra and two periods, at the yield and collapse state. However the Eqn. (8.4.2) could be transformed in terms of peak ground acceleration by assuming a constant dynamic acceleration amplification (see Elnashai at al. (1995)), as follows:

$$q = \frac{a_g^c}{a_g^y} \tag{8.4.3}$$

where: a_g^c : the peak ground acceleration for the ground motion intensity corresponding to the collapse state (see Fig. 8.4.1);

a_g^y : the peak ground acceleration for the ground motion intensity corresponding to the yield state (see Fig. 8.4.1).

Eqn. (8.4.3) could be employed to determine the actual behaviour factors, q, of a range of steel moment resisting frames under various recorded or artificial ground motions by applying some software inelastic dynamic analysis. It should be noted, that the values of the structural behaviour factor, q_{code}, specified by the Eurocode 8 represent the lower bounds of actual ductility supply of the structure. Therefore it should be expected that $q_{code} \leq q$.

Following the classification given by Mazzolani & Piluso (1996), the existing methods for the q-factor evaluation can be grouped in three main categories as follows:

(1) *Methods based on the inelastic response of single degree of freedom systems (SDOF).* Those methods are relatively simple but they require that the structures satisfy the conditions like "global collapse mechanism" and "structural regularity". Moreover the global mechanism considered under equivalent horizontal static forces increasing proportionally may not occur during a real dynamic response.
(2) *Methods based on energy approach.* These methods do not require the conditions of structural regularity and collapse by global mechanism to be satisfied. An energy method is discussed in 8.2.
(3) *Methods based on the inelastic dynamic analysis of multidegree of freedom systems (MDOF).* Those methods require non-linear dynamic time history analysis. The well known approach of Ballio and Setti (see Balio (1985)) is one of the leading in the literature.

Recently Aribert and Grecea (1997) proposed a new method, which is suitable for determination of the q-factor value, for a given structure and for any type of acceleration time history.

In order to evaluate the behavior factor of frame structures, criteria for defining the yielding state and collapse state have to be previously established. In both cases, several approaches, based upon both local and global criteria, may be defined. Besides, it is to be considered that the structural performance of moment resisting steel frames may be strongly influenced by the adopted beam-to-column joint typology. Therefore, aiming at assessing the influence of the main connection behavioural parameters in the current study, with reference to a typical structural configuration, four different q-factor evaluation approaches are considered and corresponding results duly compared. These approaches are as follows:

- Local ductility method, based on the inelastic analysis of MDOF;
- Drift method, based on the inelastic analysis of MDOF;
- Modified Akiyama-Kato method or Energy method (see 8.2);
- Aribert and Grecea method;

STUDIED FRAMES

Two different type of frames are studied as follows:

1) The 3bay x 3bay, 5-storey steel building previously designed by Mazzolani & Piluso (1997) and analysed by Fulop (1998), is used as a case study. The analysed frame and corresponding member sizes, which are obtained from a design procedure carried out according to Eurocode 8 (1994) and Eurocode 3 (1990), are presented in Fig. 8.4.2. This frame is noted here as "Mazzolani & Piluso frame".
2) The six B-type frames, which have been previously described (see 8.2). Those frames are noted here as "B-type frames".

Several solutions in terms of beam-to-column joint behaviour are considered for the case of "Mazzolani & Piluso frame". The structure with rigid joints is refereed as "RIG". The structure with semi-rigid connections located only in the middle bay of the frame is refereed as "Smn", where **m** is a key number for the connection stiffness and **n** is a key number for the connection strength. The case with pinned connections in the middle bay is called "PIN". The structure with semi-rigid connections in all beam-to-column joints is refereed as "Tmn". In particular, in all analysed cases, according to Eurocode 3 (1994) classification system, **m**=1 is referred to a rigid joint, while **m**=3 represents a semi-rigid joint with stiffness $0.8S_{j,ini}$ (see table 8.4.1). Similarly, **n**=2 is referred to a connection whose strength is equal to the one of the connected member (M_{con}=M_b), while **n**=4 corresponds to a partial strength connection with M_{con}=0.6M_b(see table 8.4.1). The rotation capacity of connections is chosen as a parameter and its influence on the global and local structural behaviour is accounted for. Five different cases based on connection plastic rotation capacity (0.015; 0.030; 0.045; 0.060; 0.075 rad) are studied for frames S14, S32, T14 and T32. Frames S14 and T14 are with partial strength rigid connections, while frames S32 and T32 are with full strength semi rigid connections. The stiffness, moment and rotation capacities of the connections for the studied cases are presented in the Table 8.4.1.

All necessary information for the case of "B-type frames" is presented in (8.2).

DRAIN-2DX software is used as an analysis tool. Beams and columns are modeled by means of plastic hinge beam-column element 02, while simple connection element 04 with elastic perfectly plastic behaviour is employed for connection modeling in the case of "Mazzolani & Piluso frame" (see Powel & Campbell (1994)). Parameters of element 04 are varied for the frames S14, S32, T14 and T32 depending on assumed connection strength and rigidity levels.

TABLE 8.4.1
STIFFNESS AND MOMENT CAPACITY OF THE CONNECTIONS

Frame	Stiffness	Moment capacity	Stiffness [kNm/m]		Moment capacity [kNm]		Rotation capacity[rad]
			IPE 400	IPE 450	IPE 400	IPE 450	
Rigid	Infinity	1.2*M_b	1.00E+20	1.00E+20	479.964	386.574	-
Semi14	1.0*$S_{j,ini}$	0.6*M_b	236180	161910	239.982	184.287	0.015 0.030 0.045 0.060 0.075
Semi32	0.6*$S_{j,ini}$	1.0*M_b	141708	97146	479.964	386.574	0.015 0.030 0.045 0.060 0.075
Tot14	1.0*$S_{j,ini}$	0.6*M_b	236180	161910	239.982	184.287	0.015 0.030 0.045 0.060 0.075
Tot32	0.6*$S_{j,ini}$	1.0*M_b	141708	97146	479.964	386.574	0.015 0.030 0.045 0.060 0.075
Pin	Zero	Zero	1.0	1.0	0.0	0.0	-

Three earthquake records were selected for this study – El Centro NS 1940, Kobe JMA NS 1995, and Tolmezzo. Some of their parameters are presented in Table 8.2.2.

Forty nonlinear time history response analyses are carried out for each structural typology and for each accelerogram. The earthquake records are normalized to 1.0g. Then the earthquake record under consideration is scaled by multiplying the acceleration values by the factor α, assuming for each time history analysis the peak ground acceleration value as follows:

$$PGA = \alpha g \tag{8.4.4}$$

where: α is the acceleration record multiplier and **g** is the gravity acceleration.

Coefficient α is varied from 0.05 up to 2.0 with step increment equal to 0.05. The integration step was chosen less than 0.01 sec. In this case force, moment and energy unbalances were in a reasonable limit.

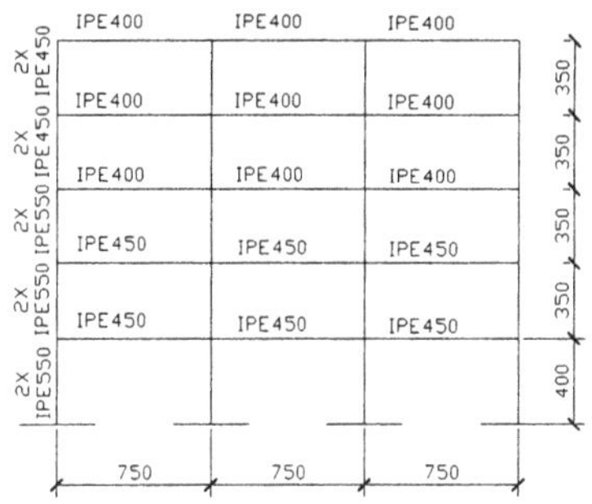

Figure 8.4.2 Structure layout

For each structural typology and for each earthquake record the multipliers α_y (corresponding to the yielding state) and α_u (corresponding to the collapse) are defined. Then, the behaviour factor q based on the local ductility criteria as well as on the drift criteria may is evaluated through the ratio:

$$q = \frac{\alpha_u}{\alpha_y} \tag{8.4.5}$$

LOCAL DUCTILITY METHOD

The yielding state is defined when the first plastic hinge in the structure is formed. The ultimate or collapse state is assumed when the cumulated rotation in the most stressed beam-column element or connection, in positive or negative direction, reaches its monotonic rotation capacity θ_{st}. The monotonic rotation capacity of beam and columns θ_{st} is calculated as an average of the values obtained by the theoretical method of Mazzolani & Piluso (1996) and the empirical method of Mitani-Makino (see Mazzolani & Piluso (1996)). In the case of "Mazzolani & Piluso frame" the monotonic capacity of connections θ_{st} is assumed as a parameter varying from 0.015 to 0.075. The assumed values for the "Mazzolani & Piluso frame" are listed in tables 8.4.2 and 8.4.3. Beam and column rotation capacity, R_{st}, (see Fig. 8.2.1 for R_{st} definition) for the six "B-type frames" are listed in table 8.2.4 (see 8.2).

TABLE 8.4.2
BEAM AND COLUMN PLASTIC ROTATION CAPACITY

Section	θ_{st}	$R_{st}=(\theta_{st}-\theta_y)/\theta_y$
IPE 450	0.0562	10.27
IPE 400	0.0596	9.67
2xIPE 550	0.0366	12.69
2xIPE 450	0.0495	13.06

TABLE 8.4.3
CONNECTION PLASTIC ROTATION CAPACITY

Frame	Connection	θ_{st}	$R_{st}=(\theta_{st}-\theta_y)/\theta_y$
Semi14 Tot14	IPE 450 - Column	0.015	7.40
		0.030	14.80
		0.045	22.20
		0.060	29.60
		0.075	37.00
	IPE 400 - Column	0.015	6.60
		0.030	13.20
		0.045	19.80
		0.060	26.40
		0.075	33.00
Semi32 Tot32	IPE 450 - Column	0.015	2.65
		0.030	5.30
		0.045	7.95
		0.060	10.60
		0.075	13.25
	IPE 400 - Column	0.015	2.40
		0.030	4.80
		0.045	7.20
		0.060	9.60
		0.075	12.00

DRIFT METHOD

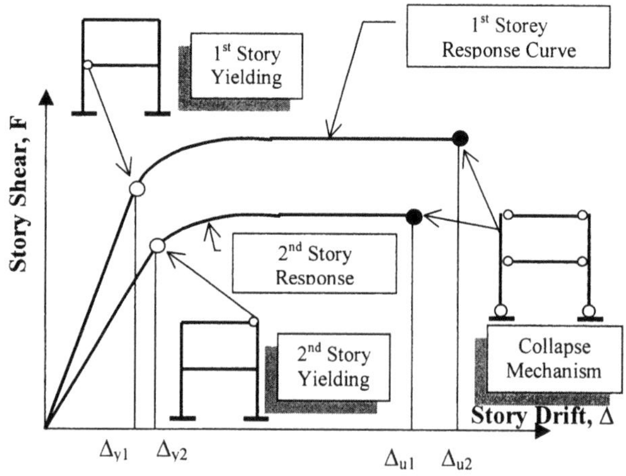

Fig. 8.4.3 Story yield and collapse definition based on push-over analysis

The assumption made herein is that the yield limit values Δ_{yi} as well as the ultimate limit values Δ_{yi} of the interstory drifts are calculated by means of a non-linear static "push-over" analysis. The vertical force distribution is fixed on the basis of story shears obtained by elastic multi-modal analysis of the structure. Interstory drift corresponding to the yielding state is defined when the first yield hinge is formed in the considered story, while the interstory drift at the ultimate state is considered when the mechanism is formed. **P-Δ** effects are considered in the analysis (see Fig.8.4.3).

MODIFIED AKIYAMA-KATO METHOD

The meaning of energy concept in the earthquake resistant design is that the safety of the structure against the severe earthquakes is assessed by comparing the structure energy dissipating capacity with the earthquake input energy. The modified Akiyama-Kato energy method is discussed in details in the 8.2.

ARIBERT & GRECEA METHOD

The third version of Aribert & Grecea (1997) method is applied herein (see 8.4)..

RESULTS OF ANALYSIS

The values of q-factors, which are calculated for all the analysed "Mazzolani & Piluso frames", by assuming a connection rotation capacity θ_{st}=0.045 and by applying the above mentioned different q-factor evaluation methods are shown on Fig. 8.4.4. The same results for the case of "B-type frame are presented in Fig. 8.4.5. However only the Energy, Local ductility and Drift methods are applied in this case. Obtained results in both cases, which have been synthesised as the average for the three records, are slightly scattered, being in the reasonable limit of 10-20%. In general, the Energy method gives the most conservative values for the behavior factor and the smallest scatter for different earthquake records.

The influence of connection ductility on the q-factors for frames with semirigid connections is presented in Fig. 8.4.6 and Fig 8.4.7. The scatter between Energy and Local ductility methods is again reasonable, even if it is more pronounced in case of frames presenting partial strength connections. Nevertheless, it is important to observe that higher behavior factor values for the S14 and T14 frame typologies do not mean a higher structural seismic performance. The Fig. 8.4.8 shows that their capacities in terms of ultimate withstanding PGA for all earthquake records are much lower in respect to the ones of the other frame typologies.

CONCLUSIONS

Several conclusions can be drawn from this study:

- the scatter of the behavior factor values, obtained by the three different methods (Local ductility, Interstory Drift and energy) for three different earthquake records (El Centro, Kobe and Tolmezzo), is in the reasonable limit for all the analysed frame typologies, where several connection strength, rigidity and ductility levels have been considered;

- the Energy method gives the smallest scatter of the behavior factor for different acceleration histories.

- the connection ductility plays an important role on the behavior factor values when the failure is ruled by connections. Evidently, the influence of connection ductility is higher for the case of partial strength connections than for full strength semi-rigid connections. Besides, such an influence is more remarkable in case the Local Ductility approach is used for q-factor evaluation rather than the one based on Energy method.

- The Aribert & Grecea method seems to be the most conservative one;

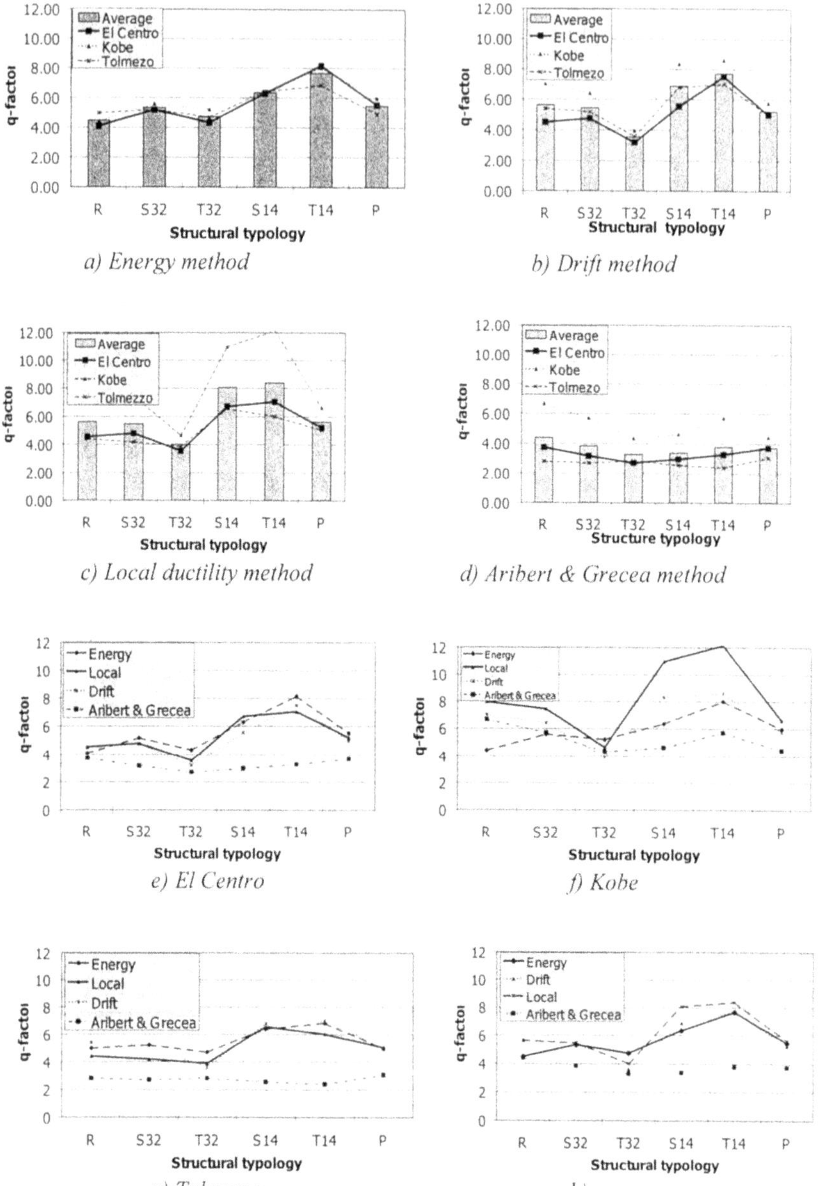

a) Energy method

b) Drift method

c) Local ductility method

d) Aribert & Grecea method

e) El Centro

f) Kobe

g) Tolmezzo

h) averrage

Fig. 8.4.4 Comparison between q-factors calculated by the different approaches for "Mazzolani and Piluso frames" (connection capacity $\theta_{st} = 0.045$)

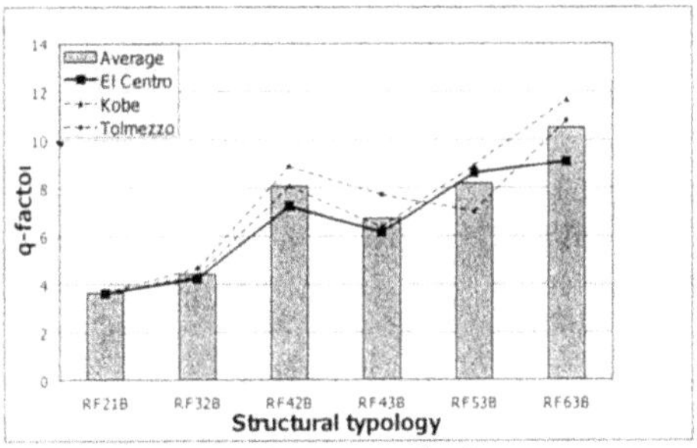

a) Energy method

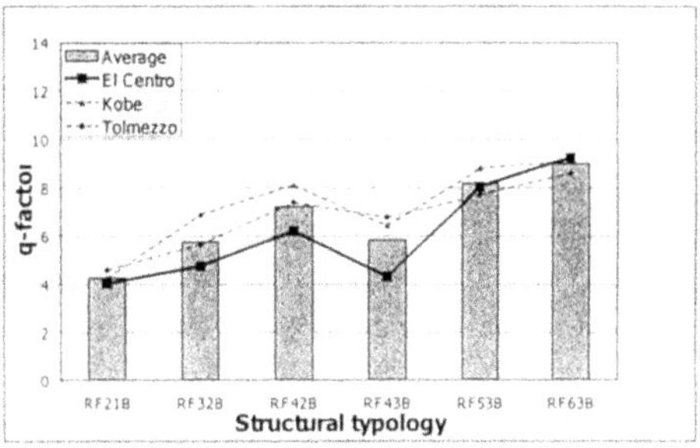

b) Drift method

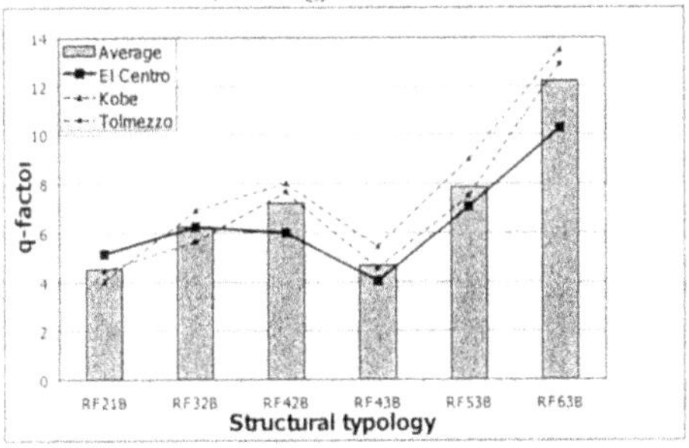

c) Local ductility method

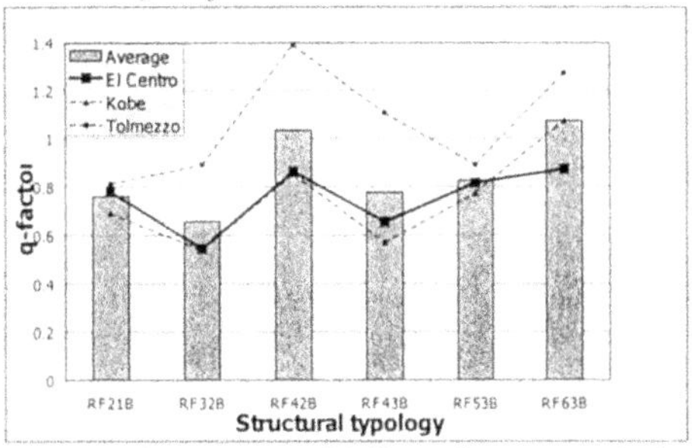

d) ultimate accelerogram multipliers (Energy method)

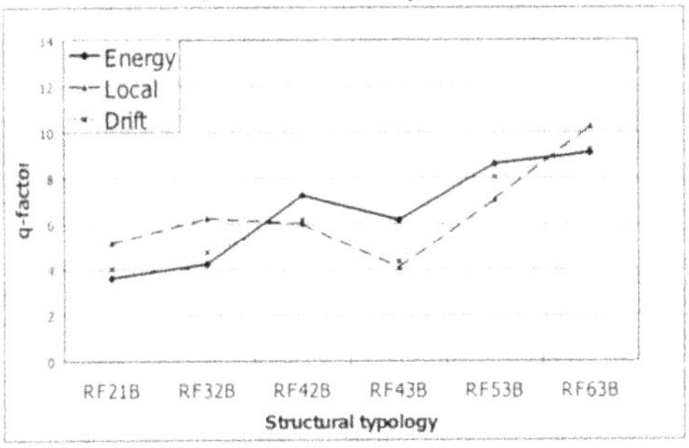

e) El Centro

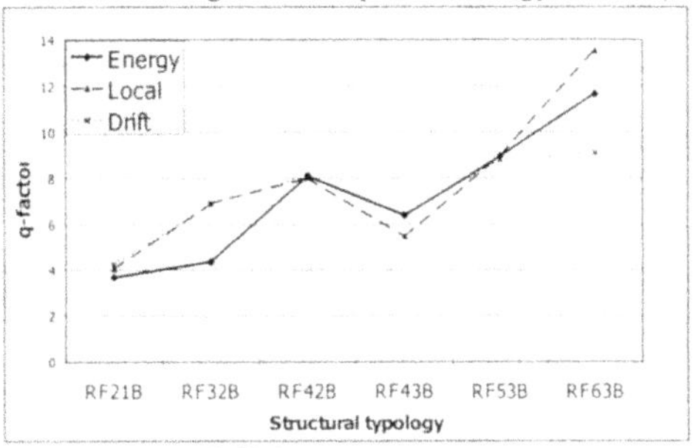

f) Kobe

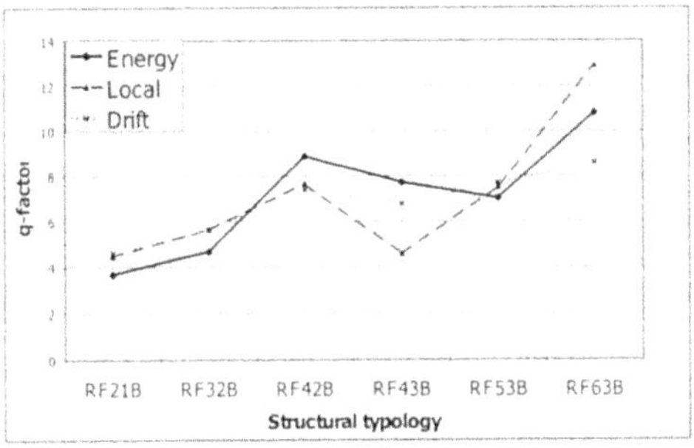

g) Tolmezzo

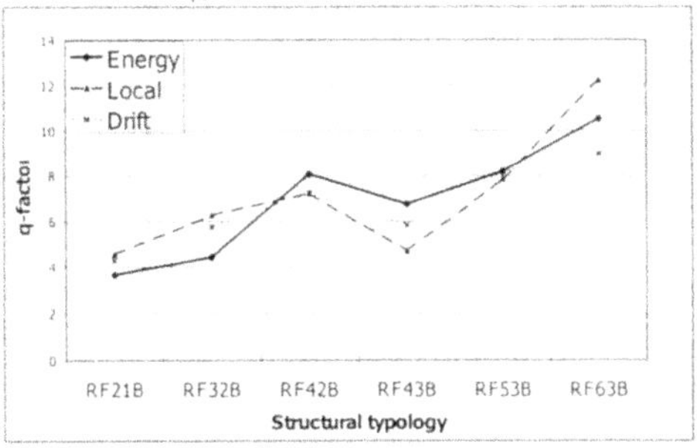

h) averrage

Fig. 8.4.5 Comparison between q-factors and ultimate accelerogram multipliers calculated by the different approaches for "B-type frames"

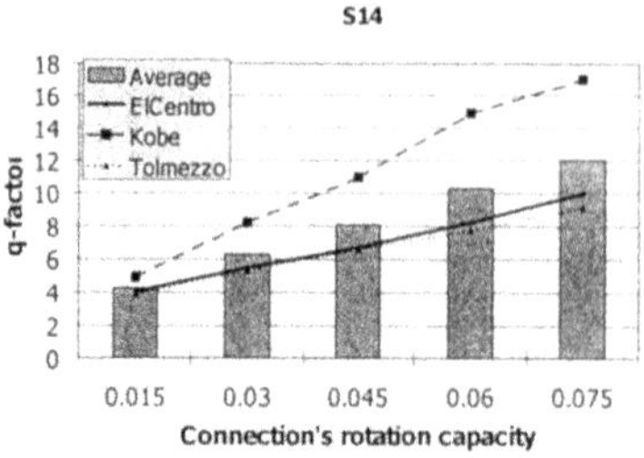

a) Local Ductility method

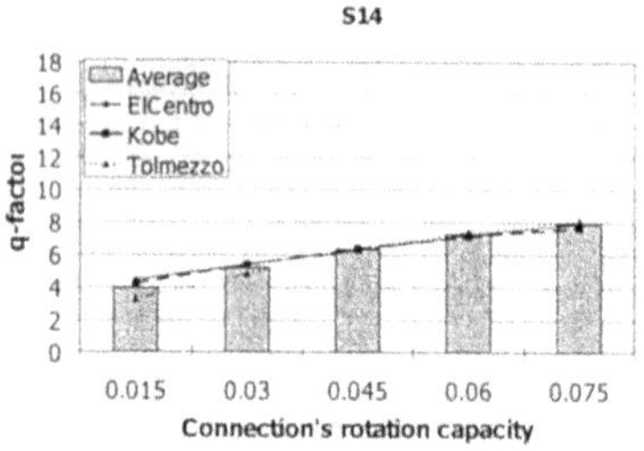

b) Energy method

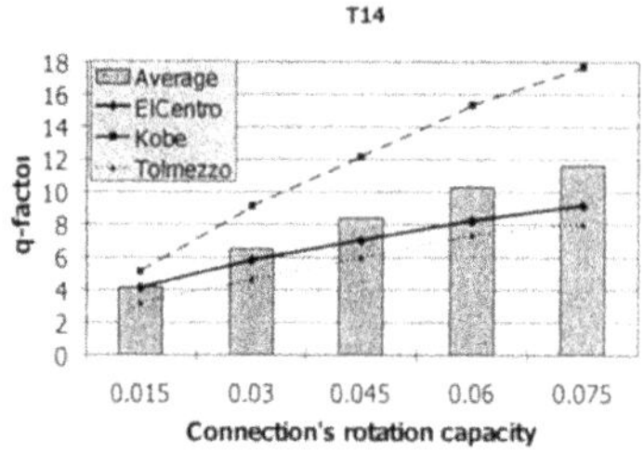

c)Local Ductility method

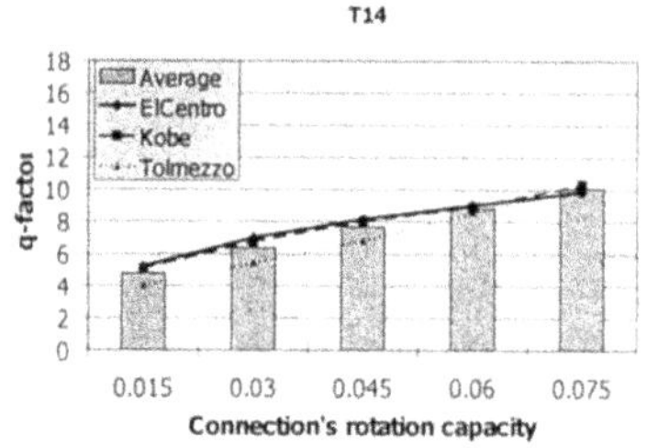

d) Energy method

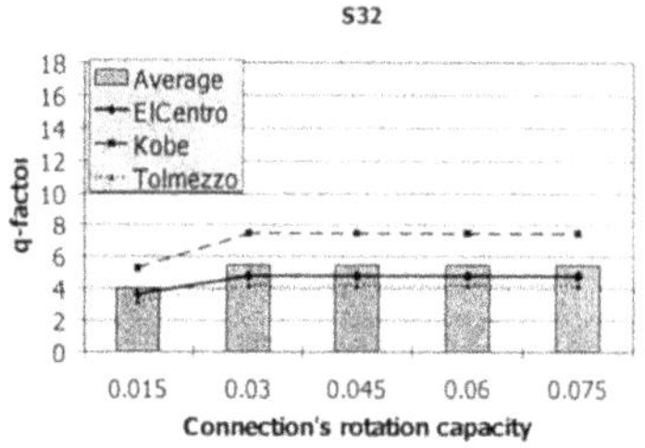

e) Local ductility method

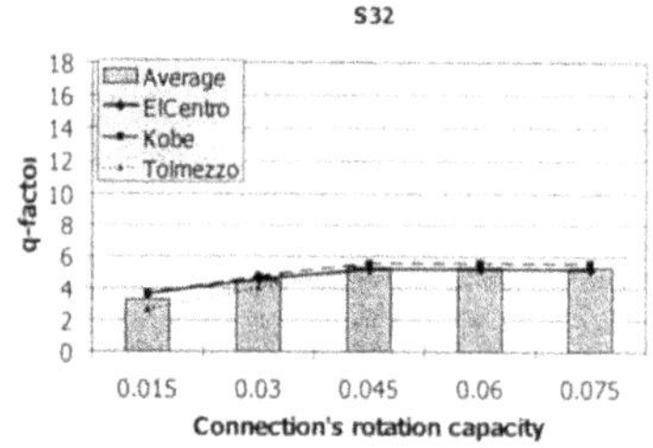

f) Energy method

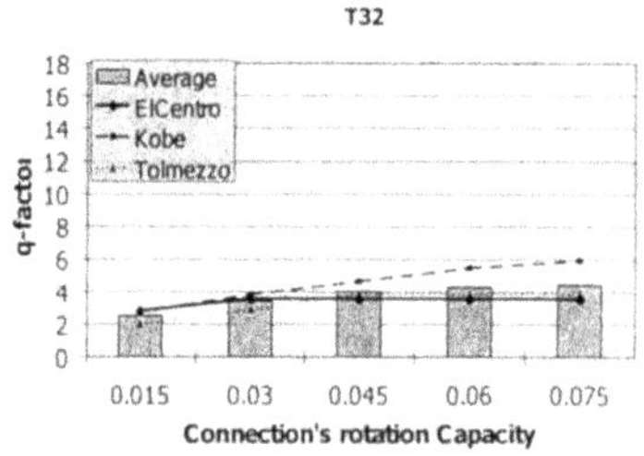

g) Local Ductility method

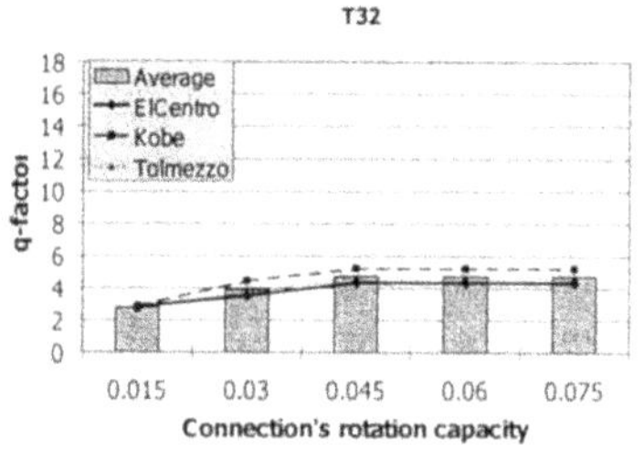

h) Energy method

Fig. 8.4.6 Connection ductility effect on the behavior factor for different earthquakes in the case of "Mazzolani and Piluso frames"

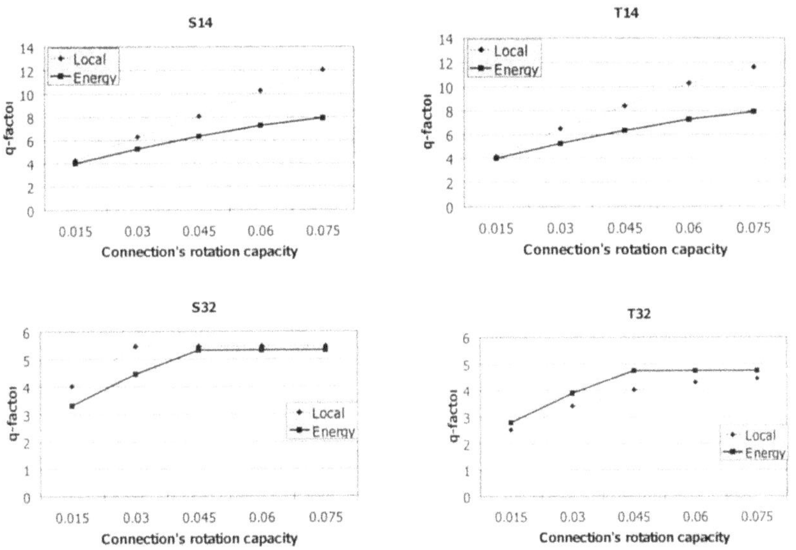

Fig. 8.4.7 Connection ductility effect on the behavior factors average for three earthquakes in the case of "Mazzolani and Piluso frames"

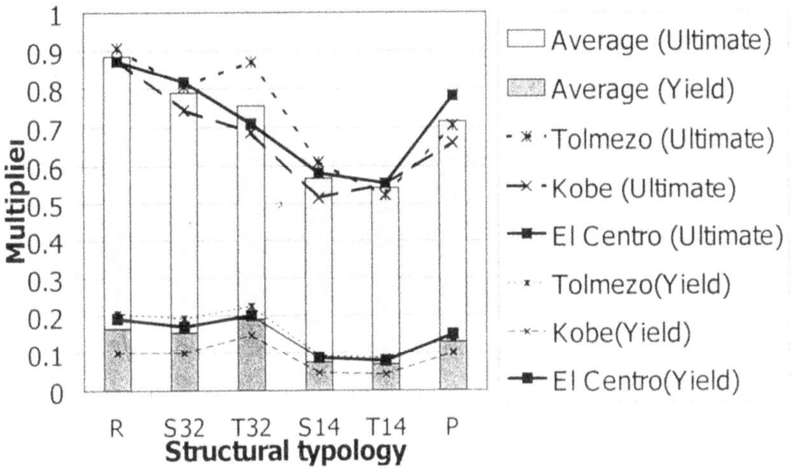

Fig. 8.4.8 Yield and ultimate accelerogram multipliers in the case of "Mazzolani and Piluso frames"

References

Aribert, J.M., and Grecea, D., (1997). "A New Method to Evaluate the q-Factor from Elastic-Plastic Dynamic Analysis and its Application to Steel Frames", Behaviour of Steel Frames in Seismic Areas", Stessa '97, Kyoto, Japan.

Balio, G., (1985). "ECCS approach for the design of steel structures against earthquakes", Symposium on Steel in Buildings, Luxemburg, IABSE-AIPC-IVBH Report, Vol. 48.

Elnashai, A.S., Broderick, B.M., and Dowling, P.J., (1995). "Earthquake-resistant composite steel/concrete structures", Journal of the Institution of Structural Engineers (The Structural Engineer), Volume 73, Number 8, UK

Fulop, L., (1998). "Seismic Performance of Moment-Resisting Frames with Semirigid Joints", Diploma Project, University of Naples Federico II, Universitatea Politehnica Timisoara, Supervisors: Prof. F. M. Mazzolani, Prof. D. Dubina; Co-ordinator: Dr. G. De Matteis.

Mazzolani, F.M. and Piluso, V., (1997). "The influence of the design configuration on the seismic response of moment-resisting frames", Proc. of Behaviour of Steel Structures in Seismic Areas (STESSA'97), Kyoto, Japan, pp. 444-453.

Mazzolani, F.M. and Piluso, V., (1996)"Theory and Design of Seismic Resistant Steel Frames", E&FN SPON.

Powell, G. H. and Campbell, S., (1994). "DRAIN-2DX. Element Description and User Guide for Element Type01, Type02, Type04, Type06 and Type15", Report No UCB/SEMM/94/08.

9

CONCLUDING REMARKS

Federico M. Mazzolani

GENERAL

'RECOS' Copernicus project, sponsored by EC, has dealt with the influence of joints on the seismic response of steel structures. Within this general subject, several special topics have been investigated, aiming at providing useful results for both engineers and researchers as well as for codification. As a whole, the current project represents one of the largest homogeneous studies developed in Europe in this field. A number of experimental tests, theoretical activities and numerical studies have been performed. Several general conclusions may be drawn. In the following they are synthesised with reference to each Chapter, which develops a specific item of the research project.

SEISMIC INPUT AND CODIFICATION

The codification of building structural analysis under seismic loading represents a very important and difficult matter. Recently, important progress in this field has been reached, allowing for designing structures by accounting for inelastic deformation capability of structural components. Nevertheless, the actual behaviour of steel buildings, as evidenced by recent earthquakes occurred all around the World, has been demonstrated to be not always correctly interpreted by present codes. Therefore, the needing of further important revisions of existing provisions has issued.

The analysis of several existing codes belonging to different geographical areas (Europe, USA and Japan) has allowed a comparison among the most important codified rules to be carried out. The American code (UBC) is certainly the most up-to-dated (1997), including important aspects arisen from the recent Northridge earthquake. In all codes the method to be chosen for performing global analysis is based on structural regularity and, sometimes, importance of the building. A specific characteristic of the Japanese code is that the ultimate limit state is investigated through a plastic analysis rather than an elastic one. As far as the definition of seismic force is concerned, analytical expressions of design spectra appear to be quite different to each other, but the shape of the curve is very similar. Seismic risk is essentially based upon the seismicity of the zone, which is defined by peak ground acceleration and type of soil. Only in very recent codes (UBC 97), the effect of seismic typology is being to be taken into account. The influence of seismic ground motion types on the structure response has been deeply investigated, considering the effect provided by both near-source and far-source, artificial and recorded, earthquake types on the response of MR frames. Due to large rapid pulses characterising ground motions generated in the vicinity of the source, the structural response of frames may be very different in relation to the peculiarities of the seismic input and therefore respect to that it may be predicted by applying code provisions. Other great efforts in this direction are therefore required, aiming at providing safer and more detailed approaches for assessing seismic risk.

DUCTILITY OF MEMBERS AND CONNECTIONS

The evaluation of available ductility of members and connections is presented in Chapter 2. This topic is very important for the seismic analysis of structures and it becomes essential when sophisticated inelastic time-history analyses are performed. In fact, in such a case, collapse of structures may be easily verified by comparing developed inelastic deformations in all structural components to the corresponding available ones. In the past, several studies have been carried out with reference to ductility of members under monotonic loading. More recently, the main efforts have been addressed to connections and, in general, to cyclic loading. Two different approaches have been herein presented to determine available ductility: (1) using collapse plastic mechanism method and (2) using a theoretical approach. In the first case, a general computer program "DUCTROT" has been developed, it being very versatile and accounting for several influencing phenomena (strain rate, cyclic loading, tensile failure), allowing members and several joint typologies to be analysed. In the second case, the approach is rather difficult but very attractive. In this project a consistent theoretical study has been devoted to the T-stub model, which is the main component in all bolted joints. Results are very satisfactory and they are available to be extended to other components and joint typologies.

CYCLIC BEHAVIOUR OF BEAM-TO-COLUMN BARE STEEL CONNECTIONS

The most appropriate and reliable method to analyse the cyclic behaviour of connections is still represented by the direct experimentation. Several full-scale tests on a number of different joint typologies have been executed within the current project. Attention has been paid to many influencing factors, trying to cover lacks in the existing literature. In particular, influence of: (a) strain rate, (b) loading asymmetry, (c) beam haunching and (d) column size have been analysed according to Table 1. Besides, some other minor influencing factors have been investigated, namely the influence of static pre-loading (e) and partial strength (f).

TABLE 1

DISTRIBUTION OF EXPERIMENTAL ACTIVITIES AMONG PARTNERS

Type of connection	Laboratory	Type of influence					
		a	*b*	*c*	*d*	*e*	*f*
Welded	Lisbon				■		
	Ljubljana	■					
	Liège	■					
	Sofia			■			
	Timisoara		■				
Bolted	Lisbon				■		
	Ljubljana	■				■	
	Liège	■					■
	Sofia						
	Timisoara		■				

The influence of strain rate on mechanical response of structures is a very important matter, it being worthy of a special regard. In fact, in same case, especially related to near-source earthquakes, due to the great velocity of the seismic action, rate loading on structural members may be much higher than the one commonly applied in laboratory quasi-static tests. On the other side, it is well known that materials exhibit higher yielding and ultimate strength as far as strain rate increases, but such increments are not the same and therefore the mechanical response of the components is undergoing some variation. The experimental evidence is essential to understand the potentiality of such an effect on the poor brittle behaviour that some steel structures exhibited during recent earthquakes. Also, it is undeniable that the available experimental results appear questionable, showing a different influence due to strain rate, as it has been shown from the test results of the Laboratory of the University of Liège. Contradictory results have been also obtained in tests performed within this project, where rigid full-strength extended end plate connections revealed a different influence of high loading velocity on cycle number to failure in relation to specific strength of adopted steel. The topic is very interesting and attractive and must be deepened in further research studies.

Influence of loading asymmetry has been investigated as well. For this scope, a number of different connection typologies have been considered at the Laboratory of the University of Timisoara. Panel zone in shear is the component that potentially may induce a strong variation of connection behaviour. As a consequence, loading asymmetry affects some response parameters of beam-to-column joints and this difference must be duly accounted for in design procedures. The presented experimental activity has been devoted to identify such an influence, essentially under a qualitative point of view.

The study on the influence of haunching analysed at the Laboratory of the University of Sofia is devoted to test the reliability of new connection details on the dissipative capacity of beam-to-column connections. This is a new trend in steel constructions, essentially promoted in U.S.A. and Japan after recent earthquakes in Northridge and in Kobe. Strengthening and weakening strategies are now being widespread elsewhere, because simple details may promote an important performance improvement in the whole behaviour. This advantage has been confirmed by the presented test results, which undeniably evidence that beam haunching is able to move the plastic hinge away from the column face, therefore conferring a high ductility and very stable hysteretic cycles to the joint. Both the two tested typologies, like tapered flange and radius cut flange, exhibited satisfactory cyclic behaviour; in addition, the latter also assures some advantage connected to the possibility to correctly predict the actual behaviour of the connection and therefore appears to be more convenient for design practice.

An important concern affecting the response of moment resisting steel frames is the ratio between beam and column strength. The influence of column size, with a fixed beam, on both welded and bolted connections has been studied at University of Lisbon. Results show that the control of such a ratio may be somewhat important in case of welded connections, whose behaviour is conditioned by the panel zone. Connections may behave as semi-rigid or rigid as a function of the column size. Also, their strength is affected by the same factor. On the contrary, the influence of column size seem to be not important in case of bolted top and seat angle connections, where the connecting elements themselves usually represent the weakest components of the whole joint.

CYCLIC BEHAVIOUR OF BEAM-TO-COLUMN STEEL-CONCRETE COMPOSITE CONNECTIONS

Steel-concrete composite connections have been investigated at INSA - Rennes Laboratory. Nowadays, composite constructions are becoming more and more popular, especially due to the improved mechanical and fire resistance respect to bare steel elements. But, the knowledge concerning the actual behaviour of beam-to-column composite joints is still incomplete. The main aspect analysed in this study is the risk of degradation of the slab and the shear connectors under cyclic loading. Tests have concerned with monotonic and cyclic response of different slab and connectors typologies, currently used in buildings. Preliminarily, push-out and push-pull tests on connectors evidenced a peculiar cyclic behaviour for such type of components, which requires the control of their ductility and fatigue resistance. Tests on the whole end plate composite joints have shown that the rotation capacity may be detrimentally affected by partial shear connectors under seismic actions as well as by other factors such as the number of ribs in the slab and the mutual position of shear connectors and reinforcement. A theoretical determination of the collapse based on the usual fatigue formulation provided by EC3 has been also proposed and applied, providing interesting results. Further activities should be devoted to analyse different types of joints in order to better state which factors can improve the rotation capacity of composite joints.

RE-ELABORATION OF EXPERIMENTAL RESULTS

Even though tests have been performed in different Laboratories, a special attention has been paid to the methodology of analysis and adopted measuring conventions. Test results are perfectly homogeneous and they may be directly compared to each other. Since the assessment of the number of cycles to failure under cyclic loading is one of the most important objective to be got by experimental tests, an unified re-elaboration of all results has been provided as well. This allowed the fatigue life endurance of tested specimens to be determined by the same procedure, which is based on a fixed level of probability of failure. Results show that the S-N lines are sensitive to specimen typology, but they may be predicted with an acceptable level of reliability.

EVALUATION OF GLOBAL SEISMIC PERFORMANCE

The evaluation of seismic resistance of steel frame buildings has been analysed though a number of numerical studies. Aiming at assessing the ductility demand for semi-rigid joint frames, analyses have been carried out at University of Timisoara, where some frames with different column cross-section sizes and types of connections (the ones tested in the Laboratory) have been investigated. As it was expected, connection typology affects the whole performance of the frame (welded joints generally behave better than the bolted ones), but an important influence is also given by the considered acceleration record. Besides, the adopted collapse criterion may influence the judgement of reliability than may be assigned to the frame. Several factors affect the numerical response of the analysed structure and they must be correctly accounted for in comparing structure performances of different building configurations.

In the same direction, the activity developed at the University of Athens has been related to the interaction between local and global ductility properties for frame structures. In principle, the task is very interesting because the demand in terms of global ductility by numerical analyses must be compared with the available ones, which is related to the level of local ductility of members and connections. Dynamic analyses have shown strong differences for different types of accelerograms, thus confirming the important rule of the acceleration record. The main trend of the performed analyses show that higher local ductility improves the structural response, but special attention should be paid to structural irregularities, especially the ones due to weak storeys, which may penalise the frame performance. Eventually, serviceability criteria may prevail over the resistance ones, restricting the possibility to exploit large inelastic deformations. As a whole, the study provides useful information and criteria on how to perform numerical analyses, but also shows that the prediction of the seismic behaviour of steel moment resisting frames is almost complicated and the application of design criteria as well as the interpretation of the corresponding results in not immediate.

FAILURE MODE AND DUCTILITY CONTROL

The ductility demand control of moment resisting frames has been deepened, by analysing some important aspects connected with both design methodology and numerical modelling.

It is commonly accepted that frames should be designed so to promote global type collapse mechanisms, in order to assure the maximum energy dissipation capacity and global ductility. Design procedures exist in case of steel frames with rigid full-strength joints. A first attempt to extend such a procedure to partial-strength semi-rigid frames has been proposed in Chapter 7.1. Results are encouraging, especially because they show that such joint typologies, in case of long span and heavily loaded beams, may be economically advantageous, without penalising the performance of the structure in terms of ductility demand.

The influence of the hysteretic behaviour of beam-to-column connection has been analysed as well. Special cyclic models have been developed and calibrated on the basis of available experimental results for typical connection typologies. In case of frame structures, the effect of joint on ductility demand has been deeply investigated aiming at identifying the main factors affecting the performance of the structure. Obviously, the results may be interpreted twofold: on one side, they emphasise the connection typologies allowing for a better seismic response of MR frames; on the other side, they are useful for assessing the susceptibility of numerical analysis results to suitably model all types of connections and, in particular, the possibility to adopt simplified hysteretic rules.

Influence of structural typology has been focussed in Chapters 7.3 and 7.4. The former is dealing with the performance of dual frame buildings, where completely pinned or semi-rigid partial strength connections are employed in either interior or exterior bays of the frame. The latter is concerned with the effect of building asymmetry. In both cases the developed tasks contribute to the understanding of the non-linear seismic response of frame building in case of non-ordinary conditions, providing the main factors having the major impact within the above influences. Besides, the way how these studies have been conducted allows some interests in the development of simplified non-linear analyses to be stated, as the one proposed by Fajfar. This kind of simplified analyses can be very helpful to correctly and reliably assess the actual behaviour of the building.

DESIGN METHODOLOGY

All the above tasks dealing with numerical analyses have shown that the interpretation of results is a difficult matter, but it is the most important for correctly assessing the structural behaviour and the effect of assumed design conditions. In particular, it is essential to be consistent in applying code provisions, assuming design criteria and evaluating structural performance. Chapter 8 analyses most of the issues connected with design methodology and evaluation of frame response, firstly providing basic definitions and relations of the most important mechanical factors and general methods used to characterise building performance. Then, the evaluation of global seismic response has been performed by means of several approaches for determining the so-called q-factor (behaviour or reduction factor). The investigation has been done comparing the existing typical methods and by proposing and developing new "ad hoc" methods, like the one of Aribert and the modification of the Kato-Akiyama energy approach. As a whole, the present study shows that for moment resisting frames, even if the definition and evaluation of behaviour factor is taken for granted by codes and researchers, this problem is worthy of further investigation and remark.

ACKNOWLEDGEMENTS

At the end of this very fruitful period of thirty months, as the convenor of the RECOS project, I would like to express my satisfaction for the brilliant obtained results. The partners of the working team, some of them are out-standing experts in this field, have been very effective in their activity, very intelligent in promoting and discussing new trends, very punctual in the respect of the deadlines during the research process, very collaborative each others in accepting also non familiar methodologies, very concerned with the importance of their tasks. So, in one word, the development of this project, thanks to the excellent composition of the working team and the corresponding results, can be indicated as a suitable model to be followed in this kind of activities, also because of the friendly atmosphere created within the group.

A very special acknowledgement of gratitude has to be extended to Gianfranco De Matteis, who co-operated in a very effective way to the development of this work in his function of technical secretary of the RECOS project.

INDEX